The Movement of Tectonic Plates

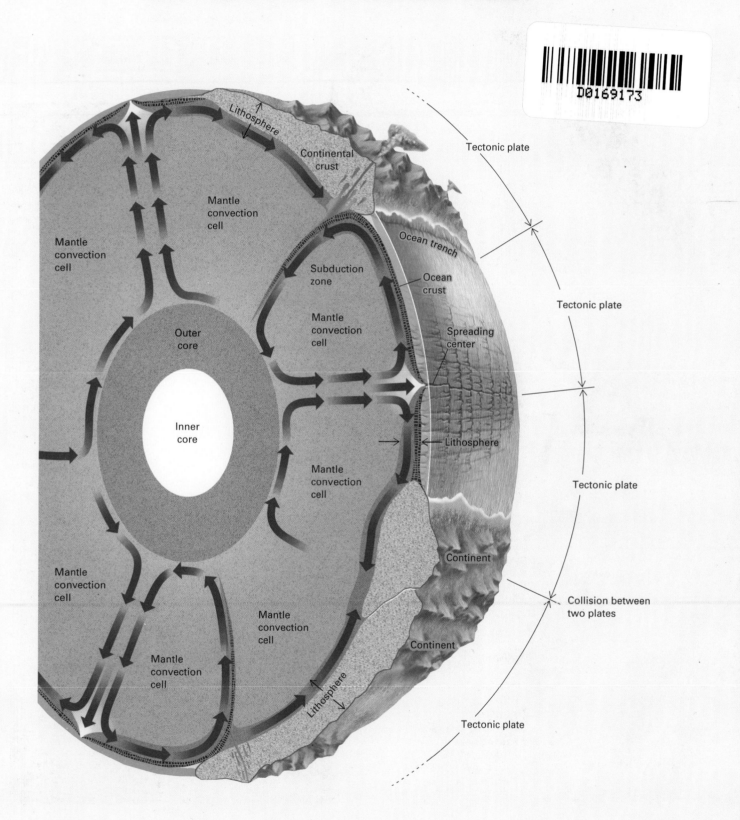

Earth Science
and the Environment

Earth Science
and the Environment

FOURTH EDITION

GRAHAM R. THOMPSON, PHD
University of Montana

JONATHAN TURK, PHD

THOMSON

™

BROOKS/COLE

Australia • Brazil • Canada • Mexico • Singapore
Spain • United Kingdom • United States

THOMSON

BROOKS/COLE

Earth Science and the Environment, fourth edition
Graham R. Thompson and Jonathan Turk

Executive Editor: *Peter Adams*
Assistant Editor: *Carol Benedict*
Editorial Assistant: *Anna Jarzab*
Technology Project Manager: *Samuel Subity*
Marketing Manager: *Kelley McAllister*
Marketing Communications Manager: *Nathaniel Bergson-Michelson*
Content Project Manager: *Belinda Krohmer*
Creative Director: *Rob Hugel*
Art Director: *Vernon Boes*
Print Buyer: *Becky Cross*
Permissions Editor: *Roberta Broyer*
Production Service: *Pre-Press Company, Inc.*
Text Designer: *Terri Wright*
Art Editor: *Lisa Torri*

Photo Researcher: *Terri Wright*
Cover Images: *Background Image – © Royalty-free Photodisc Red/ Getty Images*
Inset globe – NASA/Goddard Space Flight Center, The SeaWiFS Project and ORBIMAGE, Scientific Visualization Studio, Landsat-7 Project, and USGS EROS Data Center.
Atmosphere – NOAA
Biosphere – Royalty-Free/Corbis
Hydrosphere – © Royalty-free Image Source Limited/Jupiter Images
Geosphere – VEER Christopher Talbot Frank/Getty Images
Title Page Image: *© Steve Bloom Images/NASA/Alamy*
Cover Printer: *CTPS*
Compositor: *Pre-Press Company, Inc.*
Printer: *CTPS*

Library of Congress Control Number: 2006926271

ISBN-13: 978-0-495-11287-7
ISBN-10: 0-495-11287-9

For more information about our products, contact us at:
Thomson Learning Academic Resource Center
1-800-423-0563
For permission to use material from this text or product, submit a request online at http://www.thomsonrights.com.
Any additional questions about permissions can be submitted by e-mail to thomsonrights@thomson.com.

About the Authors

Gray Thompson is Professor of Geology at The University of Montana, where he teaches Introductory Geology, Mineralogy, Summer Field Mapping, and graduate courses in Clay Mineralogy and Shale Petrology. He has published more than 20 research papers in international journals, mostly co-authored with his students. He is also a mountaineer and professional guide with first ascents, many with Jon Turk, of peaks and routes in the Rocky Mountains, Alaska, the Yukon, Baffin Island, the Alps, the Karakoram, and the Himalayas. He has authored many articles published in international climbing magazines and journals, and has been the subject of other articles in these publications. Many of the photographs in this text were taken by Thompson and Turk on their climbing trips and expeditions over the past fifteen years.

Jon Turk is a chemist, professional geoscience writer, and adventurer. He received his Ph.D in 1971, and later that year Jon co-authored the first environmental science college textbook in the country. In the 32 years since then, Jon has continued his career as a science writer by publishing 23 environmental and geoscience texts. Jon's love of unspoiled environments and his fascination for the wild places on this planet have also led to a distinguished career as an adventurer. He has kayaked around Cape Horn as well as the 3,000 miles between Japan and Alaska, crossed the western Gobi of Mongolia, unsupported, on a mountain bike, and was the first to ascend Lamo-she Peak (6,070 meters) in the eastern Himalayas with co-author Gray Thompson. He has written numerous magazine articles about his expeditions as well as adventure/travel books, including *In the Wake of the Jomon: Stone Age Mariners and a Voyage Across the Pacific* (McGraw Hill 2005).

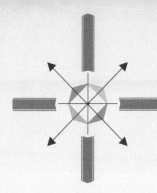

Contents Overview

Contents

UNIT 1 EARTH MATERIALS AND TIME 19

UNIT 2 INTERNAL PROCESSES 127

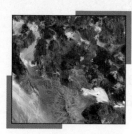

UNIT 6 ASTRONOMY 557

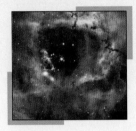

22 Motions in the Heavens 558

23 Planets and Their Moons 581

24 Stars, Space, and Galaxies 607

APPENDIX

Preface

In 1979 James Lovelock published his landmark book, *Gaia: A New Look at Life on Earth* (Oxford University Press), which proposed that plants and animals contribute to the evolution of Earth's atmosphere. Furthermore, Lovelock proposed that Earth could be likened to a living organism, whose environment is regulated by homeostatic interactions between earth, air, water, and life. While the Gaia hypothesis was initially accepted by many in the environmental community, it was criticized and debated within the scientific community. Although the debate continues, scientists have accumulated a huge amount of data over the past 27 years to show that earth, air, water, and living organisms interact in complex ways and that these interactions profoundly affect Earth's environment.

The authors of this book, Gray Thompson and Jon Turk, have been involved in environmental and earth science research and education for nearly four decades. We have carefully watched the unfolding of the Gaia debate and read volumes of literature describing the mechanisms of current and paleo planetary change. In this fourth edition of *Earth Science and the Environment*, we stress the interrelationship among Earth's four systems in greater detail than in any of our previous textbooks.

In March 2000, a Connecticut-sized iceberg broke off the Ross Ice Shelf in Antarctica and floated northward to melt in warmer waters. We ask, Why did the ice break free? Was it a consequence of normal glacial movement and changing seasons, or was it caused by global climate change? Will the breakup of the Ross Ice Shelf affect distant coastlines and ecosystems?

Every year, volcanic eruptions blast fiery lava, dark ash, and billowing steam into the sky. Sometimes nearby houses are buried and unlucky residents are killed. Beyond the immediate tragedy, we again ask questions about cause and consequence: What Earth processes produced the eruption? Will the dark cloud of suspended ash shadow the sun and alter regional or global climate?

In attempting to answer these and similar questions, we learn that Earth's systems are interrelated. For example, deep forces within Earth generate volcanic eruptions. But an eruption also emits the greenhouse gas carbon dioxide that contributes to global warming and that, in turn, may accelerate the breakup of Antarctic ice. Following the chain of events one step further, extensive ice shelf collapse causes sea level rise, which alters global climate even more.

Turning our focus from the present into deep geologic time, abundant evidence reveals that just about anywhere on our planet, the rock, the landscape, the climate, and the living organisms have changed throughout Earth's history. Fossil ferns and dinosaur bones in Connecticut tell us that this temperate region was once warm and tropical. Millions of years after the dinosaurs became extinct, continental glaciers bulldozed across the landscape leaving huge mounds of rock and sediment.

Earth science is a study of the world around us: the rocks beneath our feet, the air we breathe, the water that exists almost everywhere in our environment, and all the living organisms that share our planet with us. Earth science also extends a view into space to look at distant planets, stars, and faraway galaxies.

Earth scientists study the mechanisms of change. How did the climate in Connecticut fluctuate from warm and humid to frigidly cold? What forces drive volcanic eruptions, and why do they concentrate in some regions but not in others? As scientists sought to answer these and other related questions, they have found that Earth is a complex system composed of smaller, interconnected and interacting systems. Whenever we discuss a process or event, we simultaneously examine the disparate systems that affect or are affected by that geological, atmospheric, biological, or oceanic change.

Earth science encompasses the entire history of our planet, from its formation 4.6 billion years ago to the present. We go back in time and ask how Earth evolved, how and when water collected to form oceans, how continents rose out of the watery wilderness, and what processes created an atmosphere favorable to life.

Over the past hundred years, technology has given the human species unprecedented power to alter Earth. As a consequence, we are engaged in a great and uncontrolled experiment to observe how our planetary systems will react to these perturbations. It is an experiment unique in scale and process. In a traditional laboratory study, scientists examine two parallel systems. One is perturbed, the other (the control) is not perturbed, and scientists observe the different behaviors of the two. But we have only one Earth. If we increase the carbon dioxide concentration of the air, our planetary systems will change and we will never know what would have happened if the atmospheric carbon dioxide concentration had remained constant.

In addition to the fact that we can't perform controlled experiments on Earth, we can't easily retreat from our mistakes. Once change has occurred, no one can say, "Woops, that didn't work; let's back up, start at the beginning, and try something different." Therefore it is vitally important that we proceed carefully as we alter the environment that sustains us. No textbook can offer answers or solutions to the complex issues that we face, but the fourth edition of *Earth Science and the Environment* provides background information needed to understand basic Earth processes and to formulate critical decisions that will define our future.

Many of Earth's systems interactions discussed in the text summarize the results of research published since the last edition was written. Thus, at the same time that we are presenting the most accurate modern picture of planetary change, we also teach that science is a living process and not a set of stale facts to be memorized. We emphasize that scientists propose models, frequently disagree with one another, and then return to the laboratory, the computer screen, or the field to gain more information and to try to answer difficult questions.

Sequence of Topics

Because different earth science courses can have many different emphases, information presented in these courses can follow a wide variety of logical sequences. In this book, we have chosen to introduce Earth's materials and geological time, and then to start from Earth's interior and work outward. Thus the book is divided into six units:

Unit 1: Earth Materials and Time
Unit 2: Internal Processes
Unit 3: Surface Processes
Unit 4: The Oceans
Unit 5: The Atmosphere
Unit 6: Astronomy

Some instructors may prefer other sequences for covering these topics. The structure of this book allows many alternative sequences.

Special Features

In previous editions of *Earth Science*, we isolated many topics and Case Histories into **Focus On** boxes that were set aside and highlighted in color. The rationale was that some topics are not necessary to the sequential development of each chapter but may be of interest to certain classes. In this edition, we felt that most of these topics are so integral to the discussions that they should be incorporated with the text and not set aside. Thus, we maintain a few **Focus On** boxes to highlight some examples, but, in most cases, we have fused the main text information with the examples.

Art and Photographs

Earth science is a visual discipline. Landscapes tell their own stories, and many processes can be illustrated with line drawings. The authors took many of the pho-

Courtesy of Graham R. Thompson/Jonathan Turk

tographs in this book during their extensive travels and field research. We have upgraded much of the line art and rendered many new pieces. We have also continued to emphasize important topics with large-format concept art called "Systems Perspective" to capture the students' attention and curiosity.

Chapter Review Material

A short summary of important chapter material is provided at the end of each chapter. Following the summary, we have included and expanded our unique art called "Earth Systems Interactions." These schematic drawings show the four spheres of earth science: geosphere, hydrosphere, atmosphere, and biosphere. Arrows connect the four spheres with labels showing how the separate systems interact. This art emphasizes the interactivity of Earth's systems and at the same time provides a graphical review of many of the discussions in the chapter.

Questions

We provide two types of end-of-chapter questions. The students can answer the **review questions** in a straightforward manner from the material in the text. Thus, these questions allow students to test themselves on how completely they have learned the material in the chapter.

At the same time, **discussion questions** challenge students to apply what they have learned to an analysis of situations not directly described in the text. These questions often have no absolute correct answers. In addition, throughout the text the captions of selected figures contain questions designed to provoke thought and discussion about important issues raised by those figures.

Glossary and Appendix

A **glossary** is provided at the end of the book. **Appendices** cover mineral classification and identification, the International System of Units (SI), the elements, rock symbols, and star maps.

ANCILLARIES

This edition is accompanied by an extensive suite of both print and media supplements.

ThomsonNOW, the first assessment-driven and student-centered tutorial created for the earth science market, focuses student study time on concept mastery and is FREE to students with each purchase of a new textbook. This Web-based student tutorial is completely flexible and can be used in any order, but it features a default three-step mastery process designed to maximize the effectiveness of the system. The first step, called "What Do I Know?" consists of a diagnostic "Pre-Test" to help students identify their specific weaknesses in any particular chapter. The results provide a "Personalized Learning Plan," based on each student's results.

The second step, "What Do I Need to Learn?" provides activities drawn from the wealth of book-specific, interactive resources like animations, movies, and tutorials that students can work with.

The third step, "What Have I Learned?" is an optional "Post-Test" to ensure that the student has mastered the concept(s) in each chapter. As with the Pre-Test, student results may be e-mailed to the instructor—helping both the instructor and the student assess progress. By providing students with a better understanding of exactly what they need to focus on, **ThomsonNOW** helps students maximize their study time, bringing them a step closer to success! **ThomsonNOW** is available at **www.thomsonedu.com.**

GIS Investigations

Do you want to integrate GIS into your earth science course? These groundbreaking guides, by Hall-Wallace et al., let even novice users tap the power of GIS to explore, manipulate, and analyze large data sets. The ArcView [r]-based guides come with all the software and data sets needed to complete the exercises.

New to This Edition

ArcGIS[r] versions of these manuals are available with online, downloadable data sets and an online "Instructor's Manual." These are available alone, bundled together, and/or bundled with *Earth Science and the Environment.*

> *Exploring Tropical Cyclones: GIS Investigations for the Earth Sciences*
> ISBN: 0-534-39147-8
> ISBN: 0-495-11543-6 - ArcGIS version
> *Exploring Water Resources: GIS Investigations for the Earth Sciences*
> ISBN: 0-534-39156-7
> ISBN: 0-495-11512-6 - ArcGIS version
> *Exploring the Dynamic Earth: GIS Investigations for the Earth Sciences*
> ISBN: 0-534-39138-9
> ISBN: 0-495-11509-6 - ArcGIS version
> *Exploring the Ocean Environment: GIS Investigations for the Earth Sciences*
> ISBN: 0-534-423507
> ISBN: 0-495-11506-1 - ArcGIS version

Online Instructor's Manual/Test Bank

This manual, written by the main text authors, contains discussion sections (highlighting the most important/interesting topics and suggesting alternative teaching sequences where appropriate), answers to discussion questions from the main text, selected readings, and a test bank consisting of multiple-choice, true/false, and fill-in-the-blank questions and answers.

Transparency Acetates

0495-113921
This set of 100 transparency acetates contains photos and images taken directly from the text.

Slides

0030214793
This set of slides contains full-color photos and images taken directly from the text.

ExamView® computerized testing

Create, deliver, and customize tests and study guides (both print and online) in minutes with this easy-to-use assessment and tutorial system. ExamView® offers both a Quick Test Wizard and an Online Test Wizard that guide you step-by-step through the process of creating tests, while the unique "WYSIWYG" capability allows you to see the test you are creating on the screen exactly as it will print or display online. You can build tests of up to 250 questions using up to 12 question types. Using ExamView's complete word processing capabilities, you can enter an unlimited number of new questions or edit existing questions.

Multimedia Manager

ISBN: 0-534-49314-9 (CD format)
This easy-to-use multimedia lecture tool allows you to quickly assemble art and database files with notes to create fluid lectures. Available on CD, the Multimedia

Manager includes a database of animations and images from earth science titles published by Brooks, Cole. The simple interface makes it easy for you to incorporate graphics, digital video, animations, and audio clips into your lectures.

Book Companion Website

www.thomsonedu.com/earthscience/thompson

ACKNOWLEDGEMENTS

The first two editions of *Earth Science and the Environment* were published under the guidance of our close personal friend, senior editor John Vondeling at Saunders College Publishing. Tragically, John developed cancer and passed away. *Earth Science and the Environment* drifted in limbo for several months until it was taken under the wing of Brooks, Cole Publishing and a new editorial team, headed by acquisitions editor Keith Dodson. It has been a delight to work with Keith and the professionals at Brooks, Cole. They have brought new vitality to the manuscript preparation and to the exciting ancillary material that accompanies the text. The team at Thomson Arts and Sciences Publishing has changed over the years, but enthusiasm for the book has remained strong. Special thanks to Peter Adams, our new acquisitions editor, and the following members of the fourth edition project team: Anna Jarzab—editorial assistant, Carol Benedict—assistant editor, Sam Subity—technology project manager, Kelley McAllister—marketing manager, Nathaniel Bergson-Michelson—executive advertising project manager, Belinda Krohmer—content project manager, and Vernon Boes—art director. They have all helped us through the complex process of moving from a rough manuscript to a final book.

We would also like to extend special thanks to numerous independent contractors who worked on the book. Wendy Pizzi at Pre-Press Company has tirelessly overseen the day-to-day details of production. Nina Maclean, our personal editorial assistant, keeps confusion at bay in the home office. Terri Wright did the photo research and book design, and Lisa Torri styled the art for the Earth Systems Interactions figures, which have been given their due prominence in this edition.

In addition to the editorial and production teams, all four editions of the manuscript have been extensively reviewed at several stages, and the numerous careful criticisms of the art and text have helped shape the book and ensure accuracy.

Reviewers of the fourth edition:

Patricia Crews, *Florida Community College of Jacksonville*

Lois Breur Krause, *Clemson University*

Veronique L. Lankar, *Pace University/Mount St. Mary's College*

Leland Timothy Long, *Georgia Institute of Technology*

Sadredin C. Moosavi, *Minnesota State University–Mankato*

Reviewers of third edition:

Jens Bischof, *Old Dominion University*

Roland H. Brady, *California State University at Fresno*

Nathalie Brandes, *University of Wisconsin, Eau Claire*

Stan Celestian, *Glendale Community College*

Marvin Cochrane, *Troy State University*

Scott T. Dreher, *Indiana State University*

Jim Durbin, *University of Southern Indiana*

Joseph Graf, *Southern Oregon University*

Chris Hansen, *Southern Adventist University*

Christopher Hooker, *Waubonsee Community College, Sugar Grove*

R. V. Krishnamurthy, *Western Michigan University*

Steven J. Maier, *Northwestern Oklahoma State University*

W. Patrick Seward, *Rogers State University*

Steve Taylor, *Western Oregon University*

Reviewers of second edition:

James Albanese, *State University of New York at Oneonta*

Sandra Brake, *Indiana State University*

Mark Evans, *University of Pittsburgh*

Bryan Gregor, *Wright State University*

Clay Harris, *Middle Tennessee State University*

Steve LaDochy, *California State University–Los Angeles*

John Madsen, *University of Delaware*

Joseph Moran, *University of Wisconsin–Green Bay*

Paul Nelson, *St. Louis Community College at Meramec*

Adele Schepige, *Western Oregon State College*

Richard Smosna, *West Virginia University*
Ronald Wasowski, *King's College*

Reviewers of first edition:

James Albanese, *State University of New York at Oneonta*
John Alberghini, *Manchester Community College*
Calvin Alexander, *University of Minnesota*
Edmund Benson, *Wayne County Community College*
Robert Brenner, *University of Iowa*
Walter Burke, *Wheelock College*
Wayne Canis, *University of Northern Alabama*
Stan Celestian, *Glendale Community College*
Edward Cook, *Tunxis Community College*
James D'Amario, *Hartford Community College*
Joanne Danielson, *Shasta College*
John Ernissee, *Clarion University*
Richard Faflak, *Valley City State University*

Joseph Gould, *St. Petersburg Junior College*
Bryan Gregor, *Wright State University*
Miriam Hill, *Indiana University–Southeast*
John Howe, *Bowling Green State University*
DelRoy Johnson, *Northwestern University*
Alan Kafka, *Boston College*
William Kohland, *Middle Tennessee State University*
Thomas Leavy, *Clarion University*
Doug Levin, *Bryant College*
Jim LoPresto, *Edinboro University of Pennsylvania*
Glenn Mason, *Indiana University–Southeast*
Joseph Moran, *University of Wisconsin–Green Bay*
Alan Morris, *University of Texas–San Antonio*
Jay Pasachoff, *Williams College*
Frank Revetta, *Potsdam College*
Laura Sanders, *Northeastern Illinois University*
Barun K. Sen Gupta, *Louisiana State University*
James Shea, *University of Wisconsin–Parkside*
Kenneth Sheppard, *East Texas State University*
Doug Sherman, *College of Lake County*

All of these people and many of their associates have worked hard and efficiently to produce the finished product.

Graham R. Thompson,
Missoula, Montana

Jonathan Turk,
Darby, Montana

CHAPTER

1

Earth Systems

© John M. Roberts/CORBIS

Although scientists have searched for living organisms on other planets, so far we have found life only on Earth. From alpine meadows, such as this landscape in Mount Rainier National Park, to deserts, to the deep sea floor, life has adapted to almost all conceivable Earthy environments.

E arth is sometimes called the water planet or blue planet because azure seas cover more than two-thirds of its surface. Earth is the only planet or moon in the Solar System with rain falling from clouds, and water that runs over the land to collect in extensive oceans. It is also the only body that we know of that supports life. ■

1.1

Flowers Bloom on Earth, Venus Boils, and Mars Freezes

Our Solar System originally evolved from a cloud of dust particles and gas rotating in the vast emptiness of space. Under the relentless pull of gravity, these particles condensed into discrete bodies that gradually formed the Sun, the planets, and their moons. Thus, all of the planets formed from the same original mixture of materials. Yet today, the planets are distinctly different from one another. To appreciate these differences, let us briefly compare Earth with its two closest neighbors, Venus and Mars.

Of all the planets, Venus most closely resembles Earth in size, density, and distance from the Sun. Consequently, astronomers once thought that Venus might be similar to Earth, and that both water and life might be found there. However, data obtained from spacecraft reveal that Venus is extremely inhospitable (◆ Figure 1.1). Any Earth-like life would quickly suffocate in the carbon dioxide–rich Venusian atmosphere. Caustic sulfuric acid clouds fill the sky there. In addition, the surface of Venus is hot enough to melt lead, and therefore hot enough to destroy the organic molecules necessary for life.

As for Mars, early in its history it must have had a temperate climate somewhat like that of Earth today. Images from spacecraft and remote rovers show extinct stream canyons, sea floors, and lake beds, indicating that flowing water must have been abundant on the planet (◆ Figure 1.2). But today, the Martian surface is frigid and dry. Mars's winter ice caps are mostly frozen carbon dioxide, commonly called dry ice. If water is present on

Mars, it lies frozen beneath the planet's surface; Mars also has almost no atmosphere.

If Earth, Venus, and Mars formed from the same materials, why are the three planets so different today? Why are we the only planet favored with great oceans, cascading waterfalls, blue skies, and abundant life? Shortly after the formation of the planets, the original atmospheres of Earth, Venus, and Mars evolved into swirling mixtures of carbon dioxide, carbon monoxide, water, ammonia, methane, and other gases. To appreciate what happened next, we need to understand the behavior of carbon dioxide and water in planetary environments.

Water can be a solid, liquid, or gas. If liquid water cools, it turns to ice; when it is heated, it evaporates to form vapor. Thus water leaves the atmosphere when it condenses or freezes, whereas it enters the atmosphere when it vaporizes. Carbon dioxide also exists in a variety of forms. At Earth's surface, carbon dioxide occurs as an atmospheric gas, dissolves in seawater, and combines with calcium and oxygen to form a type of rock called limestone. (Carbon dioxide also exists as a liquid and a solid, but not in Earth's surface environment.) Carbon dioxide gas and water vapor are both greenhouse gases—they absorb infrared radiation and warm the atmosphere.

Because Venus is closer to the Sun than Earth is, it receives more solar radiation and was originally a bit warmer than Earth is now. Due to the higher temperature on Venus, water vapor either never condensed, or if it did, it quickly evaporated again. Because there were no seas for carbon dioxide to dissolve into, most of the carbon dioxide also remained in the atmosphere. Both of these greenhouse gases further heated up Venus's atmosphere. Heat released more water and carbon dioxide into the atmosphere and, subsequently, these gases trapped more heat. The temperature spiraled higher and higher.

In contrast, Earth, being farther from the Sun, was cool enough so that the water vapor condensed to form vast oceans. Large amounts of carbon dioxide then dis-

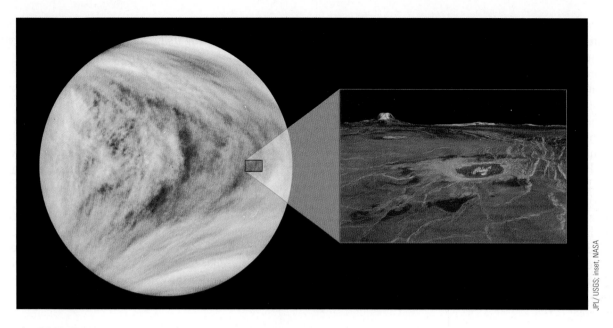

◆ **FIGURE 1.1** Before astronomers could peer through the Venusian cloud cover or measure its temperature and composition, they speculated that Venus may harbor life. Today we know that the atmosphere is so hot and corrosive that living organisms could not possibly exist.

solved in the seas or reacted to form limestone. Thus, large quantities of these two greenhouse gases were removed from the atmosphere. Fortunately for us, the temperature stabilized in a range favorable for the existence of liquid water and for the emergence and evolution of life.

Mars is a little farther from the Sun than Earth is, and consequently it receives a little less solar energy. A Martian atmosphere rich in carbon dioxide and water vapor probably evolved between 4.0 and 3.5 billion years ago. Both of these greenhouse gases absorbed the Sun's heat, producing a temperate climate despite the planet's greater distance from the Sun. Rain fell from clouds, rivers flowed, wind blew over shallow seas. Perhaps life evolved

in this favorable environment. But because Mars is further from the Sun than Earth is, its atmosphere cooled more than Earth's. This small initial cooling caused more water vapor to condense and more carbon dioxide to dissolve into the seas, lowering the amounts of greenhouse gases in the Martian atmosphere. The temperature spiraled downward. Today, the Martian surface occasionally becomes as warm as an autumn afternoon on Earth (20°C), but more commonly, the temperature is below freezing. Temperatures as low as –140°C have been recorded. At this extreme, carbon dioxide freezes into dry ice.

In summary, Venus is closer to the Sun and receives more solar radiation than either Earth or Mars. This

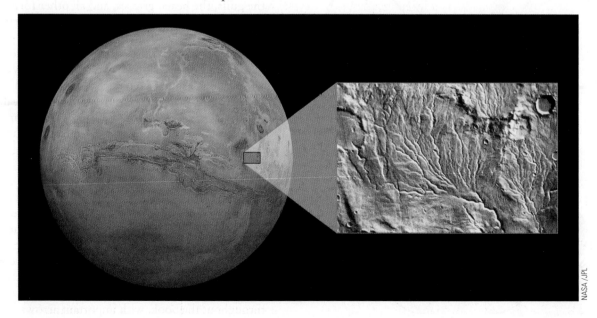

◆ **FIGURE 1.2** Water once flowed over the surface of Mars, eroding canyons and depositing sediment. However, today the planet is frigid and dry.

small amount of extra solar energy initiated changes that concentrated greenhouse gases in the atmosphere. In turn, these greenhouse gases heated up the planet much more than can be accounted for by the greater amount of solar energy received by Venus. Of the three planets, Mars is the farthest from the Sun and receives the least solar energy. Here the cool initial temperatures triggered different atmospheric changes that resulted in the loss of nearly all greenhouse gases from the Martian atmosphere—and the temperature plummeted.

On Earth, a temperate climate persisted. Life evolved. Then, over the course of a few billion years, an atmosphere rich in nitrogen and oxygen gradually replaced the atmosphere that was rich in carbon dioxide and water vapor. Today, geochemical and biological processes maintain a delicate balance of greenhouse gases so our global climate remains hospitable for life.

This example highlights the major themes summarized here:

- Planets are active, dynamic, and in a continuous state of change. On Earth, a primordial poisonous atmosphere evolved into one favorable for the existence of water and the emergence of life. At the same time, continents formed in a global ocean and barren rock weathered into soil. Life

formed, probably in the seas, and migrated onto land. During its history, Earth's temperature has oscillated frequently and dramatically.

- Planetary changes are driven by complex interactions among atmosphere, seas, and solid ground. On Earth, living organisms also alter the planet that supports them. For example, plant roots help weather rock into soil, and soil supports plants. During photosynthesis, plants produce oxygen, which plants and animals need for respiration. In many cases, small, seemingly insignificant perturbations can trigger a chain of events that will magnify the initial perturbation far beyond its initial impact. These examples illustrate how small differences in solar energy led to dramatic differences in the modern planetary environments of Earth, Venus, and Mars.

1.2
The Earth's Four Spheres

Imagine walking along a rocky coast as a storm blows in from the sea. Wind whips the ocean into whitecaps, gulls wheel overhead, and waves crash onto shore. Before you have time to escape, blowing spray has soaked your clothes. A hard rain begins as you scramble over the rocks to your car. During this adventure, you have observed the four major spheres of Earth. The rocks and soil underfoot are the surface of the **geosphere**, or the solid Earth. The rain and sea are parts of the **hydrosphere**, the watery part of our planet. The gaseous layer above the Earth's surface is the **atmosphere**. Finally, you, the gulls, the beach grasses, and all other forms of life in the sea, on land, and in the air are parts of the **biosphere**, the realm of organisms.

You can readily observe that the atmosphere is in motion because clouds waft across the sky and wind blows against your face. Animals, and to a lesser extent plants, also move. Flowing streams, crashing waves, and falling rain are all familiar examples of motion in the hydrosphere. Although it is less apparent, the geosphere is also moving and dynamic. Vast masses of subterranean rock flow vertically and horizontally within the planet's interior, continents drift, and mountains rise and then erode into sediment. Throughout this book we will study many of these phenomena to learn what energy forces set matter in motion and how the motion affects the planet we live on. ◁▷ Figure 1.3 shows schematically all the possible interactions among the spheres. This figure will be repeated throughout the book, with important arrows highlighted to emphasize interactions discussed in the accompanying text.

◁▷ **FIGURE 1.3** All of Earth's cycles and spheres are interconnected.

Figure 1.4 shows that the geosphere is by far the largest of the four spheres. The Earth's radius is about 6,400 kilometers, 1½ times the distance from New York to Los Angeles. Despite this great size, nearly all of our direct contact with Earth occurs at or very near its surface. The deepest well penetrates only about 12 kilometers, ⅟₅₃₃ of the total distance to the center. The oceans make up most of the hydrosphere, and the central ocean floor averages 5 kilometers deep. Most of Earth's atmosphere lies within 30 kilometers of the surface, and the biosphere is a thin shell about 15 kilometers thick.

The Geosphere

Our Solar System coalesced from a frigid cloud of dust and gas rotating slowly in space. The Sun formed as gravity pulled material toward the swirling center. At the same time, rotational forces spun material in the outer cloud into a thin disk. When the turbulence of the initial accretion subsided, small grains stuck together to form fist-sized masses. These planetary seeds then accreted to form rocky clumps, which grew to form larger bodies called planetesimals, 100 to 1,000 kilometers in diameter.

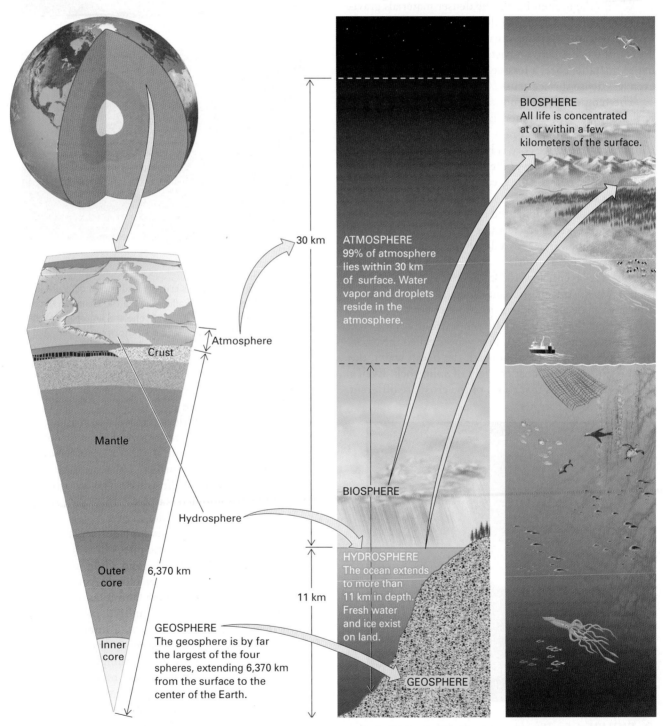

FIGURE 1.4 Most of Earth is solid rock surrounded by the hydrosphere, the biosphere, and the atmosphere.

Finally, the planetesimals collected to form the planets. This process was completed about 4.6 billion years ago.

As the Earth coalesced, the rocky chunks and planetesimals were accelerated by gravity so that they slammed together at high speeds. Particles heat up when they collide, so the early Earth warmed as it formed. Later, asteroids, comets, and more planetesimals crashed into the surface, generating additional heat. At the same time, radioactive decay heated the Earth's interior. As a result of all three of these processes, our planet became so hot that all or most of it melted soon after it formed.

Within the molten Earth, the denser materials gravitated toward the center, while the less dense materials floated toward the top, creating a layered structure. Today, the geosphere consists of three major layers: a dense, metallic **core**, a less dense rocky **mantle**, and an even less dense surface **crust** (◈ Figure 1.5).

The temperature of modern Earth becomes higher with depth, so at its center, Earth is 6,000°C, as hot as the Sun's surface. The core is composed mainly of iron and nickel. The outer core is liquid—a subterranean sea of molten metal. However the inner core, although hotter yet, is solid because the great pressure compresses the metal to a solid state.

The mantle surrounds the core and lies beneath the crust. The physical characteristics of the mantle vary with depth. Near the surface, to a depth of about 100 kilometers, the outermost mantle is cool, strong, and hard. In contrast, the layer below 100 kilometers is so hot that the rock is weak, soft, plastic, and flows slowly—like cold honey. Even deeper in Earth, pressure overwhelms temperature and the mantle rock becomes strong again.

The outermost layer is a thin veneer called the crust. Below a layer of soil and beneath the ocean water, the crust is composed almost entirely of solid rock. Even a casual observer sees that rocks of the crust are different

A

B

◈ **FIGURE 1.6** Earth's crust is made up of different kinds of rock. **(A)** The granite of Baffin Island in the Canadian Arctic is gray, hard, and strong. **(B)** The sandstone and limestone of the Utah desert are red, crumbly, and show horizontal layering.

from one another: some are soft, others hard, and they come in many colors (◈ Figure 1.6).

The uppermost mantle is relatively cool and its pressure is relatively low. Both factors combine to produce hard, strong rock similar to that of the crust. The outer part of Earth, including both the uppermost mantle and the crust, makes up the **lithosphere**. The lithosphere averages about 100 kilometers thick.

According to a theory developed in the 1960s, the lithosphere is broken into seven major segments called **tectonic plates**. These tectonic plates float on the weak, plastic, mantle rock beneath, and glide across Earth (◈ Figure 1.7). For example, North America is currently drifting toward China about as fast as your fingernail grows. In a few hundred million years—almost incomprehensibly long on a human time scale but brief when compared with planetary history—Asia and North America may collide, crumpling the edges of the continents and building a giant mountain range. In later chapters we will learn how plate tectonic theory explains earthquakes, volcanic eruptions, the formation of mountain ranges, and many other processes and events that have created our modern Earth and environment.

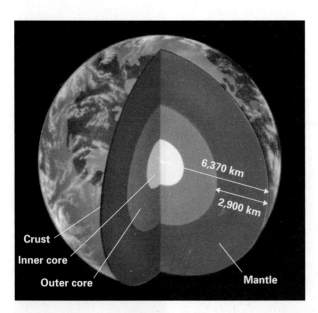

◈ **FIGURE 1.5** The geosphere is divided into three major layers: the crust, mantle, and core.

CHAPTER 1 • Earth Systems

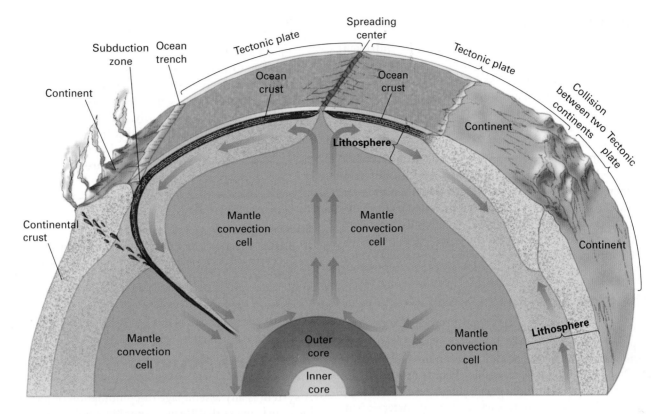

ThomsonNOW™ ◆▷ **ACTIVE FIGURE 1.7** The lithosphere is composed of the crust and the uppermost mantle. It is a 100-kilometer-thick layer of strong rock that floats on the underlying plastic mantle. The lithosphere is broken into seven major segments, called tectonic plates, that glide horizontally over the plastic mantle at rates of a few centimeters per year. In the drawing, the thickness of the mantle and the lithosphere are exaggerated to show detail.

The Hydrosphere

The hydrosphere includes all of Earth's water, which circulates among oceans, continents, glaciers, and the atmosphere. Oceans cover 71 percent of Earth and contain 97.5 percent of its water. Ocean currents transport heat across vast distances, altering global climate.

About 1.8 percent of Earth's water is frozen in glaciers. Although glaciers cover about 10 percent of Earth's land surface today, they covered much greater portions of the globe as recently as 18,000 years ago.

Only about 0.64 percent of Earth's total water exists on the continents as a liquid. Although this is a small proportion, fresh water is essential to life on Earth. Lakes, rivers, and clear, sparkling streams are the most visible reservoirs of continental water, but they constitute only 0.01 percent of Earth's water. In contrast, **ground water**, which saturates rock and soil of the upper few kilometers of the geosphere, is much more abundant and accounts for 0.63 percent of Earth's water. A minuscule amount, 0.001 percent, exists in the atmosphere, but this water is so mobile that it profoundly affects both the weather and climate of our planet (◆▷ Figure 1.8).

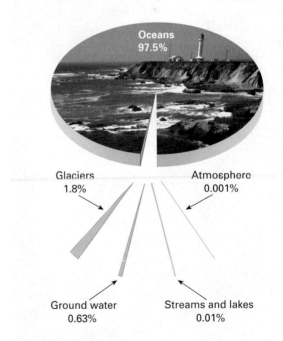

◆▷ **FIGURE 1.8** The oceans contain most of Earth's surface water. Most fresh water is frozen into glaciers. Most available fresh water is stored underground as ground water.

INTERACTIVE QUESTION: *How would a doubling of the amount of water in the atmosphere affect the oceans? How would it affect weather and climate?*

1.2   The Earth's Four Spheres   7

The Atmosphere

The atmosphere is a mixture of gases, mostly nitrogen and oxygen, with smaller amounts of argon, carbon dioxide, and other gases. It is held to Earth by gravity and thins rapidly with altitude. Ninety-nine percent is concentrated in the first 30 kilometers, but a few traces remain even 10,000 kilometers above Earth's surface.

The atmosphere supports life because animals need oxygen, and plants need both carbon dioxide and oxygen. In addition, the atmosphere supports life indirectly by regulating climate. Air acts both as a filter and as a blanket, retaining heat at night and shielding us from direct solar radiation during the day. Wind transports heat from the equator toward the poles, cooling equatorial regions and warming temperate and polar zones (◇▶ Figure 1.9).

The Biosphere

The biosphere is the zone that life inhabits. It includes the uppermost geosphere, the hydrosphere, and the lower parts of the atmosphere. Sea life concentrates near the surface, where sunlight is available. Plants also grow on Earth's surface, with roots penetrating a few meters into the soil. Animals live on the surface, fly a kilometer or two above it, or burrow a few meters underground. Large populations of bacteria live in rock to depths of as much as four kilometers, some organisms live on the ocean floor, and a few windblown microorganisms drift at heights of

◇▶ **FIGURE 1.9** Winds often drive clouds into great swirls. At the same time, global winds carry warm air to cooler regions and cool air to equatorial parts of Earth. The flow of heat makes the climates of both regions more favorable to humans.

Courtesy NASA

10 kilometers or more. But even at these extremes, the biosphere is a very thin layer at Earth's surface.

Plants and animals are clearly affected by Earth's environment. Organisms breathe air, require water, and thrive in a relatively narrow temperature range. Terrestrial organisms ultimately depend on soil, which is part of the geosphere. Less obviously, plants and animals also alter and form the environment that they live in. For example, living organisms contributed to the evolution of the modern atmosphere. We will discuss these interactions throughout the book.

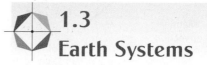

ThomsonNOW™

CLICK Thomson Interactive to work through an activity on the Earth's layers and plate locations through Plate Tectonics.

◈ 1.3 Earth Systems

A **system** is any assemblage or combination of interacting components that form a complex whole. For example, the human body is a system composed of stomach, bones, nerves, and many other organs. Each organ is discrete, yet all the organs interact to produce a living human. Blood nurtures the stomach; the stomach helps provide energy to maintain the blood.

Systems are driven by the flow of matter and energy. Thus a person ingests food, which contains both matter and chemical energy, and inhales oxygen. Waste products are released through urine, feces, sweat, and exhaled breath. Some energy is used for respiration and motion, and the remainder is released as heat.

A single bacterium is a system, just as a person is. But the human body contains hundreds of millions of bacteria that are essential to the functioning of metabolic processes, such as digestion. Thus a larger system (the human body) is composed of many smaller ones, including bacteria.

In addition, humans interact with their environment. Imagine a Stone-Age hunter-gatherer living in a small valley. All the plants and animals in this valley are components of an **ecosystem**, which is defined as a complex community of organisms and its environment, functioning as an ecological unit in nature. Therefore, to understand the human system, we must study smaller systems (bacteria) that exist within the body and also must appreciate how humans interact with larger ecosystems. But we're not finished yet. Individual ecosystems interact with climate systems, ocean currents, and other Earth systems.

Thus:

- The size of systems varies dramatically.
- Large systems contain numerous smaller systems.
- Systems interact with one another in complex ways.

As we have learned, Earth is composed of four major systems: geosphere, hydrosphere, atmosphere, and biosphere. Each of these large systems is further subdivided into a great many interacting smaller ones. For example, a single volcanic eruption is part of a system. Energy from deep within Earth melts rock, magma rises, and gases escape from deep earth layers and react chemically with surface materials. But this volcanic eruption is driven by mantle convection and movements of tectonic plates that are components of Earth's interior, also a system. At the same time, ash spewed skyward from the volcanic eruption cools the earth and becomes a component of the climate system. Looking at the eruption from yet another perspective, a volcanic mountain alters drainage patterns and erosion rates and is thus part of the hydrologic system. In this book, we will study systems of all sizes and illustrate many of the complex interactions among them.

Earth's surface systems—the atmosphere, hydrosphere, and biosphere—are ultimately powered by the Sun. Wind is powered by uneven solar heating of the atmosphere, ocean waves are driven by the wind, and ocean currents move in response to wind or temperature differences. Luckily for us, Earth receives a continuous influx of solar energy, and it will continue to receive this energy for another 5 billion years or so.

In contrast, Earth's interior is powered by the decay of radioactive elements and by residual heat from the primordial coalescence of the planet. We will discuss these sources further in later chapters.

Several energy and material cycles are fundamental to our study of Earth systems. These include the rock cycle (Chapter 3), the hydrological cycle or water cycle (Chapter 11), the carbon cycle (Chapter 17), and the nitrogen cycle (Chapter 17). During the course of these cycles, matter is always conserved—however, matter continuously and commonly changes form.

For example, ice melts to form water, and if more heat is added, water vaporizes to a gas. The process occurs in reverse when heat is removed. Thus, in the hydrological cycle, water evaporates from the ocean into the atmosphere, falls to Earth as rain or snow, and eventually flows back to the oceans. Chemical reactions also change the physical properties of matter. Common table salt is a solid, crystalline mineral. If you drop a crystal of salt in water, it dissolves, thus altering the properties of both the salt and the water. Or, as another example, carbon dioxide gas reacts with the element calcium to form both liquid solutions and solid rock, limestone. In turn, carbon in either of these forms reacts to form carbon dioxide gas and other products.

Because matter exists in so many different chemical and physical forms, most materials occur in all four of Earth's major spheres: geosphere, hydrosphere, atmosphere, and biosphere. Water, for example, is chemically bound into clays and other minerals, as a component of the geosphere. It is the primary constituent of the hydrosphere. It exists in the atmosphere as vapor and clouds. Water is an essential part of all living organisms. Or as another example, thick layers of salt occur as sedimentary rocks; large quantities of salt are dissolved in the oceans; salt aerosols are suspended in the atmosphere, and salt is an essential component of life. Thus, all the spheres continuously exchange matter and energy. In our study of Earth systems, we categorize the four separate spheres and numerous material cycles independently, but we also recognize that Earth materials and processes are all part of one integrated system (◆ Figure 1.3).

1.4
Time and Rates of Change in Earth Science

James Hutton was a gentleman farmer who lived in Scotland in the late 1700s. Although trained as a physician, he never practiced medicine and instead turned to geology. Hutton observed that a certain type of rock, called sandstone, is composed of sand grains cemented together (◆ Figure 1.10). He also noted that rocks in the Scottish Highlands slowly decompose into sand and that streams carry sand into the lowlands. He inferred that sandstone is composed of sand grains that originated by the erosion of ancient cliffs and mountains.

Hutton tried to deduce how much time was required to form a thick bed of sandstone. He studied sand grains slowly breaking away from rock outcrops. He watched sand bouncing down streambeds. Finally he traveled to beaches and river deltas where sand was accumulating. By estimating the time needed for thick layers of sand to accumulate on beaches, Hutton concluded that sandstone must be much older than human history. Hutton,

◆ **FIGURE 1.10** Hutton observed that a sandstone outcrop, like this cliff rising above the Escalante River, Utah, is composed of tiny round sand grains cemented together.

overwhelmed by the apparent magnitude of geological time, wrote:

> On us who saw these phenomena for the first time, the impression will not easily be forgotten. . . . We felt ourselves necessarily carried back to the time . . . when the sandstone before us was only beginning to be deposited, in the shape of sand and mud, from the waters of an ancient ocean. . . . The mind seemed to grow giddy by looking so far into the abyss of time. . . .
>
> We find no vestige of a beginning, no prospect of an end.

Hutton had no way to measure the magnitude of geologic time. However, modern geologists have learned that certain radioactive materials in rocks act as clocks to record the passage of time. Using these clocks and other clues embedded in Earth's crust, the moon, and in meteorites fallen from the Solar System, geologists estimate that Earth formed 4.6 billion years ago.

The primordial planet was vastly different from our modern world. There was no crust as we know it today, no oceans, and the diffuse atmosphere was vastly different from the modern one. There were no living organisms.

No one knows exactly when or how the first living organisms evolved, but we know that life existed at least as early as 3.8 billion years ago, 800 million years after the planet formed. For the following 3.3 billion years, life evolved slowly, and although some multicellular organisms developed, most of the biosphere consisted of single-celled organisms. Organisms rapidly became more complex, abundant, and varied about 544 million years ago. The dinosaurs flourished between 225 million and 65 million years ago.

Homo sapiens and our direct ancestors have been on Earth for 5 to 7 million years, only about 0.1 percent of the planet's history. In his book *Basin and Range*, John McPhee offers a metaphor for the magnitude of geologic time. If the history of Earth were represented by the old English measure of a yard—the distance from the king's nose to the end of his outstretched hand—all of human history could be erased by a single stroke of a file on his middle fingernail. ◆ Figure 1.11 summarizes Earth history in graphical form. Geologists routinely talk about events that occurred millions or even billions of years ago. If you live on the East Coast of North America, you may be familiar with the Appalachian Mountains, which started rising from a coastal plain nearly half a billion years ago. If you live in Seattle, the land you are standing on was an island in the Pacific between 80 and 100 million years ago, when dinosaurs walked Earth.

There are two significant consequences of the vast span of geologic time:

1. Events that occur slowly become significant. If a continent moves a few centimeters a year, the movement makes no noticeable alteration of Earth systems over decades or centuries. But over hundreds of millions of years, the effects are significant.
2. Improbable events occur regularly. The chances are great that a large meteorite won't crash into Earth tomorrow, or next year, or during the next century. But during the past 500 million years, several catastrophic impacts have occurred and they will probably occur sometime in the future.

Recall that James Hutton deduced that sandstone forms when rocks slowly decompose to sand, the sand is transported to lowland regions, and the grains cement together. This process occurs step by step—over many years. Hutton's conclusions led him to formulate a principle now known as **gradualism** or **uniformitarianism**. The principle states that geologic change occurs over long periods of time, by a sequence of almost imperceptible events. Hutton surmised that geologic processes operating today also operated in the past. Thus scientists can explain events that occurred in the past by observing slow changes occurring today. Sometimes this idea is summarized in the statement, "The present is the key to the past."

However, not all geologic change is gradual. William Whewell, another early geologist, argued that geologic change was sometimes rapid. He wrote that the geologic past may have "consisted of epochs of paroxysmal and catastrophic action, interposed between periods of comparative tranquility." Earthquakes and volcanoes are examples of catastrophic events, but Whewell additionally argued that occasionally huge catastrophes alter the course of Earth history. He couldn't give an example because they happen so infrequently that none had occurred within human history.

Today, geologists know that both Hutton's uniformitarianism and Whewell's **catastrophism** are correct. Over the great expanses of geologic time, slow, uniform processes alter Earth. In addition, infrequent catastrophic events radically modify the path of slow change.

Gradual Change in Earth History

Tectonic plates creep slowly across Earth's surface. Since the first steam engine was built 200 years ago, North America has migrated 8 meters westward, a distance a sprinter can run in 1 second. Thus, the movement of tectonic plates is too slow to be observed without sensitive instruments. However, over geologic time, this movement alters the shapes of ocean basins, forms lofty mountains and plateaus, generates earthquakes and volcanic eruptions, and affects our planet in many other ways.

Catastrophic Change in Earth History

Chances are small that the river flowing through your city will flood this spring, but if you live to be 100 years old, you will probably see a catastrophic flood. In fact, de-

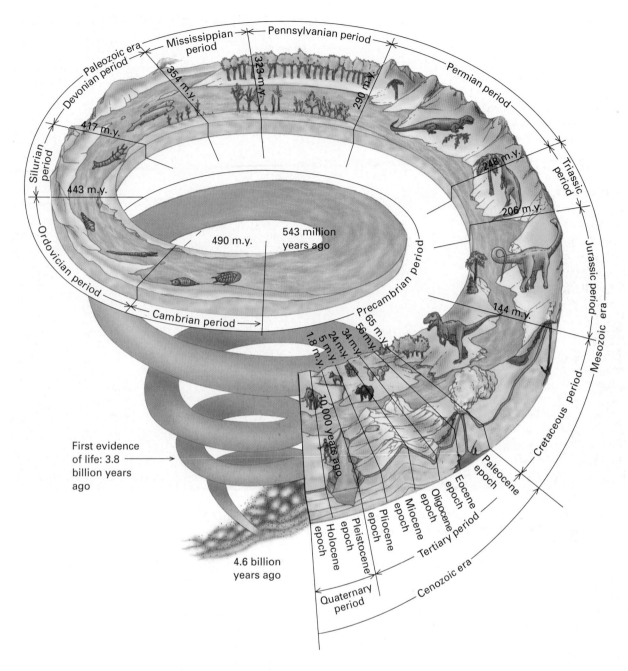

◆ **FIGURE 1.11** The geologic time scale is shown as a spiral to indicate the great length of time before multicellular organisms became abundant, about 543 million years ago.

structive floods occur every year, somewhere on the planet. In the 12 months prior to writing this book, three catastrophic geological events caused significant loss of human life: the tsunami in Southeast Asia (Chapter 7), Hurricane Katrina (Chapters 11 and 19), and the Pakistani/Kashmir earthquake (Chapter 7). But even these catastrophes are miniscule compared with events that have occurred in the past.

When geologists study the 4.6 billion years of Earth history, they find abundant evidence of catastrophic events that are highly improbable in a human lifetime or even in human history. For example, clues preserved in rock and sediment indicate that giant meteorites have smashed into our planet, vaporizing portions of the crust and spreading dense dust clouds over the sky (◆ Figure 1.12). Geologists have suggested that some meteorite impacts (much larger than that shown in Figure 1.12) have almost instantaneously driven millions of species into extinction. As another example, when the continental glaciers receded at the end of the last Ice Age, huge floods

◆ **FIGURE 1.12**   Fifty thousand years ago, a meteorite crashed into the northern Arizona desert, creating Meteor Crater. Catastrophic events have helped shape Earth history.

occurred that were massively larger than any floods in historic times. In addition, periodically throughout Earth history, catastrophic volcanic eruptions changed environmental conditions for life around the globe.

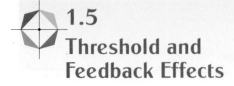

1.5
Threshold and Feedback Effects

Threshold Effects

Imagine that some process gradually warmed the atmosphere. Imagine further that the summer temperature of coastal Greenland glaciers was –1.0°C. If gradual change warmed the atmosphere by 0.2°C, the glaciers would not be affected appreciably because ice melts at 0°C and the summer temperature would be 0.8°C, still below freezing. But now imagine that gradual warming brought summer temperature to 0.2°C. At this point, the system is at a threshold. Another small increment of warming would raise the temperature to the melting point of ice. As soon as ice starts to melt, large changes

occur rapidly: glaciers recede and global sea level rises. A **threshold effect** occurs when the environment initially changes slowly (or not at all) in response to a small perturbation, but after the threshold is crossed, an additional small perturbation causes rapid and dramatic change.

If we don't understand a system thoroughly, it is easy to be deluded into a false sense of complacency. Imagine, in the case just described, that we didn't know the melting point of ice. If we observed that several years of incremental warming had no effect on the system, we might conclude that additional warming might also have no effect. But, in fact, the next small increment, which crosses the threshold, melts the ice.

Populations of plants, animals, and people also respond to thresholds. In 1912 scientists released 4 male and 21 female reindeer onto St. Paul Island in the Bering Sea. There were no predators on the island and hunting pressure was low, so the reindeer population increased dramatically. By 1938 approximately 2,000 reindeer grazed peacefully on the tundra vegetation of St. Paul Island. To an untrained eye, the introduction was successful because the reindeer were healthy and multiplying. But, the reindeer herd was overgrazing the available pasture—eating more food than was growing—heading toward a threshold. Everything was fine as long as the reindeer could move across the island to find new sources of food, but when the entire island became overgrazed, suddenly there was nothing to eat—and the reindeer starved. Between 1938 and 1950 almost the entire herd starved to death, until only eight emaciated animals remained (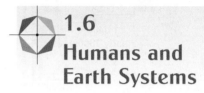 Figure 1.13).

A **feedback mechanism** occurs when a small perturbation affects another component of Earth systems, which amplifies the original effect, which perturbs the system even more, which leads to an even greater effect. The evolution of the Venusian climate is a perfect example of a feedback mechanism: Solar energy evaporated surface water. The water vapor trapped more heat in the atmosphere. The temperature rose and more water evaporated. When the water evaporated, dissolved carbon dioxide escaped into the atmosphere. This gaseous carbon dioxide trapped even more heat. The temperature rose further.

Scientists have observed many examples of feedback mechanisms in modern earth science. For example, in recent years, industrial emissions have increased the concentration of carbon dioxide in the atmosphere and the mean global atmospheric temperature has risen. In two separate studies published in September 2005, scientists showed that: (1) Higher air temperatures in England caused soil bacteria to flourish. The bacteria consumed more food, and respiration increased, releasing even more carbon dioxide into the air. (2) In a second study, extremely hot weather in France during the summer of 2003 led to a reduction of plant photosynthesis, with a consequent reduction of the uptake of carbon dioxide by plants. Thus atmospheric carbon dioxide levels increased.

In both studies, emissions of industrial carbon dioxide led to warming, but warming caused natural systems to release even more carbon dioxide, which presumably leads to additional warming. No one fully understands the eventual ramifications of this feedback loop (discussed further in Chapter 21).

Threshold and feedback mechanisms remind us that Earth systems often change in ways that are difficult to predict. These complexities provide the challenge and fascination of earth science.

1.6
Humans and Earth Systems

Ten thousand years ago, a mere eyeblink in Earth history, human population was low and technology was limited. Sparsely scattered bands hunted and gathered for their food, built simple dwellings, and warmed themselves with small fires. Human impact was not appreciably different from that of many other species on the planet. But with the invention of agriculture and the growth of cities, problems arose in some cultures. In his recent book, *Collapse*, Jared Diamond documents several case histories where people settled in a region, built a society, and consumed resources faster than they were naturally replaced. Then, like the reindeer on St. Lawrence Island, populations crashed. Thus at the height of the Mayan civilization between the third and the ninth century, the central cities supported from 1,800 to 2,600 people per square mile. (In comparison, Los Angeles County averaged 2,345 people per square mile in 2000.) Then, due to a variety of human and environmental factors, the cities were abandoned, the population crashed, and an estimated 90 to 95 percent of the Mayans died. In a similar manner the Anasazi civilizations in the American southwest disappeared in the thirteenth century and Viking colonies starved to death in Greenland. All three of these human tragedies were exacerbated by climate change and human-to-human interactions, such as war. Nevertheless, Diamond points out that after thresholds were reached, the populations didn't just decline back to a sustainable level, they collapsed catastrophically.

In mid-2005 there were 6.5 billion people on Earth (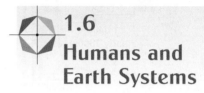 Figure 1.14). Farms, pasture lands, and cities covered 40 percent of the land area of the planet (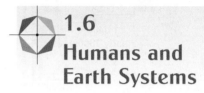 Figure 1.15). We diverted half of all the fresh water on the continents and controlled 50 percent of the total net terrestrial biological productivity. As the land and its waters have been domesticated and manipulated, other animal and plant species have perished. The species extinction

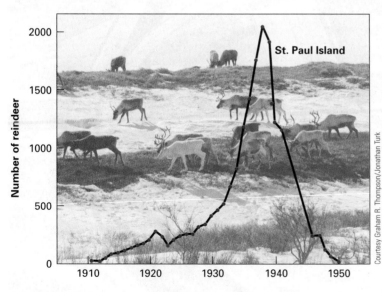

Courtesy Graham R. Thompson/Jonathan Turk

◆ **FIGURE 1.13** When reindeer were introduced onto predator-free St. Paul Island, the population rose dramatically for 30 years. Then when the food supply was exhausted, the population crashed, leaving only eight emaciated survivors.

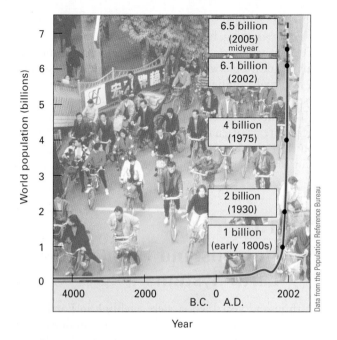

Data from the Population Reference Bureau

6.5 billion
(2005)
midyear

6.1 billion
(2002)

4 billion
(1975)

2 billion
(1930)

1 billion
(early 1800s)

World population (billions)

4000   2000   0   2002
B.C.   A.D.

Year

◈ **FIGURE 1.14** The human population has increased rapidly since the 1700s. For most of human history, there were fewer than one-half billion people on Earth. In mid-2005, 6.5 billion people inhabited our planet.

INTERACTIVE QUESTION: *Approximately how many years did it take for the human population to double from 2 billion to 4 billion? If that doubling rate were to continue, how many people would live on Earth in the year 2500? 3000?*

rate today is 1,000 times more rapid than it was before the Industrial Revolution. In fact, species extinction hasn't been this rapid since a giant meteorite smashed into Earth's surface 65 million years ago, leading to the demise of the dinosaurs and many other species.

Emissions from our fires, factories, electricity-generating plants, and transportation machines have raised the atmospheric carbon dioxide concentration to its highest level in 420,000 years. Since the Industrial Revolution, Earth's mean annual temperature has increased by 0.6°C.

Roughly 20 percent of the human population lives in wealthy industrialized nations where food, shelter, clean drinking water, efficient sewage disposal, and advanced medical attention are all readily available. At the same time, 1.5 billion people in the impoverished regions of the world survive on less than one dollar per day. One eighth to one half of the human population on this planet is malnourished, and 14 million infants or young children starve to death every year. If you were a member of a family in the Himalayan region of northern India, where you spent hours every day gathering scant firewood or cow dung for fuel, where meals consisted of an endless repetition of barley gruel and peas, where you brought your sheep and goats into the house during the winter so you could all huddle together for warmth, your view of the planet would be different from what it is now.

Courtesy of Graham R. Thompson/Jonathan Turk

◈ **FIGURE 1.15** This cloverleaf near Denver, Colorado, is but one example of the global conversion of natural ecosystems into developed landscapes.

FOCUS ON

Hypothesis, Theory, and Law

On an afternoon field trip, you may find several different types of rocks or watch a river flow by. But you can never see the rocks or river as they existed in the past or as they will exist in the future. Yet a geologist might explain how the rocks formed millions or even a few billion years ago and might predict how the river valley will change with time.

Scientists not only study events that they have never observed and never will observe, but they also study objects that can never be seen, touched, or felt. In this book we describe the center of Earth, 6,400 kilometers beneath our feet, and distant galaxies, billions of light-years away, even though no one has ever visited these places and no one ever will.

Much of science is built on inferences about events and objects outside the realm of direct experience. An inference is a conclusion based on available information, thought, and reason. How certain are we that a conclusion of this type is correct?

Scientists develop an understanding of the natural world according to a set of guidelines known as the **scientific method**, which involves three basic steps: (1) observation, (2) forming a preliminary conclusion or hypothesis, and (3) testing the hypothesis and developing a theory.

Observation

All modern science is based on observation. Recall that James Hutton observed that ocean currents and rivers transported and deposited sand. He observed that the sand accumulates slowly, layer by layer. Then he saw cliffs of layered sandstone hundreds of meters high. Observations of this kind are the starting point of science.

Forming a Hypothesis

A scientist tries to organize observations to recognize patterns. Hutton noted that the sand layers deposited along the coast look just like the layers of sand in the sandstone cliffs. Starting with these observed facts, he then concluded that the thick layers of sandstone had been deposited in an ancient ocean or river delta. He further inferred that, because layers of sand are deposited slowly, the thick layers of sandstone must have accumulated over a long time. Finally, he took one more step of logic to hypothesize that Earth is old.

A **hypothesis** is a tentative explanation built on strong supporting evidence. Once a scientist or group of scientists proposes a hypothesis, others test it by comparison with additional observations and experiments. If the hypothesis explains some of the facts but not all of them, it must be altered, or if it cannot be changed satisfactorily, it must be discarded and a new hypothesis developed.

Testing the Hypothesis and Forming a Theory

If a hypothesis explains new observations as they accumulate and is not substantively contradicted, more and more people come to accept it. As the tentative nature of a hypothesis is gradually dispelled, it is elevated to a **theory**. Theories differ widely in form and content, but all obey four fundamental criteria:

1. A theory must be based on a series of confirmed observations or experimental results.
2. A theory must explain all relevant observations or experimental results.
3. A theory must not contradict any relevant observations or other scientific principles.
4. A theory must be internally consistent. Thus, it must be built in a logical manner so that the conclusions do not contradict any of the original premises.

For example, the theory of plate tectonics states that the outer layer of Earth is broken into a number of plates that move horizontally relative to one another. As you will see in Chapter 6, this theory is supported by many observations and seems to have no major inconsistencies.

Many theories cannot be absolutely proven. For example, even though scientists are just about certain that their image of atomic structure is correct, no one has or ever will watch an individual electron travel in its orbit. Therefore, our interpretation of atomic structure is called atomic theory.

In some instances, observations are so universal that there is no doubt of their verity. A **law** is a statement of how events always occur under given conditions. It is considered to be factual and correct. For example, the law of gravity states that all objects are attracted to one another in direct proportion to their masses. We cannot conceive of any contradiction to this principle, and none has been observed. Hence, the principle is called a law. Modern science is in such constant flux and our universe is so complex, that very few laws are formulated anymore.

> **FOCUS QUESTION**
> Obtain a copy of a news article in a weekly news magazine. Underline the facts with one color pencil and the author's/authors' opinions with another. Did the author/authors follow the rules for the scientific method outlined here to reach a conclusion?

Yet regardless of whether we are rich or poor, our survival is interconnected with the flow of energy and materials among Earth systems. We affect Earth, and at the same time Earth affects us; it is the most intimate and permanent of marriages.

Predictions for the future are, by their very nature, uncertain. More than 200 years ago, in 1798, the Reverend Thomas Malthus argued that the human population was growing exponentially while food production was increasing much more slowly. As a result, humankind was facing famine, "misery and vice." But technological advancements led to an exponential increase in food production. Today food is plentiful in many regions, while at the same time serious shortages exist, especially in parts of Africa.

Many scientists are deeply concerned about our future: John Holdren, professor at the John F. Kennedy School of Government, argues that the problem is

> not that we are running out of energy, food, or water but that we are running out of environment—that is, running out of the capacity of air, water, soil and biota to absorb, without intolerable consequences for human well-being, the effects of resource extraction, transportation, transformation, and use.

Three broad trends are currently facing humanity:

1. The human population is large and continuing to increase—but the rate of increase is slowing. According to one prediction, population will stabilize at 7 billion by 2020. But other demographers predict that 14 billion people will walk the earth by 2100.

2. At the same time, extreme poverty is decreasing. In India and China, with one-third of the world's population, economic output and standards of living are rising rapidly. When people have money, they consume more resources.

3. As both population and per-capita consumption rise, pollution and pressure on Earth's resources continue to rise. Table 1.1 lists 12 major environmental impacts facing human populations today.

Professor Diamond, quoted earlier, argues that we are in a "horse race" with the environment. One horse, representing increased human consumption and the resulting environmental degradation and resource depletion, is galloping toward an unsustainable society with a possible catastrophic collapse. The other horse, powered by improved technology, increased awareness, and declining rates of population growth, leads us toward a happy, healthy, sustainable society. Which horse will win?

This book will not provide answers—because there are none—but it will provide a basis for understanding Earth processes so you can evaluate and appreciate issues concerning our stewardship of the planet.

Table 1.1   Major Environmental Problems Facing Humanity Today

1. Destruction of natural habitats such as forests, wetlands, prairies, and coral reefs at an accelerating rate.
2. Loss of ocean fisheries.
3. Loss of biodiversity through destruction of wild species.
4. Soil erosion.
5. Depletion of fossil fuels and firewood resources.
6. Water shortages.
7. Humans are approaching the ceiling where we are projected to be using most of the world's photosynthetic capacity by the middle of this century.
8. Toxic chemicals.
9. Destruction of ecosystems by the introduction of invasive species.
10. Changes in the composition of the atmosphere: greenhouse effect and depletion of ozone layer.
11. Increase in human population.
12. Increases in wealth in less-developed countries are dramatically increasing the human per-capita resource consumption.

Reprinted by permission from Jared Diamond, *Collapse: How Societies Choose to Fail or Succeed* (New York: Viking, 2005).

SUMMARY

On Venus, a runaway greenhouse effect produced torridly hot temperatures; on Mars, a reverse greenhouse effect plunged the planet into a deep freeze, while temperate conditions have prevailed on Earth.

Earth consists of four spheres: The **geosphere** is composed of a dense, hot, central **core**, surrounded by a large **mantle** that comprises 80 percent of Earth's volume, with a thin, rigid **crust**. The crust and mantle are composed of rock while the core is composed of iron and nickel. The **hydrosphere** is mostly ocean water. Most of Earth's freshwater is locked in glaciers. Most of the liquid freshwater lies in **ground water** reservoirs; streams, lakes, and rivers account for only 0.01 percent of the planet's water. The **atmosphere** is a mixture of gases, mostly nitrogen and oxygen. Earth's atmosphere supports life and regulates climate. The **biosphere** is the thin zone that life inhabits.

A **system** is composed of interrelated, interacting components.

Earth is about 4.6 billion years old; life formed at least 3.8 billion years ago; abundant multicellular life evolved about 544 million years ago, and hominids have been on this planet for a mere 5 to 7 million years. The principle of gradualism or **uniformitarianism** states that geologic change occurs over a long period of time by a sequence of almost imperceptible events. In contrast, **catastrophism** postulates that geologic change occurs mainly during infrequent catastrophic events. A **threshold effect** occurs when the environment initially changes slowly (or not at all) in response to a small perturbation, but after the threshold is crossed, an additional small perturbation causes rapid and dramatic change. A **feedback mechanism** occurs when a small initial perturbation affects another component of Earth systems, which amplifies the original effect, which perturbs the system even more, and which leads to an even greater effect.

Due to our sheer numbers and technological prowess, humans have become a significant engine of change in Earth systems.

Key Terms

geosphere 4	lithosphere 6	catastrophism 10
hydrosphere 4	tectonic plates 6	threshold effect 12
atmosphere 4	ground water 7	feedback mechanism 13
biosphere 4	system 8	scientific method 15
core 6	ecosystem 8	hypothesis 15
mantle 6	gradualism 10	theory 15
crust 6	uniformitarianism 10	law 15

For Review

1. Explain how water vapor and carbon dioxide influenced the evolution of the climate and atmosphere on Venus, Earth, and Mars.

2. List and briefly describe each of Earth's four spheres.

3. List the three major layers of Earth. Which is/are composed of rock, which is/are metallic? Which is the largest; which is the thinnest?

4. List six types of reservoirs that collectively contain most of Earth's water.

5. What is ground water? Where in the hydrosphere is it located?

6. What two gases compose most of Earth's atmosphere?

7. How thick is Earth's atmosphere?

8. Briefly discuss the size and extent of the biosphere.

9. Define a system and explain why a systems approach is useful in earth science.

10. Briefly explain the statement: "Matter can be recycled, but energy cannot."

11. Summarize the first and second laws of thermodynamics.

12. How old is Earth? When did life first evolve? How long have humans and their direct ancestors been on this planet?

13. Compare and contrast uniformitarianism and catastrophism. Give an example of each type of geologic change.

14. Briefly explain threshold and feedback effects.

15. Briefly outline the magnitude of human impact on the planet.

For Discussion

1. What cautionary warnings can you deduce from your knowledge of the evolution of climate and atmospheric composition on Venus, Earth, and Mars?

2. Only 0.64 percent of Earth's water is fresh and liquid—the rest is salty seawater or is frozen in glaciers. What are the environmental implications of such a small proportion of fresh water?

3. List five ways that organisms, including humans, change Earth. What kinds of Earth processes are unaffected by humans and other organisms?

4. Twelve generic types of interactions are possible among the four spheres. These are:

   (a) Changes in the solid Earth perturb the hydrosphere.

   (b) Changes in the hydrosphere perturb the solid Earth.

   (c) Changes in the solid Earth perturb the atmosphere.

   (d) Changes in the atmosphere perturb the solid Earth.

   (e) Changes in the solid Earth perturb the biosphere.

   (f) Changes in the biosphere perturb the solid Earth.

   (g) Changes in the hydrosphere perturb the atmosphere.

   (h) Changes in the atmosphere perturb the hydrosphere.

   (i) Changes in the hydrosphere perturb the biosphere.

   (j) Changes in the biosphere perturb the hydrosphere.

   (k) Changes in the atmosphere perturb the biosphere.

   (l) Changes in the biosphere perturb the atmosphere.

   Write an example to illustrate each possibility.

5. A windmill produces energy from the wind, which is free. Is this a contradiction of the second law of thermodynamics? Explain.

6. Give an example of a threshold or feedback effect in science, politics, human relationships, or any field you can think of.

UNIT 1

EARTH MATERIALS AND TIME

CHAPTER
2

Minerals

Jose Manuel Sanchis Calvete/CORBIS

Excellent quartz crystals are fairly abundant and easy to find. Quartz makes up about 12 percent of the Earth's crust.

ermiculite is a commercially valuable mineral that resembles black mica. Gold miners discovered a large deposit of vermiculite in Libby, Montana, in 1881. In 1919, Edward Alley discovered that the mineral expands—or "pops"—when heated, to form a lightweight, fireproof, and inexpensive material that is widely used for packing, insulation, and as a soil additive. The popping process converts water in the vermiculite crystals to steam, and the little steam explosions expand the mineral so that it looks like tiny accordions and is so light that the expanded crystals will float on water (◆ Figure 2.1). In the 1920s, the Zonolite Company formed and began mining Libby vermiculite. In 1963, W. R. Grace & Company, one of the nation's largest chemical companies, bought the Zonolite mine and operated it until closing it in 1990. While in operation, the mine produced up to 80 percent of the world's supply of vermiculite. Unfortunately, the vermiculite from the Libby mine is contaminated with naturally occurring asbestos—a mineral that is a known carcinogen, or cancer-causing substance, and a cause of other serious lung diseases.

The mine itself is located approximately 6 miles from the city of Libby (◆ Figure 2.2). During the mine's years of operation, workers loaded the vermiculite onto trucks that carried it into Libby for processing. In town, two separate expansion (or "popping") facilities were in operation at different times over the years. During processing, the plants heated the vermiculite to 350°C (about 600°F) to expand the vermiculite crystals. During the heating and expansion process, asbestos fibers entered the air. One of the popping plants was located next to a baseball field and near children's playgrounds. Waste materials from the mining and processing were used for landfill and surface covering on playgrounds and ball fields in town. While airborne asbestos levels in Libby are now low, during the mining period, asbestos levels were much higher. In fact, ambient air concentrations of up to 15 times the current Occupational Safety and Health Administration limits were reported for downtown Libby between 1960, when airborne asbestos levels were first monitored in Libby, and when the plant closed in 1990.

A **B**

◆ **FIGURE 2.1**   **(A)** Vermiculite before popping resembles the black mica called biotite. **(B)** The popping process creates steam explosions that expand the vermiculite so that it resembles tiny accordions and is light enough to float on water.

◁▷ **FIGURE 2.2** The Zonolite vermiculite mine is located about 6 miles from Libby, Montana. Asbestos contained in the vermiculite has caused cancer and other lung diseases among the miners and residents of Libby, where the vermiculite was processed.

United States Environmental Protection Agency

For a 20-year period beginning in the 1980s, the human death rate from asbestosis (a lung disease caused by inhaling asbestos) among Libby residents was approximately 40 to 60 times higher than that of average U.S. cities. Deaths from lung and related cancers were also higher than national averages. Since 1960, 192 deaths and 375 related illnesses in Libby, with a total population of only about 2,600 people, have been blamed on the asbestos contamination.

Federal health officials are now trying to determine whether the vermiculite mined in Libby and shipped across the country has spread asbestos contamination to other areas. They are particularly concerned that the problem has spread to towns where the vermiculite was shipped for processing into soil additives and home insulation. The Environmental Protection Agency (EPA) has identified 240 sites in 40 states where the vermiculite was shipped from Libby. At least 22 of those sites require some kind of cleanup, which the EPA will oversee. Facing huge lawsuits from sick and dying miners, their families, and from other asbestos-sickened residents of Libby, W. R. Grace & Company declared bankruptcy in 2002.

In February 2005, W. R. Grace and seven of its current or former top officials were indicted on charges that they knowingly put Libby workers and residents in danger through exposure to vermiculite ore contaminated with asbestos from the company's mine.

Vermiculite and asbestos are minerals. Minerals are the building blocks of rock, and therefore the geosphere is composed of minerals. Minerals make up the rocks beneath your feet, the soil that supports plants, and the deep rock of Earth's mantle. Any thorough study of Earth must include an understanding of minerals. But it is not sufficient to study minerals isolated from the rest of the planet. This book emphasizes system interactions among the four spheres: geosphere, atmosphere, hydrosphere, and biosphere. In the example just cited, minerals of the geosphere caused harm to humans, who are part of the biosphere. Other systems interactions occur whenever minerals, rock, or soil interact with air, water, or living organisms—in other words, very frequently.

Most minerals in their natural settings are harmless to humans and other species. In addition, many resources that are essential to modern life are produced from rocks and minerals: iron, aluminum, gold and all other metals, concrete, fertilizers, coal, and

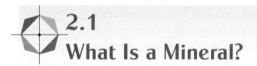

nuclear fuels. Both oil and natural gas are recovered from reservoirs in rock. The discoveries of methods to extract and use these Earth materials profoundly altered the course of human history. The Stone Age, the Bronze Age, and the Iron Age are historical periods named for the rock and mineral resources that dominated the development of civilization during those times. The Industrial Revolution occurred when humans discovered how to convert the energy stored in coal into useful work. The automobile age occurred because humans found vast oil reservoirs in shallow rocks of Earth's crust and built engines to burn the refined fuels. The computer age relies on the movement of electrons through wafer-thin slices of silicon, germanium, and other materials that are also crystallized from minerals.

However, some minerals become harmful to humans and animals when crushed to dust during mining, quarrying, road building, and other activities. Some dissolve to release chemical elements such as lead, sulfur, and arsenic that are poisonous to plants and animals, including humans. A few minerals, like asbestos, are potent environmental hazards because of other properties. In this chapter, we will consider the nature of minerals and then return to a discussion of minerals as natural resources and as environmental hazards. ■

2.1
What Is a Mineral?

Pick up any rock and look at it carefully. You will probably see small, differently colored specks, such as those in granite (◊▶ Figure 2.3). Each speck is a mineral. In this photo, the white grains are the mineral feldspar, the black ones are biotite, and the glassy-gray ones are quartz. A rock is a mixture of minerals. Most rocks contain two to five abundant minerals plus minor amounts of several others. A few rocks are made of only one mineral.

Although we can define minerals as the building blocks of rocks, such a definition does not tell us much about the nature of minerals. More precisely, a **mineral** is a naturally occurring inorganic solid with a definite chemical composition and a crystalline structure. *Chemical composition and crystalline structure are the two most important properties of a mineral:* They distinguish any mineral from all others. Before discussing them, however, let's briefly consider the other properties of minerals that this definition describes.

Natural occurrence

A synthetic diamond can be identical to a natural one, but it is not a true mineral, according to our definition, because a mineral must form by natural processes. Like diamonds, most gems that occur naturally also can be manufactured by industrial processes. Natural gems are more highly valued than manufactured ones. For this

reason, jewelers should always tell their customers whether a gem is natural or artificial, by prefacing the name of a manufactured gem with the term *synthetic*.

Inorganic

Organic substances are composed mostly of carbon that is chemically bonded to hydrogen. Inorganic compounds do not contain carbon–hydrogen bonds. Although organic compounds can be produced in laboratories and by industrial processes, plants and animals create most of Earth's organic material. The tissue of most plants and animals is organic. When these organisms die, they decompose to form other organic substances. Both coal and oil form by the decay of plants and animals and are not minerals because of their organic properties. In addition, oil is not a mineral because it is not solid and has neither a crystalline structure nor a definite chemical composition.

Some material that organisms produce is not organic. For example, limestone, one of the most common sedimentary rocks, is usually composed of the shells of dead corals, clams, and similar marine organisms. Shells, in turn, are made of the mineral calcite. Although organisms produce calcite, the calcite is a mineral: an inorganic solid that formed naturally and that has a definite chemical composition and crystalline structure.

Solid

All minerals are solids. Thus, ice is a mineral, but neither water nor water vapor is a mineral.

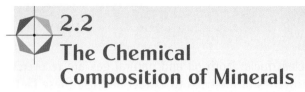

2.2
The Chemical Composition of Minerals

Minerals are the fundamental building blocks of rocks, but what are minerals made of? Minerals and all other Earth materials are composed of chemical elements. An **element** is a fundamental component of matter and cannot be broken into simpler particles by ordinary chemical processes. Most common minerals consist of a small number—usually two to five—of different chemical elements.

A total of 88 elements occur naturally in Earth's crust. However, 8 elements—oxygen, silicon, aluminum, iron, calcium, magnesium, potassium, and sodium—make up more than 98 percent of the crust (Table 2.1). Each element is represented by a one- or two-letter symbol, such as O for oxygen and Si for silicon. In nature, most chemical elements have either a positive or negative electrical charge. For example, oxygen has a charge of negative 2 (−2) and silicon has a charge of plus 4 (+4). An atom with an electrical charge, whether it is positive or negative, is called an **ion**. Ions with opposite charges are attracted to each other like the positive end of a magnet attracts the negative end. See the Focus On box for a discussion of atoms, ions, and chemical bonds in minerals.

Recall that a mineral has a definite chemical composition. A substance with a definite chemical composition is made up of chemical elements that are combined in definite proportions. Therefore, the composition can be expressed as a chemical formula, which is written by combining the symbols of the individual elements. A few minerals, such as gold and silver, consist of only a single element. Their chemical formulas, respectively, are Au (the symbol for gold) and Ag (the symbol for silver).

Feldspar   Biotite   Quartz

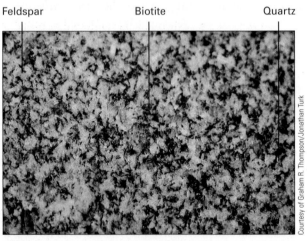

◆ **FIGURE 2.3** **(A)** The Bugaboo Mountains of British Columbia are made of granite. **(B)** Each of the differently colored grains in this close-up photo of granite is a different mineral. The white grains are feldspar, the black ones are biotite, and the glassy-gray ones are quartz. The granite sample shown here is about 4 centimeters across.

ThomsonNOW

CLICK Thomson Interactive to work through an activity on mineral formation through Mineral Labs.

Table 2.1	The Eight Most Abundant Chemical Elements in the Earth's Crust*			
Element	Symbol	Weight Percent	Atom Percent	Volume Percent†
Oxygen	O	46.60	62.55	93.8
Silicon	Si	27.72	21.22	0.9
Aluminum	Al	8.13	6.47	0.5
Iron	Fe	5.00	1.92	0.4
Calcium	Ca	3.63	1.94	1.0
Sodium	Na	2.83	2.64	1.3
Potassium	K	2.59	1.42	1.8
Magnesium	Mg	2.09	1.84	0.3
Total		98.59	100.00	100.00

*Abundances are given in percentages by weight, by numbers of atoms, and by volume.
†These numbers will vary somewhat as a function of the ionic radii chosen for the calculations.

SOURCE: From *Principles of Geochemistry* by Brian Mason and Carleton B. Moore. © 1982 by John Wiley & Sons, Inc.

Most minerals, however, are made up of two to five essential elements. For example, the formula of quartz is SiO_2: it consists of one atom of silicon (Si) for every two of oxygen (O). Quartz from anywhere in the Universe has that exact composition. If it had a different composition, it would be some other mineral. The compositions of some minerals, such as quartz, do not vary by even a fraction of a percent. The compositions of other minerals vary slightly, but the variations are limited.

The 88 elements that occur naturally in Earth's crust can combine in many ways to form many different minerals. In fact, more than 3,500 minerals are known. However, the 8 abundant elements commonly combine in only a few ways. As a result, only 9 **rock-forming minerals** (or mineral "groups") make up most rocks of Earth's crust. We will describe the rock-forming minerals in Section 2.5.

ThomsonNOW™

CLICK Thomson Interactive to work through an activity on atomic behavior.

2.3

Crystals: The Crystalline Nature of Minerals

Every mineral has a crystalline structure, and therefore every mineral is a crystal. A **crystal** is any substance whose atoms are arranged in a regular, periodically repeated pattern. The mineral halite (common table salt) has the composition NaCl: one sodium ion (Na^+) for every chlorine ion (Cl^-). ◆ Figure 2.4A is an exploded view of the ions in halite. ◆ Figure 2.4B is more realistic, showing the ions in contact. In both sketches the sodium and chlorine ions alternate in orderly rows and columns intersecting at right angles. This arrangement is the **crystalline structure** of halite.

Think of a familiar object with an orderly, repetitive pattern, such as a brick wall. The rectangular bricks repeat themselves throughout the wall. As a result, the whole wall also has the shape of a rectangle or some modification of a rectangle. In every crystal, a small group of atoms, like a single brick in a wall, repeats itself over and over. This small group of atoms is called a **unit cell**. The unit cell for halite is shown in Figure 2.4A. If you compare Figures 2.4A and 2.4B, you will notice that the simple halite unit cell repeats throughout the halite crystal.

Most minerals initially form as tiny crystals that grow as layer after layer of atoms (or ions) are added to their surfaces. A halite crystal might grow, for example, as salty seawater evaporates from a tidal pool. At first, a tiny grain might form, similar to the sketch of halite in Figure 2.4B. This model shows a halite crystal containing 125 atoms; it would be only about one millionth of a millimeter long on each side, far too small to see with the naked eye. As evaporation continued, more and more sodium and chlorine ions would precipitate onto the faces of the growing crystal. When other minerals form as molten volcanic lava cools, atoms also bond in concentric layers to the growing crystals.

The shape of a well-formed crystal like that of halite in ◆ Figure 2.4C is determined by the shape of the unit cell and the manner in which the crystal grows. For example, it is obvious from ◆ Figure 2.5A that the stacking of small cubic unit cells can produce a large cubic crystal. ◆ Figure 2.5B shows that different stacking of the same cubes can produce an eight-sided crystal, called an octahedron. Halite can crystallize as a cube or as an

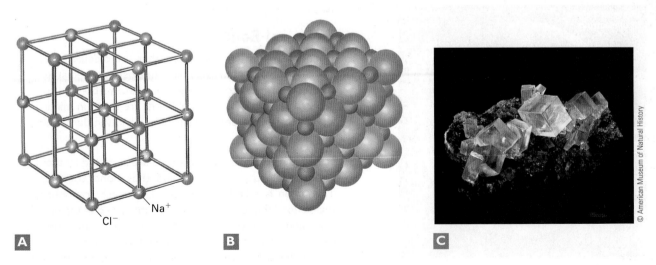

© American Museum of Natural History

A **B** **C**

◆ **FIGURE 2.4** (**A and B**) The orderly arrangement of sodium and chlorine ions in halite. (**C**) Halite crystals. The crystal model in (A) is exploded so that you can see into it; the ions are actually closely packed, as in (B). Note that ions in (A) and (B) form a cube, and the crystals in (C) are also cubes.

FOCUS ON

Elements, Atoms, and Chemical Bonds

This Focus On box is designed to serve as an introduction to chemical concepts for a reader who has not studied chemistry, or as a review for someone who is familiar with chemistry.

To understand the chemical compositions of minerals, let's first consider the nature of the chemical elements that make up Earth. An **atom** is the basic unit of an element. An atom is tiny; the diameter of the average atom is about 10^{-10} meters (1/10,000,000,000 meters). A penny contains about 1.5×10^{22} (15,000,000,000,000, 000,000,000) atoms. An atom consists of an even smaller, very dense, positively charged center called a **nucleus**, surrounded by negatively charged electrons

An **electron** is a tiny particle that orbits the nucleus, but not in a clearly defined path like that of Earth around the Sun. Scientists usually portray electron orbits as a cloud of negative charge surrounding the nucleus.

The nucleus is made up of several kinds of particles; the two largest are positively charged **protons** and uncharged **neutrons**. A neutral atom contains equal numbers of protons and electrons. The positive and negative charges balance each other so that a neutral atom has no overall electrical charge.

Electrons concentrate in shells around the nucleus. Each shell can hold a certain number of electrons. An atom is most stable when its outermost shell is completely filled with electrons. But in their neutral states,

most atoms do not have a filled outer shell. Some atoms may fill their outer shells by acquiring extra electrons until the shell becomes full. Alternatively, an atom may give up electrons until the outermost shell becomes empty. In this case, the next lower shell, which is full, becomes the outermost shell. Thus, an atom can become stable with a full outer shell by either acquiring or releasing the proper number of electrons.

When an atom loses one or more electrons, its protons outnumber its electrons and it therefore has a positive charge. If an atom gains one or more extra electrons, it is negatively charged. A charged atom is called an **ion**.

A positively charged ion is a **cation**. In nature, all of the abundant elements, except oxygen, release electrons to become cations. For example, each potassium atom (K) loses one electron to form a cation with a charge of +1 (K^+). Each silicon atom loses four electrons, forming a cation with a +4 charge, Si^{4+}. In contrast, oxygen gains two extra electrons to acquire a −2 charge, O^{2-}. Atoms with negative charges are called **anions**.

Atoms and ions rarely exist independently. Instead, they unite to form **compounds**. The forces that hold atoms and ions together to form compounds are called **chemical bonds**.

Most minerals are compounds. When ions bond together to form a mineral, they do so in proportions so that the total number of negative charges exactly balances the total number of positive charges. Thus, minerals are always electrically neutral. For example, the mineral quartz contains one (4+) silicon cation for every two (2−) oxygen anions.

Chemical Bonds

Chemical bonds join atoms together. A molecule is the smallest particle of matter that can exist in a free state; it can be a single atom, or a group of atoms bonded together. Four types of chemical bonds are found in minerals: ionic, covalent, and metallic bonds, and van der Waals forces.

IONIC BONDS. Cations and anions are attracted by their opposite electronic charges and thus bond together. This union is called an ionic bond. An ionic compound (made up of two or more ions) is neutral because the positive and negative charges balance each other. For example, when sodium and chlorine form an ionic bond, the sodium atom loses one electron to become a cation and chlorine gains one to become an anion. When they combine, the +1 charge balances the −1 charge.

Concentric shells of electrons

(Negative charge)

(Positive charge)

Nucleus

Electrons concentrate in layers, or shells, around the nucleus of an atom.

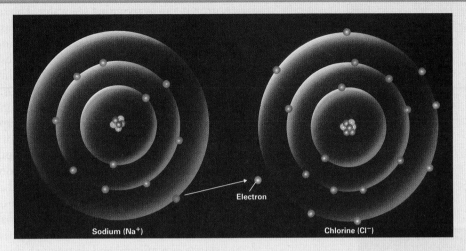

When sodium and chlorine atoms combine, sodium loses one electron, becoming a cation, Na^+. Chlorine acquires the electron to become an anion, Cl^-.

COVALENT BONDS. A covalent bond develops when two or more atoms share their electrons to produce the effect of filled outer electron shells. For example, carbon needs four electrons to fill its outermost shell. It can achieve this by forming four covalent bonds with four adjacent carbon atoms. It "gains" four electrons by sharing one with another carbon atom at each of the four bonds. Diamond consists of a three-dimensional network of carbon atoms bonded into a network of tetrahedra, similar to the silicate framework structure of quartz. The strength and homogeneity of the bonds throughout the crystal make diamond the hardest of all minerals.

In most minerals, the bonds between atoms are partly covalent and partly ionic. The combined characteristics of the different bond types determine the physical properties of those minerals.

METALLIC BONDS. In a metallic bond, the outer electrons are free; that is, they are not associated with particular atoms. The metal atoms sit in a "sea" of outer-level electrons that are free to move from one atom to another. That arrangement allows the nuclei to pack together as closely as possible, resulting in the characteristic high density of metals and metallic minerals, such as pyrite. Because the electrons are free to move through the entire crystal, metallic minerals are excellent conductors of electricity and heat.

VAN DER WAALS FORCES. Weak electrical forces called van der Waals forces also bond molecules together. These weak bonds result from an uneven distribution of electrons around individual molecules, so that one portion of a molecule may have a greater density of negative charge while another portion has a partial positive charge. Because van der Waals forces are weak, minerals in which these bonds are important, such as talc and graphite, tend to be soft and cleave easily along planes of van der Waals bonds.

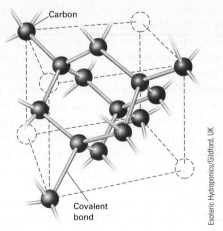

Carbon atoms in diamond form a tetrahedral network similar to that of quartz.

Esoteric Hydroponics/Glidford, UK

> **FOCUS QUESTION**
> Why do some minerals, such as native gold, silver, and graphite, conduct electricity, whereas others, such as quartz and feldspar, do not? Discuss relationships among other physical properties of minerals and the types of chemical bonds found in these minerals.

ThomsonNOW

CLICK Thomson Interactive to work through an activity on the definition of atoms and crystals.

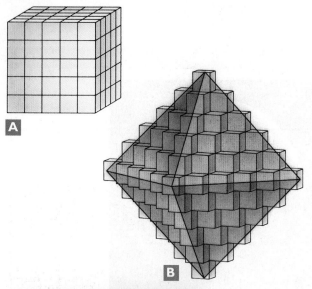

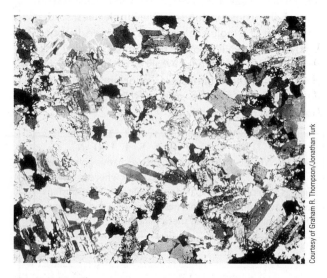

FIGURE 2.5 Both a cubic crystal **(A)** and an octahedron **(B)** can form by different kinds of stacking of identical cubes. The same cubes can also stack to form other shapes **(C)**.

A. E. Seaman Mineral Museum, Michigan Technological University/G.W. Robinson photograph

Courtesy of Graham R. Thompson/Jonathan Turk

FIGURE 2.6 A photomicrograph of a thin slice of granite. When crystals grow simultaneously, they commonly interlock and show no characteristic habit. To make this photo, a thin slice of granite was cut with a diamond saw, glued to a microscope slide, and ground to a thickness of 0.02 mm. Most minerals are transparent when such thin slices are viewed through a microscope.

octahedron. All minerals consist of unit cells stacked face to face as in halite, but not all unit cells are cubic.

A **crystal face** is a flat surface that develops if a crystal grows freely in an uncrowded environment, such as halite growing in evaporating sea water. The sample of halite in Figure 2.4C has well-developed crystal faces. When a mineral grows freely like this, it commonly forms symmetrical crystal with perfectly flat faces that reflect light like a mirror. In nature, however, mineral grains often impede the growth of adjacent crystals. For this reason, minerals rarely show perfect development of crystal faces. Figure 2.6 is a photomicrograph (a photo taken through a microscope) of a thin slice of granite in which the crystals fit like pieces of a jigsaw puzzle. This interlocking texture developed because some crystals grew around others as molten magma cooled and solidified.

To summarize the characteristics of minerals, each mineral has a definite chemical composition and a crystalline structure. Those two properties define a substance as a mineral and distinguish one mineral from all others. In addition, minerals form naturally and are inorganic solids.

2.4
Physical Properties of Minerals

How does a geologist identify a mineral in the field? Chemical composition and crystal structure distinguish each mineral from all others. For example, halite always consists of sodium and chlorine in a one-to-one ratio, with the atoms arranged in a cubic fashion. But if you pick up a crystal of halite, you cannot see the atoms. You could identify a sample of halite by measuring its chemical composition and crystal structure by laboratory procedures, but such analyses are expensive and time consuming. Instead, geologists commonly identify minerals by visual recognition, and confirm the identification with simple tests.

Most minerals have distinctive appearances. Once you become familiar with common minerals, you will recognize them just as you recognize any familiar object. For example, an apple just looks like an apple, even though apples come in many colors and shapes. In the same way, quartz looks like quartz to a geologist. The color and shape of quartz may vary from sample to sample, but it still looks like quartz. Some minerals, however, look like others, so that their physical properties must be examined further to make a correct identification. For example, halite can look like calcite, quartz, and several other minerals, but halite tastes salty and scratches easily with a knife blade, two characteristics that distinguish it from the other minerals. Geologists commonly use properties such as crystal habit, cleavage, hardness, color, and luster to identify minerals.

Crystal Habit

Crystal habit is the characteristic shape of a mineral and the manner in which aggregates of crystals grow. If a crystal grows freely, it develops a characteristic shape that the arrangement of its atoms controls, as in the cubes of halite shown in ◆ Figure 2.4C. ◆ Figure 2.7 shows three common minerals with different crystal habits. Some minerals occur in more than one habit. For example, ◆ Figure 2.8A shows quartz with a prismatic (pencil-shaped) habit, and ◆ Figure 2.8B shows massive quartz.

Cleavage

Cleavage is the tendency of some minerals to break along flat surfaces. The surfaces are planes of weak bonds in the crystal. Some minerals, such as mica, have one set of parallel cleavage planes (◆ Figure 2.9). Others have two, three, or even four different sets, as shown in ◆ Figure 2.10. Some minerals, like the micas, have excellent cleavage.

◆ **FIGURE 2.8** **(A)** Prismatic quartz grows as elongated crystals. **(B)** Massive quartz shows no characteristic shape.

You can peel sheet after sheet from a mica crystal as if you were peeling layers from an onion. Others have poor cleavage. Many minerals, such as quartz, have no cleavage at all because they have no planes of weak bonds. The number of cleavage planes, the quality of cleavage, and the angles between cleavage planes all help in mineral identification.

A flat surface created by cleavage and a flat crystal face can appear identical. However, a cleavage surface is duplicated when a crystal is broken, whereas a crystal face is not. So, if you are unsure of which type of flat surface you are looking at, break the sample with a hammer, unless, of course, you want to save it.

Fracture

Fracture is the manner in which a mineral breaks other than along planes of cleavage. Many minerals fracture into characteristic shapes. Conchoidal fracture creates smooth, curved surfaces (◆ Figure 2.11). It is characteristic of quartz and olivine. Some minerals break into splintery or fibrous fragments. Most fracture into irregular shapes.

Hardness

Hardness is the resistance of a mineral to scratching and is one of the most commonly used properties for identifying a mineral. It is easily measured and is a fundamental property of each mineral because it is controlled by the bond strength between the atoms in the mineral. Geologists commonly gauge hardness by attempting to scratch a mineral with a knife or other object of known hardness. If the blade scratches the mineral, the mineral is softer than the knife. If the knife cannot scratch the mineral, the mineral is harder.

To measure hardness more accurately, geologists use a scale based on 10 minerals, numbered 1 through 10

◆ **FIGURE 2.7** **(A)** Equant garnet crystals have about the same dimensions in all directions. **(B)** Asbestos is fibrous. **(C)** Kyanite forms bladed crystals.

◁▷ **FIGURE 2.9** Mica has a single, perfect cleavage plane. This large crystal is the variety of mica called muscovite.

(Table 2.2). Each mineral is harder than those with lower numbers on the scale, so 10 (diamond) is the hardest and 1 (talc) is the softest. The scale is known as the **Mohs hardness scale**, after F. Mohs, the Austrian mineralogist who developed it in the early nineteenth century.

The Mohs hardness scale shows that a mineral scratched by quartz but not by orthoclase has a hardness between 6 and 7 (Table 2.2). Because the minerals of the Mohs scale are not always handy, it is useful to know the hardness values of common materials. A fingernail has a hardness of slightly more than 2, a pocketknife blade slightly more than 5, window glass about 5.5, and a steel file about 6.5. If you practice with a knife and the minerals of the Mohs scale, you can develop a "feel" for minerals with hardnesses of 5 and under by how easily the blade scratches them.

When testing hardness, it is important to determine whether the mineral has actually been scratched by the object or whether the object has simply left a trail of its own powder on the surface of the mineral. To check, simply rub away the powder trail and feel the surface of the mineral with your fingernail for the groove of the scratch. Fresh, unweathered mineral surfaces must be used in hardness measurements because weathering often produces a soft rind on minerals.

Specific Gravity

Specific gravity is the weight of a substance relative to that of an equal volume of water. If a mineral weighs 2.5 times as much as an equal volume of water, its specific gravity is 2.5. You can estimate a mineral's specific gravity simply by hefting a sample in your hand. If you practice with known minerals, you can develop a feel for specific gravity. Most common minerals have specific gravities of about 2.7. Metals have much greater specific gravities; for example, gold has the highest specific gravity of all minerals, 19. Lead is 11.3, silver is 10.5, and copper is 8.9.

Color

Color is the most obvious property of a mineral, and it is often used in identification. But color can be unreliable because small amounts of chemical impurities and imperfections in crystal structure can dramatically alter color. For example, corundum (Al_2O_3) is normally a

◁▷ **FIGURE 2.10** Some minerals have more than one cleavage plane. **(A)** Feldspar has two cleavages intersecting at right angles. **(B)** Calcite has three sets of cleavage planes. Because the three sets do not intersect at right angles, these cleaved crystals have the appearance of deformed boxes. **(C)** Fluorite has four cleavage planes. Each cleavage face has a parallel counterpart on the opposite side of the crystal, and thus, perfectly cleaved fluorite forms a double pyramid.

Breck P. Kent

◆ **FIGURE 2.11** Quartz typically fractures along smoothly curved surfaces, called conchoidal fractures. This sample is smoky quartz.

American Museum of Natural History

◆ **FIGURE 2.12** Pyrite, or fool's gold, has a metallic luster. In this sample, a well-formed pyrite crystal rests on other minerals.

Table 2.2   Minerals of the Mohs Hardness Scale

Minerals of Mohs Scale	Common Objects
1. Talc	
2. Gypsum	Fingernail
3. Calcite	Copper penny
4. Fluorite	
5. Apatite	Knife blade
	Window glass
6. Orthoclase	Steel file
7. Quartz	
8. Topaz	
9. Corundum	
10. Diamond	

cloudy, translucent, brown or blue mineral. Addition of a small amount of chromium can convert corundum to the beautiful, clear, red gem known as ruby. A small quantity of iron or titanium turns corundum into the striking blue gem called sapphire. Quartz occurs in many colors, including white, clear, black, purple, and red, as a result of tiny amounts of impurities and minor defects in the perfect ordering of atoms.

Streak

Streak is the color of the fine powder of a mineral. It is observed by rubbing the mineral across a piece of unglazed porcelain known as a "streak plate." Many minerals leave a streak of powder with a diagnostic color on the plate. Streak is commonly more reliable than the color of the mineral itself for identification.

For example, the mineral hematite (iron oxide) can occur both as a dull red mineral or as a shiny black form that closely resembles black mica—but both types leave the same red powder on a streak plate.

Luster

Luster is the manner in which a mineral reflects light. A mineral with a metallic look, irrespective of color, has a metallic luster. Pyrite is a yellowish mineral with a metallic luster (◆ Figure 2.12). As a result, it looks like gold and is commonly called fool's gold. The luster of non-metallic minerals is usually described by self-explanatory words such as *glassy*, *pearly*, *earthy*, and *resinous*.

Other Properties

Properties such as reaction to acid, magnetism, radioactivity, fluorescence, and phosphorescence can be characteristic of specific minerals. Calcite and some other carbonate minerals dissolve rapidly in acid, releasing visible bubbles of carbon dioxide gas. Minerals containing radioactive elements such as uranium emit radioactivity that can be detected with a scintillometer. Fluorescent materials emit visible light when they are exposed to ultraviolet light. Phosphorescent minerals continue to emit light after the external stimulus ceases.

2.5
Mineral Classes and the Rock-Forming Minerals

Geologists classify minerals according to their chemical elements (Table 2.3). For example, the **silicates** all contain silicon and oxygen; the **carbonates** contain carbon and oxygen; and the **sulfides** contain sulfur. The **native elements** are a small class of minerals, including pure gold and silver, which consist of only a single element. Although more than 3,500 minerals are known in Earth's

Table 2.3 Important Mineral Groups

Group	Member	Formula	Economic Use
Oxides	Hematite	Fe_2O_3	Ore of iron
	Magnetite	Fe_3O_4	Ore of iron
	Corundum	Al_2O_3	Gemstone, abrasive
	Ice	H_2O	Solid form of water
	Chromite	$FeCr_2O_4$	Ore of chromium
Sulfides	Galena	PbS	Ore of lead
	Sphalerite	Zns	Ore of zinc
	Pyrite	FeS_2	Fool's gold
	Chalcopyrite	$CuFeS_2$	Ore of Copper
	Bornite	Cu_5FeS_4	Ore of copper
	Cinnabar	HgS	Ore of mercury
Sulfates	Gypsum	$CaSO_4 \cdot 2H_2O$	Plaster
	Anhydrite	$CaSO_4$	Plaster
	Barite	$BaSO_4$	Drilling mud
Native elements	Gold	Au	Electronics, jewelry
	Copper	Cu	Electronics
	Diamond	C	Gemstone, abrasive
	Sulfur	S	Sulfa drugs, chemicals
	Graphite	C	Pencil lead, dry lubricant
	Silver	Ag	Jewelry, photography
	Platinum	Pt	Catalyst
Halides	Halite	$NaCl$	Common salt
	Fluorite	CaF_2	Used in steel making
	Sylvite	KCl	Fertilizer
Carbonates	Calcite	$CaCO_3$	Portland cement
	Dolomite	$CaMg(CO_3)_2$	Portland cement
	Aragonite	$CaCO_3$	Portland cement
Hydroxides	Limonite	$FeO(OH) \cdot nH_2O$	Ore of iron, pigments
	Bauxite	$Al(OH)_3 \cdot nH_2O$	Ore of aluminum
Phosphates	Apatite	$Ca_5(F,Cl,OH)(PO_4)_3$	Fertilizer
	Turquoise	$CuAl_6(PO_4)_4(OH)_8 \cdot 4H_2O$	Gemstone
Silicates	(Silicate minerals make up 92 percent of the Earth's crust. Figure 2.15 summarizes the rock-forming minerals.)		

crust, only a small number—between 50 and 100—are common or valuable, and only 9 "rock-forming" minerals make up most of the crust. These nine are important to geologists simply because they are the most common and abundant minerals. Seven of the rock-forming minerals are silicate minerals. The other two, calcite and dolomite, are carbonates. (These mineral groups, and others, are summarized in Figures 2.14, 2.15, and 2.16.)

Silicates

Silicates make up about 92 percent of Earth's crust. They are so abundant for two reasons. First, silicon and oxygen are the two most plentiful elements in the crust. Second, silicon and oxygen combine readily.

To understand silicate minerals, remember four principles:

1. Every silicon atom in Earth's crust surrounds itself with four oxygen atoms. The bonds between silicon and its four oxygens are very strong.

2. The silicon atom and its four oxygen atoms form a pyramid-shaped structure, called the **silicate tetrahedron**, with silicon in the center and oxygens at the four corners (◆ Figure 2.13). The silicate tetrahedron is the fundamental building block of all silicate minerals.

3. Most silicate tetrahedra combine with additional elements. Those elements are mostly the other common cations (in the Focus On Box in this

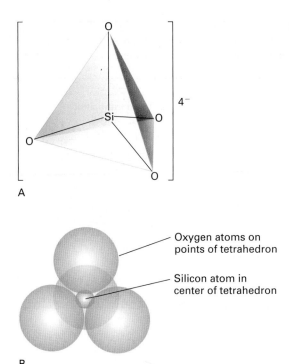

A

B

◈ **FIGURE 2.13** The silicate tetrahedron consists of one silicon atom surrounded by four oxygen atoms. It is the fundamental building block of all silicate minerals. **(A)** A schematic representation. **(B)** A proportionally accurate model.

chapter, we explain that atoms with electrical charges are called "ions"; a positively charged ion is a "cation," and a negatively charged ion is an "anion") in Earth's crust: aluminum, iron, calcium, potassium, sodium, and magnesium. Quartz is the only common silicate that contains only silicon and oxygen.

4. Silicate tetrahedra commonly link together by sharing oxygen atoms to form chains, sheets, or three-dimensional networks as shown in ◈ Figure 2.14.

Rock-Forming Silicates

The silicate minerals fall into five groups, based on five different ways in which tetrahedra link together. Each group contains at least one of the rock-forming minerals. ◈ Figures 2.14 and 2.15 display these groups as well as the most common silicate minerals in each group.

◈ Figure 2.16 shows that feldspar alone makes up about half of Earth's crust. Feldspars are divided into two main groups: plagioclase contains calcium, and alkali feldspar contains potassium or sodium (called alkali elements) instead of calcium. Orthoclase is a common alkali feldspar and is abundant in granite and other rocks of the continents. Rocks of the oceanic crust contain about 50 percent plagioclase. Plagioclase and orthoclase often look alike and can be difficult to tell apart.

Quartz comprises 12 percent of crustal rocks. It is widespread and abundant in most continental rocks but rare in oceanic crust and the mantle. Quartz is the only common silicate mineral that contains no cations other than silicon; it is pure SiO_2. It has a ratio of one 4+ silicon for every two 2− oxygens, so the positive and negative charges neutralize each other perfectly without the addition of other cations.

Pyroxene makes up another 11 percent of the crust. The basalt rock of ocean crust is made up almost entirely of two minerals, plagioclase feldspar and pyroxene. Basalt and granite are described in Chapters 3 and 8.

Basalt commonly contains 1 or 2 percent olivine, but olivine is otherwise uncommon in Earth's crust. However, it makes up a large proportion of mantle rocks, and thus is an abundant and important mineral.

Amphibole, mica, and clay are the other rock-forming silicate minerals. Amphiboles and micas are common in granite and many other types of continental rocks but are rare in oceanic crust. Clay forms when rain and atmospheric gases chemically decompose rocks of all types, and thus clay is common in soils. Clay minerals make up the rock called shale, which is the most abundant of all sedimentary rocks and is described in Chapter 3. Clay is the main raw material in the manufacture of ceramics such as plates and teacups.

Carbonates

Carbonate minerals are much less common than silicates in Earth's crust, but they are important rock-forming minerals because they form sedimentary rocks that cover large regions of every continent, as described in the following chapter. The shells and other hard parts of most marine organisms such as clams, oysters, and corals are made of carbonate minerals. These shells and skeletons accumulate to form the abundant sedimentary rock called limestone, described in Chapter 3. Calcite ($CaCO_3$) forms limestone, and dolomite($CaMg(CO_3)_2$), makes up a similar rock also called dolomite, or dolostone (◈ Figure 2.15). Limestone is mined as a raw ingredient of cement.

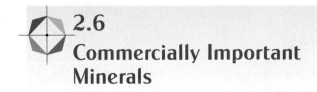

2.6
Commercially Important Minerals

Many minerals are commercially important even though they are not abundant. Our industrial society depends on metals, such as iron, copper, lead, zinc, gold, and silver. **Ore minerals** are minerals from which metals or other elements can be profitably recovered. A few, such as native gold and native silver, are composed of a single element. However, most metals are chemically bonded to other elements. Iron is commonly bonded to oxygen.

Class	Arrangement of SiO₄ tetrahedron	Unit composition	Mineral examples
(A) Independent tetrahedra	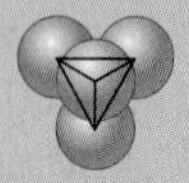	$(SiO_4)^{4-}$	Olivine: The composition varies between Mg_2SiO_4 and Fe_2SiO_4
(B) Single chains		$(SiO_3)^{2-}$	Pyroxene: The most common pyroxene is augite, $Ca(Mg, Fe, Al)(Al, Si)_2O_6$
(C) Double chains		$(Si_4O_{11})^{6-}$	Amphibole: The most common amphibole is hornblende, $NaCa_2(Mg, Fe, Al)_5(Si, Al)_8O_{22}(OH)_2$
(D) Sheet silicates	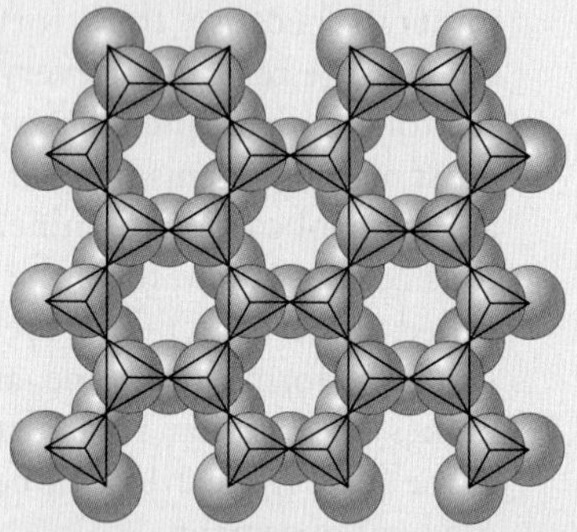	$(Si_2O_5)^{2-}$	Mica, clay minerals, chlorite, e.g.: muscovite, $KAl_2(Si_3Al)O_{10}(OH)_2$
(E) Framework silicates	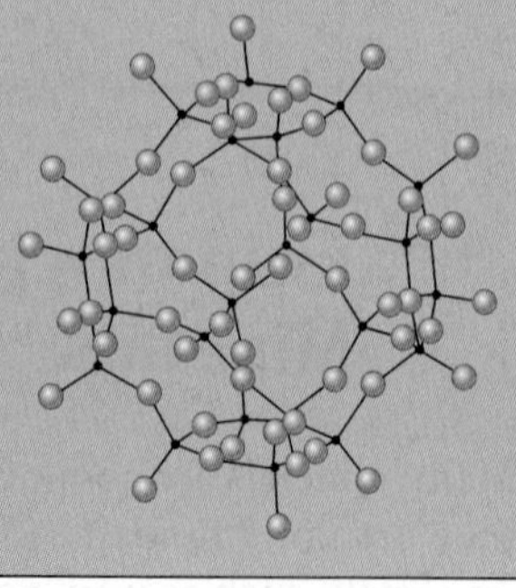	SiO_2	Quartz: SiO_2 Feldspar: As an example, potassium feldspar is $KAlSi_3O_8$

ThomsonNOW ◀▶ **ACTIVE FIGURE 2.14** The five silicate structures are based on sharing of oxygen atoms among silicate tetrahedra. **(A)** Independent tetrahedra share no oxygen atoms. **(B)** In single chains, each tetrahedron shares two oxygens with adjacent tetrahedra, forming a chain. **(C)** A double chain is a pair of single chains that are cross-linked by additional oxygen sharing. **(D)** In the sheet silicates, each tetrahedron shares three oxygens with adjacent tetrahedra. **(E)** A three-dimensional silicate framework shares all four oxygens of each tetrahedron. INTERACTIVE QUESTION: *Discuss why mica has one perfect cleavage plane. Why does quartz have no cleavage?*

◇ **FIGURE 2.15** The rock-forming minerals. **(A)** Olivine. **(B)** Pyroxene. **(C)** Amphibole. **(D)** Black biotite is one common type of mica. White muscovite (FIGURE 2.9) is the other. **(E)** Clay. **(F)** Feldspar, represented here by orthoclase feldspar. **(G)** Quartz. **(H)** Calcite. **(I)** Dolomite.

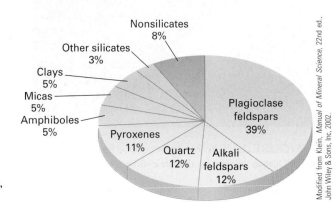

◇ **FIGURE 2.16** The silicate minerals compose 92 percent of Earth's crust. Feldspar alone makes up about 50 percent of the crust, and pyroxene and quartz constitute another 23 percent.

Modified from Klein, *Manual of Mineral Science,* 22nd ed., John Wiley & Sons, Inc, 2002.

◆ **FIGURE 2.17** Galena is the most important ore of lead.

◆ **FIGURE 2.19** Sapphire is one of the most costly precious gems.

Copper, lead, and zinc are commonly bonded to sulfur to form sulfide ore minerals (◆ Figure 2.17).

Industrial minerals are commercially important although they are not considered "ore" because they are mined for purposes other than the extraction of metals. Halite is mined for table salt, and gypsum is mined for plaster and sheetrock. Apatite and other phosphorus minerals are sources of the phosphate fertilizers crucial to modern agriculture. Limestone is the raw material of cement. Native sulfur, used to manufacture sulfuric acid, insecticides, fertilizer, and rubber, is mined from the craters of dormant and active volcanoes, where it is deposited from gases emanating from the vents (◆ Figure 2.18).

A **gem** is a mineral that is prized primarily for its beauty, although some gems, such as diamonds, are also used industrially. Depending on its value, a gem can be either precious or semiprecious. Precious gems include diamond, emerald, ruby, and sapphire (◆ Figure 2.19). Several varieties of quartz, including amethyst, agate, jasper, and tiger's eye, are semiprecious gems. Garnet, olivine, topaz, turquoise, and many other minerals sometimes occur as attractive semiprecious gems (◆ Figure 2.20).

◆ **FIGURE 2.20** Topaz, a colorless crystal, is a popular semiprecious gem.

When you look at a lofty mountain or a steep cliff, you might not immediately think about the tiny mineral grains that form the rocks. Yet minerals are the building blocks of Earth. In addition, some minerals provide the basic resources for our industrial civilization.

As we will see throughout this book, Earth's surface and the human environment change as minerals react with the air and water, with other minerals, and with living organisms.

◆ **FIGURE 2.18** Yellow native sulfur is forming today in the vent of Ollague volcano on the Chile–Bolivia border.

2.7
Harmful and Dangerous Rocks and Minerals

Most rocks and minerals in their natural states are harmless to humans and other organisms. That is not surprising because all of us evolved among the minerals and rocks that make up Earth, and species would not survive if they were poisoned by their surroundings. A few rocks and minerals, however, are harmful and dangerous. Asbestos is one such mineral; it is a powerful carcinogen—a cancer-causing substance. Some rocks and minerals emit radon gas, a radioactive carcinogen. Other rocks and minerals contain toxic elements such as mercury, lead, arsenic, or sulfur.

In nature, most environmentally hazardous rocks and minerals are buried beneath the surface, where they are unavailable to plants and animals. Natural weathering and erosion expose them so slowly that they do little harm. Most minerals that contain toxic metals or other elements such as lead, mercury, and arsenic, are also relatively insoluble. In their natural environments they weather so slowly that they release the toxic materials in low concentrations. However, if pollution controls are inadequate, mining, milling, and smelting can concentrate and release hazardous natural materials at greatly accelerated rates, poisoning humans and other organisms.

Silicosis and Black Lung

Most common minerals such as feldspar and quartz are harmless in their natural states in solid rock. However, if these minerals are ground to dust, they can enter the lungs and cause serious and even fatal inflammation and scarring of the lung tissue, called silicosis. In advanced cases, the lungs become inflamed and may fill with fluid, causing severe breathing difficulty and low blood oxygen levels. On-the-job exposure to silica dust can occur in mining, stone cutting, quarrying, building and road construction, working with abrasives, sand blasting, and other occupations and hobbies. Intense exposure to silica may produce silicosis in a year or less, but it usually takes at least 10 or 15 years of exposure before symptoms develop. Silicosis has become less common since the Occupational Safety and Health Administration (OSHA) instituted regulations requiring the use of protective equipment that limits the amount of silica dust inhaled. Coal dust, inhaled by coal miners in large quantities before federal laws required dust suppression in coal mines, has similar effects and causes a disease called black lung.

Asbestos, Asbestosis, and Cancer

Asbestos is an industrial name for a group of minerals that crystallize as long, thin fibers (◆ Figure 2.21). One type is a sheet silicate mineral called chrysotile, which has a crystal structure and composition similar to those of the micas and which forms tangled, curly fibers. The other type includes four similar varieties of amphibole that crystallize as straight, sharply pointed needles. Chrysotile has been the more valuable and commonly used type of asbestos because the fibers are flexible, tough, and can be woven into fabric.

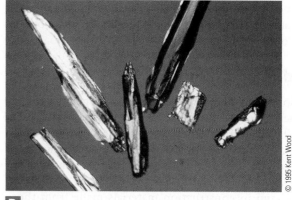

◆ **FIGURE 2.21** **(A)** One form of asbestos occurs as long, curly fibers. **(B)** The other occurs as short, sharply pointed needles. Both types cause cancer.

Asbestos is commercially valuable because it is flameproof, chemically inert, and extremely strong. For example, chrysotile fibers are eight times stronger than a steel wire of equivalent diameter. Asbestos has been used to manufacture brake linings, fireproof clothing, insulation, shingles, tile, pipe, and gaskets but now is allowed only in brake pads, shingles, and pipe.

In the early 1900s, asbestos miners and others who worked with asbestos learned that prolonged exposure to the fibers caused asbestosis. Later, in the 1950s and 1960s, studies showed that asbestos also causes lung cancer and other forms of cancer. One reason that so much time passed before scientists recognized the cancer-causing properties of asbestos is that the disease commonly does not develop until decades after exposure to asbestos.

Although it is not clear how asbestos fibers cause cancer, it seems that the shapes of the crystals play an important role. Statistical studies also show that the sharp, pointed amphibole asbestos is a more potent carcinogen than the flexible chrysotile fibers. In response to growing awareness of the health effects of asbestos, the Environmental Protection Agency (EPA) banned its use in building construction in 1978. However, the ban did not address the issue of what should be done with the asbestos already installed. In 1986, Congress passed a ruling called the Asbestos Hazard Emergency Response Act, requiring that all schools be inspected for asbestos. Public response has resulted in hasty programs to remove asbestos from schools and other buildings at a cost of billions of dollars. But what is the real level of hazard?

Most asbestos in buildings is the less potent chrysotile, woven into cloth or glued into a tight matrix, and often the surface has been further stabilized by painting. Therefore, the fibers are not free to blow around. The levels of airborne asbestos in most buildings are no higher than those in outdoor air. Some scientists argue that asbestos insulation poses no health danger if left alone, but when the material is removed, it is disturbed and asbestos dust escapes. Not only are workers endangered, but airborne asbestos persists in the building for months after completion of the project.

Radon and Cancer

Radon is one of a series of elements formed by the radioactive decay of uranium. Uranium occurs naturally in small concentrations in several minerals and in all types of rock, but it concentrates in two abundant rocks, granite and shale, that are described in the following chapter. It is also found in soil that formed from granite and shale, and in construction materials made from those rocks, such as concrete and concrete blocks. Radon is itself radioactive, and it decays into other radioactive elements. Because of their radioactivity, radon and its decay products are carcinogenic, and because it is a gas, we breathe radon into our lungs.

Radon seeps from the ground into homes and other buildings, where it concentrates in indoor air and causes an estimated 5,000 to 20,000 cancer deaths per year among Americans. The risk of dying from radon-caused lung cancer in the United States is about 0.4 percent over a lifetime, much greater than the risk of dying from cancer caused by asbestos, pesticides, or other air pollutants and nearly as high as the risk of dying in an auto accident, from a fall, or in a fire at home.

Americans are not all exposed to equal amounts of radon. Some homes contain very low concentrations of the gas; others have high concentrations. The variations in concentration are due to two factors: geology and home ventilation. Geology is important because some types of rocks, such as granite and shale, contain high concentrations of uranium and radon, and others contain relatively little.

As radon forms by slow radioactive decay of uranium in bedrock, soil, or construction materials, it seeps into the basement of a home and circulates throughout the house. Radon concentrations are highest in poorly ventilated homes built on granite or shale, on soil derived from these rocks, and in homes constructed with concrete and concrete blocks containing these types of rocks. The highest home radon concentrations ever measured were found in houses built on the Reading Prong,

a uranium-rich body of granite extending from Reading, Pennsylvania, through northern New Jersey and into New York. The air in one home in this area contained 700 times more radon than the EPA "action level"—the concentration at which the Environmental Protection Agency recommends that corrective measures be taken to reduce the amount of radon in indoor air.

People should ask two questions in regard to radon hazards: "What is the radon concentration in my home?" and "If it is high, what can be done about it?" Because radon is radioactive, it can be measured with a simple detector available at most hardware stores or from local government agencies for about $25. If the detector indicates excessive radon, three types of solutions can be implemented. The first is to extend a ventilation duct from the basement to the outside of the house. This solution prevents basement air from circulating through the house and also prevents radon from accumulating in the basement. The second solution is to ventilate the house so that indoor air is continually refreshed. However, this method allows hot air to escape and thus increases heating bills. A third solution is to pump outside air into the house to keep indoor air at a slightly higher pressure than the outside air. This positive pressure prevents gas from seeping from soil or bedrock into the basement.

It is impossible to avoid exposure to radon completely because it is everywhere, in outdoor air as well as in homes and other buildings. But it is relatively easy and inexpensive to minimize exposure and thus avoid a significant cause of cancer.

Acid Mine Drainage and Heavy Metals Contamination

Sulfide ore minerals, described in Section 2.6, are combinations of metals such as lead, zinc, copper, cadmium, mercury, and other heavy metals with sulfur. Some contain arsenic or other toxic elements in addition to the metals. These minerals are mined for their metals, which are essential to modern industrial societies.

However, mining and refining these minerals can create serious air and water pollution problems. When sulfide ore minerals are mined or refined without adequate pollution control, sulfur escapes into streams, ground water, and the atmosphere, where it forms hydrogen sulfide and sulfuric acid. Waterborne sulfides poison aquatic organisms and atmospheric sulfur compounds contribute to acid precipitation (◆ Figure 2.22).

Other toxic elements such as lead, mercury, cadmium, and arsenic can escape from mine wastes and smelters into the atmosphere and water. In addition, smelting and refining converts some poisonous elements, such as mercury, lead, and arsenic, from relatively insoluble forms to highly soluble, and therefore highly toxic, forms. Most of the sulfur and toxic metals can be removed by pollution-control devices, which are required by law in the United States. These topics are discussed further in Chapters 5, 12, and 17.

◆ **FIGURE 2.22** Acid mine drainage poisons aquatic life and stains stream beds and banks. Potosi, Bolivia.

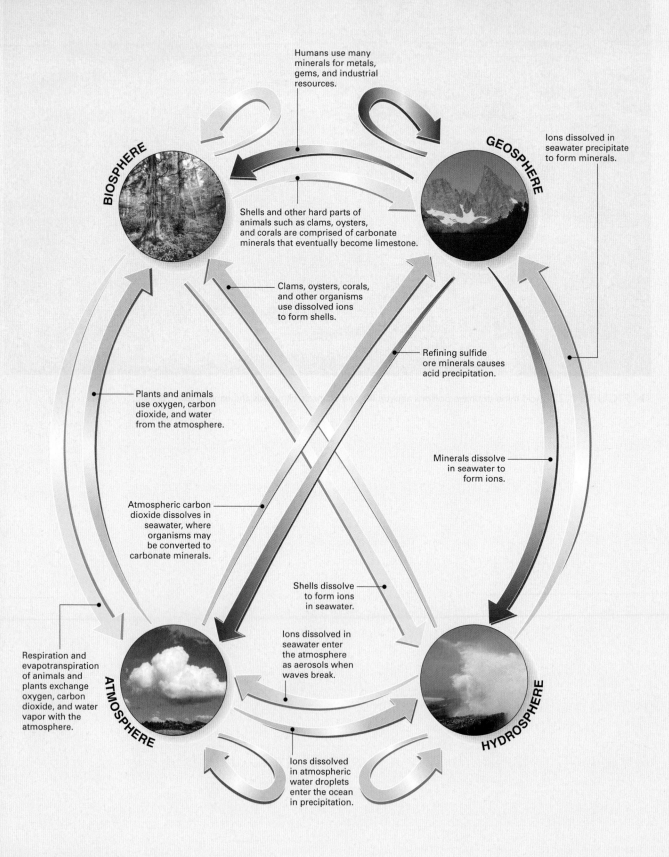

Humans use many minerals for metals, gems, and industrial resources.

BIOSPHERE

GEOSPHERE

Ions dissolved in seawater precipitate to form minerals.

Shells and other hard parts of animals such as clams, oysters, and corals are comprised of carbonate minerals that eventually become limestone.

Clams, oysters, corals, and other organisms use dissolved ions to form shells.

Refining sulfide ore minerals causes acid precipitation.

Plants and animals use oxygen, carbon dioxide, and water from the atmosphere.

Minerals dissolve in seawater to form ions.

Atmospheric carbon dioxide dissolves in seawater, where organisms may be converted to carbonate minerals.

Shells dissolve to form ions in seawater.

Respiration and evapotranspiration of animals and plants exchange oxygen, carbon dioxide, and water vapor with the atmosphere.

ATMOSPHERE

Ions dissolved in seawater enter the atmosphere as aerosols when waves break.

HYDROSPHERE

Ions dissolved in atmospheric water droplets enter the ocean in precipitation.

ALTHOUGH MINERALS MAKE UP the solid Earth—the geosphere—the elements that compose minerals cycle continuously among all four of the major Earth systems. For example, rain and soil moisture decompose rocks, slowly dissolving minerals and transferring their ions from the geosphere to the hydrosphere. Streams then carry the dissolved ions to the sea, where marine animals such as clams and oysters extract some of the ions from sea water to form their shells, incorporating the materials into the biosphere. Waves may whip the sea surface into foamy breakers, and wind carries some of the dissolved elements in the sea foam into the atmosphere. Rain then returns the ions to the land, where they may crystallize to form soil minerals, or back to the sea, where they may precipitate as evaporate minerals. Humans mine and refine some minerals for fertilizers critical to modern agriculture, transferring the materials from the geosphere to the biosphere and the hydrosphere.

In these ways, even the minerals of the seemingly permanent geosphere make their ways through the four major Earth realms. Exchanges of materials among the geosphere, the hydrosphere, the biosphere, and the atmosphere supply necessary nutrients to soil, plants, and animals, including humans, and provide necessary resources for life on Earth. ■

SUMMARY

Minerals are the substances that make up rocks. A mineral is a naturally occurring, inorganic solid with a definite **chemical composition** and a **crystalline structure**. Each mineral consists of chemical **elements** bonded together in definite proportions, so that its chemical composition can be given as a **chemical formula**. The crystalline structure of a mineral is the orderly, periodically repeated arrangement of its atoms. A **unit cell** is a small structural and compositional module that repeats itself throughout a crystal. The shape of a crystal is determined by the shape and arrangement of its unit cells. Every mineral is distinguished from others by its chemical composition and crystal structure.

Most common minerals are easily recognized and identified visually. Identification is aided by observing a few physical properties, including **crystal habit**, **cleavage**, **fracture**, **hardness**, **specific gravity**, **color**, **streak**, and **luster**.

Although more than 3,500 minerals are known in Earth's crust, only the nine rock-forming minerals are abundant in most rocks. They are feldspar, quartz, pyroxene, amphibole, mica, clay, olivine, calcite, and dolomite. The first seven on this list are silicates; their structures and compositions are based on the **silicate tetrahedron**, in which a silicon atom is surrounded by four oxygen atoms to form a pyramid-shaped structure. Silicate tetrahedra link together by sharing oxygens to form the basic structures of the silicate minerals. The silicates are the most abundant minerals because silicon and oxygen are the two most abundant elements in Earth's crust and bond together readily to form the silicate tetrahedron. Two **carbonate minerals**, calcite and dolomite, are also sufficiently abundant to be called rock-forming minerals. **Ore minerals**, **industrial minerals**, and **gems** are important for economic reasons. Most minerals and rocks are environmentally safe in their natural states, but some can release environmentally hazardous materials when they are mined, milled, or smelted.

Key Terms

mineral   23
element   24
ion   24
rock-forming mineral   25
crystal   25
crystalline structure   25
unit cell   25
atom   26
nucleus   26
electron   26
proton   26
neutron   26

cation   26
anion   26
compound   26
chemical bond   26
crystal face   28
crystal habit   29
cleavage   29
fracture   29
hardness   29
Mohs hardness scale   30
specific gravity   30
color   30

streak   31
luster   31
silicate   31
carbonate   31
sulfide   31
native elements   31
silicate tetrahedron   32
ore mineral   33
industrial mineral   36
gem   36

The Rock-Forming Minerals

feldspar	amphibole	olivine
quartz	mica	calcite
pyroxene	clay minerals	dolomite

For Review

1. What properties distinguish minerals from other substances?

2. Explain why oil and coal are not minerals.

3. What does the chemical formula for quartz, SiO_2, tell you about its chemical composition? What does $KAlSi_3O_8$ tell you about orthoclase feldspar?

4. What is an atom? An ion? A cation? An anion? What roles do they play in minerals?

5. What is a chemical bond? What role do chemical bonds play in minerals?

6. Every mineral has a crystalline structure. What does this mean?

7. What are the factors that control the shape of a well-formed crystal?

8. What is a crystal face?

9. What conditions allow minerals to grow well-formed crystals? What conditions prevent their growth?

10. List and explain minerals' physical properties that are most useful for identification.

11. Why do some minerals have cleavage and others do not? Why do some minerals have more than one set of cleavage planes?

12. Why is color often an unreliable property for mineral identification?

13. List the rock-forming minerals. Why are they called *rock-forming*? Which are silicates? Why are so many of them silicates?

14. Draw a three-dimensional view of a single silicate tetrahedron. Draw the five arrangements of tetrahedra found in the rock-forming silicate minerals. How many oxygen ions are shared between adjacent tetrahedra in each of the five configurations?

15. Make a table with two columns. List the basic silicate structures in the left column. In the right column, list one or more examples of rock-forming minerals for each structure.

16. Explain how mining can release harmful or poisonous materials from rocks that were benign in their natural environment.

For Discussion

1. Diamond and graphite are two minerals with identical chemical compositions, pure carbon (C). Diamond is the hardest of all minerals, and graphite is one of the softest. If their compositions are identical, why do they have such profound differences in physical properties?

2. List the eight most abundant chemical elements in Earth's crust. Are any unfamiliar to you? List familiar elements that are not among the eight. Why are they familiar?

3. Table 2.1 shows that silicon and oxygen together make up nearly 75 percent, by weight, of Earth's crust. But silicate minerals make up more than 95 percent of the crust. Explain the apparent discrepancy.

4. Quartz is SiO_2. Why does no mineral exist with the composition SiO_3?

5. If you were given a crystal of diamond and another of quartz, how would you identify which is diamond?

6. Would you expect minerals found on the Moon, Mars, or Venus to be different from those of Earth's crust? Explain your answer.

7. Explain how mining, crushing, or refining can change a benign mineral into a poison.

CHAPTER

3

Rocks

Courtesy Graham R. Thompson/Jonathan Turk

Climber rappels on the Grand Teton, Grand Teton National Park, Wyoming.

Earth is solid rock to a depth of 2,900 kilometers, where the mantle meets the liquid outer core. Even casual observation reveals that rocks are not all alike. The great peaks of the Sierra Nevada in California are hard, strong granite. The red cliffs of the Utah desert are soft sandstone. The top of Mount Everest is limestone containing clamshells and the remains of other small marine animals.

The marine fossils of Mount Everest tell us that this limestone formed in the sea. What forces lifted the rock to the highest point of the Himalayas? Where did the vast amounts of sand in the Utah sandstone come from? How did the granite of the Sierra Nevada form?

All of these questions ask about the processes that formed the rocks and changed them throughout geologic history. In this chapter we will study rocks: how they form and what they are made of. In later chapters we will use our understanding of rocks to interpret Earth's history. ■

ThomsonNOW™ Throughout this chapter, the ThomsonNOW logo indicates an opportunity for online self-study, which:
- Assesses your understanding of important concepts and provides a personalized study plan
- Links you to animations and simulations to help you study
- Helps to test your knowledge of material and prepare for exams

Visit ThomsonNOW at www.thomsonedu.com/login to access these resources.

3.1 Rocks and the Rock Cycle

A rock is a solid aggregate of one or more minerals. Geologists group rocks into three categories on the basis of how the rocks form: igneous rocks, sedimentary rocks, and metamorphic rocks.

Earth's interior is hot and dynamic. The high temperature that exists within a few hundred kilometers of the surface can melt solid rock to form a molten liquid called **magma**, which then rises slowly toward Earth's surface. You can read an explanation of how magma forms in Section 1 of Chapter 8. As it rises, the magma cools to become solid rock again (◆ Figure 3.1). An **igneous rock** forms when magma solidifies.

All rocks are hard and seem permanent and unchanging over a human lifetime. But this apparent permanence is an illusion created by our short observational time frame. Over geologic time, water and air attack rocks of all kinds at Earth's surface to decompose, or weather, them. Thus, over time **weathering** breaks rock down to smaller particles. **Sediment** is a term for gravel, sand, clay, and all other particles weathered and eroded from rock. Sand on a beach and mud on a lake bottom are examples of sediment. Weathering processes also form ions such as sodium and calcium dissolved in ground water and streams. Streams, wind, glaciers, and gravity then erode the sediment and dissolved ions and carry them downhill to deposit them at lower elevations. A **sedimentary rock** forms when sediment becomes cemented or compacted into solid rock (◆ Figure 3.2). Marine organisms such as clams, oysters, and corals extract dissolved calcium from seawater. They combine the calcium with carbon dioxide that dissolves in seawater from Earth's atmosphere, and use it to form their shells and hard parts. After the organisms die, those shells and coral remains accumulate and become the sedimentary rock called limestone. Thus, limestone, one of the most common sedimentary rocks, forms by direct interactions among the geosphere, the hydrosphere, the biosphere, and the atmosphere.

A **metamorphic rock** forms when heat, pressure, or hot water alter any preexisting rock. For example, Earth's crust may slowly sink, forming a depression that may be hundreds of kilometers in diameter and thousands of meters deep. Sediment accumulates in the depression, becomes cemented, and forms sedimentary rock. As more sediment accumulates, it buries the deepest layers under a huge weight. The deep burial raises the temperature and pressure on the rocks and alters both the minerals and the texture to transform the sedimentary rock to a metamorphic rock (◆ Figure 3.3). This has been happening for the past 150 million years where the Mississippi River has carried billions of tons of mud and sand to the northern part of the Gulf of Mexico, depositing a pile of sediment that now measures more than 12 kilometers in thickness in the deepest part of the basin.

No rock is permanent over geologic time; instead, all rocks change slowly from one of the three rock types to another. This continuous process is called the **rock cycle** (◆ Figure 3.4). In the example of metamorphism described previously, sediment has accumulated beneath the Mississippi Delta to form sedimentary rock. As the basin sank, new sediment buried the sedimentary rock to greater depths where rising temperature and pressure have converted it to metamorphic rock. If the temperature rises sufficiently at some future time, the metamorphic rock will melt to form magma. The magma will then rise and solidify to become igneous rock. Millions of years later, movement of Earth's crust might raise the igneous rock to the surface, where it will weather to form sediment. Rain and streams will then wash the sediment into a new basin, renewing the cycle.

◆ **FIGURE 3.1** A fiery lava flow approaches the sea on the island of Hawaii. January 28, 2005.

◆ **FIGURE 3.2** Sedimentary layers of sandstone form steep cliffs above the San Juan River in Utah.

The rock cycle can follow many different paths. For example, weathering may turn a metamorphic rock to sediment, which then becomes cemented and forms a sedimentary rock. An igneous rock may be metamorphosed. The rock cycle simply expresses the idea that rocks are not permanent but change over geologic time.

The Rock Cycle and Earth Systems Interactions

The rock cycle illustrates several types of Earth systems interactions—interactions among rocks, the atmosphere, the biosphere, and the hydrosphere. Rain and air, aided by acids and other chemicals secreted by plants, decompose solid rocks to form large amounts of clay and other sediment. During these processes, water and atmospheric gases react chemically to become incorporated into the clay. Thus, these processes transfer water and air from the atmosphere and hydrosphere to the solid minerals of the geosphere. More rain then washes the clay and other sediment into streams, which carry it to a sedimentary basin such as the Mississippi River delta on the edge of a continent. Recall from Chapter 1 that solar energy drives the hydrological cycle. Thus, sunlight evaporates mois-

◆ **FIGURE 3.3** Metamorphic rocks are commonly contorted as a result of movements of the Earth's crust that deform them.

Solidification

Igneous rock

Weathering

Sediment

Lithification

Magma

Sedimentary rock

Melting

Metamorphic rock

Metamorphism

◆ **FIGURE 3.4** The rock cycle shows that rocks change over geologic time. The arrows show paths that rocks can follow as they change.

ture to form rain, which, in turn, feeds flowing streams. But these same processes are also part of the rock cycle, illustrating once again that all Earth processes interact.

The rock cycle is also driven by Earth's internal heat. For example, when a sedimentary basin sinks under the weight of additional sediment, the deeper layers become heated and metamorphosed by Earth's heat. The same heat may melt the rocks to produce magma. But then, the magma rises upward and perhaps even erupts onto Earth's surface from a volcano. In this way, heat is transferred from Earth's interior to the atmosphere.

More importantly for Earth's surface environment, however, is the fact that volcanic eruptions release large amounts of carbon dioxide into the atmosphere. As mentioned in Chapter 1, carbon dioxide is a greenhouse gas that absorbs infrared radiation and warms the atmosphere. The greatest known episodes of volcanic eruptions in Earth history released enough carbon dioxide to raise average global atmospheric temperatures by approximately 10°C. These events and processes are described in Chapters 8, 17, and 21.

Throughout this chapter, we will emphasize these and other interactions among the atmosphere, biosphere, hydrosphere, and rocks of the geosphere to illustrate the point that those systems continuously exchange both energy and material so that Earth functions as a single, integrated system.

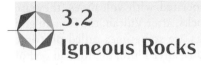

ThomsonNOW™

CLICK ThomsonNOW Interactive to work through an activity on the building blocks of rocks through rock labs and rock cycles.

3.2
Igneous Rocks

Magma: The Source of Igneous Rocks

If you drilled a well deep into the crust, you would find that Earth temperature rises about 30°C for every

◆ **FIGURE 3.5** The peaks of Sam Ford Fiord, Baffin Island, are composed of granite.

Table 3.1	Igneous Rock Textures Based on Grain Size
Grain Size	**Name of Texture**
No mineral grains (obsidian)	Glassy
Too fine to see with naked eye	Very fine grained
Up to 1 millimeter	Fine-grained
1–5 millimeters	Medium-grained
More than 5 millimeters	Coarse-grained
Relatively large grains in a finer-grained matrix	Porphyry

kilometer of depth. Below the crust, temperature continues to rise, but not as rapidly. In the mantle between depths of 100 and 350 kilometers, the temperature is so high that rocks in some areas melt to form magma. The temperature of magma varies from about 600 to 1,400°C, depending on its chemical composition and the depth at which it forms. As a comparison, an iron bar turns red-hot at about 600°C and melts at slightly over 1,500°C.

When rock melts, the resulting magma expands by about 10 percent, so it is less dense than the rock around it and therefore rises as it forms—much as a hot air balloon ascends in the atmosphere. When magma rises, it enters the cooler environment near Earth's surface, where it solidifies to form solid igneous rock.

As explained in Chapters 1 and 6, Earth melted shortly after it formed. Therefore, the early crust was quite different from our modern crust and consisted entirely of igneous rocks. Later geological activity modified the original igneous crust to form sedimentary, metamorphic, and younger igneous rocks. However, about 95 percent of Earth's crust is still igneous rock or metamorphosed igneous rock. Even though today a relatively thin layer of sedimentary rock buries much of this igneous foundation, igneous rocks are easy to view because they make up some of the world's most spectacular mountains (◆ Figure 3.5).

Types of Igneous Rocks

Magma can either rise all the way through the crust to erupt onto Earth's surface to form an **extrusive igneous rock,** or it can solidify within the crust below the surface as an **intrusive igneous rock.** Because extrusive rocks are so commonly associated with volcanoes, they are also called **volcanic rocks,** after Vulcan, the Roman god of fire. Intrusive rocks are sometimes called **plutonic rocks,** or simply **plutons,** after Pluto, the Roman god of the underworld.

When magma solidifies, it usually crystallizes to form minerals. The **texture** of a rock refers to the size, shape, and arrangement of its mineral grains, or crystals

(Table 3.1). Some igneous rocks consist of mineral grains that are too small to be seen with the naked eye; others are made up of thumb-sized, or even larger, crystals.

Extrusive (Volcanic) Rocks

Lava is fluid magma that flows from a crack or a volcano onto Earth's surface. The term also refers to rock that forms when lava cools and becomes solid. After lava erupts onto the relatively cool Earth's surface, it solidifies rapidly—perhaps over a few days or years. Crystals form but do not have much time to grow. As a result, many volcanic rocks have fine-grained textures, with crystals too small to be seen with the naked eye. **Basalt** is a common very fine grained volcanic rock (◆ Figure 3.6).

In unusual circumstances, molten lava may solidify within a few hours of erupting. Because the magma hardens so quickly, the atoms have no time to align themselves to form crystals. As a result, the atoms are arranged in a random chaotic pattern. Glass is a solid composed of such a random pattern of atoms. Volcanic glass is called **obsidian** (◆ Figure 3.7).

If magma rises slowly through the crust before erupting, some crystals may grow while most of the magma remains molten. If this mixture of magma and crystals then erupts onto the surface, the magma solidifies quickly, forming **porphyry,** a rock with the large

◆ **FIGURE 3.6** Basalt is a very fine-grained volcanic rock. The holes were gas bubbles that were preserved as the magma solidified in southeastern Idaho.

Geoffrey Sutton

◆ **FIGURE 3.7** Obsidian is natural volcanic glass. It contains no crystals. This sample is about 6 inches wide.

Geoffrey Sutton

◆ **FIGURE 3.8** Porphyry is an igneous rock containing large crystals embedded in a fine-grained matrix. This rock is about 4 inches across and is a rhyolite porphyry with large, pink feldspar phenocrysts.

crystals, called **phenocrysts**, embedded in a fine-grained matrix (◆ Figure 3.8).

Intrusive (Plutonic) Rocks

When magma solidifies *within* the crust, the overlying rock insulates the magma like a thick blanket. The magma then crystallizes slowly and the crystals grow over hundreds of thousands, or even millions, of years. As a result, most plutonic rocks are medium to coarse grained. **Granite,** the most abundant rock in continental crust, is a medium- or coarse-grained plutonic rock. The crystals in granite are clearly visible. Many are a millimeter or so across, although some crystals may be much larger.

Naming and Identifying Igneous Rocks

Geologists use both the minerals and texture to name igneous rocks. For example, any medium- or coarse-grained igneous rock consisting mostly of feldspar and

quartz is called granite. **Rhyolite** also consists mostly of feldspar and quartz but is very fine grained (◆ Figure 3.9). The same magma that solidifies slowly within the crust to form granite can also erupt onto Earth's surface to form rhyolite.

Like granite and rhyolite, most common igneous rocks are classified in pairs, each member of a pair containing the same minerals but having a different texture. The texture depends mainly on whether the rock is volcanic (fine grained) or plutonic (coarse grained). ◆ Figure 3.10 shows the names, minerals, and textures of common igneous rocks. Figure 3.10 shows that granite and rhyolite contain large amounts of *fel*dspar and *si*lica, and so are called **felsic** rocks. Basalt and **gabbro** are called **mafic** rocks because of their high

A

B

Geoffrey Sutton

◆ **FIGURE 3.9** Both of these rocks are about 5 inches across. Although granite **(A)** and rhyolite **(B)** contain the same minerals, granite is coarse grained because it cooled slowly, but rhyolite is very fine grained because it cooled rapidly.

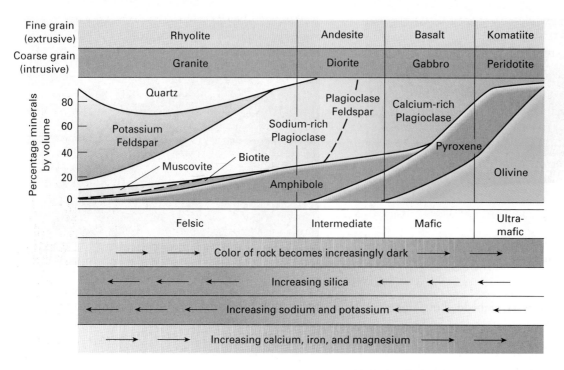

| Fine grain (extrusive) | Rhyolite | | Andesite | Basalt | Komatiite |
| Coarse grain (intrusive) | Granite | | Diorite | Gabbro | Peridotite |

| Felsic | | Intermediate | Mafic | Ultra-mafic |

Color of rock becomes increasingly dark ⟶

Increasing silica

Increasing sodium and potassium ⟵

Increasing calcium, iron, and magnesium ⟶

◆ **FIGURE 3.10** The names of common igneous rocks are based on the minerals and texture of a rock. In this figure, a mineral's abundance in a rock is proportional to the thickness of its colored band beneath the rock name. If a rock has a fine-grained texture, its name is found in the top row of rock names; if it has a coarse-grained texture, its name is in the second row. Color and chemical trends are shown in the lowest four rows.

magnesium and iron contents (ferrum is the Latin word for iron). Rocks with especially high magnesium and iron concentrations are called **ultramafic**. Rocks with compositions between those of granite and basalt are called **intermediate rocks.**

Once you learn to identify the rock-forming minerals, it is easy to name a coarse-grained plutonic rock using ◆ Figure 3.10 because the minerals are large enough to see. It is more difficult to name many volcanic rocks because the minerals are too small to identify. A field geologist often uses color to name a volcanic rock. Rhyolite is usually light in color: white, tan, red, and pink are common. Many **andesites** are gray or green, and basalt is commonly black. The minerals in many volcanic rocks cannot be identified even with a microscope because of their tiny crystal sizes. In this case, definitive identification is based on chemical and X-ray diffraction analyses carried out in the laboratory.

Common Igneous Rocks

Granite and Rhyolite

Granite contains mostly feldspar and quartz. Small amounts of black biotite or hornblende often give it a speckled appearance. Granite (and metamorphosed granite) is the most common rock in continental crust. It is found nearly everywhere beneath the relatively thin veneer of sedimentary rocks and soil that covers most of

the continents. Geologists often call this rock **basement rock** because it makes up the foundation of a continent.[1] Granite is hard and resistant to weathering; it forms steep, sheer cliffs in many of the world's great mountain ranges. Mountaineers prize granite cliffs for the steepness and strength of the rock (◆ Figure 3.11).

As granitic magma (magma with the chemical composition of granite) rises through Earth's crust, some of it may erupt from a volcano to form rhyolite, while the remainder solidifies beneath the volcano, forming granite. Most obsidian forms from magma with a granitic (rhyolitic) composition.

Basalt and Gabbro

Basalt consists of approximately equal amounts of plagioclase feldspar and pyroxene. It makes up most of the oceanic crust, as well as huge basalt plateaus on continents (◆ Figure 3.12). Gabbro is the plutonic counter-

---

1. Geologists commonly use the terms *basement rock, bedrock, parent rock,* and *country rock.* Basement rock is the igneous and metamorphic rock that lies beneath the thin layer of sediment and sedimentary rocks covering much of Earth's surface, and thus it forms the basement of the crust. Bedrock is the solid rock that lies beneath soil or unconsolidated sediments. It can be igneous, metamorphic, or sedimentary. Parent rock is any original rock before it is changed by metamorphism or any other geologic process. The rock enclosing or cut by an igneous intrusion or by a mineral deposit is called country rock.

◆ **FIGURE 3.11** The authors on Inugsuin Point Buttress, a granite wall on Baffin Island.

◆ **FIGURE 3.12** The stack of about 20 Columbia River Basalt lava flows show distinct layering in the canyon of the Grande Ronde River, Washington state, USA. Each of the layers is a separate flow.

part of basalt; it is mineralogically identical but consists of larger crystals. Gabbro is uncommon at Earth's surface, although it is abundant in deeper parts of oceanic crust, where basaltic magma crystallizes slowly.

Andesite and Diorite

Andesite is a volcanic rock intermediate in composition between basalt and granite. It is commonly gray or green and consists of plagioclase feldspar and dark minerals (usually biotite, amphibole, or pyroxene). It is named for the Andes Mountains, the volcanic chain on the western edge of South America, where it is abundant. Because it is volcanic, andesite is typically very fine grained. **Diorite** is the plutonic equivalent of andesite. It forms from the same magma as andesite and, consequently, often underlies andesitic mountain chains such as the Andes.

Peridotite and Komatiite

Peridotite is an ultramafic igneous rock that makes up most of the upper mantle but is rare in Earth's crust. It is coarse grained and composed of olivine and small amounts of pyroxene, amphibole, or mica, but no feldspar. Komatiite is the fine-grained, extrusive, equivalent of peridotite. Geologists think that it was the material of the earliest crust that formed as the molten planet cooled more than 4 billion years ago. Only a few traces of this primordial crust are known to exist today.

3.3 Sedimentary Rocks

Over geologic time, the atmosphere, biosphere, and hydrosphere weather rock and convert it to clay, sand, gravel, and ions dissolved in water. As we will learn in Chapter 10, both chemical and physical processes weather rocks and form particles such as sand and gravel, and dissolved ions. Flowing water, wind, gravity, and glaciers erode the decomposed rock, transport it downslope, and deposit it on the seacoast or in lakes and river valleys. With time, the loose, unconsolidated sediment becomes compacted and cemented, or **lithified**, and forms sedimentary rock. Rivers carry dissolved ions into the sea, where they accumulate in the oceans. Marine organisms such as clams, oysters, and corals may then use these ions to form shells and other hard parts, which accumulate after the organisms die and then form limestone. Sedimentary rocks make up only about 5 percent of Earth's crust. However, because they form on Earth's surface, sedimentary rocks are widely spread in a thin veneer over underlying igneous and metamorphic rocks. As a result, they cover about 75 percent of continents. The formation of sediment, and in many cases its lithification, all involve complex systems interactions among Earth's spheres. As you read the sections that follow, pay special attention to the interplay among earth, air, water, and life that forms the rocks at Earth's surface.

Sedimentary rocks are broadly divided into four categories:

1. **Clastic sedimentary rocks** are composed of particles of weathered rocks, such as sand grains and pebbles, called clasts, which have been transported, deposited, and lithified. Clastic rocks make up more than 85 percent of all sedimentary rocks (◆ Figure 3.13). This category includes sandstone, siltstone, and shale.

2. **Organic sedimentary rocks** consist of the lithified remains of plants or animals. Coal is an organic sedimentary rock made up of decomposed and compacted plant remains.

Cooling and Crystallization of Magma: Bowen's Experiment

Early in the 1900s, a geologist named N. L. Bowen discovered that when magma cools and crystallizes to form an igneous rock, all the minerals do not crystallize at the same time and temperature. Those that crystallize first, at the highest temperature, contain the lowest percentage of silica. Minerals with higher silica contents crystallize later, at lower temperatures.

Bowen conducted his experiments by placing finely ground rock samples into gold tubes. He used gold because it is chemically inert and does not affect chemical reactions.

Bowen welded the tubes closed to form a tight seal and placed them into a strong, hollow, steel cylinder with a screw cap, called a bomb. He then heated the bomb until the powdered rock melted, creating an artificial magma. Finally, he cooled the mixture slightly and let it sit long enough, often for a few months, for minerals to begin crystallizing from the melt.

At that point, his bombs contained a mixture of crystals and the melt. He then recorded the temperature of each bomb and plunged it into a cold liquid. The rapid cooling preserved the crystals and solidified the melt as glass. By extracting the sample from the bomb and examining it with a microscope, he was able to identify the minerals that had formed from the magma. He allowed his artificial magmas to begin forming crystals at several temperatures before quick-cooling them. By identifying the minerals that had formed at each temperature, he was able to determine the order in which minerals crystallize from a cooling magma.

When he started with a basaltic magma, Bowen found that both olivine and calcium-rich plagioclase crystallized first. Both minerals are low in silica relative to the magma. As the magma continued to cool, crystallization of additional minerals followed one path if the original crystals separated from the magma as they formed, and a different path if the crystals remained in contact with the melt. From these experiments, Bowen made the following generalizations regarding crystallization of basaltic magma.

Crystals Separating from Magma

In a natural magma chamber, the olivine and calcium-rich plagioclase that form first may become separated from the magma, perhaps because the denser crystals settle to the bottom of the chamber. This separation removes them from contact with the magma, and they are unable to react with it further. The separation of olivine and plagioclase enriches the remaining melt in silica. Consequently, as the magma cools further, minerals richer in silica form. The final bit of melt to solidify has a granitic composition; potassium feldspar, sodium plagioclase, and quartz are the last minerals to crystallize. As a result of this process, a cooling basaltic magma produces a large amount of basalt, but also a small amount of granite. This process is called fractional crystallization.

Crystals Mixing with the Magma

Alternatively, currents in a natural magma chamber may prevent crystals from separating from the magma. In this case, the crystals remain in contact with the magma and react with it as it cools. Although it may seem counterintuitive, crystals of olivine that form at high temperature then dissolve back into the melt as it cools. The dissolving olivine reacts with the melt to form pyroxene. Simultaneously, the calcium-rich plagioclase reacts with the melt to form plagioclase with less calcium and aluminum but more sodium and silica. When the magma solidifies, the resulting basalt consists of pyroxene, intermediate plagioclase, and a small amount of olivine.

Bowen also experimented with artificial magmas of intermediate and granitic compositions. He discovered that as those melts cooled, pyroxene reacted with the melt to form amphibole. The amphibole then reacted to form biotite. Plagioclase continued to react with the melt to become progressively enriched in sodium and silica and depleted in calcium and aluminum. As the last of the magma crystallized, potassium feldspar, muscovite, and quartz formed.

Bowen's Reaction Series

Bowen summarized his experimental results in a now famous Y-shaped figure known as Bowen's reaction series. It is called a reaction series to emphasize that as magma cools, minerals form at high temperatures and then react with the magma to form other minerals. Minerals of both arms of the Y contain progressively more silica with decreasing temperature.

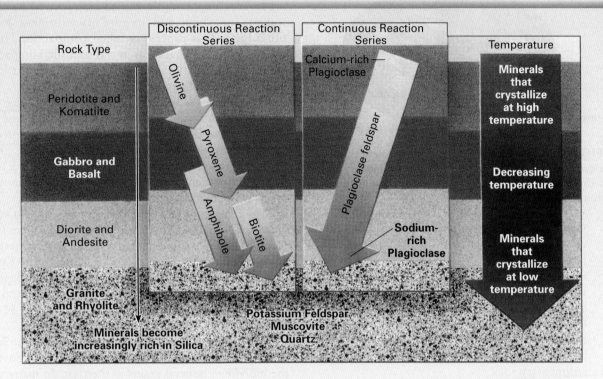

◆ **FIGURE 1** Bowen's reaction series shows the order in which minerals crystallize from a cooling magma and then react with the magma as it cools to form minerals lower on the diagram.

The right side of Bowen's reaction series describes reactions between plagioclase feldspar and cooling magma. Calcium-rich plagioclase forms first, and then reacts continuously with the melt to become progressively enriched in sodium and silica and depleted in calcium and aluminum. Thus, this arm of the Y is known as a "continuous reaction series."

The left side of the Y shows that as temperature decreases, olivine reacts with the melt to form pyroxene. Pyroxene then dissolves back into the magma as amphibole forms. The reactions continue until biotite forms as amphibole dissolves. This arm of the Y is called a "discontinuous reaction series" because each of the minerals has a different crystal structure, and the mineral reactions occur in a stepwise manner rather than continuously.

Minerals of the discontinuous series share more oxygen among silica tetrahedra as temperature decreases. Olivine is an independent tetrahedra mineral. Pyroxene is a single-chain silicate. Amphibole is a double-chain silicate, and biotite is a sheet silicate.

The first mineral to crystallize from a cooling magma depends on the magma composition. The crystallization sequence shown in the figure above is for basaltic magma. In a more silica-rich magma, pyroxene might form first instead of olivine. As the magma cooled, however, the same general reaction sequence would occur.

The rock names in ◆ Figure 1 are placed on the same level as the minerals that make up each of those rocks. Bowen's reaction series shows that basalt forms at relatively high temperatures, andesite at intermediate temperatures, and granite at relatively low temperatures.

FOCUS QUESTION

How did Bowen's experiments explain how the process of partial melting, described in the text, works? Why does partial melting produce magma that is more silica-rich and iron- and magnesium-poor than the rock that is undergoing partial melting?

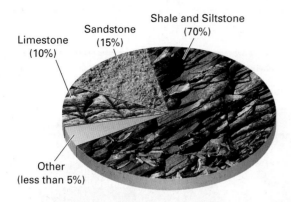

FIGURE 3.13 Sandstone, siltstone, and shale are clastic rocks that make up more than 85 percent of all sedimentary rocks. Limestone and some "other" sedimentary rocks make up less than 15 percent.

3. **Chemical sedimentary rocks** form by direct precipitation of minerals from solution. Rock salt, for example, forms when salt precipitates from evaporating seawater or saline lake water.

4. **Bioclastic sedimentary rocks** are composed of broken shell fragments and similar remains of living organisms. The fragments are clastic, but they are of a biological origin. Most limestone formed from broken shells and is a bioclastic sedimentary rock.

Clastic Sedimentary Rocks

Clastic sediment is called gravel, sand, silt, or clay, in order of decreasing particle size (Table 3.2). As clastic particles ranging in size from coarse silt to boulders tumble downstream, their sharp edges are worn off and they become rounded (◈ Figure 3.14). Finer silt and clay do not round effectively because they are so small and light that water, and even wind, cushion them as they bounce along.

If you fill a measuring cup with sand or other clastic sediment, you can still add a substantial amount of water that occupies empty space, or **pore space**, among the sand grains (◈ Figure 3.15A). Commonly, sand and similar sediment have about 20 to 40 percent pore space.

As more sediment accumulates, the weight of overlying layers compresses the buried sediment. Some of the water is forced out, and the pore space shrinks (◈ Figure 3.15B). This process is called **compaction**. If the grains have platy shapes, as in clay and silt, compaction alone may lithify the sediment as the platy grains interlock like pieces of a puzzle.

As sediment is buried, water circulates through the pore space. This water commonly contains dissolved ions that precipitate in the pore spaces. When the ions precipitate, they form a cement that binds the clastic grains firmly together to form a hard rock (◈ Figure 3.15C). Calcite, quartz, and iron oxides are the most common cements in sedimentary rocks.

The time required for lithification of sediment varies greatly. In some heavily irrigated areas of southern California, calcite has precipitated from irrigation water to cement soils within only a few decades. In the Rocky Mountains, calcite has cemented some glacial deposits that are less than 20,000 years old. In contrast, sand and gravel deposited in southwestern Montana about 30 million years ago can still be dug with a hand shovel.

Conglomerate is lithified gravel (◈ Figure 3.16). Each clast in a conglomerate is usually much larger than the individual mineral grains in the clast. In many con-

Table 3.2	Sizes and Names of Sedimentary Particles and Clastic	
Diameter (mm)	**Sediment**	**Clastic Sedimentary Rock**
256– 64– 2–	Boulders Cobbles Pebbles	Gravel (rubble)	Conglomerate (rounded particles) or breccia (angular particles)
$\frac{1}{16}$–	Sand	Sandstone
$\frac{1}{256}$–	Silt Clay	Mud	Siltstone Claystone or shale } Mudstone

Courtesy of Graham R. Thompson/Jonathan Turk

FIGURE 3.14 Boulders and cobbles collide as they move downstream. The collisions abrade the sharp edges, producing rounded rocks.

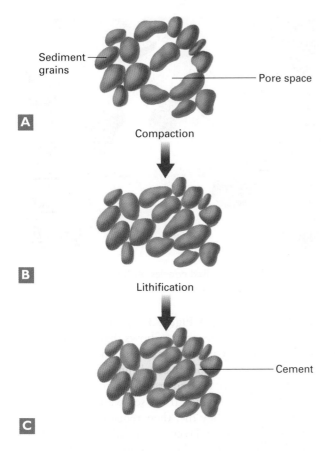

Sediment grains

Pore space

A

Compaction

B

Lithification

Cement

C

◈ **FIGURE 3.15** **(A)** Pore space is the open space between sediment grains. **(B)** Compaction squashes the grains together, reducing the pore space and lithifying the sediment by interlocking the grains. **(C)** Cement fills the remaining pore space, lithifying the sediment by gluing the grains together.

Courtesy of Graham R. Thompson/Jonathan Turk

◈ **FIGURE 3.16** Conglomerate is lithified gravel. The cobbles were rounded as they bounced downstream, before they were deposited and lithified. The large cobbles in this photo are fist-sized.

glomerates, the clasts may be fist sized or even larger. Therefore, the clasts retain most of the characteristics of the parent rock. If enough is known about the geology of an area where conglomerate is found, it may be possible to identify exactly where the clasts originated. For example, a granite clast probably came from nearby granite bedrock.

The next time you walk along a gravelly streambed, look carefully at the gravel. You will probably see sand or silt trapped among the larger clasts. In a similar way, most conglomerates contain fine sediment among the large clasts.

Sandstone consists of lithified sand grains (◈ Figure 3.17). Granite, the most common rock in continents, is

A

Courtesy of Graham R. Thompson/Jonathan Turk

B

Courtesy of Graham R. Thompson/Jonathan Turk

◈ **FIGURE 3.17** Sandstone is lithified sand. **(A)** A sandstone cliff above the Colorado River, Canyonlands, Utah. **(B)** A close-up photo shows the well-rounded sand grains.

3.3 Sedimentary Rocks

55

◆ **FIGURE 3.18** Shale near Drummond, Montana. The large fragments are about 10 centimeters across.

◆ **FIGURE 3.19** Red nodules of chert in light-colored limestone. The nodules are fist-sized.

composed largely of quartz and feldspar. When granite and similar rocks weather, the feldspar reacts chemically with air and water to form tiny clay crystals. In contrast, quartz is resistant to chemical weathering and the unaltered quartz grains simply accumulate as sand in soil. Eventually, rain erodes the sand and streams carry it toward the seacoast. As the sand grains bounce along on the bottom of a stream, they collide together, wearing away the sharp edges and becoming rounded. Finally, the rounded sand grains accumulate on beaches. Over time, they become compacted and lithified to form sandstone. Most beach sands and most sandstones consist predominantly of rounded quartz grains.

Shale (◆ Figure 3.18) consists mostly of tiny clay minerals and lesser amounts of quartz. Shale typically splits easily along very fine layering called **fissility**. The fissility of shale results from the parallel orientation of the platy clay particles.

Siltstone is lithified silt. The main component of most siltstones is quartz, although clays are commonly present. Siltstones often show layering but lack the fine fissility of shales because of their lower clay content.

◆ Figure 3.13 shows that shale and siltstone make up 70 percent of all clastic sedimentary rocks. Their abundance reflects the vast quantity of clay produced by weathering. Shale is usually gray to black due to the presence of partially decayed remains of plants and animals commonly deposited with clay-rich sediment. This organic material in shales is the source of most oil and natural gas. (The formation of oil and gas from this organic material is discussed in Chapter 5.)

Organic Sedimentary Rocks

Organic sedimentary rocks form by lithification of the remains of plants and animals. **Chert** is a rock composed of pure silica. It occurs as sedimentary beds and as irregularly shaped lumps called "nodules" in other sedimentary rocks (◆ Figure 3.19). Microscopic examination of bedded chert often shows that it is made up of the remains of tiny marine organisms whose skeletons are composed of silica rather than calcium carbonate. Bedded chert forms as tiny marine organisms that float near the sea surface extract silica from sea water to form their skeletons. When they die, their remains sink to the sea floor to accumulate in layers, eventually to form chert. In contrast, some nodular chert appears to form by precipitation from silica-rich ground water, most often in limestone. Nodular chert, then, forms as a result of interactions between the hydrosphere and the geosphere. Chert was one of the earliest geologic resources. Flint, a dark gray to black variety, was frequently used for arrowheads, spear points, scrapers, and other tools chipped to hold a fine edge.

When plants die, their remains usually decompose by reaction with oxygen. However, in warm swamps and in other environments where plant growth is rapid, dead plants accumulate so rapidly that the oxygen is used up long before the decay process is complete. The partially decayed plant remains form **peat**. As peat is buried and compacted by overlying sediments, it converts to **coal**, a hard, black, combustible rock. Coal formation is described in Chapter 5.

Chemical Sedimentary Rocks

Some common elements in rocks and minerals, such as calcium, sodium, potassium, and magnesium, dissolve during chemical weathering and are carried by ground water and streams to lakes or to the ocean. Streams that carry the salts to the ocean drain most lakes. However, some lakes, such as the Great Salt Lake in Utah, are landlocked. Streams flow into the lake, but no streams exit. As a result, water escapes only by evaporation. When water evaporates, salts remain behind and the lake water steadily becomes saltier. **Evaporites** are rocks that form when evaporation concentrates dissolved ions to the point at which they precipitate

◆ FIGURE 3.20 An evaporating lake precipitated thick salt deposits on the Salar de Uyuni, Bolivia. The salt polygons are about 2 meters across.

from solution (◆ Figure 3.20). The same process can occur if ocean water is trapped in coastal or inland basins, where it can no longer mix with the open sea.

The most common minerals found in evaporite deposits are gypsum ($CaSO_4 \cdot 2H_2O$) and halite (NaCl). (The $\cdot 2H_2O$ in the chemical formula of gypsum means the mineral structure incorporates water.) Gypsum is used in plaster and wallboard, and halite is common salt. Evaporites form important economic deposits, but they compose only a small proportion of all sedimentary rocks.

Bioclastic Sedimentary Rocks

Carbonate rocks are primarily made up of the carbonate minerals calcite and dolomite, described in the previous chapter. Calcite-rich rocks are called **limestone**, whereas rocks rich in dolomite are referred to as **dolomite**. Many geologists use the term *dolostone* for the rock name to distinguish it from the mineral dolomite.

Recall that the chemical formula for calcite is $CaCO_3$, a compound of the calcium cation, Ca^{2+}, and the carbonate anion ($CO_3{}^{2-}$). Dissolved calcium ions are released into water during the chemical weathering of calcium-bearing rocks. The carbonate anion forms when atmospheric carbon dioxide gas dissolves in water. Seawater contains large quantities of both dissolved calcium and carbon dioxide. Clams, oysters, corals, some types of algae, and a variety of other marine organisms convert dissolved calcium and carbonate ions to shells and other hard body parts. When these organisms die, waves and ocean currents break the shells into small fragments. A rock formed by lithification of such sediment is called bioclastic limestone, indicating that it forms by both biological and clastic processes. Most limestones are bioclastic. The bits and pieces of shells appear as fossils in the rock (◆ Figure 3.21).

A

B

◆ FIGURE 3.21 Most limestone is lithified shell fragments and other remains of marine organisms. **(A)** Mount Robson in British Columbia, Canada, is composed of limestone. **(B)** A close-up of shell fragments in limestone.

3.3 Sedimentary Rocks 57

Carbonate Rocks and Global Climate

Carbon dioxide is a greenhouse gas that traps heat in the atmosphere. Limestone is a hard, solid rock. Although it may seem counterintuitive, limestone rock is formed largely from carbon dioxide gas. Atmospheric carbon dioxide dissolves in seawater and then combines with dissolved calcium to form limestone rock, so formation of limestone removes carbon dioxide from both the seas and the air.

In Chapter 1 we mentioned briefly that Earth's outer shell is broken into several segments called tectonic plates. The plates float on the weak, plastic mantle below and glide across Earth, moving continents and oceans. You will learn in Chapter 6 that after a tectonic plate moves thousands of kilometers across Earth's surface, it eventually sinks deep into the mantle. In some instances, a sinking plate may carry limestone into the mantle, removing large amounts of carbon dioxide from Earth's surface and sequestering it in the deep mantle. In other cases, limestone on a sinking tectonic plate may become heated to the point that the carbon dioxide is released directly into the atmosphere in volcanic eruptions. In these ways the greenhouse gas is cycled between surface limestone, deep mantle rocks, and the atmosphere. Thus interactions between limestone and the carbon dioxide in the atmosphere and that dissolved in the oceans are important determinants of global climate. These system interactions are described in Chapter 21, "Climate Change."

Sedimentary Structures

Nearly all sedimentary rocks contain **sedimentary structures**—features that developed during or shortly after deposition of the sediment. These structures help us understand how the sediment was transported and deposited. Because sedimentary rocks form at Earth's surface, sedimentary structures and other features of sedimentary rocks also contain clues about environmental conditions at Earth's surface when the rocks formed.

The most obvious and common sedimentary structure is **bedding**, or stratification—layering that develops as sediment is deposited (◆ Figure 3.23). Bedding forms because sediment accumulates layer by layer. Nearly all sedimentary beds were originally horizontal because most sediment accumulates on nearly level surfaces.

Cross-bedding consists of small beds lying at an angle to the main sedimentary layers (◆ Figure 3.24A). Cross-bedding forms in many environments where wind or water transports and deposits sediment. For example, wind heaps sand into parallel ridges called dunes and flowing water forms similar features called sand waves. ◆ Figure 3.24B shows that cross-beds are the layers formed by sand grains tumbling down the steep, downstream face of a dune or sand wave. Cross-bedding is common in sands deposited by wind, streams, and ocean currents and by waves on beaches.

◆ **FIGURE 3.22** The Niobrara chalk of western Kansas consists of the remains of tiny marine organisms.

David Schwimmer

Organisms that form limestone thrive and multiply in warm, shallow seas that are heated by the sun. Therefore, bioclastic limestone typically forms in shallow water along coastlines at low and middle latitudes. It also forms on continents where rising sea level floods land with shallow seas. Limestone makes up many of the world's great mountains; the summit of Mount Everest is made of limestone containing marine fossils. Leonardo da Vinci puzzled over the presence of fossils on mountaintops and was perhaps the first person to propose that Earth processes actively raise rocks from the sea floor to the tops of mountains, stating, "The silent fossils on snowy windswept summits remind us of the earth's active nature."

Coquina is bioclastic limestone consisting wholly of coarse shell fragments cemented together. **Chalk** is a very-fine-grained, soft, white bioclastic limestone made of the shells and skeletons of microorganisms that float near the surface of the oceans. When they die, their remains sink to the bottom and accumulate to form chalk. The pale-yellow chalks of Kansas, the off-white chalks of Texas, and the gray chalks of Alabama remind us that all of these areas once lay beneath the sea (◆ Figure 3.22).

◇ **FIGURE 3.23** Sedimentary bedding shows clearly in the walls of the Grand Canyon.

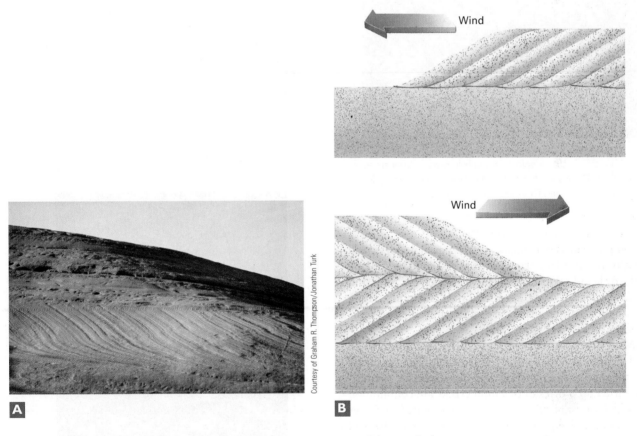

◇ **FIGURE 3.24** **(A)** Cross-bedding preserved in lithified ancient sand dunes in Arches National Park, Utah. The outcrop shown is about 5 meters high. **(B)** The development of cross-bedding as the prevailing wind direction changes.

INTERACTIVE QUESTION: *In what direction was the wind blowing when the dune in (A) formed?*

◆ **FIGURE 3.25** Ripple marks in billion-year-old siltstone in eastern Utah. The ripples formed when waves and currents piled silt into the small ridges. This outcrop is about 3 meters high.

Ripple marks are small, nearly parallel ridges and troughs that are also formed in sand and mud by moving water or wind. You can see them in mud puddles and shallow streams. They are like dunes and sand waves, but smaller. Ripple marks are often preserved in sedimentary rocks (◆ Figure 3.25).

Mud cracks are polygonal cracks that form when mud shrinks as it dries (◆ Figure 3.26). They indicate that the mud accumulated in shallow water that periodically dried up. For example, mud cracks are common on intertidal mud flats where sediment is flooded by water at high tide and exposed at low tide. The cracks often fill with sediment carried in by the next high tide and are commonly well preserved in rocks.

Occasionally, very delicate sedimentary structures are preserved in rocks. Geologists have found imprints of raindrops that fell on a muddy surface about 1 billion years ago (◆ Figure 3.27) and imprints of salt crystals that formed as a puddle of saltwater evaporated. Like mud cracks, raindrop and salt imprints show that the mud

◆ **FIGURE 3.26** Mud cracks form when wet mud dries and shrinks. The polygons are about 10 centimeters across.

must have been deposited in shallow water that intermittently dried up.

Fossils are any remains or traces of a plant or animal preserved in rock—any evidence of past life. Fossils include remains of shells, bones, or teeth; whole bodies preserved in amber or ice; and a variety of tracks, burrows, and chemical remains. Fossils are discussed further in Chapter 4.

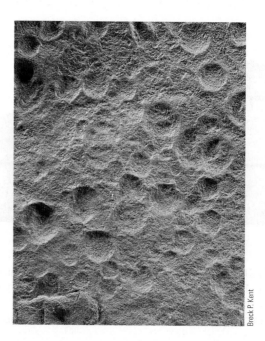

◆ **FIGURE 3.27** Delicate raindrop imprints formed by rain that fell on a mud flat about a billion years ago. Each drop imprint is about 5 millimeters in diameter.

3.4 Metamorphic Rocks

A potter forms a delicate vase from moist clay. She places the soft piece in a kiln and slowly heats it to 1,000°C. As temperature rises, the clay minerals decompose. Atoms from the clay then recombine to form new minerals that make the vase strong and hard. The breakdown of the clay minerals, growth of new minerals, and hardening of the vase all occur without melting of the solid materials.

Metamorphism (from the Greek words for "changing form") is the process by which rising temperature and pressure, or changing chemical conditions, transform rocks and minerals. Metamorphism occurs in solid rock, like the transformations in the vase as the potter fires it in her kiln. Small amounts of water and other fluids speed up the metamorphic mineral reactions, but the rock remains solid as it changes. Metamorphism can change any type of parent rock: sedimentary, igneous, or even another metamorphic rock.

A mineral that does not decompose or change in other ways, no matter how much time passes, is a "stable" mineral. Millions of years ago, weathering processes may have formed the clay minerals that today's potter used to create her vase. They were stable and had remained unchanged since they formed.

A stable mineral can become unstable when environmental conditions change. Three types of environmental change affect mineral stability and cause metamorphism: rising temperature, rising pressure, and changing chemical composition (usually caused by an influx of hot water). For example, when the potter put the clay in her kiln and raised the temperature, the clay minerals decomposed because they became unstable at the higher temperature. The atoms from the clay then recombined to form new minerals that were stable at the higher temperature. Like the clay, every mineral is stable only within a certain temperature range. In a similar manner, each mineral is stable only within a certain pressure range.

In addition, a mineral is stable only in a certain chemical environment. If hot water seeping through bedrock carries new chemicals to a rock, those chemicals may react with the original minerals to form different minerals that are stable in the new chemical environment. If hot water dissolves chemical components from a rock, new minerals may form for the same reason. Thus, water from the hydrosphere interacts with rock of the geosphere (at high temperatures) to create new rocks, showing once again that Earth's spheres, while separate in one sense, interact continuously to shape our planet.

Metamorphism occurs because each mineral is stable only within a certain range of temperature, pressure, and chemical environment. If temperature or pressure rises above that range, or if chemicals are added or removed from the rock, the rock's original minerals may decompose and their components recombine to form new minerals that are stable under the new conditions.

Metamorphic Grade

The **metamorphic grade** of a rock is the intensity of metamorphism that formed the rock. Temperature is the most important factor in metamorphism, and therefore grade closely reflects the temperature of metamorphism. Because temperature increases with depth in Earth, a general relationship exists between depth and metamorphic grade (◆ Figure 3.28). Low-grade metamorphism occurs at shallow depths, less than 10 kilometers beneath the surface, where temperature is no higher than 300 to 400°C. High-grade conditions are found deep within continental crust and in the upper mantle, 40 to 55 kilometers below Earth's surface. The temperature here is 600 to 800°C, close to the melting point of rock. High-grade conditions can develop at shallower depths, however, in rocks adjacent to hot magma. For example, today metamorphic rocks are forming beneath Yellowstone Park, where hot magma lies close to Earth's surface.

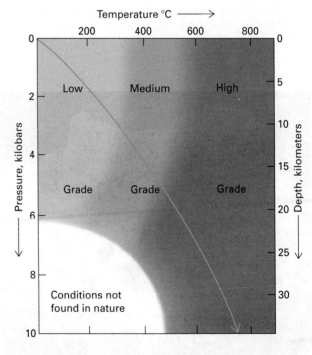

◆ **FIGURE 3.28** Metamorphic grade increases with depth because temperature and pressure rise with depth. The blue arrow traces the path of increasing temperature and pressure with depth in a normal part of the crust.

INTERACTIVE QUESTIONS: *In a normal part of the crust, what are the temperature and pressure at a depth of 25 kilometers? What grade of metamorphic rocks would you expect to exist there?*

Metamorphic Changes

Metamorphism commonly alters both the texture and mineral content of a rock.

Textural Changes

As a rock undergoes metamorphism, some mineral grains grow larger and others shrink. The shapes of the grains may also change. For example, fossils give the fossiliferous limestone shown in ◀▶ Figure 3.21B its texture. Both the fossils and the cement between them are made of small calcite crystals. If the limestone is heated, some of the calcite grains grow larger at the expense of others. In the process, the fossiliferous texture is destroyed. Metamorphism transforms limestone into a metamorphic rock called **marble** (◀▶ Figure 3.29). Like the fossiliferous limestone, the marble is composed of calcite, but the texture is now one of large interlocking grains, and the fossils have vanished.

Micas are common metamorphic minerals that form as many different parent rocks undergo metamorphism. Recall from Chapter 2 that micas are shaped like pie plates. When metamorphism occurs without **deformation**, the micas grow with random orientations, like pie plates flying through the air (◀▶ Figure 3.30A). However, when tectonic force squeezes rocks as they are heated during metamorphism, the rock deforms into folds. When rocks are folded as mica crystals are growing, the micas grow with their flat surfaces perpendicular to the direction of squeezing. This parallel alignment of micas (and other minerals) produces the metamorphic layering called **foliation** (◀▶ Figure 3.30B). Metamorphic rocks break easily along the foliation planes. This parallel fracture pattern is called **slaty cleavage** (◀▶ Figure 3.31). In most cases, slaty cleavage cuts across original sedimentary bedding.

The foliation layers range from a fraction of a millimeter to a meter or more thick. Metamorphic foliation can resemble sedimentary bedding but is different in origin. Foliation results from alignment and segregation of metamorphic minerals during metamorphism, and forms at right angles to the forces acting on the rocks. Sedimentary bedding develops because sediments are deposited in a layer-by-layer process.

Mineralogical Changes

As a general rule, when a parent rock contains only one mineral, metamorphism transforms the rock into one composed of the same mineral, but with a coarser texture. Limestone converting to marble is one example of this generalization; both rocks consist of the mineral calcite, but their respective textures are very different. Another example is the metamorphism of quartz sandstone to **quartzite**, a rock composed of recrystallized quartz grains.

In contrast, metamorphism of a parent rock containing several minerals usually forms a rock with new and different minerals *and* a new texture. For example, a typical shale contains large amounts of clay, as well as quartz and feldspar. When heated, some of those minerals decompose, and their atoms recombine to form new minerals such as mica, garnet, and a different kind of feldspar. ◀▶ Figure 3.32 shows a rock called gneiss that

A

B

◀▶ **FIGURE 3.29** Metamorphism has destroyed the fossiliferous texture of the limestone in Figure 3.29A and replaced it with the large, interlocking calcite grains of marble shown in Figure 3.29B. The large calcite crystals in 3.29B are one to two centimeters in diameter.

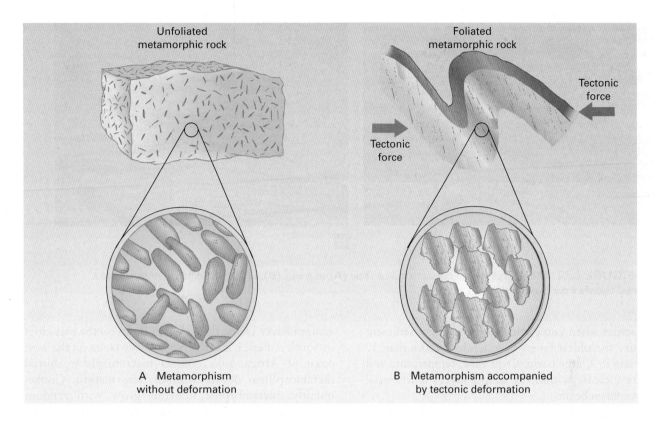

Unfoliated metamorphic rock

Foliated metamorphic rock

Tectonic force

Tectonic force

A Metamorphism without deformation

B Metamorphism accompanied by tectonic deformation

◆ **FIGURE 3.30** **(A)** When metamorphism occurs without deformation, platy micas grow with random orientations. **(B)** When deformation accompanies metamorphism, platy micas orient perpendicular to the force squeezing the rocks, forming foliated metamorphic rocks. The wavy lines in B represent original sedimentary layers that have been folded by the deformation.

formed when metamorphism altered both the texture and minerals of shale. If migrating fluids alter the chemical composition of a rock, new minerals invariably form.

Types of Metamorphism and Metamorphic Rocks

Recall that rising temperature, rising pressure, and changing chemical environment cause metamorphism.

Courtesy of Graham R. Thompson/Jonathan Turk

◆ **FIGURE 3.31** These metamorphic rocks in Vermont show well-developed slaty cleavage; they are fractured along foliation planes created by parallel alignment of micas.

In addition, deformation resulting from movement of Earth's crust develops foliation and thus strongly affects the texture of a metamorphic rock. Four different geologic processes create these changes.

Contact Metamorphism

Contact metamorphism occurs where hot magma intrudes cooler rock of any type—sedimentary, metamorphic, or igneous. The highest grade metamorphic rocks form at the contact, closest to the magma. Lower grade rocks develop farther out. A metamorphic halo around a pluton can range in width from less than a meter to hundreds of meters, depending on the size and temperature of the intrusion and the effects of water or other fluids (◆ Figure 3.33). Contact metamorphism commonly occurs without deformation. As a result, the metamorphic minerals grow with random orientations—like the pie plates flying through the air—and the rocks develop no metamorphic layering.

Burial Metamorphism

Burial metamorphism results from burial of rocks in a sedimentary basin. A large river like the Mississippi carries massive amounts of sediment to the ocean every year, where it accumulates on a delta. Over tens or even hundreds of millions of years, the weight of the sediment becomes so great that the entire region sinks, just as a

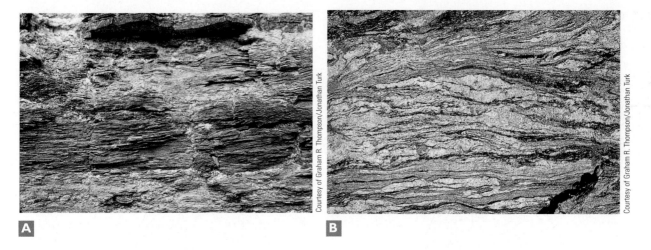

◇ **FIGURE 3.32** Metamorphism can convert a typical shale **(A)** to gneiss **(B)**, a rock with different minerals and a different texture from those of shale.

canoe settles when you climb into it. Younger sediment may bury the oldest layers to a depth of more than 10 kilometers in a large basin. Over time, temperature and pressure increase within the deeper layers until burial metamorphism begins.

Burial metamorphism is occurring today in the sediments underlying many large deltas, including the Mis-

sissippi River Delta, the Amazon Basin on the east coast of South America, and the Niger River Delta on the west coast of Africa. Like contact metamorphism, burial metamorphism occurs without deformation. Consequently, metamorphic minerals grow with random orientations and, like contact metamorphic rocks, burial metamorphic rocks are unfoliated.

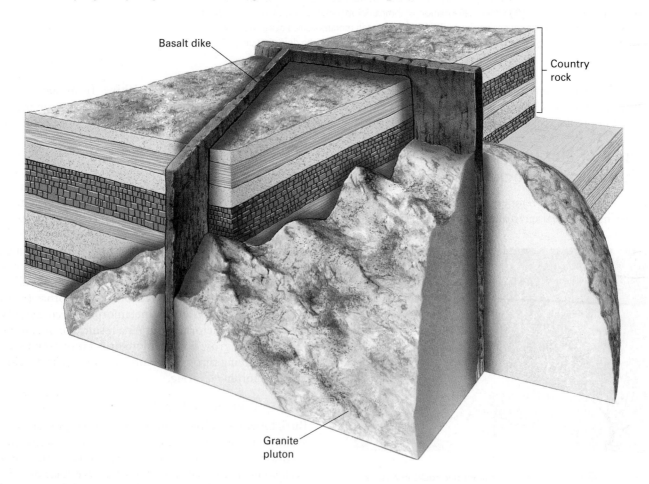

◇ **FIGURE 3.33** A halo of contact metamorphism, shown in red, surrounds a pluton. The later intrusion of the basalt dike metamorphosed both the pluton and the sedimentary rock, creating a second metamorphic halo.

Courtesy of Graham R. Thompson/Jonathan Turk

◆ **FIGURE 3.34** Light-colored granite has intruded folded metamorphic rocks in the Sierra Nevada, near Bishop, California.

Regional Dynamothermal Metamorphism

Regional dynamothermal metamorphism occurs where major crustal movements build mountains and deform rocks. The term *dynamothermal* simply indicates that the rocks are being deformed and heated at the same time. It is the most common and widespread type of metamorphism and affects broad regions of Earth's crust.

Magma rises and heats large portions of the crust in places where tectonic plates converge (see Chapter 6). The high temperatures cause new metamorphic minerals to form throughout the region. The rising magma also deforms the hot, plastic country rock as it forces its way upward. At the same time, the movements of the crust squeeze and deform rocks. As a result of all these processes acting together, regionally metamorphosed rocks are strongly foliated and are typically associated with mountains and igneous rocks (◆ Figure 3.34). Regional dynamothermal metamorphism produces zones of foliated metamorphic rocks tens to hundreds of kilometers across.

Shale is the most abundant type of sedimentary rock and consists of clay minerals, quartz, and feldspar. The mineral grains are too small to be seen with the naked eye and can barely be seen with a microscope. Shale undergoes a typical sequence of changes as metamorphic grade increases.

◆ Figure 3.35 shows the temperatures at which metamorphic minerals are stable as shale undergoes metamorphism. Thus, it shows the sequence in which minerals appear, and then decompose, as metamorphic grade increases. As regional metamorphism begins, clay minerals break down and mica and chlorite replace them (◆ Figure 3.36A). These new, platy minerals grow perpendicular to the direction of squeezing. As a result, the rock develops slaty cleavage and is called **slate** (◆ Figure 3.36B). With rising temperature and continued deformation, the micas and chlorite grow larger, and wavy or wrinkled surfaces replace the flat, slaty cleavage, giving **phyllite** a silky appearance (◆ Figure 3.36C).

As temperature continues to rise, the mica and chlorite grow large enough to be seen by the naked eye, and foliation becomes very well developed. Rock of this type is called **schist** (◆ Figure 3.36D). Schist forms approximately at the transition from low to intermediate metamorphic grades.

At high metamorphic grades, light- and dark-colored minerals often separate into bands that are thicker than the layers of schist, to form a rock called **gneiss** (pronounced *nice*) (◆ Figure 3.36E). At the highest metamorphic grade, the rock begins to melt, forming small veins of granitic magma. When metamorphism wanes and the rock cools, the magma veins solidify to form **migmatite**, a mixture of igneous and metamorphic rock (◆ Figure 3.36F).

Quartz sandstone and limestone transform to foliated quartzite and foliated marble, respectively, during regional metamorphism.

Hydrothermal Metamorphism

Water is a chemically active fluid; it attacks and dissolves rocks and minerals. If the water is hot, it attacks minerals even more rapidly. **Hydrothermal metamorphism** (also called hydrothermal alteration and metasomatism) occurs when hot water and ions dissolved in the hot water react with a rock to change its chemical composition and minerals. Circulating ground water—the water that saturates soil and bedrock—causes most hydrothermal alteration. Cold ground water sinks through bedrock fractures to depths of a few kilometers, where it is heated by the deep, hot rocks or in some cases by a hot, shallow pluton. Upon heating, the water expands and rises back toward the surface through other fractures (◆ Figure 3.37). As it rises, it alters the country rock through which it flows. In some hydrothermal environments, water reacts with sulfur minerals to form sulfuric acid, making the solution even more corrosive.

Hydrothermal metamorphism is yet another example of interactions between different Earth systems. In this case, water from the hydrosphere reacts with solid rock to change the chemical and physical characteristics of the rock, and in many cases creates valuable ore deposits used by humans.

Most rocks and magma contain low concentrations of metals such as gold, silver, copper, lead, and zinc. For example, gold makes up 0.0000002 percent of average crustal rock, while copper makes up 0.0058 percent and lead 0.0001 percent. Although the metals are present in very low concentrations, hydrothermal solutions sweep slowly through vast volumes of country rock, dissolving and accumulating the metals as they go. The solutions then deposit the dissolved metals, where they encounter changes in temperature, pressure, or chemical environment (◆ Figure 3.38). In this way, hydrothermal solutions scavenge and concentrate metals from average crustal rocks and then deposit them locally to form ore. Hydrothermal ore deposits are discussed further in Chapter 5.

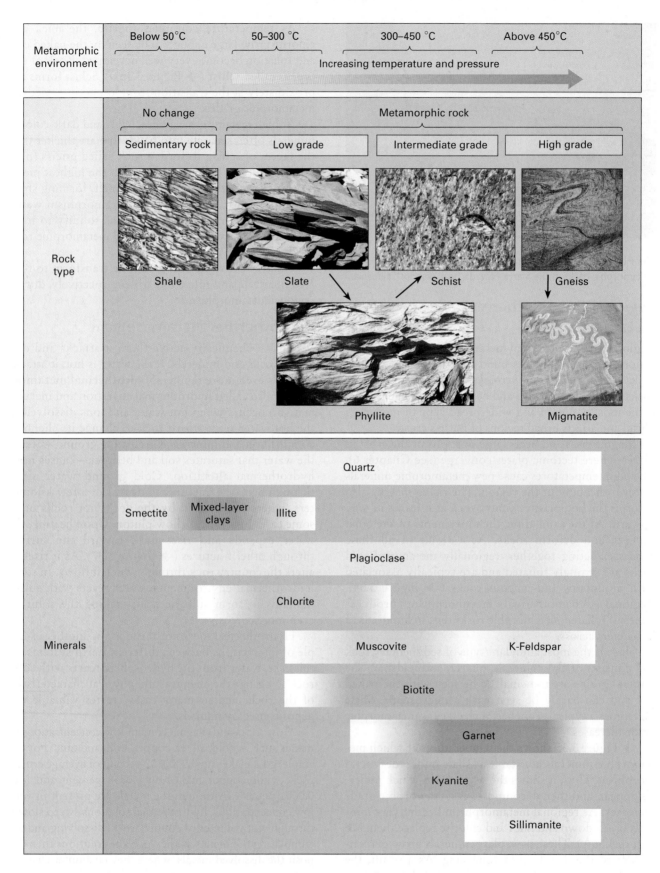

◇ **FIGURE 3.35** Shale undergoes changes both in texture and minerals as metamorphic grade increases. The lower part of the figure shows the stability ranges of common metamorphic minerals.

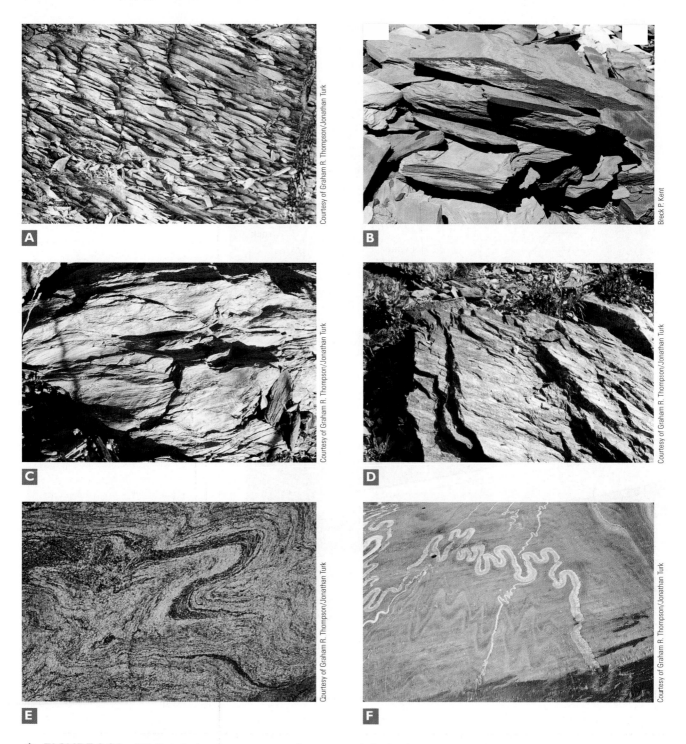

◇ FIGURE 3.36 **(A)** Shale is the most common sedimentary rock. Regional metamorphism progressively converts shale to slate **(B)**, phyllite **(C)**, schist **(D)**, and gneiss **(E)**. Migmatite **(F)** forms when gneiss begins to melt.

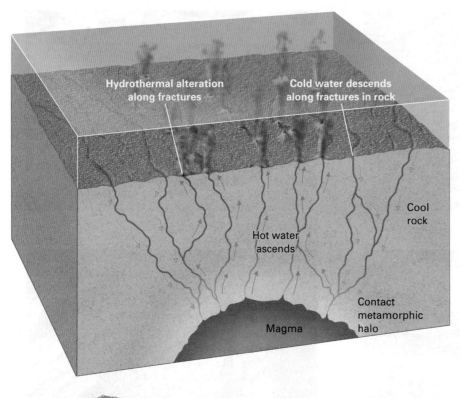

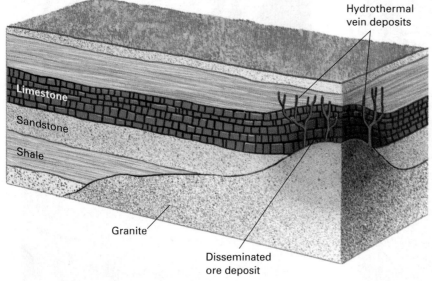

FIGURE 3.37 Ground water descending through fractured rock is heated by magma and rises through other cracks, causing hydrothermal metamorphism in nearby rock.

FIGURE 3.38 Hydrothermal ore deposits form when hot water dissolves metals from country rock and deposits them in fractures and surrounding country rock.

EARTH SYSTEMS INTERACTIONS

THE ROCK CYCLE illustrates interactions among all four Earth Systems: atmospheric gases combine with rain and ground water to form weak acids that decompose and dissolve rocks of all types—igneous, sedimentary, and metamorphic. Dissolved calcium and other ions then flow to the seas in streams and ground water where clams, oysters, corals, and many other marine organisms of our biosphere use the ions to form their shells and other hard parts. When the organisms die, their shells accumulate and become lithified to form limestone.

As another example, large volcanic eruptions emit 0.4% by weight as much carbon dioxide into the atmosphere as they erupt lava onto Earth's surface. Carbon dioxide is a greenhouse gas that can cause atmospheric warming and affect plants and animals. Other volcanic gases such as sulfur dioxide and volcanic ash can block solar radiation to cause atmospheric cooling. Thus, Earth materials cycle continuously among all four Earth Systems, and those cycles profoundly affect our environment. ■

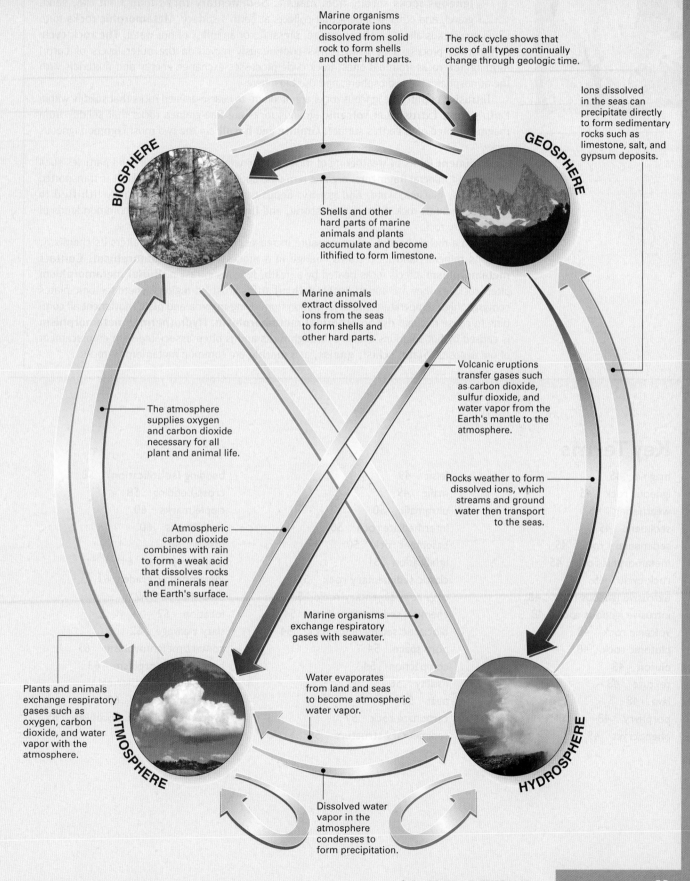

Marine organisms incorporate ions dissolved from solid rock to form shells and other hard parts.

The rock cycle shows that rocks of all types continually change through geologic time.

Ions dissolved in the seas can precipitate directly to form sedimentary rocks such as limestone, salt, and gypsum deposits.

BIOSPHERE

GEOSPHERE

Shells and other hard parts of marine animals and plants accumulate and become lithified to form limestone.

Marine animals extract dissolved ions from the seas to form shells and other hard parts.

Volcanic eruptions transfer gases such as carbon dioxide, sulfur dioxide, and water vapor from the Earth's mantle to the atmosphere.

The atmosphere supplies oxygen and carbon dioxide necessary for all plant and animal life.

Rocks weather to form dissolved ions, which streams and ground water then transport to the seas.

Atmospheric carbon dioxide combines with rain to form a weak acid that dissolves rocks and minerals near the Earth's surface.

Marine organisms exchange respiratory gases with seawater.

Plants and animals exchange respiratory gases such as oxygen, carbon dioxide, and water vapor with the atmosphere.

Water evaporates from land and seas to become atmospheric water vapor.

ATMOSPHERE

HYDROSPHERE

Dissolved water vapor in the atmosphere condenses to form precipitation.

3.4 Metamorphic Rocks

SUMMARY

Geologists divide rocks into three groups, depending upon how the rocks formed. **Igneous rocks** solidify from magma. **Sedimentary rocks** form from clay, sand, gravel, and other sediment that collects at Earth's surface. **Metamorphic rocks** form when any rock is altered by temperature, pressure, or an influx of hot water. **The rock cycle** summarizes processes by which rocks continuously recycle in the outer layers of Earth, forming new rocks from old ones. Rock cycle processes exchange energy and materials with the atmosphere, the hydrosphere, and the biosphere.

Intrusive or **plutonic** igneous rocks are medium- to coarse-grained rocks that solidify within Earth's crust. **Extrusive** or **volcanic** igneous rocks are fine-grained rocks that solidify from magma erupted onto Earth's surface. **Granite** and **basalt** are the two most common igneous rocks.

Sediment forms by weathering of rocks and minerals. It includes all solid particles such as rock and mineral fragments, organic remains, and precipitated minerals. It is transported by streams, glaciers, wind, and gravity; deposited in layers; and eventually **lithified** to form sedimentary rock. **Shale, sandstone,** and **limestone** are the most common kinds of sedimentary rock.

When a rock is heated, when pressure increases, or when hot water alters its chemistry, both its minerals and its textures change in a process called **metamorphism. Contact metamorphism** affects rocks heated by a nearby igneous intrusion. **Burial metamorphism** alters rocks as they are buried deeply within Earth's crust. In regions where tectonic plates converge, high temperature and deformation from rising magma and plate movement all combine to cause regional **dynamothermal metamorphism. Hydrothermal metamorphism** is caused by hot solutions soaking through rocks and is often associated with emplacement of ore deposits. **Slate, schist, gneiss,** and **marble** are common metamorphic rocks.

Key Terms

magma   45
igneous rock   45
weathering   45
sediment   45
sedimentary rock   45
metamorphic rock   45
rock cycle   45
extrusive igneous rock   48
intrusive igneous rock   48
volcanic rock   48
plutonic rock   48
pluton   48
texture   48
lava   48
porphyry   48
phenocryst   49

felsic   49
mafic   49
ultramafic   50
intermediate rock   50
basement rock   50
lithification   51
clastic sedimentary rock   51
organic sedimentary rock   51
chemical sedimentary rock   54
bioclastic sedimentary rock   54
pore space   54
compaction   54
fissility   56
peat   56
carbonate rock   57
sedimentary structure   58

bedding (stratification)   58
cross-bedding   58
ripple marks   60
mud cracks   60
fossils   60
metamorphism   61
metamorphic grade   61
deformation   62
foliation   62
slaty cleavage   62
contact metamorphism   63
burial metamorphism   63
regional dynamothermal
   metamorphism   65
hydrothermal metamorphism   65

Common Igneous Rocks

Felsic	Intermediate	Mafic	Ultramafic
obsidian	andesite	basalt	peridotite
granite	diorite	gabbro	rhyolite

Common Sedimentary Rocks

Clastic	Bioclastic	Organic	Chemical
conglomerate	limestone	chert	evaporite
sandstone	dolomite	coal	
siltstone	(dolostone)		
shale	coquina		
	chalk		

Common Metamorphic Rocks

Unfoliated	Foliated
marble	slate
quartzite	phyllite
	schist
	gneiss
	migmatite

For Review

1. Explain what the rock cycle tells us about Earth processes.
2. Describe ways in which rock cycle processes exchange energy and materials with the atmosphere, the hydrosphere, and the biosphere.
3. What are the three main kinds of rock in Earth's crust?
4. How do the three main types of rock differ from each other?
5. What is magma? Where does it originate?
6. Describe how igneous rocks are classified and named.
7. What are the two most common kinds of igneous rock?
8. Describe and explain the differences between a plutonic and a volcanic rock.
9. What is sediment? How does it form?
10. How do sedimentary grains become rounded?
11. Where in your own area would you look for rounded sediment?
12. Describe how sediment becomes lithified.
13. What are the differences among shale, sandstone, and limestone?
14. Explain why almost all sedimentary rocks are layered, or bedded.
15. How does cross-bedding form?
16. What is metamorphism? What factors cause metamorphism?
17. What kinds of changes occur in a rock as it is metamorphosed?
18. What is metamorphic foliation? How does it differ from sedimentary bedding?
19. How do contact metamorphism and regional metamorphism differ, and how are they similar?

For Discussion

1. Magma usually begins to rise toward Earth's surface as soon as it forms. It rarely accumulates as large pools in the upper mantle or lower crust, where it originates. Why?
2. In the San Juan Mountains of Colorado, parts of the range are made of granite, and other parts are volcanic rocks. Explain why these two types of igneous rock are likely to be found together.
3. Suppose you were given a fist-sized sample of igneous rock. How would you tell whether it is volcanic or plutonic in origin?
4. How would you tell whether another rock sample is igneous, sedimentary, or metamorphic in origin?
5. Why is shale the most common sedimentary rock?

6. One sedimentary rock is composed of rounded grains, and the grains in another are angular. What can you tell about the history of the two rocks from these observations?

7. What do mud cracks and raindrop imprints in shale tell you about the water depth in which the mud accumulated?

8. How would you distinguish between contact metamorphic rocks and regional metamorphic rocks in the field?

9. What happens to bedding when sedimentary rocks undergo regional metamorphism?

10. How can granite form during metamorphism?

11. Explain how the geosphere, the atmosphere, the hydrosphere, and the biosphere all contribute to the rock cycle.

ThomsonNOW™

Assess your understanding of this chapter's topics with additional comprehensive interactivities at **www.thomsonedu.com/login**, which also has current and up-to-date web links, additional readings, and exercises.

CHAPTER 4

Geologic Time: A Story in the Rocks

Dirk Wiersma / Photo Researchers, Inc.

A fossil of an ammonoid, a marine animal about 2 centimeters in diameter. The original shell and soft tissues have been replaced by calcite and other minerals. Ammonoids lived between 500 million years ago, and 65 million years ago.

James Hutton, one of the earliest geologists, realized that sandstone forms as sand slowly accumulates and is compressed into solid rock. He also saw that sand and other sediment accumulate layer by layer, so that the deepest layers of sedimentary rocks are the oldest and the layers become younger toward Earth's surface. Early geologists also found ancient shells, bones, and other fossils in sandstone and in other kinds of sedimentary rocks. They realized that if the rocks formed with younger layers above older ones, and if fossils were preserved as the rock formed, then fossils in deep layers must be older than the fossils in the upper layers. Therefore, scientists can trace the history of life on Earth by studying fossils embedded in successively younger rock layers. This history is now summarized in a geologic column and time scale, shown later in this chapter. ■

4.1

Earth Rocks, Earth History, and Mass Extinctions

In many locations, nineteenth-century geologists found thick sequences of rock layers with abundant fossils that we now know represent millions, or hundreds of millions, of years of Earth history. But younger layers above the fossil-rich rocks contained few fossils or none at all. Looking even higher in the sequence of rock layers, they found abundant fossils again, but of organisms very different from those in the older rocks. Even more surprising, fossils of many of the most abundant animals and plants in lower layers were never seen again in younger rocks. Thus, those organisms simply disappeared from the face of Earth forever.

This fossil record suggests that sudden, catastrophic events had abruptly decimated life on Earth, and that new life forms emerged following these **mass extinctions.** However, scientists did not immediately accept the idea that catastrophic events had caused sudden mass extinctions. Instead, they suggested that the rock record might not be complete. Maybe some rocks had been destroyed by erosion, or perhaps sedimentary rocks had not formed for a long period of time. According to this reasoning, the part of the record that was lost could have contained evidence for the gradual decline of the species that became extinct, and for the slow emergence of new species. Following conventional wisdom of the time, they concluded that the extinctions of old life forms and emergence of new ones occurred slowly, as a result of gradually changing conditions.

But if pieces of the rock record were missing in some locations, they must exist in other places on Earth. Geo-

logists searched for rocks to fill in the gaps and provide evidence for gradual extinctions, but they did not find them. Eventually, the evidence for near instantaneous extinctions became compelling. From studies of this type, scientists learned that at least six times in Earth history, catastrophic extinctions decimated life on Earth (◆ Figure 4.1).

The most dramatic extinction occurred 248 million years ago, at the end of the Permian Period. At that time, 90 percent of all species in the oceans suddenly died out. On land, about two thirds of reptile and amphibian species and 30 percent of insect species vanished.

One modern scientist exclaimed that this extinction event was "the closest life has come to complete extermination since its origin."[1]

The death of most life forms at the end of Permian time left huge ecological voids in the biosphere. Ocean ecosystems changed as new organisms emerged in an environment free of predators and competition. On land, dinosaurs and many other terrestrial animals and plants emerged and proliferated.

One hundred and sixty million years later, another catastrophic extinction wiped out one fourth of all species, including the dinosaurs. This mass extinction occurred 65 million years ago. Small mammals survived this disaster, facing a new world free of the efficient predators that had hunted them. The number of new mammal species increased rapidly after this extinction event, eventually leading to the evolution of humans.

What kinds of catastrophic events caused these sudden extinctions? Why would species that had survived for tens of millions of years instantaneously disappear? Consider three hypotheses for mass extinctions. Each of these hypotheses involves large-scale systems

1. Douglas Erwin, *The Great Paleozoic Crisis* (New York: Columbia University Press, 1993), 187.

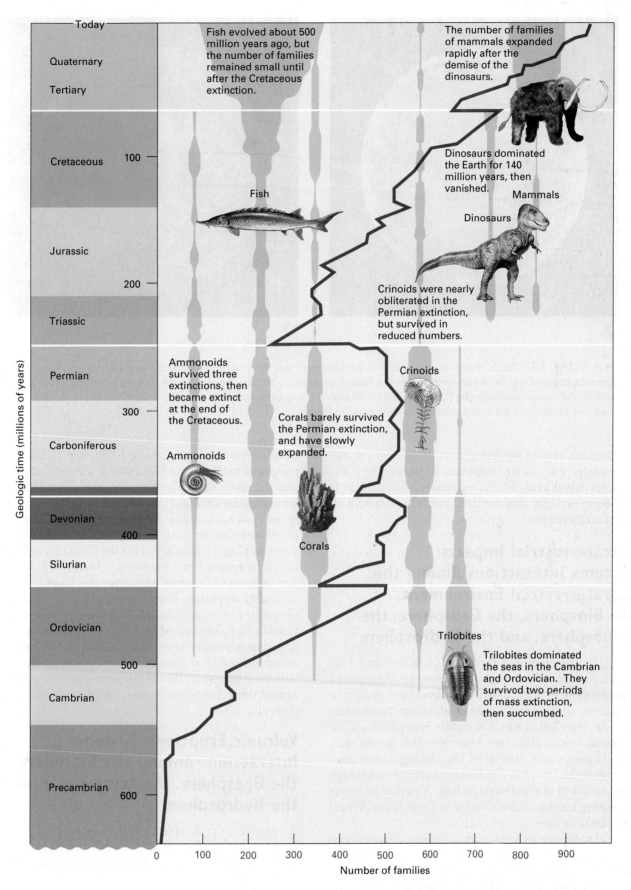

◆ FIGURE 4.1 The red line shows that the number of families of organisms, indicated by the thickness of the blue shaded areas, has varied throughout Earth history. Sudden leftward shifts of the red line indicate mass extinctions. The six mass extinctions described in the text occurred in late Cambrian, the end of Ordovician, late Devonian, the end of Permian, late Triassic, and the end of Cretaceous times. The numbers in the left column indicate time in millions of years ago.

4.1   Earth Rocks, Earth History, and Mass Extinctions

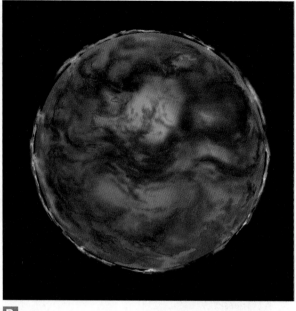

◆ **FIGURE 4.2**   Today, most scientists agree with the hypothesis that the Cretaceous extinction was caused by a giant meteorite that slammed into Earth just north of Mexico's Yucatán peninsula. In this artist's rendition, **(A)** a large meteorite struck just north of the Yucatán peninsula. **(B)** The impact pulverized both the meteorite and rocks of Earth's crust, creating a thick, dark cloud that blocked out the sunlight, causing Earth's surface to freeze.

interactions among the biosphere, the geosphere, the atmosphere, and the hydrosphere. In the first hypothesis, described next, a fifth, extraterrestrial, system is seen as the trigger that initiated radical changes in all four Earth systems.

Extraterrestrial Impacts: Systems Interactions Among the Extraterrestrial Environment, the Biosphere, the Geosphere, the Atmosphere, and the Hydrosphere

In 1977, the father-and-son team of Walter and Luis Alvarez were studying rocks that formed at the time the dinosaurs and so many other life forms became extinct. In one layer, they found abundant dinosaur fossils. Just above it, they found very few fossils of any kind and no dinosaur fossils. Between these two rock layers, they found a thin, sooty, clay layer. They brought samples of the clay to the laboratory and found that it contained high concentrations of the element iridium. This discovery was surprising because iridium is rare in Earth rocks. Where did it come from?

Although rare in Earth's crust, iridium is more abundant in meteorites. Alvarez and Alvarez suggested that 65 million years ago, a meteorite 10 kilometers in diameter hit the Earth with energy equal to 10,000 times that of today's entire global nuclear arsenal. The collision vaporized both the meteorite and Earth's crust at the point of impact, forming a plume of hot dust and gas that ig-

nited fires around the planet (◆ Figure 4.2). Soot from the global wildfires and iridium-rich meteorite dust rose into the upper atmosphere, circling the globe. The thick, dark cloud blocked out the Sun and halted photosynthesis for as much as a year. Surface waters froze, and many plants and animals died. This event, called the terminal Cretaceous extinction, killed off the dinosaurs and many other life forms. Then the sooty, iridium-rich dust settled to Earth to form the distinctive clay layer.

Other scientists have found evidence of a huge meteorite crater that formed 65 million years ago in the Caribbean Sea north of the Yucatán peninsula. Most scientists now agree with the meteorite impact hypothesis for the terminal Cretaceous extinction. Some scientists have suggested that an extraterrestrial impact also caused the Permian extinction, but the evidence is less compelling.[2]

Volcanic Eruptions: Systems Interactions Among the Biosphere, the Geosphere, the Atmosphere, and the Hydrosphere

A volcanic eruption ejects gas and fine volcanic ash into the atmosphere. The ash acts as an umbrella that reflects sunlight back into space and thus causes climatic cooling.

---

2. Douglas Erwin, *The Great Paleozoic Crisis* (New York: Columbia University Press, 1993).

One volcanic gas, sulfur dioxide, forms small particles called aerosols that also reflect sunlight and cool Earth.

Carbon dioxide is another gas emitted by volcanoes. It is a greenhouse gas that can cause warming of Earth's atmosphere. Thus, volcanoes erupt materials capable of causing both atmospheric cooling and warming. In many eruptions, the cooling overwhelms the warming effects. This cooling can occur rapidly—within a few weeks or months of the eruption. The relationships between volcanic eruptions and climatic cooling and warming are discussed further in Chapters 8 and 21.

Scientists have noted that some mass extinctions coincided with unusually high rates of volcanic activity. For example, massive flood basalts erupted onto Earth's surface in Siberia 248 million years ago—at the same time that the Permian extinction occurred. According to one hypothesis, explosive volcanic eruptions blasted massive amounts of volcanic ash and sulfur aerosols into the air. The ash and aerosols spread like a pall throughout the upper atmosphere, blocking out the Sun. Earth's atmospheric and oceanic temperatures plummeted, plants withered, and animals starved or froze to death. If this hypothesis is correct, the Permian extinction (and perhaps others) resulted from a classic Earth systems interaction, where a volcanic eruption occurring within the geosphere altered the composition of the atmosphere, with profound consequences for atmospheric and marine temperatures, and for life.

Supercontinents and Earth's Carbon Dioxide Budget: Systems Interactions Among the Biosphere, the Geosphere, the Atmosphere, and the Hydrosphere

In the modern oceans, cold polar seawater sinks to the sea floor and flows as a deep ocean current toward the equator. This current transports oxygen to the deep ocean basins and forces the deep water to rise near the equator, where it is warmed. The current thus mixes both heat and gases throughout the oceans and the atmosphere. You can read more about ocean currents in Chapter 16.

Ocean currents are partially controlled by the positions of the continents. Recall from Chapter 1 that the continents slowly drift across the globe. Several times in Earth history, all the continents have joined together to form one giant supercontinent. One such supercontinent assembled during Permian time (◀▶ Figure 4.3 A). Computer calculations indicate that the assembly of all continents into a single land mass, and the consequent development of a single global ocean, prevented mixing between the surface water and the ocean depths.[3]

---

3. A. H. Knoll et al., "Comparative Earth History and Late Permian Mass Extinction," *Science* 273 (July 26, 1996): 452 ff.

Surface marine organisms absorb atmospheric carbon dioxide and use it to produce organic tissue. These organisms then die and settle to the sea floor. This process removes carbon dioxide from the atmosphere, converts the carbon to organic tissue, and transports it to the sea floor. Organisms living near the deep sea floor then consume the fallen litter and release carbon dioxide into the deep ocean water. Much of this gas dissolves in seawater and is held there by the pressure of overlying water. Modern ocean currents mix deep and shallow seawater, returning the carbon dioxide to the atmosphere.

However, because there was little vertical mixing in the Permian oceans, carbon accumulated in ever-increasing concentration in the deep oceans (◀▶ Figure 4.3 B). This process continued to remove carbon dioxide from the atmosphere and stored it in the deep oceans during late Permian time. But carbon dioxide is a greenhouse gas that absorbs heat and warms the atmosphere. As carbon dioxide was removed from the air and stored in deep ocean water, the atmosphere cooled. Some scientists have suggested that continental ice sheets formed as a result of the global cooling. Sea level fell as water evaporated from the oceans and accumulated on continental glaciers, exposing continental shelves and stressing populations of shallow-dwelling marine organisms.

The cool air cooled the surface of the sea. When water cools, it becomes denser. Eventually a threshold was reached at which polar surface water became denser than the deep water. At this point the cool, dense surface water sank. The sinking surface waters forced the carbon dioxide–rich deep water to rise to the sea surface, where it rapidly released massive amounts of carbon dioxide into the atmosphere (◀▶ Figure 4.3 C). According to this hypothesis, the carbon dioxide asphyxiated life both in the seas and on the continents, killing most life on Earth. Later, new organisms evolved when the carbon dioxide levels fell.

Each of these three hypotheses involves large-scale interactions among the atmosphere, the hydrosphere, the biosphere, and the geosphere, and the first also includes an extraterrestrial factor. The third hypothesis, if correct, is a classic example of both Earth systems feedback mechanisms and a threshold effect. The feedback process occurred as tectonic plate movement altered the shapes of the ocean basins. Ocean currents changed when the basins changed. The alteration in currents, in turn, changed atmospheric composition and unleashed a threshold event that rapidly killed most living organisms on Earth.

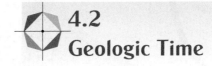

4.2 Geologic Time

While most of us think of time in terms of days or years, Earth scientists commonly refer to events that happened

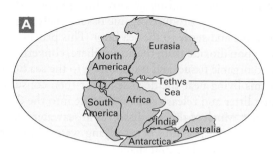

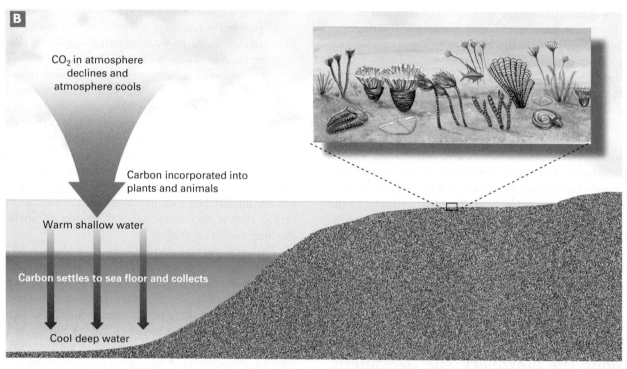

◈ **FIGURE 4.3** **(A)** The assembly of all continents into the supercontinent Pangaea during Permian time prevented vertical mixing of ocean water. **(B)** Near-surface aquatic plants absorbed atmospheric carbon dioxide and incorporated the carbon into organic tissue. This carbon collected on the sea floor. As a result, the carbon dioxide concentration in the air declined and the atmosphere cooled. **(C)** When the atmosphere cooled sufficiently, polar surface water became dense enough to sink, stirring up the deep ocean water and releasing vast amounts of carbon dioxide back toward the surface. This carbon dioxide asphyxiated both marine and terrestrial organisms.

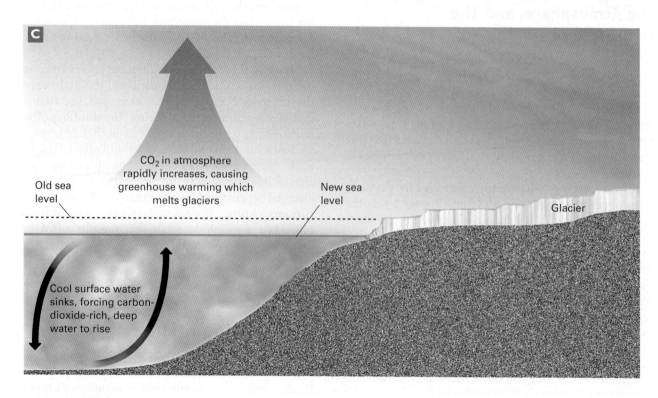

millions or billions of years ago. In Chapter 1 you learned that Earth is approximately 4.6 billion years old. Yet humans and our human-like ancestors have existed for only 5 to 7 million years, and recorded history is only a few thousand years old. How do scientists measure the ages of rocks and events that occurred millions or billions of years ago, long before recorded history or even human existence?

Scientists measure geologic time in two ways. **Relative age** measurement refers only to the order in which events occurred. Determination of relative age is based on a simple principle: In order for an event to affect a rock, the rock must exist first. Thus, the rock must be older than the event. This principle seems obvious, yet it is the basis of much geologic work. As you learned in Chapter 3, sediment normally accumulates in horizontal layers. However, the sedimentary rocks in ◆ Figure 4.4 are folded and tilted. We deduce that the folding and tilting occurred after the sediment accumulated. The order in which rocks and geologic features formed can usually be interpreted by such observation and logic.

Absolute age is age in years. Dinosaurs became extinct 65 million years ago. The terminal Permian extinction occurred 248 million years ago. The Teton Range in Wyoming began rising 6 million years ago.

Absolute age tells us both the order in which events occurred and the amount of time that has passed since they occurred.

4.3 Relative Geologic Time

Absolute age measurements have become common only since the second half of the twentieth century. Prior to that time, geologists used field observations to determine relative ages. Even today, with sophisticated laboratory processes available, most field geologists routinely use relative age measurement. Geologists use a combination of common sense and a few simple principles to determine the order in which rocks formed and changed over time.

The principle of original horizontality is based on the observation that sediment usually accumulates in horizontal layers (◆ Figure 4.5A). If sedimentary rocks lie at an angle, as in ◆ Figure 4.5B, or if they are folded as in Figure 4.4, we can infer that tectonic forces tilted or folded them after they formed.

◆ **FIGURE 4.4** Limestone formed 500 million years ago in a shallow sea. Four hundred million years later, tectonic forces uplifted and folded it in the Canadian Rockies, Alberta.

◈ **FIGURE 4.5** **(A)** The principle of original horizontality tells us that most sedimentary rocks are deposited with horizontal bedding (San Juan River, Utah). When we see tilted rocks **(B)**, we infer that they were tilted after they were deposited (Connecticut).

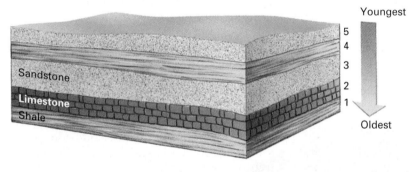

◈ **FIGURE 4.6** In a sequence of sedimentary beds, the oldest bed (labeled 1) is the lowest, and the youngest (labeled 5) is on top.

INTERACTIVE QUESTION: *Imagine that the horizontal beds in this figure were tilted so they stood vertically. How could you tell which was the oldest and which was the youngest?*

◈ **FIGURE 4.7** A basalt dike cutting across sandstone in the walls of Grand Canyon. The dike must be younger than the sandstone.

The principle of superposition states that sedimentary rocks become younger from bottom to top (as long as tectonic forces have not turned them upside down or thrust an older layer over a younger one). This is because younger layers of sediment always accumulate on top of older layers. In ◈ Figure 4.6, the sedimentary layers become progressively younger in the order 1, 2, 3, 4, and 5.

The principle of crosscutting relationships is based on the obvious fact that a rock must first exist before anything can happen to it. ◈ Figure 4.7 shows a basalt dike cutting across layers of sandstone. Clearly, the dike must be younger than the sandstone. ◈ Figure 4.8 shows sedimentary rocks intruded by three granite dikes. Dike 2 cuts dike 1, and dike 3 cuts dike 2, so dike 2 is younger than 1, and dike 3 is the youngest. The sedimentary rocks must be older than all of the dikes.

The theory of **evolution** states that life forms have changed throughout geologic time. Species emerge, persist for a while, and then become extinct. **Fossils** are the remains and other traces of prehistoric life that are preserved in sedimentary rocks. They are useful in determining relative ages of rocks because different animals and plants lived at different times in Earth history. For example, trilobites lived from 544 million to about 248 million years ago and the first dinosaurs appeared about 248 million years ago.

The principle of superposition tells us that sedimentary rock layers become younger from bottom to top. If the rocks formed over a long time, different

◆ **FIGURE 4.8** Three granite dikes cutting sedimentary rocks. The dikes become younger in the order 1, 2, 3. The sedimentary rocks must be older than all three dikes.

Fossils also allow geologists to interpret ancient environments because most organisms thrive in specific environments and do not survive in others. For example, corals live in clear, warm, shallow seas. Yet today, fossil corals are abundant in rocks of many cold mountain environments such as the Canadian Rockies and the summit of Mount Everest. The fossil corals tell us that the rocks that are exposed in these mountains once lay submerged beneath clear, warm, shallow seas. Today, geologists understand that tectonic processes raised these marine rocks and their fossils to form some of the world's highest peaks.

fossils appear and then vanish from bottom to top in the same order in which the organisms evolved and then became extinct. Rocks containing dinosaur bones must be younger than those containing trilobite remains.

The principle of faunal succession states that species succeeded one another through time in a definite and recognizable order and that the relative ages of sedimentary rocks can therefore be recognized from their fossils.

Interpreting Geologic History from Fossils

Paleontologists study fossils to understand the history of life and evolution. The oldest known fossils are traces of bacteria-like organisms that lived about 3.5 billion years ago. A much younger fossil consists of the frozen and mummified remains of a Bronze Age man recently found in a glacier near the Italian-Austrian border (◆ Figure 4.9).

4.4 Unconformities and Correlation

The 2-kilometer-high walls of Grand Canyon are composed of sedimentary rocks lying on older igneous and metamorphic rocks (◆ Figure 4.10). Their ages range from about 200 million years to nearly 2 billion years. The principle of superposition tells us that the deepest rocks are the oldest, and the rocks become younger as we climb up the canyon walls. However, no principle assures us that the rocks formed continuously from 2 billion to 200 million years ago. The rock record may be incomplete. Suppose that no sediment were deposited for a period of time, or erosion removed some sedimentary layers before younger layers accumulated. In either case a gap would exist in the rock record. We know that any rock layer is younger than the layer below it, but without more information we do not know how much younger.

Layers of sedimentary rocks are **conformable** if they were deposited without interruption. An **unconformity** represents an interruption in deposition, usually of long duration. During the interval when no sediment was deposited, some rock layers may have been eroded. Thus, an unconformity represents a long time interval for which no geologic record exists in that place. The lost record may involve hundreds of millions of years.

Several types of unconformities exist. In a **disconformity**, the sedimentary layers above and below the

© Corbis/CORBIS SYGMA

◆ **FIGURE 4.9** The mummified remains of this Bronze Age hunter were recently discovered preserved by glacial ice near the Austrian-Italian border.

Nonconformity

◆ **FIGURE 4.10** The 2-kilometer-high walls of Grand Canyon are composed of sedimentary rocks lying on older igneous and metamorphic rocks. Their ages range from 2 billion years to 200 million years old. The two-billion-year-old schist in the lower part of the photo lies beneath the horizontally bedded, younger sedimentary rocks.

unconformity are parallel (◆ Figures 4.11 and 4.12). A disconformity may be difficult to recognize unless an obvious soil layer or erosional surface developed. However, geologists identify disconformities by determining the ages of rocks using methods based on fossils and absolute dating, described later in this chapter.

In an **angular unconformity**, tectonic activity tilted older sedimentary rock layers, and erosion planed them off, before younger sediment accumulated (◆ Figures 4.13 and 4.14). A **nonconformity** is an unconformity in which sedimentary rocks lie on igneous or metamorphic rocks. The nonconformity

shown in ◆ Figure 4.10 represents a time gap of about 1 billion years.

Ideally geologists would like to develop a continuous history for each region of Earth by interpreting rocks that formed in that place throughout geologic time. However, there is no single place on Earth where rocks formed and were preserved continuously. Erosion has removed some rock layers and at times no rocks formed. Consequently, the rock record in any one place is full of unconformities, or, historical gaps. To assemble as complete and continuous a record as possible, geologists combine evidence from many

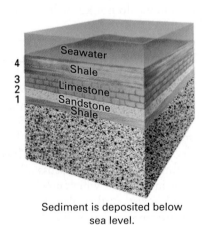

4
3
2
1

Seawater

Shale

Limestone

Sandstone

Shale

Sediment is deposited below sea level.

Eroded surface

2
1

Rocks are exposed above sea level and layers 4 and 3 are removed by erosion.

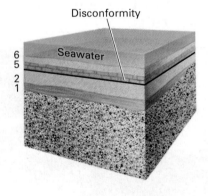

Disconformity

6
5

Seawater

2
1

Rocks subside below sea level and layers 5 and 6 are deposited on the eroded surface.

ThomsonNOW ◆ **ACTIVE FIGURE 4.11**

A disconformity is created by uplift and erosion, followed by deposition of additional layers of sediment. The disconformity represents a gap in the rock record.

CHAPTER 4 • Geologic Time: A Story in the Rocks

FIGURE 4.12 A disconformity separates parallel layers of sandstone and overlying conglomerate in Wyoming. Some sandstone layers were eroded away before the conglomerate was deposited.

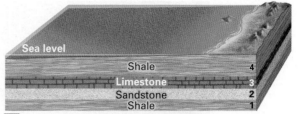

A Sediment is deposited below sea level

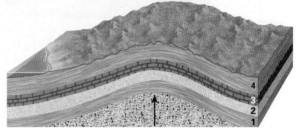

B Rocks are uplifted and tilted

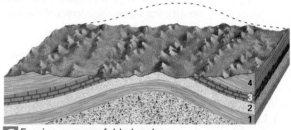

C Erosion exposes folded rocks

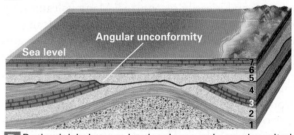

D Rocks sink below sea level and new rocks are deposited

FIGURE 4.13 An angular unconformity develops when older sedimentary rocks are folded and eroded before younger sediment accumulates.

INTERACTIVE QUESTION: *Draw a sequence in which older sedimentary rocks are tilted and eroded before younger sediment accumulates.*

localities. To do this, rocks of the same age from different localities must be matched in a process called **correlation** (◆ Figure 4.15). But how do we correlate rocks of the same age over great distances?

If you follow a single sedimentary bed from one place to another, then it is clearly the same layer in both places. But this approach is impractical over long distances and where rocks are not exposed. Another problem arises when correlation is based on continuity of a sedimentary layer. When rocks are correlated for the purpose of building a geologic time scale, geologists want to show that certain rocks all formed at the same time. Suppose that you are attempting to trace a beach sandstone that formed as sea level rose. The beach would have migrated inland over time and, as a result, the sandstone becomes younger in a landward direction (◆ Figure 4.16).

Over a distance short enough to be covered in an hour or so of walking, the age difference may be unimportant. But if you traced a similar beach sand for hundreds of kilometers, its age may vary by millions of years. Thus, there are two kinds of correlation: *Time correlation* means age equivalence, but *lithologic correlation* means continuity of a rock unit, such as the sandstone. The two are not always the same because some rock units, such as the sandstone, were deposited at different times in different places. To construct a record of Earth history and a geologic time scale, Earth

scientists use several types of evidence to correlate rocks of the same age.

Index fossils accurately indicate the ages of sedimentary rocks. An index fossil is produced by an organism that (1) is abundantly preserved in rocks, (2) was geographically widespread, (3) existed as a species or genus for only a relatively short time, and (4) is easily identified in the field. Floating or swimming marine animals make the best index fossils because they spread rapidly and widely throughout the seas. The shorter the time span that a species existed, the more precisely the index fossil reflects the age of a rock.

In many cases, the presence of a single type of index fossil is sufficient to establish the age of a rock. More

Angular
unconformity

◆ **FIGURE 4.14** An angular unconformity near Capitol Reef National Park, Utah. The dark line shows the bottom of the angular unconformity.

commonly, an assemblage of several fossils is used to date and correlate rocks. ◆ Figure 4.17 shows an example of how index fossils and fossil assemblages are used in correlation.

A **key bed** is a thin, widespread sedimentary layer that was deposited rapidly and synchronously over a wide area and is easily recognized. Many volcanic eruptions eject great volumes of fine, glassy volcanic ash into the atmosphere. Wind carries the ash over great distances before it settles. Some historic ash clouds have encircled the globe. When the ash settles, the glass rapidly crystallizes to form a pale clay layer that is incorporated into sedimentary rocks. Such volcanic eruptions occur at a precise point in time, so the ash is the same age everywhere. The thin, sooty, iridium-rich clay layer deposited 65 million years ago from debris of a giant meteorite impact (described at the beginning of this chapter) is a classic example of a key bed. How does a geologist measure the absolute age of an event that occurred before calendars and even before

4.5
Absolute Geologic Time

humans evolved to keep calendars? Think of how a calendar measures time. Earth rotates about its axis at a constant rate, once a day. Thus, each time the Sun rises, you know that a day has passed and you check it off on your calendar. If you mark off each day as the Sun rises, you record the passage of time. To know how many days have passed since you started keeping time, you just count the checkmarks. Absolute age measurement depends on two factors: a process that occurs at a constant rate (Earth rotates once every 24 hours) and some way to keep a cumulative record of that process (mark-

ing the calendar each time the Sun rises). Measurement of time with a calendar, a clock, an hourglass, or any other device depends on these two factors.

Geologists have found a natural process that occurs at a constant rate and accumulates its own record: it is the radioactive decay of elements that are present in many rocks. Thus, many rocks have built-in calendars. We must understand radioactivity to read the calendars.

Recall from Chapter 2 that an atom consists of a small, dense nucleus surrounded by a cloud of electrons. A nucleus consists of positively charged protons and neutral particles called neutrons. All atoms of any given element have the same number of protons in the nucleus. However, the number of neutrons may vary. **Isotopes** are atoms of the same element with different numbers of neutrons. For example, all isotopes of potassium-40 contain 19 protons and 21 neutrons. Potassium-39 has 19 protons but only 20 neutrons.

Many isotopes are stable, meaning that they do not change with time. If you studied a sample of potassium-39 for 10 billion years, all the atoms would remain unchanged. Other isotopes are unstable or radioactive. Given time, their nuclei spontaneously decay.[4] Potassium-40 decays naturally to form two other isotopes, argon-40 and calcium-40 (◆ Figure 4.18).

A radioactive isotope such as potassium-40 is known as a parent isotope. An isotope created by radioactivity,

---

4. Radioactive nuclei decay by three processes. In electron capture, the nucleus captures an electron. The electron combines with a proton to create a neutron. The decay of potassium-40 to argon-40 is an example of this process. In beta emission, a neutron ejects an electron from the nucleus, and the neutron becomes a proton. The ejected electron is called a beta particle. The decay of potassium-40 to calcium-40 is an example of this process. In a third type of decay, called alpha emission, a nucleus emits an alpha particle consisting of two protons and two neutrons. Uranium-238 emits eight alpha particles and six beta particles in a sequence of steps to become lead-206.

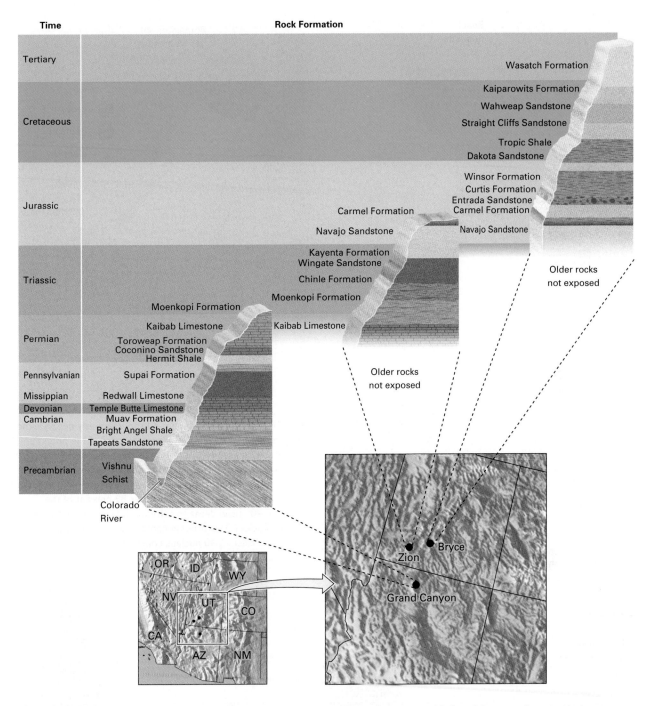

Time | **Rock Formation**

Tertiary — Wasatch Formation

Cretaceous — Kaiparowits Formation, Wahweap Sandstone, Straight Cliffs Sandstone, Tropic Shale, Dakota Sandstone

Jurassic — Winsor Formation, Curtis Formation, Entrada Sandstone, Carmel Formation, Navajo Sandstone, Carmel Formation, Navajo Sandstone

Older rocks not exposed

Triassic — Kayenta Formation, Wingate Sandstone, Chinle Formation, Moenkopi Formation, Moenkopi Formation

Permian — Kaibab Limestone, Kaibab Limestone, Toroweap Formation, Coconino Sandstone, Hermit Shale

Older rocks not exposed

Pennsylvanian — Supai Formation

Missippian — Redwall Limestone

Devonian — Temple Butte Limestone

Cambrian — Muav Formation, Bright Angel Shale, Tapeats Sandstone

Precambrian — Vishnu Schist

Colorado River

OR ID WY NV UT CO CA AZ NM

Zion · Bryce · Grand Canyon

◆ **FIGURE 4.15** Correlation of sedimentary rocks from three different locations in Utah and Arizona allows geologists to develop a more complete picture of geologic history in the region, despite the fact that each location contains only rocks of limited ages. The rock units such as the Tapeats Sandstone and the Moenkopi Formation, are formations. A formation is an easily identified rock unit that can be shown on cross sections such as these.

such as argon-40 or calcium-40, is called a daughter isotope.

Many common elements, such as potassium, consist of a mixture of radioactive and nonradioactive isotopes. With time, the radioactive isotopes decay but the nonradioactive ones do not. Some elements, such as uranium, consist only of radioactive isotopes. The amount of uranium on Earth slowly decreases as it decomposes to other elements, such as lead.

Radioactivity and Half-Life

If you watch a single atom of potassium-40, when will it decompose? This question cannot be answered because any particular potassium-40 atom may or may not decompose at any time. But recall from Chapter 2 that even a small sample of a material contains a huge number of atoms. Each atom has a certain probability of decaying at any time. Averaged over time, half of the atoms

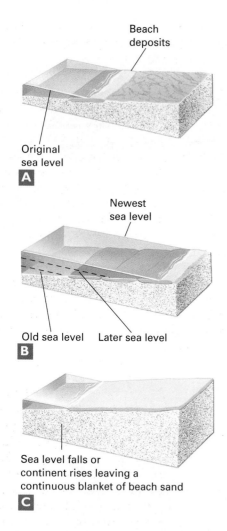

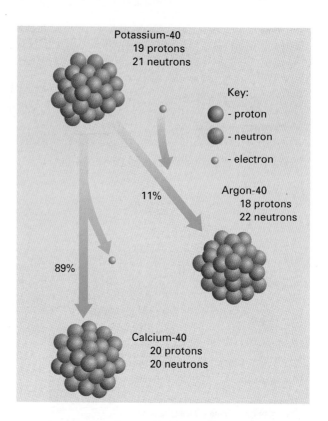

FIGURE 4.16 As sea level rises (**A** and **B**) or falls (**C**) slowly through time, a beach migrates laterally, forming a single sand layer that is of different ages in different localities.

FIGURE 4.18 Potassium-40 spontaneously decays either to argon-40 or to calcium-40. Eleven percent of the potassium-40 atoms decay to argon-40. In this case, the potassium-40 nucleus captures an electron. The electron combines with a proton to produce a neutron. This process creates an argon-40 nucleus with 18 protons and 22 neutrons. The other 89 percent of the potassium-40 nuclei decay to calcium-40 by expelling an electron from the nucleus. This process converts a neutron to a proton, producing a calcium-40 nucleus containing 20 protons and 20 neutrons.

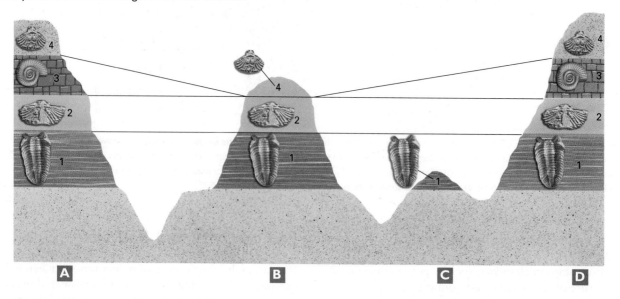

FIGURE 4.17 Index fossils demonstrate age equivalency of sedimentary rocks from widely separated localities. Sedimentary beds containing the same index fossils are interpreted to be of the same age. The fossils show that at locality B, layer 3 is missing because layer 4 directly lies on top of layer 2. Either layer 3 was deposited and then eroded away before 4 was deposited at locality B, or layer 3 was never deposited here. At locality C, all layers above 1 are missing, either because of erosion or because they were never deposited.

CHAPTER 4 • Geologic Time: A Story in the Rocks

in any sample of potassium-40 will decompose in 1.3 billion years. The **half-life** is the time it takes for half of the atoms in a radioactive sample to decompose. The half-life of potassium-40 is 1.3 billion years. Therefore, if 1 gram of potassium-40 were placed in a container, 0.5 gram would remain after 1.3 billion years, 0.25 gram after 2.6 billion years, and so on. Each radioactive isotope has its own half-life; some half-lives are fractions of a second and others are measured in billions of years.

Radiometric Dating

Two aspects of radioactivity are essential to the calendars in rocks. First, the half-life of a radioactive isotope is constant. It is easily measured in the laboratory and is unaffected by geologic or chemical processes, so radioactive decay occurs at a known, constant rate. Second, as a parent isotope decays, its daughter accumulates in the rock. The longer the rock exists, the more daughter isotope accumulates. The accumulation of a daughter isotope is analogous to marking off days on a calendar. Because radioactive isotopes are widely distributed throughout Earth, many rocks have built-in calendars that allow us to measure their ages. **Radiometric**

dating is the process of determining the absolute ages of rocks, minerals, and fossils by measuring the relative amounts of parent and daughter isotopes.

◆▶ Figure 4.19 shows the relationships between age and relative amounts of parent and daughter isotopes. At the end of one half-life, 50 percent of the parent atoms have decayed to daughter. At the end of two half-lives, the mixture is 25 percent parent and 75 percent daughter. To determine the age of a rock, a geologist measures the proportions of parent and daughter isotopes in a sample and compares the ratio to a similar graph. Consider a hypothetical parent–daughter pair having a half-life of 1 million years. If we determine that a rock contains a mixture of 25 percent parent isotope and 75 percent daughter, ◆▶ Figure 4.19 shows that the age is two half-lives, or 2 million years.

If the half-life of a radioactive isotope is short, that isotope gives accurate ages for young materials. For example, carbon-14 has a half-life of 5,730 years. Carbon-14 dating gives accurate ages for materials younger than 50,000 years. It is useless for older materials because by 50,000 years virtually all the carbon-14 has decayed, and no additional change can be measured.

Isotopes with long half-lives give good ages for old rocks, but not enough daughter accumulates in young rocks to be measured. For example, rubidium-87 has a half-life of 47 billion years. In a geologically short period of time—10 million years or less—so little of its daughter has accumulated that it is impossible to measure accurately. Therefore, rubidium-87 is not useful for rocks younger than about 10 million years. The six radioactive isotopes that are most commonly used for dating are summarized in Table 4.1.

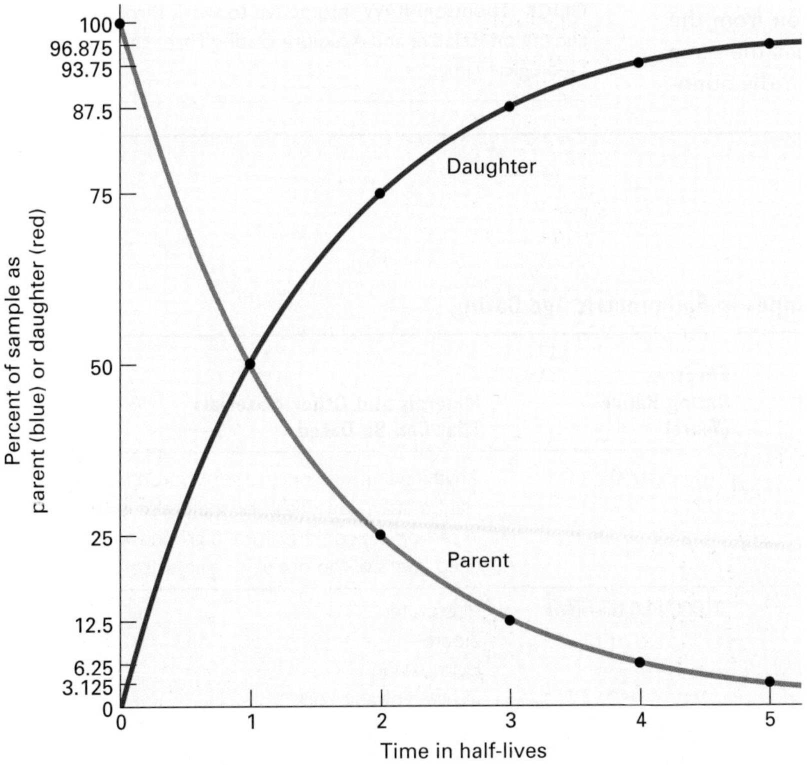

What Does a Radiometric Date Mean?

Biotite—black mica—is rich in potassium, so the potassium-40/argon-40 (abbreviated K/Ar) parent–daughter pair can be used successfully. Suppose that we collect a fresh sample of biotite-bearing granite, separate out a few grams of biotite, and obtain a K/Ar date of 100 million years. What happened 100 million years ago to start the K/Ar "calendar"?

The granite started out as molten magma, which slowly solidified below Earth's surface. But at this point, the granite was still buried several kilometers below the surface and was still hot. Later, perhaps over millions of years, the granite cooled slowly as tectonic forces pushed it toward the surface, where we collected our sample. So what does our

◆▶ **FIGURE 4.19** As a radioactive parent isotope decays to a daughter, the proportion of parent decreases (blue line) and the amount of daughter increases (red line). The half-life is the amount of time required for half of the parent to decay to daughter. At time zero, when the radiometric calendar starts, a sample is 100 percent parent. At the end of one half-life, 50 percent of the parent has converted to daughter. At the end of two half-lives, 25 percent of the sample is parent and 75 percent is daughter. Thus, by measuring the proportions of parent and daughter in a rock, the rock's age in half-lives can be obtained. Because the half-lives of all radioactive isotopes are well known, it is simple to convert age in half-lives to age in years.

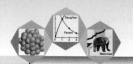

Carbon-14 Dating

Carbon-14 dating differs from the other parent–daughter pairs shown in Table 4.1 in two ways. First, carbon-14 has a half-life of only 5,730 years, in contrast to the millions or billions of years of other isotopes used for radiometric dating.

Second, accumulation of the daughter, nitrogen-14, cannot be measured. Nitrogen-14 is the most common isotope of nitrogen, and it is impossible to distinguish nitrogen-14 produced by radioactive decay from other nitrogen-14 in an object being dated.

The abundant stable isotope of carbon is carbon-12. Carbon-14 is continuously created in the atmosphere as cosmic radiation bombards nitrogen-14, converting it to carbon-14. The carbon-14 then decays back to nitrogen-14. Because it is continuously created and because it decays at a constant rate, the ratio of carbon-14 to carbon-12 in the atmosphere remains nearly constant. While an organism is alive, it absorbs carbon from the atmosphere. Therefore, the organism contains the same ratio of carbon-14 to carbon-12 as found in the atmo-

sphere. However, when the organism dies, it stops absorbing new carbon. Therefore, the proportion of carbon-14 in the remains of the organism begins to diminish at death as the carbon-14 decays. Thus, geologists make carbon-14 age determinations by measuring the ratio of carbon-14 to carbon-12 in organic material. As time passes, the proportion of carbon-14 steadily decreases. By the time an organism has been dead for 50,000 years, so little carbon-14 remains that measurement is very difficult.

> **FOCUS QUESTION**
> Explain why carbon-14 dating would not be useful to date Mesozoic fossils.

ThomsonNOW

CLICK ThomsonNOW Interactive to work through an activity on Relative and Absolute Dating Through Geological Time.

Table 4.1 The Most Commonly Used Isotopes in Radiometric Age Dating

Isotopes		Half-Life of Parent (Years)	Effective Dating Range (Years)	Minerals and Other Materials That Can Be Dated
Parent	**Daughter**			
Carbon-14	Nitrogen-14	5730 ± 30	0100–50,000	Anything that was once alive: wood, other plant matter, bone, flesh, or shells; also, carbon in carbon dioxide dissolved in ground water, deep layers of the ocean, or glacier ice
Potassium-40	Argon-40 Calcium-40	1.3 billion	50,000–4.6 billion	Muscovite Biotite Hornblende Whole volcanic rock
Uranium-238	Lead-206	4.5 billion	10 million–4.6 billion	Zircon Uraninite and pitchblende
Uranium-235 Thorium-232	Lead-207 Lead-208	710 million 14 billion		
Rubidium-87	Strontium-87	47 billion	10 million–4.6 billion	Muscovite Biotite Potassium feldspar Whole metamorphic or igneous rock

100-million-year radiometric date tell us—The age of formation of the original granite magma? The time when the magma became solid? The time of uplift?

We can measure time because potassium-40 decays to argon-40, which accumulates in the biotite crystal. As more time elapses, more argon-40 accumulates. Therefore, the biotite calendar does not start recording time until argon atoms begin to accumulate in biotite.

Argon is an inert gas that is trapped in biotite as the argon forms. But if the biotite is heated above a certain temperature, the argon escapes. Biotite retains argon only when it cools below 350°C. Thus, the 100-million-year age date on the granite is the time that has passed since the granite cooled below 350°C, probably corresponding to cooling that occurred during uplift and erosion.

Two additional conditions must be met for a radiometric date to be accurate: First, when a mineral or rock cools, no original daughter isotope is trapped in the mineral or rock. Second, once the clock starts, no parent or daughter isotopes are added or removed from the rock or mineral. Metamorphism, weathering, and circulating fluids can add or remove parent or daughter isotopes. Sound geologic reasoning and discretion must be applied when choosing materials for radiometric dating and when interpreting radiometric age dates.

4.6
The Geologic Column and Time Scale

As mentioned earlier, no single locality exists on Earth where a complete sequence of rocks formed continuously throughout geologic time. However, geologists have correlated rocks from many localities around the world to create the **geologic column** and **geologic time scale**, which is a nearly complete composite record of geologic time (◆ Figure 4.20). The worldwide geologic column is frequently revised as geologic mapping continues.

Just as a year is subdivided into months, months into weeks, and weeks into days, geologic time is split into smaller intervals. The units are named, just as months and days are. The largest time units are **eons**, which are divided into **eras**. Eras are subdivided, in turn, into **periods**, which are further subdivided into **epochs**.

The geologic column and time scale were originally constructed on the basis of relative age determinations. When geologists developed radiometric dating, they added absolute ages to the column and time scale. Geologists commonly use the time scale to date rocks in the field. Imagine that you are studying sedimentary rocks in the walls of Grand Canyon. If you find an index fossil or a key bed that has already been radiometrically dated by other scientists, you know the age of the rock you are studying, and you do not need to send the sample to a laboratory for radiometric dating.

The Earliest Eons of Geologic Time: Precambrian Time

The earliest eons—the **Hadean**, **Archean**, and **Proterozoic** Eons—often are not subdivided at all, even though together they constitute a time interval of 4 billion years, almost eight times as long as the **Phanerozoic** Eon. Why is Phanerozoic time subdivided so finely? Or, conversely, what prevents subdivision of the earlier eons?

The Hadean Eon (Greek for "beneath the Earth") is the earliest time in Earth history and ranges from the planet's origin 4.6 billion years ago to 3.8 billion years ago. Geologic time units are based largely on fossils found in rocks. However, only a few Earth rocks are known that formed during the Hadean Eon and no fossils of Hadean age are known. It may be that erosion or metamorphism destroyed traces of Hadean life, or that Hadean time preceded the evolution of life. In any case, Hadean time is not amenable to subdivision based on fossils.

Most rocks of the Archean Eon (Greek for "ancient") are igneous or metamorphic, although a few Archean sedimentary rocks are preserved. Some contain microscopic fossils of single-celled organisms. Although the exact date is uncertain, the fossil record indicates that life began on Earth between 3.2 and 3.5 billion years ago. Fossils are neither numerous nor well preserved enough to permit much fine tuning of Archean time.

Diverse groups of fossils have been found in sedimentary rocks of the Proterozoic Eon (Greek for "earlier life"). The most complex are multicellular and have different kinds of cells arranged into tissues and organs. Some of these organisms look so much like modern jellyfish, corals, and worms that some paleontologists view them as ancestors of organisms alive today. However, other paleontologists believe that the resemblance is only superficial. A few types of Proterozoic shell-bearing organisms have been identified, but shelled organisms did not become abundant until the Paleozoic Era.

Commonly, the Hadean, Archean, and Proterozoic Eons are collectively referred to by the informal term **Precambrian**, because they preceded the Cambrian Period, when fossil remains first became very abundant.

The Phanerozoic Eon

Sedimentary rocks of the Phanerozoic Eon contain abundant fossils (*phaneros* is Greek for "evident"). Four changes occurred at the beginning of Phanerozoic time that greatly improved the fossil record:

1. The number of species with shells and skeletons dramatically increased.

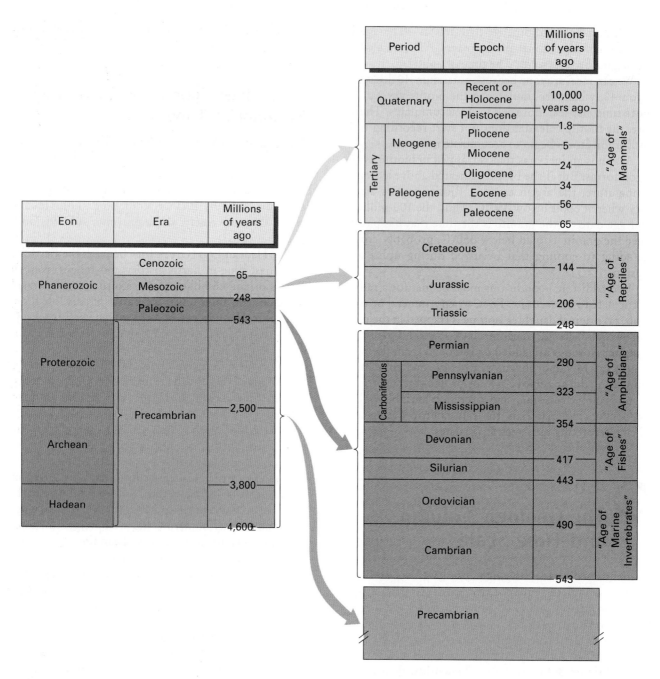

◈ **FIGURE 4.20** The geologic column and geologic time scale is a record of geologic time divided into eons, eras, periods, and epochs. Time is given in millions of years (for example, 1,000 stands for 1,000 million, which is one billion). The table is not drawn to scale. We know relatively little about events that occurred during the early part of Earth's history. Therefore, the first 4 billion years are given relatively little space on this chart, while the more recent Phanerozoic Eon, which spans only 543 million years, receives proportionally more space.

2. The total number of individual organisms preserved as fossils increased greatly.
3. The total number of species preserved as fossils increased greatly.
4. The average sizes of individual organisms increased.

Shells and skeletons are much more easily preserved than plant remains and soft body tissues. Thus, in rocks of earliest Phanerozoic time and younger, the most abundant fossils are the hard, tough shells and skeletons.

Subdivision of the Phanerozoic Eon into three eras is based on the most common types of life during each era. Sedimentary rocks formed during the **Paleozoic Era** (Greek for "old life") contain fossils of early life forms, such as invertebrates, fishes, amphibians, reptiles, ferns, and cone-bearing trees (◈ Figure 4.21). The Paleozoic Era ended abruptly about 248 million years ago when the terminal Permian mass extinction wiped out 90 percent of all marine species, two-thirds of reptile and amphibian species, and 30 percent of insect species.

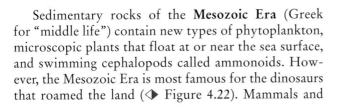
FIGURE 4.21 In Paleozoic time, the land was inhabited by reptiles and amphibians and covered by ferns, ginkgoes, and conifers.

FIGURE 4.22 In this reconstruction of Mesozoic home life, a duckbill dinosaur *(Maiasauras)* nurtures babies in their nest 100 million years ago in Montana.

Sedimentary rocks of the **Mesozoic Era** (Greek for "middle life") contain new types of phytoplankton, microscopic plants that float at or near the sea surface, and swimming cephalopods called ammonoids. However, the Mesozoic Era is most famous for the dinosaurs that roamed the land (◇ Figure 4.22). Mammals and

flowering plants also evolved during this era. The Mesozoic Era ended 65 million years ago with another mass extinction that wiped out the dinosaurs.

During the **Cenozoic Era** (Greek for "recent life"), mammals and grasses became abundant (◇ Figure 4.23). Humans have evolved and lived wholly in the Cenozoic Era.

FIGURE 4.23 During Cenozoic time, grasses proliferated and animals adapted to the new food source.

The eras of Phanerozoic time are subdivided into periods, the time unit most commonly used by geologists. Some of the periods are named after special characteristics of the rocks formed during that period. For example, the Cretaceous Period is named from the Latin word for chalk (*creta*), after chalk beds of this age in Africa, North America, and Europe. Other periods are named for the geographic localities where rocks of that age were first described. For example, the Jurassic Period is named for the Jura Mountains of France and Switzerland. The Cambrian Period is named for Cambria, the Roman name for Wales, where rocks of this age were first studied.

In addition to the abundance of fossils, another reason that details of Phanerozoic time are better known than those of Precambrian time is that many of the older rocks have been metamorphosed, deformed, and eroded. It is simple probability that the older a rock is, the greater the chance that metamorphism or erosion has destroyed the rock or the features that record Earth's history.

EARTH SYSTEMS INTERACTIONS

AT THE BEGINNING OF this chapter, we described three hypotheses that scientists have formulated to explain mass extinctions. All three hypotheses involve global-scale interactions among the atmosphere, the hydrosphere, the biosphere, and the geosphere. One of the hypotheses invokes an extraterrestrial factor—an asteroid impact—as a trigger that initiated major changes in, and interactions among, all four Earth systems. Each hypothesis provides numerous examples of how Earth systems interact to create and change our environment, as well as demonstrating the roles that Earth systems feedback processes and threshold effects play in Earth systems interactions. ■

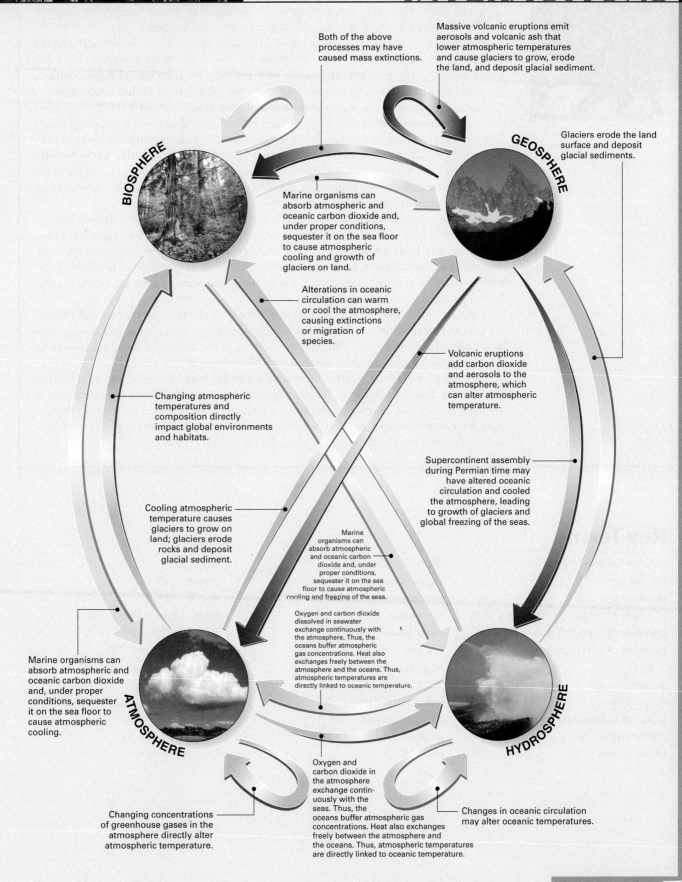

Both of the above processes may have caused mass extinctions.

Massive volcanic eruptions emit aerosols and volcanic ash that lower atmospheric temperatures and cause glaciers to grow, erode the land, and deposit glacial sediment.

BIOSPHERE

GEOSPHERE

Glaciers erode the land surface and deposit glacial sediments.

Marine organisms can absorb atmospheric and oceanic carbon dioxide and, under proper conditions, sequester it on the sea floor to cause atmospheric cooling and growth of glaciers on land.

Alterations in oceanic circulation can warm or cool the atmosphere, causing extinctions or migration of species.

Volcanic eruptions add carbon dioxide and aerosols to the atmosphere, which can alter atmospheric temperature.

Changing atmospheric temperatures and composition directly impact global environments and habitats.

Supercontinent assembly during Permian time may have altered oceanic circulation and cooled the atmosphere, leading to growth of glaciers and global freezing of the seas.

Cooling atmospheric temperature causes glaciers to grow on land; glaciers erode rocks and deposit glacial sediment.

Marine organisms can absorb atmospheric and oceanic carbon dioxide and, under proper conditions, sequester it on the sea floor to cause atmospheric cooling and freezing of the seas.

Oxygen and carbon dioxide dissolved in seawater exchange continuously with the atmosphere. Thus, the oceans buffer atmospheric gas concentrations. Heat also exchanges freely between the atmosphere and the oceans. Thus, atmospheric temperatures are directly linked to oceanic temperature.

Marine organisms can absorb atmospheric and oceanic carbon dioxide and, under proper conditions, sequester it on the sea floor to cause atmospheric cooling.

ATMOSPHERE

HYDROSPHERE

Oxygen and carbon dioxide in the atmosphere exchange continuously with the seas. Thus, the oceans buffer atmospheric gas concentrations. Heat also exchanges freely between the atmosphere and the oceans. Thus, atmospheric temperatures are directly linked to oceanic temperature.

Changing concentrations of greenhouse gases in the atmosphere directly alter atmospheric temperature.

Changes in oceanic circulation may alter oceanic temperatures.

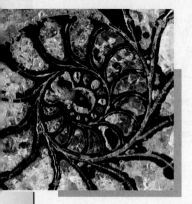

SUMMARY

At least five catastrophic mass extinctions have occurred in geologic history. The greatest, 248 million years ago, annihilated 90 percent of all marine species, two-thirds of reptile and amphibian species, and 30 percent of insect species.

Determinations of relative time are based on geologic relationships among rocks and the evolution of life forms through time. The criteria for relative dating are summarized in a few simple principles: the **principle of original horizontality**, the **principle of superposition**, the **principle of crosscutting relationships**, and **the principle of faunal succession**.

Layers of sedimentary rock are **conformable** if they were deposited without major interruptions. An **unconformity** represents a major interruption of deposition and a significant time gap between formation of successive layers of rock. In a **disconformity**, layers of sedimentary rock on either side of the unconformity are parallel. An **angular unconformity** forms when lower layers of rock are tilted and partially eroded prior to deposition of the upper beds. In a **nonconformity**, sedimentary layers lie on top of an erosion surface developed on igneous or metamorphic rocks.

Fossils are used to date rocks according to the **principle of faunal succession**. Correlation is the demonstration of equivalency of rocks that are geographically separated. **Index fossils** and **key beds** are important tools in **time correlation**, the demonstration that sedimentary rocks from different geographic localities formed at the same time. Worldwide correlation of rocks of all ages has resulted in the **geologic column**, a composite record of rocks formed throughout the history of Earth.

Absolute time is measured by **radiometric dating**, which relies on the fact that radioactive **parent isotopes** decay to form **daughter isotopes** at a fixed, known rate as expressed by the **half-life** of the isotope. The cumulative effects of the radioactive decay process can be determined because the daughter isotopes accumulate in rocks and minerals.

The major units of the geologic time scale are **eons, eras, periods,** and **epochs**. The **Phanerozoic Eon** is finely and accurately subdivided because sedimentary rocks deposited at this time are often well preserved and they contain abundant well-preserved fossils. In contrast, **Precambrian** rocks and time are only coarsely subdivided because fossils are scarce and poorly preserved and the rocks are often altered.

Key Terms

For Review

1. Describe the two ways of measuring geologic time. How do they differ?

2. Give an example of how the principle of original horizontality might be used to determine the order of events affecting a sequence of folded sedimentary rocks.

3. How does the principle of superposition allow us to determine the relative ages of a sequence of unfolded sedimentary rocks?

4. Explain the principle of crosscutting relationships and how it can be used to determine age relationships among sedimentary rocks.

5. Explain a conformable relationship in sedimentary rocks.

6. What geologic events are recorded by an angular unconformity? A disconformity? A nonconformity? What can be inferred about the timing of each set of events?

7. Discuss the principle of faunal succession and the use of index fossils in time correlation.

8. What are the two types of correlation of rock units? How do they differ?

9. What tools or principles are most commonly used in correlation?

10. What is radioactivity?

11. What is a stable isotope? An unstable isotope?

12. What is the relationship between parent and daughter isotopes?

13. What is meant by the half-life of a radioactive isotope? How is the half-life used in radiometric age dating?

14. Why are some radioactive isotopes useful for measuring relatively young ages, whereas others are useful for measuring older ages?

15. What geologic event is actually measured by a radiometric age determination of an igneous rock or mineral?

16. Why is the Phanerozoic Eon separated into so many subdivisions in contrast to much longer Precambrian time, which has few subdivisions?

17. List major events that occurred during the Hadean, Archean, and Proterozoic Eons.

18. List four changes that occurred at the beginning of Phanerozoic time that greatly improved the fossil record.

19. Briefly discuss the three major eras of the Phanerozoic Eon.

20. Describe three hypotheses that explain mass extinctions.

For Discussion

1. Suppose that you landed on the Moon and were able to travel in a vehicle that could carry you over the lunar surface to see a wide variety of rocks, but that you had no laboratory equipment to work with. What principles and tools could you use to determine relative ages of Moon rocks?

2. Imagine that one species lived between 544 and 505 million years ago and another lived between 520 and 248 million years ago. What can you say about the age of a rock that contains fossils of both species?

3. Imagine that someone handed you a sample of sedimentary rock containing abundant fossils. What could you tell about its age if you did not use radiometric dating and you did not know where it was collected from? What additional information could you determine if you studied the outcrop that it came from?

4. What geologic events are represented by a potassium–argon age from flakes of biotite in (a) granite, (b) biotite schist, and (c) sandstone?

5. Suppose you were using the potassium/argon (K/Ar) method to measure the age of biotite in granite. What would be the effect on the age measurement if the biotite had been slowly leaking small amounts of argon since it crystallized?

6. On rare occasions, layered sedimentary rocks are overturned so that the oldest rocks are on the top and the youngest are buried most deeply. Discuss two ways of determining whether such an event had occurred.

7. Devise two metaphors for the length of geologic time in addition to the metaphor used in Chapter 1. Locate some of the most important time boundaries in your analogy.

8. Discuss how sedimentary rocks could record major changes in temperature or rainfall in an area.

CHAPTER 5

Geologic Resources

Courtesy of Graham R. Thompson/Jonathan Turk

An offshore oil platform in the Gulf of Mexico.

Since human-like creatures emerged 5 to 7million years ago, our use of geologic resources has become increasingly sophisticated. Early hominids used sticks and rocks as simple weapons and tools. Later prehistoric people used flint and obsidian to make more effective weapons and tools, and used natural pigments to create elegant art on cave walls (◀▶ Figure 5.1). About 8,000 B.C., people learned to shape and fire clay to make pottery. Archaeologists have found copper ornaments in Turkey dating from 6500 B.C.; 1,500 years later, Mesopotamian farmers used copper farm implements. Today, geologic resources provide the silicon chip that operates your computer, the titanium valves in a space probe, and the gasoline that powers your car.

We use two types of geologic resources: **mineral resources** and **energy resources**. Mineral resources include all useful rocks and minerals. As we will see in the sections that follow, many mineral resources are concentrated by processes that involve interactions among rock of the geosphere, atmospheric gases, and water from the hydrosphere. In turn, humans have used the mineral resources to create the industrial world that has altered our planet. The primary energy resources of the twenty-first century are coal, petroleum, and natural gas; all formed from the decayed remains of prehistoric plants and animals that have been altered by Earth systems processes. ■

5.1
Mineral Resources

Mineral resources include both metal **ore** and **nonmetallic mineral resources**. Ore is rock sufficiently enriched in one or more minerals to be mined profitably (◀▶ Figure 5.2). Geologists usually use the term *ore* to refer to metallic mineral deposits, and the term is commonly accompanied by the name of the metal—for example, *iron ore* or *silver ore*.

A nonmetallic resource is any useful rock or mineral that is not a metal, such as salt, building stone, sand, and gravel. When we think about "striking it rich" from mining, we usually think of gold. However, more money has been made mining sand and gravel than by mining gold. For example, in the United States in the year 2004, sand and gravel produced $6.2 billion in revenue, but gold produced only $3.2 billion. Sand and gravel are mined from stream and glacial deposits, sand dunes, and beaches. In turn, these materials are mixed with portland cement to make concrete. Portland cement is made by heating a mixture of crushed limestone and clay. Reinforced with steel, concrete is used to build roads, bridges, and buildings. Thus reinforced concrete is one of the basic building materials of the modern world. In addition, many buildings are faced with stone—usually granite or limestone, although marble, slate, sandstone, and other rocks used for building are also mined from quarries cut into bedrock (◀▶ Figure 5.3).

Table 5.1 lists important metals and other elements obtained from mineral resources. About 40 metals are commercially important. Some, such as iron, lead, copper, aluminum, silver, and gold, are familiar. Others, such as vanadium, titanium, and tellurium, are less well known but vital to industry.

All mineral resources are nonrenewable: we use them up at a much faster rate than natural processes create them, although many can be recycled.

5.2
Ore and Ore Deposits

If you pick up any rock and send it to a laboratory for analysis, the report will probably show that the rock contains measurable amounts of iron, gold, silver, aluminum, and other valuable metals. However, the concentrations of these metals are so low in most rocks that the extraction cost would be much greater than the income gained by selling the metals. In certain locations, however, geologic processes have enriched metals many times above their normal concentrations. Table 5.2 shows that the concentration of a metal in ore may exceed its average abundance in ordinary rock by a factor of more than 100,000.

Successful exploration for new ore deposits requires an understanding of the processes that concentrate metals to form ore. For example, platinum concentrates

Erich Lessing/Art Resource, NY

◆ **FIGURE 5.1** A cave painting, created between the fifth and second millennium B.C., showing a bison. Altamira Cave, Santillana de Mar, Spain.

in certain types of igneous rocks. Therefore, if you were exploring for platinum, you would focus on those rocks rather than on sandstone or limestone. The following sections describe some of the more common ore-forming processes.

With the exception of magmatic processes, which occur deep within the crust, the natural processes

Montana Historical Society

◆ **FIGURE 5.2** In the early 1900s, miners extracted gold, copper, and other metal ores from underground mines such as this one 600 meters below the surface in Butte, Montana.

(listed next) that concentrate ore minerals all involve interactions between rocks and minerals of the geosphere with water from the hydrosphere. Weathering and sedimentary processes also involve rainwater and other climatic processes that occur within the atmosphere.

Magmatic Processes

Magmatic processes form mineral deposits as liquid magma solidifies to form an igneous rock. These processes create metal ores as well as some gems and nonmetallic mineral deposits such as valuable sulfur deposits and building stone.

Some large bodies of igneous rock, particularly those of mafic (basaltic) composition, solidify in layers. Each layer contains different minerals and is of a different chemical composition than is found in adjacent layers. Some of the layers may contain rich ore deposits. The layering can develop by at least two processes:

1. Recall from Chapter 3 that cooling magma does not solidify all at once. Instead, higher-temperature minerals crystallize first, and lower-temperature minerals form later as the magma cools and temperature drops. Most minerals are denser than magma. Consequently, early formed crystals may sink to the bottom of a magma chamber in a process called **crystal settling** (◆ Figure 5.4). In some instances, ore minerals crystallize with other early formed minerals and accumulate in layers near the bottom of a pluton.

◀▷ FIGURE 5.3 A Chinese quarryman splits a large granite block with a sledgehammer. After he splits the rock, the circular saws in the background cut it into thin slabs for floors and walls.

Table 5.1   Some Important Elements and Their Uses

Mineral	Type	Some Uses
Aluminum (Al)	Metal	Structural materials (airplanes, automobiles), packaging (beverage cans), fireworks
Borax ($Na_2B_4O_7$)	Nonmetal	Diverse manufacturing uses—glass, enamel, artificial gems, soaps, antiseptics
Chromium (Cr)	Metal	Chrome plate, pigments, steel alloys (tools, jet engines, bearings)
Cobalt (Co)	Metal	Pigments, alloys (jet engines, tool bits), medicine, varnishes
Copper (Cu)	Metal	Alloy ingredient in gold jewelry, silverware, brass, and bronze; electrical wiring, pipes, cooking utensils
Gold (Au)	Metal	Jewelry, money, dentistry, alloys, specialty electronics
Gravel	Nonmetal	Concrete (buildings, roads)
Gypsum ($CaSO_4 - 2H_2O$)	Nonmetal	Plaster of Paris, wallboard, soil treatments
Iron (Fe)	Metal	Basic ingredient of steel (buildings, machinery)
Lead (Pb)	Metal	Pipes, solder, battery electrodes, pigments
Magnesium (Mg)	Metal	Alloys (aircraft), firecrackers, bombs, flashbulbs
Manganese (Mn)	Metal	Steel, alloys (steamship propellers, gears), batteries, chemicals
Mercury (Hg)	Liquid metal	Thermometers, barometers, dental inlays, electric switches, street lamps, medicine
Molybdenum (Mo)	Metal	Steel alloys, lamp filaments, boiler plates, rifle barrels
Nickel (Ni)	Metal	Money, alloys, metal plating
Phosphorus (P)	Nonmetal	Medicine, fertilizers, detergents
Platinum (Pt)	Metal	Jewelry, delicate instruments, electrical equipment, cancer chemotherapy, industrial catalyst
Potassium (K)*	Nonmetal	Salts used in fertilizers, soaps, glass, photography, medicine, explosives, matches, gunpowder
Common salt (NaCl)	Nonmetal	Food additive
Sand (largely SiO_2)	Nonmetal	Glass, concrete (buildings, roads)
Silicon (Si)	Nonmetal	Electronics, solar cells, ceramics, silicones
Silver (Ag)	Metal	Jewelry, silverware, photography, alloys
Sulfur (S)	Nonmetal	Insecticides, rubber tires, paint, matches, papermaking, photography, rayon, medicine, explosives
Tin (Sn)	Metal	Cans and containers, alloys, solder, utensils
Titanium (Ti)	Metal	Paints; manufacture of aircraft, satellites, and chemical equipment
Tungsten (W)	Metal	High-temperature applications, lightbulb filaments, dentistry
Zinc (Zn)	Metal	Brass, metal coatings, electrodes in batteries, medicine (zinc salts)

*Potassium, which is very reactive chemically, is never found free in nature; it is always combined with other elements.

Table 5.2 Comparison of Concentrations of Specific Elements in Earth's Crust with Concentrations Needed to Operate a Commercial Mine

Element	Natural Concentration in Crust (% by Weight)	Concentration Required to Operate a Commercial Mine (% by Weight)	Enrichment Factor
Aluminum	8	24–32	3–4
Iron	5.8	40	6–7
Copper	0.0058	0.46–0.58	80–100
Nickel	0.0072	1.08	150
Zinc	0.0082	2.46	300
Uranium	0.00016	0.19	1,200
Lead	0.00010	0.2	2,000
Gold	0.0000002	0.0008	4,000
Mercury	0.000002	0.2	100,000

2. Some large bodies of mafic magma crystallize from the bottom upward. Thus, early formed ore minerals become concentrated near the base of the pluton.

The largest ore deposits found in mafic-layered plutons are the rich chromium and platinum reserves of South Africa's Bushveld intrusion. The pluton is about 375 by 300 kilometers in area—roughly the size of the state of Maine—and about 7 kilometers thick. The Bushveld deposits contain more than 20 billion tons of chromium and more than 10 billion grams of platinum, the greatest reserves in any known deposit on Earth. The world's largest known nickel deposit occurs in a layered mafic pluton at Sudbury, Ontario, and rich platinum ores are mined from layered plutons in southern Montana and Noril'sk, Russia.

Hydrothermal Processes

Hydrothermal processes are probably responsible for the formation of more ore deposits, and a larger total quantity of ore, than all other processes combined. To form a hydrothermal ore deposit, hot water (hence the roots *hydro* for water and *thermal* for hot) dissolves metals from rock or magma. The metal-bearing solutions then seep through cracks or through permeable rock, where they precipitate to form an ore deposit.

Hydrothermal water comes from three sources:

1. Granitic magma contains more dissolved water than solid granite rock. Thus, the magma gives off hydrothermal water as it solidifies. For this reason, hydrothermal ore deposits are commonly associated with granite and similar igneous rocks.
2. Ground water can seep into the crust, where it is heated and forms a hydrothermal solution. This is particularly true in volcanic areas where hot rock or magma heat ground water at shallow depths. For this reason, hydrothermal ore deposits are also common in volcanic regions.
3. In the oceans, hot, young basalt near the Mid-Oceanic Ridge heats seawater as it seeps into cracks in the sea floor.

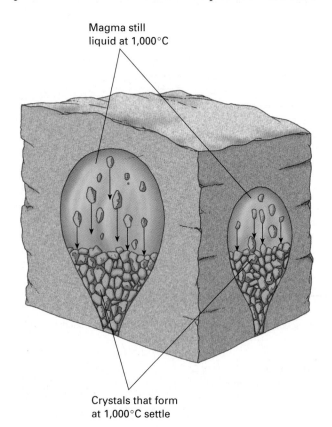

Magma still liquid at 1,000°C

Crystals that form at 1,000°C settle

◇ **FIGURE 5.4** The first crystals to form settle and concentrate near the bottom of a magma chamber. The crystals are magnified for clarity. In actuality the crystals may be a few millimeters in diameter, whereas the magma chamber may be many kilometers in width.

Although water by itself is capable of dissolving minerals, most hydrothermal waters also contain dissolved salts, which greatly increase its ability to dissolve minerals. Therefore, hot, salty, hydrothermal water is a very powerful solvent, capable of dissolving and transporting metals.

Table 5.2 shows that tiny amounts of all metals are found in average rocks of the Earth's crust. For example, gold makes up 0.0000002 percent of the crust, while copper makes up 0.0058 percent and lead 0.0001 percent. Although the metals are present in very low concentrations in country rock, hydrothermal solutions percolate through vast volumes of rock, dissolving and accumulating the metals. The solutions then deposit the metals, where they encounter changes in temperature, pressure, or chemical environment (◆ Figure 5.5). In this way, hydrothermal solutions scavenge metals from large volumes of normal crustal rocks and then deposit them locally to form ore.

A **hydrothermal vein** deposit forms when dissolved metals precipitate in a fracture in rock. Ore veins range from less than a millimeter to several meters in width. A single vein can yield several million dollars worth of gold or silver. The same hydrothermal solutions may also soak into pores in country rock near the vein to create a large but much-less-concentrated **disseminated ore** deposit. Because they commonly form from the same solutions, rich ore veins and disseminated deposits are often found together. The history of many mining districts is one in which early miners dug shafts and tunnels to follow the rich veins. After the veins were exhausted, later miners used huge power shovels to extract low-grade ore from disseminated deposits surrounding the veins.

Disseminated copper deposits, with ore veins, are abundant along the entire western margin of North and South America (◆ Figure 5.6). They are most commonly associated with large granitic plutons. Both the plutons and the copper ore formed as a result of subduction that occurred as tectonic plates carrying North and South America migrated westward to converge with oceanic plates, as described in the next chapter. Other metals, including lead, zinc, molybdenum, gold, and silver, are found with the copper ore. Examples of such deposits occur at Butte, Montana; Bingham, Utah; Morenci, Arizona; and Ely, Nevada.

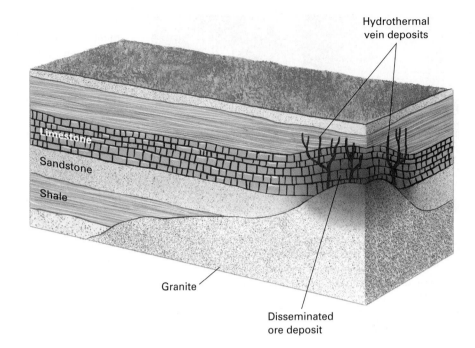

◆ **FIGURE 5.5** Hot water scavenges metals from country rock and deposits metallic minerals in veins that fill fractures in bedrock. It also deposits low-grade disseminated metal ore in large volumes of rock surrounding the veins.

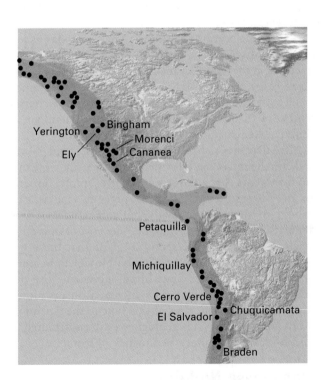

◆ **FIGURE 5.6** Rich deposits of copper and other metals exist throughout the Cordillera of North, Central, and South America. These deposits formed along modern or ancient tectonic plate boundaries. Dots indicate mines. A few of the larger mines are named.

◀▶ **FIGURE 5.7** A black smoker spouts from the East Pacific rise. Seawater becomes hot as it circulates through the hot basalt near a spreading center or submarine volcano. The hot water dissolves sulfur, iron, zinc, copper, and other ions from the basalt. The ions then precipitate as "smoke," consisting of tiny metallic mineral grains, when the hot solution meets cold ocean water. The hot, nutrient-rich water sustains thriving plant and animal communities.

In volcanically active regions of the sea floor, near the Mid-Oceanic Ridge and submarine volcanoes, seawater circulates through the hot, fractured oceanic crust. The hot seawater dissolves metals from the rocks and then, as it rises through the upper layers of oceanic crust, cools and precipitates the metals to form **submarine hydrothermal ore** deposits.

The metal-bearing solutions can be seen today as jets of black water, called **black smokers**, spouting from fractures in the Mid-Oceanic Ridge (◀▶ Figure 5.7). The black color is caused by precipitation of fine-grained metal sulfide minerals as the solutions cool upon contact with seawater. The precipitating metals accumulate as chimneylike structures near the hot-water vent.

Ore deposits rich in copper, lead, zinc, gold, silver, and other metals have formed in submarine volcanic environments throughout time. For example, on land an ore deposit containing 2.5 percent zinc is rich enough to mine commercially, but some undersea zinc deposits contain 55 percent zinc. The cost of operating machinery beneath the sea is so great that these deposits cannot be mined profitably. However, in some places tectonic forces have lifted submarine ore deposits to Earth's surface. The ancient Romans mined copper, lead, and zinc ores of this type in the Apennine Mountains of Italy. This geologic wealth contributed to the political and military ascendancy of Rome.

Manganese Nodules

About 25 to 50 percent of the Pacific Ocean floor is covered with golf ball– to bowling ball–sized **manganese nodules** (◀▶ Figure 5.8). A typical nodule contains 20 to 30 percent manganese, 6 percent iron, about 1 percent

each of copper and nickel, and lesser amounts of other metals. (Much of the remaining 60 to 70 percent consists of oxygen and other anions chemically bonded to the metals.) The metals are probably added to seawater by volcanic activity at the Mid-Oceanic Ridge, perhaps by the black smokers. Chemical reactions between seawater and sea-floor sediment precipitate the dissolved metals to form the nodules.

The nodules grow by about 10 layers of atoms per year, which amounts to 3 millimeters per million years. Curiously, they are found only on the surface of, but never within, the sediment on the ocean floor. Because sediment accumulates much faster than the nodules grow, why doesn't it bury the nodules as they form? Photographs show that animals churn up sea-floor sediment. Worms burrow into it, and other animals pile sediment against the nodules to build shelters. Some geologists suggest that these activities constantly lift the nodules onto the surface.

A trillion or more tons of manganese nodules lie on the sea floor. They contain several valuable industrial metals that could be harvested without drilling or blasting. One can imagine undersea television cameras locating the nodules and giant vacuums scooping them up and lifting them to a ship. But, because the sea floor is a difficult environment in which to operate complex machinery, harvest of manganese nodules is not profitable at the present time.

Sedimentary Processes

Placer Deposits

Gold is denser than any other mineral. Therefore, if you swirl a mixture of water, gold dust, and sand in a gold

◀▶ **FIGURE 5.8** Manganese nodules cover many parts of the sea floor in both the Pacific and Atlantic Oceans. These lie at a depth of 5,500 meters in the northeastern Atlantic.

FOCUS ON

The 1872 Mining Law

The 1872 Mining Law governs the mining of metal ore on public land in the United States. In the mid-1800s, when pioneers were exploring the western frontier, the intent of the law was to encourage the exploration and development of America's natural resources and to foster economic growth. The law states that any individual or corporation can obtain mineral rights to metal ore on public land. Miners need only prove that they can make profits by mining the deposits and must complete a small amount of development work on the claims each year. Miners pay no royalties or other fees to the government or public for the ore extracted. In some cases, corporations or citizens holding claims can buy the land for as little as $2.50 per acre, thus acquiring permanent ownership of public land.

Conservationists and other public interest groups have attacked the 1872 law, claiming that it is one of the biggest giveaway programs in the United States. For example, in 1994 the federal government sold 2,000 acres near Elko, Nevada, to American Barrick Resources, Inc., for $9,765. Geologists estimate that the property contains $10 billion worth of gold, although profits will be much less because of mining expenses. In another instance, a mining company bought 160 acres near Keystone, Colorado. They operated a mine on a portion of the land and sold the rest for ski condominiums for $11,000 per acre.

In response, defenders of the 1872 law argue that mineral exploration entails a great deal of risk and expense because the probability of finding new ore is low. In addition, mining companies must invest large sums of money to locate and develop an ore deposit before any profit is earned. The potential rewards, as guaranteed by the 1872 law, must be great to make exploration attractive to investors. Defenders also argue that the law has worked well for more than a century and has produced one of the world's most productive and efficient mining industries.

In response to public criticism, the U.S. government declared a moratorium in 1995 on transferring ownership to claim holders, although all other provisions of the 1872 law remained in effect. However, in October 2005 the House Resources Committee passed a budget package with a concealed provision that would end the ban on the sale of public land to claim holders and mining companies. Environmental groups once again criticized the proposal, saying it lets mining companies buy Western land at fire-sale prices as part of a budget plan to raise funds and cut federal spending. This bill was defeated mid-December 2005.

> **FOCUS QUESTION**
> How does the public benefit from mining of resources on public land?

---

pan, the gold falls to the bottom first (◆ Figure 5.9). Differential settling also occurs in nature. Many streams carry silt, sand, and gravel with an occasional small grain of gold. The gold settles first when the current slows down. Over years, currents agitate the sediment and the dense gold works its way into cracks and crevices in the streambed. Thus grains of gold concentrate in gravel and bedrock cracks in the streambed, forming a **placer** deposit (◆ Figure 5.10). Most of the prospectors who rushed to California in the Gold Rush of 1849 searched for placer deposits.

Precipitates

Ground water dissolves minerals as it seeps through soil and bedrock. In most environments, ground water eventually flows into streams and then to the sea. Some of the dissolved ions, such as sodium and chloride, make seawater salty. In deserts, however, playa lakes develop with no outlet to the ocean. Water flows into the lakes but can escape only by evaporation. As the water evaporates, the dissolved salts concentrate until they precipitate to form **evaporite** deposits.

You can perform a simple demonstration of evaporation and precipitation. Fill a bowl with warm water and add a few teaspoons of table salt. The salt dissolves and you see only a clear liquid. Set the bowl aside for a few days until the water evaporates. The salt precipitates and encrusts the sides and bottom of the bowl.

Evaporite deposits formed in desert lakes include table salt, borax, sodium sulfate, and sodium carbonate. These salts are used in the production of paper, soap, and medicines, and for the tanning of leather.

◆ **FIGURE 5.9** Jeffery Embrey panning for gold near his cabin in Park City, Montana, in 1898.

Several times during the past 500 million years, shallow seas covered large regions of North America and all other continents. At times, those seas were so poorly connected to the open oceans that water did not circulate freely between them and the oceans. Consequently, evaporation concentrated the dissolved salts until they precipitated as marine evaporites. Periodically, storms flushed new seawater from the open ocean into the shallow seas, providing a new supply of salt. Thick marine evaporite beds, formed in this way, underlie nearly 30 percent of North America. Table salt, gypsum (used to manufacture plaster and sheetrock), and potassium salts (used in fertilizer) are mined extensively from these deposits.

Most of the world's supply of iron is mined from sedimentary rocks called **banded iron formations**. These iron-rich rocks precipitated from the seas between 2.6 and 1.9 billion years ago, as a result of rising atmospheric oxygen concentrations. The processes that formed these important ores are closely related to the development of life on Earth and are described in Chapter 17.

Weathering Processes

In environments with high rainfall, the abundant water dissolves and removes most of the soluble ions from soil and rock near Earth's surface. This process leaves the

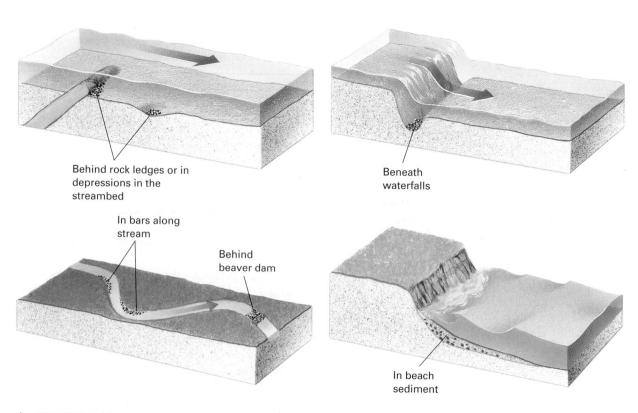

Behind rock ledges or in depressions in the streambed

Beneath waterfalls

In bars along stream

Behind beaver dam

In beach sediment

◆ **FIGURE 5.10** Placer deposits form where water currents slow down and deposit high-density minerals.

◆ **FIGURE 5.11** This spheroidal texture is typical of bauxite, which is aluminum ore formed by intense tropical weathering of aluminum-rich rocks.

relatively insoluble ions in the soil to form **residual ore** deposits. Both aluminum and iron have very low solubilities in water. **Bauxite**, the principal source of aluminum, forms as a residual deposit, and in some instances iron also concentrates enough to become ore. Most bauxite deposits form in warm, rainy, tropical, or subtropical environments where chemical weathering occurs rapidly. Thus bauxite ores are common in Jamaica, Cuba, Guinea, Australia, and parts of the southeastern United States (◆ Figure 5.11). Some bauxite deposits are found today in regions with dry, cool climates. Most of them, however, formed when the regions had a warm, wet climate, and they reflect climatic change since their origins.

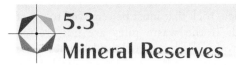

5.3
Mineral Reserves

Mineral reserves are the known amount of ore in the ground. The term can refer to the amount of ore remaining in a particular mine, or it can be used on a global or national scale. Mining depletes mineral reserves by decreasing the amount of ore remaining in the ground, but reserves may increase in two ways. First, geologists may discover new mineral deposits, thereby adding to the known amount of ore. Second, subeconomic mineral deposits—those in which the metal is not sufficiently concentrated to be mined at a profit—can become profitable if the price of that metal increases or if improvements in mining or refining technology reduce extraction costs.

Consider an example of the changing nature of reserves. In 1966 geologists estimated that global reserves

of iron were about 5 billion tons.[1] At that time, world consumption of iron was about 280 million tons per year. Assuming that consumption continued at the 1966 rate, the global iron reserves identified in 1966 would have been exhausted in 18 years, and we would have run out of iron ore in 1984. But iron ore is still plentiful and cheap today because new and inexpensive methods of processing lower-grade iron ore were developed. Thus, large deposits that were subeconomic in 1966, and therefore not counted as reserves, are now ore.

The Geopolitics of Metal Resources

Earth's mineral resources are unevenly distributed, and no single nation is self-sufficient in all minerals. For example, almost two-thirds of the world's molybdenum reserves, and more than one-third of the lead reserves, are located in the United States. More than half of the aluminum reserves are found in Australia and Guinea. The United States uses about 30 percent of all aluminum produced in the world, yet it has no large bauxite deposits. Zambia and the Democratic Republic of Congo supply half of the world's cobalt, although neither nation uses the metal for its own industry.

Five nations—the United States, Russia, South Africa, Canada, and Australia—supply most of the mineral resources used by modern societies. Many other nations have few mineral resources. For example, Japan has almost no metal or fuel reserves; despite its modern economy and high productivity, it relies entirely on imports for metals and fuel.

Developed nations consume most of Earth's mineral resources. While the United States, Japan, Germany, and Russia have traditionally been the largest consumers of these resources, China's industrial base is now growing so rapidly that it may outpace the developed world in consumption.

Currently, the United States depends on 25 other countries for more than half of its mineral resources. Some must be imported because we have no reserves of our own. We do have reserves of others, but we consume them more rapidly than we can mine them, or we can buy them more cheaply than we can mine them.

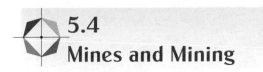

5.4
Mines and Mining

Miners extract both ore and coal (described in the following section) from underground mines and surface mines. A large **underground mine** may consist of tens of kilometers of interconnected passages that commonly follow ore veins or coal seams (◆ Figure 5.12). The lowest

---

1. B. Mason, *Principles of Geochemistry,* 3rd ed. (New York: John Wiley, 1966), Appendix III.

FIGURE 5.12 Machinery extracts coal from an underground coal mine.

FIGURE 5.14 A huge power shovel dwarfs a person standing inside the Navajo Strip Mine in New Mexico.

levels may be several kilometers deep. In contrast, a **surface mine** is a hole excavated into Earth's surface. The largest human-created hole is the open-pit copper mine at Bingham Canyon, Utah. It is 4 kilometers in diameter and 0.8 kilometers deep. The mine produced 263,700 tons of copper in 2004 and smaller amounts of gold, silver, and molybdenum (◆ Figure 5.13). Most modern coal mining is done by large power shovels that extract coal from huge surface mines (◆ Figure 5.14).

In the United States, the Surface Mining Control and Reclamation Act requires that mining companies restore mined land so that it can be used for the same purposes for which it was used before mining began. In addition,

a tax is levied to reclaim land that was mined before the law was enacted. Enforcement and compliance of environmental laws waxes and wanes with the political climate in Washington. Yet environmental awareness has increased dramatically over the past generation, and, overall, mining operations are less polluting today than they were when your parents were in their teens. One of the big challenges for the future is to clean up old mines that were operated under lax or nonexistent environmental regulations of the past. In the United States, more than 6,000 unrestored coal and metal surface mines cover an area of about 90,000 square kilometers, slightly smaller than the state of Virginia. This figure does not include abandoned sand and gravel mines and rock quarries, which probably account for an even larger area.

Although underground mines do not directly disturb the land surface, some abandoned mines collapse, and occasionally buildings have fallen into the holes (◆ Figure 5.15). Over 800,000 hectares (2 million acres) of land in central Appalachia have settled into underground coal mine shafts.

Mining of both metal ores and coal also creates huge piles of waste rock, rock that must be removed to get at the ore or coal. If the waste piles are not treated properly, rain erodes the loose rock and leaches toxic elements such as arsenic, sulfur, and heavy metals from the piles, choking the streams with sediment and contaminating both stream water and ground water. These environmental effects are discussed in Chapter 12.

FIGURE 5.13 The Bingham Canyon, Utah, open-pit copper mine is the largest human-created hole on Earth. It is 4 kilometers in diameter and 0.8 kilometer deep.

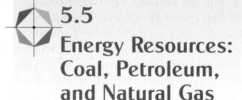

5.5
Energy Resources: Coal, Petroleum, and Natural Gas

Coal, petroleum, and natural gas are called **fossil fuels** because they formed from the remains of plants and animals. Fossil fuels are not only nonrenewable but also

◆ **FIGURE 5.15** This house tilted and broke in half as it sank into an abandoned underground coal mine.

unrecyclable. When a lump of coal or a liter of oil is burned, the energy dissipates and is, for all practical purposes, lost. Thus our fossil fuel supply inexorably diminishes.

Coal

Coal is a combustible rock composed mainly of carbon. Humans began using coal before they used petroleum and natural gas because coal is easily mined and can be burned without refining.

Coal-fired electric generating plants burn about 92 percent of the coal consumed in the United States. The remainder is used to make steel or to produce steam in factories. Although it is easily mined and abundant in many parts of the world, coal emits air pollutants that can be removed only with expensive control devices.

Large quantities of coal formed worldwide during the Carboniferous Period, between 360 and 286 million years ago, and later in Cretaceous and Paleocene times, when warm, humid swamps covered broad areas of low-lying land. Coal is probably forming today in some places, such as in the Ganges Delta in India, but the process is much slower than the rate at which we are consuming coal reserves. As shown in ◆ Figure 5.16, widespread availability of this fuel is projected at least until the year 2200.

When plants die in forests and grasslands, organisms consume some of the plant litter, and chemical reactions with oxygen and water decompose the remainder. As a result, little organic matter accumulates except in the topsoil. In some warm swamps, however, plants grow and die so rapidly that newly fallen vegetation quickly buries older plant remains. The new layers prevent atmospheric oxygen from penetrating into the deeper layers, and decomposition stops before it is complete, leaving brown, partially decayed plant matter called **peat**. Commonly, peat is then buried by mud deposited in the swamp.

Plant matter is composed mainly of carbon, hydrogen, and oxygen and contains large amounts of water. During burial, rising pressure expels the water and chemical reactions release most of the hydrogen and oxygen. As a result, the proportion of carbon increases until coal forms (◆ Figure 5.17). The grade of coal, and the heat that can be recovered by burning coal, varies considerably depending on the carbon content (Table 5.3).

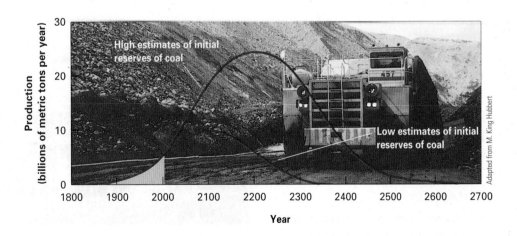

◆ **FIGURE 5.16** Past and predicted global coal supplies based on two estimates of reserves. Shaded area shows coal already consumed.

A Litter falls to floor of stagnant swamp

B Debris accumulates, barrier forms, decay is incomplete

C Sediment accumulates, organic matter is converted to peat

D Peat is lithified to coal

◈ **FIGURE 5.17** Peat and coal form as sediment buries organic litter in a swamp.

Table 5.3 Classification of Coal by Grade, Heat Value, and Carbon Content

Type	Color	Water (%)	Other Volatiles and Non-combustible Compounds (%)	Carbon (%)	Heat Value (BTU/lb)
Peat	Brown	75	10	15	3,000–5,000
Lignite	Dark Brown	45	20	35	7,000
Bituminous (soft coal)	Black	5–15	20–30	55–75	12,000
Anthracite (hard coal)	Black	4	1	95	14,000

Petroleum

The first commercial oil well was drilled in the United States in 1859, ushering in a new energy age. Crude oil, as it is pumped from the ground, is a gooey, viscous, dark liquid made up of thousands of chemical compounds. It is then refined to produce propane, gasoline, heating oil, and other fuels (◈ Figure 5.18). Petroleum also is used to manufacture plastics, nylon, and other useful materials.

Formation of Petroleum

Streams carry organic matter from decaying land plants and animals to the sea and to some large lakes, and deposit it with mud in shallow coastal waters. Marine plants and animals die and settle to the sea floor, adding

◈ **FIGURE 5.18** An oil refinery converts crude oil into useful products such as gasoline.

more organic matter to the mud. Over millions of years, younger sediment buries this organic-rich mud to depths of a few kilometers, where rising temperature and pressure convert the mud to shale. At the same time, the elevated temperature and pressure convert the organic matter to liquid **petroleum** that is dispersed throughout the rock (◇ Figure 5.19). The activity of bacteria may enhance the process. Typically, petroleum forms in the temperature range of 50 to 100°C.

The shale or other sedimentary rock in which oil originally forms is called the **source rock**. Oil dispersed in shale cannot be pumped from an oil well because shale is relatively impermeable; that is, liquids do not flow through it rapidly. But under favorable conditions, petroleum migrates slowly to a nearby layer of permeable rock—usually sandstone or limestone—where it can flow readily. Because petroleum is less dense than water or rock, it then rises through the permeable rock until it is trapped within the rock or escapes onto Earth's surface.

Many **oil traps** form where a layer of impermeable rock such as shale prevents the petroleum from rising further. Oil or gas then accumulates in a petroleum **reservoir**. Folds and faults create several types of oil traps (◇ Figure 5.20). In some regions, large, lightbulb-shaped bodies of salt have flowed upward through solid rocks to form salt domes. The rising salt folded the surrounding rock to form an oil trap (◇ Fig. 5.20D). The salt originated as a sedimentary bed of marine evaporite, and it rose because salt is less dense than the surrounding rocks. An oil reservoir is not an underground pool or lake of oil. It consists of oil-saturated permeable rock that is like an oil-soaked sponge.

Geologic activity can destroy an oil reservoir as well as create one. A fault may fracture the reservoir rock, or tectonic forces may uplift the reservoir and expose it to erosion. In either case, the petroleum escapes once the trap is destroyed. Sixty percent of all oil wells are found in relatively young rocks that formed during the Cenozoic Era. Undoubtedly, much petroleum that had formed in older Mesozoic and Paleozoic rocks escaped long ago and decomposed at Earth's surface.

Petroleum Extraction, Transport, and Refining

To extract petroleum, an oil company drills a well into a reservoir. After the hole has been bored, the expensive drill rig is removed and replaced by a pumper that slowly extracts the petroleum. Fifty years ago, many reservoirs lay near the surface and oil was easily pumped from shallow wells. But these reserves have been depleted, and modern oil wells are often a few kilometers or more deep.

In the past, oil wells were drilled vertically into the reservoir. Oil then flowed through the reservoir to the well, where pumps raised it to the surface. The amount of oil that reached the well was limited by the permeability of the reservoir rock and by the viscosity of the oil.

On average, more than half of the oil in a reservoir is too viscous to be pumped to the surface by conventional techniques. This oil is left behind after an oil field has

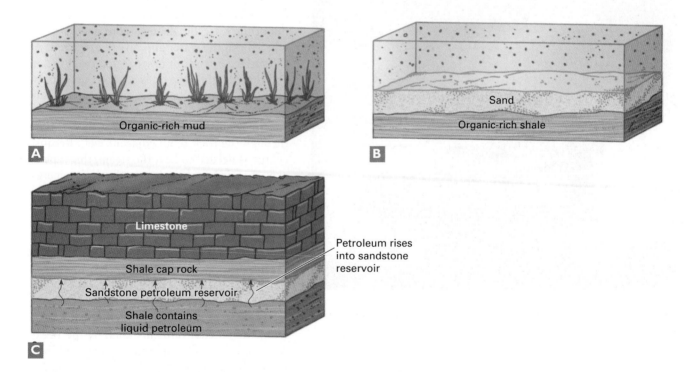

◇ **FIGURE 5.19** **(A)** Organic matter from land and sea settles to the sea floor and mixes with mud. **(B)** Younger sediment buries this organic-rich mud. Rising temperature and pressure convert the mud to shale, and the organic matter to petroleum. **(C)** The petroleum is trapped in the reservoir by an impermeable cap rock.

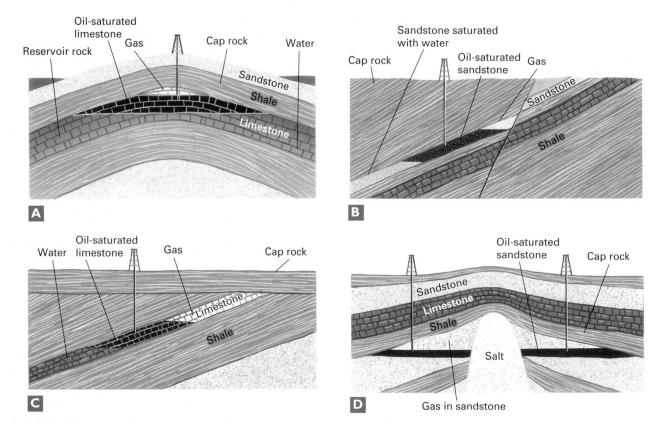

◈ FIGURE 5.20 Four types of oil traps. **(A)** Petroleum rises into permeable limestone capped by impermeable shale in a structural dome. **(B)** A trap forms where a fault has moved impermeable shale against permeable sandstone. **(C)** Horizontally bedded shale traps oil in tilted limestone. **(D)** Sedimentary salt rises and deforms overlying strata to create a trap.

◈ FIGURE 5.21 An offshore, oil-drilling platform extracts oil from the continental shelf beneath a shallow sea.

"gone dry." Recently, oil companies have developed methods of drilling horizontally through reservoirs, allowing access to vast amounts of oil left by earlier wells. Additional oil can be extracted by **secondary** and **tertiary recovery techniques**. In one simple secondary

process, water is pumped into one well, called the "injection well." The pressurized water floods the reservoir, driving oil to nearby wells, where both the water and oil are extracted. At the surface, the water is separated from the oil and reused, while the oil is sent to the refinery. One tertiary process forces superheated steam into the injection well. The steam heats the oil and makes it less viscous so that it can flow through the rock to an adjacent well. Because energy is needed to heat the steam, this type of extraction is not always cost effective or energy efficient. Another tertiary process pumps detergent into the reservoir. The detergent dissolves the remaining oil and carries it to an adjacent well, where the petroleum is then recovered and the detergent recycled.

Because an oil well occupies only a few hundred square meters of land, most cause relatively little environmental damage. However, oil companies are now extracting petroleum from fragile environments such as the ocean floor and the Arctic tundra. To obtain oil from the sea floor, engineers build platforms on pilings driven into the ocean floor and mount drill rigs on these steel islands (◈ Figure 5.21). Despite great care, accidents occur during the drilling and extraction of oil. When accidents occur at sea, millions of barrels of oil can spread throughout the waters, poisoning marine life and disrupting marine ecosystems. Significant oil spills have occurred in virtually all offshore drilling areas.

Natural Gas

Natural gas forms in source rock or an oil reservoir when crude oil is heated above 100°C during burial. (Natural gas is mostly methane, CH_4, an organic molecule consisting of a single carbon atom bonded to four hydrogen atoms.) Consequently, many oil fields contain a mixture of oil, with natural gas floating above the heavier liquid petroleum. In other instances, the lighter, more mobile gas escaped into the atmosphere or was trapped in a separate underground reservoir.

Natural gas is used without refining for home heating, cooking, and to fuel large electrical generating plants. Because natural gas contains few impurities, it releases no sulfur or other pollutants when it burns, although, as with all fossil fuels, combustion of natural gas releases carbon dioxide, a greenhouse gas. This fuel has a higher net energy yield, produces fewer pollutants, and is less expensive to produce than petroleum. At current consumption rates, global natural gas supplies will last for 80 to 200 years.

Although most commercial natural gas is produced from petroleum fields, 7 percent of current U.S. gas production comes from coal seams, where both natural Earth heat and microbial activity slowly convert buried coal to **coal bed methane**. The coal bed methane reserves in the United States are estimated to be more than 700 trillion cubic feet (Tcf), although less than 100 Tcf may be economically recoverable.

Most coal beds have a high capacity to store water in small voids in the coal itself. As natural processes convert coal to methane, the gas dissolves in the coal bed ground water, where pressure of overlying rock keeps the methane in solution. Natural gas companies drill thousands of wells into the coal beds and pump the ground water to the surface, decreasing the pressure on the remaining water in the coal bed. The decreased pressure allows the methane to separate from the water. It is then piped to the surface, where it is compressed and sent to market.

Because they store so much water, coal beds are important ground water reservoirs for farmers and ranchers, especially in the arid and semi-arid western United States, where extensive coal bed methane development is now occurring. Here, coal bed methane development has two serious impacts on regional agriculture and ecosystems. The extraction of so much water from the coal beds has depleted essential aquifers and lowered the water table over large areas. Secondly, coal bed water is commonly salty. After it is pumped to the

surface, the saltwater can poison streams and soils, rendering them useless for agriculture and wildlife. State and federal regulations on water extraction and disposal methods attempt to minimize these impacts.

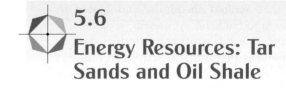

5.6
Energy Resources: Tar Sands and Oil Shale

In 2004, 84 percent of energy used in the United States came from petroleum, coal, and natural gas, which traditionally have been the cheapest fuels. However, the price of crude petroleum jumped from less than $40 to nearly $60 per barrel, a 50 percent increase, between 2004 and 2005.

In the paragraphs that follow, we discuss heavy oil and renewable resources. In the past, these energy sources have been more expensive than the three traditional fossil fuels. However, the cost of producing these alternative energies has been decreasing at the same time that the costs of traditional fuels have been increasing. As a result, the world is on the threshold of a major restructuring of the global economy of energy. Many fuel sources that were uneconomical even a year ago are now cheaper than the conventional sources.

Tar Sands

In some regions, large sand deposits are permeated with heavy oil and an oil-like substance called **bitumen**, which are too thick to be pumped. The richest **tar sands** exist in Alberta (Canada), Utah, and Venezuela.

◆ **FIGURE 5.22** Heavy equipment mines tar sands, which are abundant in Alberta, Canada.

In Alberta alone, tar sands contain an estimated 1 trillion barrels of petroleum. About 10 percent of this fuel is shallow enough to be surface mined (◆ Figure 5.22). Tar sands are dug up and heated with steam to make the heavy oil and bitumen fluid enough to separate from the sand. The oil and bitumen are then treated chemically and heated to convert them to crude oil. At present, several companies mine tar sands profitably, producing more than 13 percent of Canada's petroleum requirements. In 2004 Syncrude Canada, the world's largest producer of crude oil from oil sands, produced more than 87 million barrels of oil at $39 per barrel, significantly below world market prices for petroleum from conventional sources. Deeper deposits, comprising the remaining 90 percent of the reserve, can be extracted using subsurface techniques similar to those discussed for secondary and tertiary recovery.

Oil Shale

Some shales and other sedimentary rocks contain a waxy, solid organic substance called **kerogen**. Kerogen is organic material that has not yet converted to oil. Kerogen-bearing rock is called **oil shale**. If oil shale is mined, mixed with water, and then heated, the kerogen converts to petroleum. In the United States, oil shales contain the energy equivalent of 2 to 5 trillion barrels of petroleum, enough to fuel the nation for 300 to 700 years at the 2006 consumption rate (◆ Figure 5.23). However, many oil shales are of such low grade that more energy is required to mine and convert the kerogen to petroleum than is generated by burning the oil. Consequently, these low-grade shales may never be used for fuel.

Water consumption is a serious problem in oil shale development. Approximately two barrels of water are needed to produce each barrel of oil from shale. Oil shale occurs most abundantly in the semiarid western United States. In this region, scarce water is also needed for agriculture, domestic use, and industry.

A recent study by the RAND Corporation estimates that there are 800 billion barrels of recoverable oil shale resources from the Green River Valley in Colorado, Utah, and Wyoming—enough to satisfy U.S. oil needs for 100 years at current consumption rates. However, recovering crude oil from shale is expensive, and the study predicts that oil prices will have to reach $70 to $95 per barrel before the process will be economical. As with the case of tar sands in Alberta, increasing world oil prices and technological advances in oil shale recovery techniques combine to make oil shale a likely source of energy in the future.[2]

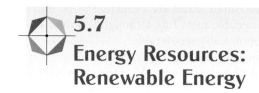

◆ **FIGURE 5.23** Secondary recovery, tar sands, and oil shale increase our petroleum reserves significantly.

5.7 Energy Resources: Renewable Energy

Solar, wind, geothermal, hydroelectric, wood, and other biomass fuels are renewable—natural processes replenish them as we use them. Although the amount of energy produced today by renewable sources is small compared to that provided by fossil and nuclear fuels, renewable resources have the potential to supply all of our energy needs.

In the previous section, we learned that tar sands have become economical as the cost of conventional petroleum has increased. Similarly, as the price of conventional fossil fuels rose in 2004 and 2005, some renewables became less expensive than conventional energy. Except for biomass fuels, renewable energy sources emit no carbon dioxide and therefore do not contribute to global warming.

Solar Energy

Current technologies allow us to use solar energy in three ways: passive solar heating, active solar heating, and electricity production by solar cells.

A passive solar house is built to absorb and store the Sun's heat directly. In active solar heating systems,

2. James T. Bartis, *Oil Shale Development in the United States* (Santa Monica, Calif.: RAND Corporation, 2005).

solar thermal collectors absorb the Sun's energy and use it to heat water. Pumps then circulate the hot water through radiators to heat a building, or the inhabitants use the hot water directly for washing and bathing. Solar thermal collectors are becoming increasingly popular worldwide, with an estimated 18 million square meters of capacity added in 2004, bringing the total capacity to nearly 150 million square meters.

A **solar cell** or photovoltaic (PV) produces electricity directly from sunlight. A modern solar cell is a semiconductor. Sunlight energizes electrons in the semiconductor so that they travel through the crystal, producing an electric current (◆ Figure 5.24). The sun bathes the earth with 86,000 trillion watts of energy at any given time, more than 6,600 times the amount currently used by humans each year.

Although solar power still accounts for less than 1 percent of world energy demand, solar energy is our most abundant resource, and photovoltaic cell production is the fastest-growing segment of the energy industry. The cost of solar cells has decreased 100-fold in the last 30 years, and global production more than doubled between 2002 and 2004. Photovoltaic arrays are now competitive with electricity costs during peak demand times in California. Photovoltaics are also cost effective for electricity needs far from existing power lines.

The 2004 production capacity of U.S. solar panels was 397 megawatts, amounting to much less than one percent of total usage. In Japan, the world leader in solar energy, the government plans to generate 10 percent of the nation's electricity with PVs by 2030.

Wind Energy

In the eight years from 1997 to 2004, production of electricity from wind in the United States increased fourfold to 6,740 megawatts, as much as that of six nuclear power plants and enough to power a city of more than 2 million (◆ Figure 5.25). Utility-scale, wind-power projects now under construction or under negotiation will add at least 5,000 megawatts of wind capacity in the United States over the next five years. Three of the largest wind farms in the world are located in California, at Altamont Pass, San Gorgonio Pass, and Tehachapi Pass. They are actually collections of dozens of individual wind farms that have several owners and turbine types, and have been built and modified over several decades. All three of these areas are seeing renewed growth; the old, small wind turbines are being replaced with much-larger, more-efficient wind turbines. Today, the smallest

◆ **FIGURE 5.24** Solar panels installed in 2005 provide energy for the U.S. Mission to the United Nations in Geneva, Switzerland. This is the largest solar energy project ever installed at a U.S. government building overseas.

◆ **FIGURE 5.25** Wind turbines generate electricity at Cowley Ridge, Alberta, Canada.

utility-scale wind turbines generate about 700 kilowatts, with some models approaching 5,000 kilowatts (5 megawatts). One megawatt of wind capacity is enough to supply 240 to 300 average American homes. Thus, the 2,100 megawatts of wind-generated electricity currently produced in California is enough to supply 500,000 to 620,000 households. Other gigantic wind farms now generate electricity in Texas, eastern Oregon, and several other states. A huge, untapped potential for wind generation exists in several midwestern and western states, where winds blow strongly and almost continuously.

The U.S. Department of Energy has announced a goal of generating 6 percent of U.S. electricity from wind by 2020—a goal consistent with the current rate of growth of wind energy nationwide. However, independent experts predict that as public demand for clean energy grows and as the cost of producing energy from the wind continues to decline, wind energy will provide an even greater portion of the nation's energy supply. Wind energy production is growing rapidly because construction of wind generators is now cheaper than building new fossil fuel–fired power plants. Wind energy is also clean and virtually limitless. Most people feel that the environmental side effects of wind farms (noise, unsightliness, or interference with bird migration patterns) are acceptable alternatives to the pollution and greenhouse gas emission involved in mining and burning coal to produce the same amount of electricity.

Worldwide, wind is the second-fastest-growing source of energy and many countries are rapidly investing in new wind farms. Germany, the world leader in wind energy, currently generates 6.6 percent of its electricity from wind power—16,629 megawatts in 2004—and Spain is not far behind with 6 percent.

Geothermal Energy

Energy extracted from Earth's internal heat is called geothermal energy. Natural hot ground water can be pumped to the surface to generate electricity, or it can be used directly to heat homes and other buildings. Alternatively, cool surface water can be pumped deep into the ground, to be heated by subterranean rock, and then circulated to the surface for use. The United States is the largest producer of geothermal electricity in the world, with a production capacity of 2,133 megawatts, an amount equivalent to the power output of two large nuclear reactors.

Hydroelectric Energy

If a river is dammed, the energy of water dropping downward through the dam can be harnessed to turn turbines that produce electricity. Hydroelectric generators supply between 15 and 20 percent of the world's electricity. They provide about 3 percent of all energy consumed in the United States, but about 8 percent of our electricity.

The United States is unlikely to increase its production of hydroelectric energy. Large dams are expensive to build, and few suitable sites remain. Environmentalists commonly oppose dam construction because the resulting reservoirs flood large areas—destroying wildlife habitat, agricultural land, towns, and migratory fish

populations. For example, the dams on the Columbia River and its tributaries are largely responsible for the demise of salmon populations in the Pacific Northwest. Undammed wild rivers and their canyons are prized for their aesthetic and recreational value.

Biomass Energy

Biomass (plant) fuels currently produce about 9,700 megawatts of energy, about one-third of that generated by nuclear power plants. Wood is the most productive of all biomass, followed by controlled garbage incineration and alcohol fuels. The use of biofuels, fuels derived from crops and agricultural wastes, is growing rapidly around the world. In general, these fuels are cleaner burning than fossil fuels and can be produced domestically in most countries, thereby creating local jobs and reducing foreign oil imports. However, production of biofuels is not always a net energy gain; in some cases, more energy is used in the production and processing of these fuels than can be extracted from them. There are two main types of biofuel: ethanol and biodiesel.

Ethanol is derived from corn, grains, and sugar and is a common additive to gasoline. World production of ethanol increased by 13.6 percent in 2004, reaching almost 33 billion liters or the equivalent of 200 million barrels of oil. Ethanol is now commonly used as a fuel additive to increase the octane level of gasoline and to add oxygen that lowers carbon monoxide emissions during the combustion process.

Biodiesel can be made from seed crops such as sunflowers or from organic wastes, including recycled cooking oil. The European Union is the world's top producer of biodiesel, producing 1.6 billion liters in 2003, a 43 percent increase over 2001 levels. Many European car manufacturers have designed their vehicles to use this fuel.

Limitations to Renewable Energy Resources

Solar cells, wind turbines, hydroelectric dams, and geothermal power plants harness nature's energy to produce electricity. Solar heaters, geothermal, and biomass resources are used to heat water or buildings. But aside from biofuels, none of these resources can be used directly to power mobile transportation systems such as cars and trucks.[3] Several methods are available to convert these energy sources for use in transportation.

Perhaps the easiest way to use electricity to transport people and goods is the old-fashioned electric train. Electric streetcars, commuter trains, and subways have been used for decades. If we build more electric mass-

---

3. Electricity can be used to charge a battery that powers a car. This option is discussed further in the text.

transit systems, and if people use them, we could shift away from our dependence on the internal-combustion engine and on petroleum. Eventually, the required electricity consumption could be supplied by renewable energy sources.

Another solution is the electric car. Battery-only and gasoline–battery hybrid cars are discussed in the following subsection, "Transportation."

Energy planners also envision a **hydrogen economy**, in which a **fuel cell** uses the chemical energy of hydrogen to produce electricity cleanly and efficiently, with water and heat as byproducts. Fuel cells can provide energy for systems as large as a power station and as small as a laptop computer. They can also power cars, trucks, trains, and other vehicles.

A fuel cell uses hydrogen and oxygen to create heat and electricity by electrochemical processes. In one such process, the cell separates hydrogen's negatively charged electron from the hydrogen nucleus, which then consists of a single positively charged proton. The electrons, in turn, combine with oxygen, which then reacts with the hydrogen proton to form water and heat energy. In some types of fuel cells, the electrons travel through an electrical circuit to reach the other side of the cell. This movement of electrons is an electrical current. Thus, fuel cells can produce both heat and an electrical current, which can then be used as power sources for transportation, electrical appliances, and most other energy-consuming equipment.

Fuel cells have several benefits over fossil-fuel and nuclear technologies now used in power plants and vehicles. They emit only water vapor. Although water vapor is a greenhouse gas, it is seen as a lesser environmental threat than carbon dioxide, and fuel cells emit no pollutants that create smog and cause health problems. Hydrogen is an abundant component of water and other common materials; the supply is nearly limitless. Thus, hydrogen and fuel cells have the potential to eliminate our dependence on petroleum imports.

Separating hydrogen from water requires electric energy, which must come from somewhere. Energy planners suggest using wind- or solar-powered electricity to dissociate water and separate the hydrogen for use in fuel cells, which can then power autos, trucks, trains, and other mobile forms of transportation. Hydrogen also can be extracted from fossil fuels, a process that consumes less electrical energy than separating the hydrogen from water—but the process consumes the fossil fuels, which are much more expensive than water.

Researchers have also genetically manipulated certain bacteria to produce more hydrogen and less oxygen during photosynthesis. They estimate that if the conversion efficiency of this process can be increased to 10 percent, a 100-mile square plot of "bioreactors" could power a fleet of 200 million

hydrogen vehicles. Over the past 35 years of writing these textbooks, we have seen many such promising ideas. Not all of them come to fruition, but they remind us of the innovative power of the human mind and of our capacity for generating creative solutions to complex problems.

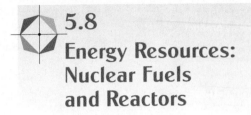

5.8
Energy Resources: Nuclear Fuels and Reactors

Nuclear fuels are radioactive isotopes used to generate electricity in nuclear reactors. Uranium is the most commonly used nuclear fuel. These energy resources, like mineral resources, are nonrenewable, although uranium is abundant.

A modern nuclear power plant uses nuclear **fission** to produce heat and generate electricity. One isotope of uranium, U-235, is the major fuel. When a U-235 nucleus is bombarded with a neutron, it breaks apart (*fission* means "splitting"). The initial reaction releases two or three neutrons. Each of these neutrons can trigger the fission of additional nuclei; hence, this type of nuclear reaction is called a **branching chain reaction** (◆ Figure 5.26). Because this fission is initiated by neutron bombardment, it is not a spontaneous process and is different from natural radioactivity.

To fuel a nuclear reactor, uranium concentrated with U-235 is compressed into small pellets. Each pellet could easily fit into your hand but contains the energy equivalent of 1 ton of coal (◆ Figure 5.27A). A column of pellets is encased in a 2-meter-long pipe, called a **fuel rod** (◆ Figure 5.27B). A typical nuclear power plant contains about 50,000 fuel rods bundled into assemblies of 200 rods each. **Control rods** made of neutron-absorbing alloys are spaced among the fuel rods. The control rods fine-tune the reactor. If the reaction speeds up because too many neutrons are striking uranium atoms, then the power-plant operator lowers the control rods to absorb more neutrons and slow down the reaction. If fission slows down because too many neutrons are absorbed, the operator raises the control rods. If an accident occurs and all internal power systems fail, the control rods fall into the reactor core and quench the fission.

The reactor core produces tremendous amounts of heat. A fluid, usually water, is pumped through the reactor core to cool it. The cooling water (which is now radioactive from exposure to the core) is then passed through a radiator, where it heats another source of water to produce steam. The steam drives a turbine, which in turn generates electricity (◆ Figure 5.28).

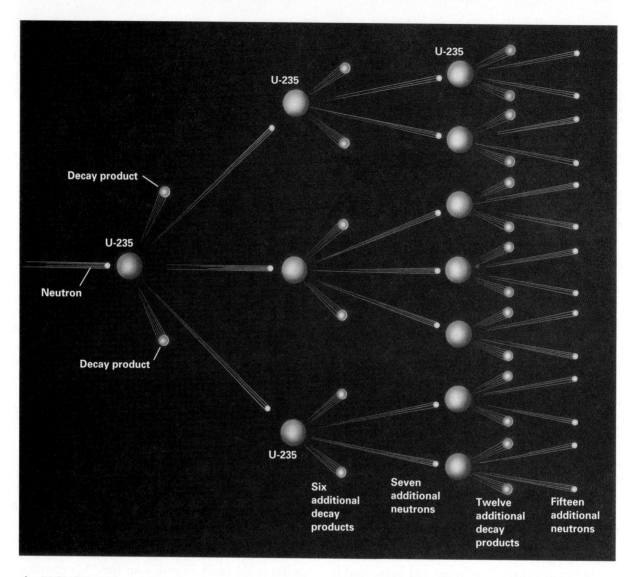

◇ **FIGURE 5.26** In a branching chain reaction, a neutron strikes a uranium-235 nucleus and the nucleus splits into two roughly equal fragments and emits two or three neutrons. These neutrons can then initiate additional reactions, which produce more neutrons. A branching chain reaction accelerates rapidly through a sample of concentrated uranium-235.

◇ **FIGURE 5.27** **(A)** Fuel pellets contain enriched uranium-235. Each pellet contains the energy equivalent of 1 ton of coal. **(B)** Fuel pellets are encased into narrow rods that are bundled together and lowered into the reactor core.

Every step in the mining, processing, and use of nuclear fuel produces radioactive wastes. The mine waste discarded during mining is radioactive. Enrichment of the ore produces additional radioactive waste. When a U-235 nucleus undergoes fission in a reactor, it splits into two useless radioactive nuclei that must be discarded. Finally, after several months in a reactor, the U-235 concentration in the fuel rods drops until the fuel pellets are no longer useful. In some countries, these pellets are reprocessed to recover U-235, but in the United States this process is not economical and the pellets are discarded as radioactive waste. In Chapter 12 we discuss problems and solutions for storing radioactive wastes.

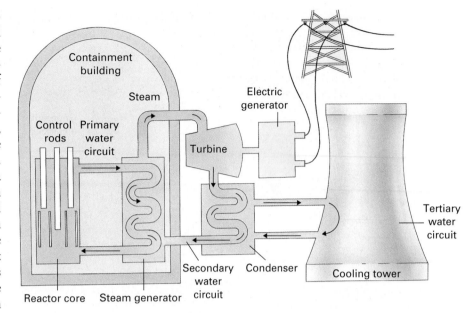

◆ **FIGURE 5.28** In a nuclear power plant, fission energy creates heat, which is used to produce steam. The steam drives a turbine, which generates electricity.

In the early 1970s, the nuclear industry was growing rapidly and many energy experts predicted that nuclear energy would dominate the generation of electric energy. Some experts even suggested that electricity would become "too cheap to meter." These predictions have not been realized. Four factors led to the decline of the nuclear power industry: (1) Construction of new reactors in the United States became so costly that electricity generated by nuclear power became more expensive than that generated by coal-fired power plants. (2) After major accidents at Three Mile Island in the United States and Chernobyl in Russia, people became concerned about safety. (3) Serious concerns remain about the safe disposal of nuclear wastes. (4) The demand for electricity has risen less than expected during the past three decades. As a result, growth of the nuclear power industry has nearly halted. After 1974, many planned nuclear power plants were canceled, and after 1981—over 25 years ago—no new orders were placed for nuclear power plants in the United States. The dramatic turnaround led to serious financial altercations; *Forbes* business magazine called the United States nuclear power program "the largest managerial disaster in U.S. business history, involving $1 trillion in wasted investment and $10 billion in direct losses to stockholders."

Elsewhere, nuclear power production has seen a 2 percent increase in generating capacity between 2003 and 2004, the highest rate of growth ever reached. Plants are currently under construction in several countries, including India, Japan, and China. While some countries are decommissioning older plants and replacing them with other forms of power, many have plans for new nuclear reactors to be built in coming years.

Today, with rising fuel prices, many U.S. policy makers are suggesting that the nuclear option be reconsidered. However, at the same time, alternative energy resources such as wind and solar are being explored aggressively.

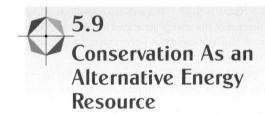

5.9
Conservation As an Alternative Energy Resource

The single quickest and most effective way to decrease energy consumption and to prolong the availability of fossil fuels is to conserve energy (◆ Figure 5.29). Policies to improve energy efficiency are more cost effective than building new power plants. Such policies help to reduce air pollution and dependence on oil imports while saving money for consumers and industry.

Energy conservation has already produced dramatic results in the United States, where energy consumption has decreased by 47 percent per dollar of gross domestic product (GDP) in the last 30 years. However, total fuel consumption continues to rise, and there is much room for continued improvement. Some energy experts have suggested that if people in industrialized nations use more efficient equipment and develop more efficient habits, these nations could conserve as much as half of the energy that they consume.

Energy use in the United States falls into three categories: buildings, industry, and transportation. Two kinds of conservation strategies can be applied in each of

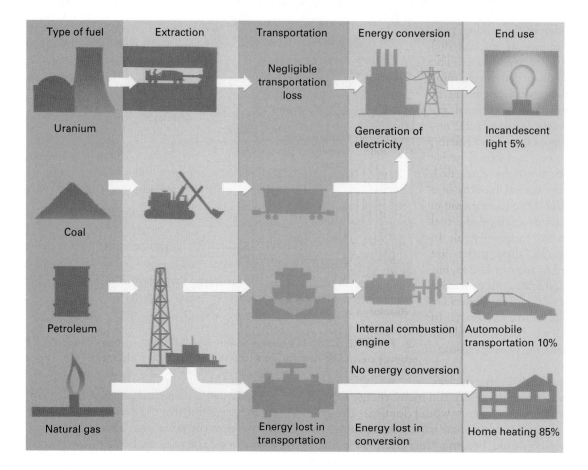

◆ **FIGURE 5.29** The end-use efficiencies of common energy-consuming systems. Only 15 percent of the energy generated in the United States performs useful tasks; 85 percent is wasted. Half of the waste is unavoidable, but half could be avoided with improved technology and conservation.

INTERACTIVE QUESTION: *Use the data in this figure to list ways in which you could conserve energy this week. Order the list from most effective (greatest energy savings) to least effective.*

those categories. Technical solutions involve switching to more efficient implements. Social solutions involve decisions to use existing energy systems more efficiently.

Technical Solutions

Buildings

Residential and commercial buildings consume about 39 percent of all the energy produced in the United States. Most of that energy is used for heating, air conditioning, and lighting. Factoring energy efficiency into home and office building design or retrofitting existing structures to reduce waste can save thousands of dollars in energy costs over the lifetime of the structure.

About 22 percent of the electricity generated in the United States is used for lighting. A fluorescent bulb uses one-fourth of the energy consumed by a comparable incandescent bulb. New solid-state technology promises to make further advances in energy-efficient lighting, with industry experts predicting that LED lights will be 9 times more efficient than incandescent bulbs in the near future. Although fluorescent bulbs cost about $5 each, they last 10 times longer than incandescent bulbs and yield an energy savings equivalent to 3 times their cost.

Switching to more efficient light sources would conserve the electricity generated by about 120 large generating plants and would save $30 billion in fuel and maintenance expenses every year. With a switch to LED lighting, these savings could be much greater.

Industry

Industry consumes another 33 percent of the energy used in the United States. In general, conservation practices are cost effective, and many companies are taking advantage of the fact that saving energy is profitable. According to the Environmental Protection Agency, average U.S. energy use per dollar of GDP declined 2.1 percent per year between 1996 and mid-2005. During that

118

CHAPTER 5 • Geologic Resources

time, energy demand increased, but 78 percent of that increased demand was met by increased efficiency. However, industry still wastes great amounts of energy.

For example, about 70 percent of the electricity consumed by industry—half of all electricity produced in the United States—drives electric motors. Most motors are inefficient because they run only at full speed and are slowed by brakes to operate at the proper speeds to perform their tasks. This approach is like driving your car with the gas pedal pressed to the floor and controlling your speed with the brakes. Replacing older electric motors with variable-speed motors would save an amount of electricity equal to that generated by 150 large power plants, but such replacement has been slow.

Transportation

About 28 percent of all energy, more than 50 percent of the oil usage in the United States, and one-third of the nation's carbon emissions are consumed transporting people and goods. Americans own one-fourth of all automobiles in the world, and drive as many miles as the rest of the world's population combined. On their way to and from work, American drivers, about two-thirds of them driving alone, consume ten percent of all oil consumed in the world. Meanwhile, the automobile markets in China and India are growing rapidly. In 2003, auto sales in China increased 60 percent, adding 4 million private cars to the streets. The global passenger car fleet is now estimated at more than 530 million vehicles.

The efficiency of auto and truck engines is about 10 percent. Thus, we can save much energy by using more efficient cars and trucks. Over the past few decades, automobile manufacturers have offered increasingly efficient mass-production automobiles. In 1996, General Motors introduced a production electric car. But consumers weren't interested and after leasing only 800 vehicles, GM abandoned the project.

A hybrid car, versions of which are now produced by Honda, Toyota, Ford, and Lexus, uses a small, fuel-efficient gasoline engine combined with an electric motor that assists the engine when accelerating. When the driver applies the brakes, the electric motor acts as a generator and charges the batteries while the car is slowing down. When a hybrid car is stopped in traffic, the engine shuts off and restarts automatically when the driver puts the car back in gear. In addition, aerodynamic body design; narrow, stiff tires; and lightweight materials increase the efficiency of hybrid cars.

Hybrids consume less gas and produce less pollution per mile than conventional gasoline engines. Current models of hybrid cars achieve fuel efficiencies ranging from 61/68 (traffic/highway) to 31/27 miles per gallon depending on make and model. The Honda Civic Hybrid gets 10 to 15 miles more per gallon than the same car with a regular gas engine. The hybrid engine of the Toyota Prius produces 90 percent fewer harmful emissions than a comparable gasoline engine.

Hybrid cars currently on the market cost from $3,500 to $6,000 more per car than comparable cars with conventional gasoline engines. Thus, if gas is priced at $2.50 per gallon, it would take the average driver (15,000 miles per year) 5 to 10 years to amortize the higher price for a hybrid. Additionally, the cost of hybrid batteries ranges from $1,000 to $3,000. Although the warranty may cover the hybrid battery, once the warranty expires, replacement can be expensive.

Using hybrids and other energy-efficient vehicles, American motorists can achieve a 50 percent or greater increase in fuel economy, which would save 3 million barrels of oil a day—about one-third of the current oil imports.

Social Solutions

Social solutions involve altering human behavior to conserve energy. Energy-conserving actions can be applied in buildings, in industry, and in transportation. Some result in inconvenience to individuals. For example, if you choose to carpool rather than drive your own car, you save fuel but inconvenience yourself by coordinating your schedule with your carpool companions. Our national fondness for large sport-utility vehicles and pickup trucks, which have poor fuel economy, has resulted in a decline in average fuel efficiency of all American cars and light trucks from 26.2 mpg in 1987 to 25.0 mpg in 2003. As explained previously, high-mileage cars are on the market, but they will only make an impact if people make the social decision to use them. People argue that this social decision comes at a cost because light vehicles make the driver and passengers more vulnerable in case of an accident, but studies have shown that lighter, more agile vehicles, with better turning capacity and more effective braking, are actually safer than heavier SUVs.

At home and in the workplace, lowering the thermostat in winter and wearing a sweater, and using less air conditioning in summer, might reduce the comfort margin but can save considerable energy. Many social solutions, however, are cost free in terms of inconvenience.

When practiced by everyone, simply turning off the lights, the television set, and other appliances when you leave the room will conserve large amounts of energy.

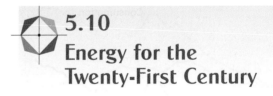

5.10
Energy for the Twenty-First Century

The United States consumes 25 percent of the world's oil yet owns only 3 percent of the known reserves. In the year 2004, fossil fuels supplied 86 percent of all energy used in the United States; oil alone accounted for 40 percent, natural gas for 23 percent, and coal for 22 percent (◆ Figure 5.30). Thus, oil is our major

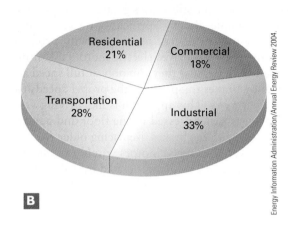

Fossil fuels = 86%
(petroleum, coal, natural gas)

Petroleum 40%
Domestic 13%
Coal 22%
Import 27%
Natural Gas 23%
Other 3%
Nuclear 8%
Renewables 6%
Hydroelectric 3%

A

End-Use Sector Shares of Total Consumption, 2004

Residential 21%
Commercial 18%
Transportation 28%
Industrial 33%

B

Energy Information Administration/Annual Energy Review 2004.

◆ **FIGURE 5.30** **(A)** In the year 2004, fossil fuels supplied 86 percent of all energy used in the United States; petroleum alone accounted for 40 percent, natural gas for 23 percent, and coal for 22 percent. **(B)** Pie graph (B) shows how energy from all sources was consumed in the United States in 2004.

source of energy. In addition, oil is the only portable energy resource currently in popular use, and thus is the main energy resource for transportation in the United States. At current rates of consumption, we have more than 200 years of domestic coal reserves, and at least several decades of natural gas reserves. Oil, however, is another story.

In 1956, M. King Hubbert, a geologist, was working at the Shell research lab in Houston, Texas. Hubbert compared U.S. domestic oil reserves with current and predicted rates of oil consumption. He then forecast that U.S. oil production would peak in the early 1970s and would thereafter decline continuously. He predicted that Americans would have to make up an ever-increasing difference between domestic oil supply and consumption by relying on larger and larger imports, or they would have to turn to other energy resources. Other experts and economists ridiculed his prediction, but in 1970 the U.S. domestic oil production reached its maximum and it has been slowly declining since then.

In 2004, the United States ranked third in global oil production. (Saudia Arabia was first; Russia second.) But because U.S. petroleum consumption was higher than it was in 1970 and production was lower, the United States imported nearly 70 percent of its petroleum (◆ Figure 5.31). Because of this high dependence on foreign oil, the U.S. energy future is intimately linked with the global one.

In 1971, Hubbert predicted that global oil production would peak between 1995 and 2000 and that the world supply of oil will be 90 percent depleted between 2020 and 2030. Twice in the 1970s, the major oil-producing nations reduced oil production and raised the price, causing economic disruptions. In 1990, Iraq invaded Kuwait and threw global petroleum markets into turmoil. Within a week, the price of petroleum skyrocketed from below $20 per barrel to $40 per barrel; it then fell again after the brief Gulf War. In 2006, as we write this edition of *Earth Science and the Environment*, the Middle East is again in turmoil and American troops are again in Iraq, once a major oil producer. In September of 2005, Hurricane Katrina struck the oil-producing regions of the U.S. Gulf Coast, reducing the Gulf Coast oil production by 90 percent and lowering national production by 25 percent. As a result of the combined effects of the Iraq war and storm damage to oil-production facilities on the U.S. Gulf Coast, the price of oil topped $70 per barrel and gasoline prices exceeded $3.50 per gallon in the spring of 2006.

In 2002 Kenneth Deffeyes, an oil geologist and professor at Princeton University, used Hubbert's methods to recalculate petroleum supply and demand. As a result, he predicted a global peak in oil production before 2010[4]. He argued that even new exploration and recovery technolo-

Overview, 1949–2004

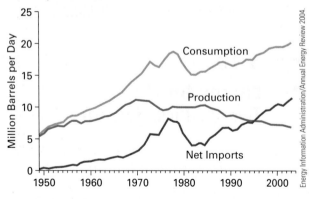

Energy Information Administration/Annual Energy Review 2004.

◆ **FIGURE 5.31** Our reliance on oil imports has increased steadily since 1971 as Hubbert predicted.

4. Kenneth Deffeyes, *Hubbert's Peak: The Impending World Oil Shortage* (Princeton, N.J.: Princeton University Press, 2002).

gies could not forestall declining worldwide oil production and increasing consumption in the very near future.

In rebuttal to this article, many energy experts argue that just as Hubbert was wrong about global oil production peaking between1995 and 2000, Deffeyes is wrong for the same reason: His estimates of reserves are too low and there is more oil than he estimated.

However, these arguments simply alter the estimate of *when* global oil supplies will fall below global demand. No one seriously suggests that Earth has an unlimited supply and that the current energy economy will persist indefinitely.

An important question, then, is, when will oil supplies fail to meet global demands? In *Twilight in the Desert—The Coming Saudi Oil Shock and the World Economy,* author Matthew R. Simmons points out that Saudi Arabia, the leading oil-producing nation in the world, is secretive about the status of its oil fields but assures its international customers that the Saudi oil fields have sufficient reserves to supply global demand for at least the following 30 years. Simmons points out, however, that nearly all Saudi oil comes from only five major oil fields, that they have been producing for several decades, and that they may be rapidly approaching a point when they can no longer continue to yield oil at current rates, contrary to Saudi assurances of a 30-year reserve. If Simmons is correct and the Saudis are misleading a world eager to believe their assurances, global supply may fall below demand in the near future.

For the past 20 years, global discoveries of oil have been less than consumption; today we consume two barrels of oil for every barrel discovered. World use of oil surged 3.4 percent in 2004, the fastest rate of increase in 16 years. A large part of this increased consumption has occurred because economies in many less-developed countries are expanding at a rapid rate. For example, in China, fossil fuel use jumped 11 percent in 2004 alone, making this country the second-largest consumer of petroleum in the world. The International Energy Agency predicts that by 2030 China's oil imports will be equal to the current U.S. levels. If these trends continue, global energy demand will increase 50 percent in 25 years.

So what will happen when global oil production drops below demand? First of all, petroleum will not just "run out" one day, with all the wells suddenly going dry. As supply dwindles and demand increases, the price of fuel will rise and production of petroleum will increase from oil shales and tar sands. Some economists and geologists have suggested that the fluctuations in gasoline prices seen in recent years are the first wave of disruptions resulting from declining global oil reserves and that much greater disturbances will occur imminently.

People will not be able to afford as much fuel as they would like, so social and technical conservation strategies, which were previously rejected, will be implemented. As a result, demand will decrease. But if this decrease is not sufficient, petroleum prices will continue to rise. Many economists predict that the price increase could be dramatic. This is an example of an economic threshold effect. As long as potential supply is greater than demand, the price of oil reflects mainly the cost of drilling, shipping, and refining. But the moment we cross the threshold where demand is greater than supply, then an auction occurs where the rich can outbid the poor.

In *Twilight in the Desert* Simmons outlines conditions where petroleum could suddenly skyrocket from the present cost of about $50 per barrel to $200 per barrel. When conventional sources become expensive, secondary and tertiary reserves and oil shale development will then become profitable. Tar sands that were buried so deeply that they were unprofitable will become economical. In World War II, when allies cut Germany's oil imports, Nazi chemists synthesized petroleum from coal. In the future, when the global price of petroleum becomes high enough, such synthesis will become profitable throughout the world. But all of these measures are simply a new face on an old monkey: fossil fuels as the overwhelming component of global energy production (◈ Figure 5.32).

◈ **FIGURE 5.32** Geologic resources have been one of the foundations of human development from the Stone Age to the space age. No one can accurately predict how society will adapt when global energy demand significantly exceeds supply. However, we can be virtually certain that the future will not mimic the past.

Is it possible to alter global energy production from a fossil-fuel economy to an economy of renewable energy resources? In 2001, the United Nations Intergovernmental Panel on Climate Change (IPCC) concluded that significant reduction of fossil fuel use is possible with renewable "technologies that exist in operation or pilot-plant stage today . . . without any drastic technological breakthroughs."[5] In other words, they suggested that if we vigorously develop all the renewable energy resources listed in Section 5.7, the global economic system could absorb a drastic decline in petroleum production without massive disruptions. A year later, 18 prominent energy experts published a rebuttal in *Science,* proposing the exact opposite conclusion. Using almost the same phrases, with the simple addition of the word *not,* they argued that, "Energy resources that can produce 100 to 300 percent of present world power consumption without fossil fuels and greenhouse emissions do not exist operationally or as pilot plants."[6] Their basic counter argument is that global energy consumption is huge and renewable sources have low power densities. Thus we do not have the available land nor could we quickly build the required infrastructure to replace fossil fuels. This debate has continued, in many forums, through 2005, as this book was written.

When experts disagree, it is difficult for lay people to evaluate the merits of the contradictory arguments. But whoever is right, it is clear that if global energy demand significantly exceeds supply, the world will fall into unprecedented economic chaos. Commerce will slip into unimaginable depression. Food supplies will diminish and food distribution will become expensive. Poor people, who are already on the edge of malnourishment, will starve. We can only hope that human ingenuity will combine with economic and political commitment to develop alternative energy resources before these catastrophes become reality.

5. B. Metz et al., Eds., *Climate Change 2001: Mitigation* (New York: Cambridge University Press, 2001), 8.

6. Martin Hoffert et al., *Science 298* (November 1, 2002).

EARTH SYSTEMS INTERACTIONS

EARTH'S MINERAL AND ENERGY resources form by processes involving multiple interactions among all four Earth systems: rock of the geosphere, atmospheric gases, organisms of the biosphere, and water of the hydrosphere. Humans have used those geologic resources to create the modern societies, cultures, and industries that make up our contemporary global and local environments, but human use of geologic resources have also altered those environments. Thus, interactions among the four Earth systems make modern life possible for humans, but at the same time these interactions have changed Earth systems in ways that are both beneficial and detrimental to humans and other organisms with whom we share the planet. ∎

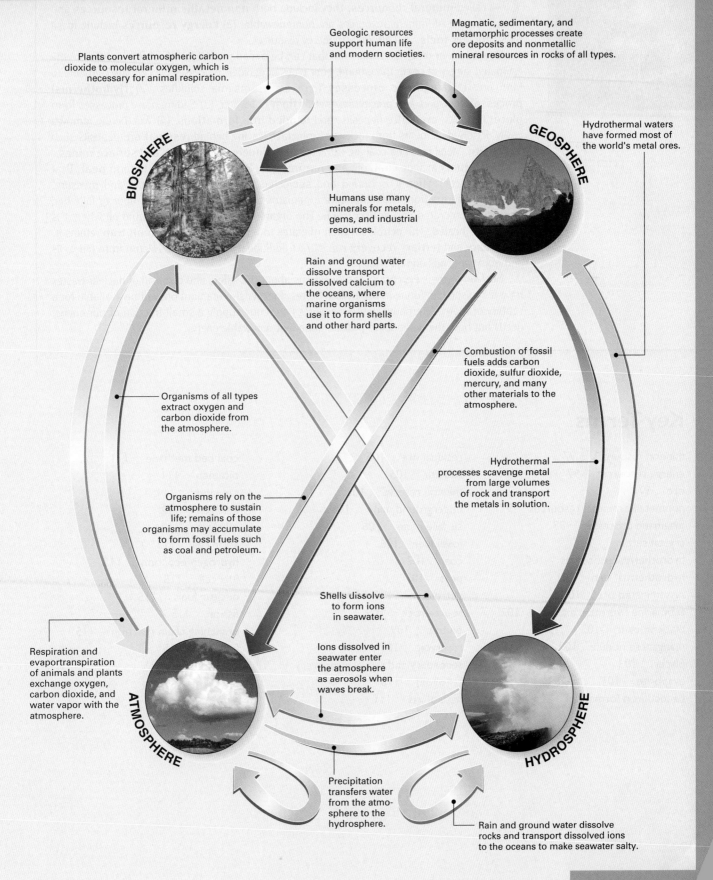

Plants convert atmospheric carbon dioxide to molecular oxygen, which is necessary for animal respiration.

Geologic resources support human life and modern societies.

Magmatic, sedimentary, and metamorphic processes create ore deposits and nonmetallic mineral resources in rocks of all types.

BIOSPHERE

GEOSPHERE

Hydrothermal waters have formed most of the world's metal ores.

Humans use many minerals for metals, gems, and industrial resources.

Rain and ground water dissolve transport dissolved calcium to the oceans, where marine organisms use it to form shells and other hard parts.

Organisms of all types extract oxygen and carbon dioxide from the atmosphere.

Combustion of fossil fuels adds carbon dioxide, sulfur dioxide, mercury, and many other materials to the atmosphere.

Hydrothermal processes scavenge metal from large volumes of rock and transport the metals in solution.

Organisms rely on the atmosphere to sustain life; remains of those organisms may accumulate to form fossil fuels such as coal and petroleum.

Shells dissolve to form ions in seawater.

Respiration and evaportranspiration of animals and plants exchange oxygen, carbon dioxide, and water vapor with the atmosphere.

Ions dissolved in seawater enter the atmosphere as aerosols when waves break.

ATMOSPHERE

HYDROSPHERE

Precipitation transfers water from the atmosphere to the hydrosphere.

Rain and ground water dissolve rocks and transport dissolved ions to the oceans to make seawater salty.

SUMMARY

Geologic resources fall into two major categories: (1) Useful rocks and minerals are called **mineral resources**; they include both **nonmetallic mineral resources** and metals. All mineral resources are nonrenewable. (2) **Energy resources** include fossil fuels, **nuclear fuels**, and alternative energy resources.

Ore is rock or other Earth material that can be mined profitably. **Mineral reserves** are the estimated supply of ore in the ground. Four types of geologic processes concentrate elements to form ore: (1) **Magmatic processes** form ore as magma solidifies. (2) **Hydrothermal processes** transport and precipitate metals from hot water. (3) Sedimentary processes form **placer** deposits, **evaporite** deposits, and **banded iron formations**. (4) Weathering removes easily dissolved elements from rocks and minerals, leaving behind **residual ore** deposits such as **bauxite**. Metal ores and coal are extracted from **underground mines** and **surface mines**.

Fossil fuels include **coal**, oil, and **natural gas**. Plant matter decays to form **peat**. Peat converts to coal when it is buried and subjected to elevated temperature and pressure. **Petroleum** forms from the remains of organisms that settle to the ocean floor or lake bed and are incorporated into **source rock**. The organic matter converts to liquid oil when it is buried and heated. The petroleum then migrates to a **reservoir**, where an **oil trap** retains it. **Secondary** and **tertiary recovery** can extract additional supplies of petroleum from old wells and from tar sands and oil shale.

Nuclear power is expensive, and questions about the safety and disposal of nuclear wastes have diminished its future in the United States. Inexpensive uranium ore will be available for a century or more. Alternative energy resources currently supply a small fraction of our energy needs but have the potential to provide abundant renewable energy.

Key Terms

mineral resources   97
energy resources   97
ore   97
nonmetallic mineral resources   97
magmatic processes   98
crystal settling   98
hydrothermal processes   100
hydrothermal vein   101
disseminated ore   101
submarine hydrothermal ore   102
black smoker   102
manganese nodule   102
placer   103
evaporite   103
banded iron formation   104

residual ore   105
bauxite   105
mineral reserve   105
underground mine   105
surface mine   106
fossil fuel   106
coal   107
peat   107
petroleum   109
source rock   109
oil trap   109
reservoir   109
secondary and tertiary recovery
    (of oil)   110
natural gas   111

coal bed methane   111
bitumen   111
tar sands   111
kerogen   112
oil shale   112
solar cell   113
hydrogen economy   115
fuel cell   115
nuclear fuels   115
fission   115
branching chain reaction   115
fuel rod   115
control rod   115

For Review

1. Describe the two major categories of geologic resources.

2. Describe the differences between nonrenewable and renewable resources. List one example of each.

3. What is ore? What are mineral reserves? Describe three factors that can change estimates of mineral reserves.

4. If most elements are widely distributed in ordinary rocks, why should we worry about running short?

5. Explain crystal settling.

6. Discuss the formation of hydrothermal ore deposits.

7. Discuss the formation of marine evaporites.

8. Explain how coal forms. Why does it form in some environments but not in others?

9. Explain the importance of source rock, reservoir rock, and oil traps in the formation of petroleum reserves.

10. Discuss two sources of petroleum that will be available after conventional wells go dry.

11. List the relative advantages and disadvantages of using coal, petroleum, and natural gas as fuels.

12. Explain how a nuclear reactor works. Discuss the behavior of neutrons, the importance of control rods, and how the heat from the reaction is harnessed to produce useful energy.

13. Discuss the status of the nuclear power industry in the United States.

14. List the alternative energy resources described in this chapter. Can you think of others?

15. How does conservation act as an alternative energy resource?

16. Describe each of the ways in which solar energy can be used.

17. Describe the unique advantages of hydrogen fuel.

For Discussion

1. What factors can make our metal reserves last longer? What factors can deplete them rapidly?

2. It is common for a single mine to contain ores of two or more metals. Discuss how geologic processes might favor concentration of two metals in a single deposit.

3. List 10 objects that you own. What resources are they made of? How long will each of the objects be used before it is discarded? Will the materials eventually be recycled or deposited in the trash? Discuss ways of conserving resources in your own life.

4. If you were searching for petroleum, would you search primarily in sedimentary rock, metamorphic rock, or igneous rock? Explain.

5. If you were a space traveler abandoned on an unknown planet in a distant solar system, what clues would you look for if you were searching for fossil fuels?

6. Is an impermeable cap rock necessary to preserve coal deposits? Why or why not?

7. Discuss problems in predicting the future availability of fossil fuel reserves. What is the value of the predictions?

8. Compare the depletion of mineral reserves with the depletion of fossil fuels. How are the two problems the same, and how are they different?

9. Discuss the environmental, economic, and political implications of the development of a solar cell that produces electricity more cheaply than conventional sources.

10. Wind energy, hydroelectric power, and energy from sea waves are sometimes called secondary forms of solar energy. Discuss how the Sun is the primary source of those alternative energy resources.

11. Discuss Earth systems interactions in the concentration of ores by: hydrothermal processes, weathering processes, precipitation, and placer processes. What Earth systems are involved? How do they interact?

12. Discuss Earth systems interactions in the formation and human use of coal, petroleum, and nuclear fuels. What Earth systems are involved? How do they interact?

UNIT 2

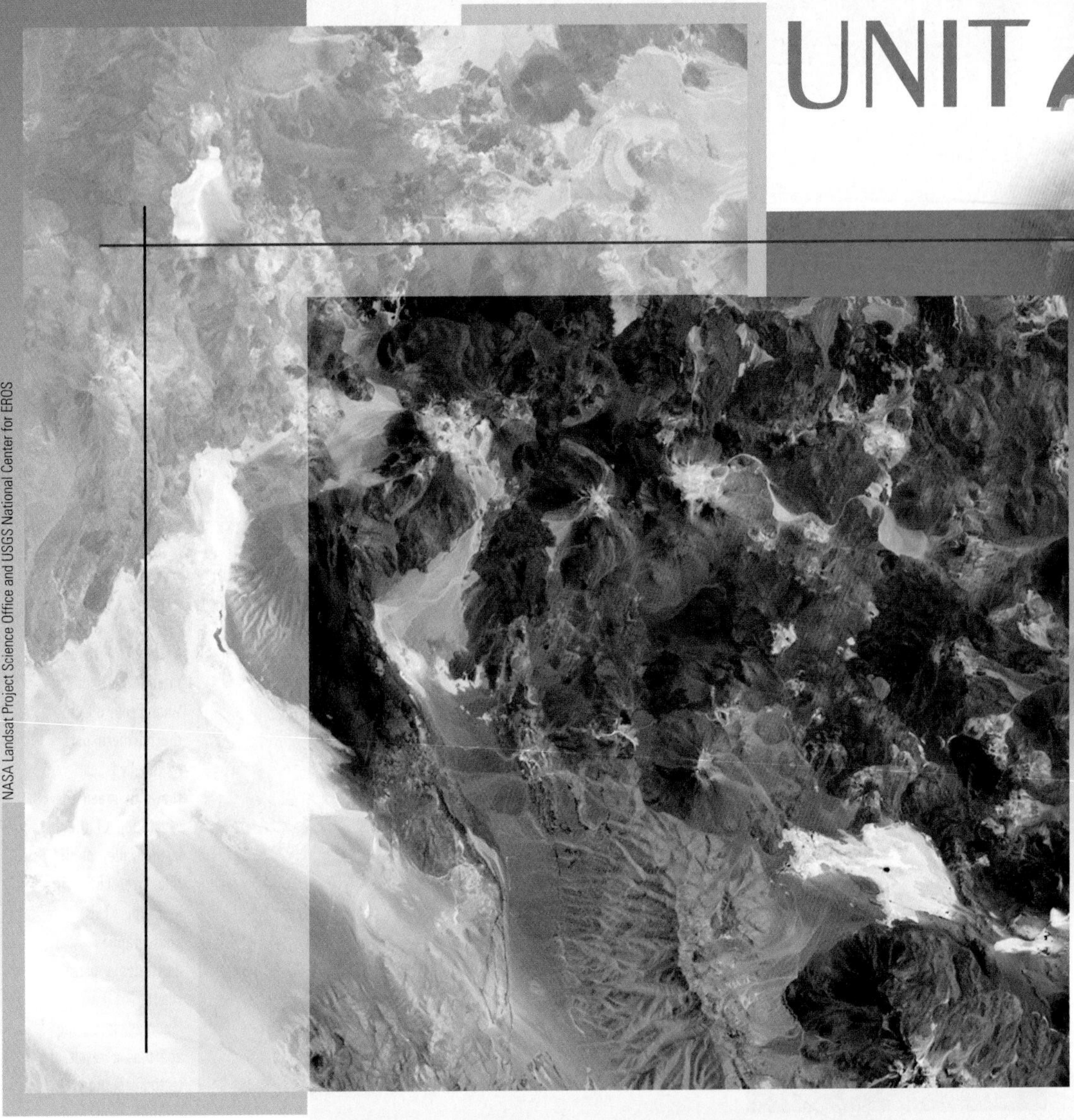

NASA Landsat Project Science Office and USGS National Center for EROS

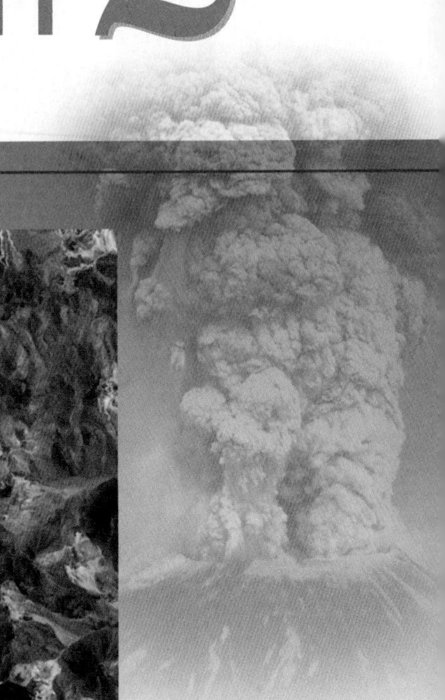

INTERNAL PROCESSES

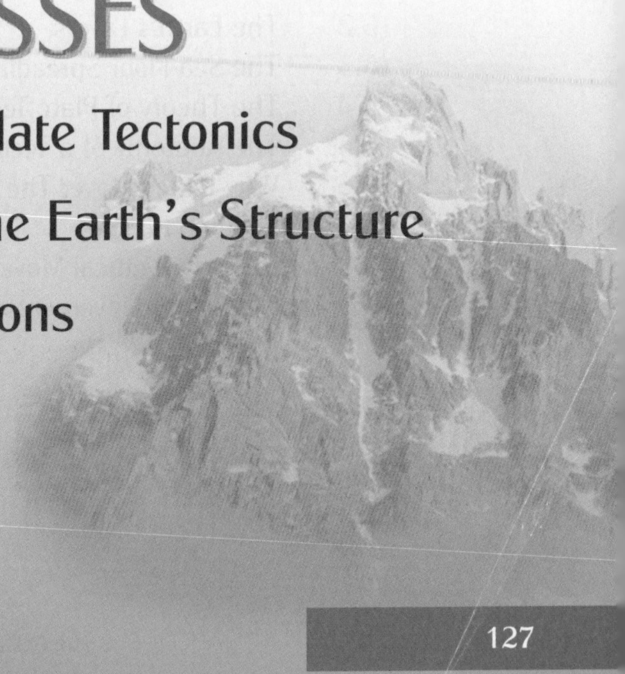

The Active Earth: Plate Tectonics

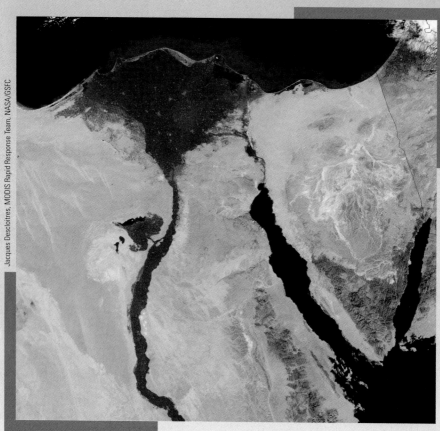

A false-color image satellite view of the Sinai Peninsula and the northern part of the Red Sea shows in green the vegetation of the fertile valley of the Nile River and delta in Egypt. This photograph illustrates a place where The Earth's landmass is pulling apart as a result of tectonic motion.

Jacques Descloitres, MODIS Rapid Response Team, NASA/GSFC

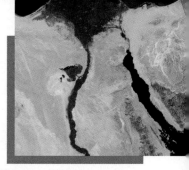

About 5 billion years ago, a ball of dust and gas, one among billions in the universe, collapsed into a slowly spinning disc. The inner portion of the disc then grew to form our Sun, while the outer parts coalesced to form planets orbiting the Sun. Our Earth is one of those planets. As particles of dust and gas were drawn together by gravity to form Earth, the colliding particles became hotter and hotter from impact energy. Over time, frozen crystals of carbon dioxide, methane, and ammonia melted. As the temperature continued to rise, water ice melted as well. The young planet continued to heat up as asteroids, comets, and other space debris crashed into its surface. As Earth formed into a cohesive sphere, the decay of radioactive isotopes within the interior generated additional heat. About 4.6 billion years ago, the planet became hot enough that it melted. Then, as the bombardment slowed down and radioactive isotopes decayed and became less abundant, much of Earth's heat radiated into space and our planet began to cool.

Today, although Earth's surface has cooled to temperatures supportive of living organisms, the interior remains hot, both from heat left over after the early melting event and from continued decay of the remaining radioactive isotopes. Consequently, Earth becomes hotter with depth, and the temperature at Earth's center is about 6,000°C—similar to that of the Sun's surface. This heat drives Earth's internal engine, creating earthquakes, volcanic eruptions, mountain building, and continual movements of the continents and ocean basins. These effects, in turn, profoundly affect our environment—Earth's atmosphere, hydrosphere, and biosphere. The Earth engine and its effects are described in the theory of **plate tectonics**, a simple theory that provides a unifying framework for understanding the way Earth works and how Earth systems interact to create our environment. The term *tectonics* is taken from the Greek *tektonikos*, meaning "construction."

Like most great scientific revolutions, the development of plate tectonics theory developed sequentially over many years, building on earlier observations, hypotheses, and theories. The story illustrates how a scientific theory evolves and how scientists rely on the work and discoveries of earlier scientists. ■

6.1

Alfred Wegener and the Origin of an Idea: The Continental Drift Hypothesis

Although the theory of plate tectonics was not developed until the 1960s, it was foreshadowed early in the twentieth century by a young German scientist named Alfred Wegener, who noticed that the African and South American coastlines on opposite sides of the Atlantic Ocean seemed to fit as if they were adjacent pieces of a jigsaw puzzle (◆ Figure 6.1). He realized that the apparent fit suggested that the continents had once been joined together and had later separated by thousands of kilometers to form the Atlantic Ocean.

Wegener was not the first to make this observation, but he was the first scientist to pursue it with additional research. Studying world maps, Wegener realized that not only did the continents on both sides of the Atlantic fit together, but other continents, when moved properly, also fit like additional pieces of the same jigsaw puzzle (◆ Figure 6.2). On his map, all the continents joined together formed one supercontinent that he called **Pangea** from the Greek root words for "all lands." The northern part of Pangea is commonly called **Laurasia** and the southern part **Gondwanaland**.

Wegener understood that the fit of the continents alone did not prove that a supercontinent had existed. He began seeking additional evidence in 1910 and continued work on the project until his death in 1930.

He mapped the locations of fossils of several species of animals and plants that could neither swim well nor fly. Fossils of the same species are now found in Antarctica, Africa, Australia, South America, and India. Why

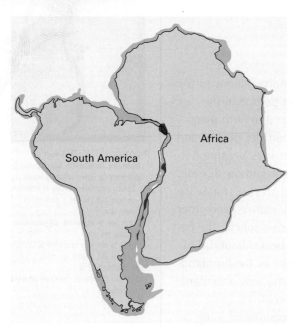

◇▶ **Figure 6.1** The African and South American coastlines appear to fit together like adjacent pieces of a jigsaw puzzle on Wegener's reconstruction map. The red areas show locations of distinctive rock types in South America and Africa. The brown regions are the continental shelves, which are the actual edges of the continents.

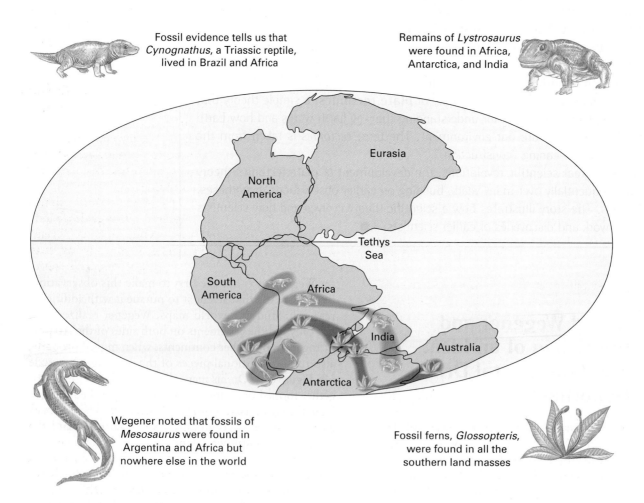

Fossil evidence tells us that *Cynognathus*, a Triassic reptile, lived in Brazil and Africa

Remains of *Lystrosaurus* were found in Africa, Antarctica, and India

Wegener noted that fossils of *Mesosaurus* were found in Argentina and Africa but nowhere else in the world

Fossil ferns, *Glossopteris*, were found in all the southern land masses

◇▶ **FIGURE 6.2** Geographic distributions of plant and animal fossils on Wegener's map indicate that a single supercontinent, called Pangea, existed from about 300 to 200 million years ago.

would the same species be found on continents separated by thousands of kilometers of ocean? When Wegener plotted the same fossil localities on his Pangea map, he found that they all lie in the same region of Pangea (◈ Figure 6.2). Wegener then suggested that each species, rather than migrating across the open ocean, had evolved and spread over that part of Pangea before the supercontinent broke apart. Certain types of sedimentary rocks form in specific climatic zones. Glaciers and gravel deposited by glacial ice, for example, form in cold climates and are therefore found at high latitudes and high altitudes. Sandstones that preserve the structures of desert sand dunes form where deserts are common, near latitudes 30° north and south. Coral reefs and coal swamps thrive in near-equatorial tropical climates. Thus each of these rocks reflects the latitude at which it formed.

Wegener plotted 250-million-year-old glacial deposits on a map showing the modern distribution of continents (◈ Figure 6.3A). Notice that glacial deposits would have formed in tropical and subtropical zones. ◈ Figure 6.3B shows the same glacial deposits, and other climate-indicating rocks, plotted on Wegener's Pangea map. Here the glaciers cluster neatly about the South Pole. The other rocks are also found in logical locations.

Wegener also noticed several instances in which an uncommon rock type or a distinctive sequence of rocks on one side of the Atlantic Ocean is identical to rocks on the other side. When he plotted the rocks on a Pangea map, those on the east side of the Atlantic were continuous with their counterparts on the west side (◈ Figure 6.1). For example, the deformed rocks of the Cape Fold belt of South Africa are similar to rocks found in the Buenos Aires province of Argentina. Plotted on a Pangea map, the two sequences of rocks appear as a single, continuous belt.

Wegener's concept of a single supercontinent that broke apart to form the modern continents is called the theory of **continental drift**. The theory was so revolutionary that skeptical scientists demanded an explanation of *how* continents could move. They wanted an explanation of the mechanism of continental drift. Wegener had concentrated on developing evidence that continents had drifted, not on what caused them to move. Finally, perhaps out of exasperation and as an afterthought to what he considered the important part of his theory, Wegener suggested two alternative possibilities: first, that continents plow their way through oceanic crust, shoving it aside as a ship plows through water; or second, that continental crust slides over oceanic crust. These suggestions turned out to be ill considered.

Physicists quickly proved that both of Wegener's mechanisms were impossible. Oceanic crust is too strong for continents to plow through it. The attempt would be like trying to push a matchstick boat through heavy tar. The boat, or the continent, would break apart.

Furthermore, frictional resistance is too great for continents to slide over oceanic crust.

These conclusions were taken by most scientists as proof that Wegener's theory of continental drift was wrong. Notice, however, that the physicists' calculations proved only that the mechanism proposed by Wegener was incorrect. They did not disprove, or even consider, the huge mass of evidence indicating that the continents were once joined together.

Alfred Wegener was an adventurer in addition to being a scientist. In 1930 the German government sent him to Greenland to build a weather station near the center of the ice cap to support commercial airline flights between Europe and North America. In the late summer and autumn of the year, Wegener and his companions freighted supplies by dog sled to the station, which was to be staffed by a single observer through the coming Arctic winter. After their last planned trip, the observer decided that the supplies were insufficient for the long winter and stated that he wouldn't stay unless he had more food and equipment.

With winter rapidly closing in, Wegener and his coworker, an Inuit named Rasmus Willimsen, set off on one more trip for additional supplies. They departed on Wegener's 50th birthday; Wegener was never seen again. Willimsen's frozen body was found the following summer by a search party.

Objections to Wegener's hypothesis of continental drift continued from all sides of the scientific community. Because Wegener had been unable to defend his hypothesis, other scientists rejected it, and continental drift was largely forgotten. During the 30-year period from 1930 to 1960, a few geologists supported the continental drift hypothesis, but most ignored it.

Much of the hypothesis of continental drift is similar to modern plate tectonics theory. Modern evidence indicates that the continents were together much as Wegener had portrayed them in his map of Pangea. Today, most geologists recognize the importance of Wegener's contributions.

6.2
The Earth's Layers

The energy released by an earthquake travels through Earth as waves. After Wegener died and his theory was mostly forgotten, geologists discovered that both the speed and direction of these waves change abruptly at certain depths as the waves pass through Earth. They soon realized that these changes reveal that Earth is a layered planet. ◈ Figure 6.4 and Table 6.1 describe the layers. It is necessary to understand Earth's layers to consider the plate tectonics theory. In Chapter 7 we discuss how changes in earthquake waves reveal Earth's layering.

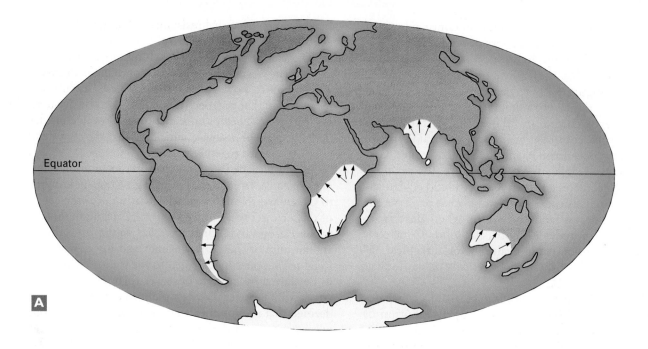

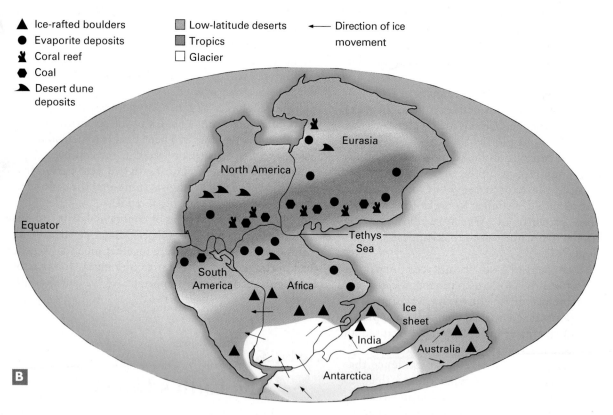

Legend:

▲ Ice-rafted boulders

● Evaporite deposits

♟ Coral reef

⬟ Coal

🝙 Desert dune deposits

▢ Low-latitude deserts

▨ Tropics

□ Glacier

← Direction of ice movement

◇ **FIGURE 6.3** **(A)** 250-million-year-old glacial deposits are displayed in white on a map showing the modern distribution of continents. The black arrows show directions of glacial movement, indicated by glacial features described in Chapter 13, "Glaciers and Ice Ages." **(B)** 300-million-year-old glacial deposits and other climate-sensitive sedimentary rocks plotted on Wegener's map of Pangea.

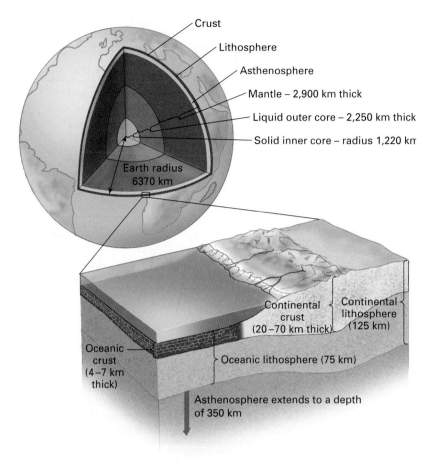

Crust
Lithosphere
Asthenosphere
Mantle – 2,900 km thick
Liquid outer core – 2,250 km thick
Solid inner core – radius 1,220 km
Earth radius 6370 km

Continental crust (20–70 km thick)
Continental lithosphere (125 km)
Oceanic crust (4–7 km thick)
Oceanic lithosphere (75 km)
Asthenosphere extends to a depth of 350 km

◆ **FIGURE 6.4** Earth is a layered planet. The insert is drawn on an expanded scale to show near-surface layering. Note that the average thickness of the lithosphere is about 75 kilometers beneath the oceans but about 125 kilometers beneath continents.

that near the mantle/core boundary, it is about 3,300°C. These changes cause the strength of mantle rock to vary with depth, and thus they create layering within the mantle.

At this point in our story, it is important to understand the effects of temperature and pressure on rocks. Most people understand that rising temperature will eventually melt a rock. Less obvious, however, is the fact that high pressure *prevents* melting. This is because rock expands by about 10 percent when it melts. High pressure makes it more difficult for a rock to expand and therefore inhibits melting. If the combined effects of temperature and pressure are close to—but just below—a rock's melting point, the rock remains solid but loses strength, so it becomes weak and plastic. Thus the rock can ooze, or flow slowly, like road tar on a hot day. At that point, if temperature rises or pressure decreases, the rock will begin to melt.

Because both temperature and pressure increase with depth in Earth, their combined effects cause changes in the physical properties of rocks with increasing depth (◆ Figure 6.5). These changes alter the strength of upper-mantle rocks with depth, and create two distinctly different layers in the upper mantle. Although the composition of the mantle is the same in the two layers, the strength of the rocks is very different.

The Crust

The crust is the outermost and thinnest layer. Because it is cool relative to the layers below, the crust consists of hard, strong rock (◆ Figure 6.4). Crust beneath the oceans differs from that of continents.

Oceanic crust is between 4 and 7 kilometers thick and is composed mostly of dark, dense basalt. In contrast, the average thickness of continental crust is about 20 to 40 kilometers, although under mountain ranges it can be as much as 70 kilometers thick. Continents are composed primarily of light-colored, less-dense granite.

The Mantle

The mantle lies directly below the crust. It is almost 2,900 kilometers thick and makes up 80 percent of Earth's volume. The mantle is composed of peridotite (described in Chapter 3), which is denser than the basalt and granite of the crust and dark in color.

Although the chemical composition is probably similar throughout the entire mantle, temperature and pressure increase with depth. ◆ Figure 6.5 shows that the temperature at the top of the mantle is near 1,000°C and

The Lithosphere

Figure 6.5 shows that the uppermost mantle is cool and its pressure is low, conditions similar to those in the crust. Both factors combine to produce hard, strong rock similar to that of the crust. The outer part of Earth, including both the uppermost mantle and the crust, make up the **lithosphere**. The lithosphere averages about 100 kilometers in thickness but varies from about 75 kilometers thick beneath ocean basins to about 125 kilometers under the continents (◆ Figure 6.4). The lithosphere, then, consists mostly of the cold, strong uppermost mantle. The crust is just a thin layer of buoyant rock embedded in the upper mantle. Some parts of the lithosphere have continents embedded in them, whereas other regions of lithosphere are topped by oceanic crust.

The Asthenosphere

At a depth varying from 75 to 125 kilometers beneath Earth's surface, the combined effects of rising temperature and pressure approach the melting point of mantle rock. As a result, the mantle abruptly loses strength and

Table 6.1   The Layers of the Earth

	Layer	Composition	Depth	Properties
Crust	Oceanic crust	Basalt	4–7 km	Cool, hard, and strong
	Continental crust	Granite	20–70 km	Cool, hard, and strong
Lithosphere	Lithosphere includes the crust and the uppermost portion of the mantle	Varies; the crust and the mantle have different compositions	75–125 km	Cool, hard, and strong
Mantle	Uppermost portion of the mantle included as part of the lithosphere			
	Asthenosphere	Entire mantle is plastic, ultramafic rock. Its mineralogy varies with depth	Extends to 350 km	Hot, weak, and 1% or 2% melted
	Remainder of upper mantle		Extends from 350 to 660 km	Hot, under great pressure, and mechanically strong
	Lower mantle		Extends from 660 to 2,900 km	High pressure forms minerals different from those of the upper mantle
Core	Outer core	Iron and nickel	Extends from 2,900 to 5,150 km	Liquid
	Inner core	Iron and nickel	Extends from 5,150 km to the center of Earth	Solid

becomes weak and plastic, and 1 to 2 percent of the rock melts, although 98 to 99 percent remains solid. The weak, plastic, and partly molten character extends to a depth of about 350 kilometers, where increasing pressure overwhelms temperature, and the rock becomes stronger again. This layer of weak mantle rock extending from about 100 to 350 kilometers deep is the **asthenosphere** (Greek for "weak layer"). The average temperature in the asthenosphere is about 1,800°C, although the temperature increases with depth as it does in other Earth layers. Pressure in the asthenosphere rises from about 35 kilobars near the top, to about 120 kilobars at the base. Two familiar examples of solid but plastic materials are Silly Putty™ and hot road tar. If you apply force to a plastic solid, it flows slowly.

When building a house, you start with a strong, hard foundation. However, Earth is not constructed in this way. The strong, hard lithosphere lies on top of the soft, weak asthenosphere. Thus, the lithosphere is not rigidly supported by the rock beneath it. Instead, the lithosphere floats on the soft plastic rock of the asthenosphere. This concept of a floating lithosphere is important to our understanding of plate tectonics and Earth's internal processes.

The Mantle Below the Asthenosphere

At the base of the asthenosphere, increasing pressure overwhelms the effect of rising temperature, and the strength of the mantle increases again. Although the mantle below 350 kilometers is stronger than the asthenosphere, it never regains the strength of the lithosphere. It is plastic and capable of flowing slowly, over geologic time. Recent evidence suggests that the lowermost mantle, at the core boundary, is partly molten[1]. The temperature profile is given in ◆ Figure 6.5.

The Core

The core is the innermost of Earth's layers. It is a sphere with a radius of about 3,470 kilometers, about the same size as Mars. Earth's core is composed largely of iron and nickel. The outer core is molten because of the high temperature in that region. Near its center, the core's temperature is nearly 7,000°C, hotter than the Sun's sur-

---

1. Quentin Williams and Edward J. Ganero, "Seismic Evidence for Partial Melt at the Base of the Earth's Mantle," *Science 273* (September 13, 1996): 1528.

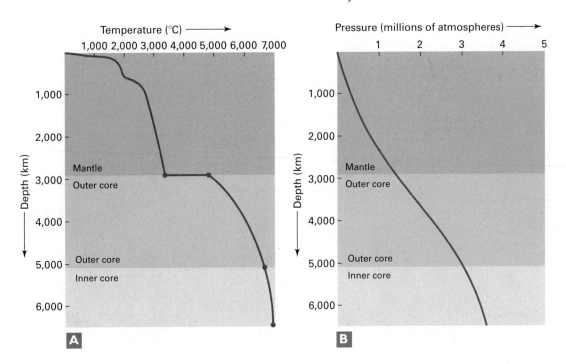

◆ **FIGURE 6.5** **(A)** Earth's internal temperature increases with depth. At the center of Earth, the temperature is close to 7,000°C, hotter than the Sun's surface. For reference, the temperature of an oven baking a chocolate cake is about 175°C, and a heated steel bar turns cherry red at 746°C. **(B)** Pressure increases more uniformly with depth than temperature.

face. The pressure is three and a half million times that of Earth's atmosphere at sea level. The extreme pressure compresses the inner core to a solid, despite the fact that it is even hotter than the molten outer core.

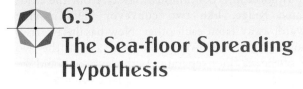

ThomsonNOW™

CLICK ThomsonNOW Interactive to work through an activity on the Earth's Layers.

6.3
The Sea-floor Spreading Hypothesis

Shortly after World War II, scientists began to explore the floors of Earth's oceans. Although these studies ultimately played a large role in the development of plate tectonics theory, they were initially undertaken for military and economic reasons. Defense strategists wanted a detailed knowledge of sea-floor topography for submarine warfare, and the same information was needed to lay undersea telephone cables. As they mapped the sea floor, oceanographers discovered the largest mountain chain on Earth, now called the **Mid-Oceanic Ridge system** (◆ Figure 6.6). One leg of this huge submarine mountain chain, called the **Mid-Atlantic Ridge**, lies directly in the middle of the Atlantic Ocean, halfway between North and South America to the west, and Europe and Africa to the east.

As you learned in Chapter 3, oceanic crust is composed mostly of basalt, an igneous rock that forms when lava cools and becomes solid. Basalt is rich in iron. As lava cools and becomes basalt, the iron-rich minerals become weak magnets that align their magnetic fields parallel to the Earth's magnetic field. Thus, basalt records the orientation of Earth's magnetic field at the time the rock cools.

In addition to mapping the topography of the sea floor, oceanographers towed devices called magnetometers behind their research vessels to detect and record magnetic patterns in the deep oceans. ◆ Figure 6.7 shows the magnetic orientations of sea-floor rocks near a part of the Mid-Atlantic Ridge southwest of Iceland. In this figure, green stripes represent basalt with a magnetic orientation parallel to Earth's current magnetic field, called **normal magnetic polarity**. The blue stripes represent rocks with magnetic orientations that are exactly opposite to the current magnetic field, called **reversed magnetic polarity**. Notice that the stripes form a symmetric pattern of normal and reversed polarity about the axis of the ridge, and that the central stripe is green, indicating that basalt at the ridge axis has a magnetic orientation parallel to that of Earth's magnetic field today.

Why do the sea-floor rocks have alternating normal and reversed polarity, and why is the pattern symmetrically distributed about the Mid-Oceanic Ridge? In the mid-1960s, three scientists—a Cambridge graduate student named Frederick Vine; his professor, Drummond Matthews; and Lawrence Morley, a Canadian working independently of the other two—proposed an explanation for these odd magnetic patterns on the sea floor.

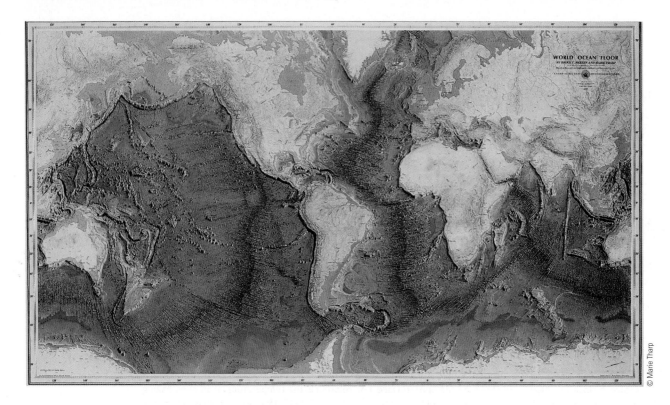

◆ **FIGURE 6.6** An artist's rendition of the sea floor. The Mid-Oceanic Ridge system is a submarine mountain chain that encircles the globe like the seam on a baseball.

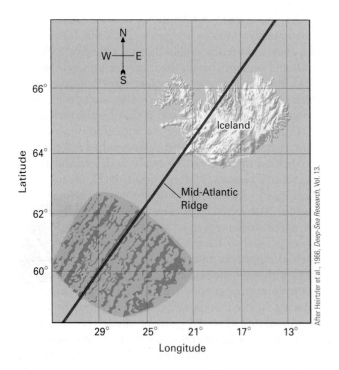

After Heirtzler et al., 1966, Deep-Sea Research, Vol. 13.

◆ **FIGURE 6.7** The Mid-Atlantic Ridge, shown in red, runs through Iceland. Magnetic orientation of sea-floor rocks near the ridge are shown in the lower-left portion of the map. The green stripes represent sea-floor rocks with normal magnetic polarity, and the blue stripes represent rocks with reversed polarity. The stripes form a symmetrical pattern of alternating normal and reversed polarity on each side of the ridge.

INTERACTIVE QUESTION: *Explain why the stripes of normal and reversed polarity are not neatly linear.*

They knew that other scientists had recently discovered that Earth's magnetic field has reversed its polarity on the average of every 500,000 years during the past 65 million years. When a **reversal** of Earth's field occurs, the north magnetic pole becomes the south magnetic pole, and vice versa. This discovery surprised most geologists, but periodic reversals of the Earth's magnetic field were well known by the early 1960s.

Vine, Matthews, and Morley suggested that the sea floor is spreading continuously away from the Mid-Oceanic Ridge, like two conveyor belts moving outward, away from each other. New basalt lava rises through cracks that form at the ridge axis as the two sides of the sea floor separate. As the lava cools and solidifies, the basalt acquires the magnetic orientation of Earth's field. Because Earth's field reverses every half-million years, each stripe represents new sea floor formed during a half-million years (◆ Figure 6.8). Thus, the sea floor and oceanic crust should become older with increasing distance from the ridge axis.

At the same time that these sea-floor magnetic patterns were detected, oceanographers discovered that the thin layer of mud that overlies sea-floor basalt in most parts of the oceans is thinnest at the Mid-Oceanic Ridge and becomes progressively thicker at greater distance from the ridge. They reasoned that if mud settles onto the sea floor at the same rate everywhere, and if the ridge is the newest part of the sea floor, the mud layer should be thinnest at the ridge. The mud should thicken with increasing distance from the ridge because oceanic crust becomes older away from the ridge axis.

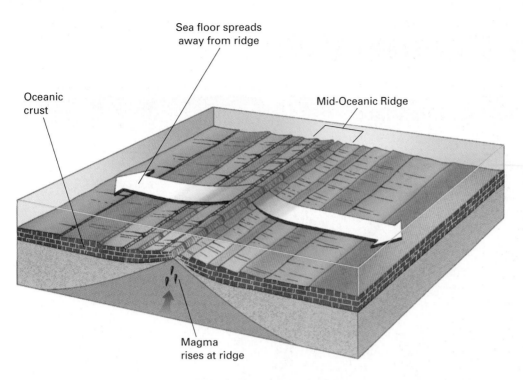

Sea floor spreads
away from ridge

Oceanic
crust

Mid-Oceanic Ridge

Magma
rises at ridge

◆ **FIGURE 6.8** As new oceanic crust cools at the Mid-Oceanic Ridge, it acquires the magnetic orientation of Earth's field. Alternating stripes of normal (green) and reversed (blue) polarity record reversals in Earth's magnetic field that occurred as the crust spread outward from the ridge.

Furthermore, the oceanographers found that fossils in the deepest layers of mud overlying basalt were very young at the ridge axis but become progressively older with increasing distance from the ridge. Again, this discovery indicated that the sea floor becomes older with increasing distance from the ridge axis.

Symmetrical magnetic patterns and similar mud age and thickness trends were quickly discovered at other parts of the Mid-Oceanic Ridge system in other ocean basins, and the hypothesis of **sea-floor spreading** was proposed as a general model for the origin of all oceanic crust. In a very few years, the sea-floor spreading hypothesis became the basis for development of the much-broader theory of plate tectonics.

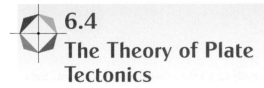

ThomsonNOW

CLICK ThomsonNOW Interactive to work through an activity on the changing Earth through geological time.

6.4 The Theory of Plate Tectonics

Like most great unifying scientific ideas, the plate tectonics theory is simple. Briefly, it states that the lithosphere is a shell of hard, strong rock about 100 kilometers thick that floats on the hot, plastic asthenosphere

(◆ Figure 6.9). The lithosphere is broken into seven large (and several smaller) segments called tectonic plates (◆ Figure 6.10). They are also called *lithospheric plates* and, simply, *plates*—the terms are interchangeable. The tectonic plates glide slowly over the asthenosphere at rates ranging from less than 1 to about 16 centimeters per year, about as fast as your fingernails grow. Continents and ocean basins make up the upper parts of the plates. As a tectonic plate glides over the asthenosphere, the continents and oceans move with it.

A **plate boundary** is a fracture that separates one plate from another. Neighboring plates can move relative to one another in three ways, shown by the insets in ◆ Figure 6.10. At a **divergent boundary**, two plates move apart from each other. At a **convergent boundary**, two plates move toward each other, and at a **transform boundary**, they slide horizontally past each other. Table 6.2 summarizes characteristics and examples of each type of plate boundary.

The great forces generated at a plate boundary build mountain ranges and cause volcanic eruptions and earthquakes. In contrast to plate boundaries, the interior portion of a plate is usually tectonically quiet because it is far from the zones where two plates interact.

Divergent Plate Boundaries

At a divergent plate boundary (also called a spreading center and a rift zone), two plates spread apart from one another (center portion of ◆ Figure 6.11). The underlying asthenosphere then oozes upward to fill the gap

Systems Perspective

The Movement of Tectonic Plates

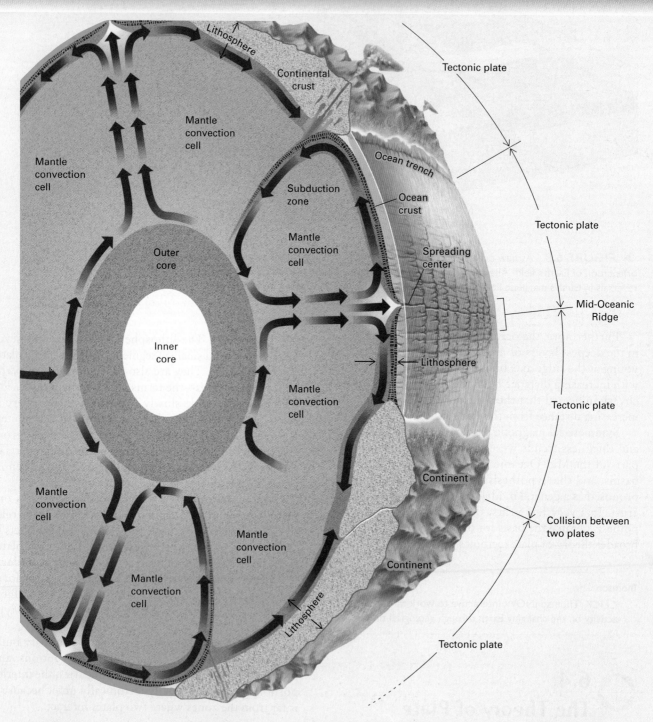

◆ **FIGURE 6.9** A cutaway view of Earth shows that the lithosphere glides horizontally across the asthenosphere. Continents and oceans ride piggyback on the lithosphere, and consequently both continents and oceans move across Earth's surface at rates of a few centimeters each year. The mantle and lithosphere circulate in huge elliptical cells that rise from the deepest mantle, then flow across Earth's surface, and finally sink back to the mantle–core boundary. In this figure, the thickness of the lithosphere is exaggerated for clarity.

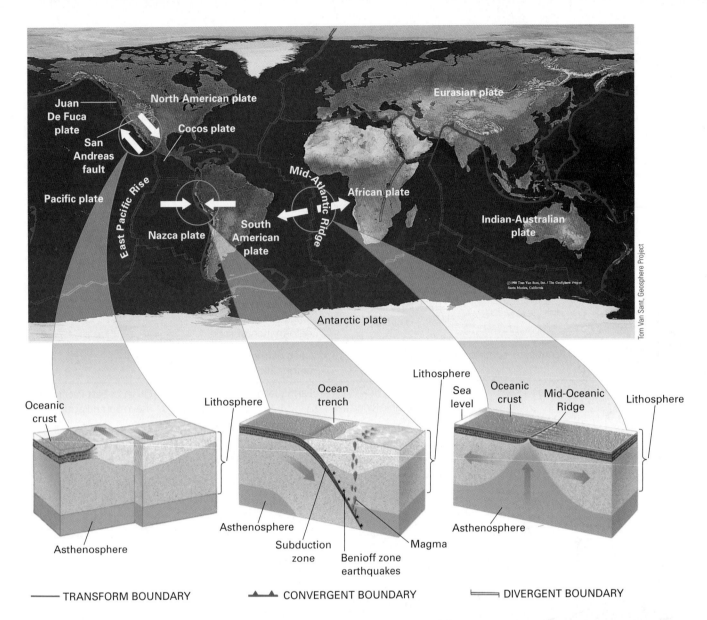

◆ FIGURE 6.10 Earth's lithosphere is broken into seven large tectonic plates, called the African, Eurasian, Indian-Australian, Antarctic, Pacific, North American, and South American plates. A few of the smaller plates are also shown. White arrows show that the plates move in different directions. The three different types of plate boundaries are shown below the map: At a transform plate boundary, rocks on opposite sides of the fracture slide horizontally past each other. Two plates come together at a convergent boundary. Two plates separate at a divergent boundary.

between the separating plates. As the asthenosphere rises between the separating plates, some of it melts to form magma.[2] Most of the magma rises to Earth's surface, where it cools to form new crust. Most of this activity occurs beneath the seas because most divergent plate boundaries lie in the ocean basins.

Both the lower lithosphere (the part beneath the crust) and the asthenosphere are parts of the mantle and thus have similar chemical compositions. The main difference between the two layers is one of mechanical strength. The cool lithosphere is strong and hard, but the hot asthenosphere is weak and plastic. As the asthenosphere rises closer to Earth's surface between two separating plates, it cools from the higher asthenosphere temperatures to the lower temperature of the lithosphere, gains mechanical strength, and, therefore, *transforms into new lithosphere*. In this way, new lithosphere continuously forms at a divergent boundary.

---

2. It seems counterintuitive that the rising, cooling asthenosphere should melt to form magma, but the melting results from decreasing pressure rather than a temperature change. This process is discussed in Chapter 8.

Table 6.2   Characteristics and Examples of Plate Boundaries

Type of Boundary	Types of Plates Involved	Topography	Geologic Events	Modern Examples
Divergent	Ocean–ocean	Mid-Oceanic Ridge	Sea floor spreading, shallow earthquakes, rising magma, volcanoes	Mid-Atlantic Ridge
	Continent–continent	Rift valley	Continents torn apart, earthquakes, rising magma, volcanoes	East African rift
Convergent	Ocean–ocean	Island arcs and ocean trenches	Subduction, deep earthquakes, rising magma, volcanoes, deformation of rocks	Western Aleutians
	Ocean–continent	Mountains and ocean trenches	Subduction, deep earthquakes, rising magma, volcanoes, deformation of rocks	Andes
	Continent–continent	Mountains	Deep earthquakes, deformation of rocks	Himalayas
Transform	Ocean–ocean	Major offset of Mid-Oceanic Ridge axis	Earthquakes	Offset of East Pacific rise in South Pacific
	Continent–continent	Small deformed mountain ranges, deformations along fault	Earthquakes, deformation of rocks	San Andreas fault

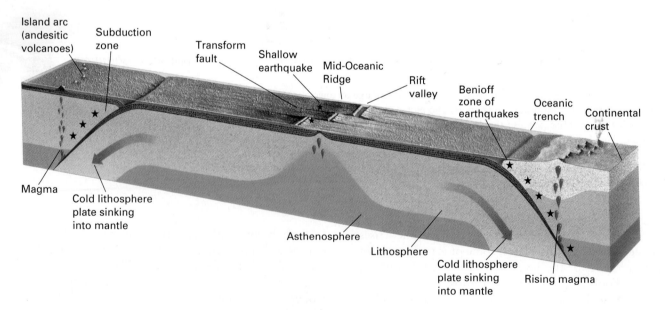

◆ **FIGURE 6.11** Lithospheric plates move away from a spreading center by gliding over the weak, plastic asthenosphere. In the center of the drawing, new lithosphere forms at a spreading center. Note that the lithosphere beneath the spreading center is only 10 or 15 kilometers thick but that it becomes thicker as the new lithosphere moves away from the spreading center and cools. At the sides of the drawing, old lithosphere sinks into the mantle at subduction zones.

At a spreading center, the rising asthenosphere is hot, weak, and plastic. Only the upper 10 to 15 kilometers cools enough to gain the strength and hardness of lithosphere rock. As a result, the lithosphere, including the crust and the upper few kilometers of mantle rock, can be as little as 10 or 15 kilometers thick at a spreading center. But as the lithosphere spreads, it cools from the top downward (center portion of ◆ Figure 6.11).

As it spreads outward and cools, the new lithosphere also thickens because the boundary between cool rock and hot rock migrates downward. Consequently, the thickness of the lithosphere increases as it moves away from the spreading center. Think of ice freezing on a pond. On a cold day, water under the ice freezes and the ice becomes thicker. The lithosphere continues to thicken until it attains a steady-state thickness of about 75 kilometers beneath an ocean basin, and as much as 125 kilometers beneath a continent.

The Mid-Oceanic Ridge: Rifting in the Oceans

New lithosphere at an oceanic spreading center is warmer than older lithosphere rock farther away from the spreading center, and therefore the new lithosphere has lower density. Consequently, it floats to a high level, forming the undersea mountain chain called the Mid-Oceanic Ridge system (◆ Figure 6.6). In a few places, such as Iceland, the ridge rises above sea level. But as lithosphere migrates away from the spreading center, it cools and becomes denser. As a result, it sinks into the soft, plastic asthenosphere like a small boat sinking into the water as people step on board. Thus, the Mid-Oceanic Ridge rises 2 to 3 kilometers above the surrounding sea floor and comes within 2 kilometers of the sea surface, but the average depth of the sea floor away from the Mid-Oceanic Ridge is about 5 kilometers (◆ Figure 6.11).

The Mid-Oceanic Ridge system encircles Earth like the seam on a baseball, forming Earth's longest mountain chain. Basaltic magma that oozes onto the sea floor at the ridge creates about 22 cubic kilometers, or 70 billion tons, of new oceanic crust each year. The Mid-Oceanic Ridge system is described further in Chapter 15.

Splitting Continents: Rifting in Continental Crust

A divergent plate boundary can rip a continent in half in a process called **continental rifting**. A rift valley develops in a continental rift zone because continental crust stretches, fractures, and sinks as it is pulled apart. Continental rifting is now taking place along the East African Rift (◆ Figure 6.12). If the rifting continues, eastern Africa will separate from the main portion of the continent, and a new ocean basin will open between the separating portions of Africa.

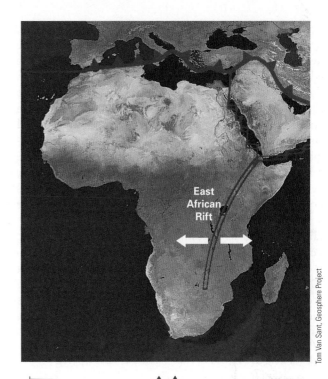

| Divergent Boundary | Convergent Boundary | Transform Boundary |

◆ **FIGURE 6.12** The continent of Africa is splitting apart along the East African Rift.

INTERACTIVE QUESTION: *Describe the modern climatic conditions and the plant and animal life in the rift. How will they change if a new ocean basin forms in the rift during the next several million years?*

Convergent Plate Boundaries

At a convergent plate boundary, two lithospheric plates move toward each other. Not all lithospheric plates are made of equally dense rock. Where two plates of different densities converge, the denser one sinks into the mantle beneath the other. This sinking process is called **subduction** and is shown in both the right and left sides of ◆ Figure 6.11. A **subduction zone** is a long, narrow belt where a lithospheric plate is sinking into the mantle. On a worldwide scale, the rate at which old lithosphere sinks into the mantle at subduction zones is equal to the rate at which new lithosphere forms at spreading centers. In this way, Earth maintains a global balance between the creation of new lithosphere and the destruction of old lithosphere.

Plate convergence can occur (1) between a plate carrying oceanic crust and another carrying continental crust, (2) between two plates carrying oceanic crust, and (3) between two plates carrying continental crust.

Convergence of Oceanic Crust with Continental Crust

Recall that oceanic crust is generally denser than continental crust. In fact, the entire lithosphere beneath the oceans is denser than continental lithosphere. When an oceanic plate converges with a continental plate, the denser oceanic plate sinks into the mantle beneath the edge of the continent. As a result, many subduction zones are located at continental margins. Today, oceanic plates are sinking beneath the western edge of South America; along the coasts of Oregon, Washington, and British Columbia; and at several other continental margins shown in ◇ Figure 6.10. When the descending plate reaches the asthenosphere, large quantities of magma are generated by processes explained in Chapter 8. Thus volcanoes are common along subduction zones. Chapter 9 describes how the Andes—a chain of volcanic mountains—formed as a result of subduction of a Pacific oceanic plate beneath the west coast of South America.

The oldest sea-floor rocks on Earth are only about 200 million years old because oceanic crust continuously recycles into the mantle at subduction zones. Rocks as old as 3.96 billion years are found on continents because subduction consumes little continental crust.

Convergence of Two Plates Carrying Oceanic Crust

Recall that newly formed oceanic lithosphere is hot, thin, and of low density, but as it spreads away from the Mid-Oceanic Ridge, it becomes older, cooler, thicker,

and denser. Thus, the density of oceanic lithosphere increases with its age. When two oceanic plates converge, the older, denser one sinks into the mantle. Oceanic subduction zones are common in the southwestern Pacific Ocean (◇ Figure 6.10). Their effects on the geology of the sea floor are described in Chapter 15.

Convergence of Two Plates Carrying Continents

If two converging plates carry continents, neither can sink deeply into the mantle because of their low densities. A continent does not normally sink into the mantle at a subduction zone for the same reasons that a log does not sink in a lake: both are of lower density than the material beneath them. In this case, the two continents collide and crumple against each other, forming a huge mountain chain. The Himalayas, the Alps, and the Appalachians all formed as results of continental collisions (◇ Figure 6.13). The formation of the Himalayas is described in Chapter 9.

Transform Plate Boundaries

A transform plate boundary forms where two plates slide horizontally past one another as they move in opposite directions (◇ Figure 6.10). California's San

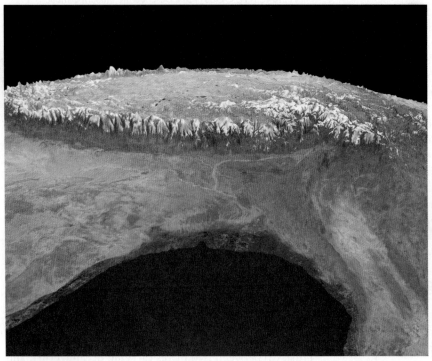

NASA/Goddard Space Flight Center Scientific Visualization Studio

◇ **FIGURE 6.13**   A collision between India and Asia created the Himalayan mountain chain. This photo shows the Himalayas with the vertical scale greatly exaggerated. It also shows the Ganges–Brahmaputra River system flowing through the Himalayas into the Bay of Bengal.

INTERACTIVE QUESTION: *How does the growth of a great mountain range affect the weather and climate of nearby regions?*

Andreas Fault is a transform boundary between the North American plate and the Pacific plate. Frequent earthquakes occur along transform plate boundaries. This type of boundary can occur in both oceans and continents.

ThomsonNOW

CLICK ThomsonNOW Interactive to work through these activities and view these animations:
- Plate Boundaries and Locations through Plate Tectonics
- Distribution of Volcanism
- Earthquakes in Space and Time, Seisminc Risk, USA through Earthquakes and Tsunamis.

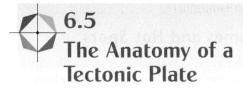

6.5
The Anatomy of a Tectonic Plate

The nature of a tectonic plate can be summarized as follows:

1. A plate is a segment of the lithosphere; thus it includes the uppermost mantle and all of the overlying crust.
2. A single plate can carry both oceanic crust and continental crust. The average thickness of a lithospheric plate covered by oceanic crust is 75 kilometers, whereas that of lithosphere covered by a continent is 125 kilometers. Lithosphere may be as little as 10 to 15 kilometers thick at an oceanic spreading center.
3. A plate is composed of hard, mechanically strong rock.
4. A plate floats on the underlying hot, plastic asthenosphere and glides horizontally over it.
5. A plate behaves like a slab of ice floating on a pond. It may flex slightly, as thin ice does when a skater goes by, allowing minor vertical movements. In general, however, each plate moves as a large, intact sheet of rock.
7. A plate margin is tectonically active. Earthquakes, mountain ranges, and volcanoes are common at plate boundaries. In contrast, the interior of a lithospheric plate is normally tectonically stable.
8. Tectonic plates move at rates that vary from less than 1 to 16 centimeters per year. Because continents and oceans make up the upper parts of the moving lithosphere, both continents and oceans migrate across Earth's surface at the same rates at which the plates move. Manhattan Island is now 9 meters farther from London than it was more than 225 years ago when the Declaration of Independence was written. Alfred Wegener was correct in saying that continents drift across Earth's surface.

ThomsonNOW

CLICK ThomsonNOW Interactive to work through an activity on the Distribution of Volcanism and Plate Boundaries Through Plate Tectionics.

6.6
Why Plates Move: The Earth As a Heat Engine

After geologists had developed the plate tectonics theory, they began to ask, *why* do the great slabs of lithosphere glide over Earth's surface? Recent research[3, 4] shows that after subduction begins, a tectonic plate sinks all the way to the core–mantle boundary, to a depth of 2,900 kilometers. At the same time, equal volumes of hot rock rise from the deep mantle to the surface beneath a spreading center, forming new lithosphere to replace that lost to subduction. The term **convection** refers to the upward and downward flow of fluid material in response to heating and cooling. The process of mantle convection continually stirs the entire mantle as rock that is hotter than its surroundings rises toward Earth's surface and old plates that are colder than their surroundings sink into the mantle. A single mantle convection cell may be thousands of kilometers across. In this way, the entire mantle–lithosphere system circulates in great cells, carrying rock from the core–mantle boundary to Earth's surface and then back into the deepest mantle (◆ Figure 6.9).

A soup pot on a hot stove illustrates the process of convection. Rising temperature causes most materials, including soup (or rock) to expand. When soup at the bottom of the pot is heated by the underlying stove, it becomes warm and expands. It then rises because it is less dense than the soup at the top. When the hot soup reaches the top of the pot, it flows along the surface until it cools and sinks (◆ Figure 6.14).

Just as the soup on the stove rests on the hot burner, the base of the mantle lies on—and is heated by—the hotter core. Thus, heat from the core, supplemented by additional heat generated by radioactivity within the mantle, drives the entire mantle and lithosphere in huge cells of convecting rock. A tectonic plate is the upper portion of a convecting cell and thus glides over the asthenosphere as a result of the convection (◆ Figure 6.9).

The upward flow of hot rock transports great quantities of heat from Earth's interior to its surface, where it radiates into space. If the mantle did not carry heat to the surface in this way, Earth's interior would be much hotter than it is today.

3. Summarized by Richard A. Kerr, "Deep Sinking Slabs Stir the Mantle," *Science 275* (January 31, 1997): 613–615.
4. Hans-Peter Bunge and Mark Richards, "The Origin of Large Scale Structure in Mantle Convection: Effects of Plate Motions and Viscosity Stratification," *Geophysical Research Letters 23* (October 15, 1996): 2987–2990.

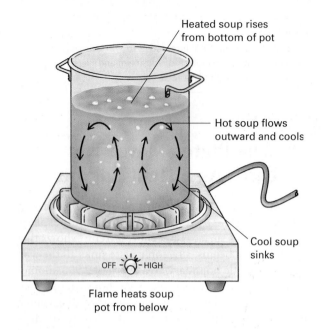

Heated soup rises
from bottom of pot

Hot soup flows
outward and cools

Cool soup
sinks

OFF -☼- HIGH

Flame heats soup
pot from below

◈ **FIGURE 6.14** Soup convects when it is heated from the bottom of the pot.

Two other processes shown in ◈ Figure 6.15 may facilitate the movement of tectonic plates. Notice that the base of the lithosphere slopes downward from a spreading center; the grade can be as steep as 8 percent, steeper than most paved highways. Calculations show that even if the slope were less steep, gravity would cause the lithosphere to slide away from a spreading center over the soft, plastic asthenosphere at a rate of a few centimeters per year.

In addition, the lithosphere becomes denser as it moves away from a spreading center and cools. Eventually, old lithosphere may become denser than the asthenosphere below. Consequently, it can no longer float on the asthenosphere and sinks into the mantle in a subduction zone, dragging the trailing portion of the plate over the asthenosphere. Both of these processes may contribute to the movement of a lithospheric plate as it glides over the asthenosphere.

Mantle Plumes and Hot Spots

In contrast to the huge, curtain-shaped mass of mantle that rises beneath a spreading center, a **mantle plume** is a relatively small rising column of plastic mantle rock that is hotter than surrounding rock. Many plumes rise from great depths in the mantle, probably because small

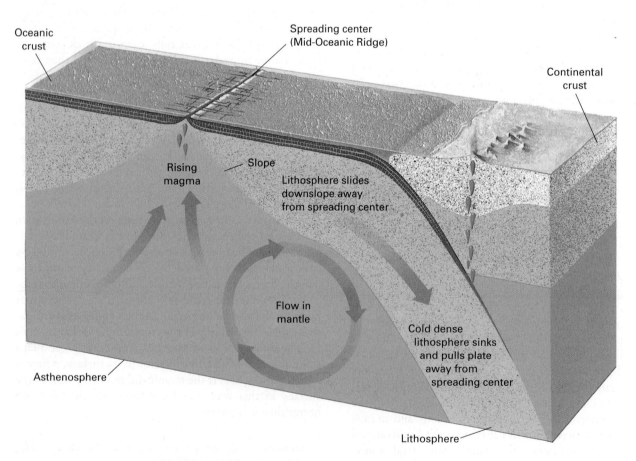

Oceanic crust

Spreading center
(Mid-Oceanic Ridge)

Continental crust

Rising magma

Slope

Lithosphere slides downslope away from spreading center

Flow in mantle

Cold dense lithosphere sinks and pulls plate away from spreading center

Asthenosphere

Lithosphere

◈ **FIGURE 6.15** New lithosphere glides downslope away from a spreading center. At the same time, the old, cool part of the plate sinks into the mantle at a subduction zone, pulling the rest of the plate along with it. (The steepness of the slope at the base of the lithosphere is exaggerated in this figure.)

zones of rock near the core–mantle boundary become hotter and more buoyant than surrounding regions of the deep mantle.

Others may form as a result of heating in shallower portions of the mantle. As pressure decreases in a rising plume, magma forms at a **hot spot** in the mantle just beneath the lithosphere and rises to erupt from volcanoes on Earth's surface. The Hawaiian Island chain is an example of a volcanic center at a hot spot. It erupts in the middle of the Pacific tectonic plate because the plume originates deep in the mantle, far from any plate boundary.

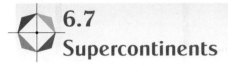

6.7
Supercontinents

Prior to 2 billion years ago, large continents as we know them today may not have existed. Instead, many—perhaps hundreds—of small masses of continental crust and island arcs similar to Japan, New Zealand, and the modern islands of the southwest Pacific Ocean dotted a global ocean basin. Then, between 2 billion and 1.8 billion years ago, tectonic plate movements swept these microcontinents together to form a single landmass called a **supercontinent**. After a few hundred million years, this supercontinent broke into fragments. The fragments then separated, each riding away from the others on its own tectonic plate. About 1 billion years ago, the fragments of continental crust reassembled, forming a second supercontinent. In turn, this continent fractured and the continental fragments reassembled into a third supercontinent about 300 million years ago, 70 million years before the appearance of dinosaurs. This third supercontinent is Alfred Wegener's Pangea, which began to break apart about 235 million years ago in late Triassic time. The tectonic plates have continued their slow movement to create the mosaic of continents and ocean basins that shape the map of the world as we know it today.

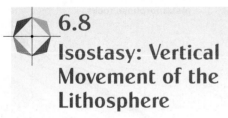

6.8
Isostasy: Vertical Movement of the Lithosphere

If you have ever used a small boat, you may have noticed that the boat settles in the water as you get into it and rises as you step out. The lithosphere behaves in a similar manner. If a large mass is added to the lithosphere, it settles and the underlying asthenosphere flows laterally away from that region to make space for the settling lithosphere.

But how is mass added or subtracted from the lithosphere? One process that adds and removes mass is the growth and melting of large glaciers. When a glacier grows, the weight of ice forces the lithosphere downward. For example, in the central portion of Greenland, a 3,000-meter-thick ice sheet has depressed the continental crust below sea level. Conversely, when a glacier melts, the continent rises—it rebounds. Geologists have discovered Ice Age beaches in Scandinavia tens of meters above modern sea level. The beaches formed when glaciers had depressed the Scandinavian crust. They now lie well above sea level because the land rose as the ice melted. The concept that the lithosphere is in floating equilibrium on the asthenosphere is called **isostasy**, and the vertical movement in response to a changing burden is called isostatic adjustment (◄▶ Figure 6.16).

The iceberg pictured in ◄▶ Figure 6.17 illustrates an additional effect of isostasy. A large iceberg has a high peak, but its base extends deeply below the water surface. The lithosphere behaves in a similar manner. Continents rise high above sea level, and the lithosphere beneath a continent has a "root" that extends as much as 125 kilometers into the asthenosphere.

In contrast, most ocean crust lies approximately 5 kilometers below sea level, and oceanic lithosphere extends only about 75 kilometers into the asthenosphere. For similar reasons, high mountain ranges have deeper roots than low plains, just as the bottom of a large iceberg is deeper than the base of a small one.

6.9
How Plate Movements Affect Earth Systems

The movements of tectonic plates generate volcanic eruptions and earthquakes, which help shape Earth's surface. They also build mountain ranges and change the global distributions of continents and oceans. Although the convecting mantle brings heat to Earth's surface, it is only $\frac{1}{5,000}$ as much as the solar heat that warms the surface and atmosphere. Thus the Sun provides almost all of the heat to warm the atmosphere and the oceans, creating weather and climate. However, tectonic activities strongly affect our environment—impacting global and regional climate, the atmosphere, hydrosphere, and biosphere—in other ways.

Volcanoes

Volcanic eruptions occur where hot magma rises to Earth's surface. Volcanic eruptions are common at both divergent and convergent plate boundaries. The processes that form magma in these tectonic environments are described in Chapter 8.

At a divergent boundary, hot asthenosphere oozes upward to fill the gap left between the two separating

Systems Perspective

Tectonic Movement Drives Many Important Components of the Rock Cycle

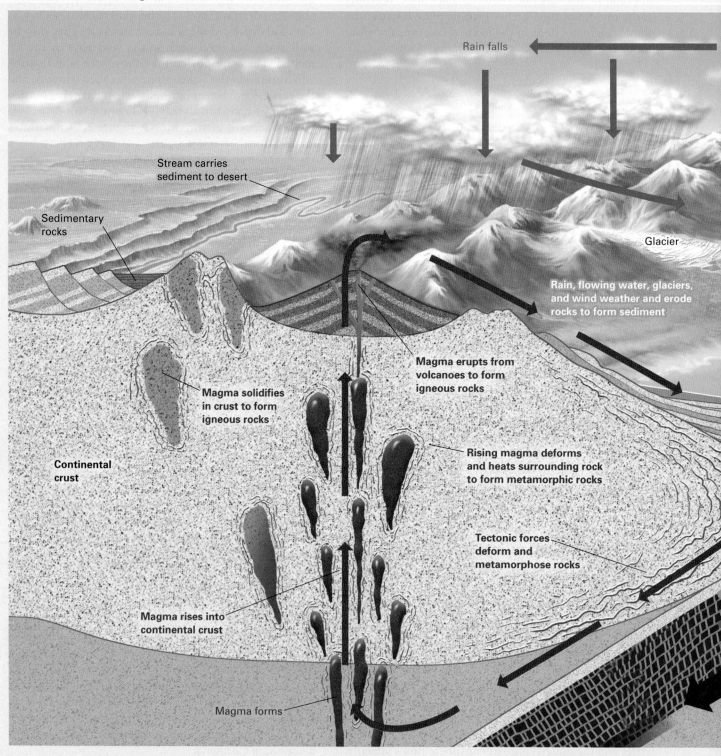

Rain falls

Stream carries
sediment to desert

Sedimentary
rocks

Glacier

Rain, flowing water, glaciers,
and wind weather and erode
rocks to form sediment

Magma solidifies
in crust to form
igneous rocks

Magma erupts from
volcanoes to form
igneous rocks

Rising magma deforms
and heats surrounding rock
to form metamorphic rocks

Continental
crust

Tectonic forces
deform and
metamorphose rocks

Magma rises into
continental crust

Magma forms

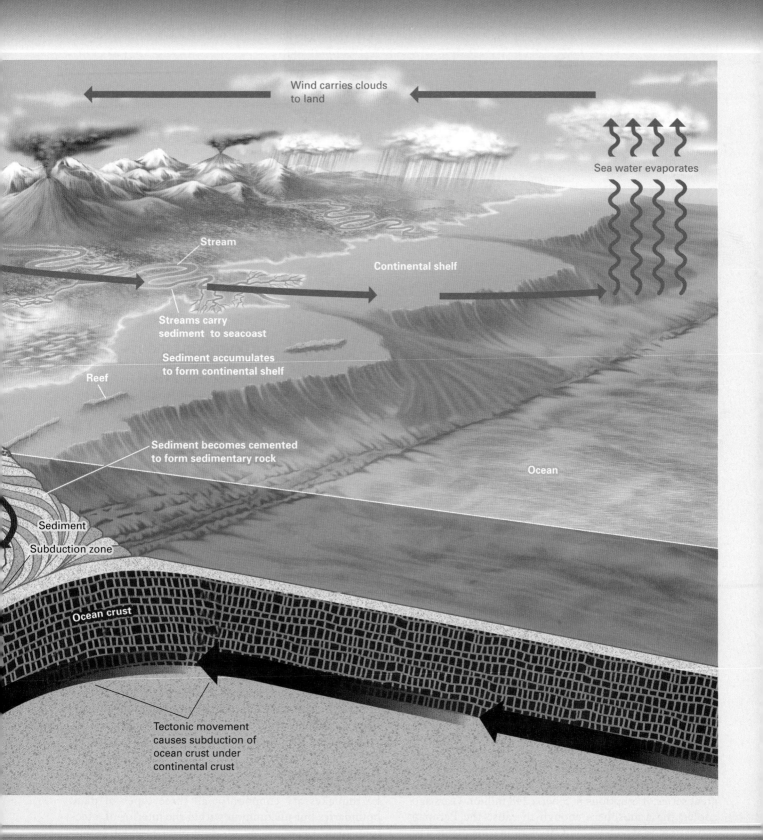

Wind carries clouds
to land

Sea water evaporates

Stream

Continental shelf

Streams carry
sediment to seacoast

Sediment accumulates
to form continental shelf

Reef

Sediment becomes cemented
to form sedimentary rock

Ocean

Sediment

Subduction zone

Ocean crust

Tectonic movement
causes subduction of
ocean crust under
continental crust

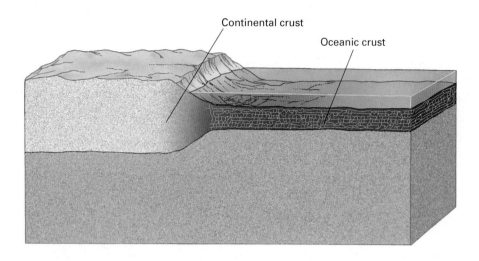

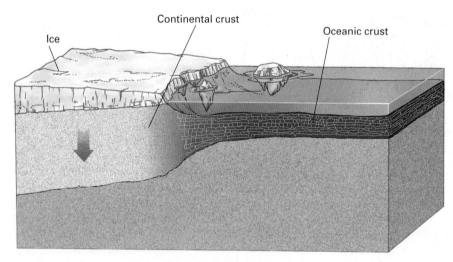

◇ **FIGURE 6.16** The weight of an ice sheet causes continental crust to sink isostatically.

INTERACTIVE QUESTION: *How might the growth of a large polar continental sheet affect coastal environments and ecosystems on nearby unglaciated coastal areas, and on an unglaciated continent near the equator?*

plates. Portions of the rising asthenosphere melt to form basaltic magma, which erupts onto Earth's surface. The Mid-Oceanic Ridge is a submarine chain of volcanoes and lava flows formed at divergent plate boundaries. Volcanoes are also common in continental rifts, such as the East African Rift.

Huge quantities of magma also form in a subduction zone and rise through the overlying lithosphere. Some solidifies within the crust, and some erupts from volcanoes on Earth's surface. Volcanoes of this type are common in the Cascade Range of Oregon, Washington, and British Columbia. They are also found in western South America, Japan, the Philippines, and near most other subduction zones (◇ Figure 6.18).

Volcanic eruptions directly alter the atmosphere to change global climate. Volcanoes emit ash and sulfur compounds that reflect sunlight and cool the atmosphere. For two years after Mount Pinatubo erupted in 1991, Earth cooled by a few tenths of a degree Celsius.

Temperature rose again in 1994 after the ash and sulfur settled out. Some scientists have suggested that a great series of eruptions in Siberia 248 million years ago cooled the atmosphere enough to cause the Permian extinction, in which 90 percent of all marine plants and animals and two-thirds of the reptiles and amphibians died in the most catastrophic mass extinction event in Earth history.

Volcanoes also emit carbon dioxide—a greenhouse gas that warms the atmosphere by absorbing infrared radiation. A gigantic series of eruptions 120 million years ago created a vast lava plateau on the sea floor off the coast of Peru. The eruptions also emitted enough carbon dioxide to warm the atmosphere by more than 10°C, profoundly altering habitats and species distributions around Earth. The warm climate created huge swamps in which dinosaurs flourished, and the abundant vegetation formed the massive Cretaceous coal deposits of North America and other continents.

Thus, volcanic eruptions can cause global warming or cooling. The net result of a volcanic eruption—warming or cooling—depends on the size of the eruption, its violence, and the proportion of solids and gases released.

Earthquakes

Earthquakes are common at all three types of plate boundaries but uncommon within the interior of a tectonic plate. Quakes concentrate at plate boundaries sim-

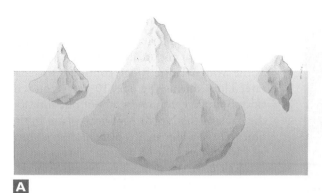

A

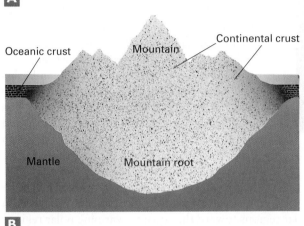

Oceanic crust Mountain Continental crust

Mantle Mountain root

B

◈ **FIGURE 6.17** **(A)** Icebergs illustrate some of the effects of isostasy. A large iceberg has a deep root and also a high peak. **(B)** In an analogous manner, continental crust extends more deeply into the mantle beneath high mountains than it does under the plains. Oceanic lithosphere is thinner and denser and consequently floats at the lower level.

ply because those boundaries are zones where one plate slips past another. The slippage is rarely smooth and continuous.

Instead, the fractures may be locked up for months to hundreds of years. Then, one plate suddenly slips a few centimeters, or even a few meters, past its neighbor. An earthquake is a vibration in rock caused by these abrupt movements.

Mountain Building

Many of the world's great mountain chains, including the Andes and parts of the mountains of western North America, formed at subduction zones. Several processes combine to build a mountain chain at a subduction zone. The great volume of magma rising into the crust adds volume to the crust. Additional crustal thickening may occur where two plates converge for the same reason that a mound of bread dough thickens when you compress it from both sides. Finally, volcanic eruptions build chains of volcanoes. This thick crust then floats upward on the soft, plastic asthenosphere to form a mountain chain.

In some cases, two continents collide at a convergent plate boundary. The collision thrusts great masses of rock upward, creating huge mountain chains such as the Himalayas, the Alps, and the Appalachians.

Great chains of volcanic mountains form at rift zones because the new, hot lithosphere floats to a high level, and large amounts of magma form in these zones. The Mid-Oceanic Ridge and the volcanoes of the East African Rift and the Rio Grande Rift of the southwestern United States are examples of such mountain chains.

David Weintraub/Photo Researchers, Inc.

◈ **FIGURE 6.18** Mount St. Helens, most recently erupted in 1980, is an active volcano in the Cascade Range of Washington near a convergent plate boundary.

Courtesy of Graham R. Thompson/Jonathan Turk

◆ **FIGURE 6.19** The snowy peaks of the Andes tower over the dry Bolivian Altiplano. Lago Nube, Cordillera Apolobamba.

Mountains are generally colder and wetter than nearby lowlands (◆ Figure 6.19). Streams run faster and cut deeper in the mountains than they do in the lowlands; landslides and glaciation are more common and vegetation and wildlife are different. Mountain ranges strongly affect global wind and precipitation patterns, as discussed in Chapters 19 and 20. Thus, plate tectonics, driven by heat from Earth's core, ultimately affects both weather and climate.

Migrating Continents and Oceans

Both continents and ocean basins migrate over Earth's surface because they are parts of the moving lithospheric plates: they simply ride piggyback on the plates. North America is now moving away from Europe at about 2.5 centimeters per year, as the Mid-Atlantic Ridge continues to separate. Thus, during the next 10 years you will ride about 25 centimeters, about one foot, on the back of the North American plate. In the time span of a human life, plate motion is slow.

However, in the 65 million years since the extinction of the dinosaurs, North America has migrated about 1,620 kilometers, about the distance from New York City to Minneapolis. As the Atlantic Ocean widens, the Pacific is shrinking. Thus, as continents move, ocean basins open and close over geologic time.

Migrations of continents and ocean basins alter both regional and global climate. Note from ◆ Figure 6.3B that 300 million years ago, North America straddled the equator, and equatorial regions of Africa lay near the South Pole. As a result, North America was hotter and Africa was colder than they are today.

Ocean currents carry warm water from the equator toward the poles, and cool water from polar regions toward the equator, warming polar regions and cooling the tropics. Similarly, winds transport heat and moisture over the globe. Movements of continents and ocean basins alter both ocean currents and wind patterns, and may convert a desert to a rainforest or a glacier-covered region to grassland. Streams and drainage patterns must respond to the altered rainfall distribution. Lakes may dry up, or new lakes form. Plants and animals may die or migrate away to be replaced by new species that are adapted to the new climatic conditions. As we will learn in Chapter 21, changes in ocean currents and wind systems also alter global climate patterns.

EARTH SYSTEMS INTERACTIONS

ABOUT 5 BILLION YEARS AGO, our Earth began to form from a cold, diffuse cloud of gas and dust. The four systems, or spheres, that we call the geosphere, the hydrosphere, the atmosphere, and the biosphere, have emerged over geologic time as the solid Earth cooled, the atmosphere and oceans formed, and life began. Each of those systems, in turn, consists of many smaller subsystems. For example, the geosphere is divided into the core, the mantle, and the crust; the biosphere is divided into the plant and animal kingdoms, each of which is further subdivided into multiple layers of phyla, class, order, family, genus, and species.

However, the division of Earth into systems, subsystems, and smaller units, is merely a matter of intellectual necessity for scientists who must organize the huge array of data and observations of the natural world into a manageable framework. That is, Earth is really an integrated whole, and the "systems" that we use to describe it are artifacts created by scientists to understand our world. ■

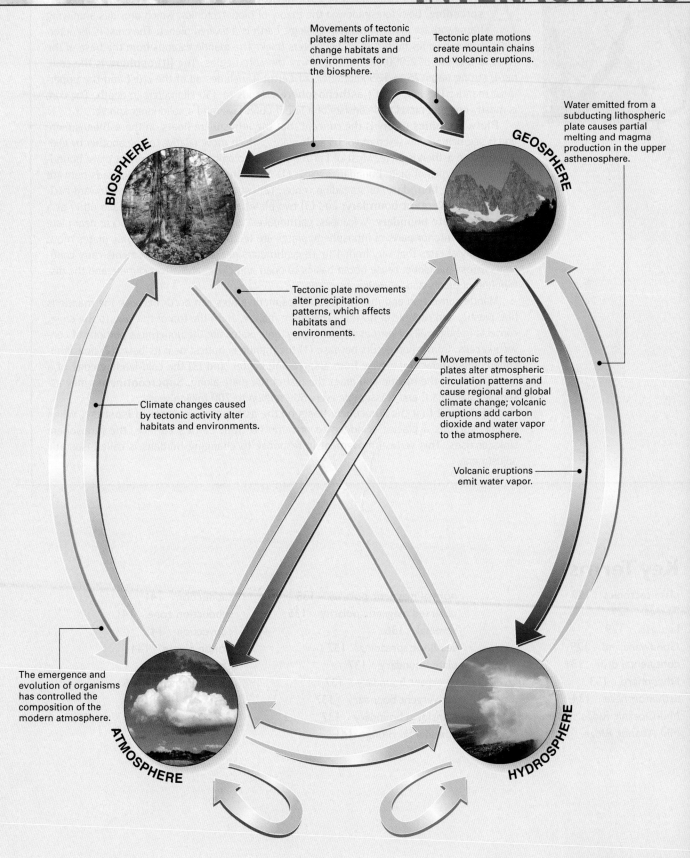

Movements of tectonic plates alter climate and change habitats and environments for the biosphere.

Tectonic plate motions create mountain chains and volcanic eruptions.

Water emitted from a subducting lithospheric plate causes partial melting and magma production in the upper asthenosphere.

BIOSPHERE

GEOSPHERE

Tectonic plate movements alter precipitation patterns, which affects habitats and environments.

Movements of tectonic plates alter atmospheric circulation patterns and cause regional and global climate change; volcanic eruptions add carbon dioxide and water vapor to the atmosphere.

Climate changes caused by tectonic activity alter habitats and environments.

Volcanic eruptions emit water vapor.

The emergence and evolution of organisms has controlled the composition of the modern atmosphere.

ATMOSPHERE

HYDROSPHERE

SUMMARY

Alfred Wegener's hypothesis of **continental drift**, and the hypothesis of **sea-floor spreading**, both foreshadowed the theory of plate tectonics, which provides a unifying framework for much of modern geology. Earth is a layered planet. The crust is its outermost layer and varies from 4 to 70 kilometers thick. The mantle extends from the base of the crust to a depth of 2,900 kilometers, where the core begins. The **lithosphere** is the cool, hard, strong outer 75 to 125 kilometers of Earth; it includes all of the crust and the uppermost mantle. The hot, plastic **asthenosphere** extends to 350 kilometers in depth. The core is mostly iron and nickel and consists of a liquid outer layer and a solid inner sphere.

Plate tectonics theory is the concept that the lithosphere floats on the asthenosphere and is segmented into seven major tectonic plates, which move relative to one another by gliding over the asthenosphere. Most of Earth's major geological activity occurs at plate boundaries. Three types of plate boundaries exist: (1) New lithosphere forms and spreads outward at a **divergent boundary**, or spreading center; (2) two lithospheric plates move toward each other at a **convergent boundary**; and (3) two plates slide horizontally past each other at a **transform plate boundary**. Volcanoes, earthquakes, and mountain building occur near plate boundaries. Interior parts of lithospheric plates are tectonically stable. Tectonic plates move horizontally at rates that vary from 1 to 16 centimeters per year. Plate movements carry continents across the globe, cause ocean basins to open and close, and affect climate and the distribution of plants and animals.

Mantle convection and movement of lithospheric plates can occur because the mantle is hot, plastic, and capable of flowing. The entire mantle, from the top of the core to the crust, convects in huge cells. Horizontally moving tectonic plates are the uppermost portions of convection cells. Convection occurs because (1) the mantle is hottest near its base, (2) new lithosphere glides downslope away from a spreading center, and (3) the cold leading edge of a plate sinks into the mantle and drags the rest of the plate along. **Supercontinents** may assemble, split apart, and reassemble every 500 million to 700 million years.

The concept that the lithosphere floats on the asthenosphere is called **isostasy**. When weight, such as a glacier, is added to or removed from Earth's surface, the lithosphere sinks or rises. This vertical movement in response to changing burdens is called isostatic adjustment.

Key Terms

plate tectonics   129
Pangea   129
Laurasia   129
Gondwanaland   129
continental drift   131
lithosphere   133
asthenosphere   134
Mid-Oceanic Ridge system   135
Mid-Atlantic Ridge   135

normal magnetic polarity   135
reversed magnetic polarity   135
reversal   136
sea-floor spreading   137
plate boundary   137
divergent boundary   137
convergent boundary   137
transform boundary   137
continental rifting   141

subduction   141
subduction zone   141
convection   143
mantle plume   144
hot spot   145
supercontinent   145
isostasy   145

For Review

1. Briefly describe Alfred Wegener's theory of continental drift. What evidence supported his ideas? How does Wegener's theory differ from the modern theory of plate tectonics?

2. Briefly describe the sea-floor spreading hypothesis and the evidence used to develop the hypothesis. How does this idea differ from Wegener's theory and the theory of plate tectonics?

3. Draw a cross-sectional view of Earth. List all the major layers and the thickness of each.

4. Describe the physical properties of each of Earth's layers.

5. Describe and explain the important differences between the lithosphere and the asthenosphere.

6. What properties of the asthenosphere allow the lithospheric plates to glide over it?

7. Describe some important differences between the crust and the mantle.

8. Describe some important differences between oceanic crust and continental crust.

9. Explain how sea-floor spreading supports the modern theory of plate tectonics.

10. How is it possible for the solid rock of the mantle to flow and convect?

11. Summarize the important aspects of the plate tectonics theory.

12. How many major tectonic plates exist? List them.

13. Describe the three types of tectonic plate boundaries.

14. Explain why tectonic plate boundaries are geologically active and the interior regions of plates are geologically stable.

15. Describe some differences between the lithosphere beneath a continent and that beneath oceanic crust.

16. Describe a reasonable model for a mechanism that causes movement of tectonic plates.

17. Why would a lithospheric plate floating on the asthenosphere suddenly begin to sink into the mantle to create a new subduction zone?

18. Describe the Mid-Oceanic Ridge.

19. Why are the oldest sea-floor rocks only about 200 million years old, whereas some continental rocks are almost 4 billion years old?

For Discussion

1. Discuss why a unifying theory, such as the plate tectonics theory, is desirable in any field of science.

2. Central Greenland lies below sea level because the crust is depressed by the ice cap. If the glacier were to melt, would Greenland remain beneath the ocean? Why or why not?

3. At a rate of 5 centimeters per year, how long would it take for a continent to drift the width of your classroom? The distance between your apartment or dormitory and your classroom? The distance from New York to London?

4. Why do most major continental mountain chains form at convergent plate boundaries? What topographic and geologic features characterize divergent and transform plate boundaries in continental crust? Where do these types of boundaries exist in continental crust today?

5. The largest mountain in the solar system is Olympus Mons, a volcano on Mars. It is 25,000 meters high, nearly three times the elevation of Mount Everest. Speculate on the factors that might permit such a large mountain on Mars.

6. The core's radius is 3,470 kilometers, and that of the mantle is 2,900 kilometers, yet the mantle contains 80 percent of Earth's volume. Explain this apparent contradiction.

7. If you built a model of Earth that is 1 meter in radius, how thick would the crust, lithosphere, asthenosphere, mantle, and core be?

8. Look at the map in Figure 6.10 and name a tectonic plate that is covered mostly by continental crust. Name one that is mostly ocean. Name two plates that are about half ocean and half continent.

9. Give an example of how tectonic activity might change a portion of the geosphere. Explain how that change could affect the hydrosphere, atmosphere, and biosphere.

ThomsonNOW

Assess your understanding of this chapter's topics with additional quizzes and comprehensive interactivities at **http://www.thomsonedu.com/login**, which also has current and up-to-date web links, additional readings, and exercises.

CHAPTER 7

Earthquakes and the Earth's Structure

AP/Wide World Photos

Villagers in Lhoknga, Indonesia, walk past a tug boat that was washed up on the island's main coastal road by the great tsunamis of December 2004.

On December 26, 2004, an earthquake in the Indian Ocean caused a tsunami that killed approximately 283,000 people in India and Southeast Asia. On October 8, 2005 an earthquake along the Indian-Pakistani border killed 79,000 (◆ Figure 7.1). Shortly after the tsunami disaster, Dr. Donald DePaolo, a geochemist at the University of California, Berkeley, wrote, "It's hard to find something uplifting about 283,000 lives being lost. But the type of geologic processes that caused the earthquake and the tsunami is an essential characteristic of the Earth. As far as we know, it doesn't occur on any other planetary body and has something very directly to do with the fact that the Earth is a habitable planet."

As we learned in Chapter 6, the Earth's tectonic plates glide slowly over the soft asthenosphere, about as fast as your fingernail grows. At plate boundaries, friction often holds the plates locked and stationary. Rocks stretch and compress, forces build—but for decades or even a century, or more, nothing happens. Then suddenly, the plates snap free and Earth shakes. Thus, an earthquake is a classic example of a threshold effect. The interiors of plates move at a steady rate, but often motion is locked at plate boundaries. Then, when the threshold is reached, sudden and cataclysmic motion—an earthquake—occurs.

While plate tectonic motion and the resulting earthquakes lead to death and destruction, we must appreciate that Earth's tectonic activity profoundly affects Earth systems and may cause the planet to be habitable. An earthquake, no matter how tragic, is one component of a complex system and a planet that sustains us.

Recall from Chapter 6 that the movement of tectonic plates:

- Formed the oceans and the atmosphere. Gases released during volcanic eruptions (another manifestation of plate movement) brought water to the surface of an otherwise dry, inhospitable Earth and have repeatedly altered the composition of the atmosphere (Chapters 6, 8, 15, 20, and 21).
- Made the continents (Chapters 6 and 8).
- Rejuvenates soils. Volcanic ash adds valuable nutrients to many soils (Chapter 8).
- Regulates global chemistry. The surface of tectonic plates that descend in a subduction zone is covered with water-saturated mud. The plates carry this mud downward, and thus transport water from the surface into the mantle. Water rises again at spreading centers. During its tenure in the interior of Earth, water exchanges numerous elements with Earth's interior, renewing vital nutrients at the surface (Chapters 6, 8, 15, 20, and 21).
- Concentrates metals. Water streaming through the seabed's hot, volcanic, spreading centers concentrated hydrothermal mineral resources that may later be bulldozed onto land by tectonic motion (Chapter 5). ■

ThomsonNOW™ Throughout this chapter, the ThomsonNOW logo indicates an opportunity for online self-study, which:
- Assesses your understanding of important concepts and provides a personalized study plan
- Links you to animations and simulations to help you study
- Helps to test your knowledge of material and prepare for exams

Visit ThomsonNOW at http://www.thomsonedu.com/login to access these resources.

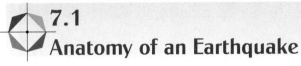

7.1
Anatomy of an Earthquake

If you squeeze a rock in your hand, it does not change shape. Place it on the ground and jump on it or kick it and, again, nothing happens. The rock appears rigid. But if you apply enough stress, rock will deform. When stress is applied to a rock, the rock can deform in one of three ways:

At first, the rock deforms elastically. If the stress is removed, the rock springs back to its original size and shape. A rubber band deforms elastically when you stretch it. The energy used to stretch the rubber band is

◆ **FIGURE 7.1** Residents and police officers search the remains of a 19-story housing complex destroyed by the Oct. 8, 2005, magnitude 7.6 earthquake in Islamabad, Pakistan.

The ground rises and falls and undulates back and forth. Buildings topple, bridges fall, roadways and pipelines snap. An **earthquake** is a sudden motion or trembling of Earth caused by the abrupt release of energy that is stored in rocks.

In Chapter 1, we used the melting of ice as an example of a threshold effect. If you warm ice from −10°C to −0.0000001°C, some of its properties, such as density, change, but the ice remains solid. However, when the ice warms the final tiny increment to a fraction of a degree above zero, a big change suddenly occurs; it melts. In an analogous manner, when Earth's crustal rocks are stretched or compressed, they slowly change shape, but then, when a critical threshold is reached, they fracture, causing an earthquake. The difference between the ice example and the earthquake is that we know the melting point of ice precisely, but, as we will discuss shortly, scientists are unable to determine precisely when and where an earthquake will occur.

Most movement of crustal rock occurs by slippage along established faults because the friction binding the two sides of the fault is weaker than the rock itself (◆ Figure 7.4). But occasionally, solid rock may fracture, creating a new fault. After the earthquake dies

stored in the elongated rubber. When the stress is removed, the rubber band springs back and releases the stored energy. In the same way, an elastically deformed rock will spring back to its original shape and release its stored elastic energy when the force is removed.

However, every rock has a limit beyond which it cannot deform elastically. Under certain conditions, an elastically deformed rock may suddenly fracture (◆ Figure 7.2). The fracture releases the elastic energy, like letting go of one end of the stretched rubber band, and the rock springs back to its original shape. When large masses of rock in Earth's crust deform and then fracture, the resultant rapid motion creates vibrations that travel through Earth and are felt as an earthquake.

Under other conditions, when its elastic limit is exceeded, a rock continues to deform, like putty. This behavior is called **plastic deformation**. A rock that has deformed plastically keeps its new shape when the stress is released and, consequently, does not store the energy used to deform it. Therefore, earthquakes do not occur when rocks deform plastically. ◆ Figure 7.3 shows a rock outcrop that was deformed plastically by forces deep within Earth's crust. Parallel rock layers were bent and twisted into an S-shaped fold.

We learned in Chapter 6 that Earth's lithosphere is broken into seven large and several smaller tectonic plates. Although tectonic plates move at rates between 1 and 16 centimeters per year (about as fast as your fingernail grows), friction prevents the plates from slipping past one another continuously. For decades, or even a century or two, the cool, strong rock near the plate boundary stretches or compresses elastically, but the edges remain locked and immobile. Potential energy builds, just like the potential energy in a stretched rubber band or a compressed spring. Then, when the strain reaches a critical value, rock snaps loose or fractures.

◆ **FIGURE 7.2** Stressed rock fractured and displaced this roadway during the Loma Prieta earthquake near San Francisco in 1989.

◆ **FIGURE 7.3** This rock deformed plastically when stressed.

away, the stored-up potential energy in the rock is released but the fault remains as a weakness in the rock. When more tectonic stress builds, the rock is more likely to move along the fault than to crack again to create a new fault. Thus earthquakes occur repeatedly along established faults. For example, the San Andreas Fault in Southern California lies along a tectonic plate boundary that has moved many times in the past and will certainly move again in the future (◆ Figure 7.5).

During an earthquake, rock moves from a few centimeters to a meter or two. Thus a single earthquake makes only small changes in the topography or geography of a region. But, over tens of millions of years, the inexorable motion of plates, accompanied by tens or hundreds of thousands of earthquakes, changes the surface of the planet.

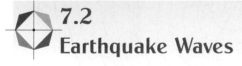

CLICK ThomsonNOW Interactive to work through an activity on Earthquakes and Space and Time Through Earthquakes and Tsunamis.

✦ 7.2
Earthquake Waves

If you have ever bought a watermelon, you know the challenge of picking out a ripe, juicy one without being able to look inside. One trick is to tap the melon gently with your knuckle. If you hear a sharp, clean sound, it is probably ripe; a dull thud indicates that it may be over-ripe and mushy. The watermelon illustrates two points that can be applied to Earth: (1) the energy of your tap travels through the melon, and (2) the nature of the melon's interior affects the quality of the sound.

A wave transmits energy from one place to another. A drumbeat travels through air as a sequence of waves, the Sun's heat travels to Earth as waves, and a tap travels through a watermelon in waves. Waves that travel through rock are called **seismic waves**. Earthquakes and explosions produce seismic waves. **Seismology** is the study of earthquakes and the nature of Earth's interior based on evidence from seismic waves.

The initial rupture point, where abrupt movement creates an earthquake, typically lies below the surface at a point called the **focus**. The point on Earth's surface directly above the focus is the **epicenter**. An earthquake produces several types of seismic waves. **Body waves** travel through Earth's interior and carry some of the energy from the focus to the surface (◆ Figure 7.6). **Surface waves** then radiate from the epicenter along Earth's surface. Although the wave mechanisms are different, surface waves undulate across the ground like the waves that ripple across the water after you throw a rock into a calm lake.

Body Waves

Two main types of body waves travel through Earth's interior. A **P wave** is a compressional elastic wave that causes alternate compression and expansion of the rock (◆ Figure 7.7). P waves are called *primary* waves because they are so fast that they are the first seismic waves to reach an observer. Consider a long spring such as the popular Slinky™ toy. If you stretch a Slinky™ and strike one end, a compressional wave travels along its length. P waves travel through air, liquid, and solid material. Next time you take a bath, immerse your head until your ears

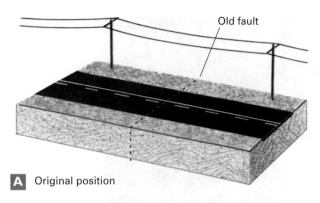

A Original position

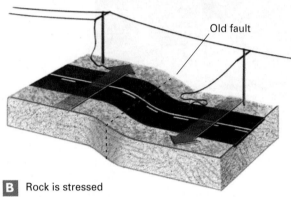

B Rock is stressed

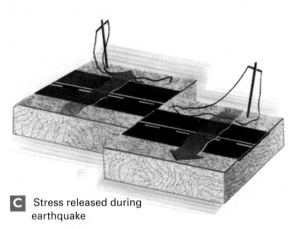

C Stress released during earthquake

◆ **FIGURE 7.4** **(A)** A road is built across an old fault. **(B)** The rock stores elastic energy when it is stressed by a tectonic force. **(C)** When the rock moves rapidly along the fault, it snaps back to its original shape, creating an earthquake.

◆ **FIGURE 7.5** California's San Andreas Fault, the source of many earthquakes, is the boundary between the Pacific plate, on the left in this photo, and the North American plate on the right.

are under water and listen as you tap the sides of the tub with your knuckles. You are hearing P waves.

P waves travel at speeds between 4 and 7 kilometers per second in Earth's crust and at about 8 kilometers per second in the uppermost mantle. For comparison, the speed of sound in air is only 0.34 kilometer per second, and the fastest jet fighters fly at about 0.85 kilometer per second.

A second type of body wave, called an **S wave**, is a shear wave. S waves arrive after P waves and are the "secondary" waves to reach an observer. An S wave can be illustrated by tying a rope to a wall, holding the opposite end, and giving the rope a sharp up-and-down jerk

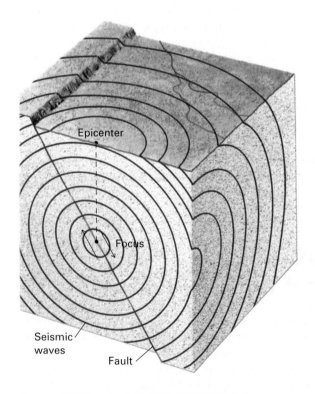

ThomsonNOW ◆ **ACTIVE FIGURE 7.6** Body waves radiate outward from the focus of an earthquake.

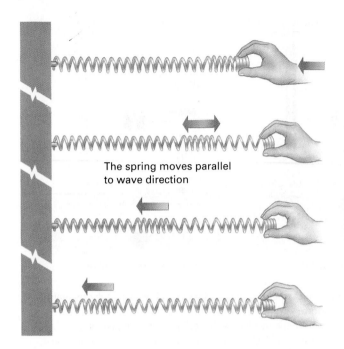

◆ **ACTIVE FIGURE 7.7** Model of a P, or compressional, wave. The spring moves parallel to the direction of wave propagation.

(◆ Figure 7.8). Although the wave travels parallel to the rope, the individual particles in the rope move at right angles to the rope length. A similar motion in an S wave produces shear stress in a rock and gives the wave its name. S waves are slower than P waves and travel at speeds between 3 and 4 kilometers per second in the crust.

Unlike P waves, S waves move only through solids. Because molecules in liquids and gases are only weakly

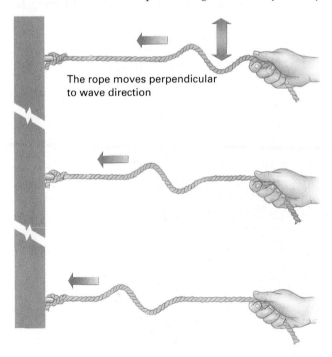

◆ **ACTIVE FIGURE 7.8** Model of an S, or shear, wave. The rope moves perpendicular to the direction of wave propagation.

bound to one another, they slip past each other and thus cannot transmit a shear wave.

Surface Waves

Surface waves travel more slowly than body waves. Two types of surface waves occur simultaneously: an up-and-down rolling motion and a side-to-side vibration. During an earthquake, Earth's surface rolls like ocean waves and writhes from side to side like a snake (◆ Figure 7.9).

Measurement of Seismic Waves

A **seismograph** is a device that records seismic waves. To understand how a seismograph works, consider the act of writing a letter while riding in an airplane. If the plane hits turbulence, inertia keeps your hand relatively stationary as the plane moves back and forth beneath it, and your handwriting becomes erratic.

Early seismographs worked on the same principle. A heavy weight was suspended from a spring. A pen attached to the weight was aimed at the zero mark on a piece of graph paper (◆ Figure 7.10). The graph paper was mounted on a rotating drum that was attached firmly to bedrock. During an earthquake, the graph paper jiggled up and down, but inertia kept the weight and its pen stationary. As a result, the paper moved up and down beneath the pen. The rotating drum recorded earthquake motion over time. This record of Earth vibration is called a **seismogram** (◆ Figure 7.11). Modern seismographs use electronic motion detectors, which transmit the signal to a computer.

Measurement of Earthquake Strength

Over the past century, geologists have devised several scales to express the strength of an earthquake. Before seismographs became widespread, geologists evaluated earthquakes on the Mercalli scale, which was based on structural damage. On the **Mercalli scale**, an earthquake that destroyed many buildings was rated as more intense than one that destroyed only a few.

This system did not accurately measure the energy released by a quake, however, because structural damage also depends on distance from the focus, the rock or soil beneath the structure, and the quality of construction. In

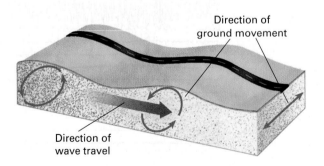

◆ **FIGURE 7.9** During an earthquake, surface waves move Earth's surface up and down and from side to side.

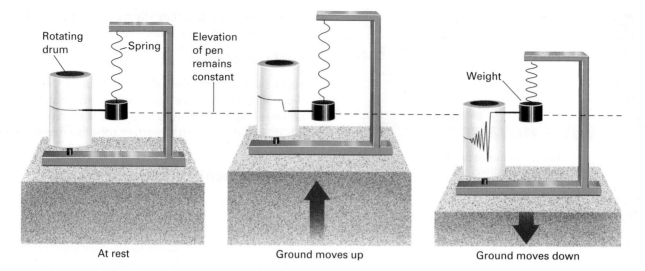

FIGURE 7.10 A seismograph records ground motion during an earthquake. In early versions of seismography, the pen draws a straight line across the rotating drum when the ground is stationary. When the ground rises abruptly during an earthquake, it carries the drum up with it. But the spring stretches, so the weight and pen hardly move. Therefore the pen marks a line lower on the drum. Conversely, when the ground sinks, the pen marks a line higher on the drum. During an earthquake the pen traces a jagged line as the drum rises and falls.

1935 Charles Richter devised the **Richter scale** to express the amount of energy released during an earthquake. Richter magnitude is calculated from the height of the largest earthquake body wave recorded on a specific type of seismograph. The Richter scale is more quantitative than earlier intensity scales, but it is not a precise measure of earthquake energy. A sharp, quick jolt would register as a high peak on a Richter seismograph, but a very large earthquake can shake the ground for a long time without generating extremely high peaks. Thus, a great earthquake can release a huge amount of energy that is not reflected in the height of a single peak and that is not adequately expressed by Richter magnitude.

Modern equipment and methods enable seismologists to measure the amount of movement and the surface area of a fault that moved during a quake. The product of these two values allows them to calculate the **moment magnitude**. Most seismologists now use moment magnitude rather than Richter magnitude because it more closely reflects the total amount of energy released during an

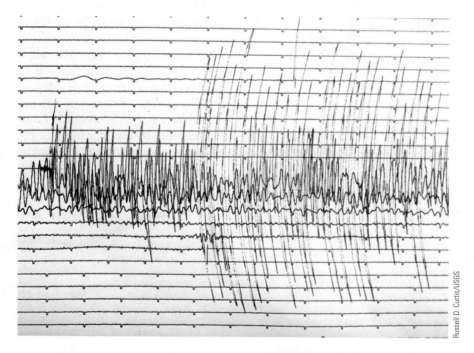

Russell D. Curtis/USGS

FIGURE 7.11 This seismogram recorded north–south ground movements during the October 1989 Loma Prieta earthquake.

CHAPTER 7 • Earthquakes and Earth's Structure

earthquake. An earthquake with a moment magnitude of 6.5 has an energy of about 10^{25} (1 followed by 25 zeros) ergs.[1] The atomic bomb dropped on the Japanese city of Hiroshima at the end of World War II released about that much energy. (Of course, atomic bombs also release intense heat and deadly radioactivity, while earthquakes only shake the ground.)

On both the moment magnitude and Richter scales, the energy of the quake increases by a factor of about 30 for each increment on the scale. Thus a magnitude-6 earthquake releases roughly 30 times more energy than a magnitude-5 earthquake.

The largest possible earthquake is determined by the strength of rocks. A strong rock can store more elastic energy before it fractures than a weak rock can. The largest earthquakes ever measured had moment magnitudes of 8.5 to 8.7, about 900 times greater than the energy released by the Hiroshima bomb.

Locating the Source of an Earthquake

If you have ever watched an electrical storm, you may have used a simple technique for estimating the distance between you and the place where the lightning strikes. After the flash of a lightning bolt, count the seconds that pass before you hear the thunder. Although the electrical discharge produces thunder and lightning simultaneously, light travels much faster than sound. Therefore, light reaches you virtually instantaneously, whereas sound travels much more slowly, at 340 meters per second. If the time interval between the flash and the thunder is 1 second, then the lightning struck 340 meters away and was very close.

The same principle is used to determine the distance from a recording station to both the epicenter and the focus of an earthquake. Recall that P waves travel faster than S waves and that surface waves are slower yet. If a seismograph is located close to an earthquake epicenter, the different waves will arrive in rapid succession for the same reason that the thunder and lightning come close together when a storm is close. On the other hand, if a seismograph is located far from the epicenter, the S waves arrive at correspondingly later times after the P waves arrive, and the surface waves are even farther behind, as shown in ◆ Figure 7.12.

Geologists use a **time-travel curve** to calculate the distance between an earthquake epicenter and a seismograph. To make a time-travel curve, a number of seismic stations at different locations record the time of arrival of seismic waves from an earthquake with a known epicenter and occurrence time. Using these data, a graph such as ◆ Figure 7.13 is drawn. This graph can then be used to measure the distance between a recording station and an earthquake whose epicenter is unknown.

Time-travel curves were first constructed from data obtained from natural earthquakes. However, scientists do not always know precisely when and where an earthquake occurred. In the 1950s and 1960s, geologists studied seismic waves from atomic bomb tests to improve the time-travel curves because they knew both the location and timing of the explosions.

◆ Figure 7.13 shows us that if the first P wave arrives 3 minutes before the first S wave, the recording station is about 1,900 kilometers from the epicenter. But this distance does not indicate whether the earthquake originated to the north, south, east, or west. To pinpoint the location of an earthquake, geologists compare data from three or more recording stations. If a seismic station in New York City records an earthquake with an epicenter 6,750 kilometers away, geologists know that the epicenter lies somewhere on a circle 6,750 kilometers from New York City (◆ Figure 7.14). The same epicenter is reported to be 2,750 kilometers from a seismic station in London and 1,700 kilometers from one in Godthab, Greenland. If one circle is drawn for each recording station, the arcs intersect at the epicenter of the quake.

7.3
Earthquakes and Tectonic Plate Boundaries

Before the plate tectonics theory was developed, geologists recognized that earthquakes occur frequently in some regions and infrequently in others, but they did not understand why. Modern geologists know that most earthquakes occur along plate boundaries, where tectonic plates diverge, converge, or slip past one another (◆ Figure 7.15). Thus, earthquakes are related to the slow movement of mantle rock that begins 2,900 kilometers beneath us at the core–mantle boundary.

Earthquakes at a Transform Plate Boundary: The San Andreas Fault Zone

The populous region from San Francisco to San Diego straddles the San Andreas Fault zone, which is a transform boundary between the Pacific plate and the North American plate (◆ Figure 7.16). The fault itself is vertical and the rocks on opposite sides move

---

1. An erg is a standard unit of energy in scientific usage. One erg is a small amount of energy; approximately 3×10^{12} ergs are needed to light a 100-watt light bulb for one hour. However, 10^{25} is a very large number and 10^{25} ergs represents a considerable amount of energy.

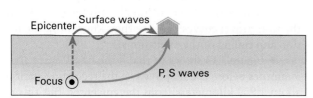

A Recording station A near focus

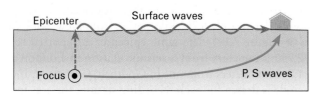

B Recording station B far from focus

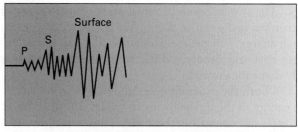

Seismogram from station A

Time

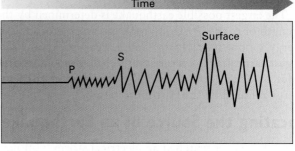

Seismogram from station B

◆ **FIGURE 7.12** The time intervals between arrivals of P, S, and surface waves at a recording station increase with distance from the focus of an earthquake.

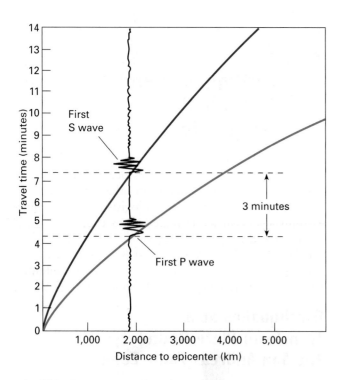

◆ **FIGURE 7.13** A time-travel curve. With this graph you can calculate the distance from a seismic station to the epicenter of an earthquake. In the example shown, a 3-minute delay between the first arrivals of P waves and S waves corresponds to an earthquake with an epicenter 1,900 kilometers from the seismic station.

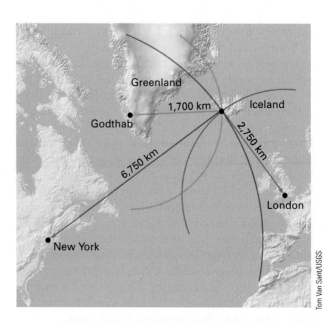

ThomsonNOW ◆ **ACTIVE FIGURE 7.14** Locating an earthquake. The distance from each of three seismic stations to the earthquake is determined from time-travel curves. The three arcs are drawn. They intersect at only one point, which is the epicenter of the earthquake.

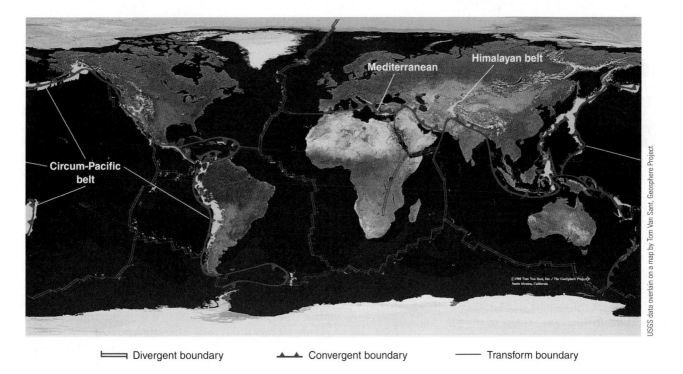

Divergent boundary ⟝⟞   ▲▲ Convergent boundary   —— Transform boundary

◆ **FIGURE 7.15** Earth's major earthquake zones coincide with tectonic plate boundaries. Each yellow dot represents a moderate or large earthquake that occurred between 1985 and 1995.

INTERACTIVE QUESTION: *What factors, besides earthquake frequency, would you use to locate a tectonic plate boundary?*

horizontally. A fault of this type is called a **strike-slip fault** (◆ Figure 7.17). Plate motion stresses rock adjacent to the fault, generating numerous smaller faults that threaten many urban areas. The San Andreas Fault and its associated faults form a broad region called the **San Andreas Fault zone.**

In the past few centuries, hundreds of thousands of earthquakes have occurred in this zone. Geologists of

the United States Geological Survey record about 10,000 earthquakes every year, although most are so weak that they can be detected only with seismographs. Severe quakes occur periodically. One shook Los Angeles in 1857, and another destroyed San Francisco in 1906 (◆ Figure 7.18). A large quake in 1989 occurred south of San Francisco; another rocked Northridge, part of metropolitan Los Angeles, in January 1994

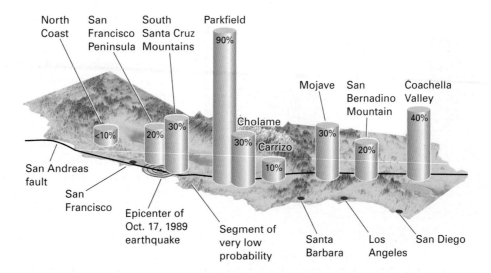

◆ **FIGURE 7.16** An earthquake hazard map of the San Andreas Fault zone of California. (Numbers equal the probability of an earthquake with a magnitude greater than 5 in the next 30 years.)

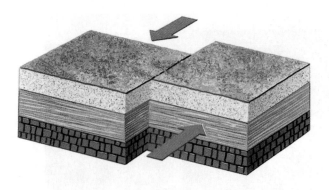

FIGURE 7.17 A strike-slip fault is vertical and the rock on opposite sides of the fracture moves horizontally.

(◀ Figure 7.19). The fact that the San Andreas Fault zone is part of a major plate boundary tells us that more earthquakes are inevitable.

The plates move past one another in three different ways along different segments of the San Andreas Fault zone:

1. Along some portions of the fault zone, rocks slip past one another at a continuous, snail-like pace called **fault creep**. The movement occurs without violent and destructive earthquakes because the rocks move continuously and slowly.

2. In other segments of the fault, the plates pass one another in a series of small hops, causing numerous small, nondamaging earthquakes.

3. Along the remaining portions of the fault, friction prevents slippage of the fault, although the plates continue to move past one another. In this case, rock near the fault deforms and stores elastic energy. Because the plates move past one another at an average of 3.5 centimeters per year,

3.5 meters of elastic deformation accumulate over a period of 100 years. When the accumulated elastic energy exceeds friction, the rock suddenly slips along the fault and snaps back to its original shape, producing a large, destructive earthquake.

Plates move slowly and silently every day. Rocks stretch, bend, and compress, unseen beneath our feet. Geologists understand the mechanisms of earthquakes and know for certain that more destructive quakes will occur along the San Andreas Fault. But no one knows when or where the "Big One" will strike.

Earthquakes at Convergent Plate Boundaries

Recall that in a subduction zone, a relatively cold, rigid lithospheric plate dives beneath another plate and slowly sinks into the mantle. In most places, the subducting plate slips past the plate above it with intermittent jerks, giving rise to numerous earthquakes. The earthquakes concentrate along the upper part of the sinking plate, where it scrapes past the opposing plate (◀ Figure 7.20). This earthquake zone is called the **Benioff zone**, after the geologist who first recognized it. Many of the world's strongest earthquakes occur in subduction zones.

The small Juan de Fuca plate, which lies off the coasts of Oregon, Washington, and southern British Columbia, is diving beneath North America at a rate of 3 to 4 centimeters per year. Thus, the region should experience subduction zone earthquakes. Yet although small earthquakes occasionally shake the Pacific Northwest, no large ones have occurred in the past 150 to 200 years.

Geologists have offered two hypotheses to explain why earthquakes are relatively uncommon in the Pacific Northwest. Subduction may be occurring slowly and continuously by fault creep. If this is the case, elastic energy would not accumulate in nearby rocks, and strong earthquakes would be unlikely. Alternatively, friction may be locking together rocks along the fault In this case, the rocks stretch or compress; thereby accumulating a huge amount of elastic energy that will be released in a giant, destructive quake sometime in the future.

Recently geologists have discovered evidence of great prehistoric earthquakes in the Pacific Northwest. A major coastal earthquake commonly creates violent sea waves, which deposit a layer of sand along the coast. Geologists have found several such sand layers, each burying a layer of peat and mud that accumulated in coastal swamps during the quiet intervals between earthquakes. In addition, they have found submarine landslide deposits lying on the deep sea floor near the coast. These deposits formed when earthquakes triggered submarine landslides that

FIGURE 7.18 The 1906 earthquake and fire destroyed most of San Francisco.

Chromosohm/Sohm/Photo Researchers

◇ **FIGURE 7.19**   On January 17, 1994, a magnitude 6.6 earthquake struck Northridge in the San Fernando Valley just north of Los Angeles, killing 55 people and causing $8 billion in property damage. This building stood on Olympic Boulevard.

carried sand and mud from the coast to the sea floor. Geologists estimate that 13 major earthquakes, separated by 300 to 900 years, struck the coast during the past 7,700 years. There is also evidence for one major historic earthquake. Oral accounts of the native inhabitants chronicle the loss of a small village in British Columbia and a significant amount of ground shaking in northern California. Thus, many geologists anticipate another

major, destructive earthquake in the Pacific Northwest during the next 600 years.

When two converging plates both carry continents, neither can sink deeply into the mantle. Instead the crust buckles up to form high mountains, such as the Himalayas. Rocks fracture or slip during this buckling process, generating frequent earthquakes. The 2005 earthquake along the India–Pakistan border, mentioned briefly in the introduction to this chapter, was generated by the slow convergence of the subcontinent of India with Asia.

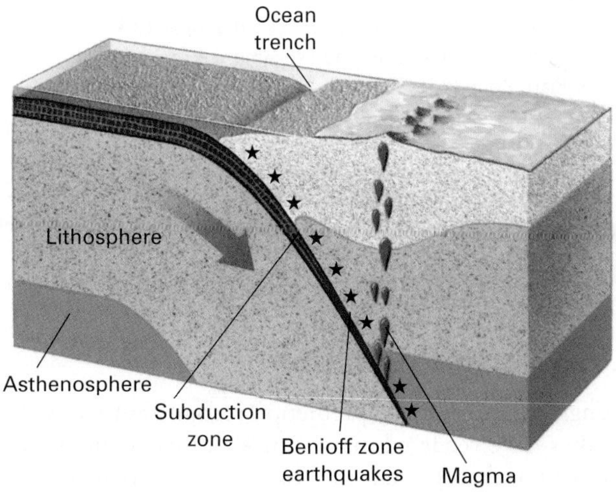

Earthquakes at Divergent Plate Boundaries

Earthquakes frequently shake the Mid-Oceanic Ridge system as a result of faults that form as the two plates separate. Blocks of oceanic crust drop downward along most mid-oceanic ridges, forming a rift valley in the center of the ridge. Only shallow earthquakes occur along the Mid-Oceanic Ridge because here the asthenosphere rises to levels as shallow as 10 to 15 kilometers below Earth's surface and is too hot and plastic to fracture.

Earthquakes in Plate Interiors

No major earthquakes have occurred in the central or eastern United States in the past 100 years, and no lithospheric plate boundaries are known in these regions. However, the largest historical earthquake sequence in the contiguous 48 states occurred near New Madrid, Missouri. In 1811 and 1812, three shocks with estimated moment magnitudes between 7.3 and 7.8 altered the

◇ **FIGURE 7.20**   A descending lithospheric plate generates magma and earthquakes in a subduction zone. Earthquake foci, shown by stars, concentrate along the upper portion of the subducting plate, called the Benioff zone.

INTERACTIVE QUESTION: *Using this illustration, show how the position of earthquake epicenters changes with the depth of the earthquake focus in a subduction zone.*

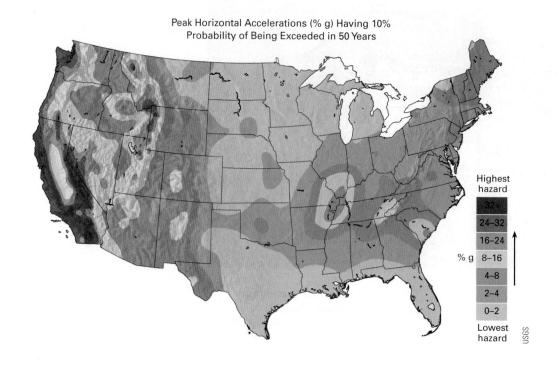

Peak Horizontal Accelerations (% g) Having 10%
Probability of Being Exceeded in 50 Years

Highest
hazard

32+
24–32
16–24
% g | 8–16
4–8
2–4
0–2

Lowest
hazard

USGS

◆ **FIGURE 7.21** This map shows earthquake hazards in the United States. The predictions are based on records of frequency and magnitude of historical earthquakes (% *g* equals percent of the acceleration of gravity).

course of the Mississippi River and rang church bells 1,500 kilometers away in Washington, D.C.

Earthquakes in plate interiors are not as well understood as those at plate boundaries, but modern research is revealing some clues. Tectonic forces stretched North America in Precambrian time. As the continent pulled apart, rock fractured to create two huge fault zones that crisscross the continent like a giant X. Although the fault zones failed to develop into a divergent plate boundary, they remain weaknesses in the lithosphere. New Madrid lies at a major intersection of the faults—the center of the X. As the North American plate glides over the asthenosphere, it may pass over irregularities, or "bumps," in that plastic zone, causing slippage and earthquakes along the deep faults. In addition to the quakes recorded in New Madrid in 1811 and 1812, major earthquakes occurred close to the years 1600, 1300, and 900.

In 2005, researchers reported in the journal *Nature* that the rocks adjacent to the New Madrid fault are currently deforming about as fast as rocks adjacent to active tectonic plate boundaries. This deformation indicates that additional earthquakes are likely in the future. [2]

2. Smalley et al., *Nature* 435, June 23, 2005, 1088 ff.

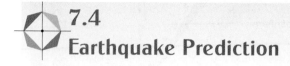

ThomsonNOW™

CLICK ThomsonNOW Interactive to work through these activities:
• Earthquakes in Space and Time and Seismic Risk Through Earthquakes and Tsunamis
• Plate Boundaries Through Plate Tectonics

7.4
Earthquake Prediction

Long-Term Prediction

Long-term earthquake prediction recognizes that earthquakes have recurred many times along existing faults and will probably occur in these same regions again. For example, in the United States, although some faults exist in plate interiors (as in the New Madrid earthquake zone), the most active faults lie along the west coast at plate boundaries (◆ Figure 7.21).

Long-term prediction tells us *where* earthquakes are likely to occur. This information is useful because it allows engineers to establish strict building codes in earthquake-prone regions (see Section 7.5 that follows). However, long-term prediction gives only a vague idea of *when* the next earthquake will strike. Near New Madrid, earthquakes have occurred separated by 400, 300, and 200 years. The last one struck in 1811. Based on an average 300-year interval, one might expect a quake in the year 2100, but such an analysis could be in error by hundreds of years.

Short-Term Prediction

Short-term predictions are forecasts that an earthquake may occur at a specific place and time. Short-term prediction depends on signals that immediately precede an earthquake.

Foreshocks are small earthquakes that precede a large quake by a few seconds to a few weeks. The cause of foreshocks can be explained by a simple analogy. If you try to break a stick by bending it slowly, you may hear a few small cracking sounds just before the final snap. If foreshocks consistently preceded major earthquakes, they would be a reliable tool for short-term prediction. However, foreshocks preceded only about half of a group of recent major earthquakes selected for study. In addition, some swarms of small shocks thought to be foreshocks were not followed by a large quake.

Another approach to short-term earthquake prediction is to measure changes in the land surface near an active fault zone. Seismologists monitor unusual Earth movements with tiltmeters and laser surveying instruments because distortions of the crust may precede a major earthquake. This method has successfully predicted some earthquakes, but in other instances predicted quakes did not occur or quakes occurred that had not been predicted.

Other types of signals can be used in short-term prediction. When rock is deformed prior to an earthquake, microscopic cracks may develop as the rock approaches its rupture point. In some cases, the cracks release radon gas previously trapped in rocks and minerals. The cracks may fill with water and cause the water levels in wells to fluctuate. Air-filled cracks do not conduct electricity as well as solid rock, so the electrical conductivity of rock decreases as microscopic cracks form.

In January 1975, Chinese geophysicists recorded swarms of foreshocks and unusual bulges near the city of Haicheng, which had a previous history of earthquakes. When the foreshocks became intense on February 1, authorities evacuated portions of the city. The evacuation was completed on the morning of February 4, and in the early evening of the same day, an earthquake destroyed houses, apartments, and factories but caused few deaths.

After that success, geologists hoped that a new era of earthquake prediction had begun. But a year later, Chinese scientists failed to predict an earthquake in the adjacent city of Tangshan. This major quake was not preceded by foreshocks, so no warning was given, and 240,000 people died. Over the past few decades, short-term prediction has not been reliable. Although some geologists continue to search for reliable indicators for short-term earthquake prediction, many other geologists believe that this goal will remain elusive.

While geologists cannot reliably predict an earthquake hours, days, or weeks before the event, it is possible to predict earth movement a few seconds before buildings begin to shake and topple. This prediction is based on two principles: (A) If a ring of seismographs is built around the outside perimeter of a city, the seismographs would detect most earthquakes before they reached the city. A warning could then be transmitted to the city electronically, because electronic signals travel faster than earthquake waves. (B) Second, P waves travel faster than either S waves or surface waves. However, P waves are less destructive than the other two types. Therefore, when seismographs detect the P waves, they can be programmed to issue an automated warning a few seconds before the more-destructive waves arrive. Combining these two approaches, Japanese scientists were able to broadcast accurate warnings 16 seconds before an August 25, 2005, magnitude 7.2 earthquake in Sendai, Japan. Sixteen seconds may not sound like much, but the warning signals, in theory, could trigger computerized control systems to shut off gas mains, high-tension electric transmission lines, sensitive equipment in chemical factories and refineries, and mass transit systems. In addition, controllers could abort airplane landings, alert physicians performing surgery, and trigger emergency signals that could alert motorists on freeways. All of these procedures could save lives and material damage. Unfortunately, in the Sendai, Japan, earthquake, while the warning system functioned flawlessly so that city planners knew that an earthquake was imminent, the computerized controls and public warning systems were not adequately in place, so there was little response to the warning. City planners continue to improve the system.

7.5 Earthquake Damage and Hazard Mitigation

Violent ground motion may toss a person to the ground and break his arm, but this motion, by itself, is seldom lethal. Most earthquake fatalities and injuries occur when falling structures crush people. Structural damage, injury, and death depend on the magnitude of the quake, its proximity to population centers, rock and soil types, topography, and the quality of construction in the region.

◀▶ **FIGURE 7.22** The 1985 Mexico City earthquake had a moment magnitude of 8.1 and killed between 8,000 and 10,000 people. Earthquake waves amplified within the soil to destroy many buildings in the city.

How Rock and Soil Influence Earthquake Damage

In many regions, bedrock lies at or near Earth's surface and buildings are anchored directly to the rock. Bedrock vibrates during an earthquake and buildings may fail if the motion is violent enough. However, most bedrock returns to its original shape when the earthquake is over, so if structures can withstand the shaking, they will survive. Thus, bedrock forms a desirable foundation in earthquake-hazard areas.

In many places, structures are built on sand, silt, or clay. Sandy sediment and soil commonly settle during an earthquake. This displacement tilts buildings, breaks pipelines and roadways, and fractures dams. To avert structural failure in such soils, engineers drive steel or concrete pilings through the sand to the bedrock below. These pilings anchor and support the structures, even if the ground beneath them settles.

Mexico City provides one example of what can happen to clay-rich soils during an earthquake. The city is built on a high plateau ringed by even higher mountains. When the Spaniards invaded central Mexico in the sixteenth century, lakes dotted the plateau and the Aztec capital lay on an island at the end of a long causeway in one of the lakes. Over the following centuries, European settlers drained the lake and built the modern city on the water-soaked, clay-rich, lake-bed sediment. On September 19, 1985, an earthquake with a moment magnitude of 8.1 struck about 500 kilometers west of the city. Seismic waves shook the wet clay beneath the city and reflected back and forth between the bedrock sides and bottom of the basin, just as waves in a bowl of Jello™ bounce off the side and bottom of the bowl. The reflections amplified the waves, which destroyed more than 500 buildings and killed between 8,000 and 10,000 people (◀▶ Figure 7.22). Meanwhile, there was comparatively little damage in Acapulco, a city much closer to the epicenter but built on bedrock.

If soil is saturated with water, the sudden shock of an earthquake can cause the grains to shift closer together, expelling some of the water. When this occurs, increased stress is transferred to the pore water, and the pore pressure may rise sufficiently to suspend the grains in the water. In this case, the soil loses its shear strength and behaves as a fluid. This process is called **liquefaction**. When soils liquefy on a hillside, the mixture of water and soil, called a slurry, flows downslope, carrying structures along with it. During the 1964 earthquake near Anchorage, Alaska, a clay-rich bluff 2.8 kilometers long, 300 meters wide, and 22 meters high liquefied. The slurry carried houses into the ocean and buried some so deeply that bodies were never recovered.

Construction Design and Earthquake Damage

On December 22, 2003, a magnitude-6.5 earthquake struck in central California, 11 kilometers from San Simeon. Two people were killed and 40 buildings severely damaged. Four days later, a magnitude-6.6 earthquake in southeastern Iran, 185 kilometers from the city of Kerman, killed 30,000 people, injured another 30,000, and destroyed 85 percent of the buildings in the city.

On August 16, 2005, a magnitude-7.2 earthquake struck 80 kilometers from Sendai, Japan. There were 70 injuries and no deaths. Two months later, a 7.6-magnitude earthquake shook a mountainous region along the India–Pakistan border, killing 79,000, injuring 65,000, and leaving 3 million people homeless (◀▶ Figure 7.23).

Most earthquake mortality occurs when buildings fall and crush the unlucky inhabitants. Structure failure is caused by a variety of factors, including soil type and distance from the epicenter. But the catastrophically tremendous differences between mortality in the poor, less-developed countries and in the rich, developed countries is due to one simple factor—quality of construction.

Some common framing materials used in buildings, such as wood and steel, bend and sway during an earthquake, but they resist failure. However, brick, stone, adobe (dried mud), and other masonry products are brittle and likely to fail during an earthquake. Although masonry can be reinforced with steel, in many regions of the world people cannot afford such structural reinforcement.

◀▶ **FIGURE 7.23** Survivors search for bodies in Muzaf-farabad, capital of Pakistani Kashmir, following the October 8, 2005, magnitude-7.6 earthquake that shook the mountainous region along the India–Pakistan border, killing 79,000 people, injuring 65,000, and leaving 3 million homeless.

Thus earthquake mortality is an example of the interface between natural systems and human ones. Forces deep within Earth, beyond human control, drive earthquake frequency and magnitude. But earthquake mortality is closely related to political and economic systems. Throughout the remainder of this book, we will study natural disasters such as volcanic eruptions, floods, and hurricanes. In each case we will repeat the same theme: We can mitigate the consequences of disaster by adapting our human systems to natural ones.

Fire

Earthquakes commonly rupture buried gas pipes and electrical wires, leading to fire, explosions, and electrocutions. Water pipes may also break, hampering firefighters. Most of the damage from the 1906 San Francisco earthquake resulted from fires.

Again, design features can reduce fire risk. A straight pipe is likely to rupture if the ground stretches and compresses, whereas a zigzag pipe may flex like an accordion and remain intact.

Landslides

Landslides are common when Earth trembles. Earthquake-related landslides are discussed in more detail in Chapter 10.

Tsunamis

When an earthquake occurs beneath the sea, part of the sea floor rises or falls. Water is displaced in response to the rock movement, forming a wave (◀▶ Figure 7.24). Sea waves produced by an earthquake are often called tidal waves, but they have nothing to do with tides. Therefore, geologists call them by their Japanese name, **tsunami**.

In the open sea, a tsunami is so flat that it is barely detectable. Typically, the crest may be only 1 to 3 meters high, and successive crests may be more than 100 to 150 kilometers apart. However, a tsunami may travel at 750 kilometers per hour. When the wave approaches the shallow water near shore, the base of the wave drags against the bottom and the water stacks up, increasing the height of the wave. The rising wall of water then flows inland. A tsunami can flood the land for as long as 5 to 10 minutes.

For 80 million years, the Indian–Australian Plate has been migrating northeast, driving into the Asian continent. Over vast millennia, as dinosaurs thrived and then became extinct, as mammals evolved, as humans arose out of the African savannah to eventually build rocket ships and computers, this slow, inexorable motion of tectonic plates has caused the Himalayas to rise.

Today the central part of the Indian–Australian Plate is subducting beneath the islands of Sumatra and Java. On December 26, 2004, approximately 1,200 kilometers of rock along the subducting sea floor slipped suddenly, with vertical movement up to 15 meters (◀▶ Figure 7.25).

A Earthquake! Sea floor drops, sea level falls with it

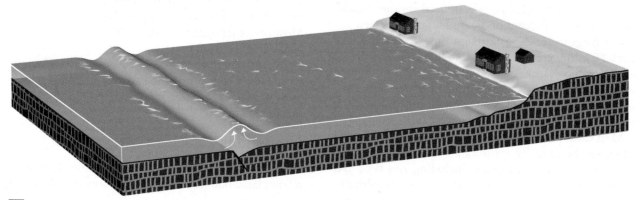

B Water rushes into low spot, and overcompensates, creating a bulge

Long shallow waves in open sea

Waves steepen and rise in shallow water

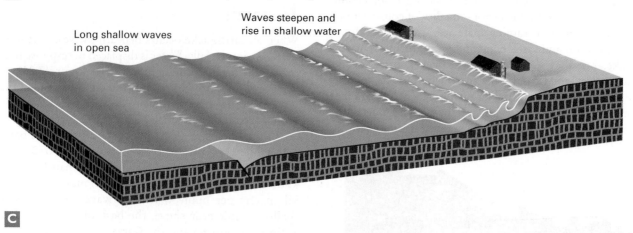

C

ThomsonNOW ◁▷ **ACTIVE FIGURE 7.24** A tsunami develops when part of the sea floor drops during an earthquake. Water rushes to fill the low spot, but the inertia of the rushing water forces too much water into the area, creating a bulge in the water surface. The long, shallow waves can build up into destructive giants when they reach shore.

The resulting earthquake of magnitude 9.0 to 9.3 lasted for almost 10 minutes and was the second-largest seismic event ever recorded, second only to the 1960 Chilean quake. To appreciate the magnitude, imagine a fault line stretching from New York to Chicago suddenly moving vertically as high as a five-story building. This tremendous displacement of rock initiated a massive tsunami that radiated in all directions, killing an estimated 283,000 people along the Indian Ocean coastlines (◁▷ Figure 7.26). Survivors reported that, moments prior to the deadly wave, coastal water retreated, exposing dry mud in ocean bays. Then the wave raced inward, rearing upward as much as 30 meters, as high as a ten-story building. Although mortality was highest in Sumatra, people died in coastal areas as far away as Port Elizabeth, South Africa, 8,000 km from the epicenter.

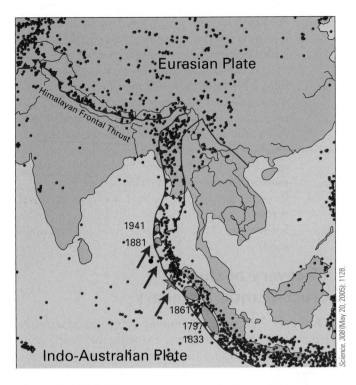

Science, 308 (May 20, 2005): 1128.

◇▶ **FIGURE 7.25** Fault map of the Indo Sumatra-Andaman earthquake region. Green star shows the center of the December 2004 Sumatra-Andaman earthquake. Red dots show earthquakes greater than magnitude 5.0 from 1965 to December 2004, and the red arrows show plate motion. The frequency of past earthquakes and the continuing motion of the plates guarantee that more earthquakes will occur in the region in the near future.

As we will learn in Chapter 16, coastlines are desirable places to live and population density is high near the sea. From our study of plate tectonics in Chapter 6, we learned that, in addition to the Sumatra area, there are active subduction zones along many of the world's coastlines, including western South America, the northwest coast of North America, most of the east coast of Asia, the Aleutians, and southwest Alaska. All are vulnerable to tsunami disasters. We can't prevent tsunamis, but we can, hopefully, mitigate loss of life. An early-warning system, similar to that discussed for earthquakes, could alert coastal inhabitants several minutes ahead of the deadly wave, giving people time to retreat to higher ground. In addition, barrier islands and coral reefs cause waves to break and dissipate their energy offshore. Thus preservation of these natural features protects populated regions along the coast. This issue will be discussed further in Chapter 16 and again in Chapter 19, with reference to the Hurricane Katrina disaster.

Cost–Benefit Analysis

As explained previously, engineers can reduce damage and loss of life by building earthquake-resistant structures in high-hazard zones. The problems lie in defining the terms *earthquake-resistant* and *high hazard* and then balancing risk assessment, benefit, and cost.

For example, engineers can construct a highway bridge to withstand a magnitude-8.0 quake with a

AP/Wide World Photos

◇▶ **FIGURE 7.26** A lone mosque stands in a destroyed coastal village near Aceh, Sumatra island, Indonesia, following the December 2004, tsunamis that killed people in 11 countries from Asia to Africa.

nearby epicenter. However, the bridge would be very expensive to build and the risk is small that a major earthquake will occur near *any particular* bridge. So city planners build bridges that will withstand a small nearby quake or a larger quake with a more distant epicenter. If a major quake strikes, bridges near the epicenter will fail, while those farther away will survive. Some people will die but construction costs will be affordable.

People cannot agree on how much protection we should buy. In a city such as Los Angeles, where a large quake is likely in our lifetimes, we must ask, how much is human life worth? In the New Madrid region, where historical frequency has been low but past quakes have been violent, we ask, how much are we willing to gamble that another large quake will not occur within the expected lifespan of a building or bridge? New York and Boston do not lie near any plate boundaries, but quakes up to magnitude 6.5 have occurred in the northeast. The problems are exacerbated in a locale like the Gujarat region of India, where people are poor and earthquakes are violent but infrequent. So how much should we spend on earthquake-resistant design in each of these regions? There are no definite answers; we can only do our best to develop budgets that balance geology, chance, and cost.

7.6 Studying Earth's Interior

Recall that Earth is composed of a thin crust, a thick mantle, and a core. The three layers are distinguished by different chemical compositions. In turn, both the mantle and core contain finer layers based on changing physical properties. Scientists have learned a remarkable amount about Earth's structure, even though the deepest well is only a 12-kilometer-deep hole in northern Russia. Scientists deduce the composition and properties of Earth's interior by studying the behavior of seismic waves. Some of the principles necessary for understanding the behavior of seismic waves are as follows:

1. In a uniform, homogeneous medium, a wave radiates outward in concentric spheres and at constant velocity.
2. The velocity of a seismic wave depends on the nature of the material that it travels through. Thus seismic waves travel at different velocities in different types of rock. In addition, within a type of rock, wave velocity varies with changing rigidity and density of that rock.

3. When a wave passes from one material to another, it refracts (bends) and sometimes reflects (bounces back). Both refraction and reflection are easily seen in light waves. If you place a pencil in a glass half filled with water, the pencil appears bent. Of course the pencil does not bend; the light rays do. Light rays slow down when they pass from air to water, and as the velocity changes, the waves refract. If you look in a mirror, the mirror reflects your image. In a similar manner, boundaries between Earth's layers refract and reflect seismic waves.
4. P waves are compressional waves and travel through gases, liquids, and solids, whereas S waves are shear waves and travel only through solids.

Discovery of the Crust–Mantle Boundary

◆ Figure 7.27 shows that some waves travel directly from the earthquake focus, through the crust, to a nearby seismograph. Others travel downward into the

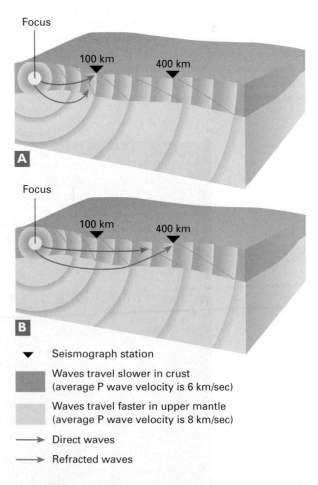

◆ **FIGURE 7.27** The travel paths of seismic waves. **(A)** The direct waves reach the closer seismic station first. **(B)** The refracted waves reach a more distant seismic station before the direct waves. Even though the refracted waves travel a longer distance, they travel at a higher speed through the mantle.

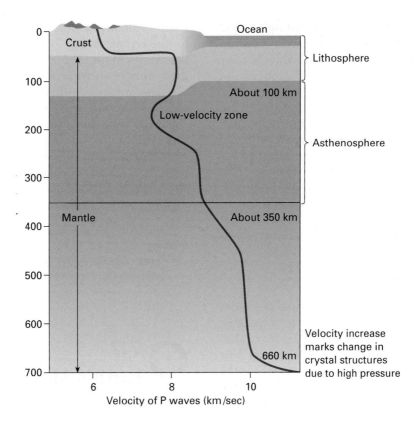

FIGURE 7.28 Velocities of P waves in the crust and the upper mantle. As a general rule, the velocity of P waves increases with depth. However, in the asthenosphere, the temperature is high enough and the pressure is low enough that rock is plastic. As a result, seismic waves slow down in this region, called the low-velocity zone. Wave velocity increases rapidly at the 660-kilometer discontinuity, the boundary between the upper and the lower mantle, probably because of a change in mineral content due to increasing pressure.

mantle and then refract back upward to the same seismograph. The route through the mantle is longer than that through the crust. However, seismic waves travel faster in the mantle than they do in the crust. Over a short distance (less than 300 kilometers), waves traveling through the crust arrive at a seismograph before those following the longer route through the mantle. However, for longer distances, the longer route through the mantle is faster because waves travel more quickly in the mantle.

The situation is analogous to the two different routes you may use to travel from your house to a friend's. The shorter route is a city street where traffic moves slowly. The longer route is an interstate highway, but you have to drive several kilometers out of your way to get to the highway and then another few kilometers from the highway to your friend's house. If your friend lives nearby, it is faster to take the city street. But if your friend lives far away, it is faster to take the longer route and make up time on the highway.

In 1909, **Andrija Mohorovičić** discovered that seismic waves from a distant earthquake traveled more rapidly than those from a nearby earthquake. By analyzing the arrival times of earthquake waves to many different seismographs, Mohorovičić identified the boundary between the crust and the mantle. Today, this boundary is called the **Mohorovičić discontinuity,** or the **Moho,** in honor of its discoverer.

The Moho lies at a depth ranging from 4 to 70 kilometers. As explained previously, oceanic crust is thinner than continental crust, and continental crust is thicker under mountain ranges than it is under plains.

The Structure of the Mantle

The mantle is almost 2,900 kilometers thick and composes about 80 percent of Earth's volume. Much of our knowledge of the composition and structure of the mantle comes from seismic data. As explained earlier, seismic waves speed up abruptly at the crust–mantle boundary (◀▶ Figure 7.28). Seismic waves slow down again when they enter the asthenosphere at a depth between 75 and 125 kilometers. The plasticity and partially melted character of the asthenosphere slow down the seismic waves. At the base of the asthenosphere 350 kilometers below the surface, seismic waves speed up again because increasing pressure overwhelms the temperature effect and the mantle becomes stronger and less plastic.

At a depth of about 660 kilometers, seismic wave velocities increase again because pressure is great enough that the minerals in the mantle recrystallize to form denser minerals. The zone where the change occurs is called the **660-kilometer discontinuity**. The base of the mantle lies at a depth of 2,900 kilometers. Recent research has indicated that the base of the mantle, at the core–mantle boundary, may be so hot that despite the tremendous pressure, rock in this region is liquid.

Discovery of the Core

Using a global array of seismographs, seismologists detect direct P and S waves up to 105° from the focus of an earthquake. Between 105° and 140° is a "shadow zone" where no direct P waves arrive at Earth's surface. This shadow zone is caused by a discontinuity, which is the mantle–core boundary. When P waves pass from the mantle into the core, they are refracted, or bent as shown in ◆ Figure 7.29. The refraction deflects the P waves away from the shadow zone.

No S waves arrive beyond 105°. Their absence in this region shows that they do not travel through the outer core. Recall that S waves are not transmitted through liquids. The failure of S waves to pass through the outer core indicates that the outer core is liquid.

Refraction patterns of P waves, shown in ◆ Figure 7.29, indicate that another boundary exists within the core. It is the boundary between the liquid outer core and the solid inner core. Although seismic waves tell us that the outer core is liquid and the inner core is solid, other evidence tells us that the core is composed of iron and nickel.

More detailed studies, conducted in the 1980s and 1990s, show that seismic waves travel at different speeds in different directions through the inner core—thus the inner core is not homogeneous. One series of measurements suggests that the inner core, lying within its liquid sheath, is rotating at a significantly faster rate than the mantle and crust. Other researchers have proposed that the solid inner core may convect just as the solid mantle does.

Density Measurements

The overall density of Earth is 5.5 grams per cubic centimeter (g/cm³), but both crust and mantle have average densities less than this value. The density of the crust ranges from 2.5 to 3.0 g/cm³, and the density of the mantle varies from 3.3 to 5.5 g/cm³. Because the mantle and crust account for slightly more than 80 percent of

Earth's volume, the core must be very dense to account for the average density of Earth. Calculations show that the density of the core must be 10 to 13 g/cm³, which is the density of many metals under high pressure.

Many meteorites are composed mainly of iron and nickel. Cosmologists think that meteorites formed at about the same time that the solar system did and that they reflect the composition of the primordial solar system. Because Earth coalesced from meteorites and similar objects, scientists believe that iron and nickel must be abundant on Earth. Therefore, they conclude that the metallic core is composed of iron and nickel.

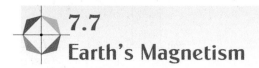

CLICK ThomsonNOW Interactive to work through an activity on Core Studies Through Earth's Layers.

7.7 Earth's Magnetism

Early navigators learned that no matter where they sailed, a needle-shaped magnet aligned itself in an approximately north–south orientation. From this observation, they learned that Earth has a magnetic north pole and a magnetic south pole (◆ Figure 7.30).

Earth's interior is too hot for a permanent magnet to exist. Instead, Earth's magnetic field is probably electromagnetic in origin. If you wrap a wire around a nail and connect the ends of the wire to a battery, the nail becomes magnetized and can pick up small iron objects. The battery causes electrons to flow through the wire, and this flow of electrical charges creates the electromagnetic field.

Most likely, Earth's magnetic field is generated within the outer core. Metals are good conductors of electricity and the metals in the outer core are liquid and very mobile. Two types of motion occur in the liquid outer core: (1) Because the outer core is much hotter at its base than at its surface, the liquid metal convects. (2) The rising and falling metals are then deflected by Earth's spin. These convecting, spinning liquid conductors are thought to generate Earth's magnetic field.

Magnetic fields are common in planets, stars, and other objects in space. Our solar system almost certainly possessed a weak magnetic field when it first formed. The flowing metals of Earth's liquid outer core amplified some of this original magnetic force to initiate Earth's magnetic field. Earth's magnetic field is approximately aligned with Earth's rotational axis because Earth's spin affects the flow of metal in the outer core.

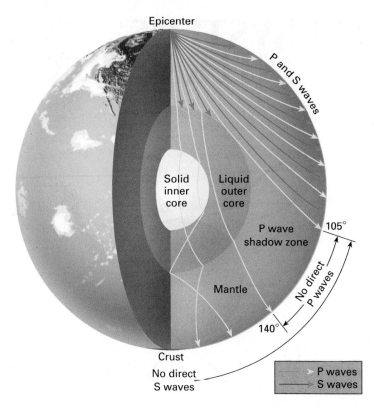

Epicenter

P and S waves

Solid inner core

Liquid outer core

P wave shadow zone

105°

Mantle

No direct P waves

140°

Crust

No direct S waves

P waves
S waves

◆ **ACTIVE FIGURE 7.29** Cross section of Earth showing paths of seismic waves. They bend gradually because of increasing pressure with depth. They also bend sharply where they cross major layer boundaries in Earth's interior.

Note that S waves do not travel through the liquid outer core, and therefore direct S waves are only observed within an arc of 105° of the epicenter.

P waves are refracted sharply at the core–mantle boundary, so there is a shadow zone of no direct P waves from 105° to 140°.

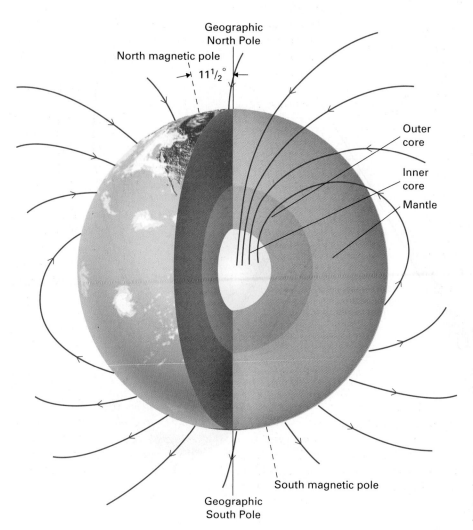

Geographic North Pole

North magnetic pole

11½°

Outer core

Inner core

Mantle

South magnetic pole

Geographic South Pole

◆ **FIGURE 7.30** The magnetic field of Earth. Note that the magnetic North Pole is offset 11.5° from the geographic pole.

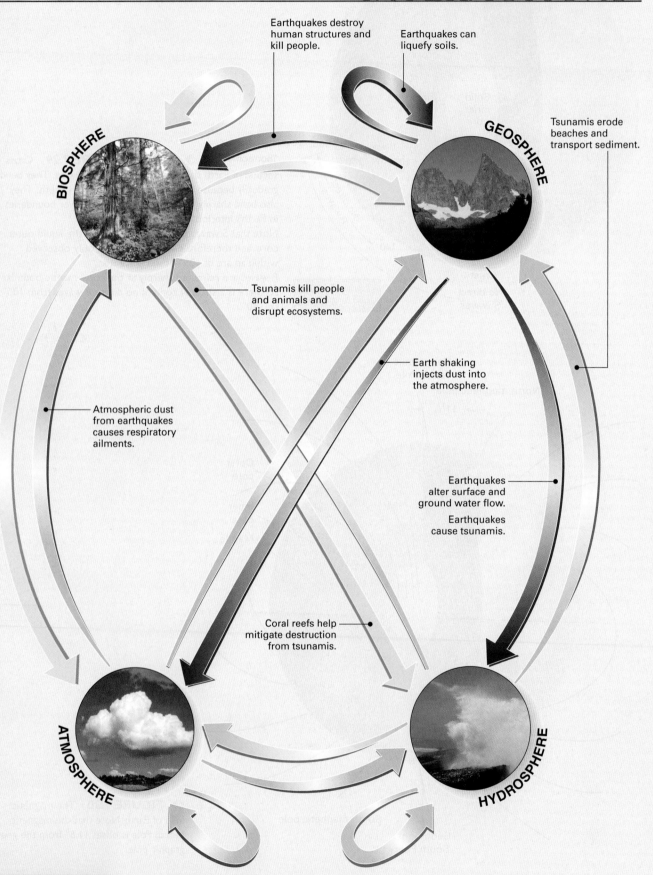

Earthquakes destroy human structures and kill people.

Earthquakes can liquefy soils.

Tsunamis erode beaches and transport sediment.

BIOSPHERE

GEOSPHERE

Tsunamis kill people and animals and disrupt ecosystems.

Earth shaking injects dust into the atmosphere.

Atmospheric dust from earthquakes causes respiratory ailments.

Earthquakes alter surface and ground water flow.

Earthquakes cause tsunamis.

Coral reefs help mitigate destruction from tsunamis.

ATMOSPHERE

HYDROSPHERE

AS DISCUSSED IN the introduction, earthquakes are one manifestation of plate tec̶
motion that shapes and alters all other Earth systems. In addition to the large-scale exa̶
listed in the introduction, a great many smaller-scale systems interactions occur. For exa̶
after the 1994 Northridge earthquake in California, doctors in the region noted an inc̶
in respiratory diseases among local residents. Shaking ground had introduced dust int̶
atmosphere (a geosphere to atmosphere interaction), which in turn affected people and̶
sumably, other animals of the biosphere. As another example, earthquakes frequently ̶
streambeds, thus affecting the hydrosphere. ■

SUMMARY

An **earthquake** is a sudden motion or trembling of Earth caused by the abrupt release of slowly accumulated energy in rocks. Most earthquakes occur along tectonic plate boundaries. Earthquakes occur either when the elastic energy accumulated in rock exceeds the friction that holds rock along a fault, or when the elastic energy exceeds the strength of the rock and the rock breaks.

Seismology is the study of earthquakes and the nature of Earth's interior based on evidence from seismic waves. An earthquake starts at the initial point of rupture, called the **focus**. The location on Earth's surface directly above the focus is the **epicenter**. Seismic waves include **body waves**, which travel through the interior of Earth, and **surface waves**, which travel on the surface. **P waves** are compressional body waves. **S waves** are body waves that travel slower than P waves. They consist of a shearing motion and travel through solids but not liquids. **Surface waves** travel more slowly than either type of body wave. Seismic waves are recorded on a **seismograph**; a record of earth vibration is called a **seismogram**. Early in this century, geologists used the **Mercalli scale** to record the extent of earthquake damage. The **Richter scale** expresses earthquake magnitudes. Modern geologists use the **moment magnitude** scale to record the energy released during an earthquake. The distance from a seismic station to an earthquake is calculated by constructing a time-travel curve, which calibrates the difference in arrival times between S and P waves. The epicenter can be located by measuring the distance from three or more seismic stations.

Earthquakes are common at all three types of plate boundaries. The **San Andreas Fault** zone is an example of a **strike-slip fault** along a transform plate boundary. Along some portions of the fault zone, rocks slip past one another at a continuous, snail-like pace called **fault creep**. In other regions, friction prevents slippage until elastic deformation builds and is eventually released in a large earthquake. Subduction zone earthquakes occur along the **Benioff zone** when the subducting plate slips suddenly. Earthquakes occur at divergent plate boundaries as blocks of lithosphere along the fault drop downward. Earthquakes occur in plate interiors along old faults.

Long-term earthquake prediction is based on the observation that most earthquakes occur on preexisting faults at tectonic plate boundaries. Short-term prediction is based on occurrences of **foreshocks**, release of radon gas, and changes in the land surface, the water table, and electrical conductivity. Earthquake damage is influenced by rock and soil type, construction design, and the likelihood of fires, landslides, and **tsunamis**.

Earth's internal structure and properties are known by studies of earthquake wave velocities and refraction and reflection of seismic waves as they pass through Earth. The **Mohorovičić discontinuity**, or the **Moho**, lies at a depth ranging from 4 to 70 kilometers. Oceanic crust is thinner than continental crust, and continental crust is thicker under mountain ranges than it is under plains. The mantle is almost 2,900 kilometers thick and composes about 80 percent of Earth's volume. The upper portion of the mantle is part of the hard and rigid lithosphere. Beneath the lithosphere, the asthenosphere is plastic and partially melted. Beneath the lithosphere, the mantle is stronger and less plastic; at the **660-kilometer discontinuity**, the mineral composition of the mantle changes. Wave and density studies show that the outer core is liquid iron and nickel and the inner core is solid iron and nickel. Flowing metal in the outer core generates Earth's magnetic field.

Key Terms

plastic deformation 156	S wave 158	fault creep 164
earthquake 156	seismograph 159	Benioff zone 164
seismic wave 157	seismogram 159	foreshock 167
seismology 157	Mercalli scale 159	liquefaction 168
focus 157	Richter scale 160	tsunami 169
epicenter 157	moment magnitude 160	Mohorovičić discontinuity (Moho)
body wave 157	time-travel curve 161	173
surface wave 157	strike-slip fault 163	660-kilometer discontinuity 174
P wave 157	San Andreas Fault zone 163	

For Review

1. Explain how energy is stored prior to, and then released during, an earthquake.

2. Give two mechanisms that can release accumulated elastic energy in rocks.

3. Why do most earthquakes occur at the boundaries between tectonic plates? Are there any exceptions?

4. Define *focus* and *epicenter*.

5. Discuss the similarities and differences among P waves, S waves, and surface waves.

6. Explain how a seismograph works. Sketch what an imaginary seismogram would look like before and during an earthquake.

7. Describe the similarities and differences between the Richter and moment magnitude scales. What is actually measured and what information is obtained?

8. Describe how the epicenter of an earthquake is located.

9. Discuss earthquake mechanisms at the three types of tectonic plate boundaries.

10. What is the Benioff zone? At what type of tectonic boundary does it occur?

11. Why do only shallow earthquakes occur along the Mid-Oceanic Ridge?

12. Discuss earthquake mechanisms at plate interiors.

For Discussion

1. Which of the following materials is elastic, which is completely rigid: tennis ball, rubber band, steel bar, rock, brick?

2. The seismograph shown in Figure 7.10 measures up-and-down earth movement. Draw a diagram for a seismograph that would record side-to-side motion. Explain how your seismograph works.

3. Using the graph in Figure 7.13 determine how far away from an earthquake you would be if the first P wave arrived 5 minutes before the first S wave.

4. Using a map of the United States, locate an earthquake that is 1,000 kilometers from Seattle, 1,300 kilometers from San Francisco, and 700 kilometers from Denver.

5. Mortality was high in the Gujarat earthquake in 1993 because the quake occurred at night when people were sleeping in their homes. However, mortality in the Northridge earthquake was low because it occurred early in the morning rather than during rush hour. Is there a contradiction in these two statements?

6. Imagine that geologists predict a major earthquake in a densely populated region. The prediction may be right or it may be wrong. City planners may heed it and evacuate the city or ignore it. The possibilities lead to four combinations of predictions and responses, which can be set out in a grid as follows:

		Will the predicted earthquake really occur?	
		Yes	No
Is the city evacuated?	Yes		
	No		

For example, the space in the upper-left corner of the grid represents the situation in which the predicted earthquake occurs and the city is evacuated. For each space in the square, outline the consequences of that sequence of events.

7. Discuss the trade-off between money and human lives when considering construction in earthquake-prone zones. Would you be willing to pay twice as much for a house that was earthquake-resistant, as compared with a normal house? Would you be willing to pay more taxes for safer highway bridges? Do you feel that different types of structures (e.g., residential homes and apartments, commercial buildings, nuclear power plants) should be built to different safety standards?

8. Argue for or against placing stringent building codes for earthquake-resistant design in the Seattle area. Would your answer be different for Boston? Houston?

9. From Figure 7.28, what is the speed of P waves at a depth of 25 kilometers? 200 kilometers? 500 kilometers?

10. Give two reasons why the Earth's magnetic field cannot be formed by a giant bar magnet within the Earth's core.

ThomsonNOW

Assess your understanding of this chapter's topics with additional comprehensive interactivities at **http://www.thomsonedu.com/login**, which also has up-to-date web links, additional readings, and exercises.

CHAPTER 8

Volcanoes and Plutons

AP/Wide World Photos

Mount Etna, Sicily, Italy, erupts during the night of March 23, 2000. The volcano has the longest record of documented eruptions among all Earth volcanoes, dating back to 1500 BC. Since then, the volcano has erupted about 200 times and has been very active in recent decades.

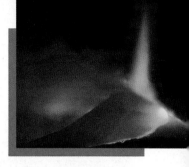

A volcanic eruption can be one of the most violent of all geologic events. Since the year 1815 when the eruption of Mount Tambora in Indonesia caused 92,000 fatalities, eruptions have killed approximately 250,000 people and have caused more than $10 billion in damage. Some violent eruptions have buried towns and cities in hot lava or volcanic ash. The 1902 eruption of Mount Pelée on the Caribbean island of Martinique buried the city of Saint Pierre in red-hot volcanic ash that killed 29,000 people. The 1980 eruption of Mount St. Helens in Washington State and the 1991 Mount Pinatubo eruption in the Philippines were similarly violent events. On June 25, 1997, the Soufrière Hills volcano on the tiny Caribbean island of Montserrat erupted to kill 19 people in the most destructive event in a five-year-long sequence of escalating volcanic crises. An ash plume rose 10 kilometers over the volcano, and volcanic ash buried 4 square kilometers of the island. Other volcanoes erupt gently. Tourists flock to Hawaii and Iceland to photograph fire fountains erupting into the sky, and geologists study advancing lava flows in relative safety (◀▶ Figure 8.1).

Volcanic eruptions can trigger other deadly events. The 1815 Tambora eruption destroyed crops and caused death by starvation of 92,000 people. The 1883 eruption of Krakatoa in the southwest Pacific Ocean generated tsunamis that killed 36,000 people. In 1985, a small eruption of Nevado del Ruiz in Colombia triggered a mud flow that buried the town of Armero, killing more than 22,000 people.

In contrast to the high visibility of a volcanic eruption, we cannot see plutonic rocks form because they crystallize within the crust, commonly at depths of 5 to 20 kilometers below the surface. However, tectonic forces have raised these deeply buried igneous rocks toward Earth's surface, and erosion has exposed them in many of the world's greatest mountain ranges.

California's Sierra Nevada, the Rockies, the Appalachians, the European Alps, and the Himalayas all contain vast quantities of plutonic rocks. ■

8.1
Magma

In Chapter 3 we stated that rocks melt in certain environments to form magma. This process is one example of the constantly changing nature of rocks described by the rock cycle. Why do rocks melt, and in what environments does magma form?

Processes that Form Magma

Recall that the asthenosphere is the layer in the upper mantle that extends from a depth of about 100 kilometers to 350 kilometers. In that layer the combined effects of temperature and pressure are such that 1 or 2 percent of the mantle rock is molten, as explained in Chapter 6. Although the majority of the asthenosphere is solid rock, it is so hot and so close to its melting point that relatively small changes can melt large volumes of rock. In order to understand volcanoes and plutonic rocks, we must first understand three processes that melt the asthenosphere to form magma: increasing temperature, decreasing pressure, and addition of water (◀▶ Figure 8.2).

Increasing Temperature

Everyone knows that a solid melts when it becomes hot enough. Butter melts in a frying pan and snow melts under the spring Sun. For similar reasons, an increase in temperature will melt a hot rock. Oddly, however, increasing temperature is the least important cause of magma formation in the asthenosphere.

Decreasing Pressure

A mineral is a solid composed of an ordered array of atoms bonded together. When a mineral melts, the atoms become disordered and move freely, taking up more space than the solid mineral. Consequently, magma occupies about 10 percent more volume than the rock that melted to form it (◀▶ Figure 8.3).

If a rock is heated to its melting point on Earth's surface, it melts readily because there is little pressure to keep it from expanding. The temperature in the asthenosphere is hot enough to melt rock, but the high

◆ **FIGURE 8.1** Two geologists retreat from a slowly advancing lava flow on the island of Hawaii.

pressure prevents the rock from expanding, so it does not melt. However, if the pressure were to decrease, large volumes of asthenosphere rocks would melt.

Melting caused by decreasing pressure is called **pressure-release melting**. In the section "Environments of Magma Formation," we will see how certain tectonic processes decrease pressure on asthenosphere rocks, thereby melting them.

Addition of Water

A wet rock generally melts at a lower temperature than an otherwise identical dry rock. Consequently, addition of water to rock that is near its melting temperature can melt the rock. Certain tectonic processes, described shortly, add water to the hot rock of the asthenosphere to form magma.

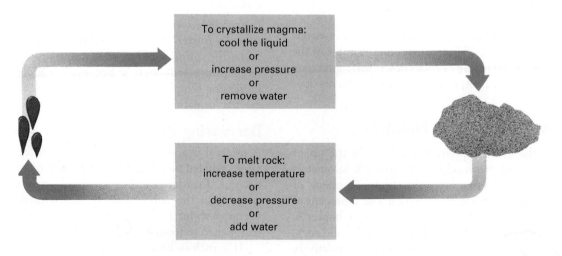

◆ **FIGURE 8.2** The lower box shows that increasing temperature, addition of water, and decreasing pressure all melt rock to form magma. The upper box shows that cooling, increasing pressure, and water loss all solidify magma to form an igneous rock.

CHAPTER 8 • Volcanoes and Plutons

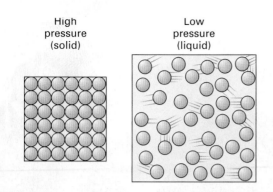

◆ **FIGURE 8.3** When most minerals melt, the volume of the resulting magma increases. As a result, high pressure favors the dense, orderly arrangement of a solid mineral and low pressure favors the random, less-dense arrangement of molecules in liquid magma.

Environments of Magma Formation

Magma forms abundantly in three tectonic environments: spreading centers, mantle plumes, and subduction zones. Let us consider each environment to see how rising temperature, decreasing pressure, and addition of water can melt the asthenosphere to create magma.

Magma Production in a Spreading Center

As lithospheric plates separate at a spreading center, hot, plastic asthenosphere oozes upward to fill the gap (◆ Figure 8.4). As the asthenosphere rises, pressure drops and pressure-release melting forms basaltic magma. (The terms *basaltic* and *granitic* refer to magmas with the chemical compositions of basalt and granite, respectively.) Because the magma is of lower density than the surrounding rock, it rises toward the surface.

Most of the world's spreading centers lie in the ocean basins, where they form the Mid-Oceanic Ridge system. The magma created at the spreading centers erupts from the ridge beneath the sea and solidifies to form new oceanic crust. The oceanic crust then spreads outward, riding atop the separating tectonic plates. Nearly all of Earth's oceanic crust is created in this way at the Mid-Oceanic Ridge system. In most places the ridge lies beneath the sea. In a few places, such as Iceland, the ridge rises above sea level and volcanic eruptions are common (◆ Figure 8.5). Some spreading centers, like the East African Rift, occur in continents, and here, too, basaltic magma erupts onto Earth's surface.

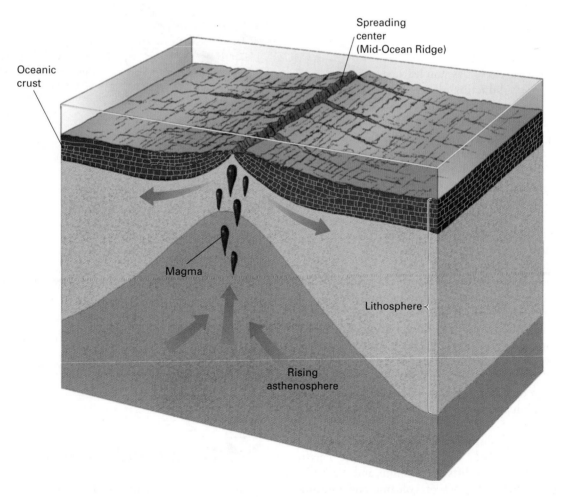

◆ **FIGURE 8.4** Pressure-release melting produces magma beneath a spreading center, where hot asthenosphere rises to fill the gap left by the two separating tectonic plates.

Magma Production in a Mantle Plume

Recall from Chapter 6 that a mantle plume is a rising column of hot, plastic rock that originates deep within the mantle. The plume rises because it is hotter than the surrounding mantle and, consequently, is less dense and more buoyant. As a plume rises, pressure-release melting just beneath the lithosphere forms magma, which then rises toward Earth's surface (◀ Figure 8.6).

Because mantle plumes and hot spots form below the lithosphere, hot spots can occur within a tectonic plate. For example, the Yellowstone hot spot, responsible for the volcanoes and hot springs in Yellowstone National Park, lies far from the nearest plate boundary. If a mantle plume rises beneath the sea, volcanic eruptions build submarine volcanoes and volcanic islands.

The Hawaiian Islands are a chain of hot spot volcanoes that formed over a long-lived mantle plume beneath the Pacific Ocean.

Magma Production in a Subduction Zone

In a subduction zone, addition of water, decreasing pressure, and heating all combine to form huge quantities of magma (◀ Figure 8.7). As you learned in Chapter 6, a subducting plate is covered by water-saturated oceanic crust. As the wet rock dives into the hot mantle, rising temperature vaporizes the water, forming steam that ascends into the hot asthenosphere directly above the sinking plate.

As the subducting plate descends, it drags plastic asthenosphere rock down with it, as shown by the elliptical arrows in ◀ Figure 8.7. Rock from deeper in the asthenosphere then flows upward to replace the sinking rock. Pressure decreases as this hot rock rises.

Friction generates heat in a subduction zone as one plate scrapes past the opposite plate. ◀ Figure 8.7 shows that addition of water, pressure release, and frictional heating combine to melt asthenosphere rocks in the zone where the subducting plate passes into the asthenosphere.

Addition of water is probably the most important factor in a subduction zone, and frictional heating is the least important.

Volcanoes and plutonic rocks are common features of a subduction zone. The volcanoes of the Pacific Northwest, the granite cliffs of Yosemite, and the Andes Mountains are all examples of volcanic and plutonic rocks formed at subduction zones. The Ring of Fire is a

◀ **FIGURE 8.5**   Iceland is one of the few places on Earth where volcanoes of the Mid-Oceanic Ridge system rise above the sea. This photo shows an eruption of Iceland's Mount Surtsey.

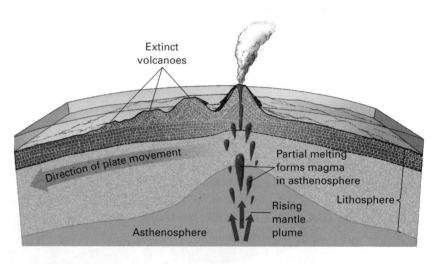

ThomsonNOW™   ◀ **ACTIVE FIGURE 8.6**   Pressure-release melting produces magma in a rising mantle plume. The magma rises to form a volcanic hot spot.

zone of active volcanoes that traces the subduction zones encircling the Pacific Ocean basin.

About 75 percent of Earth's active volcanoes (exclusive of the submarine volcanoes at the Mid-Oceanic Ridge) lie in the Ring of Fire (◀ Figure 8.8).

ThomsonNOW™

CLICK ThomsonNOW Interactive to work through an activity on Rock Lab

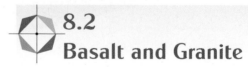

8.2 Basalt and Granite

Basalt and granite are the most common igneous rocks. Basalt makes up most of the oceanic crust, and granite is

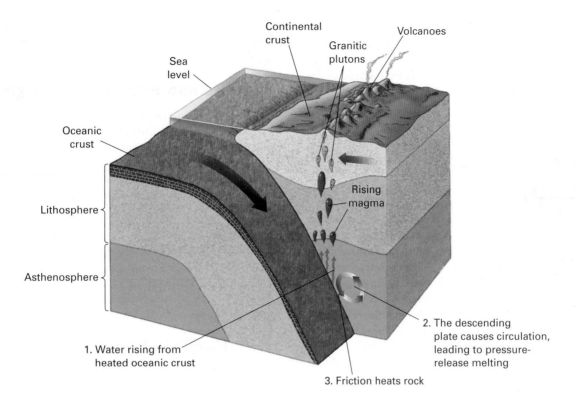

◈ **FIGURE 8.7** Three processes melt the asthenosphere to form magma at a subduction zone: (1) Steam rises from wet oceanic crust on top of the subducting plate; (2) circulation in the asthenosphere decreases pressure on hot mantle rock; (3) friction heats rocks in the subduction zone.

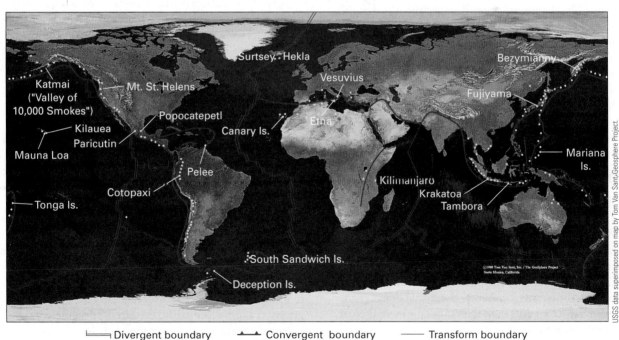

◈ **FIGURE 8.8** Seventy-five percent of Earth's active volcanoes (yellow dots) lie in the Ring of Fire, a chain of subduction zones at convergent plate boundaries (heavy red lines with teeth) that encircles the Pacific Ocean.

the most abundant rock in continents. Because of their abundance, it is interesting to consider how basalt and granite form.

Recall that basaltic magma forms by melting of the asthenosphere. But the asthenosphere is peridotite. Basalt and peridotite are quite different in composition: Peridotite contains about 40 percent silica, but basalt contains about 50 percent. Peridotite contains considerably more iron and magnesium than basalt. How does peridotite melt to create basaltic magma? Why does the magma have a composition different from that of the rock that melted to produce it?

Any pure substance, such as ice, has a definite melting point. Ice melts at exactly 0°C. In addition, ice melts to form water, which has exactly the same composition as the ice, pure H_2O. A rock does not behave in this way because it is a mixture of several minerals. It is a general rule that a mixture of two or more minerals will begin to melt at a temperature lower than the melting point of any one of the minerals in its pure state. In one set of experiments designed to simulate melting of mantle peridotite, for example (remember that peridotite consists mainly of olivine and pyroxene, with small amounts of calcium feldspar), pure olivine melted at 1,890°C, pure pyroxene melted at 1,390°C, and pure calcium feldspar melted at 1,550°C; but peridotite rock consisting of all three minerals began to melt at 1,270°C. Furthermore, the composition of the first bit of melt is usually richer in silica than the rock that is undergoing this process of **partial melting**.

Thus, when mantle peridotite begins to melt, the magma is of basaltic composition—about 20 percent richer in silica than peridotite. Because the new basaltic magma is less dense than the peridotite rock, the magma begins to rise toward Earth's surface. This process is called partial melting because only a small amount of the original peridotite melts to form basaltic magma, leaving silica-depleted peridotite in the asthenosphere. ◆ Figures 8.4 and 8.6 show that basaltic magma forms in this way at a spreading center and in a mantle plume. ◆ Figure 8.7 shows that basaltic magma also forms in a subduction zone.

Granite and Granitic Magma

Granite contains more silica than basalt and therefore melts at a lower temperature—typically between 700°C and 900°C (see the Focus On Box titled "Cooling and Crystallization of Magma: Bowen's Experiment" in Chapter 3 for a discussion of this process). Thus basaltic magma is hot enough to melt granitic continental crust. In some tectonic environments, basaltic magma forms beneath a continent and then rises into continental crust. Because the lower continental crust is hot, even a small amount of basaltic magma melts large amounts of lower continental crust to form granitic magma. Typically, the granitic magma rises a short distance and then solidifies within the crust to form plutonic rocks. Most granitic plutons solidify at depths between 5 kilometers and about 20 kilometers.

Granite forms by this process in a subduction zone at a continental margin, a continental rift zone, and a mantle plume rising beneath a continent.

Andesite and Intermediate Magma

Igneous rocks of intermediate composition, such as andesite and diorite, form by processes similar to those that generate granitic magma. Their magmas contain less silica than granite, either because they form by melting of continental crust that is lower in silica or because basaltic magma has mixed with the granitic magma.

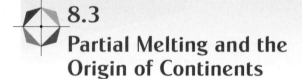

ThomsonNOW

CLICK ThomsonNOW Interactive to work through an activity on Rock Lab.

8.3
Partial Melting and the Origin of Continents

Recall from earlier discussions that Earth melted shortly after its formation about 4.6 billion years ago. Magma at the surface then cooled to form the earliest crust. From the evidence of a few traces of old crust combined with computer models, geologists surmise that the first crust was lava with the composition of peridotite, called komatiite.

Our explanation of the formation of granitic magma by melting of granitic continental crust leaves us with two interesting questions: When did granitic continents form? If the early crust and mantle were composed of peridotite, how did granitic continental crust evolve at Earth's surface?

When Did Continents Form?

The 3.96-billion-year-old Acasta gneiss in Canada's Northwest Territories is Earth's oldest known rock. It is metamorphosed granitic rock, similar to modern continental crust, and implies that at least some granitic crust had formed by early Precambrian time.

Geologists have found grains of a mineral called zircon in a sandstone in western Australia. The zircon gives radiometric dates of 4.2 billion years, although the sandstone is younger. Zircon commonly forms in granite. Geologists infer that the very old zircon initially formed in granite, which later weathered and released the zircon grains as sand. Eventually, the zircon became part of the younger sandstone. The presence of these zircon grains suggests that granitic rocks existed 4.2 billion years ago. Geologists have also found granitic rocks nearly as old as the Acasta gneiss and the Australian zircon grains in Greenland and Labrador.

These dates tell us that some granitic continental crust probably existed by 4.2 billion years ago.

Partial Melting and the Origin of Granitic Continents

Recall from Chapter 3 that the mantle, oceanic crust, and continental crust have different compositions. These differences are reviewed in the table below.

If the earliest crust had a composition similar to that of the mantle, we must describe how oceanic and continental crust evolved from mantle rocks. In the previous section we explained that partial melting produces magma that contains a higher proportion of silica than the rock from which the melt formed. Geologists believe that the earliest crust was komatiite lava, formed from the melted mantle as Earth cooled from the surface downward after its early melting event. Later, partial melting of this primordial crust formed a basaltic crust that was richer in silica. Then, partial melting of the basalt probably formed intermediate rocks such as andesite, which underwent another partial melting to form the silica-rich granitic continents. The process of partial melting may explain how the silica-rich continents evolved in steps from the silica-poor mantle. But what tectonic processes caused the sequence of melting steps?

In modern times, magma forms in three geologic environments: spreading centers, subduction zones, and mantle plumes. Similar magma-forming environments may have existed in early Precambrian time, but geologists are uncertain which were most important.

Some observations imply that Archean tectonics were similar to modern horizontal plate movements, and most magma formed at spreading centers and subduction zones. Other evidence indicates that horizontal plate motion was minor and that vertical mantle plumes dominated early Precambrian tectonics.

Mantle (Peridotite)	Oceanic Crust (Basalt)	Continental Crust (Granite)
High magnesium and iron content; low silica	Less magnesium and iron and more silica compared with mantle rock	Low magnesium and iron; high silica

Horizontal Tectonics

According to one hypothesis, heat-driven convection currents in a hot, active mantle initiated horizontal plate movement in the early Precambrian crust. The dense primordial crust dove into the mantle in subduction zones, where partial melting created basaltic magma. As a result, the crust gradually became basaltic.

At a later date, the earliest continental crust formed by partial melting of basaltic crust in a new generation of subduction zones. In the next chapter, you will see that island arcs form today in the same manner. Therefore, the first continents probably consisted of small granitic or andesitic blobs—like island arcs—surrounded by basaltic crust (◆ Figure 8.9A). Gradually, continued plate movement led to further subduction and isolated islands coalesced to form microcontinents. In the process, sediments accumulated and were metamorphosed during the collisions (◆ Figure 8.9B).

Vertical Mantle Plume Tectonics

Several researchers have proposed, instead, that mantle plumes dominated early Precambrian tectonics. Upwellings of mantle rock, perhaps rising from the mantle–core boundary, led to partial melting of portions of the upper mantle. This magma then solidified to form basaltic crust. Continued partial melting eventually formed granitic continental crust (◆ Figure 8.10).

As evidence for both models accumulates, some geologists suggest that both mechanisms were important. According to this hypothesis, mantle plumes formed thick basaltic oceanic plateaus, which then oozed outward to initiate horizontal motion. This motion caused subduction and another melting episode that generated continental crust by partial melting of the basalt plateaus.

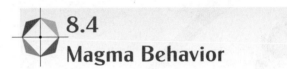

8.4 Magma Behavior

Once magma forms, it rises toward Earth's surface because it is less dense than surrounding rock. As it rises, two changes occur. Magma cools as it enters shallower and cooler levels of Earth, and pressure drops because the weight of overlying rock decreases. ◆ Figure 8.2 shows that cooling tends to solidify magma but that decreasing pressure tends to keep it liquid.

So, does magma solidify or remain liquid as it rises toward Earth's surface? The answer depends on the type of magma. Basaltic magma commonly rises to the surface to erupt from a volcano or to flow onto the sea floor at the Mid-Oceanic Ridge. In contrast, granitic magma usually solidifies within the crust.

The contrasting behavior of granitic and basaltic magmas is a consequence of their different compositions. Granitic magma contains about 70 percent silica, whereas the silica content of basaltic magma is only about 50 percent. In addition, granitic magma generally contains up to 10 percent water, whereas basaltic magma contains only 1 to 2 percent water. These differences are summarized in the following table.

Typical Granitic Magma	Typical Basaltic Magma
70% silica	50% silica
Up to 10% water	1–2% water

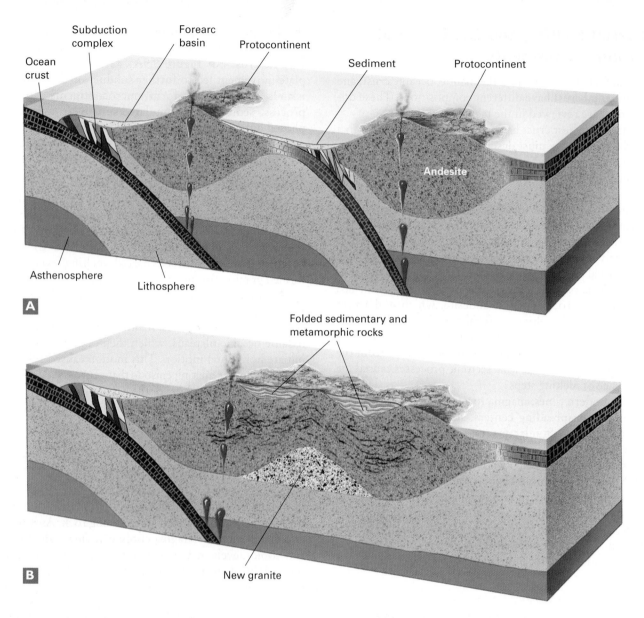

Asthenosphere
Lithosphere

A

B

◆ **FIGURE 8.9** **(A)** According to one model, early continents formed in subduction zones. **(B)** As plates continued to move, small continents coalesced. Sediments that had eroded from the original islands were folded, and uplifted. Some were subjected to so much heat and pressure that they became metamorphic rocks. New granite formed from partial melting of the crust at the subduction zone.

Effects of Silica on Magma Behavior

In the silicate minerals, silicate tetrahedra link together to form the chains, sheets, and framework structures described in Chapter 2. Silicate tetrahedra link together in a similar manner in magma. They form long chains and similar structures if silica is abundant in the magma, but shorter chains if less silica is present.

Because of its higher silica content, granitic magma contains longer chains than does basaltic magma. The long chains become tangled, making the magma stiff, or viscous. It rises slowly because of its viscosity and has ample time to solidify within the crust before reaching the surface. In contrast, basaltic magma, with its shorter silicate chains, is less viscous and flows easily.

Because of its fluidity, it rises rapidly to erupt at Earth's surface.

Effects of Water on Magma Behavior

A second, and more important, difference between the two magmas is that granitic magma contains more water than basaltic magma does. Water lowers the temperature at which magma solidifies. If dry granitic magma solidifies at 700°C, the same magma with 10 percent water may not become solid until the temperature drops below 600°C.

Water tends to escape as steam from hot magma. But deep in the crust where granitic magma forms, high pressure prevents the water from escaping. As the magma rises, pressure decreases and water escapes. Because the

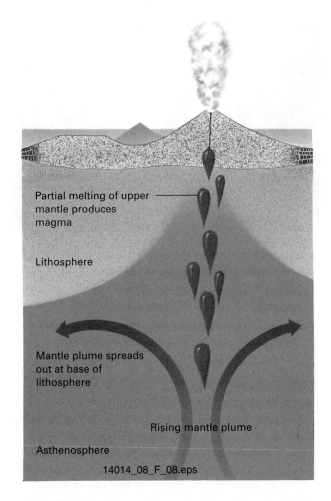

Partial melting of upper mantle produces magma

Lithosphere

Mantle plume spreads out at base of lithosphere

Rising mantle plume

Asthenosphere

14014_08_F_08.eps

◆ **FIGURE 8.10** Another hypothesis contends that early continental crust formed over rising mantle plumes, in a process called "plume tectonics."

magma loses water, its solidification temperature rises, causing it to crystallize. Water loss causes rising granitic magma to solidify within the crust.

Because basaltic magmas have only 1 to 2 percent water to begin with, water loss is relatively unimportant. As a result, rising basaltic magma usually remains liquid all the way to Earth's surface, and basalt volcanoes are common.

ThomsonNOW™

CLICK ThomsonNOW Interactive to work through an activity on Rock Lab.

 **8.5 Plutons**

In most cases, granitic magma solidifies within Earth's crust to form a large mass of granite called a **pluton** (◆ Figure 8.11). Many granite plutons measure tens of kilometers in diameter. How can such a large mass of viscous magma rise through solid rock?

If you place oil and water in a jar, screw the lid on, and shake the jar, oil droplets disperse throughout the water. When you set the jar down, the droplets coalesce to form larger bubbles, which rise toward the surface, easily displacing the water as they ascend. Granitic magma rises in a similar way except that the process is slower because it rises through solid rock. Granitic magma forms near the base of continental crust, where surrounding rock is hot and plastic. As the magma rises, it pushes aside the plastic country rock, which then slowly flows back to fill in behind the rising bubble.

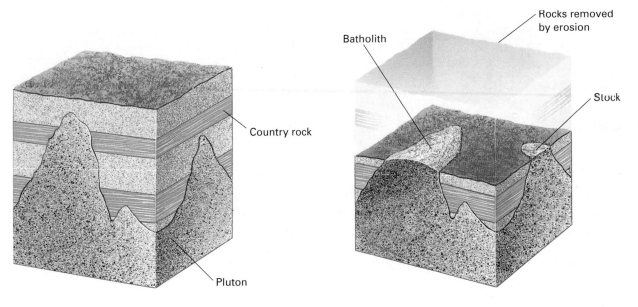

◆ **FIGURE 8.11** A pluton is any intrusive igneous rock. A batholith is a pluton with more than 100 square kilometers exposed at Earth's surface. A stock is similar to a batholith but has a smaller surface area.

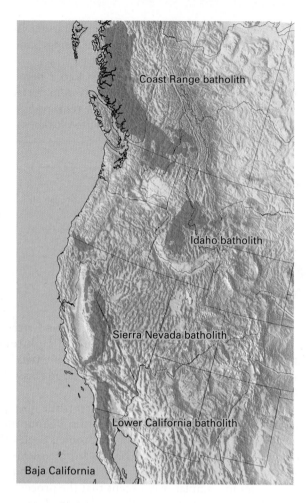

◁▷ **FIGURE 8.12** The large batholiths in western North America form high mountain ranges.

◁▷ **FIGURE 8.13** Granite plutons make up most of California's Sierra Nevada in Yosemite National Park.

After a pluton forms, tectonic forces may push it upward, and erosion may expose parts of it at Earth's surface. A **batholith** is a pluton exposed over more than 100 square kilometers of Earth's surface. An average batholith is about 10 kilometers thick, although a large one may be 20 kilometers thick. A **stock** is similar to a batholith but is exposed over less than 100 square kilometers.

◁▷ Figure 8.12 shows the major batholiths of western North America. Many mountain ranges, such as California's Sierra Nevada, contain large granite batholiths. A batholith is commonly composed of numerous smaller plutons intruded sequentially over millions of years. For example, the Sierra Nevada batholith contains about 100 individual plutons, most of which were emplaced

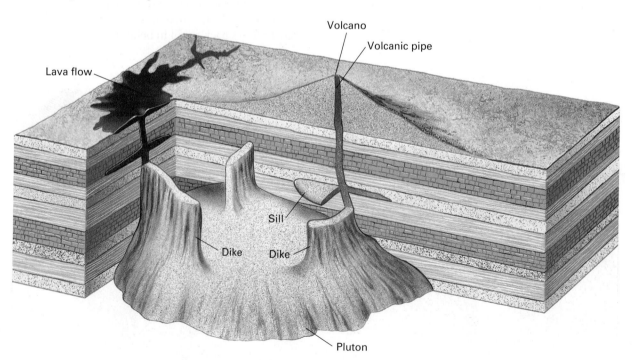

◁▷ **FIGURE 8.14** A large magma body may crystallize within the crust to form a pluton. Some of the magma may rise to the surface to form volcanoes and lava flows; some intrudes country rock to form dikes and sills.

◆ **FIGURE 8.15** Several granite dikes have been intruded into older country rock.

over a period of 50 million years. The formation of this complex batholith ended about 80 million years ago (◆ Figure 8.13).

A large body of magma pushes country rock aside as it rises. In contrast, a smaller mass of magma may flow into a fracture or between layers in country rock. A **dike** is a tabular, or sheetlike, intrusive rock that forms when magma oozes into a fracture (◆ Figure 8.14).

Dikes cut across sedimentary layers or other features in country rock and range from less than a centimeter to more than a kilometer thick (◆ Figure 8.15). A dike is commonly more resistant to weathering than surrounding rock is. As the country rock erodes, the dike is left standing on the surface (◆ Figure 8.16).

Magma that oozes between layers of country rock forms a sheetlike rock parallel to the layering, called a **sill** (◆ Figure 8.17). Like dikes, sills vary in thickness from less than a centimeter to more than a kilometer and may extend for tens of kilometers in length and width.

◆ **FIGURE 8.16** A large dike near Shiprock, New Mexico, has been left standing after softer sandstone country rock eroded.

Sill

◆ **FIGURE 8.17** A black basalt sill has been injected between layers of sandstone. Grand Canyon, Arizona.

8.6
Volcanoes

The material erupted from volcanoes creates a wide variety of rocks and landforms, including lava plateaus and several types of volcanoes. Many islands, including the Hawaiian Islands, Iceland, and most islands of the southwestern Pacific Ocean, were built entirely by volcanic eruptions.

Lava and Pyroclastic Rocks

Lava is fluid magma that flows onto Earth's surface; the word also describes the rock that forms when the magma solidifies. Lava with low viscosity may continue to flow as it cools and stiffens, forming smooth, glassy-surfaced, wrinkled, or "ropy" ridges. This type of lava is called **pahoehoe** (◈▶ Figure 8.18). If the viscosity of lava is higher, its surface may partially solidify as it flows. The solid crust breaks up as the deeper, molten lava continues to move, forming **aa** lava, with a jagged, rubbled, broken surface. When lava cools, escaping gases such as water and carbon dioxide form bubbles in the lava. If the lava solidifies before the gas escapes, the bubbles are preserved as holes in the rock (◈▶ Figure 8.19).

Hot lava shrinks as it cools and solidifies. The shrinkage pulls the rock apart, forming cracks that grow as the rock continues to cool. In Hawaii, geologists have observed this phenomenon while watching fresh lava cool: When a solid crust measuring only 0.5 centimeter thick had formed on the surface of the glowing liquid, five- or six-sided cracks developed. As the lava continued to cool and solidify, the cracks grew downward through the flow. Such cracks, called **columnar joints**, are regularly spaced and intersect to form five- or six-sided columns (◀▶ Figure 8.20).

If a volcano erupts explosively, it may eject both liquid magma and solid rock fragments. A rock formed from this material is called a **pyroclastic rock** (from *pyro*, meaning "fire," and *clastic*, meaning "particles"). The smallest particles, called **volcanic ash**, consist of tiny fragments of glass that formed when liquid magma exploded into the air. **Cinders** are volcanic fragments that vary in size from 2 to 64 millimeters.

Fissure Eruptions and Lava Plateaus

The gentlest type of volcanic eruption occurs when magma is so fluid that it oozes from cracks in the land surface called **fissures** and flows over the land like water. Basaltic magma commonly erupts in this manner because of its low viscosity. Fissures and fissure eruptions vary greatly in scale. In some cases, lava pours from small cracks on the flank of a volcano. Fissure flows of this type are common on Hawaiian and Icelandic volcanoes.

In other cases, however, fissures extend for tens or hundreds of kilometers and pour thousands of cubic kilometers of lava onto Earth's surface. A fissure eruption of this type creates a **flood basalt**, which covers the landscape like a flood. It is common for many such fissure eruptions to occur in rapid succession and to create a **lava plateau**, or basalt plateau, covering thousands of square kilometers.

The Columbia River plateau in eastern Washington, northern Oregon, and western Idaho is a lava plateau containing 350,000 cubic kilometers of basalt. The lava is up to 3,000 meters thick and covers 200,000 square kilometers (◀▶ Figure 8.21). It formed about 15 million years

◈ **FIGURE 8.18** A car buried in pahoehoe lava, Hawaii.

◈ **FIGURE 8.19** Aa lava showing vesicles, gas bubbles frozen into the flow, in Shoshone, Idaho.

FIGURE 8.20 **(A)** Columnar joints at Devil's Postpile National Monument. **(B)** A view from the top, where glaciers have polished the columns.

FIGURE 8.21 **(A)** The Columbia River basalt plateau covers much of Washington, Oregon, and Idaho. **(B)** Columbia River basalt in eastern Washington. Each layer is a separate lava flow.

INTERACTIVE QUESTION: *What effects might the eruption of this basaltic lava have had on global climate and on the biosphere?*

ago as basaltic magma oozed from long fissures in Earth's surface. The individual flows are between 15 and 100 meters thick. An even larger lava plateau in Siberia contains one million cubic kilometers of basaltic lava that poured onto Earth's surface 248 million years ago.[1]

Similar large lava plateaus occur in western India, northern Australia, Iceland, Brazil, Argentina, and Antarctica.

Volcanoes

If lava is too viscous to spread out as a flood, it builds a hill or mountain called a **volcano**. Volcanoes differ widely in shape, structure, and size (Table 8.1). Lava and rock fragments commonly erupt from an opening called a **vent** located in a **crater**, a bowl-like depression at the

1. A. R. Basu et al., "High −He3 Plume Origin and Temporal–Spatial Evolution of the Siberian Flood Basalts," *Science* 269 (August 11, 1995): 5225.

Table 8.1   Characteristics of Different Types of Volcanoes

Type of Volcano	Form of Volcano	Size	Type of Magma	Style of Activity	Examples
Basalt plateau	Flat to gentle slope	100,000–1,000,000 km² in area; 1–3 km thick	Basalt	Gentle eruption from long fissures	Columbia River plateau
Shield volcano	Slightly sloped, 6° to 12°	Up to 9,000 m high	Basalt	Gentle, some fire fountains	Hawaii
Cinder cone	Moderate slope	100–400 m high	Basalt or andesite	Ejections of pyroclastic material	Parícutin, Mexico
Composite volcano	Alternate layers of flows and pyroclastics	100–3,500 m high	Variety of types of magmas and ash	Often violent	Vesuvius, Mount St. Helens, Aconcagua
Caldera	Cataclysmic explosion leaving a circular depression called a caldera	Less than 40 km in diameter	Granite	Very violent	Yellowstone, San Juan Mountains

summit of the volcano (◈ Figure 8.22). As mentioned previously, lava or pyroclastic material may also erupt from a fissure on the flanks of the volcano.

Shield Volcanoes

Fluid basaltic magma often builds a gently sloping mountain called a **shield volcano** (◈ Figure 8.23). The sides of a shield volcano generally slope away from the vent at angles between 6° and 12° from horizontal. Although their slopes are gentle, shield volcanoes can be enormous. The height of Mauna Kea volcano in Hawaii, measured from its true base on the sea floor to its top, exceeds the height of Mount Everest.

Although shield volcanoes, such as those of Hawaii and Iceland, erupt regularly, the eruptions are normally gentle and rarely life threatening. Lava flows occasionally overrun homes and villages, but the flows advance slowly enough to give people time to evacuate.

Cinder Cones

A **cinder cone** is a small volcano composed of pyroclastic fragments. A cinder cone forms when large amounts of gas accumulate in rising magma. When the gas pressure builds sufficiently, the entire mass erupts explosively, hurling cinders, ash, and molten magma into the air. The particles then fall back around the vent to accumulate as a small mountain of pyroclastic debris. A cinder cone is usually active for only a short time because once the gas escapes, the driving force behind the eruption is gone.

As the name implies, a cinder cone is symmetrical. It also can be steep (about 30°), especially near the vent where ash and cinders pile up (◈ Figure 8.24). Most are less than 300 meters high, although a large one can be up to 700 meters high. A cinder cone erodes easily and quickly because the pyroclastic fragments are not cemented together.

Courtesy of Graham R. Thompson/Jonathan Turk

◈ **FIGURE 8.22**   Steam rises from vents in the crater of Marum volcano, Vanuatu.

FIGURE 8.23 Mount Skjoldbreidier in Iceland shows the typical low-angle slopes of a shield volcano.

FIGURE 8.24 **(A)** These cinder cones in southern Bolivia are composed of loose pyroclastic fragments **(B)**.

About 350 kilometers west of Mexico City, numerous extinct cinder cones are scattered over a broad plain. Prior to 1943, a small hole, one or two meters in diameter, existed on the plain. The hole had been there for as long as anyone could remember, and people grew corn just a few meters away. In February 1943, as two farmers were preparing their field for planting, smoke and sulfurous gases rose from the hole. As night fell, hot, glowing rocks flew skyward, creating spectacular arcing flares like a giant fireworks display. By morning, a 40-meter-high cinder cone had grown where the hole had been. For the next five days, pyroclastic material erupted 1,000 meters into the sky and the cone grew to 100 meters in height. After a few months, a fissure opened at the base of the cone, extruding lava that buried the town of San Juan Parangaricutiro. Two years later the cone had grown to a height of 400 meters. After nine years, the eruptions ended, and today the volcano, called El Parícutin, is dormant.

Composite Cones

A **composite cone**, sometimes called a **stratovolcano**, forms over a long period of time from alternating lava flows and pyroclastic eruptions. The hard lava covers the loose pyroclastic material and protects it from erosion (◆ Figure 8.25).

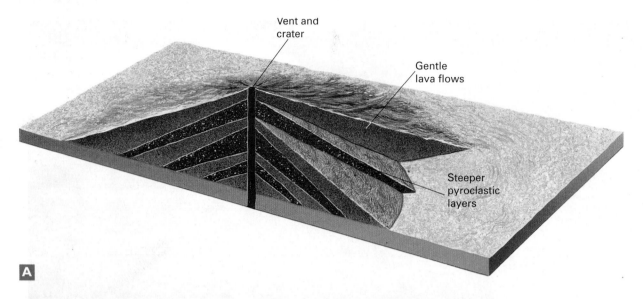

Vent and crater

Gentle lava flows

Steeper pyroclastic layers

A

Chuck Pefly/Getty Images

B

◈ **FIGURE 8.25** **(A)** A composite cone consists of alternating layers of lava and loose pyroclastic material. **(B)** Mount Rainier rises behind Seattle's skyline.

Many of the highest mountains of the Andes and some of the most spectacular mountains of western North America are composite cones. Repeated eruption is a trademark of a composite volcano. Mount St. Helens erupted dozens of times in the 4,500 years preceding its most recent eruption in 1980. Mount Rainier, also in Washington, has been dormant in recent times but could become active again and threaten nearby populated regions.

ThomsonNOW™

CLICK ThomsonNOW Interactive to work through activities on:
• Magma Chemistry and Explosivity
• Volcano Watch, USA through Volcanic Landforms

 8.7

Volcanic Explosions: Ash-Flow Tuffs and Calderas

Although granitic magma usually solidifies within the crust, under certain conditions it rises to Earth's surface, where it erupts violently. The granitic magmas that rise to the surface probably contain only a few percent water, like basaltic magma. These magmas reach the surface because, like basaltic magmas, they have little water to lose.

As the magma ascends, decreasing pressure allows the small amount of dissolved water to form a frothy, pressurized mixture of gas and liquid magma that may be as hot as 900°C (◈ Figure 8.26A). As the mixture rises to within

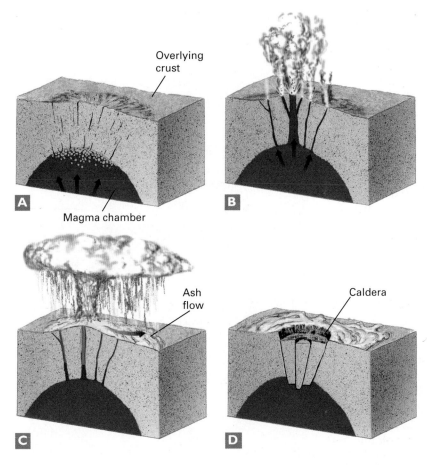

Overlying crust

Magma chamber

A

B

Ash flow

Caldera

C

D

◆ **FIGURE 8.26** **(A)** When granitic magma rises to within a few kilometers of Earth's surface, it stretches and fractures overlying rock. Gas separates from the magma and rises to the upper part of the magma body. **(B)** The gas-rich magma explodes through fractures, rising as a vertical column of hot ash, rock fragments, and gas. **(C)** When the gas is used up, the column collapses and spreads outward as a high-speed ash flow. **(D)** Because so much material has erupted from the top of the magma chamber, the roof collapses to form a caldera.

INTERACTIVE QUESTION: *How might a large caldera eruption alter local ecosystems? How might it alter global climate?*

a few kilometers of Earth's surface, it fractures overlying rocks and explodes skyward through the fractures (◆ Figure 8.26B). As an analogy, think of a bottle of beer or soda pop. When the cap is on and the contents are under pressure, carbon dioxide gas is dissolved in the liquid. When you remove the cap, pressure decreases and bubbles rise to the surface. If conditions are favorable, the frothy mixture erupts through the bottleneck.

A large and violent eruption can blast a column of pyroclastic material 10 or 12 kilometers into the sky, and the column might be several kilometers in diameter. A cloud of fine ash may rise even higher—into the upper atmosphere. The force of material streaming out of the magma chamber can hold the column up for hours or even days. Recent eruptions of Mount Pinatubo on the Philippines and the Soufrière Hills volcano on the Caribbean island of Montserrat blasted ash columns high into the atmosphere and held them up for hours.

After an eruption, upper layers of the remaining magma are depleted in dissolved water and the explosive potential is low. However, deeper parts of the magma continue to release water vapor, which rises and builds pressure again to begin another cycle of eruption. Time intervals between successive eruptions vary from a few thousand to about half a million years.

Ash Flows

When most of the gas has escaped from the upper layers of magma, the eruption ends. The airborne column of ash, rock, and gas then falls back to

Courtesy of Graham R. Thompson/Jonathan Turk

Geoffrey Sutton

◆ **FIGURE 8.27** **(A)** Ash-flow tuff forms the spectacular cliffs of Smith Rock State Park near Bend, Oregon. **(B)** Ash-flow tuff forms when an ash flow comes to a stop. The fragments in the tuff are pieces of rock that were carried along with the volcanic ash and gas.

Earth's surface, spreading over the land and rushing down stream valleys (◆ Figure 8.26C). Such a flow is called an **ash flow**, or *nuèe ardente*, a French term for "glowing cloud." Small ash flows move at speeds up to 200 kilometers per hour. Large flows have traveled distances exceeding 100 kilometers. The 2,000-year-old Taupo flow on New Zealand's South Island raced down a stream valley with such speed and inertia that it leaped over a 700-meter-high ridge as it crossed from one valley into another.

When an ash flow stops, most of the gas escapes into the atmosphere, leaving behind a chaotic mixture of volcanic ash and rock fragments called **ash-flow tuff** (◆ Figure 8.27). The largest known ash-flow tuff from a single eruption is located in the San Juan Mountains of southwestern Colorado and has a volume greater than 3,000 cubic kilometers. Another of comparable size lies in southern Nevada.

Calderas

After the gas-charged magma erupts, the roof of the magma chamber collapses into the space that the magma had filled (◆ Figure 8.26D). Because most magma bodies are circular, when viewed from above, the collapsing roof forms a circular depression called a **caldera**. A large caldera may be 40 kilometers in diameter and have walls as much as a kilometer high. Some calderas fill up with volcanic debris as the ash column collapses; others maintain the circular depression and steep walls (◆ Figure 8.28). We usually think of volcanic landforms as mountain peaks, but the topographic depression of a caldera is an exception. ◆ Figure 8.29 shows that calderas, ash-flow tuffs, and related rocks occur over a large part of western North America. Consider two well-known examples.

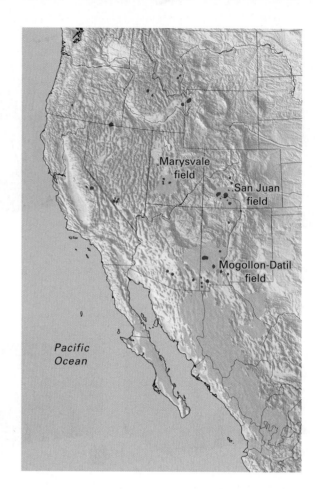

◆ **FIGURE 8.29** Calderas (red dots) and ash-flow tuffs (orange areas) are abundant in western North America.

Yellowstone National Park and the Long Valley Caldera

Yellowstone National Park in Wyoming and Montana is the oldest national park in the United States. Its geology consists of three large overlapping calderas and the ash-flow tuffs that erupted from them (◆ Figure 8.30). The oldest eruption took place 1.9 million years ago and ejected 2,500 cubic kilometers of pyroclastic material. This is enough ash to bury all of Yellowstone Park to a depth of 250 meters. The next major eruption occurred 1.3 million years ago and produced about 400 cubic kilometers of ash. The most recent, 0.6 million years ago, ejected 1,000 cubic kilometers of ash and other debris and produced the Yellowstone caldera in the center of the park.

Intervals of 0.6 to 0.7 million years separate the three Yellowstone eruptions; 0.6 million years have passed since the most recent one. The park's geysers and hot springs are formed by hot magma beneath Yellowstone, and numerous small earthquakes indicate that the magma is moving. Geologists would not be surprised if another eruption occurred at any time.

A geologic environment similar to that of Yellowstone is found near Yosemite National Park in eastern California. Here the 170-cubic-kilometer Bishop Tuff erupted

◆ **FIGURE 8.28** Crater Lake, Oregon, is a caldera caused by a volcanic eruption that occurred about 6,850 years ago. The caldera is about 10 km in diameter. The eruption released about 50 cubic km of magma to the surface. It was one of the largest eruptions in the last 10,000 years.

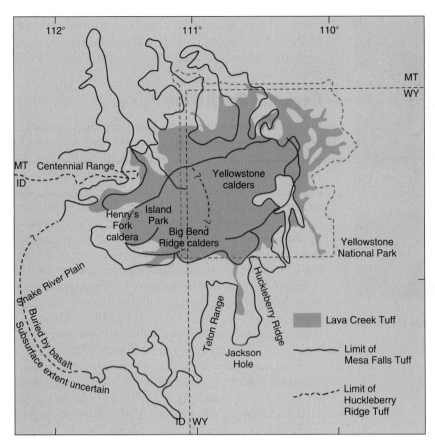

◆ **FIGURE 8.30** This map shows the locations of the three Yellowstone Calderas and the associated ash-flow tuff deposits.

from the Long Valley caldera 0.7 million years ago. Although only one major eruption has occurred, seismic monitoring indicates that magma lies beneath Mammoth Mountain, a popular California ski area, on the southwest edge of the Long Valley caldera. Unusual amounts of carbon dioxide—a common volcanic gas—have escaped from the caldera since 1994. As in the case of Yellowstone, another eruption would not surprise most geologists.

It is difficult to appreciate the potential effects of a Yellowstone or Long Valley type of eruption on the modern world. No event even of the magnitude of the smaller Long Valley eruption has occurred in recorded history. Measured by the amount of pyroclastic material erupted, the 1980 Mount St. Helens eruption was 2,500 times smaller than the first Yellowstone event, and 170 times smaller than the eruption of the Bishop Tuff. Yet the Mount St. Helens eruption killed 63 people, caused hundreds of millions of dollars in damage, and disrupted people's lives hundreds of miles from the volcano. It seems inevitable that an eruption on the scale of Yellowstone, or even of the Long Valley caldera, especially if it occurred in a populous region, would kill many thousands of people, entomb towns and cities beneath meters of red-hot ash, change the courses of rivers and streams, and destroy transportation systems and agricultural land. It would probably raise a dust cloud in the upper atmo-

sphere that would darken the Sun over the entire planet for months or years, cooling the atmosphere, altering global climate, and changing global ecosystems. It is obvious that plants and animals have survived many such events in geologic history, but that is little consolation for the individuals who eventually must experience another such volcanic crisis.

The Destruction of Pompeii

In AD 79, Mount Vesuvius erupted and destroyed the Roman cities of Pompeii, Herculaneum, and several neighboring villages near what is now Naples, Italy. Prior to that eruption, the volcano had been inactive for about 700 years, so long that farmers had cultivated vineyards on the sides of the mountain all the way to the summit. During the eruption, an ash flow streamed down the flanks of the volcano, burying the cities and towns under 5 to 8 meters of hot ash. When archaeologists located and excavated Pompeii 17 centuries later, they found molds of inhabitants trapped by the ash flow as they attempted to flee or find shelter (◆ Figure 8.31). Some of the molds appear to preserve facial expressions of terror. After the AD 79 eruption, Mount Vesuvius returned to relative quiescence but became active again in 1631. It was frequently active from 1631 to 1944, and in this century it erupted in 1906, 1929, and 1944.

Mount Vesuvius is a stratovolcano that formed over the past 25,000 years by many eruptions that varied from gently flowing lava to the explosion that buried Pompeii. A recent study of seismic velocities beneath the volcano showed that seismic waves suddenly slow from 6 to 2.7 kilometers per second at a depth of 10 kilometers. The sudden velocity decrease suggests that molten magma still exists at that depth.[2] Because stratovolcanoes erupt frequently and remain active for long periods, and because magma underlies the volcano, geologists consider Mount Vesuvius a high risk for future eruptions.

ThomsonNOW™

CLICK ThomsonNOW Interactive to work through activities on:
- Magma Chemistry and Explosivity
- Volcano Watch, USA through Volcanic Landforms

2. A. Zollo et al., "Seismic Evidence for a Low-Velocity Zone in the Upper Crust Beneath Mount Vesuvius," *Science* 274 (October 25, 1996): 592–594.

◆ **FIGURE 8.31** Archaeologists uncover molds of Pompeiians killed in the AD 79 eruption of Mount Vesuvius.

8.8
Risk Assessment: Predicting Volcanic Eruptions

Table 8.2 summarizes the major known volcanic disasters since AD 1500. The potential for such disasters in the future makes a volcanic eruption one of the greatest of geologic hazards. It also makes risk assessment and prediction of volcanic eruptions an important part of modern science.

Approximately 1,300 active volcanoes are recognized globally, and 5,564 eruptions have occurred in the past 10,000 years. These figures do not include the numerous submarine volcanoes of the Mid-Oceanic Ridge system. Many volcanoes have erupted recently, and we are certain that others will erupt soon. How can geologists predict an eruption and reduce the risk of a volcanic disaster?

Regional Prediction

Volcanoes concentrate near subduction zones, spreading centers, and hot spots over mantle plumes but are rare in other places. Therefore, the first step in assessing the volcanic hazard of an area is to understand its tectonic environment. Western Washington and Oregon are near a subduction zone and in a region likely to experience future volcanic activity. Kansas and Nebraska are not.

Furthermore, the potential violence of a volcanic eruption is related to the geologic environment of the volcano. If an active volcano lies on continental crust, the eruptions may be violent because granitic magma may form. In contrast, if the region lies on oceanic crust, the eruptions may be gentle because basaltic volcanism is more likely. Violent eruptions are likely in western Washington and Oregon, but less so on Hawaii or Iceland.

Risk assessment is based both on frequency of past eruptions and on potential violence. However, regional prediction only estimates probabilities and cannot be used to determine when a particular volcano will erupt or the intensity of a particular eruption.

Short-Term Prediction

In contrast to regional predictions, short-term predictions attempt to forecast the specific time and place of an impending eruption. They are based on instruments that monitor an active volcano to detect signals that the volcano is about to erupt. The signals include changes in the shape of the mountain and surrounding land, earthquake swarms indicating movement of magma beneath the mountain, increased emissions of ash or gas, increasing temperatures of nearby hot springs, and any other signs that magma is approaching the surface.

Table 8.2   Some Notable Volcanic Disasters Since the Year AD 1500 Involving 5,000 or More Fatalities

| | | | Primary Cause of Death and Number of Deaths | | | | |
Volcano	Country	Year	Pyroclastic Flow	Debris Flow	Lava Flow	Posteruption Starvation	Tsunami
Kelut	Indonesia	1586		10,000			
Vesuvius	Italy	1631	18,000				
Etna	Italy	1669			10,000		
Lakagigar	Iceland	1783				9,340	
Unzen	Japan	1792					15,190
Tambora	Indonesia	1815	12,000			80,000	
Krakatoa	Indonesia	1883					36,420
Pelée	Martinique	1902	29,000				
Santa Maria	Guatemala	1902	6,000				
Kelut	Indonesia	1919		5,110			
Nevado del Ruiz	Colombia	1985		>22,000			

In 1978, two U.S. Geological Survey (USGS) geologists, Dwight Crandall and Don Mullineaux, noted that Mount St. Helens had erupted more frequently and violently during the past 4,500 years than any other volcano in the contiguous 48 states. They predicted that the volcano would erupt again before the end of the century.

In March 1980, about two months before the great May eruption, puffs of steam and volcanic ash rose from the crater of Mount St. Helens, and swarms of earthquakes occurred beneath the mountain. This activity convinced other USGS geologists that Crandall and Mullineaux's prediction was correct. In response, they installed networks of seismographs, tiltmeters, and surveying instruments on and around the mountain.

In the spring of 1980, the geologists warned government agencies and the public that Mount St. Helens showed signs of an impending eruption. The U.S. Forest Service and local law enforcement officers quickly evacuated the area surrounding the mountain, averting a much larger tragedy (◁▷ Figure 8.32). Using similar kinds of information, geologists predicted the 1991 Mount Pinatubo eruption in the Philippines, saving many lives.

Although the June 25, 1997, eruption of the Soufrière Hills volcano on the Caribbean island of Montserrat killed 19 people and destroyed many homes and farms,

◁▷ **FIGURE 8.32** United States Geological Survey geologists accurately predicted the May 1980, eruption of Mount St. Helens.

predictions of the eruption by the Montserrat Volcano Observatory saved many additional lives (◁▷ Figure 8.33). The island was long known to harbor an active volcano, and the observatory was established to monitor the volcano and to predict eruptions. In January 1992, observatory seismographs recorded swarms of earthquakes in the

A

B

◁▷ **FIGURE 8.33** **(A)** The June 25, 1997, eruption of the Soufrière Hills volcano on the Carribean island of Montserrat looms over nearby villages. **(B)** Pyroclastic flows buried and devastated several communities, killing an estimated 19 people. Warnings from the Montserrat Volcano Observatory led to evacuations that saved many other lives.

INTERACTIVE QUESTION: *Look up the description of this eruption on the website of the Montserrat Volcano Observatory, and discuss the economic and health effects of this eruption on the inhabitants of Montserrat.*

southern part of the island. The quakes continued intermittently through the summer of 1995, when the volcano began to emit smoke, steam, and ash. Then, on August 21, 1995, an eruption blanketed the town of Plymouth in a thick ash cloud and darkened the sky for about 15 minutes. Government authorities ordered the first evacuation of southern Montserrat. From September through November, people returned to their homes as observatory scientists noted continued earthquake activity, ash and steam eruptions, and the swelling of a dome in the volcano's crater. Surveyors detected increasing deformation of the land around the volcano, indicating that magma was continuing to rise. On December 1, southern Montserrat was evacuated for a second time.

In the beginning of January 1996, some people again returned to their homes despite continued volcanic activity. Then, on April 6, the first in a series of pyroclastic flows erupted. Some flowed all the way to the sea. The southern part of the island was evacuated for a third time. Through the summer, small eruptions continued, swelling of the land increased, and earthquakes continually shook the island. Some people returned again to their homes and farms, but many, warned both by the obvious continued activity and by observatory scientists, stayed away.

The observatory recorded continued swelling of the dome, swarms of earthquakes, and many large and small eruptions, including pyroclastic flows, through the rest of 1996 and the first half of 1997. On June 25, 1997, major pyroclastic flows reached to within 50 meters of the airport. Surges and flows devastated several communities, killing 19 people.

Although some people returned to homes and farms during lulls in the long eruptive sequence, there is little doubt that the warnings issued by the observatory and evacuations ordered by the government saved many lives.

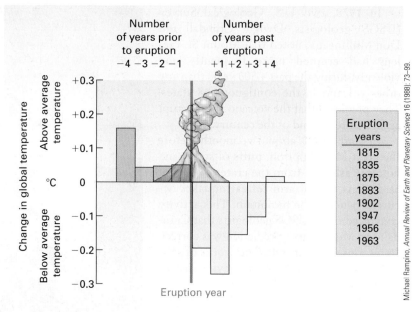

Michael Rampino, *Annual Review of Earth and Planetary Science* 16 (1988): 73–99.

◆ FIGURE 8.34 A plot of temperatures in the Northern Hemisphere shows that atmospheric cooling follows major volcanic eruptions.

8.9
Volcanic Eruptions and Global Climate

We have learned that magma production, the first step in the development of a volcanic eruption, occurs in the upper mantle. However, in another example of a systems interaction, an eruption can profoundly affect the atmosphere, climate, and living organisms. For example, the eruptions of Mount Pinatubo in 1991 produced the greatest ash and sulfur clouds in the latter half of the twentieth century. Satellite measurements show that the total solar radiation reaching Earth's surface declined by 2 to 4 percent after the Pinatubo eruptions. The following two years, 1992 and 1993, were a few tenths of a degree Celsius cooler than the temperatures of the previous decade. Temperatures rose again in 1994, after the ash and sulfur settled out.

A plot of global temperatures before and after eight recent major volcanic eruptions shows a correlation between global cooling and volcanic eruptions (◆ Figure 8.34). The correlation substantiates meteorological models showing that high-altitude dust reflects sunlight and cools the atmosphere.

Historic eruptions have been minuscule compared with some in the more distant past. About 248 million years ago, at the end of the Permian Period, 90 percent of all marine species and two-thirds of reptile and amphibian species died suddenly in the most catastrophic mass extinction in Earth history. This extinction event coincided with a massive volcanic eruption in Siberia that disgorged a million cubic kilometers of flood basalt onto Earth's surface to form a great lava plateau. The eruption must have released massive amounts of ash and sulfur compounds into the upper atmosphere, leading to cooler global climates. Many geologists think that Earth cooled enough to cause or at least contribute to the mass extinction. However, as discussed earlier, other scientists have proposed alternative mechanisms for the extinction.

Large reservoirs of carbon are stored in Earth's mantle, where basaltic magma forms. Modern basalt eruptions emit about 0.44 percent by weight as much carbon dioxide as they do lava. Thus huge amounts of carbon dioxide enter the atmosphere during the eruption of a large basalt plateau. Carbon dioxide is a greenhouse gas that causes warming of Earth's atmosphere.

About 120 million years ago, a massive upwelling of mantle rock created the mid-Cretaceous superplume, which erupted to form a vast submarine lava plateau beneath the Pacific Ocean off the west coast of South America. These eruptions also released large quantities of carbon dioxide. Mid-Cretaceous fossils indicate that Earth's climate was several degrees warmer than at present. Dinosaurs flourished in lush swamps, and huge coal deposits formed. Climate models show that this volcanic episode probably warmed Earth's climate by 12 to 15°C during mid-Cretaceous time.

Note that one mechanism links volcanic eruptions with global cooling and, paradoxically, the other associates eruptions with warming. A dark cloud cools the air immediately. Increased atmospheric carbon dioxide warms the air over longer time periods. Some volcanic eruptions may emit more ash and less carbon dioxide, whereas others may emit a higher proportion of gases and less ash. In either case, volcanic activity can profoundly alter climate.

We have learned that movements of mantle rock deep within Earth drive most volcanic activity. However, recent research indicates that changes in climate and sea level may affect volcanic activity. In one study, scientists noted a statistical correlation between sea level change and explosive volcanic activity in the Mediterranean.[3] The authors of the study postulated that changes in sea level alter the pressure on magma that is slowly rising through the crust. If sea level rises, pressure increases on magma beneath an island volcano and the magma is less likely to erupt. If sea level falls, pressure decreases and an eruption becomes more likely. In Chapter 16 we explain how sea level rises and falls in response to climate changes.

Therefore, if this hypothesis is correct, climate changes and fluctuations in sea level affect the frequency of volcanic eruptions. Volcanic eruptions, in turn, affect climate. This circle of cause and effect reminds us how complex and interactive Earth systems really are.

3. W. J. McGuire et al., "Correlation between Role of Sea-Level Change and Frequency of Explosive Volcanism in the Mediterranean," *Nature* 389 (October 2, 1997): 473.

EARTH SYSTEMS INTERACTIONS

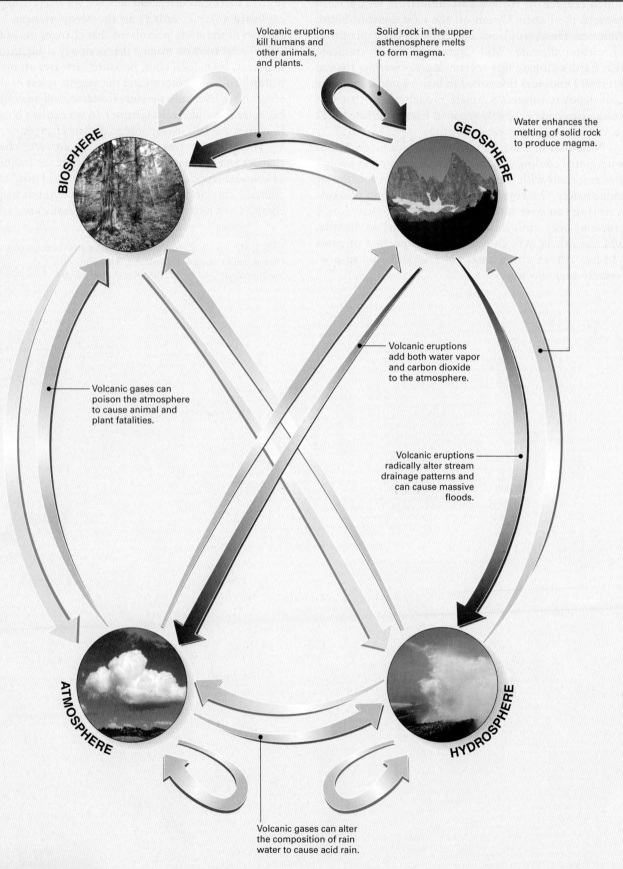

Volcanic eruptions kill humans and other animals, and plants.

Solid rock in the upper asthenosphere melts to form magma.

BIOSPHERE

GEOSPHERE

Water enhances the melting of solid rock to produce magma.

Volcanic eruptions add both water vapor and carbon dioxide to the atmosphere.

Volcanic gases can poison the atmosphere to cause animal and plant fatalities.

Volcanic eruptions radically alter stream drainage patterns and can cause massive floods.

ATMOSPHERE

HYDROSPHERE

Volcanic gases can alter the composition of rain water to cause acid rain.

EARTH SYSTEMS INTERACTIONS

SUMMARY

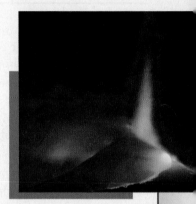

Rocks of the asthenosphere partially melt to produce basaltic magma as a result of three processes: rising temperature, **pressure-release melting**, and addition of water. These processes occur beneath spreading centers, in mantle plumes, and in subduction zones to form both volcanoes and plutons. Basaltic magma forms by **partial melting** of mantle peridotite. Granitic magma forms by a two-step process, in which basaltic magma forms by any of the previously mentioned processes and then rises into and melts granitic rocks of the lower continental crust. Earth's earliest continents were probably formed by partial melting of the original peridotite crust to form basalt, and then by further partial melting of the basalt to form andesite and then granite.

Basaltic magma usually erupts in a relatively gentle manner onto Earth's surface from a **volcano**. In contrast, granitic magma typically solidifies within Earth's crust. When granitic magma does erupt onto the surface, it often does so violently. These contrasts in behavior of the two types of magma are caused by differences in silica and water content.

Any intrusive mass of igneous rock is a **pluton**. A **batholith** is a pluton with more than 100 square kilometers of exposure at Earth's surface. A **dike** and a **sill** are both sheetlike plutons. Dikes cut across layering in country rock, and sills run parallel to layering.

Magma may flow onto Earth's surface as lava or may erupt explosively as **pyroclastic** material. Fluid lava forms **lava plateaus** and **shield volcanoes**. A pyroclastic eruption may form a cinder cone. Alternating eruptions of fluid lava and pyroclastic material from the same vent create a **composite cone**. When granitic magma rises to Earth's surface, it may erupt explosively, forming **ash-flow tuffs** and **calderas**.

Volcanic eruptions are common near a subduction zone, near a spreading center, and at a hot spot over a mantle plume, but are rare in other environments. Eruptions on a continent are often violent, whereas those in oceanic crust are gentle. Such observations form the basis of regional predictions of volcanic hazards. Short-term predictions are made on the basis of earthquakes caused by magma movements, swelling of a volcano, increased emissions of gas and ash from a vent, and other signs that magma is approaching the surface. Large volcanic eruptive episodes are associated with global climate change.

Key Terms

pressure-release melting   182
partial melting   186
pluton   189
batholith   190
stock   190
dike   191
sill   191
lava   192
pahoehoe   192

aa   192
columnar joint   192
pyroclastic rock   192
volcanic ash   192
cinders   192
fissure   192
flood basalt   192
lava plateau   192
volcano   193

vent   193
crater   193
shield volcano   194
cinder cone   194
composite cone   195
stratovolcano   195
ash flow   198
ash-flow tuff   198
caldera   198

For Review

1. Describe several different ways in which volcanoes and volcanic eruptions can threaten human life and destroy property.

2. Describe three processes that generate magma in the asthenosphere.

3. Describe magma formation in a spreading center, a hot spot, and a subduction zone.

4. Describe the origin of granitic magma.

5. How much silica does average granitic magma contain? How much does basaltic magma contain?

6. How much water does average granitic magma contain? How much does basaltic magma contain?

7. Why does magma rise soon after it forms?

8. Explain why basaltic magma and granitic magma behave differently as they rise toward Earth's surface.

9. Many rocks and even entire mountain ranges at Earth's surface are composed of granite. Does this observation imply that granite forms at the surface?

10. Explain the difference between a dike and a sill.

11. How do a shield volcano, a cinder cone, and a composite cone differ from one another? How are they similar?

12. How does a composite cone form?

13. How does a caldera form?

14. Explain why additional eruptions in Yellowstone Park seem likely. Describe what such an eruption might be like.

For Discussion

1. How and why does pressure affect the melting point of rock and, conversely, the solidification temperature of magma? How does the explanation differ for basaltic and granitic magma?

2. Why does water play an important role in magma generation in subduction zones, but not in the other two major environments of magma generation?

3. How could you distinguish between a sill exposed by erosion and a lava flow?

4. Imagine that you detect a volcanic eruption on a distant planet but have no other data. What conclusions could you draw from this single bit of information? What types of information would you search for to expand your knowledge of the geology of the planet?

5. Explain why some volcanoes have steep, precipitous faces, but why many do not.

6. Compare and contrast the danger of living 5 kilometers from Yellowstone National Park with the danger of living an equal distance from Mount St. Helens. Would your answer differ for people who live 50 kilometers or for those who live 500 kilometers from the two regions?

7. Use long-term prediction methods to evaluate the volcanic hazards in the vicinity of your college or university.

8. Discuss some possible consequences of a large caldera eruption in modern times. What is the probability that such an event will occur?

9. We explained how volcanic eruptions can influence climate and how climate can influence volcanic eruptions. Discuss the roles of the biosphere, the hydrosphere, the atmosphere, and the geosphere in these interactions.

The Snake River and Teton Range, Grand Teton National Park, Wyoming, USA. The highest peak in the photo is the Grand Teton.

© Royalty-Free/Corbis

Mountains form some of the most majestic landscapes on Earth. People find peace in the high mountain air and quiet valleys. But mountains also display nature's power. Fifty million years ago, India collided with southern Asia, folding the rocks and thrusting the Himalayas skyward. Today, storms swirl among the peaks, creating rain and snow that slowly erode the rocks. Glaciers and streams scour deep valleys into the bedrock. In the first part of this chapter, we will study how rocks behave when tectonic forces stress them. Then we will learn how tectonic processes formed Earth's mountain ranges. ■

ThomsonNOW™ Throughout this chapter, the ThomsonNOW logo indicates an opportunity for online self-study, which:
- Assesses your understanding of important concepts and provides a personalized study plan
- Links you to animations and simulations to help you study
- Helps to test your knowledge of material and prepare for exams
Visit ThomsonNOW at http://www.thomsonedu.com/login to access these resources.

9.1 Folds and Faults: Geologic Structures

How Rocks Respond to Tectonic Stress

Continents move about as fast as your fingernail grows. This movement is too slow to feel or sense in any human way. Yet, try to imagine the forces involved as continent- or ocean-sized slabs of rock slowly grind past one another or collide head on. In the past two chapters, we studied how tectonic activity generates earthquakes and volcanic eruptions. In this chapter we will consider how rock crumples and buckles under tectonic forces.

Stress is a force exerted against an object. When a rock is stressed, the rock may deform elastically or plastically, or it may simply break by brittle fracture, as described in Chapter 7. Several factors control how a rock responds to stress:

1. The nature of the rock. Think of a quartz crystal, a gold nugget, and a rubber ball. If you strike quartz with a hammer, it shatters. That is, it fails by brittle fracture. In contrast, if you strike the gold nugget, it deforms in a plastic manner—it flattens and stays flat. If you hit the rubber ball, it deforms elastically and rebounds immediately, sending the hammer flying back at you.

   Initially, all rocks react to stress by deforming elastically by a slight amount. Near Earth's surface, where temperature and pressure are low, different types of rocks behave differently with continuing stress. Granite and quartzite tend to fracture in a brittle manner. Other rocks, such

as shale, limestone, and marble, tend to deform plastically.

2. Temperature. The higher the temperature, the greater the tendency of a rock to deform in a plastic manner. It is difficult to bend an iron bar at room temperature, but if the bar is heated to a red-hot temperature, it becomes plastic and bends easily.

3. Pressure. High pressure favors plastic behavior. During burial, both temperature and pressure increase. Both factors promote plastic deformation, so deeply buried rocks have a greater tendency to bend and flow than do shallow rocks.

4. Time. Stress applied slowly, rather than suddenly, also favors plastic behavior. Marble park benches in New York City have sagged plastically under their own weight within 100 years. In contrast, rapidly applied stress, such as the blow of a hammer, causes brittle fracture in a marble bench.

Geologic Structures

Tectonic movement creates tremendous stress near plate boundaries; this stress deforms rocks. A **geologic structure** is any feature produced by rock deformation. Tectonic stress creates three types of structures: folds, faults, and joints.

Folds

A **fold** is a bend in rock. Some folded rocks display little or no fracturing, indicating that the rocks deformed in a plastic manner. In other cases, folding occurs by a combination of plastic deformation and brittle fracture (◆ Figure 9.1). Folds formed in this manner exhibit many tiny fractures.

FIGURE 9.1 This rock on the shore of the Nahanni River, Northwest Territories, Canada, folded plastically and then fractured.

If you hold a sheet of clay between your hands and squeeze your hands together, the clay deforms into a sequence of folds (◆ Figure 9.2). This demonstration illustrates three characteristics of folds:

1. Folding usually results from compression. For example, tightly folded rocks in the Himalayas indicate that the region was squeezed as tectonic plates converged (◆ Figure 9.3).
2. Folding always shortens the horizontal distances in rock. Notice in ◆ Figure 9.4 that the distance between the two points A and A′ is shorter in the folded rock than it was before folding.
3. A fold usually occurs as part of a group of many similar folds.

◆ Figure 9.5 summarizes the characteristics of five common types of folds. The figure shows that a fold arching upward is called an **anticline** and one arching downward is a **syncline**. The sides of a fold are called the **limbs**. Even though an anticline is structurally a high point in a fold, anticlines do not necessarily form topographic ridges. Conversely, synclines do not always form valleys. Landforms are created by combinations of

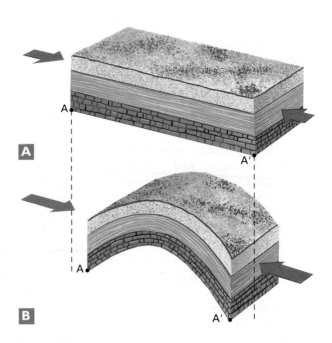

FIGURE 9.3 Folded sedimentary rocks in Himalayas reflect tectonic compression. Zanskar River Canyon, India.

FIGURE 9.2 Clay deforms into folds when compressed.

FIGURE 9.4 **(A)** Horizontally layered sedimentary rocks **(B)** fold in response to compressive stress. Notice that points A and A′ are closer after folding.

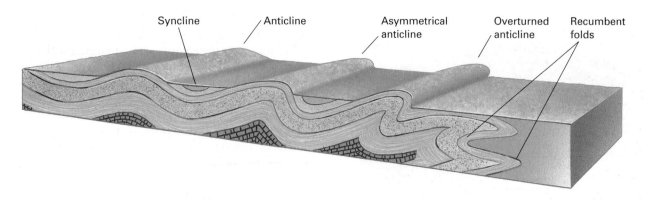

Syncline   Anticline   Asymmetrical   Overturned   Recumbent
                          anticline   anticline   folds

◆ **FIGURE 9.5**   Five common types of folds. Folds can be symmetrical, as shown on the left, or asymmetrical, as shown in the center. If a fold has tilted beyond the perpendicular, it is overturned.

tectonic and surface processes. In ◆ Figure 9.6, a syncline lies beneath the peak and an anticline forms a saddle between two peaks simply because some portions of the folded rock eroded faster than others.

A circular or elliptical anticlinal structure is called a **dome**. A dome resembles an inverted cereal bowl. Sedimentary layering dips away from the top of a dome in all directions (◆ Figure 9.7). A similarly shaped syncline is called a **basin**. Domes and basins can be small structures only a few kilometers in diameter or less. Alternatively, a large basin or dome can be 500 kilometers in diameter. These huge structures form by the sinking or rising of continental crust in response to vertical movements of the underlying mantle. The Black Hills of South Dakota are a

Courtesy of Graham R. Thompson/Jonathan Turk

◆ **FIGURE 9.6**   On this peak in the Canadian Rockies, the syncline lies beneath the summit and the anticline forms the low point on the ridge.

INTERACTIVE QUESTION: *Give a plausible explanation for these relationships.*

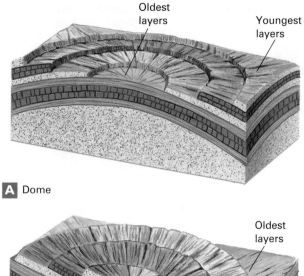

A Dome

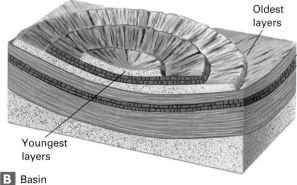

B Basin

◆ **FIGURE 9.7** **(A)** Sedimentary layering dips away from a dome in all directions. **(B)** Layering dips toward the center of a basin.

large structural dome. The Michigan basin covers much of the state of Michigan, and the Williston basin covers much of eastern Montana, northeastern Wyoming, the western Dakotas, and southern Alberta and Saskatchewan.

Faults

A **fault** is a fracture along which rock on one side has moved relative to rock on the other side (◆ Figure 9.8). **Slip** is the distance that rocks on opposite sides of a fault have moved. Movement along a fault may be gradual, or the rock may move suddenly, generating an earthquake. Some faults are a single fracture in rock; most, however, consist of numerous closely spaced fractures within a **fault zone** (◆ Figure 9.9). Rock may slide hundreds of meters or many kilometers along a large fault zone.

Rock moves repeatedly along many faults and fault zones for two reasons: (1) Tectonic forces commonly persist in the same place over long periods of time (for example, at a tectonic plate boundary). (2) Once a fault forms, it is easier for rock to move again along the same fracture than for a new fracture to develop nearby.

Hydrothermal solutions often precipitate rich ore veins in a fault zone. Miners then dig shafts and tunnels along veins to get the ore. Many faults are not vertical

but dip into Earth at an angle. Therefore, many veins have an upper and a lower side. Miners refer to the side that hangs over their heads as the **hanging wall** and the side they walk on is the **footwall** (◆ Figure 9.10). These names are commonly used to describe both ore veins and faults.

A **normal fault** forms where tectonic movement stretches Earth's crust, pulling it apart. As the crust stretches and fractures, the hanging wall moves down relative to the footwall. Notice that the horizontal distance between points on opposite sides of the fault, such as A and A′ in ◆ Figure 9.10, is greater after normal faulting occurs.

◆ Figure 9.11A shows a wedge-shaped block of rock called a **graben** dropped downward between a pair of normal faults. The word *graben* comes from the German word for "a long depression, ditch, or valley." (Think of a large block of rock settling downward to form a valley.) If tectonic forces stretch the crust over a large area, many normal faults may develop, allowing numerous grabens to settle downward between the faults, as in the Basin and Range Province in Nevada (◆ Figure 9.11B). The blocks of rock between the down-dropped grabens then appear to have moved upward relative to the grabens; they are called **horsts**.

Normal faults, grabens, and horsts are common where the crust is rifting at a spreading center, such as the Mid-Oceanic Ridge and the East African Rift zone, and where tectonic forces stretch a single plate, as in the Basin and Range Province of Utah, Nevada, and adjacent parts of western North America. They reflect extension of Earth's crust.

In a region where compressive tectonic forces squeeze the crust, geologic structures must accommodate crustal shortening. A fold accomplishes shortening.

Alternatively, compressive forces may fracture the rock to produce a **reverse fault** (◆ Figure 9.12). In a reverse fault, the hanging wall has moved up relative to the footwall, as tectonic force squeezed the rock horizontally. In the figure, the distance between points A and A′ is shortened by the faulting.

A **thrust fault** is a special type of reverse fault that is nearly horizontal (◆ Figure 9.13). In some thrust faults, the rocks of the hanging wall have moved many kilometers over the footwall. For example, all of the rocks of Glacier National Park in northwestern Montana slid 50 to 100 kilometers eastward along a thrust fault to their present location. This thrust is one of many that formed from about 180 to 45 million years ago as compressive tectonic forces built the mountains of western North America. Most of those thrusts moved large slabs of rock, some even larger than that of Glacier Park, from west to east in a zone extending from Alaska to Mexico.

A **strike-slip fault** is one in which the fracture is vertical, or nearly so, and rocks on opposite sides of the fracture move horizontally past each other (◆ Figure 9.14). A transform plate boundary is a strike-slip fault.

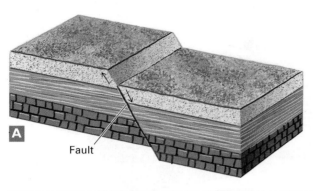

Fault

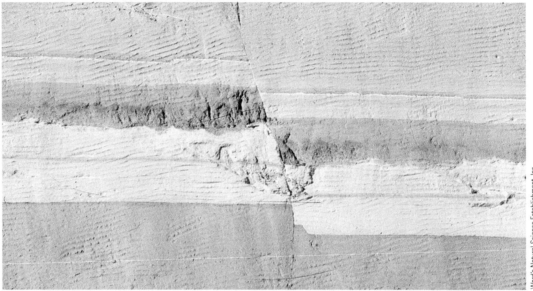

B

◈ **FIGURE 9.8** **(A)** A fault is a fracture along which rock on one side has moved relative to rock on the other side. **(B)** A small fault has dropped the right side of these volcanic ash layers downward about 60 centimeters relative to the left side.

Ward's Natural Science Establishment, Inc.

◈ **FIGURE 9.9** Faults with a large slip commonly move along numerous closely spaced fractures, forming a fault zone.

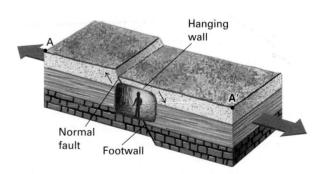

Hanging wall

Normal fault

Footwall

◈ **FIGURE 9.10** A normal fault accommodates extension of the crust. The upper side of a fault is called the hanging wall, and the lower side is called the footwall.

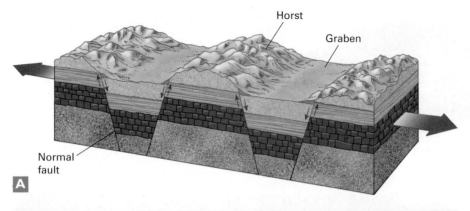

Horst

Graben

Normal fault

A

TSADO/NASA/Tom Stack and Assoc.

B

◇ **FIGURE 9.11** **(A)** Horsts and grabens commonly form where tectonic forces stretch the crust over a broad area. **(B)** The Basin and Range Province in Nevada is composed of parallel ridges, which are horsts, and valleys, which are grabens. The valleys have no external drainage and are filling with sediment. In spring, water from melting snow drains into the valleys, creating lakes. The water evaporates by midsummer, leaving dry lake beds that appear as light-tan areas in this satellite photograph.

As explained previously, the famous San Andreas Fault zone is a zone of strike-slip faults that form the boundary between the Pacific plate and the North American plate.

Joints

A **joint** is a fracture in rock and is therefore similar to a fault, except that in a joint, rocks on either side of the fracture have not moved. We discussed columnar joints in basalt in Chapter 8. Tectonic forces also create joints

(◇ Figure 9.15). Most rocks near Earth's surface are jointed, but joints become less abundant with depth because rocks become more plastic and less prone to fracturing at deeper levels in the crust.

Joints and faults are important in engineering, mining, and quarrying because they are planes of weakness in otherwise strong rock. Dams constructed in jointed rock often leak, not because the dams themselves have holes but because water seeps into the joints and flows around the dam through the fractures. You can

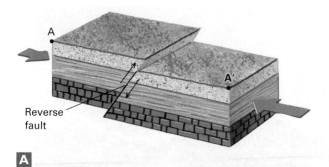

A

B

◈ **FIGURE 9.12** **(A)** A reverse fault accommodates crustal shortening when the crust is compressed. **(B)** A small reverse fault in Zion National Park, Utah.

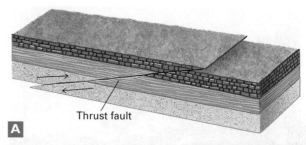

A

B

◈ **FIGURE 9.13** **(A)** A thrust fault is a low-angle reverse fault. **(B)** A small thrust fault near Flagstaff, Arizona.

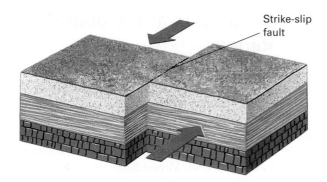

◈ **FIGURE 9.14** A strike-slip fault is nearly vertical, but movement along the fault is horizontal.

◈ **FIGURE 9.15** Joints such as these in sandstone along the Escalante River in Utah are fractures along which the rock has not slipped.

9.1   Folds and Faults: Geologic Structures   215

◆ **FIGURE 9.16** Wildly folded sedimentary rocks form the Nuptse-Lhotse wall, seen from an elevation of 7,600 meters on Mount Everest.

commonly see seepage caused by such leaks in canyon walls downstream from a dam.

Folds, Faults, and Plate Boundaries

Each of the three types of plate boundaries produces different tectonic stresses and therefore different kinds of folds and faults. Tectonic plates drift apart at a divergent boundary (the Mid-Oceanic Ridge system and continental rifts), stretching adjacent rock and producing normal faults and grabens but little folding of rocks.

At a transform boundary, friction often holds rock together as the plates gradually slip past each other. The resultant stress may fold, fault, and uplift nearby rocks. Forces of this type have formed the San Gabriel Mountains along the San Andreas Fault zone, as well as mountain ranges north of the Himalayas.

Near a convergent plate boundary, compression commonly produces large regions of folds, reverse faults, and thrust faults. Folds and thrust faults are common in the mountains of western North America, the Appalachian Mountains, the Alps, and the Himalayas, all of which formed at convergent boundaries (◆ Figure 9.16).

Although plate convergence commonly creates horizontal compression, in some instances crustal extension and normal faulting also occur at a convergent plate boundary. The Himalayas are an intensely folded mountain chain that formed by the collision between India and southern Asia. Yet, normal faults are common in parts of the Himalayas. We will describe how this occurs in Section 9.5.

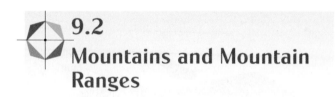

9.2
Mountains and Mountain Ranges

Tectonic forces have created mountains at each of the three types of tectonic plate boundaries. As you learned in Chapter 6, the world's largest mountain chain, the Mid-Oceanic Ridge system, formed at divergent plate boundaries beneath the ocean. Mountains also rise at divergent plate boundaries on land. Mount Kilimanjaro and Mount Kenya, two volcanic peaks near the equator, lie along the East African Rift. Other ranges, such as the San Gabriel Mountains of California, formed at transform plate boundaries. However, the volcanic island arcs of the southwestern Pacific Ocean, Alaska's Aleutian Islands, and the great continental mountain chains, including the Andes, Appalachians, Alps, Himalayas, and Rockies, all rose near convergent plate boundaries. Folding and faulting of rocks, earthquakes, volcanic eruptions, intrusion of plutons, and metamorphism all occur at a convergent plate boundary. The term **orogeny** refers to the process of mountain building and includes all of these activities.

Because plate boundaries are linear, mountains most commonly occur as long, linear, or slightly curved ranges and chains. For example, the Andes extend in a narrow band along the west coast of South America, and the Appalachians form a gently curving uplift along the east coast of North America.

Most continental mountain ranges rise because the crust becomes thicker where tectonic plates converge. Several processes thicken the crust:

1. In a subduction zone, the descending slab generates magma, which rises to cool within the overlying crust to form plutons, or erupts onto the surface to form volcanic peaks. Both the plutons and volcanic rocks thicken the continental crust by adding volumes of new material to it.
2. Magmatic activity heats the lithosphere above a subduction zone, causing it to expand and become thicker.
3. In a region where two continents collide, such as the modern Himalayas, one continent may be forced beneath the other. This process, called underthrusting, can double the thickness of continental crust in the collision zone (◆ Figure 9.17).
4. Compressive forces squeeze the crust horizontally to increase its thickness. These compressive forces are important in both subduction zones and continent–continent collisions.

As the lithosphere thickens, it also rises isostatically. As the mountain chain grows higher and heavier, eventually the underlying rocks cannot support the weight of the mountains. The crust and underlying lithosphere then spread outward beneath the mountains. As an analogy, consider pouring cold honey onto a table top. At first, the honey piles up into a high, steep mound, but soon it begins to flow outward under its own weight, lowering the top of the mound.

At the same time, streams, glaciers, and landslides erode the peaks as they rise, carrying the sediment into

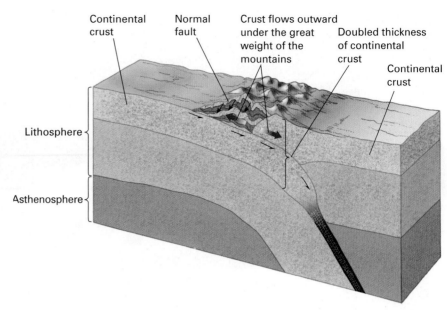

◆ **FIGURE 9.17** Several factors affect the height of a mountain range. Today the Himalayas are being uplifted by continued underthrusting. At the same time, erosion removes the tops of the peaks, and the sides of the range slip downward along normal faults. As these processes remove weight from the range, the mountains rise isostatically.

INTERACTIVE QUESTION: *Would you expect the height of a mountain range to rise and then decline at constant rates?*

adjacent valleys. Initially, when the mountains erode, they become lighter and rise isostatically, just as a canoe rises when you step out of it. Eventually, erosion wins over isostatic rebound. The Appalachians are an old range where erosion is now wearing away the remains of peaks that may once have been the size of the Himalayas.

With this background, let's look at mountain building in three types of convergent plate boundaries.

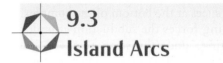

9.3
Island Arcs

An **island arc** is a volcanic mountain chain that forms where two plates carrying oceanic crust converge. The convergence causes the older, colder, and denser plate to sink into the mantle, creating a subduction zone and an oceanic trench. Magma forms in the subduction zone and rises to build submarine volcanoes. These volcanoes may eventually grow above sea level, creating an arc-shaped volcanic island chain next to the trench.

A layer of sediment a half kilometer or more thick commonly covers the basaltic crust of the deep sea floor. As the two plates converge, some of the sediment is scraped from the subducting slab and jammed against

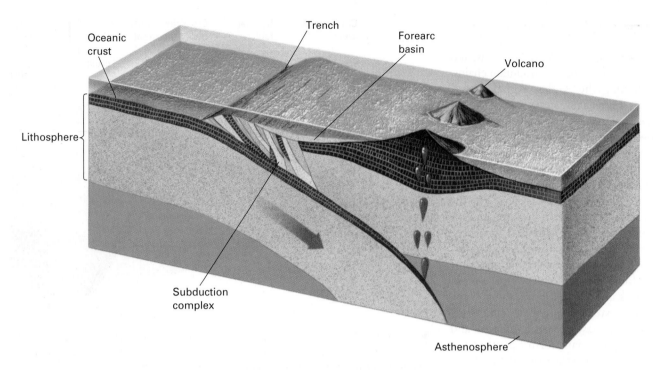

Oceanic crust

Trench

Forearc basin

Volcano

Lithosphere

Subduction complex

Asthenosphere

◆ **FIGURE 9.18** A subduction complex contains slices of oceanic crust and upper mantle scraped from the top of a subducting plate.

the inner wall (the wall toward the island arc) of the trench. Occasionally, slices of basalt from the oceanic crust, and even pieces of the upper mantle, are scraped off and mixed in with the sea-floor sediment. The process is like a bulldozer scraping soil from bedrock and occasionally knocking off a chunk of bedrock along with the soil. This scraping and compression folds and fractures the sediment and rock. The rocks added to the island arc in this way are called a **subduction complex** (◆ Figure 9.18).

Growth of the subduction complex occurs by addition of the newest slices at the bottom of the complex. This **underthrusting** forces the subduction complex upward. In addition, underthrusting thickens the crust, leading to isostatic uplift of the subduction complex. At the same time, compressive forces fold the upper crust, forming a sedimentary basin called a **forearc basin** between the subduction complex and the island arc. This process is similar to holding a flexible notebook horizontally between your hands. If you move your hands closer together, the middle of the notebook bends downward to form a topographic depression analogous to a forearc basin. The forearc basin fills with sediment eroded from the volcanic islands.

Island arcs are abundant in the Pacific Ocean, where convergence of oceanic plates is common. The western Aleutian Islands and most of the island chains of the southwestern Pacific are island arcs (◆ Figure 9.19).

9.4 The Andes: Subduction at a Continental Margin

The Andes are the world's second-highest mountain chain, with 49 peaks above 6,000 meters, nearly 20,000 feet (◆ Figure 9.20). The highest peak is Aconcagua, at 6,962 meters. The Andes rise from the Pacific coast of South America, starting nearly at sea level. Igneous rocks make up most of the Andes, although the chain also contains folded sedimentary rocks, especially in the eastern foothills.

In early Jurassic time, about 190 million years ago, the lithospheric plate that included South America started moving westward. To accommodate the westward motion, oceanic lithosphere began to sink into the mantle beneath the west coast of South America. Thus a subduction zone had formed by early Cretaceous time, 140 million years ago (◆ Figure 9.21A).

By 130 million years ago, the sinking plate was generating vast amounts of basaltic magma (◆ Figure 9.21B). Some of this magma rose to the surface to erupt from volcanoes. Most of it, however, melted portions of the lower continental crust of South America to form andesitic and granitic magma. As a result, both volcanoes and plutons formed along the entire length of western South America. As the oceanic plate sank beneath

Jacques Descloitres, MODIS, NASA/GSFC

◇▶ **FIGURE 9.19** Alaska's Aleutian Islands are a volcanic island arc forming the northern portion of the Ring of Fire, a zone of volcanic activity surrounding the Pacific Ocean. Beneath the sea on the south side of the islands is the Aleutian trench, a deep depression in the sea floor caused by subduction of the Pacific Plate as it sinks beneath the Aleutians.

Courtesy of Graham R. Thompson/Jonathan Turk

◇▶ **FIGURE 9.20** The Cordillera Apolobamba in Bolivia rises over 6,000 meters.

9.4   The Andes: Subduction at a Continental Margin

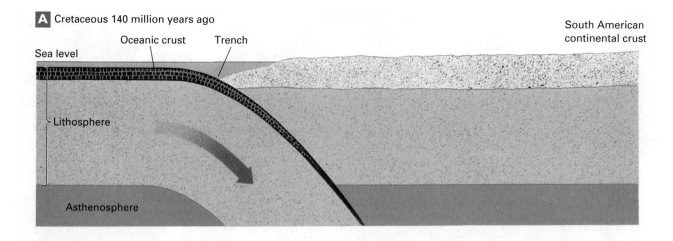

A Cretaceous 140 million years ago

Oceanic crust Trench

South American
continental crust

Sea level

Lithosphere

Asthenosphere

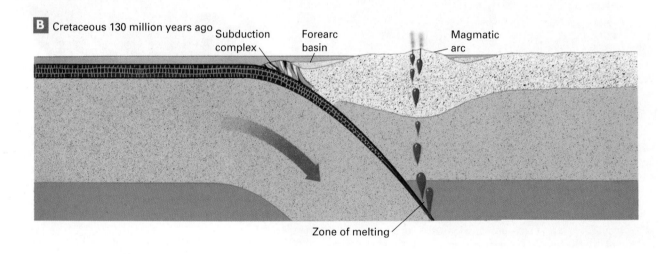

B Cretaceous 130 million years ago

Subduction
complex

Forearc
basin

Magmatic
arc

Zone of melting

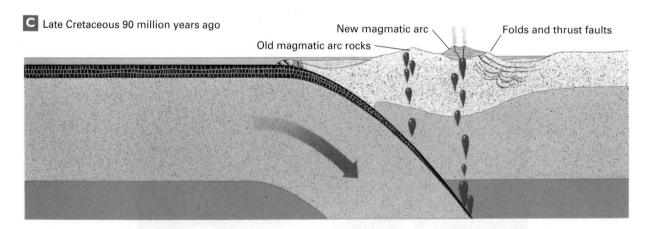

C Late Cretaceous 90 million years ago

New magmatic arc

Old magmatic arc rocks

Folds and thrust faults

◆ **FIGURE 9.21** Development of the Andes, seen in cross section looking northward. **(A)** As the South American lithospheric plate moved westward in early Cretaceous time, about 140 million years ago, a subduction zone and a trench formed at the west coast of the continent. **(B)** By 130 million years ago, igneous activity began and a subduction complex and forearc basin formed. **(C)** By 90 million years ago, the trench and region of igneous activity had both migrated eastward. Old volcanoes became dormant and new ones formed to the east.

INTERACTIVE QUESTION: *From the plate map in Chapter 6, would you expect the Andes to be impacted by continent–continent collision, as the Himalayas were? If your answer is yes, give a plausible time frame. If your answer is no, why not?*

the continent, slices of sea-floor mud and rock were scraped from the subducting plate, forming a subduction complex similar to that of an island arc.

The rising magma heated and thickened the crust beneath the Andes, causing it to rise isostatically and form great peaks. At the same time, the magmatic arc shifted several kilometers eastward. When the peaks became sufficiently high and heavy, the weak, soft rock oozed outward under its own weight. This spreading formed a great belt of thrust faults and folds along the east side of the Andes (◆ Figure 9.21C).

The Andes, then, are a mountain chain consisting predominantly of igneous rocks formed by subduction at a continental margin. The chain also contains extensive sedimentary rocks on both sides of the mountains; those rocks formed from the sediment eroded from the rising peaks. The Andes are a good general example of subduction at a continental margin, and this type of plate margin is called an **Andean margin**.

9.5 The Himalayas: A Collision between Continents

The world's highest mountain chain, the Himalayas, separates China from India and includes the world's highest peaks (◆ Figure 9.22). If you were to stand on the southern edge of the Tibetan Plateau and look southward, you would see the high peaks of the Himalayas. Beyond this great mountain chain lie the rainforests and hot, dry plains of India. If you had been able to stand in the same place 200 million years ago and look southward, you would have seen only ocean. At that time, India was located south of the equator, separated from Tibet by thousands of kilometers of open ocean. The Himalayas had not yet begun to rise (◆ Figure 9.23).

Formation of an Andean-Type Margin

By 80 million years ago, a triangular piece of lithosphere that included present-day India had split off from a large mass of continental crust near the South Pole (◆ Figure 9.24A). It began drifting northward toward Asia at a high speed—geologically speaking—perhaps as fast as 20 centimeters per year. As this Indian plate started to move, its leading edge, consisting of oceanic crust, sank beneath Asia's southern margin. This subduction formed an Andean-type continental margin along the edge of Tibet; volcanoes erupted and granite plutons rose into the continent (◆ Figure 9.24B).

◆ **FIGURE 9.22** The Himalayas rose when a plate carrying oceanic crust converged with a plate carrying continental crust, followed by a continent–continent collision. Lamo-she, Eastern Tibet.

Continent–Continent Collision

By 40 million years ago, subduction had consumed all of the oceanic lithosphere between India and Asia and the two continents collided (◆ Figure 9.25A). Because both are continental crust, neither could sink deeply into the mantle. Igneous activity then ceased because subduction stopped (◆ Figure 9.25B). The collision did not stop the northward movement of India, but it did slow it down to about 5 centimeters per year.

As India continued to move northward, it began to underthrust beneath Tibet, doubling the thickness of continental crust in the region (◆ Figure 9.25C). Thick piles of sediment that had accumulated on India's northern continental shelf were scraped from harder basement rock as India slid beneath Tibet. These

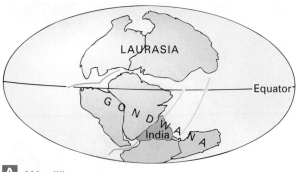

A 200 million years ago

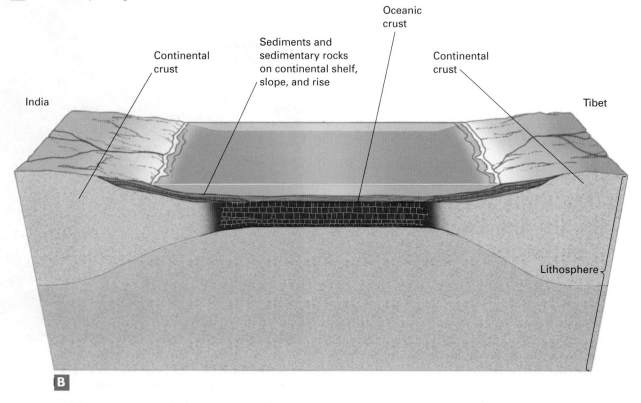

B

◆ **FIGURE 9.23** **(A)** Two hundred million years ago, India was part of a large continent located near the South Pole. **(B)** At that time India, southern Asia, and the intervening ocean basin were parts of the same lithospheric plate. (In this figure, the amount of oceanic crust between Indian and Asian continental crust is abbreviated to fit the diagram.)

sediments were pushed into great folds and thrust faults (◆ Figures 9.25 B and C).

At the same time, India crushed Tibet and wedged China out of the way along huge strike-slip faults. India has pushed southern Tibet 1,500 to 2,000 kilometers northward since the beginning of the collision. These compressional forces have created major mountain ranges and basins north of the Himalayas.

The Himalayas Today

Today, the Himalayas contain igneous, sedimentary, and metamorphic rocks. Many of the sedimentary rocks contain fossils of shallow-dwelling marine organisms that lived in the shallow sea of the Indian conti-

nental shelf. Plutonic and volcanic Himalayan rocks formed when the range was an Andean margin. Rocks of all types were metamorphosed by the tremendous stresses and heat generated during subduction and continent–continent collision.

The underthrusting of India beneath Tibet and the squashing of Tibet have doubled the thickness of continental crust and lithosphere under the Himalayas and the Tibetan Plateau to the north. Consequently, the region floats isostatically at high elevation (◆ Figure 9.25C). Even the valleys lie at elevations of 3,000 to 4,000 meters, and the Tibetan Plateau has an average elevation of 4,000 to 5,000 meters. One reason that the Himalayas contain all of Earth's highest peaks is simply that the entire plateau lies at such a high elevation. From the

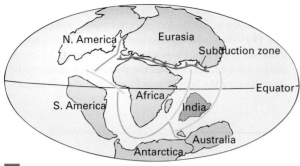

A 80 million years ago

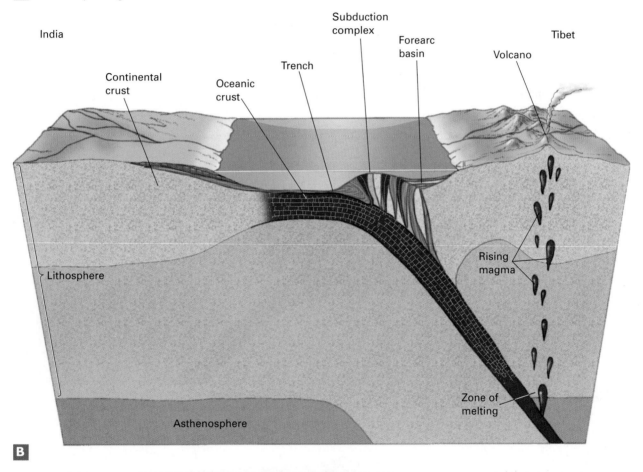

B

◆ **FIGURE 9.24** **(A)** By 80 million years ago, India was migrating northward, approaching the equator. **(B)** When India began moving northward, the plate broke and subduction began at the southern margin of Asia. By 80 million years ago, an oceanic trench and subduction complex had formed. Volcanoes erupted and granite plutons formed in the region now called Tibet. (Again, the amount of oceanic crust is abbreviated.)

valley floor to the summit, Mount Everest is actually smaller than Alaska's Denali (Mount McKinley), North America's highest peak. Mount Everest rises about 3,300 meters from base to summit, whereas Denali rises about 4,200 meters. The difference in elevation of the two peaks lies in the fact that the base of Mount Everest is at about 5,500 meters, but Denali's base is at 2,000 meters.

As the Himalayas rose, they became too heavy to support their own weight. The crust beneath the range then spread outward. This stretching created normal faults in the high mountains. As the crust spread away from the highest parts of the range, it compressed rocks near the margins of the chain. As a result, extensional normal faults formed in the mountains, while compressional thrust faults developed in the foothills. At the same time, tectonic forces resulting from the continued northward movement of India pushed the mountains upward. These processes continue today and no one knows when India will stop its northward movement or

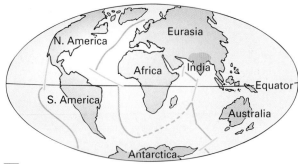

◆ **FIGURE 9.25** **(A)** By 40 million years ago, India had moved 4,000 to 5,000 kilometers northward and collided with Asia. **(B)** When India collided with Tibet, the leading edge of India was underthrust beneath southern Tibet. **(C)** Continued plate movement has doubled the thickness of continental crust, creating the high Tibetan Plateau and the Himalayas.

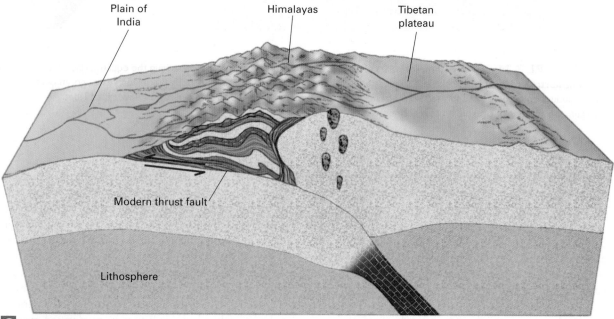

Volcanism ceases when India begins to underthrust

Folds and thrust faults in sedimentary rocks

India

Tibet

B 40 million years ago

Plain of India

Himalayas

Tibetan plateau

Modern thrust fault

Lithosphere

C Today

CHAPTER 9 • Mountains

how high the mountains will become. However, when the uplift ends, erosion will eventually lower the lofty peaks to rolling hills.

The Himalayan chain is only one example of a mountain chain built by a collision between two continents. The Appalachian Mountains formed when eastern North America collided with Europe, Africa, and South America between 470 and 250 million years ago.

The European Alps formed during repeated collisions between northern Africa and southern Europe beginning about 30 million years ago. The Urals, which separate Europe from Asia, formed by a similar process about 250 million years ago.

9.6
Mountains and Earth Systems

Tectonic forces that originate deep within Earth lift mountain ranges. However, once mountains rise, they interact with the hydrosphere, atmosphere, and biosphere. Many of these interactions are described in Chapter 6.

Air rises as it flows across a mountain range. As we will learn in Chapter 19, moisture condenses from rising air to produce rain or snow. Rain forms streams that race down steep hillsides, eroding gullies and canyons, while snow may accumulate to form glaciers that scour soil and bedrock. Thus mountains promote precipitation, which erodes mountains.

Recent research has shown that the rising of the Himalayas coincided with global cooling. Some scientists have suggested that the rising mountains caused this cooling trend, although many other factors may have contributed to the cooling trend. In Chapter 6, we described how a rising mountain chain affects local climate by altering wind and precipitation patterns and by causing development of glaciers. In Chapter 21, we will discuss how the rising of the Himalayas may have affected global climate.

Mountains affect plant and animal habitats. Cool climates prevail at high elevations, even at the equator. Mean annual temperature changes as much over 1,000 meters of elevation as it does over a distance of 1,000 kilometers of latitude. As a result, plant and animal

Courtesy of Graham R. Thompson/Jonathan Turk

◆ **FIGURE 9.26** As the human population in mountainous regions increases, people have cut forests on steep hillsides to plant crops. This terraced hillside in Nepal collapsed during heavy monsoon rainfall.

communities change rapidly with elevation. For example, if you climb one of Colorado's 14,000-foot mountains, you reach alpine tundra near the summit. But if you traveled north from Colorado at low elevations, you would not reach similar tundra until you arrived at the Canadian Barren Lands, 3,000 kilometers away.

In many mountainous regions, human populations have increased dramatically over the past few decades. As farmland has become scarce, people have cut forests and cultivated steep slopes. Unfortunately, when the protective forest and native grass are removed, soil erosion increases dramatically. Mud washes into streams and landslides carry soil downslope (◀◆ Figure 9.26). In this manner, human activity increases erosion rates. For example, in mountainous regions of Ethiopia, soil loss of 40 tons per hectare[1] has caused annual decrease in crop production of 1 to 2 percent per year. If current rates continue, the land will be barren in 50 to 100 years.

About half of the world's humans rely on water that falls in mountains and is often stored in forest soils or as snow or glacial ice. When people cut forests and replace them with farmland, however, rainwater and snowmelt are not retained efficiently. Thus stream flow increases dramatically in the wet season and declines abnormally during dry times. Global warming, discussed in Chapter 21, has led to dramatic melting of glaciers and early spring melting of winter snowfields. Thus runoff patterns are changing. These alterations in the hydrological cycle increases the risk of alternating periods of flood and drought, threatening downstream communities.

---

1. A hectare equals approximately 2.47 acres.

EARTH SYSTEMS INTERACTIONS

TECTONIC PROCESSES CHANGE THE geosphere by moving continents and raising great mountain chains. Shifts in continental positions and mountains, in turn, alter atmospheric circulation and precipitation patterns to cause regional and, in some cases, global climate changes. Those changes disrupt old environments and habitats of the biosphere, and create new ones. Thus, tectonic processes that directly alter the geosphere, indirectly cause major changes in the biosphere, atmosphere, and hydrosphere. ■

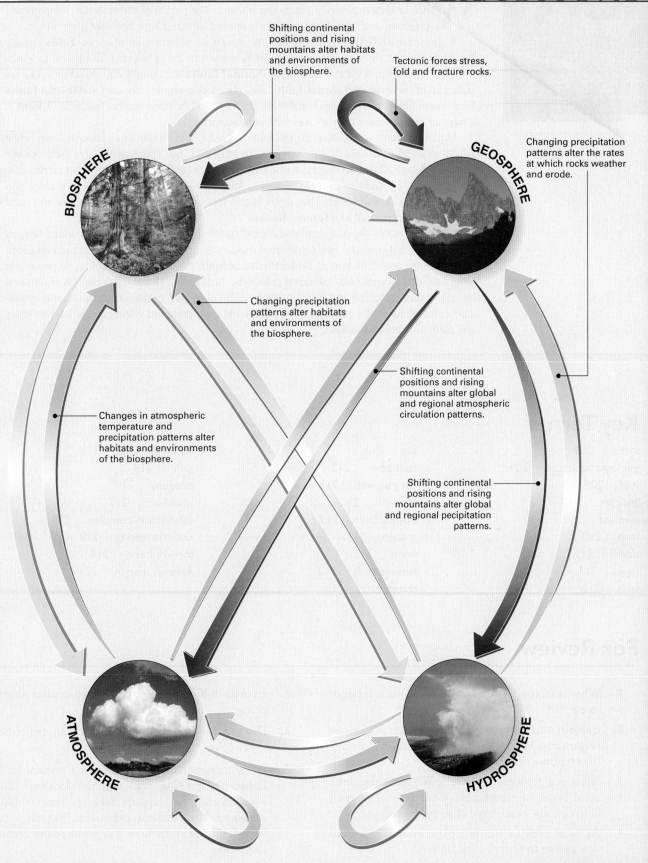

Shifting continental positions and rising mountains alter habitats and environments of the biosphere.

Tectonic forces stress, fold and fracture rocks.

Changing precipitation patterns alter the rates at which rocks weather and erode.

BIOSPHERE

GEOSPHERE

Changing precipitation patterns alter habitats and environments of the biosphere.

Shifting continental positions and rising mountains alter global and regional atmospheric circulation patterns.

Changes in atmospheric temperature and precipitation patterns alter habitats and environments of the biosphere.

Shifting continental positions and rising mountains alter global and regional precipitation patterns.

ATMOSPHERE

HYDROSPHERE

SUMMARY

When **stress** is applied to rocks, the rocks can deform in an elastic or a plastic manner, or they may rupture by brittle fracture. The nature of the material, temperature, pressure, and rate at which stress is applied all affect rock behavior under stress.

A **geologic structure** is any feature produced by deformation of rocks. **Folds** usually form when rocks are compressed. A **fault** is a fracture along which rock on one side has moved relative to rock on the other side. **Normal faults** are usually caused when rocks are pulled apart; **reverse** and **thrust faults** are caused by compression; and **strike-slip faults** form where blocks of crust slip horizontally past each other along vertical fractures. A **joint** is a fracture in rock where the rock has not been displaced.

Mountains form when the crust thickens and rises isostatically. They become lower when crustal rocks flow outward or are worn away by erosion. If two converging plates carry oceanic crust, a volcanic **island arc** forms. If one plate carries oceanic crust and the other carries continental crust, an **Andean margin** develops. Andean margins are dominated by granitic plutons and andesitic volcanoes. They also contain rocks of a **subduction complex** and sedimentary rocks deposited in a **forearc basin**.

When two plates carrying continental crust converge, an Andean margin develops first, as oceanic crust between the two continental masses is subducted. Later, when the two continents collide, one continent is **underthrust** beneath the other. The geology of mountain ranges formed by continent–continent collisions, such as the Himalayan chain, is dominated by vast regions of folded and thrust-faulted sedimentary and metamorphic rocks and by earlier-formed plutonic and volcanic rocks. Mountains exert profound effects on climate, weather, and plant and animal habitats.

Key Terms

stress 209	slip 212	strike-slip fault 212
geologic structure 209	fault zone 212	joint 214
fold 209	hanging wall 212	orogeny 217
anticline 210	footwall 212	island arc 217
syncline 210	normal fault 212	subduction complex 218
limb 210	graben 212	underthrusting 218
dome 211	horst 212	forearc basin 218
basin 211	reverse fault 212	Andean margin 221
fault 212	thrust fault 212	

For Review

1. What is tectonic stress? Explain the main types of stress.

2. Explain the different ways in which rocks can respond to tectonic stress. What factors control the response of rocks to stress?

3. What is a geologic structure? What are the three main types of structures? What type(s) of rock behavior does each type of structure reflect?

4. At what type of tectonic plate boundary would you expect to find normal faults?

5. Explain why folds accommodate crustal shortening.

6. Draw a cross-sectional sketch of an anticline–syncline pair and label the limbs.

7. Draw a cross-sectional sketch of a normal fault. Label the hanging wall and the footwall. Use your sketch to explain how a normal fault accommodates crustal extension. Sketch a reverse fault and show how it accommodates crustal shortening.

8. Explain the similarities and differences between a fault and a joint.

9. In what tectonic environment would you expect to find a strike-slip fault, a normal fault, and a thrust fault?

10. What mountain chain has formed at a divergent plate boundary? What are the main differences between this chain and those developed at convergent boundaries? Explain the differences.

11. Explain why erosion initially causes a mountain range to rise and then eventually causes the peak heights to decrease.

12. Describe the similarities and differences between an island arc and the Andes. Why do the differences exist?

13. Describe the similarities and differences between the Andes and the Himalayan chain. Why do the differences exist?

14. Draw a cross-sectional sketch of an Andean-type plate margin to a depth of several hundred kilometers.

15. Draw a sequence of cross-sectional sketches showing the evolution of a Himalayan-type plate margin. Why does this type of boundary start out as an Andean-type margin?

For Discussion

1. Discuss the relationships among types of lithospheric plate boundaries, predominant tectonic stress at each type of plate boundary, and the main types of geologic structures you might expect to find in each environment.

2. Why are thrust faults, reverse faults, and folds commonly found together?

3. Why do most major continental mountain chains form at convergent plate boundaries? What topographic and geologic features characterize divergent and transform plate boundaries in continental crust? Where do these types of boundaries exist in continental crust today?

4. Explain why extensional forces act on mountains rising in a tectonically compressional environment.

5. Explain why many mountains contain sedimentary rocks even though subduction leads to magma formation and the formation of igneous rocks.

6. Give a plausible explanation for the formation of the Ural Mountains, which lie in an inland portion of Asia.

7. Compare and explain the similarities and differences between the Andes and the Himalayan chain. How would the Himalayas, at their stage of development about 60 million years ago, have compared with the modern Andes?

8. Where would you be most likely to find large quantities of igneous rocks in the Himalayan chain—in the northern parts of the chain near Tibet, or southward, near India? Discuss why.

9. Geologists have measured some Himalayan peaks as now rising as fast as 1 centimeter per year. Compare this rate with the average rate of tectonic uplift. Offer a plausible explanation for any differences.

10. Where would you be most likely to find very old rocks—the sea floor, at the base of a growing mountain range, or within the central portion of a continent?

11. Explain how human development in mountains might endanger mountain residents and, at the same time, impact downstream settlements.

12. Explain how mountains affect climate and how climate impacts mountains. Discuss the relative time frames of these interactions.

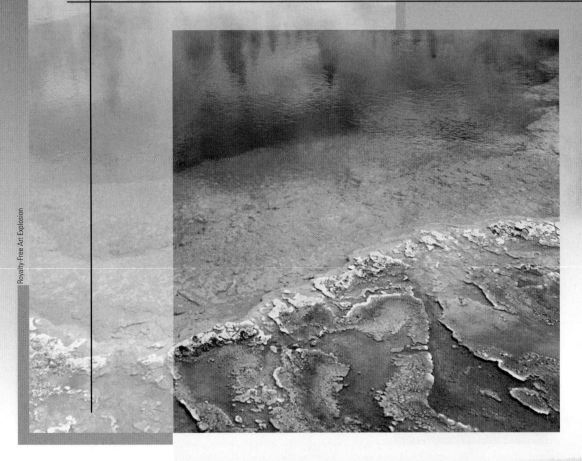

UNIT 3

SURFACE PROCESSES

CHAPTER 10

Weathering, Soil, and Erosion

© Royalty-Free/Corbis

Storm waves and sea currents have eroded the foundations of houses in Pacifica, California.

As the solar system formed, swarms of meteorites, comets, and asteroids crashed into the planets and their moons, forming huge impact craters on their surfaces. The earliest craters on Earth's surface quickly disappeared because our planet was so hot that plastic and molten rock oozed inward to fill the craters, much as soft mud will flow back in to fill a hole made by a falling rock. As the crust thickened and cooled, a second wave of bombardment formed a new set of craters. Today, however, none of these ancient craters remain on Earth's surface, but the Moon is pockmarked with craters of all ages. Why have the craters vanished from Earth but been preserved on the Moon?

Tectonic processes such as mountain building, volcanic eruptions, and earthquakes continually change the face of our planet and renew Earth's surface over geologic time. In addition, Earth is sufficiently massive to have retained its atmosphere and water. The atmosphere and hydrosphere then weather and erode the surface rocks of the geosphere. The combination of tectonic activity and erosion has eliminated all traces of early craters from Earth's surface. In contrast, the smaller Moon has lost most of its heat, so tectonic activity is nonexistent. In addition, the Moon's gravitational force is too weak to have retained an atmosphere or water to erode its surface. As a result, the Moon's ancient craters have persisted for billions of years. The combination of weathering, erosion, and tectonic activity continuously changes Earth's surface. ∎

ThomsonNOW Throughout this chapter, the ThomsonNOW logo indicates an opportunity for online self-study, which:
• Assesses your understanding of important concepts and provides a personalized study plan
• Links you to animations and simulations to help you study
• Helps to test your knowledge of material and prepare for exams
Visit ThomsonNOW at http://www.thomsonedu.com/login to access these resources.

10.1 Weathering and Erosion

Return to your childhood haunts and you will see that the rock outcrops in woodlands or parks have not changed. Over geologic time, however, air and water attack rocks and break them down at Earth's surface. The processes that decompose rocks and convert them to loose gravel, sand, clay, and soil are called **weathering** (◀▶ Figure 10.1).

Weathering involves little or no movement of the decomposed rocks and minerals. The weathered material simply accumulates where it forms. However, loose soil and other weathered material offer little resistance to rain or wind and are easily eroded. Thus **erosion** processes soon pick up and carry off weathered rocks and minerals. Rain, running water, wind, glaciers, and gravity are the most common agents of erosion. These agents may then transport the weathered material great distances and finally deposit it as layers of sediment at Earth's surface.

Weathering, erosion, transport, and deposition typically occur in an orderly sequence. For example, water freezes in a crack in granite, weathering the rock and loosening a grain of quartz. A hard rain erodes the grain and washes it into a stream. The stream then transports the quartz to the seashore and deposits it as a grain of sand on a beach (◀▶ Figure 10.2).

Weathering occurs by both mechanical and chemical processes. **Mechanical weathering** (also called **physical weathering**) reduces solid rock to small fragments but does not alter the chemical composition of rocks and minerals. Think of crushing a rock with a hammer; the fragments are no different from the parent rock except that they are smaller.

◀▶ **FIGURE 10.1** Weathering processes are breaking this boulder into smaller fragments. Erosion has not removed the fragments, so the original shape of the boulder is still visible.

Courtesy of Graham R. Thompson/Jonathan Turk

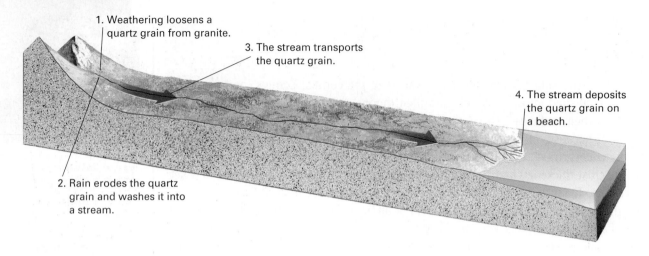

1. Weathering loosens a quartz grain from granite.

2. Rain erodes the quartz grain and washes it into a stream.

3. The stream transports the quartz grain.

4. The stream deposits the quartz grain on a beach.

◆ **FIGURE 10.2** Weathering loosens a quartz grain from granite. Rain erodes the grain, washing it into a stream. The stream transports it and finally deposits the grain on a beach.

In contrast, **chemical weathering** occurs when air and water chemically react with rock to alter its composition and mineral content. Chemical weathering is similar to rusting of a steel knife blade in that the final products differ both physically and chemically from the starting material.

10.2 Mechanical Weathering

Five processes cause mechanical weathering: pressure-release fracturing, frost wedging, abrasion, organic activity, and thermal expansion and contraction. Two additional processes—salt cracking and hydrolysis-expansion—result from combinations of mechanical and chemical processes.

Pressure-Release Fracturing

Many igneous and metamorphic rocks form deep below Earth's surface. Imagine, for example, that a granitic pluton solidifies from magma at a depth of 15 kilometers. At that depth, the pressure from the weight of overlying rock is about 5,000 times that at Earth's surface. Over millennia, tectonic forces may raise the granite to form a mountain range. As the granite rises, the overlying rock erodes and the pressure decreases. When the pressure diminishes, the rock expands, but because the rock is now cool and brittle, it fractures as it

expands. This process is called **pressure-release fracturing**. Many igneous and metamorphic rocks that formed at depth, but now lie at Earth's surface, have fractured in this manner (◆ Figure 10.3).

Frost Wedging

Although pressure release fracturing occurs by processes involving only the geosphere, all other forms of weathering involve interactions between rocks of the geosphere and the hydrosphere, atmosphere, and biosphere.

Water expands by 7 to 8 percent when it freezes. If water accumulates in a crack and then freezes, the ice expands in a process called **frost wedging**. In spring and fall in a temperate climate, water freezes at night and thaws during the day. Ice pushes rock apart but at the same time

Courtesy of Graham R. Thompson/Jonathan Turk

◆ **FIGURE 10.3** Pressure-release fracturing contributed to the fracturing of this granite in California's Sierra Nevada.

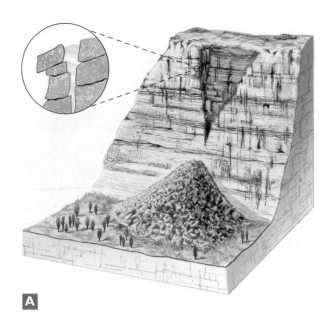

Courtesy of Graham R. Thompson/Jonathan Turk

◈ **FIGURE 10.4** **(A)** Frost wedging dislodges rocks from cliffs and creates talus slopes. **(B)** Frost wedging produced this talus cone in the Valley of the Ten Peaks, Canadian Rockies.

cements it together. When the ice melts, rock fragments tumble from a steep cliff. Experienced mountaineers try to travel in the early morning before ice melts. Large piles of loose angular rocks, called **talus**, lie beneath many cliffs (◈ Figure 10.4). These rocks fell from the cliffs mainly as a result of frost wedging.

Abrasion

Rocks, grains of sand, and silt collide with one another when currents or waves carry them along a stream or beach. During these collisions, their sharp edges and corners wear away and the particles become rounded.

The mechanical wearing of rocks by friction and impact is called **abrasion** (◈ Figure 10.5). Note that water itself is not abrasive—it is the collisions among rock, sand, and silt that round the rocks.

Wind hurls sand and other small particles against rocks, sandblasting unusual shapes (◈ Figure 10.6). Glaciers also abrade rock as they drag rocks, sand, and silt across bedrock.

Courtesy of Graham R. Thompson/Jonathan Turk

◈ **FIGURE 10.5** Abrasion rounded these rocks in a streambed in Yellowstone National Park, Wyoming.

Courtesy of Graham R. Thompson/Jonathan Turk

◈ **FIGURE 10.6** Wind abrasion selectively weathered and eroded the base of this rock in Lago Poopo, Bolivia, because windblown sand moves mostly near the ground surface. The base of the pedestal is about a meter in diameter.

Organic Activity

If soil collects in a crack in bedrock, a seed may fall there and sprout. The roots work their way into the crack, expand, and may eventually widen the crack (◀▶ Figure 10.7). City dwellers often see the results of this type of **organic activity** where tree roots raise and crack concrete sidewalks.

Thermal Expansion and Contraction

Rocks at Earth's surface are exposed to daily and yearly cycles of heating and cooling. They expand when they are heated and contract when they cool. When temperature changes rapidly, the surface of a rock heats or cools faster than its interior, and as a result, the surface expands or contracts faster than the interior. The forces generated by this **thermal expansion** and **contraction** may fracture the rock.

In mountains or deserts at midlatitudes, temperature may fluctuate from –5°C to +25°C during a spring day. Is this 30° difference sufficient to fracture rocks? The answer is uncertain. In one laboratory experiment, scientists heated and cooled granite repeatedly by more than 100°C and no fractures formed. These results imply that normal temperature changes might not be an important cause of mechanical weathering. However, the rocks used in the experiment were small and the experiment was carried out over a brief time. Perhaps thermal expansion and contraction are more significant in large outcrops, or perhaps daily heating-cooling cycles repeated over hundreds of thousands of years may promote fracturing.

In contrast to small daily or annual temperature changes, fire heats rock by hundreds of degrees. If you line a campfire with granite stones, the rocks commonly

◀▶ **FIGURE 10.8** Fire causes rocks to spall, breaking in concentric shells. Yosemite National Park, 2006.

break as you cook your dinner. In a similar manner, forest fires or brush fires occur commonly in many ecosystems and are an important agent of mechanical weathering (◀▶ Figure 10.8).

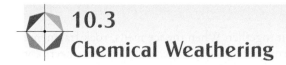

10.3
Chemical Weathering

Rock is durable over a human lifetime. Over geologic time, however, air and water chemically attack rocks near Earth's surface. The most important processes of chemical weathering are dissolution, hydrolysis, and oxidation. Water, acids and bases, and oxygen cause these processes to decompose rocks.

Dissolution

We are all familiar with the fact that some minerals dissolve readily in water while others do not. If you put a crystal of halite (rock salt, or table salt) in water, the crystals rapidly dissolve to form a solution. The process is called **dissolution**. Halite dissolves so rapidly and completely in water that it is rare in moist, natural environments. On the other hand, if you drop a crystal of quartz into pure water, only a tiny amount will dissolve while nearly all of the crystal will remain intact.

To understand how water dissolves a mineral, think of an atom on the surface of a crystal. It is held in place because it is attracted to the other atoms in the crystal by the electrical forces called chemical bonds. At the same time, electrical attractions to the outside environment pull the atom away from the crystal. The result is like a tug-of-war. If the bonds between the atom and the crystal are stronger than the attraction of the atom to its outside environment, the crystal remains intact. If outside

◀▶ **FIGURE 10.7** As this tree grew from a crack in bedrock, its growing roots widened the crack.

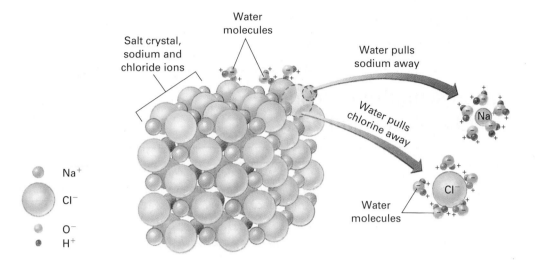

Salt crystal, sodium and chloride ions

Water molecules

Water pulls sodium away

Water pulls chlorine away

Na

Water molecules

Cl⁻

Na⁺

Cl⁻

O⁻

H⁺

◆ **FIGURE 10.9** Halite dissolves in water because the attractions between the water molecules and the sodium and chloride ions are greater than the strength of the chemical bonds in the crystal.
(Note that although the oxygen atoms are labeled − and the hydrogen atoms are labeled +, these are partial charge separations only.)

attractions are stronger, they pull the atom away from the crystal and the mineral dissolves.

Water (H_2O) is a polar molecule. The oxygen atom has a great affinity for electrons, which move away from the hydrogen atoms to concentrate around the oxygen. The two hydrogen atoms are then left with a slight positive charge because their electrons have moved closer to the oxygen atom. When a crystal of halite is dropped in a glass of water, the negatively charged oxygen atoms in the water pull the positively charged sodium ions away from the halite, while the positively charged hydrogen atoms remove the negatively charged chlorine ions (◆ Figure 10.9).

Rocks and minerals dissolve more rapidly when water is acidic or basic. An acidic solution contains a high concentration of hydrogen ions (H^+), whereas a basic solution contains a high concentration of hydroxyl ions (OH^-). Acids and bases dissolve most minerals more effectively than pure water because they provide more electrically charged hydrogen and hydroxyl ions to pull atoms out of crystals. For example, limestone is made of the mineral calcite ($CaCO_3$). Calcite barely dissolves in pure water but is quite soluble in acid. If you place a drop of strong acid on limestone, bubbles of carbon dioxide gas instantly form as the calcite dissolves.

Water found in nature is never pure. Atmospheric carbon dioxide dissolves in raindrops and reacts to form a weak acid called carbonic acid (see the Focus On box for a description of this process). As a result, even the purest rainwater falling in the Arctic or on remote mountains is slightly acidic. Air pollution can make rain even more acidic.

Thus water—especially if it is acidic or basic—dissolves ions from soil and bedrock and carries the dissolved material away. Ground water also dissolves rock to produce spectacular caverns in limestone (◆ Figure 10.10). Cave formation is discussed further in Chapter 11.

Courtesy Graham R. Thompson/Jonathan Turk

◆ **FIGURE 10.10** Caverns form when ground water dissolves limestone.

FOCUS ON

Representative Reactions in Chemical Weathering

Dissolution of Calcite

Calcite, the mineral that comprises limestone and marble, weathers in natural environments in a three-step process. In the first two steps, water reacts with carbon dioxide in the air to produce carbonic acid, a weak acid:

$$CO_2 + H_2O \;\text{Step 1}\; H_2CO_3 \;\text{Step 2}\; H^+ + HCO_3^-$$

Carbon dioxide   Water   Carbonic acid   Hydrogen ion   Bicarbonate ion

In the third step, calcite dissolves in the carbonic acid solution.

$$CaCO_3 + H^+ \;\text{Step 3}\; Ca^2 + HCO_3^-$$

Calcite   Hydrogen ion   Calcium ion   Bicarbonate ion

Hydrolysis

This is an example of a reaction in which feldspar hydrolyzes to clay.

$$2KAlSi_3O_8 + 2H^+ + H_2O \longrightarrow Al_2Si_2O_5(OH)_4 + 2K^+ + 4SiO_2$$

Orthoclase feldspar   Hydrogen ion   Water   Kaolinite, a clay mineral   Potassium ion   Silica

Hydrolysis

In **hydrolysis**, water reacts with a mineral to form a new mineral with the water as part of its crystal structure. Most common minerals weather by hydrolysis. For example, feldspar, the most abundant mineral in Earth's crust, weathers by hydrolysis to form clay (see the Focus On box "Representative Reactions in Chemical Weathering").

Quartz is the only rock-forming silicate mineral that does not weather to form clay. Quartz is resistant to chemical weathering because it dissolves extremely slowly and only to a small extent. When granite weathers, the feldspar and other minerals decompose to form clay but the unaltered quartz grains fall free from the rock. Hydrolysis has so deeply weathered some granites that quartz grains can be pried out with a fingernail at depths of several meters (◀ Figure 10.11). The rock looks like granite but has the consistency of sand.

Because quartz is so tough and resistant to both mechanical and chemical weathering, it is the primary component of sand. Much of this sand is transported

Courtesy of Graham R. Thompson/Jonathan Turk

◆ **FIGURE 10.11** Coarse grains of quartz and feldspar accumulate directly over weathered granite. The lens cap in the middle illustrates scale.

to streams and ultimately to the sea-coast, where it concentrates on beaches and eventually forms sandstone.

Oxidation

Many elements react with atmospheric oxygen, O_2. Iron rusts when it reacts with water and oxygen. Rusting is an example of a more general process called **oxidation**.[1] Iron is abundant in many minerals; if the iron in such a mineral oxidizes, the mineral decomposes.

Many valuable metals such as iron, copper, lead, and zinc occur as sulfide minerals in ore deposits. When they oxidize during weathering, the sulfur reacts to form sulfuric acid, a strong acid. The sulfuric acid washes into streams and ground water, where it may harm aquatic organisms. Many natural ore deposits generate sulfuric acid when they weather, and the reaction may be accelerated when ore is dug up and exposed at a mine site. This problem was discussed in Chapter 5.

Chemical and Mechanical Weathering Acting Together

Chemical and mechanical weathering work together, often on the same rock at the same time. After mechanical processes fracture a rock, water and air seep into the cracks to initiate chemical weathering.

In environments where ground water is salty, saltwater seeps through pores and cracks in bedrock. When the water evaporates, the dissolved salts crystallize. The growing crystals exert tremendous forces that loosen mineral grains and widen cracks in a process called **salt cracking**. Thus salt chemically precipitates in rock, and the growing salt crystals mechanically loosen mineral grains and break the rock apart.

Many sea cliffs show pits and depressions caused by salt cracking because spray from the breaking waves brings the salt to the rock. Salt cracking is also common in deserts, where surface water and ground water often contain dissolved salts ( Figure 10.12).

Granite commonly fractures by **exfoliation**, a process in which large plates or shells split away like the layers of an onion (◆ Figure 10.13). The plates may be only 10 or 20 centimeters thick near the surface, but they thicken with depth. Because exfoliation fractures

◆ **FIGURE 10.12** Crystallizing salt loosened sand grains to form this depression in sandstone in Cedar Mesa, Utah. The white patches are salt crystals.

are usually absent below a depth of 50 to 100 meters, they seem to be a result of exposure of the granite at Earth's surface.

Exfoliation is frequently explained as a form of pressure-release fracturing. However, many geologists suggest that hydrolysis-expansion may be the main cause of exfoliation. During hydrolysis, feldspars and other silicate minerals react with water to form clay. As a result of the addition of water, clays have a greater volume than the original minerals have. Thus a chemical reaction (hydrolysis) forms clay, and the mechanical expansion that occurs as the clay forms may cause exfoliation. This explanation is compatible with the observation that exfoliation concentrates near Earth's surface because water and chemical weathering are most abundant close to the surface.

10.4 Soil

Bedrock breaks into smaller fragments as it weathers, and much of it decomposes to clay and sand. Therefore, on most land surfaces, a thin layer of loose rock fragments, clay, and sand overlies bedrock (◆ Figure 10.14). This material is called **regolith**. Soil scientists define soil as upper layers of regolith that support plants, although many earth scientists and engineers use the terms *soil* and *regolith* interchangeably.

Components of Soil

Soil is a mixture of mineral grains, organic material, water, and gas. The mineral grains include clay, silt, sand, and rock fragments. Clay is so fine grained and closely

---

1. Oxidation is properly defined as the loss of electrons from a compound or element during a chemical reaction. In the weathering of common minerals, this usually occurs when the mineral reacts with molecular oxygen.

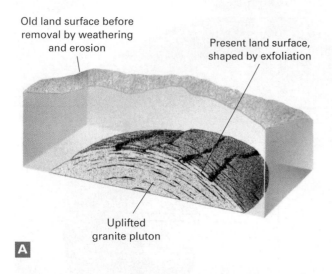

Old land surface before removal by weathering and erosion

Present land surface, shaped by exfoliation

Uplifted granite pluton

A

B

◆ **FIGURE 10.13** **(A)** Exfoliation occurs when concentric rock layers fracture and become detached from a granite outcrop. **(B)** Exfoliation has fractured this granite in Pinkham Notch, New Hampshire.

packed that water, and even air, do not flow through a clay-rich soil readily, so plants growing in clay soils suffer from lack of oxygen. In contrast, water and air flow easily through sandy soil. The most fertile soil is **loam**, a mixture of sand, clay, silt, and generous amounts of organic matter.

Minerals and organic matter contain nutrients necessary for plant growth (Table 10.1). If you walk through a forest or prairie, you can find bits of leaves, stems, and flowers on the soil surface. This material is called **litter**. When litter decomposes sufficiently that you can no longer determine the origin of individual pieces, it becomes **humus**. Humus is an essential component of most fertile soils. Humus soaks up so much moisture that humus-rich soil swells after a rain and shrinks during dry spells. This alternate shrinking and swelling loosens the soil, allowing roots to grow into it easily.

◆ **FIGURE 10.14** A very thin, dark-colored layer of soil overlies light-colored limestone to support this forest in Canada's Northwest Territories. Most plant life is supported by a soil layer only a few centimeters to a few meters thick.

A rich layer of humus also insulates the soil from excessive heat and cold and reduces water loss by evaporation. Humus also retains soil nutrients and makes them available to plants.

In intensive agriculture, farmers commonly plow the soil and leave it exposed for weeks or months. Humus oxidizes in air and decomposes. At the same time, rain dissolves soil nutrients and carries them away. Farmers replace the lost nutrients with chemical fertilizers but rarely replenish the humus. As a result, much of the soil's ability to absorb and regulate water and nutrients is lost. When rainwater flows over the surface, it transports soil particles, excess fertilizer, and pesticide residues, polluting streams and ground water.

Soil Horizons

A typical well-developed soil consists of several layers called **soil horizons**. The uppermost layer is called the **O horizon**, named for its organic component. This layer consists mostly of litter and humus, with a small proportion of minerals (◆ Figure 10.15). The next layer down, called the **A horizon**, is a mixture of humus, sand, silt, and clay. The combined O and A horizons are called **topsoil**. A kilogram of average fertile

Table 10.1 Nutrients Essential for Plant Growth

Macronutrients from air and water		Micronutrients from soil
Carbon ⎱ Oxygen ⎰ Hydrogen	The basic building blocks of all organic tissue	Iron Copper Manganese
Macronutrients from soil		Boron
Nitrogen	Most critical element for growth of plant proteins	Zinc Chlorine
Phosphorus	Important for metabolism and development of cell membranes	Molybdenum
Potassium	Maintains plant cell permeability	
Calcium	Important for plant cell walls	
Magnesium	Important for production of chlorophyll	
Sulfur	Required for synthesis of plant vitamins and proteins	

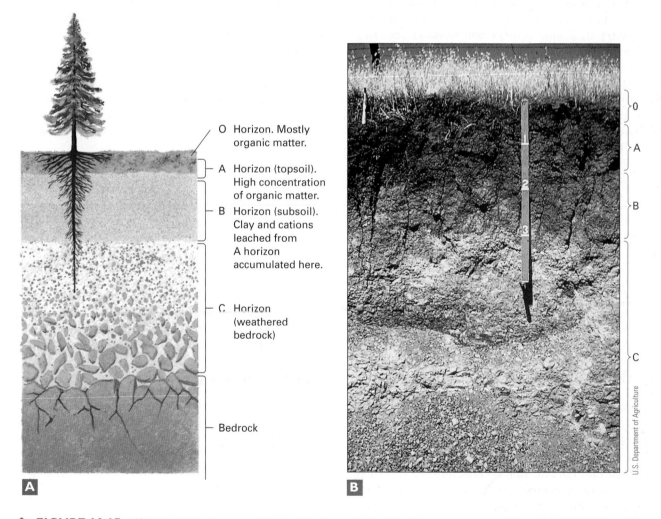

O Horizon. Mostly organic matter.

A Horizon (topsoil). High concentration of organic matter.

B Horizon (subsoil). Clay and cations leached from A horizon accumulated here.

C Horizon (weathered bedrock)

Bedrock

A

B

U.S. Department of Agriculture

◈ **FIGURE 10.15** **(A)** A typical soil consists of several horizons, distinguished by color, texture, and chemistry. **(B)** In this soil, the darkest uppermost layer is the O horizon; the A horizon is less dark; the whiter layer is the B horizon. The B horizon grades downward into the weathered bedrock of the C horizon. The scale is in feet.

FOCUS ON

Soil Erosion and Agriculture

In nature, soil erodes approximately as rapidly as it forms. However, improper farming, livestock grazing, and logging can accelerate erosion. Plowing removes plant cover that protects soil. Logging removes forest cover, and the machinery breaks up the protective litter layer. Similarly, intensive grazing strips away protective plants. Rain, wind, and gravity then erode the exposed soil easily and rapidly. Meanwhile, soil continues to form by weathering at its usual slow, natural pace. Thus increased rates of erosion caused by agriculture can lead to net soil loss. In addition, erosion contaminates waterways with silt as well as with herbicides and pesticides.

Soil is a natural resource that can be difficult or expensive to renew once it has become degraded. Intense pressures on the world's soil resources have resulted in loss of productivity and biodiversity and increasing desertification of once-fertile lands. When farmers use proper conservation measures, soil can be preserved indefinitely or even be improved. Some regions of Europe and China have supported continuous agriculture for centuries without soil damage. However, in recent years, marginal lands on hillsides, in tropical rainforests, and along the edges of deserts have been brought under cultivation. These regions are particularly vulnerable to soil deterioration, and today, soil is being lost at an alarming rate throughout the world.

The Worldwatch Institute estimates that about 0.4 billion tons of topsoil are created globally by weathering each year. However, soil is eroding more rapidly than it is forming on about 35 percent of the world's croplands. About 25 billion tons are lost every year. The soil lost annually would fill a train of freight cars long enough to encircle our planet 150 times.

In the United States, approximately one third of the topsoil that existed when the first European settlers arrived has been lost. However, as a result of increasing awareness of soil erosion and improved farming and conservation practices, soil erosion rates declined by almost 40 percent between 1982 and 2001. At present, erosion is continuing in the United States at an average rate of about 1.9 billion tons per year, costing $30 billion to $40 billion annually in lost productivity.

In China, 3.6 million square kilometers, more than one-third of the land, is affected by soil erosion, threatening the country's ability to continue to provide food and water for its 1.3 billion people. In 2004, 1.6 billion tons of topsoil were lost to erosion, equivalent to a 1-centimeter layer covering 125,000 square kilometers. The Chinese government is working on strategies to rehabilitate land that has been damaged or threatened by poor management.

Fertile cropland is also lost when it is developed for roads, cities, and suburban housing. In the United

topsoil contains about 30 percent by weight organic matter, including approximately 2 trillion bacteria, 400 million fungi, 50 million algae, 30 million protozoa, and thousands of larger organisms such as insects, worms, nematodes, and mites.

The third layer, the **B horizon** or subsoil, is a transitional zone between topsoil and weathered parent rock below. Roots and other organic material grow in the B horizon, but the total amount of organic matter is low. The lowest layer, called the **C horizon**, consists of partially weathered rock that grades into unweathered parent rock. This zone contains little organic matter.

When rain falls on soil, it sinks into the O and A horizons, weathering minerals, forming clay, and carrying dissolved ions to lower levels. This downward movement of water and dissolved ions is called **leaching**. The A horizon is sandy because water also carries the clay downward but leaves the sand behind.

Water, dissolved ions, and clay from the A horizon accumulate in the B horizon. This layer retains moisture because of its high clay content. Although moisture retention may be beneficial, if too much clay accumulates, the B horizon creates a dense, waterlogged soil.

Soil-Forming Factors

Why are some soils rich and others poor, some sandy and others loamy? Six factors control soil characteristics: parent rock, climate, rates of plant growth and decay, slope aspect and steepness, time, and transport of soil materials.

Parent Rock

The texture and composition of soil depend partly on its parent rock. For example, when granite decomposes, the feldspar converts to clay and the rock releases quartz as

States, 28,000 square kilometers of prime farmland, an area twice as large as the state of Connecticut, were urbanized between 1982 and 1997. Today, much of the fruits and vegetables grown in the United States are raised in rapidly urbanizing areas, mostly in California and Florida.

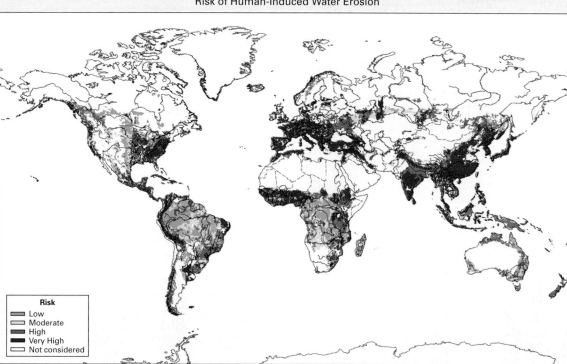

Risk of Human-Induced Water Erosion

Risk
Low
Moderate
High
Very High
Not considered

sand grains. If the clay leaches into the B horizon, a sandy soil forms. In contrast, because basalt contains no quartz, soil formed from basalt is likely to be rich in clay and contain only small amounts of sand. Nutrient abundance also depends in part on the parent rock. For example, pure quartz sandstone contains no nutrients, and soil formed on it must get its nutrients from outside sources.

Climate

Rainfall and temperature affect soil formation. Rain seeps downward through soil, but several other factors pull the water back upward. Roots suck soil water toward the surface, and water near the surface evaporates. In addition, water is electrically attracted to soil particles. This attraction initiates a process called **capillary action**, which also draws water upward toward the soil surface.

During a rainstorm, water seeps through the A horizon, dissolving soluble ions such as calcium, magnesium, potassium, and sodium. In arid and semiarid regions, rainstorms typically are of short duration and little rain falls. Consequently, when the rain stops, capillary action and plant roots then draw most of the water back up toward the surface, where it evaporates or is taken up by plants. As the water escapes, many of its dissolved ions precipitate in the B horizon, encrusting the soil with salts. A soil of this type is a **pedocal** (◆ Figure 10.16A). This process often deposits enough calcium carbonate in the form of the mineral calcite to form a hard cement called **caliche** in the soil. The Greek word *pedon* means soil; *pedocal* is a composite of *pedon* and the first three letters of *calcite*. In the Imperial Valley in California, irrigation water contains high concentrations of calcium carbonate. A thick, continuous layer of caliche forms in the soil as the water evaporates. To continue growing

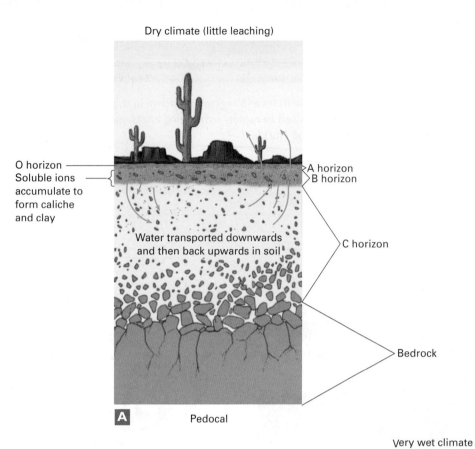

Dry climate (little leaching)

O horizon
Soluble ions
accumulate to
form caliche
and clay

A horizon
B horizon

C horizon

Bedrock

Water transported downwards
and then back upwards in soil

A   Pedocal

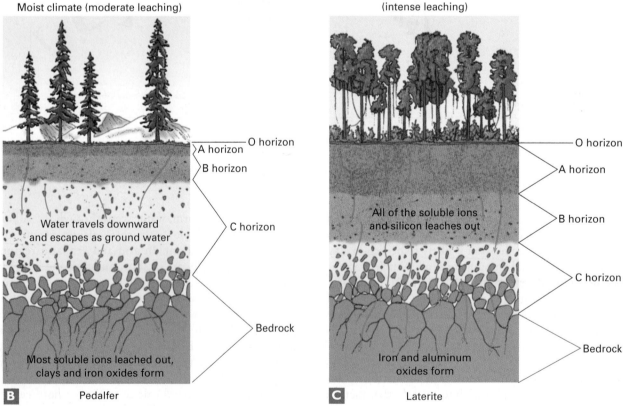

Moist climate (moderate leaching)

O horizon
A horizon
B horizon

C horizon

Bedrock

Water travels downward
and escapes as ground water

Most soluble ions leached out,
clays and iron oxides form

B   Pedalfer

Very wet climate
(intense leaching)

O horizon

A horizon

B horizon

C horizon

Bedrock

All of the soluble ions
and silicon leaches out

Iron and aluminum
oxides form

C   Laterite

◈ **FIGURE 10.16** **(A)** Pedocals, **(B)** pedalfers, and **(C)** laterites are different types of soils that form in different climates.

INTERACTIVE QUESTION: *Why is the A horizon in example A so much thinner than the A horizons in examples B and C?*

Courtesy of Graham R. Thompson/Jonathan Turk

◆ **FIGURE 10.17** Salts have poisoned this Wyoming soil. Saline water seeps into the depression and evaporates to deposit white salt crystals on the ground and on the fence posts.

crops, farmers must then rip this layer apart with heavy machinery.

Because nutrients concentrate when water evaporates, many pedocals are fertile if irrigation water is available. However, salts often concentrate so much that they become toxic to plants (◆ Figure 10.17). As mentioned previously, all streams contain small concentrations of dissolved salts. If arid or semiarid soils are intensively irrigated, salts can accumulate until plants cannot grow. This process is called **salinization**. Some historians argue that salinization destroyed croplands and thereby contributed to the decline of many ancient civilizations, such as the Babylonian Empire.

In a wet climate, rainstorms are of longer duration and more rain falls. As a result, water seeps downward through the soil to the water table, leaching soluble ions from both the A and B horizons. The less-soluble elements, such as aluminum, iron, and some silicon, remain behind, accumulating in the B horizon to form a soil type called a **pedalfer** (◆ Figure 10.16B). The subsoil in a pedalfer is commonly rich in clay, which is mostly aluminum and silicon and has the reddish color of iron oxide; hence the prefix *ped* is followed by the chemical symbols *Al*, for aluminum, and *Fe*, for iron.

In regions of high temperature and very high rainfall, such as a tropical rainforest, so much water seeps through the soil that it leaches away nearly all the soluble cations. Only very insoluble aluminum and iron minerals remain (◆ Figure 10.16C). Soil of this type is called a **laterite**. Laterites are often colored rust-red by iron oxide (◆ Figure 10.18). A highly aluminous laterite, called **bauxite**, is the world's main type of aluminum ore.

Courtesy of Graham R. Thompson/Jonathan Turk

◆ **FIGURE 10.18** This aluminum-rich Georgia laterite formed when water leached away the more soluble ions.

A

B

◆ **FIGURE 10.19** **(A)** Lateritic tropical soils, such as those in Vanuatu in the southwest Pacific Ocean, often support lush growth but contain few nutrients and little humus. **(B)** Arctic soil of Baffin Island, Canada, supports sparse vegetation and contains little organic matter.

The second important component of climate, average annual temperature, affects soil formation in two ways. First, chemical reactions proceed more rapidly in warm temperatures than in cold ones. Second, plant growth and decay are temperature dependent, as discussed next.

Rates of Plant Growth and Decay

In the tropics, plants grow and decay rapidly and growing plants quickly absorb the nutrients released by decaying plants. Heavy rainfall leaches nutrients from the soil, creating a laterite. As a result, little humus accumulates and few nutrients are stored in the soil. Thus even though the tropical rainforests support great populations of plants and animals, many of these forests are anchored on poor, laterite soils (◆ Figure 10.19A). The plants depend on a rapid cycle of growth, death, and decay. Conversely, the Arctic is so cold that plant growth and decay are slow. Therefore litter and humus form slowly and Arctic soils also contain little organic matter (◆ Figure 10.19B).

The most fertile soils are those of temperate prairies and forests. There, large amounts of plant litter drop to the ground in the autumn, but decay is slow during the winter, and plant growth during the growing season is not fast enough to extract all the nutrients from the soil. As a result, thick layers of nutrient-rich humus accumulate in temperate climates. ·

Slope Aspect and Steepness

Aspect is the orientation of a slope with respect to the Sun. In the semiarid regions in the Northern Hemisphere, thick soils and dense forests cover the cool, shady north slopes of hills, but thin soils and grass dominate hot, dry, southern exposures (◆ Figure 10.20). The reason for this difference is that in the Northern Hemisphere more water evaporates from the hot, sunny, southern slopes. Therefore, fewer plants grow, weathering occurs slowly, and

soil development is retarded. Plants grow more abundantly on the moister northern slopes and more rapid weathering forms thicker soils.

In general, hillsides have thin soils and valleys are covered by thicker soil, because soil erodes from hills and accumulates in valleys. When hilly regions were first settled and farmed, people naturally planted their crops in the valley bottoms, where the soil was rich and water was abundant. Recently, as population has expanded, farmers have moved to the thinner, less stable, hillside soils.

Time

Chemical weathering occurs slowly in most environments, and time is therefore an important factor in

◆ **FIGURE 10.20** In semiarid climatic regions, thick forests cover the cool, shady north slopes of hills (right), but grass and sparse trees dominate hot, dry southern exposures (left).

determining the extent of weathering. Recall that most minerals weather to clay. In geologically young soils, weathering may be incomplete and the soils may contain many partly weathered mineral fragments. As a result, young soils are often sandy or gravelly. As soils mature, weathering continues and the clay content increases.

Soil Transport

By studying recent lava flows, scientists have determined how quickly plants return to an area after it has been covered by hard, solid rock. In many cases, plants appear when a lava flow is only a few years old, even before weathering has formed soil. Closer scrutiny shows that the plants have rooted in tiny amounts of soil that wind or water transported from nearby areas.

In many of the world's richest agricultural areas, most of the soil was transported from elsewhere. Streams deposit sediment, wind deposits dust, and soil slides downslope from mountainsides into valleys. These foreign materials mix with locally formed soil, changing its composition and texture. The soils of river flood plains and deltas, and the rich windblown soils of central China and the American Great Plains, are examples of transported soils.

10.5
Erosion

Weathering decomposes bedrock, and plants add organic material to the regolith to create soil at Earth's surface. However, soil does not accumulate and thicken throughout geologic time. If it did, Earth would be covered by a mantle of soil hundreds or thousands of meters thick, and rocks would not exist at Earth's surface. Instead, interactions with flowing water, wind, and glaciers erode soil as it forms (◆ Figure 10.21). In addition, some weathered material simply slides downhill under the influence of gravity. In fact, all forms of erosion combine to remove soil about as fast as it forms. For this reason, soil is usually only a few meters thick or less in most parts of the world.

Once soil erodes, the clay, sand, and gravel begin a long journey as they are carried downhill by the same agents that eroded them: streams, glaciers, wind, and gravity. On their journey they may come to rest in a streambed, a sand dune, or a lake bed, but those environments are only temporary stops. Eventually, they erode again and are carried downhill until, finally, they are deposited where the land meets the sea. Some of the sediment accumulates on deltas, and coastal marine currents redistribute some of it along the shore. Eventually, younger sediment may bury older layers until they become lithified to form sedimentary rocks.

Erosion and transport of sediment by streams, glaciers, and wind are the subjects of Chapters 11, 13, and 14. In the last three sections of this chapter, we will discuss landslides: erosion by gravity.

10.6
Landslides

Mass wasting is the downslope movement of earth material, primarily caused by gravity. The word **landslide** is a general term for mass wasting and for the landforms created by mass wasting.

Although gravity acts constantly on all slopes, the strength of the rock and soil usually hold the slope in place. In some places, however, natural processes or human activity may destabilize a slope and cause mass wasting. For example, a stream can erode the base of a hillside, undercutting it until it slides. Rain, melting snow, or a leaking irrigation ditch can add weight and lubricate soil, causing it to slide downslope. Mass wasting occurs naturally in all hilly or mountainous terrain. Steep slopes are especially vulnerable, and landslide scars are common in the mountains.

In recent years, the global human population has increased dramatically. As the most-desirable land has become overpopulated, large numbers of people have moved to mountains once considered too harsh for homes and farms. In wealthier nations, people have moved into the hills to escape congested cities. As a result, permanent settlements have grown in previously uninhabited, steep terrain. Many of these slopes are naturally unstable. Construction and agriculture have destabilized others.

Courtesy of Graham R. Thompson/Jonathan Turk

◆ **FIGURE 10.21** Rain and flowing water erode soil as rapidly as it forms on these slopes in Arches National Monument.

The Hubbard Brook Experimental Forest

Ecologists have studied water and nutrient cycling in the Hubbard Brook Experimental Forest in New Hampshire. That particular ecosystem consists of a series of small valleys, each drained by a single creek. The bedrock beneath the valleys is impermeable, and water cannot escape by seeping underground. Researchers built concrete dams across several of the creeks and anchored the dams on bedrock. They then measured the amount of water flowing from each valley and the amounts of sediment and dissolved nutrients carried by each creek.

Figure 1 shows that in a healthy forest only 25 percent of the rainwater soaks into the soil or escapes as runoff; 75 percent returns to the atmosphere through transpiration and evaporation. Thus a forest recycles much of its water into the atmosphere.

To study the effects of vegetation on this ecosystem, scientists cut all the trees and shrubs in one Hubbard

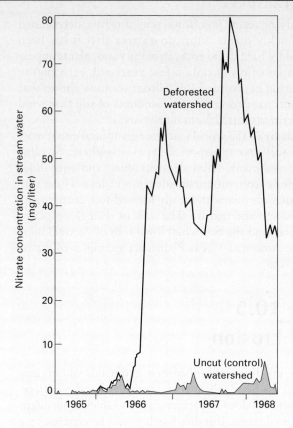

◆ FIGURE 2 The Hubbard Brook experiment showed that a completely devegetated forest valley lost its soil nutrients much faster than an adjacent, undisturbed watershed.

Brook valley and sprayed the soil with herbicides to prevent regrowth.

With no plants to absorb water, most of the rainwater flowed downslope quickly. Stream flow increased by 40 percent over that measured in an adjacent undisturbed valley. In addition, the amount of dissolved nutrients and sediment carried by the stream draining the devegetated valley increased.

The scientists at Hubbard Brook learned that the soils of the deforested valley lost nutrients six to eight times faster than the adjacent undisturbed watershed did (Figure 2). The great increase in water runoff carried large amounts of nutrients from the soil that otherwise would have been conserved by the vegetation. Scientists have recorded similar results for forests, grasslands, and wetlands. These experiments demonstrate the interrelationships between soil and vegetation. Plants maintain healthy soil, which, in turn, sustains them.

◆ FIGURE 1 A natural forest returns about 75 percent of rainfall to the atmosphere.

Every year, small landslides destroy homes and farmland. Occasionally, an enormous landslide buries a town or city, killing thousands of people. Landslides cause billions of dollars in damage every year, a cost approximately equal to the damage caused by earthquakes over a 20-year period. In many instances, losses occur because the average person does not recognize dangers that are obvious to most geologists.

Consider four examples of recent landslides that have affected humans:

1. A movie star builds a mansion on the edge of a picturesque California cliff. After a few years, the cliff collapses and the house slides into the valley (◀▶ Figure 10.22A).
2. A ditch carrying irrigation water across a hillside in Montana leaks water into the ground. After years of seepage, the muddy soil slides downslope and piles against a house at the bottom of the hill (◀▶ Figure 10.22B).
3. Excavations for roads and high-rise buildings undercut the base of a steep hillside in Hong Kong. Suddenly, the slope slides, destroying everything in its path (◀▶ Figure 10.22C).
4. On January 10, 2005, a landslide destroys 36 houses and kills 10 people in the community of La Conchita on the Southern California coast (◀▶ Figure 10.22D). The high bluff looming over La Conchita has produced numerous landslides described in historical accounts dating back to 1865, and the unstable nature of the slope was well-known by 1924 when the land was subdivided and developed. The portion of the slope that failed in 2005 was part of a mass of rock and soil that had moved during an earlier and larger landslide that had occurred in 1995.

A
J. T. Gill, USGS

B
Courtesy of Graham R. Thompson/Jonathan Turk

C
Hong Kong Government Information Services

D
© AP/Wide World Photos

◀▶ **FIGURE 10.22** Landslides cause billions of dollars in damage every year. **(A)** A few days after this photo was taken, the corner of the house hanging over the gully fell in. **(B)** A landslide, triggered by a leaking irrigation ditch, threatens a house in Darby, Montana. **(C)** A landslide destroyed several expensive buildings in Hong Kong. **(D)** This aerial photo taken on Jan. 11, 2005, shows the landslide that buried homes in La Conchita, California, south of Santa Barbara on the previous day. The landslide killed 10 people and injured more than a dozen.

Why do landslides occur? Is it possible to avoid or predict such catastrophes to reduce property damage and loss of life?

Why Do Landslides Occur?

Imagine that you are a geological consultant on a construction project. The developers want to build a road at the base of a hill, and they wonder whether landslides will threaten the road. What factors should you consider?

Steepness of the Slope

Obviously, the steepness of a slope is a factor in mass wasting. If frost wedging dislodges a rock from a steep cliff, the rock tumbles to the valley below. However, a similar rock is less likely to roll down a gentle hillside.

Type of Rock and Orientation of Rock Layers

If sedimentary rock layers dip in the same direction as a slope, the upper layers may slide over a layer of weak rock. Imagine a hill underlain by shale, sandstone, and limestone oriented so that their bedding lies parallel to

the slope, as shown in ◆ Figure 10.23A. If the base of the hill is undercut (◆ Figure 10.23B), the upper layers of sandstone and limestone may slide over the weak shale. In contrast, if the rock layers dip at an angle to the hillside, the slope may be stable even if it is undercut (◆ Figures 10.23C and 10.23D).

Several processes can reduce the stability of a slope. A stream or ocean waves can erode its base. Road building and excavation can also destabilize it. Therefore, a geologist or engineer must consider not only a slope's stability before construction but how the project might alter its stability.

The Nature of Unconsolidated Materials

The **angle of repose** is the maximum slope or steepness at which loose material remains stable. If the slope becomes steeper than the angle of repose, the material slides. The angle of repose varies for different types of material. Rocks commonly tumble from a cliff to collect at the base as angular blocks of talus. The angular blocks interlock and jam together. As a result, talus typically has a steep angle of repose, up to 45°. In contrast, rounded sand grains do not interlock and therefore have a lower angle of repose, about 30° to 35° (◆ Figure 10.24).

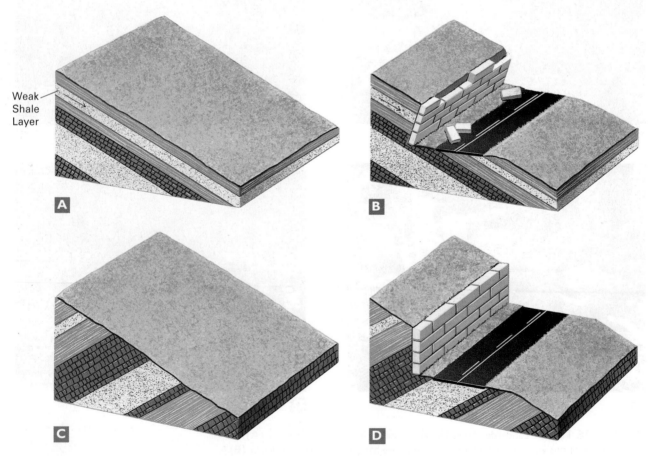

Weak Shale Layer

A

B

C

D

◆ **FIGURE 10.23** **(A)** Sedimentary rock layers dip parallel to this slope. **(B)** If a road cut undermines the slope, the dipping rock provides a good sliding surface, and the slope may fail. **(C)** Sedimentary rock layers dip at an angle to this slope. **(D)** The slope may remain stable even if it is undermined.

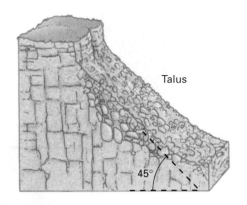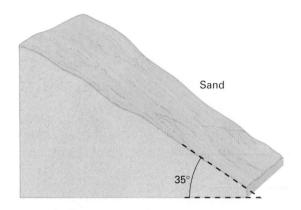

Talus

Sand

45°

35°

◈ **FIGURE 10.24** The angle of repose is the maximum slope that can be maintained by a specific material.

Water and Vegetation

To understand how water affects slope stability, think of a sand castle. Even a novice sand-castle builder knows that sand must be moistened to build steep walls and towers (◈ Figure 10.25). But too much water causes the walls to collapse. Small amounts of water bind sand grains together because the electrical charges of water molecules attract the grains. However, excess water lubricates the sand and adds weight to a slope.[2] When some soils become water saturated, they flow downslope, just as the sand castle collapses. In addition, if water collects on impermeable clay or shale, it may provide a weak, slippery layer so that overlying rock or soil can move easily.

Roots hold soil together and plants absorb water; therefore, a vegetated slope is more stable than a similar bare one. Many forested slopes that were stable for centuries slid when the trees were removed during logging, agriculture, or construction.

Landslides are common in deserts and regions with intermittent rainfall. For example, Southern California has dry summers and occasional heavy winter rain. Vegetation is sparse because of summer drought and wildfires. When winter rains fall, bare hillsides often become saturated and slide. Mass wasting occurs for similar reasons during infrequent but intense storms in deserts.

Earthquakes and Volcanoes

An earthquake may cause a landslide by shaking an unstable slope, causing it to move. A volcanic eruption may

Courtesy of Graham R. Thompson/Jonathan Turk

◈ **FIGURE 10.25** The angle of repose depends on both the type of material and its water content. Dry sand forms low mounds, but if you moisten the sand, you can build steep, delicate towers.

melt snow and ice near the top of a volcano; the water then soaks into the slope to release a landslide.

Thus mass wasting is common in earthquake-prone regions and in volcanically active areas. Two of the landslide disasters described in Section 10.8 resulted from a volcanic eruption and an earthquake.

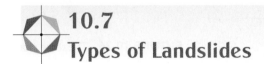

10.7
Types of Landslides

A landslide can occur slowly or rapidly. In some cases, rocks fall freely down the face of a steep mountain. In other instances, rock or soil creeps downslope so slowly that the movement may be unnoticed by a casual observer.

Landslides fall into three categories: flow, slide, and fall (◈ Figure 10.26). To understand these categories,

---

2. Water fills pores in loose material and is under pressure of overlying material. Excess water raises that pressure, and it is actually the increased pore pressure that lowers slope stability.

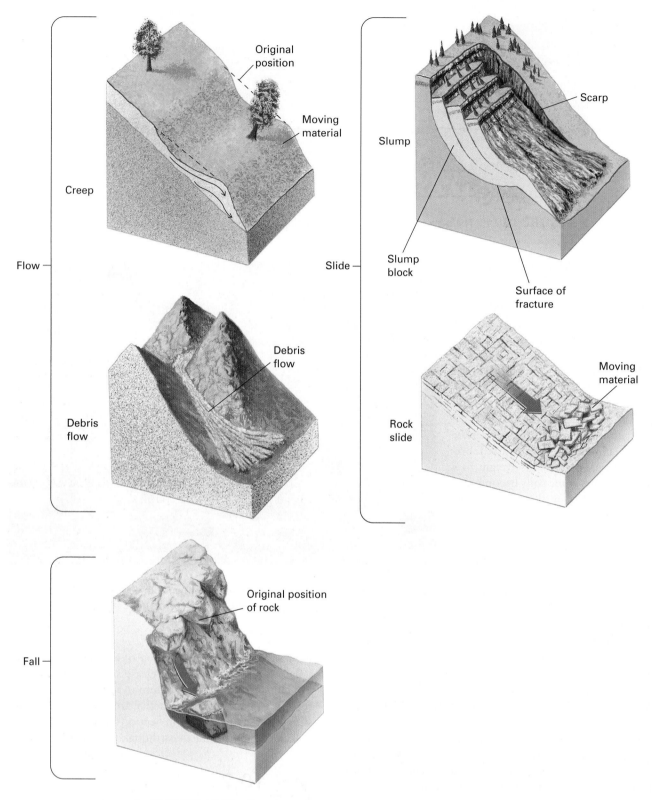

◇ FIGURE 10.26 The three categories of mass wasting are flow, slide, and fall.

think again of a sand castle. Sand that is saturated with water flows down the face of the structure. During **flow**, loose, unconsolidated soil or sediment moves as a fluid. Some slopes flow slowly—at a speed of 1 centimeter per year or less. In contrast, mud with high water content can flow as rapidly as water falling over a waterfall.

If you undermine the base of a sand castle, the wall may fracture and a segment of the wall may slide downward. Movement of a coherent block of material along a fracture is called **slide**. Slide is usually faster than flow, but it still may take several seconds for the block to slide down the face of the castle.

Table 10.2 Categories of Mass Wasting

Type of Movement	Description	Subcategory	Description	Comments
Flow	Individual particles move downslope independently of one another, not as a consolidated mass. Typically occurs in loose, unconsolidated regolith.	Creep	Slow, visually imperceptible movement.	Trees on creep slopes are tilted down-hill and develop pistol-butt shape.
		Debris flow	More than half the particles larger than sand size; rate of movement varies from less than 1m/year to 100km/hr or more.	Common in arid regions with intermittent heavy rainfall, or can be triggered by volcanic eruption.
		Earth flow and mudflow	Movement of fine-grained particles with large amounts of water.	
Slide	Material moves as discrete blocks; can occur in regolith or bedrock.	Slump	Downward slipping of a block of Earth material, usually with a backward rotation on a concave surface.	Trees on slump blocks remain rooted and are tilted up-hill.
		Rockslide	Usually rapid movement of a newly detached segment of bedrock.	
Fall	Materials fall freely in air; typically occurs in bedrock.	—	—	Occurs only on steep cliffs.

If you take a huge handful of sand out of the bottom of the castle, the whole tower topples. This rapid, free-falling motion is called **fall**. Fall is the most rapid type of mass wasting. In extreme cases, like the face of a steep cliff, rock can fall at a speed dictated solely by the force of gravity and air resistance.

Table 10.2 outlines the characteristics of flow, slide, and fall. Details of these three types of mass wasting are explained in the following sections.

Flow

As the name implies, **creep** is the slow, downhill flow of rock or soil under the influence of gravity. A creeping slope typically moves at a rate of about 1 centimeter per year, although wet soil can creep more rapidly. During creep, the shallow soil or rock layers move more rapidly than deeper material moves. As a result, anything with roots or a foundation tilts downhill (◀ Figure 10.27).

Trees have a natural tendency to grow straight upward. As a result, when soil creep tilts a growing tree, the tree develops a J-shaped curve in its trunk, called pistol butt (◀ Figure 10.28). If you ever contemplate buying hillside land for a home site, examine the trees. If they have pistol-butt bases, the slope is probably creeping, and creeping soil may tear a building apart.

If heavy rain falls on unvegetated soil, the water can saturate the soil to form a slurry of mud and rocks called a **mudflow**. A slurry is a mixture of water and

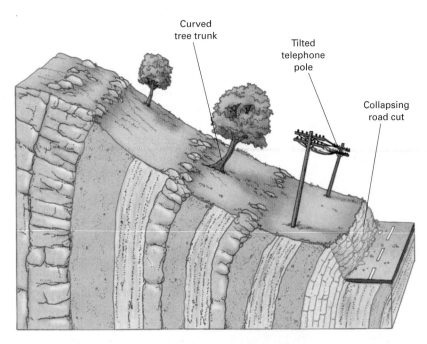

Curved tree trunk

Tilted telephone pole

Collapsing road cut

◆ **FIGURE 10.27** During creep, the land surface moves more rapidly than deeper layers, so objects embedded in rock or soil tilt downhill.

solid particles that flows as a liquid. Wet concrete is a familiar example of a slurry. It flows easily and is routinely poured or pumped from a truck.

The advancing front of a mudflow often forms tongue-shaped lobes (◊ Figure 10.29). A slow-moving mudflow travels at a rate of about 1 meter per year, but others can move as fast as a car speeding along an interstate highway. A mudflow can pick up boulders and automobiles and destroy houses, filling them with mud or even dislodging them from their foundations.

Slide

In some cases, a large block of rock or soil, or sometimes an entire mountainside, breaks away and slides downslope as a coherent mass or as a few intact blocks. Two types of slides occur: slump and rockslide.

A **slump** occurs when blocks of material slide downhill over a gently curved fracture in rock or regolith (◊ Figure 10.30). Trees remain rooted in the moving blocks. However, because the blocks rotate on the concave fracture, trees on the slumping blocks are tilted backward. Thus you can distinguish slump from creep because slump tilts trees uphill, whereas creep tilts them downhill. At the lower end of a large slump, the blocks often break apart and pile up to form a jumbled, hummocky topography.

It is useful to identify slump because it often recurs in the same place or on nearby slopes. Therefore, a slope that shows evidence of past slump is not a good place to build a house.

During a **rockslide** (or rock avalanche), bedrock slides downslope over a fracture plane. Characteristically, the rock breaks up as it moves and a turbulent mass of rubble tumbles down the hillside. In a large rockslide, the falling debris traps and compresses air beneath and among the tumbling blocks. The compressed air reduces friction and allows some rockslides to attain speeds of 500 kilometers per hour. The same mechanism allows a snow or ice avalanche to cover a great distance at high speed.

Fall

If a rock dislodges from a steep cliff, it falls rapidly under the influence of gravity. Several processes commonly detach rocks from cliffs. Recall from our discussion of weathering that frost wedging can dislodge rocks from cliffs and cause rockfall. Rockfall also occurs when ocean waves or a stream undercuts a cliff (◊ Figure 10.31).

◊ **FIGURE 10.28** If a hillside creeps as a tree grows, the tree develops pistol butt.

◊ **FIGURE 10.29** The 1980 eruption of Mount St. Helens melted large quantities of glacial ice near the summit of the peak. The meltwater mixed with soil to create lobe-shaped mudflows.

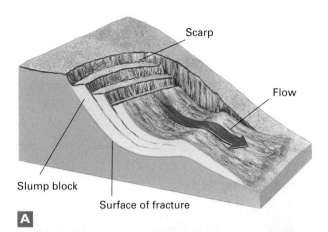

◆ **FIGURE 10.30** **(A)** In slump, blocks of soil or rock remain intact as they move downslope. **(B)** Trees tilt back into the hillside on this slump along the Quesnell River, British Columbia.

Courtesy of Graham R. Thompson/Jonathan Turk

10.8
Three Historic Landslides

The 1925 Rock Avalanche near Kelly, Wyoming

The mountainside above the Gros Ventre River near Kelly, Wyoming, was composed of a layer of sandstone resting on shale, which in turn was supported by a thick bed of limestone (◆ Figure 10.32). The rocks dipped 15° to 20° toward the river and parallel to the slope. Over time, the Gros Ventre River had undercut the sandstone, leaving the slope above the river unsupported.

In the spring of 1925, snowmelt and heavy rains seeped into the ground, saturating the soil and bedrock

and increasing their weight. The water collected on the shale, forming a slippery surface. Finally, in a catastrophic threshold failure, the sandstone layer broke loose and slid over the shale. In a few moments, approximately 38 million cubic meters of rock tumbled into the valley. The sandstone crumbled into blocks that formed a 70-meter-high natural dam across the Gros Ventre River. Two years later, the lake overflowed the dam, washing it out and creating a flood downstream that killed several people.

The Madison River Slide, Montana

In August 1959, a moderate-sized earthquake jolted the area just west of Yellowstone National Park. This region is sparsely populated, and most of the buildings in the area are wood-frame structures that can withstand quakes. As a result, the earthquake itself caused little property damage and no loss of life. However, the quake triggered a massive rockslide from the top of Red Mountain, which lay directly above a U.S. Forest Service campground on the banks of the Madison River. About 30 million cubic meters of rock broke loose and slid into the valley below, burying the campground and killing 26 people. Compressed air escaping from the slide created intense winds that lifted a car off the ground and carried it through the air and deposited it in a stand of trees more than 10 meters away. The slide's momentum carried the rocks more than 100 meters up the mountain on the opposite side of the valley. The debris dammed the Madison River, forming a lake that was later named Quake Lake. ◆ Figure 10.33 shows the debris and some of the damage caused by this slide.

Mount St. Helens: Volcanoes and Landslides in Washington State

Several volcanoes in western Washington State have been active in recent geologic history. The 1980 eruption of Mount St. Helens blew away the entire north side of the mountain (Chapter 8). The heat of the eruption melted glaciers and snowfields near the summit, and the water mixed with volcanic ash and soil to create mammoth mudflows, one of which is shown in ◆ Figure 10.29.

Are catastrophic landslides likely elsewhere in Washington State? Unfortunately, the answer is yes. Mount Baker and Mount Rainier are active, glacier-covered volcanoes that lie near Seattle (◆ Figure 10.34). A large eruption could melt the glaciers on either mountain to create flows similar to those that devastated the valleys below Mount St. Helens in 1980. Mount Baker lies 20 kilometers upriver from the town of Glacier, Washington, and less than 50 kilometers upriver from the city of Bellingham, which has a population of more than 50,000. Mount Rainier is situated upstream from numerous small towns. We can imagine an eruption on either peak initiating a mudflow that could bury nearby towns.

We are not predicting an eruption of Mount Baker or Mount Rainier; we are simply stating that both scenarios are geologically plausible. What should be done in

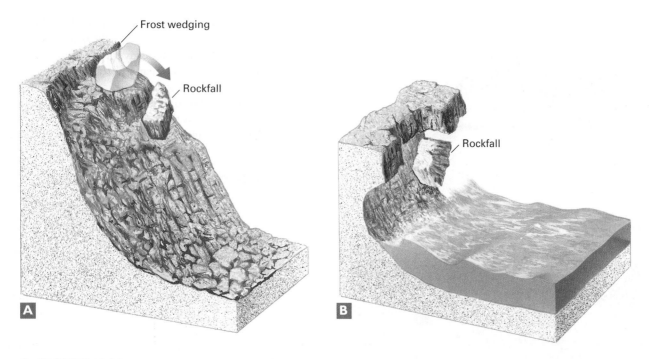

◆ **FIGURE 10.31** **(A)** Rockfall commonly occurs in spring or fall when freezing water dislodges rocks from cliffs. **(B)** Undercutting of cliffs by waves, streams, or construction can also cause rockfall.

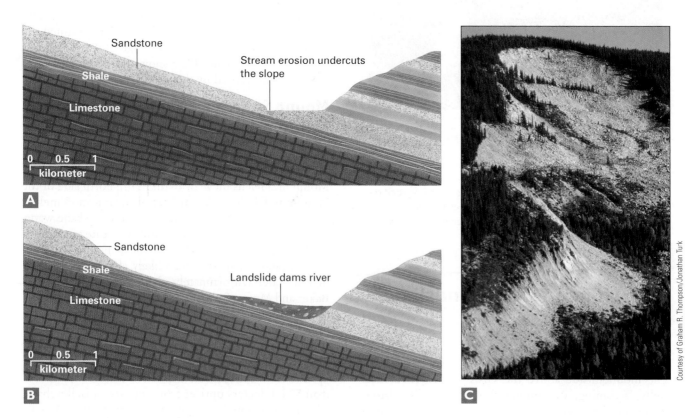

◆ **FIGURE 10.32** A profile of the Gros Ventre hillside **(A)** before and **(B)** after the slide. **(C)** About 38 million cubic meters of rock and soil broke loose and slid downhill during the Gros Ventre slide.

◆ **FIGURE 10.33** This landslide near Yellowstone Park buried a campground, killing 26 people.

◆ **FIGURE 10.34** A wisp of steam rises near the summit of Mount Baker, at the bottom of the photograph.

response to the hazard? It is impractical to move an entire city. Therefore, the only alternative is to monitor the mountains continuously and hope that they will not erupt or, if one does, that it will provide enough warning that urban areas can be evacuated in time.

One of the most important tools in evaluating landslide hazards is to understand that landslides commonly occur in the same area as earlier landslides because the geologic conditions that cause mass wasting tend to be constant over a large area and for long periods of time. Thus, if a hillside has slumped, nearby hills may also be vulnerable to mass wasting. In addition, landslides and mudflows commonly follow the paths of previous slides and flows. If an old mudflow lies in a stream valley, future flows may follow the same valley.

Many towns were founded decades or centuries ago, before geologic disasters were understood. Often the choice of a town site was not dictated by geologic considerations but by factors related to agriculture, commerce, or industry—such as proximity to rivers and ocean harbors and the quality of the farmland. Once a city is established, it is virtually impossible to move it. Furthermore, geologists' warnings that a disaster might occur are often ignored.

Awareness and avoidance are the most effective defenses against mass wasting. Geologists evaluate landslide probability by combining data on soil and bedrock stability, slope angle, climate, and history of slope failure in the area. They include evaluations of the probability of a triggering event, such as a volcanic eruption or earthquake. Building codes then regulate or prohibit construction in unstable areas. For example, according to the United States Uniform Building Code, a building cannot be constructed on a sandy slope steeper than 27°, even though the angle of repose of sand is 30° to 35°. Thus the law leaves a safety margin of 3° to 8°. Architects can obtain permission to build on more precipitous slopes if they anchor the foundation to stable rock.

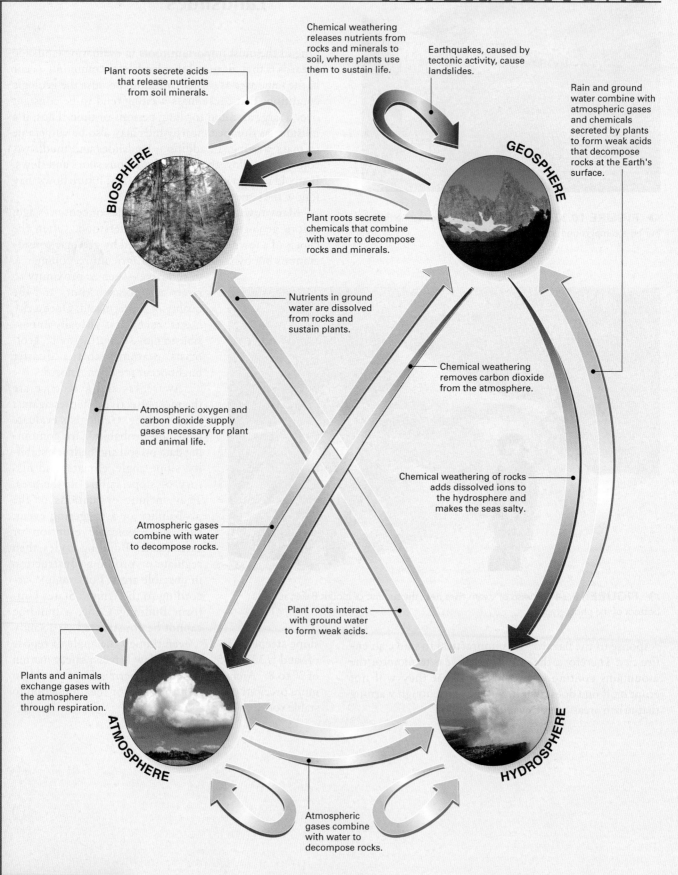

Chemical weathering releases nutrients from rocks and minerals to soil, where plants use them to sustain life.

Earthquakes, caused by tectonic activity, cause landslides.

Plant roots secrete acids that release nutrients from soil minerals.

Rain and ground water combine with atmospheric gases and chemicals secreted by plants to form weak acids that decompose rocks at the Earth's surface.

BIOSPHERE

GEOSPHERE

Plant roots secrete chemicals that combine with water to decompose rocks and minerals.

Nutrients in ground water are dissolved from rocks and sustain plants.

Chemical weathering removes carbon dioxide from the atmosphere.

Atmospheric oxygen and carbon dioxide supply gases necessary for plant and animal life.

Chemical weathering of rocks adds dissolved ions to the hydrosphere and makes the seas salty.

Atmospheric gases combine with water to decompose rocks.

Plant roots interact with ground water to form weak acids.

Plants and animals exchange gases with the atmosphere through respiration.

ATMOSPHERE

HYDROSPHERE

Atmospheric gases combine with water to decompose rocks.

RAIN AND GROUND WATER of the hydrosphere combine with atmospheric gases and chemicals secreted by plants to form weak acids that continually attack and decompose rocks and minerals at Earth's surface. These processes, in turn, create soil that supports plant and animal life on land. Erosion and transport of soils by flowing water, landslides, glaciers, and wind leads to the formation of sedimentary rocks. Thus, the weathering, soil-forming, and erosion cycle directly involves interactions among all four of the major Earth systems. ■

EARTH SYSTEMS INTERACTIONS

SUMMARY

Weathering is the decomposition and disintegration of rocks and minerals at Earth's surface. **Erosion** is the removal of weathered rock or soil by flowing water, wind, glaciers, or gravity. After being eroded from the immediate environment, rock or soil may be transported large distances and eventually deposited.

Mechanical weathering can occur by pressure-release fracturing, frost wedging, abrasion, organic activity, and thermal expansion and contraction.

Chemical weathering occurs when chemical reactions decompose minerals. A few minerals dissolve readily in water. Acids and bases enhance the solubility of minerals. Rainwater is slightly acidic due to reactions between water and atmospheric carbon dioxide.

During hydrolysis water reacts with a mineral to form new minerals. Hydrolysis of feldspar and other common minerals, except quartz, is a form of chemical weathering. Oxidation is the reaction with oxygen to decompose minerals.

Chemical weathering and mechanical weathering often operate together. For example, solutions seeping into cracks may cause rocks to expand by growth of salts or hydrolysis. Hydrolysis may also fracture granite to form exfoliation slabs.

Soil is the layer of weathered material overlying bedrock. Sand, silt, clay, and humus are commonly found in soil. Water leaches soluble ions downward through the soil. Clays are also transported downward by water. The uppermost layer of soil, called the O horizon, consists mainly of litter and humus. The amount of organic matter decreases downward. Leaching removes dissolved ions and clay from the A horizon and deposits them in the B horizon.

Six major factors control soil characteristics: parent rock, climate, rates of plant growth and decay, slope aspect and steepness, time, and transport.

In dry climates, pedocals form. In pedocals, leached ions precipitate in the B horizon, where they accumulate and may form caliche. In moist climates, pedalfer soils develop. In these regions, soluble ions are removed from the soil, leaving high concentrations of less-soluble aluminum and iron. Laterite soils form in very moist climates, where all of the more-soluble ions are removed.

Mass wasting is the downhill movement of rock and soil under the influence of gravity. The stability of a slope and the severity of a landslide depend on (1) steepness of the slope, (2) orientation and type of rock layers, (3) nature of unconsolidated materials, (4) water and vegetation, and (5) earthquakes or volcanic eruptions.

Landslides fall into three categories: flow, slide, and fall. During flow, a mixture of rock, soil, and water moves as a viscous fluid. Creep is a slow type of flow that occurs at a rate of about 1 centimeter per year. A mudflow is a fluid mass of sediment and water. Slide is the movement of a coherent mass of material. Slump is a type of slide in which the moving mass slips on a concave fracture. In a rockslide, a newly detached segment of bedrock slides along a tilted bedding plane or fracture. Fall occurs when particles fall or tumble down a steep cliff.

Earthquakes and volcanic eruptions trigger devastating mass wasting. Proper planning and engineering can avert much damage to human habitation.

Key Terms

For Review

1. Explain the differences among weathering, erosion, transport, and deposition.

2. Explain the differences between mechanical weathering and chemical weathering.

3. List five processes that cause mechanical weathering.

4. What is a talus slope? What conditions favor the formation of talus slopes?

5. List three processes that cause chemical weathering.

6. What is hydrolysis? What happens when granitic rocks undergo hydrolysis? What minerals react? What are the reaction products?

7. What are the components of healthy soil? What is the function of each component?

8. Characterize the four major horizons of a mature soil.

9. List six factors that control soil characteristics and briefly discuss each one.

10. Imagine that soil forms on granite in two regions, one wet and the other dry. Will the soil in the two regions be the same or different? Explain.

11. Explain how soils formed from granite will change with time.

12. What are laterite soils? How are they formed? Why are they unsuitable for agriculture?

13. List and describe each of the factors that control slope stability.

14. What is the angle of repose? Why is the angle of repose different for different types of materials?

15. How does vegetation affect slope stability?

16. Why is mass wasting common in deserts and semiarid lands?

17. How do volcanic eruptions cause landslides?

18. How do earthquakes cause landslides?

19. Discuss the differences among flow, slide, and fall. Give examples of each.

20. Compare and contrast creep and slump. What does a pistol-butt tree trunk tell you about slope stability?

21. Explain how trees are tilted but not killed by slump.

22. How do landslides reach and destroy towns and villages many kilometers from the steep slopes where the slides originate?

For Discussion

1. What process is responsible for each of the following observations or phenomena—is the process a mechanical or chemical change?

   (a) A board is sawn in half.

   (b) A board is burned.

   (c) A cave is formed when water seeps through a limestone formation.

   (d) Calcite is formed when mineral-rich water is released from a hot underground spring.

   (e) Meter-thick sheets of granite peel off a newly exposed pluton.

   (f) Rock fall is more common in the mountains of a temperate region in the spring than in midsummer.

2. Most substances contract when they freeze, but water expands. How would weathering be affected if water contracted instead of expanded when it froze?

3. Discuss the similarities and differences between salt cracking and frost wedging.

4. What types of weathering would predominate on the following fictitious planets? Defend your conclusions. (a) Planet X has a dense atmosphere composed of nitrogen, oxygen, and water vapor with no carbon dioxide. Temperatures range from a low of 10°C in the winter to 75°C in the summer. Windstorms are common. No living organisms have evolved. (b) The atmosphere of Planet Y consists mainly of nitrogen and oxygen, with smaller concentrations of carbon dioxide and water vapor. Temperatures range from a low of 260°C in the polar regions in the winter to 135°C in the tropics. Windstorms are common. A lush blanket of vegetation covers most of the land surfaces.

5. The Arctic regions are cold most of the year and summers are short there. Thus decomposition of organic matter is slow. In contrast, decay is much more rapid in the temperate regions. How does this difference affect the fertility of the soils?

6. The Moon is considerably less massive than Earth, and therefore its gravitational force is less. It has no atmosphere and therefore no rainfall. The interior of the Moon is cool, and thus it is geologically inactive. Would you expect mass wasting to be a common or an uncommon event in mountainous areas of the Moon? Defend your answer.

7. Explain how wildfires affect slope stability and mass wasting.

8. What types of mass wasting (if any) would be likely to occur in each of the following environments? (a) A very gradual (2 percent) slope in a heavily vegetated tropical rainforest. (b) A steep hillside composed of alternating layers of conglomerate, shale, and sandstone, in a region that experiences distinct dry and rainy seasons. The dip of the rock layers is parallel to the slope. (c) A hillside similar to that of b, in which the rock layers are oriented perpendicular to the slope. (d) A steep hillside composed of clay in a rainy environment in an active earthquake zone.

9. Identify a hillside in your city or town that might be unstable. Using as much data as you can collect, discuss the magnitude of the potential danger. Would the landslide be likely to affect human habitation?

10. Explain how the mass wasting triggered by earthquakes and volcanoes can have more serious effects than the earthquake or volcano itself. Is this always the case?

11. Develop a strategy for minimizing loss of life from mass wasting if Mount Baker should show signs of an impending eruption similar to the signs shown by Mount St. Helens in the spring of 1980.

ThomsonNOW™

Assess your understanding of this chapter's topics with additional comprehensive interactivities at **http://www.thomsonedu.com/login**, which also has up-to-date web links, additional readings, and exercises.

CHAPTER 11

Fresh Water: Streams, Lakes, Ground Water, and Wetlands

Courtesy of Graham R. Thompson/Jonathan Turk

Spring snow melt rushes down Lloyd Creek in the Bitterroot Mountains of Montana.

Think of a pleasing, relaxing landscape, and you will probably envision water: A gentle surf lapping against white coral sand with palm trees in the background; a gurgling brook dropping over moss-covered rocks beneath a canopy of oak and maple; or a sunrise burning through wispy fog on a mirror-smooth lake. Oceans cover two-thirds of Earth, and water also plays a seminal role in many systems interactions on the continents. We have already learned that water is an important agent in the rock cycle. The previous chapter, "Weathering, Soil, and Erosion," was also largely about water: as a corrosive chemical, as a physical force when it freezes, and as a medium that transports rock and sediment. Most of the remaining topics of this book, excluding the astronomy chapters, are about water. ■

11.1
The Water Cycle

The **hydrologic cycle**, or the water cycle, describes the constant circulation of water among the four spheres: the hydrosphere (or watery part of the planet), the geosphere (the land), the biosphere, and the atmosphere. About 1.3 billion cubic kilometers of water exist at Earth's surface. Of this huge quantity, 97.5 percent is salty seawater, and another 1.8 percent is frozen into the great ice caps of Antarctica and Greenland. Thus, although the hydrosphere contains a great amount of water, only 0.64 percent is fresh and available in streams, ground water, lakes, and wetlands (◆ Figure 11.1).

As explained in Chapter 1, water evaporates from the seas and from land to form water vapor in the atmosphere. This vapor eventually condenses and falls back to the surface as rain or snow. Most precipitation lands on the ocean, partly because the oceans cover most of the planet. The precipitation that falls on the continents follows four paths (◆ Figure 11.2):

1. Surface water flowing to the sea in streams and rivers is called **runoff**. This water may stop temporarily in a lake or wetland, but eventually it flows to the oceans.
2. Some water seeps into the ground—the geosphere—to become part of a vast, subterranean reservoir known as **ground water.** Although surface water is more conspicuous, 60 times more water is stored as ground water than in all streams, lakes, and wetlands combined. Ground water seeps through bedrock and soil toward the sea, although it flows much more slowly than surface water.
3. Most of the remainder of water that falls onto land evaporates back into the atmosphere. Water also evaporates directly from plants as they breathe, in a process called **transpiration**.
4. A small amount of water is also incorporated into the biosphere as plant and animal tissue when plants grow.

The hydrologic cycle not only describes the movement of water, it also describes a primary mechanism for the movement of energy from one part of the globe to another. Ocean currents transport huge quantities of heat from the equator toward the poles, thus cooling the equator and warming the higher latitudes. Evaporation

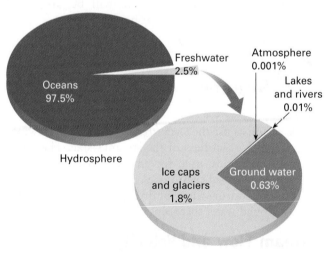

◆ **FIGURE 11.1**   Ninety-seven and a half percent of Earth's water is in the oceans, 1.8 percent is locked up in glaciers, 0.63 percent is ground water, 0.01 percent is in lakes and rivers, and 0.001 percent is in the atmosphere.
INTERACTIVE QUESTION: *Calculate the percentage of freshwater found in glaciers, ground water, lakes and rivers, and the atmosphere.*

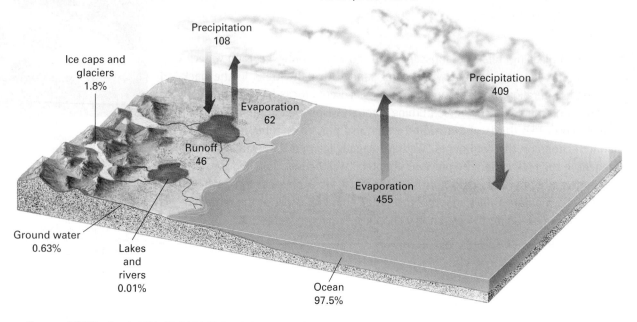

Atmosphere 0.001%

Precipitation
108

Ice caps and
glaciers
1.8%

Evaporation
62

Runoff
46

Precipitation
409

Evaporation
455

Ground water
0.63%

Lakes
and
rivers
0.01%

Ocean
97.5%

ThomsonNOW ◀▶ **ACTIVE FIGURE 11.2** The hydrologic cycle shows that water circulates constantly among the sea, the atmosphere, and the land. Whole numbers indicate thousands of cubic kilometers of water transferred each year. Percentages show proportions of total global water in different portions of Earth's surface.

is a cooling process, whereas condensation releases heat. Water vapor is a greenhouse gas, thereby warming the atmosphere, but clouds and sparkling glaciers reflect sunlight back out to space and thereby cool Earth. There are so many feedback cycles, many with opposite effects, that one scientist wrote, "Water acts as the Venetian blind of our planet, as its central heating system, and as its refrigerator, all at the same time."[1] This chapter chronicles the movement of surface water. Other components of the hydrologic cycle will be discussed throughout the remainder of the book.

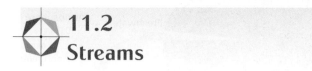 **11.2
Streams**

Earth scientists use the term *stream* for all water flowing in a channel, regardless of the stream's size. The term *river* is commonly used for any large stream fed by smaller ones, called **tributaries**. Most streams run year round, even during times of drought, because they are fed by ground water that seeps into the streambed.

Stream Flow and Velocity

Three factors control stream velocity:

Gradient

Gradient is the steepness of a stream. Obviously, if all other factors are equal, water flows more rapidly down a

1. Heike Langenberg, in the editorial introducing a special section, "Climate and Water," *Nature* 419 (September 12, 2002): 187.

steep slope than a gradual one. A tumbling mountain stream may drop 40 meters or more per kilometer, whereas the lower Mississippi River has a gradient of only 0.1 meters per kilometer.

Discharge

Discharge is the amount of water flowing down a stream. It is expressed as the volume of water flowing past a point per unit time, usually in cubic meters per second (m^3/sec). The velocity of a stream increases when its discharge increases. Thus, a stream flows faster during flood, even though its gradient is unchanged.

When Huckleberry Finn floated down the Mississippi on a raft, he probably drifted downstream at about 7 to 9 kilometers per hour (2 to 2.5 meters per second). However, a modern rafter on one of the small mountain tributaries of the Mississippi may drift downstream at only 5.4 kilometers per hour (1.5 meters per second). Thus, although it is counterintuitive, a large, lazy-appearing river may flow more rapidly than a small, steep mountain stream.

The largest river in the world is the Amazon, with an average discharge of 150,000 m^3/sec. In contrast, the Mississippi, the largest river in North America, has an average discharge of about 17,500 m^3/sec, approximately one-ninth that of the Amazon.

A stream's discharge can change dramatically from month to month or even during a single day. For example, the Selway River, a mountain stream in Idaho, has a discharge of 100 to 130 m^3/sec during early summer, when mountain snow is melting rapidly. During the dry season in late summer, the discharge drops to about

20 m³/sec (◆ Figure 11.3). A desert stream may dry up completely during summer but become the site of a flash flood during a sudden thunderstorm.

Channel characteristics

Channel characteristics refer to the shape and roughness of a stream channel. The floor of the channel is called the **bed**, and the sides of the channel are the **banks**. Friction between flowing water and the stream channel slows current velocity. Consequently, water flows more slowly near the banks than near the center of a stream. If you paddle a canoe down a straight stream channel, you move faster when you stay away from the banks. The amount of friction depends on the roughness and shape of the channel. Boulders on the banks or in the streambed increase friction and slow a stream down, whereas the water flows more rapidly if the bed and banks are smooth.

Stream Erosion and Sediment Transport

Streams shape Earth's surface by eroding soil and bedrock. The flowing water carries the eroded sediment grain by grain toward the sea. A stream may deposit some of the sediment on its flood plain, forming a level valley bottom, while it carries the remainder to the seacoast, where the sediment accumulates to form deltas and sandy beaches.

Stream erosion and sediment transport depend on a stream's energy. A rapidly flowing stream has more energy to erode and transport sediment than a slow stream

of the same size. The **competence** of a stream is a measure of the largest particle it can carry. A fast-flowing stream can transport cobbles and even boulders in addition to small particles. A slow stream carries only silt and clay.

The **capacity** of a stream is the total amount of sediment it can carry past a point in a given amount of time. Capacity is proportional to both current speed and discharge. Thus a large, fast stream has a greater capacity than a small, slow one. Because the ability of a stream to erode and carry sediment is proportional to both velocity and discharge, most erosion and sediment transport occur during the few days each year when the stream is in flood. Relatively little erosion and sediment transport occur during the remainder of the year. To see this effect for yourself, look at any stream during low water. It will most likely be clear, indicating little erosion or sediment transport. Look at the same stream later, when it is flooding. It will probably be muddy and dark, indicating that the stream is eroding its bed and banks and transporting the sediment.

After a stream erodes soil or bedrock, it transports the sediment downstream in three ways. Ions dissolved in water are called **dissolved load**. A stream's ability to carry dissolved ions depends mostly on its discharge and its chemistry, not its velocity. Thus even the still waters of a lake or ocean contain dissolved substances; that is why the sea and some lakes are salty. Although dissolved ions are invisible, they comprise more than half of the total sediment load carried by some rivers. More commonly, dissolved ions make up less than 20 percent of the total sediment load of streams.

If you place loamy soil in a jar of water and shake it up, the sand grains settle quickly. But the smaller silt and clay particles remain suspended in the water as **suspended load**, giving it a cloudy appearance. Clay and silt are small enough that even the slight turbulence of a slow stream keeps them in suspension. A rapidly flowing stream can carry sand in suspension.

During a flood, when stream energy is highest, the rushing water can roll boulders and cobbles along the bottom as **bed load**. Sand also moves in this way, but if the stream velocity is sufficient, sand grains bounce and hop along the streambed like millions of tiny marbles.

The world's two muddiest rivers—the Yellow River in China and the Ganges River in India—each carry more than 1.5 billion tons of sediment to the ocean every year. Prior to 1955 when dams and flood control retained sediment, the sediment load of the Mississippi River was about 450 million tons per year (◆ Figure 11.4). Most streams carry the greatest proportion of

◆ **FIGURE 11.3** The 1988 hydrograph for the Selway River in Idaho shows that the discharge varied from 125 m³/second in the spring to 20 m³/second in the summer. The sharp peaks reflect high discharge during periods of rapid snowmelt.
INTERACTIVE QUESTION: *Would you expect the hydrograph to look the same every year? If so, why? If not, how would it differ?*

Bed load
30 million
metric tons

Suspended
load
300 million
metric tons

Dissolved
load
120 million
metric tons

◆ **FIGURE 11.4**   The Mississippi River carries the greatest proportion of its sediment as suspended load. Numbers indicate sediment load per year.

sediment in suspension, less in solution, and the smallest proportion as bed load.

Downcutting and Base Level

A stream erodes downward into its bed and laterally against its banks. Downward erosion is called **downcutting** (◆ Figure 11.5). The **base level** of a stream is the deepest level to which it can erode its bed. Most streams cannot erode below sea level, which is called the ultimate base level. This concept is straightforward. Water can only flow downhill. If a stream were to cut its way down to sea level, it would stop flowing and hence would no longer erode its bed.

In addition to ultimate base level, a stream may have a number of local, or temporary, base levels. For example, a stream stops flowing where it enters a lake. It then stops eroding its channel because it has reached a temporary base level (◆ Figure 11.6). A layer of rock that resists erosion may also establish a temporary base level because it flattens the stream gradient. Thus, the stream slows down and erosion decreases. The top of a waterfall is a temporary base level commonly established by resistant rock. Niagara Falls is formed by a resistant layer of dolomite overlying softer shale. Although the dolomite resists erosion at the top of the falls, the turbulent falling water erodes the underlying shale. When its foundation is undermined, the dolomite cap collapses and the falls migrate upstream. As a result, Niagara Falls has retreated 11 kilometers upstream since its formation about 9,000 years ago (◆ Figure 11.7). Thus the falls retreat a little more than one meter per year. The erosion

◆ **FIGURE 11.5**   Deer Creek, a tributary of Grand Canyon, has downcut its channel into solid sandstone.

◆ **FIGURE 11.6** A steep mountain stream flowing from a glacier in the Canadian Rockies reaches a temporary base level and stops flowing where it enters a lake.

rate is rapid because the water generates considerable energy as it tumbles over the falls.

A stream like that in ◆ Figure 11.8A erodes rapidly in the steep places where its energy is high, and deposits sediment in the low-gradient stretches where it flows more slowly. Over time, erosion and deposition smooth out the irregularities in the gradient. The resulting **graded**

stream has a smooth, concave profile (◆ Figure 11.8B). Once a stream becomes graded, there is no net erosion or deposition and the stream profile no longer changes. An idealized graded stream such as this does not actually exist in nature, but many streams come close.

Sinuosity of a Stream Channel

A steep mountain stream usually downcuts rapidly compared with the rate of lateral erosion. As a result, it cuts a relatively straight channel with a steep-sided, V-shaped valley (◆ Figure 11.9). The stream maintains its relatively straight path because it flows with enough energy to erode and transport any material that slumps into its channel.

In contrast, a low-gradient stream is less able to erode downward into its bed. Much of the stream energy is directed against the banks, causing **lateral erosion**. Lateral erosion undercuts the valley sides and widens a stream valley. Most low-gradient streams flow in a series of bends, called **meanders** (◆ Figure 11.10A and B). A meandering stream wanders back and forth across its flood plain, forming a wide valley with a flat bottom.

As a stream flows into a meander bend, the inertia of the moving water tends to keep the water moving in a straight path. Consequently, most of the current flows to the outside bank of the meander. As a result, both the velocity and channel depth are greatest near the outside of the bend, and the stream erodes its outside bank. At the same time, sediment is deposited in the slower water on the inside of the meander to form a **point bar**.

◆ **FIGURE 11.7** Niagara Falls has eroded 11 kilometers upstream in the last 9,000 years and continues to erode today.

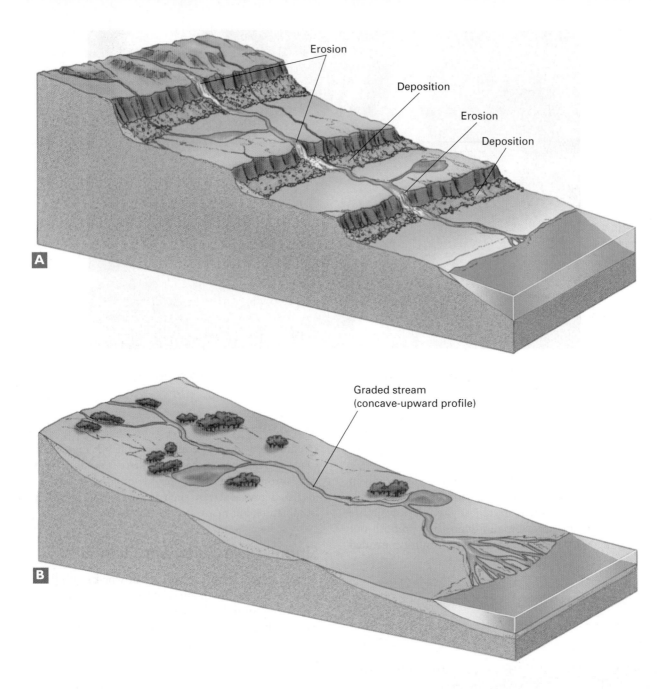

Erosion

Deposition

Erosion

Deposition

A

Graded stream
(concave-upward profile)

B

◆ **FIGURE 11.8** **(A)** An ungraded stream has many temporary base levels. **(B)** With time, the stream smooths out the irregularities to develop a graded profile.

Because a meandering stream erodes the outside banks of its meanders, and deposits sediment on the insides of the bends, the meanders migrate slowly down the flood plain as the stream constantly modifies its channel. Occasionally, an **oxbow lake** forms where the stream cuts across the neck of a meander and isolates an old meander loop (◆ Figures 11.10C and D).

Most streams flow in a single channel. In contrast, a **braided stream** flows in many shallow, interconnecting channels (◆ Figure 11.11). A braided stream forms where more sediment is supplied to a stream than it can

carry. The excess sediment accumulates in the channel, filling it and forcing the stream to overflow its banks and erode new channels. As a result, a braided stream flows simultaneously in several channels and shifts back and forth across its flood plain.

Braided streams are common in both deserts and glacial environments because both produce abundant sediment. A desert yields large amounts of sediment because it has little or no vegetation to prevent erosion. Glaciers grind bedrock into fine sediment, which is carried by streams flowing from the melting ice.

◆ **FIGURE 11.9** A steep mountain stream eroded a V-shaped valley into soft shale in the Canadian Rockies.

Drainage Basins

Only a dozen or so major rivers flow into the sea along the coastlines of the United States (◆ Figure 11.12). Each is fed by a number of tributaries, which are in turn fed by smaller tributaries. Mountain ranges or plateaus separate adjacent river systems. The region drained by a single river is called a **drainage basin**. For example, the Rocky Mountains separate the Colorado and Columbia drainage basins to the west from the Mississippi and Rio Grande basins to the east. Together, those four river systems drain more than three-fourths of the United States.

11.3 Stream Erosion and Mountains: How Landscapes Evolve

According to a model popular in the first half of this century, streams erode Earth's surface and create landforms in an orderly sequence (◆ Figure 11.13). At first, they cut steep, V-shaped valleys into mountains. Over time, the streams erode the mountains away and widen the valleys into broad flood plains. Eventually, the entire landscape flattens, forming a large, featureless plain. However, if this were the only mechanism affecting Earth's surface during its 4.6-billion-year history, all landforms would have eroded to a flat plain.

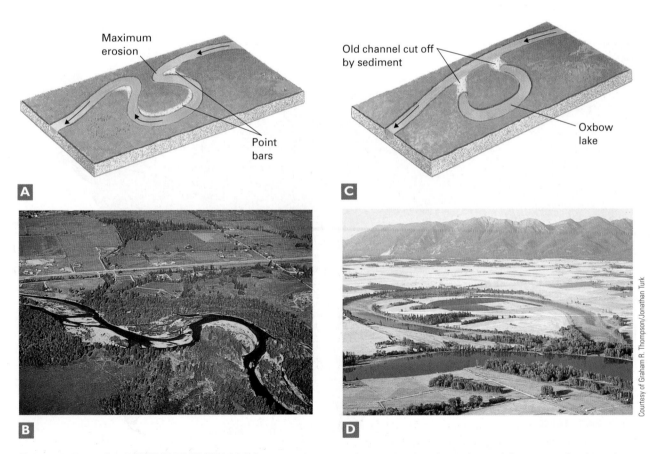

ThomsonNOW ◆ **ACTIVE FIGURE 11.10** **(A)** A stream erodes the outsides of meanders and deposits sand and gravel on the inside bends to form point bars. **(B)** Meanders and point bars form the channel of the Bitterroot River, Montana. **(C)** Over time, a stream may erode through the neck of a meander to form an oxbow lake. **(D)** This oxbow lake formed in the Flathead River, Montana.

◈ **FIGURE 11.11** Joe Creek in Canada's Yukon Territory is heavily braided because glaciers provide more sediment than the stream can carry.

Why, then, do mountains, valleys, and high plateaus still exist?

The model tells only half the story. Streams do continuously erode the landscape, flattening mountains and widening flood plains. But at the same time, tectonic activity may uplift the land and interrupt the simple,

idealized sequence. In this way, Earth's hydrosphere and atmosphere work together with tectonic processes in the geosphere to create landforms.

Consider the Himalayas. Today they are a rapidly rising mountain range as a result of tectonic processes and isostatic uplift described in Chapter 9. As the mountains rise, landslides, glaciers, and running water erode the peaks. The water flows into streams, which cut deep, V-shaped valleys through the range and transport vast quantities of sediment to the sea. Modern streams carry an annual average of 1,000 tons of sediment per square kilometer of land surface area from the Himalayas. This erosion rate, averaged over the entire Himalayan Range, amounts to a lowering of the land surface by 0.5 millimeters each year. But the average rate of tectonic and isostatic uplift of the Himalayas is about 5 millimeters per year, 10 times faster than the erosion rate.

Sometime in the future, tectonic uplift will cease. But erosion will continue to remove rock and sediment from the mountains. At first, this erosion will not lower the mountains by much because the range will continue to rise isostatically. Eventually, erosion will dominate over the isostatic uplift. Then, streams, glaciers, and landslides will wear the Himalayas down to low, rounded mountains that will probably resemble the modern Appalachians.

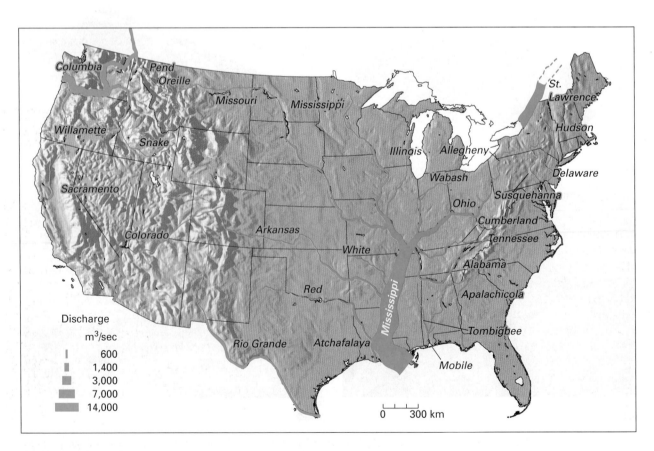

◈ **FIGURE 11.12** Most of the surface water in the United States flows to the sea from approximately a dozen major streams.

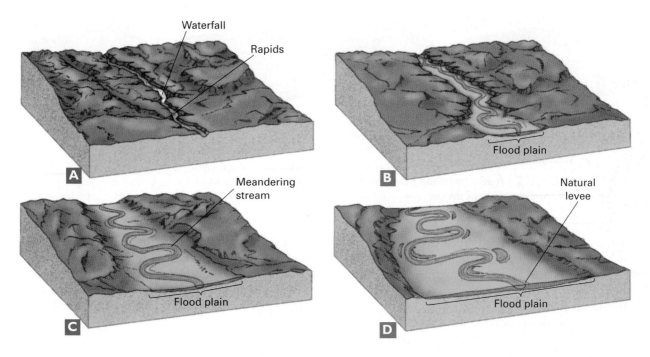

11.4
Stream Deposition

When streams slow down they deposit their bed load first. If water slows sufficiently, the suspended load may also fall to the bottom, but the dissolved load remains in the water until the chemical environment changes. New landforms are created when rivers deposit sediment.

If a steep mountain stream flows onto a flat plain, its gradient and velocity decrease abruptly. As a result, it deposits most of its sediment in a fan-shaped mound called an **alluvial fan**. Alluvial fans are common in many arid and semiarid mountainous regions (◆ Figure 11.14).

A stream also slows abruptly where it enters the still water of a lake or ocean. The sediment settles out to form a nearly flat landform called a **delta**. Part of the delta lies above water level, and the remainder of it lies slightly below water level. Deltas are commonly fan shaped, resembling the Greek letter delta (Δ).

Both deltas and alluvial fans change rapidly. The stream abandons sediment-choked channels, while new channels develop, as in a braided stream. As a result, a stream feeding a delta or fan splits into many channels called **distributaries**. A large delta may spread out in this manner until it covers thousands of square kilometers (◆ Figure 11.15). Most fans, however, are much smaller, covering a fraction of a square kilometer to a few square kilometers.

◆ **FIGURE 11.14**   This alluvial fan in Death Valley formed where a steep mountain stream deposited most of its sediment as it entered the flat valley. A road runs across the lower part of the fan.

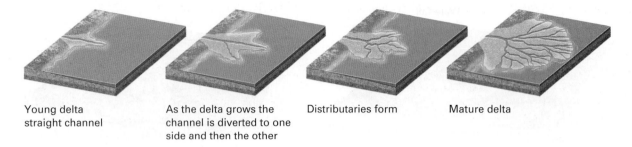

Young delta
straight channel

As the delta grows the
channel is diverted to one
side and then the other

Distributaries form

Mature delta

◆ **FIGURE 11.15** A delta forms and grows with time where a stream deposits its sediment as it flows into a lake or the sea.

Some sediment is carried beyond the delta and deposited on the sea floor, creating a **submarine delta.** ◆ Figure 11.16 shows that the submarine deltas created by the Himalayas cover an area significantly greater than the Indian subcontinent.

Even though deltas cover only a small fraction of Earth's total land surface, they are environments that include each of Earth's four spheres. The delta is composed of solid sediment, but it is a watery zone where branched distributaries encounter the ocean. Because the land is fertile, with easy access to river and ocean, delta ecosystems are rich in natural plant and animal life—but the fertile soils and easy access to transportation systems

◆ **FIGURE 11.16** The Indus River and the Ganges-Bramaphutra River have transported millions of cubic kilometers of sediment eroded from the Himalayas to the Indian Ocean. Note that the combined area for the Bengal and Indus fans is larger than the entire subcontinent of India!

make deltas desirable places for human habitation. Delta land is barely above sea level and river level, so it is certain to be flooded during atmospheric disturbances such as heavy rains and hurricanes. We will discuss some of these systems interactions in the sections and chapters that follow.

11.5
Floods

When rainfall is heavy or snowmelt is rapid, more water flows down a stream than the channel can hold, creating a **flood**. During a flood, the stream overflows onto lowlying adjacent land called the **flood plain**. Massive floods occur somewhere in the world every year. In late August 2002, the Elbe River in eastern Europe peaked at a record 9.39 meters above flood stage, 0.6 meters higher than the previous record, recorded in 1845. One hundred people died, and the total damages were expected to exceed $20 billion.

Although we normally think of floods as destructive events, flood plain ecosystems depend on floods. For example, cottonwood tree seeds only germinate after a flood, and waterfowl depend on flood plain wetlands. Many species of fish gradually lose out to stronger competitors during normal flows but have adapted better to floods, so their populations increase as a result of flooding. Thus species diversity is maintained by alternate periods of flooding and normal flow. Deltas are created and enlarged by floods. At normal times, when rivers are confined within their banks, the flowing water transports sediment out to sea and deposits it on the ocean floor. But during floods, river water rises above the stream banks and covers the delta land. When the flood waters slow down, they deposit sediment, thus enlarging the delta.

In some cases, flooding can directly benefit humans. Frequent small floods dredge bigger channels, which reduce the severity of large floods. Flooding streams carry large sediment loads and deposit them on flood plains to form fertile soil. So, paradoxically, the same floods that cause death and disaster create the rich soils that make

CHAPTER 11 • Fresh Water: Streams, Lakes, Ground Water, and Wetlands

flood plains and deltas so attractive for farming and human habitation.

But at the same time, floods are the costliest natural disasters on the planet—simply because people choose to live in harm's way. Many riverbank cities originally grew as ports, to take advantage of the easy transportation afforded by rivers. In addition, flood plains and deltas provide rich soils for farms, flat land for roads and buildings, and access to abundant water for industry and agriculture.

While flooding is a natural event, flood frequency and severity are often augmented by logging, farming, and urbanization. Recall from Chapter 10 that when all the vegetation was cut or killed in a New Hampshire forest, stream flow increased 40 percent. In a similar manner, when forests are cut and replaced by farms, stream flow may increase. Similar conditions develop when prairies are plowed and farmed. Nearly all rainwater runs off heavily paved urban areas directly into nearby streams.

Flood Control and the 1993 Mississippi River Floods

During the late spring and summer of 1993, heavy rain soaked the upper Midwest. Thirteen centimeters of rain fell in already saturated central Iowa in a single day. In mid-July, 2.5 centimeters of rain fell in 6 minutes in Papillion, Nebraska. As a result of the intense rainfall over such a large area, the Mississippi River and its tributaries flooded. In Fargo, North Dakota, the Red River, fed by a daylong downpour, rose 1.2 meters in 6 hours, flooding the town and backing up sewage into homes and the Dakota Hospital. In St. Louis, Missouri, the Mississippi crested 14 meters above normal and 1 meter above the highest previously recorded flood level. At its peak, the flood inundated nearly 44,000 square kilometers in a dozen states. Damage to homes and businesses on the flood plain reached $16 billion. Forty-five people died.

Artificial Levees and Channels

An **artificial levee** is a wall built along the banks of a stream to prevent rising water from spilling out of the stream channel onto the flood plain. In the past 70 years, the U.S. Army Corps of Engineers has spent billions of dollars building 11,000 kilometers of levees along the banks of the Mississippi and its tributaries. During the 1993 Mississippi River flood, levees held the waters back in some places, preventing $19 billion in property damage. However, in other places, the levees failed (◀▶ Figure 11.17). Unfortunately, levees also create conditions that may increase both flood intensity and property damage.

◀▶ **FIGURE 11.17** Levees are effective in containing rivers during small floods. However, large floods often breach the levees and the water spills onto the surrounding flood plain.

One factor is entirely human—the protection promised by levees encourages people to build in the flood plain. In the absence of a levee, people might decide to build on high ground, safe from floods. But when levees are built, people are more likely to construct homes and businesses in harm's way. After the 1993 Mississippi River flood, old levees were repaired and new ones constructed. People immediately took advantage of the presumed safety and built 28,000 new homes

and 27 square kilometers of stores and factories on land that was flooded in 1993. But levees have failed in the past, so these structures are vulnerable.

Levees also may cause much greater floods in the future, and they can cause higher floods along nearby reaches of a river. In the absence of levees, when a stream floods, it deposits mud and sand on the flood plain (◀ Figure 11.18A). When artificial levees are built, the stream cannot overflow during small floods, so it deposits

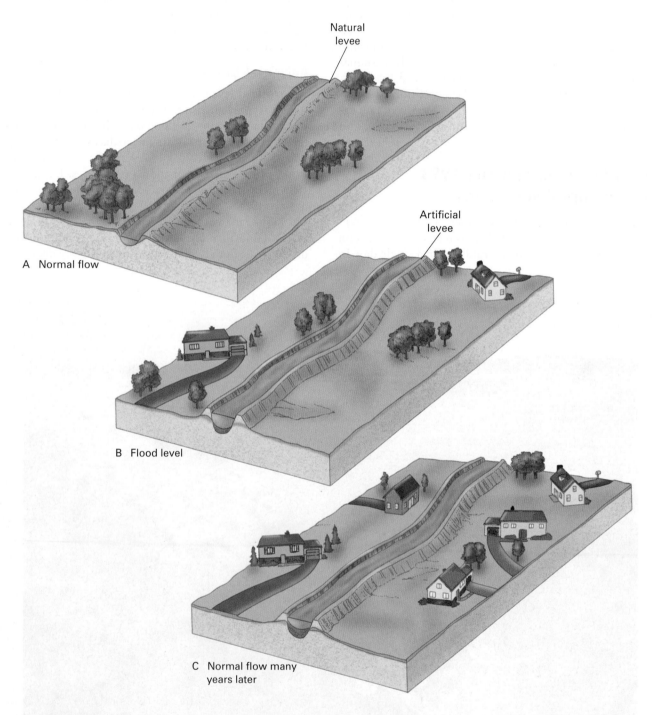

◀ FIGURE 11.18 **(A)** In the natural state, a flooding stream carries sediment from the stream channel onto the flood plain. **(B)** Artificial levees cause sediment to accumulate in a stream channel. **(C)** Eventually, the channel rises above the flood plain, creating the potential for a disastrous flood.

the sediment in its channel, raising the level of the streambed. (◆ Figure 11.18B). After several small floods, the entire stream may rise above its flood plain, contained only by the levees (◆ Figure 11.18C). This configuration creates the potential for a truly disastrous flood because if the levee should be breached during a large flood, the entire stream then flows out of its channel and onto the flood plain. As a result of levee building and channel sedimentation, portions of the Yellow River in China now lie 10 meters above its flood plain, and the Mississippi River lies above adjacent parts of New Orleans, where it flows through the city. Thus, levees may solve flooding problems in the short term, but in a longer time frame they may cause even larger and more destructive floods.

Levees may also cause higher flood levels upstream. A flooding river spreads horizontally over its flood plain, temporarily forming a wide path of flowing water. But where levees constrict the river into its narrow channel, they form a partial dam, causing the waters to rise to even higher flood levels upstream (◆ Figure 11.19). Artificial levees and channels contributed to the severity of the 2002 Elbe flood, discussed at the beginning of this section.

Engineers have tried to solve the problem of channel sedimentation by dredging **artificial channels** across me-

anders. When a stream is straightened, its gradient and velocity increase and it scours more sediment from its channel. This solution, however, also has its drawbacks. A straightened stream is shorter than a meandering one, and consequently the total volume of its channel is reduced. Therefore, the channel cannot contain as much excess water, and flooding is likely to increase both upstream and downstream from the straightened reach of the stream.

Flood Control, the Mississippi River Delta, and Hurricane Katrina

In August 2005, Hurricane Katrina drove storm waters over the levees protecting New Orleans, flooding the city, chasing 1.3 million people from their homes, and causing approximately $200 billion in damages (◆ Figure 11.20). This disaster will be discussed in Chapter 19 in the coverage of extreme weather events, but we also discuss it here because the problem was exacerbated by decades of flood control practices along the Mississippi. Recall that repeated small floods build a delta. If levees reduce the frequency of small floods, less sediment is deposited and the delta is not built up as rapidly.

In addition, dams trap sediment upstream and reduce the total sediment load of the river. In a recent communication, Dr. Ned Andrews of the USGS explained that sediment load of the Mississippi has decreased by 200 million tons per year, compared with pre-1955 levels. Multiplying by 50 years, there is a cumulative deficit of approximately 10 billion tons of sediment, which amounts to approximately 6.7 cubic kilometers of material that would have been deposited on the delta under natural conditions but has not been deposited due to dam construction.

◆ **FIGURE 11.19** Levees force a flooding river into a restricted channel, forming a partial dam that raises the flood level upstream from the restriction.

◆ **FIGURE 11.20** In August 2005, U.S. Coast Guard personnel search for survivors in the aftermath of Hurricane Katrina.

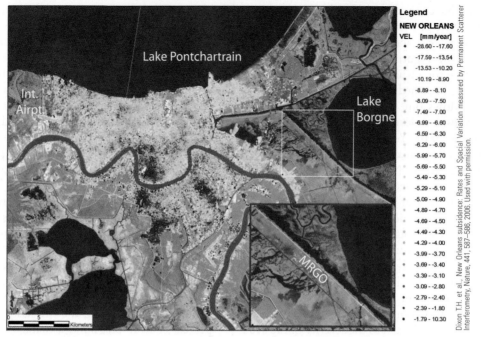

Legend
NEW ORLEANS
VEL [mm/year]
* -28.60 - -17.60
* -17.59 - -13.54
* -13.53 - -10.20
* -10.19 - -8.90
* -8.89 - -8.10
* -8.09 - -7.50
* -7.49 - -7.00
* -6.99 - -6.60
* -6.59 - -6.30
* -6.29 - -6.00
* -5.99 - -5.70
* -5.69 - -5.50
* -5.49 - -5.30
* -5.29 - -5.10
* -5.09 - -4.90
* -4.89 - -4.70
* -4.69 - -4.50
* -4.49 - -4.30
* -4.29 - -4.00
* -3.99 - -3.70
* -3.69 - -3.40
* -3.39 - -3.10
* -3.09 - -2.80
* -2.79 - -2.40
* -2.39 - -1.80
* -1.79 - 10.30

Dixon T.H. et al., New Orleans subsidence: Rates and Spacial Variation measured by Permanent Scatterer Interferometry, Nature, 441, 587–586, 2006. Used with permission.

◆ **FIGURE 11.21** The city of New Orleans and the Mississippi River appear in the center of this figure showing the subsidence rates in the area in millimeters per year.

In a natural system, a delta is built up by sediment deposition but eroded by ocean waves and currents. The Mississippi River delta grew for 200 million years because deposition was greater than erosion. But, in the last 50 years, deposition rates have decreased because of flood control and dams, while erosion rates have actually increased because urban and agricultural development has destroyed natural plant communities that normally hold the soil.

Another problem affecting the delta arises from development of abundant oil reserves beneath the delta. As the oil is removed, the surface of the delta sinks, or subsides (◆ Figure 11.21). Because delta land is only marginally above sea level at best, even small amounts of subsidence can cause the land to sink below sea level.

As a result of all these processes, the delta has shrunk. Between 1930 and 2005, nearly 5,000 square kilometers of the Mississippi delta have sunk below sea level or washed into the sea. Ten years ago, the authors of this textbook, Gray Thompson and Jon Turk, wrote, "If current rates of erosion continue, the sea will . . . flood New Orleans . . . causing severe economic losses."[2] Tragically, this dire prediction has occurred.

The flooding of New Orleans in 2005 is an example of both a complex systems interaction and a threshold event. While Hurricane Katrina triggered the flood, human interference and development within a complex natural river/delta system set the stage. The delta subsided and eroded slowly over 75 years. Natural buffers were removed, with no immediate ill effects. Then a catastrophic storm pushed waters inland. The levees failed, causing a huge human disaster.

Flood Plain Management

As we have shown, in many cases attempts at controlling floods either do not work or they shift the problem to a different time or place. An alternative approach to flood control is to abandon some flood control projects and let the river spill out onto its flood plain. Of course the question is, What land should be allowed to flood? Every farmer and homeowner on the river wants to maintain the levees that protect his or her land. Currently, federal and state governments are establishing wildlife reserves in some flood plains. Because no development is allowed in these reserves, they will flood during the next high water. However, a complete river management plan involves complex political and economic considerations.

11.6
Lakes

Lakes and lake shores are attractive places to live and play. Clean, sparkling water, abundant wildlife, beautiful scenery, aquatic recreation, and fresh breezes all

2. Jonathan Turk and Graham Thompson, *Environmental Geoscience,* Saunders College Publishing, Philadelphia, PA, 1995, 428.

◆ FIGURE 11.22 An alpine lake in Montana's Beartooth Mountains.

◆ FIGURE 11.23 Gravel deposited by a glacier forms a dam to create this mountain lake in the Sierra Nevada.

come to mind when we think of going to the lake. Despite their great value, lakes are fragile and ephemeral. Modern, post–Ice Age humans live in a special time in Earth's history, when Earth's surface is dotted with beautiful lakes.

The Life Cycle of a Lake

A **lake** is a large, inland body of standing water that occupies a depression in the land surface (◆ Figure 11.22). Streams flowing into the lake carry sediment, which fills the depression in a relatively short time, geologically speaking. Soon the lake becomes a swamp, and with time the swamp fills with more sediment and vegetation and becomes a meadow or forest with a stream flowing through it.

If most lakes fill quickly with sediment, why are they so abundant today? Most lakes exist in places that were covered by glaciers during the latest ice age. About 18,000 years ago, great continental ice sheets extended well south of the Canadian border, and mountain glaciers scoured the alpine valleys as far south as New Mexico and Arizona. Similar ice sheets and alpine glaciers existed in higher latitudes of the Southern Hemisphere. We are just now emerging from that glacial episode.

The glaciers created lakes in several ways. Flowing ice eroded numerous depressions in the land surface, which then filled with water. The Finger Lakes of upper New York State and the Great Lakes are examples of large lakes occupying glacially scoured depressions.

The glaciers also deposited huge amounts of sediment as they melted and retreated. Some of these great piles of glacial debris formed dams across stream valleys. When the glaciers melted, streams flowed down the valleys but were blocked by the dams. Many modern lakes occupy glacially dammed valleys (◆ Figure 11.23). In addition, large blocks of ice may be left behind as a glacier recedes. When these ice blocks melt, they leave depressions, which then fill with water, forming a **kettle lake**.

Most of these glacial lakes formed within the past 10,000 to 20,000 years, and sediment is rapidly filling them. Many smaller lakes have already become swamps. In the next few hundred to few thousand years, many of the remaining lakes will fill with mud. The largest, such as the Great Lakes, may continue to exist for tens of thousands of years. But the life spans of lakes such as these are limited, and it will take another glacial episode to replace them.

Lakes also form by nonglacial means. A volcanic eruption can create a crater that fills with water to form a lake, such as Crater Lake in Oregon. Oxbow lakes form in abandoned river channels. Other lakes, such as Lake Okeechobee of the Florida Everglades, form in flat lands with shallow ground water. These types of lakes, too, fill with sediment and, as a result, have limited lives.

A few lakes, however, form in ways that extend their lives far beyond that of a normal lake. For example, Russia's Lake Baikal is a large, deep lake lying in a depression created by an active fault. Although rivers pour sediment into the lake, movement of the fault repeatedly deepens the basin. As a result, the lake has existed for more than a million years, so long that indigenous species of seals, other animals, and fish have evolved in its ecosystem.

Nutrient Balance in Lakes

When plants and animals die in a lake, their bodies settle to the bottom, carrying essential nutrients with them. If the water is deep enough, the nutrients accumulate below the zone where there is enough sunlight for plant growth. Thus in a deep lake, sunlight is available near the surface, but nutrients are abundant only on the bottom. Plankton (small, free-floating organisms) grow poorly on the surface due to the lack of nutrients, and bottom-rooted plants cannot grow due to lack of sunlight. Thus, the lakes contain low concentrations of nitrates, phosphates, and other critical nutrients that sustain aquatic food webs. The purity of the water gives the lakes a deep blue color that we associate with a clean, healthy lake. Such a lake is called **oligotrophic**, meaning "poorly nourished" (◄► Figure 11.24). Oligotrophic lakes have low productivities, meaning that they sustain relatively few living organisms, although a lake of this type is attractive for recreation and typically contains a few huge trout or similar game fish.

As a lake fills with sediment, however, it becomes shallower and sunlight reaches more and more of the lake bottom. The sunlight allows bottom-rooted plants to grow. As the plants die and rot, their litter adds nutrients to the lake water. Plankton increase in numbers, as do fish and other organisms. The lake becomes so productive that its surface may become covered with a green scum of plankton or a dense mat of rooted plants. The litter contributes to the sediment filling the lake, and eventually the lake becomes a swamp. A lake of this kind with a high nutrient supply is called a **eutrophic** lake (◄► Figure 11.25). Eutrophication occurs naturally as part of the life cycle of a lake. However, addition of nutrients in the form of sewage and other kinds of pollution has greatly accelerated the eutrophication of many lakes.

Temperature Layering and Turnover in Lakes

If you have ever dived into a deep lake on a summer day, you probably discovered that the top meter or so of lake water can be much warmer than deeper water. This occurs because sunshine warms the upper layer of water, making it less dense than the cooler, deeper water. The warm, less-dense water floats on the cooler, denser water. The boundary between the warm and cool layers is called the **thermocline** (◄► Figure 11.26).

In temperate climates, colder autumn weather cools the surface water to a temperature below that of deeper water, so that the surface water becomes more dense than the deeper water. Consequently it sinks, mixing the surface and deep waters and equalizing the water temperature throughout the lake. This process is called fall **turnover**. In the winter, ice floats on the surface and temperature layering develops again. In spring, as ice melts on the lake, surface water again becomes more dense than deep lake waters, and spring turnover occurs. As summer comes, the lake again develops thermal layering.

Turnover in temperate lakes illustrates an important Earth systems interaction among the atmosphere, the hydrosphere, and the biosphere. During summer and winter when the lake water is layered, bottom-dwelling organisms may use up most or all of the oxygen in deep waters. At the same time, surface organisms may deplete surface waters of dissolved nutrients.

However, surface water is rich in oxygen because it is in contact with the atmosphere, and deep water may be rich in nutrients because it is in contact with bottom sediment. Turnover enriches deep water in oxygen and, at the same time, supplies nutrients to the surface water. The latter effect often becomes evident in the form of an algal bloom—a sudden and obvious increase in the amount of floating green algae on a lake's surface—in spring and fall.

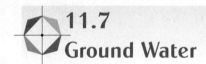

11.7
Ground Water

Two thousand years ago, residents of the coastal desert in Syria, near the town of Latakia, rowed 4 kilometers into the salty Mediterranean, dove to the ocean floor, and collected fresh drinking water in goatskin containers. This fresh water originated from an undersea spring supplied by vast reserves of subterranean **ground water** that saturates Earth's crust in a zone between a few meters and a few kilometers below the surface.

If you drill a hole into the ground in most places, a few meters to a hundred or more meters deep, its bottom fills with water after a short time, usually within a few minutes to a few days. The water appears even if no rain falls and no streams flow nearby. The water that seeps into the hole is a more common manifestation of our ground water reserves.

Ground water is exploited by digging wells and pumping the water to the surface. It provides drinking

Oligotrophic lake

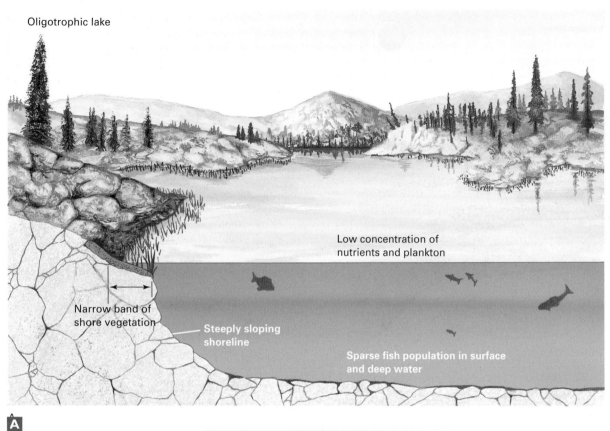

Low concentration of nutrients and plankton

Narrow band of shore vegetation

Steeply sloping shoreline

Sparse fish population in surface and deep water

A

◈ **FIGURE 11.24** **(A)** An oligotrophic, or nutrient-poor, lake contains pure water and few organisms. **(B)** An oligotrophic lake with a low nutrient content; few plants or fish; and clear, blue water in the Dry Ranges, Alberta, Canada.

B

water for more than half of the population of North America and is a major source of water for irrigation and industry. However, deep wells and high-speed pumps now extract ground water more rapidly than natural processes replace it in many parts of the central and western United States. In addition, industrial, agricultural, and domestic contaminants seep into ground water in many parts of the world. Such pollution is often difficult to detect and expensive to clean up. Ground water use and pollution are discussed in Chapter 12.

Characteristics of Ground Water

Ground water fills small cracks and voids in soil and bedrock. The proportional volume of these open spaces is called **porosity**. Sand and gravel typically have high

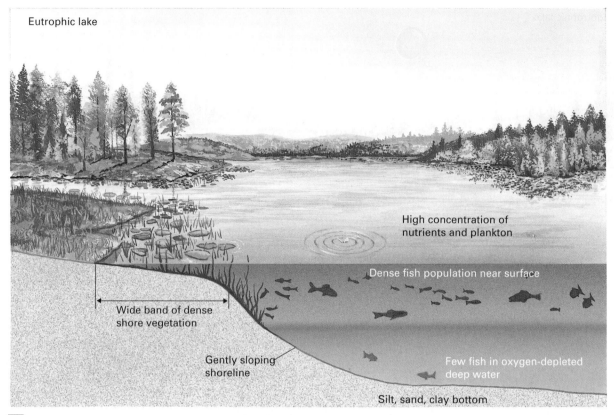

Eutrophic lake

High concentration of nutrients and plankton

Dense fish population near surface

Wide band of dense shore vegetation

Gently sloping shoreline

Few fish in oxygen-depleted deep water

Silt, sand, clay bottom

A

Courtesy of Graham R. Thompson/Jonathan Turk

B

◇ **FIGURE 11.25**

(A) A eutrophic, or nutrient-rich, lake contains many plants and other organisms. **(B)** A mat of vegetation covers the surface of a eutrophic lake near Penticton, British Columbia.

INTERACTIVE QUESTION: *Which type of lake, oligotrophic or eutrophic would be more desirable for humans? Discuss and defend your answer.*

porosities—40 percent or more. Mud can have a porosity of 90 percent or more. It has such a high porosity because the tiny clay particles are electrically attracted to water, and consequently, clay-rich mud absorbs a very high proportion of water. Most rocks have lower porosities than loose sediment. Sandstone and conglomerate can have 5 to 30 percent porosity. Shale typically has a porosity of less than 10 percent. Igneous and metamorphic rocks have very low porosities unless they are fractured.

Porosity indicates the amount of water that rock or soil can hold. In contrast, **permeability** is the ability of rock or soil to transmit water (or any other fluid). Water

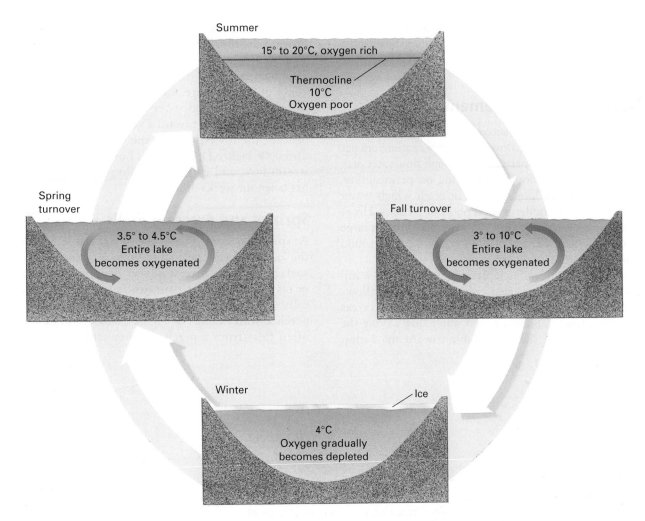

Summer
15° to 20°C, oxygen rich
Thermocline
10°C
Oxygen poor

Spring turnover
3.5° to 4.5°C
Entire lake becomes oxygenated

Fall turnover
3° to 10°C
Entire lake becomes oxygenated

Winter
Ice
4°C
Oxygen gradually becomes depleted

◇▶ **FIGURE 11.26** Lakes in temperate climates develop temperature layering in both summer and winter. As a result, bottom waters become depleted in oxygen. In fall and spring, water temperature becomes constant throughout the lake and turnover brings new supplies of oxygen to the deep waters.

can flow rapidly through material with high permeability. Most materials with high porosity also have high permeability. Sand and sandstone have numerous, relatively large, well-connected pores that allow the water to flow through the material. However, if the pores are very small, as in clay and shale, electrical attractions between water and soil particles slow the passage of water. Clay typically has a high porosity, but because its pores are so small and the electrical attractions slow the passage of water, it commonly has a very low permeability and transmits water slowly.

The Water Table and Aquifers

When rain falls, much of it soaks into the ground. Water does not descend into the crust indefinitely, however. Below a depth of a few kilometers, the pressure from overlying rock closes the pores, making bedrock both nonporous and impermeable. Water accumulates on this

impermeable barrier, filling pores in the rock and soil above it. This completely wet layer of soil and bedrock above the barrier is called the **zone of saturation**. The **water table** is the top of the zone of saturation (◇▶ Figure 11.27). The unsaturated zone, or **zone of aeration**, lies above the water table. In this layer, the rock or soil may be moist but is not saturated.

If you dig into the unsaturated zone, the hole does not fill with water. However, if you dig below the water table into the zone of saturation, you have dug a **well**, and the water level in a well is at the level of the water table. During a wet season, rain seeps into the ground to **recharge** the ground water, and the water table rises. During a dry season, the water table falls. Thus, the water level in most wells fluctuates with the seasons.

An **aquifer** is any body of rock or soil that can yield economically significant quantities of water. An aquifer must be both porous and permeable so that water flows into a well to replenish water that is pumped out. Sand

and gravel, sandstone, limestone, and highly fractured bedrock of any kind make excellent aquifers. Shale, clay, and unfractured igneous and metamorphic rocks are poor aquifers.

Ground Water Movement

Nearly all ground water seeps slowly through bedrock and soil. Ground water flows at about 4 centimeters per day (about 15 meters per year), although flow rates may be much faster or slower depending on permeability. Most aquifers are like sponges through which water seeps, rather than underground pools or streams. However, ground water can flow very rapidly through large fractures in bedrock, and in a few regions, underground rivers flow through caverns.

In general, the water table is higher beneath a hill than it is beneath an adjacent valley. Ground water flows from zones where the water table is highest toward areas where it is lowest. Some ground water flows along the sloping surface of the water table toward the valley.

Much of the ground water, however, flows downward beneath the hill. This occurs because water pressure is greatest beneath the highest part of the water table. Ground water flows from zones of high pressure toward zones of low pressure. Consequently, the pressure difference forces ground water to flow downward beneath a hill, then laterally toward a valley, and finally upward beneath the lowest part of the valley where a stream flows (◀▶ Figure 11.27). This is how ground water feeds stream flow, and why streams flow even when no rain has fallen for weeks or months.

Springs and Artesian Wells

A **spring** occurs where the water table intersects the land surface and water flows or seeps onto the surface. In some places, a layer of impermeable rock or clay lies above the main water table, creating a locally saturated zone, the top of which is called a **perched water table** (◀▶ Figure 11.28). Hillside springs often flow from a perched water table. Springs also

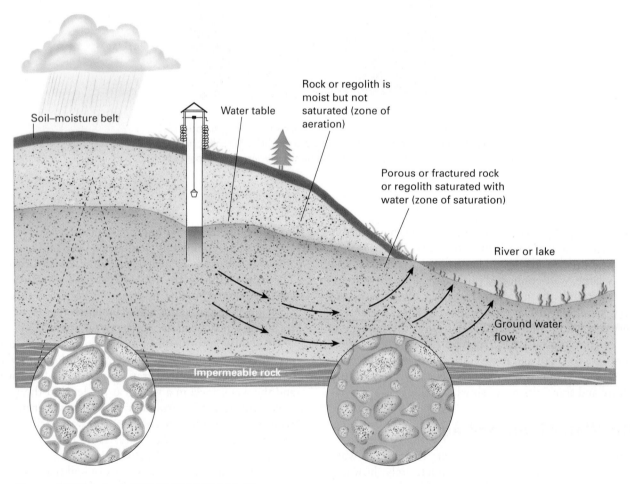

Soil–moisture belt

Water table

Rock or regolith is moist but not saturated (zone of aeration)

Porous or fractured rock or regolith saturated with water (zone of saturation)

River or lake

Ground water flow

Impermeable rock

ThomsonNOW ◀▶ **ACTIVE FIGURE 11.27** The water table is the top of the zone of saturation near Earth's surface. It intersects the land surface at lakes and streams and is the level of standing water in a well.

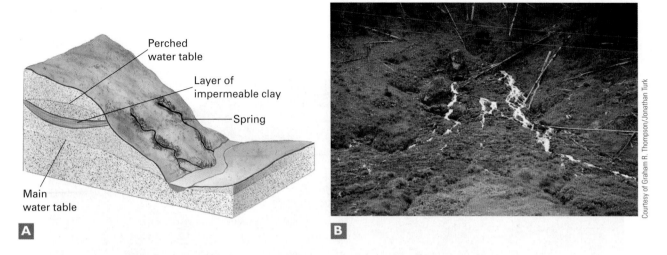

A

B

Courtesy of Graham R. Thompson/Jonathan Turk

◆ **FIGURE 11.28** **(A)** Springs can form where a perched water table intersects a hillside. **(B)** Water flows from a spring on a hillside in British Columbia.

occur where fractured bedrock or cavern systems intersect the land surface.

◆ Figure 11.29 shows a tilted layer of permeable sandstone sandwiched between two layers of impermeable shale. An inclined aquifer, such as the sandstone layer, bounded top and bottom by impermeable rock is

an **artesian aquifer**. Water in the lower part of the aquifer is under pressure from the weight of water above. Therefore, if a well is drilled through the shale and into the sandstone, water rises in the well without being pumped. A well of this kind is called an **artesian well**. If pressure is sufficient, the water spurts out onto the land.

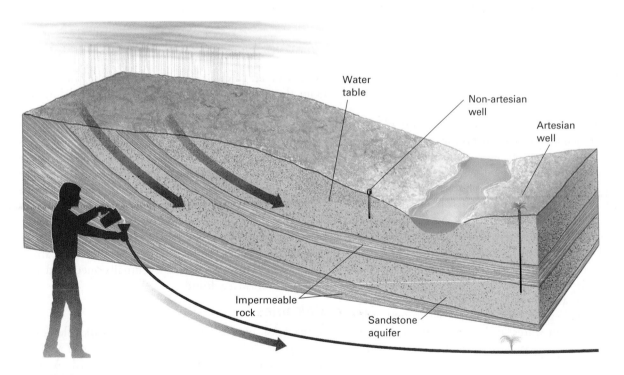

ThomsonNOW ◆ **ACTIVE FIGURE 11.29** An artesian aquifer forms where a tilted layer of permeable rock, such as sandstone, lies sandwiched between layers of impermeable rock, such as shale. Water rises in an artesian well without being pumped. A hose with a hole (inset) shows why an artesian well flows spontaneously.

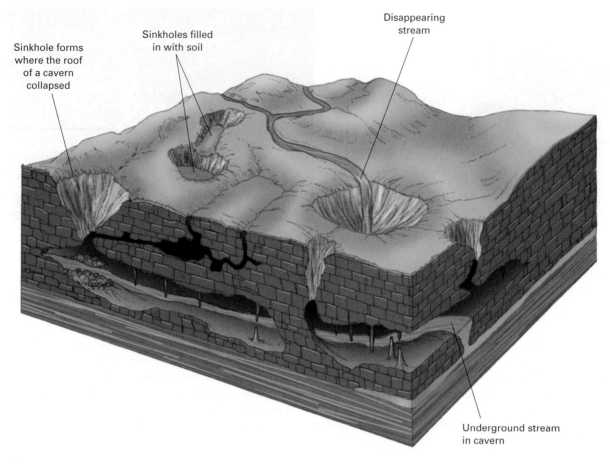

Sinkhole forms where the roof of a cavern collapsed

Sinkholes filled in with soil

Disappearing stream

Underground stream in cavern

◆ **FIGURE 11.30** Sinkholes and caverns are characteristic of karst topography. Streams commonly disappear into sinkholes and flow through the caverns to emerge elsewhere.

Caverns, Sinkholes, and Karst Topography

Just as streams erode valleys and form flood plains, ground water also creates landforms (◆ Figure 11.30). Recall that rainwater reacts with atmospheric carbon dioxide to become slightly acidic and capable of dissolving limestone. A **cavern** forms when acidic water seeps into cracks in limestone, dissolving the rock and enlarging the cracks. Mammoth Cave in Kentucky and Carlsbad Caverns in New Mexico are two famous caverns formed in this way.

Although caverns form when limestone dissolves, most caverns also contain features formed by deposition of calcite. When a solution of water, dissolved calcite, and carbon dioxide percolates through the ground, it is under pressure from water in the cracks above it. If a drop of this solution seeps into the ceiling of a cavern, the pressure decreases because the drop comes in contact with the air. The high humidity of the cave prevents the water from evaporating rapidly, but the lowered pressure allows some of the carbon dioxide to escape as a gas. When the carbon dioxide escapes, the drop becomes less acidic. This decrease in acidity causes some of

the dissolved calcite to precipitate as the water drips from the ceiling. Over time, a beautiful and intricate **stalactite** grows to hang icicle-like from the ceiling of the cave (◆ Figure 11.31).

Only a portion of the dissolved calcite precipitates as the drop seeps from the ceiling. When the drop falls to the floor, it spatters and releases more carbon dioxide. The acidity of the drop decreases further, and another minute amount of calcite precipitates. Thus, a **stalagmite** builds from the floor upward to complement the stalactite. Because stalagmites are formed by splashing water, they tend to be broader than stalactites. As the two features continue to grow, they may eventually meet and fuse together to form a **column**.

Sinkholes

If the roof of a cavern collapses, a **sinkhole** forms on Earth's surface. A sinkhole can also form as limestone dissolves from the surface downward. A well-documented sinkhole formed in May 1981 in Winter Park, Florida. During the initial collapse, a three-bedroom house, half a swimming pool, and six Porsches in a dealer's lot all fell into the underground cavern. Within a few days, the sinkhole was 200 meters wide and

◆ **FIGURE 11.31** Stalactites and stalagmites form as calcite precipitates in a limestone cavern. Luray Caverns, Virginia.

50 meters deep, and it had devoured additional buildings and roads (◆ Figure 11.32).

Although sinkholes form naturally, human activities can accelerate the process. The Winter Park sinkhole formed when the water table dropped, removing support for the ceiling of the cavern. The water table fell as a result of a severe drought augmented by excessive removal of ground water by humans.

Karst Topography

Karst topography forms in broad regions underlain by limestone and other readily soluble rocks. Caverns and sinkholes are common features of karst topography. Surface streams often pour into sinkholes and disappear into caverns. In the area around Mammoth Cave in Kentucky, streams are given names such as Sinking Creek, an indication of their fate. The word *karst* is derived from a region in Croatia where this type of topography is well developed. Karst landscapes are found in many parts of the world.

11.8
Hot Springs, Geysers, and Geothermal Energy

At numerous locations throughout the world, hot water naturally flows to the surface to produce **hot springs**. Ground water can be heated in three ways:

1. Earth's temperature increases by about 30°C per kilometer of depth in the upper portion of the

◆ **FIGURE 11.32** This sinkhole in Winter Park, Florida, collapsed suddenly in May 1981, swallowing several houses and a Porsche agency.

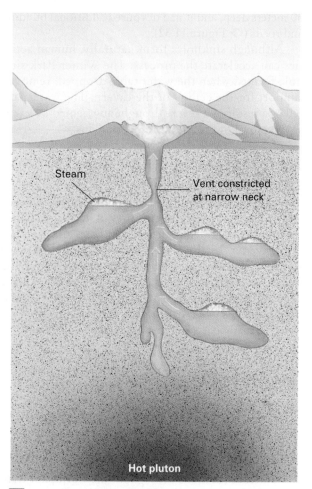

A

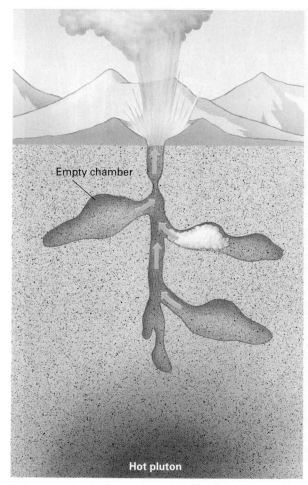

B

◇ **FIGURE 11.33** Before a geyser erupts **(A)** ground water seeps into underground chambers and is heated by hot igneous rock. Foam constricts the geyser's neck, trapping steam and raising pressure. **(B)** When the pressure exceeds the strength of the blockage, the constriction blows out. Then the hot ground water flashes into vapor and the geyser erupts.

crust. Therefore, if ground water descends through cracks to depths of 2 to 3 kilometers, it is heated by 60°C to 90°C. The hot water or steam then rises because it is less dense than cold water. However, it is unusual for fissures to descend so deep into Earth, and this type of hot spring is uncommon.

2. In regions of recent volcanism, magma or hot igneous rock may remain near the surface and can heat ground water at relatively shallow depths. Hot springs heated in this way are common throughout western North and South America because these regions have been magmatically active in the recent past and remain so today. Shallow magma heats the hot springs and geysers of Yellowstone National Park.

3. Many hot springs have the odor of rotten eggs from small amounts of hydrogen sulfide (H_2S) dis-

solved in the hot water. The water in these springs is heated by chemical reactions. Sulfide minerals, such as pyrite (FeS_2), react chemically with water to produce hydrogen sulfide and heat. The hydrogen sulfide rises with the heated ground water and gives it the strong odor.

Most hot springs bubble gently to the surface from cracks in bedrock. However, **geysers** violently erupt hot water and steam. Geysers generally form over open cracks and channels in hot underground rock. Before a geyser erupts, ground water seeps into the cracks and is heated by the rock (◇ Figure 11.33). Gradually, steam bubbles form and start to rise, just as they do in a heated teakettle. If part of the channel is constricted, the bubbles accumulate and form a temporary barrier that allows the steam pressure below to increase. The rising pressure forces some of the bubbles upward

◆ **FIGURE 11.34** An eruption of Riverside Geyser in Upper Geyser Basin, Yellowstone National Park, Wyoming.

past the constriction and short bursts of steam and water spurt from the geyser. This lowers the steam pressure at the constriction, causing the hot water to vaporize, blowing steam and hot water skyward (◆ Figure 11.34).

The most famous geyser in North America is Old Faithful in Yellowstone National Park, which erupts on the average of once every 65 minutes. Old Faithful is not as regular as people like to believe; the intervals between eruptions vary from about 30 to 95 minutes.

Hot ground water can be used to drive turbines and generate electricity, or it can be used directly to heat homes and other buildings. Energy extracted from Earth's heat is called **geothermal energy**. In the United States, a total of 70 geothermal plants in California, Hawaii, Utah, and Nevada have a generating capacity of 2,800 megawatts, enough to supply over two million people with electricity and equivalent to the power output of 25 million barrels of oil. However, this amount of energy is minuscule compared with the potential of geothermal energy.

11.9
Wetlands

Wetlands are known across North America as swamps, bogs, marshes, sloughs, mudflats, and flood plains. They are regions that are water soaked or flooded for part or all of the year. Some wetlands are wet only during exceptionally wet years and may be dry for several years at a time.

Wetland ecosystems vary so greatly that the concept of a wetland defies a simple definition. Wetlands share certain properties, however: The ground is wet for at least part of the time; the soils reflect anaerobic (lacking oxygen) conditions; and the vegetation consists of plants such as red maple, cattails, bulrushes, mangroves, and other species adapted to periodic flooding or water saturation. North American wetlands include all stream flood plains, frozen arctic tundra, warm Louisiana swamps, coastal Florida mangrove swamps, boggy mountain meadows of the Rockies, and the immense swamps of interior Alaska (◆ Figure 11.35).

A

B

◆ **FIGURE 11.35** North American wetlands extend from **(A)** the Everglades to **(B)** the immense Alaskan swamps.

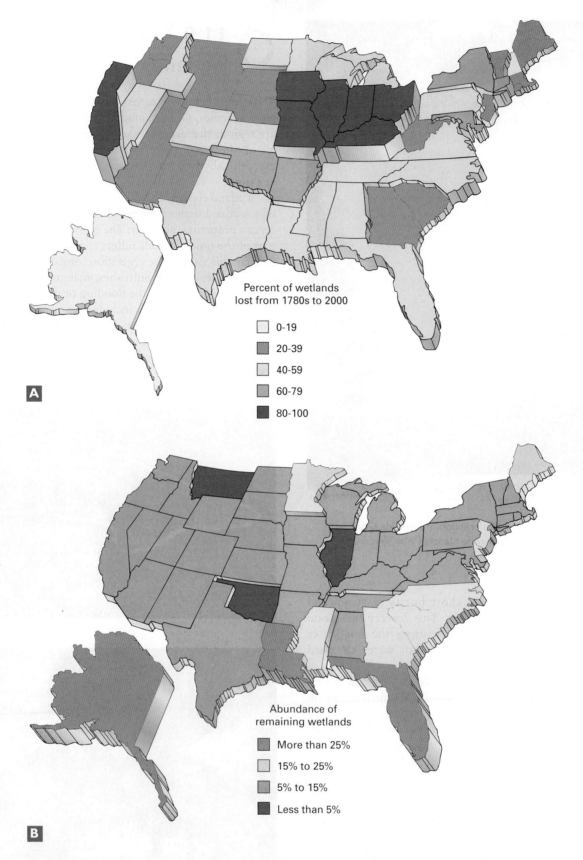

Percent of wetlands lost from 1780s to 2000

☐ 0-19
■ 20-39
☐ 40-59
■ 60-79
■ 80-100

A

Abundance of remaining wetlands

■ More than 25%
☐ 15% to 25%
■ 5% to 15%
■ Less than 5%

B

◈ **FIGURE 11.36** **(A)** The proportion of wetlands lost, by state, in the United States in the past 200 years. **(B)** The abundance of remaining wetlands in the United States. Percentages indicate the proportional area of each state still covered by wetlands.

Wetlands are among the most biologically productive environments on Earth. Two-thirds of the Atlantic fish and shellfish consumed by humans rely on coastal wetlands for at least part of their life cycles. One-third of the endangered species of both plants and animals in the United States also depend on wetlands for survival. More than 400 of the 800 species of protected migratory birds and one-third of all resident bird species feed, breed, and rest in wetlands.

When European settlers first arrived in North America, 87 million hectares of wetlands existed (exclusive of those in Alaska). Americans have long viewed wetlands as mosquito-infested, malarial swamps occupying land that can be farmed or otherwise developed if drained or filled. In the mid-1800s, the federal government passed legislation known as the Swamp Land Acts, which established an official policy to fill and drain wetlands to convert them to agricultural uses wherever possible.

Over the past century, farmers, ranchers, and developers drained or filled more than half of the original wetlands. California and several upper-midwestern states have lost more than 80 percent of their wetlands (◆ Figure 11.36A). Wetlands now make up between 6 and 9 percent of the lower 48 states and as much as 60 percent, or about 80 million hectares, of Alaska (◆ Figure 11.36B). Currently, between about 120,000 and 200,000 hectares of wetlands are destroyed each year.

Government policy and private practice have ignored the beneficial qualities of wetlands. Aquatic organisms consume many pollutants and degrade them to harmless by-products. Because these organisms abound in wetlands, the ecosystems are natural sewage treatment systems. Wetlands also mitigate flooding by absorbing excess water that might otherwise overrun towns and farms. Wild ducks and geese and other migratory birds depend on wetlands for breeding, food, and cover. The importance of wetlands in water purification, flood control, and wildlife habitat did not become widely recognized until the 1960s. At present, the focus of federal and state laws has changed from destruction of wetlands to their protection and preservation. But it has proved more difficult to reverse practice than policy, and wetland losses continue today.

EARTH SYSTEMS INTERACTIONS

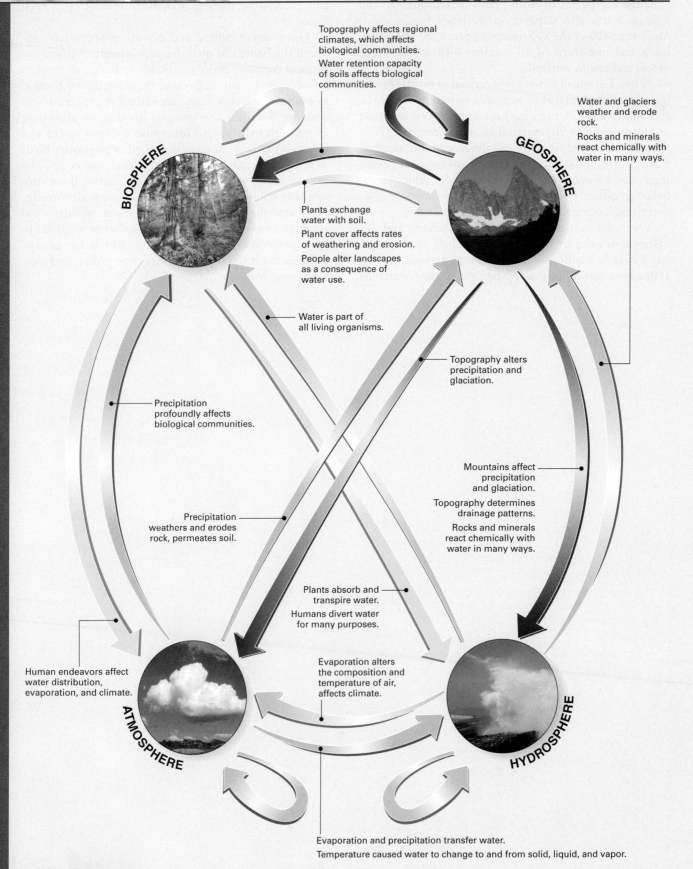

Topography affects regional climates, which affects biological communities.

Water retention capacity of soils affects biological communities.

BIOSPHERE

GEOSPHERE

Water and glaciers weather and erode rock.

Rocks and minerals react chemically with water in many ways.

Plants exchange water with soil.

Plant cover affects rates of weathering and erosion.

People alter landscapes as a consequence of water use.

Water is part of all living organisms.

Topography alters precipitation and glaciation.

Precipitation profoundly affects biological communities.

Mountains affect precipitation and glaciation.

Topography determines drainage patterns.

Rocks and minerals react chemically with water in many ways.

Precipitation weathers and erodes rock, permeates soil.

Plants absorb and transpire water.

Humans divert water for many purposes.

Human endeavors affect water distribution, evaporation, and climate.

Evaporation alters the composition and temperature of air, affects climate.

ATMOSPHERE

HYDROSPHERE

Evaporation and precipitation transfer water.

Temperature caused water to change to and from solid, liquid, and vapor.

PICK UP A HANDFUL of fertile garden soil and squeeze it between your fingers. It feels moist and exudes a small odor, largely from the myriads of microscopic organisms that live in the clod of dirt. Atmospheric gases permeate the pores between the sediment grains. Thus you are holding in your hands components of all four of Earth's spheres.

Water permeates rocks of the geosphere (to a depth of a few kilometers below the surface) and reacts with atmospheric gases and chemicals secreted by plant roots to decompose rocks and minerals. It evaporates into the atmosphere and falls as rain and snow. All living organisms are composed primarily of water. Therefore, the hydrosphere interacts with the Earth's geosphere, biosphere, and atmosphere in so many ways that it would be impossible to list them all. Step outside and record all the forms of water that you can see or that exist unseen in your immediate environment: clouds, vapor, rivers, streams, lakes, ocean, water incorporated chemically into rocks or that exists in soil, water in your blood and cells. Throughout this text, we often find it helpful to mentally dissociate a system into its component parts, but in the real world the system is one integrated whole. ∎

SUMMARY

Only about 0.63 percent of Earth's water is fresh. The rest is salty seawater and glacial ice. The constant circulation of water among the sea, land, the biosphere, and the atmosphere is called the **hydrologic cycle**, or the **water cycle**. Most of the water that evaporates from the oceans returns to the oceans as precipitation. The precipitation that falls on the continents returns to the sea via **runoff** and moving ground water, or it returns to the atmosphere via evaporation and **transpiration**. The hydrologic cycle not only describes the movement of water, it is also a primary mechanism for the movement of energy from one part of the globe to another.

A **stream** is any body of water flowing in a channel. The term *river* is commonly used for any large stream fed by smaller ones called **tributaries**. The velocity of a stream is determined by its **gradient** and **discharge**. Channel characteristics refer to the shape and roughness of a stream channel. The floor of the channel is called the bed, and the sides of the channel are the banks. Streams shape Earth's surface by eroding soil and bedrock. Stream erosion and sediment transport depend on **competence** and **capacity**. Streams transport sediment as **dissolved load, suspended load**, and **bed load**.

Downcutting, lateral erosion, and **mass wasting** combine to form a stream valley. **Base level** is the lowest elevation to which a stream can erode its bed; it is usually sea level. A lake or resistant rock can form a local, or temporary, base level. A graded stream has a smooth, concave profile. Mountain streams downcut rapidly and form V-shaped valleys, whereas lower-gradient streams form wider valleys and **meanders** by lateral erosion and mass wasting. A stream erodes its outside bank and deposits sediment on the inside, to form a **point bar**. An **oxbow lake** forms where the stream cuts across the neck of a meander. An **alluvial fan** forms where a steep stream flows onto a flat plain, and a **delta** forms where a stream enters a lake or ocean.

A **drainage basin** is the region drained by a single stream. Streams erode the land surface to flatten rugged topography, but tectonic activity and isostatic rebound uplift the land at the same time that erosion levels it.

A **flood** occurs when a stream overflows its banks and flows over its **flood plain**. **Artificial levees** and channels can contain small floods but they may exacerbate large ones.

Lakes are short-lived landforms because streams fill them with sediment. Recent glaciers created many modern lakes; as a result, we live in an unusual time of abundant lakes. The life history of a lake commonly involves a progression from an **oligotrophic** lake to a **eutrophic** lake to a swamp and, finally, to a flat meadow or forest. Temperate lakes normally develop temperature layering across a boundary called the **thermocline** during the summer and winter, but these lakes experience **turnover** twice a year, which mixes the lake water and supplies oxygen to deep waters and nutrients to surface waters.

Much of the rain that falls on land seeps into soil and bedrock to become **ground water**. **Porosity** is the proportion of rock or soil that consists of open space. **Permeability** is the ability of soil or bedrock to transmit water. Ground water saturates the upper few kilometers of soil and bedrock to a level called the **water table**, which separates the **zone of saturation** from the **zone of aeration**. An **aquifer** is a body of rock that can yield economically significant quantities of water. Aquifers are porous and permeable.

Most ground water moves slowly, about 4 centimeters per day. **Springs** occur where the water table intersects the surface of the land and may lie above a perched water table. Dipping layers of permeable and impermeable rock can produce an **artesian aquifer**.

Caverns form where ground water dissolves limestone. **Stalactites, stalagmites**, and **columns** form from precipitating limestone. A **sinkhole** forms when the roof of a limestone cavern collapses. **Karst** topography, with numerous caves, sinkholes, and subterranean streams, is characteristic of limestone regions. **Hot springs** and **geysers** develop when hot ground water rises to the surface. Hot ground water has been tapped to produce **geothermal energy**.

Wetlands are among the most biologically productive environments on Earth. In addition, they are natural water purification systems, and they mitigate flood effects by absorbing floodwaters. Despite these facts, government and private efforts have eliminated more than half of the wetlands that existed in the lower 48 states when European settlers first arrived.

Key Terms

For Review

1. In which physical state (solid, liquid, or vapor) does most of Earth's free water exist? Which physical state accounts for the least?

2. Describe the movement of water through the hydrologic cycle. Explain briefly how energy is also cycled by the same processes.

3. Describe the factors that determine the velocity of stream flow and how those factors interact.

4. Describe the factors that determine how much sediment a stream can transport.

5. Give two examples of natural features that create temporary base levels. Why are they temporary?

6. Draw a profile of a graded stream and an ungraded stream.

7. Explain how a stream forms and shapes a valley.

8. Explain how an oxbow lake forms.

9. In what type of terrain would you be likely to find a V-shaped valley? Where would you be likely to find a meandering stream with a broad flood plain?

10. Why are most lakes short-lived landforms?

11. What geological conditions create a long-lived lake?

12. Describe the differences between an oligotrophic lake and a eutrophic lake.

13. What is a thermocline? Under what conditions does a thermocline form in a lake?

14. What conditions cause mixing, or turnover, of surface and deep waters in a temperate lake?

15. (a) Draw a cross section of soil and shallow bedrock, showing the zone of saturation, water table, and zone of aeration. (b) Explain each of the preceding terms.

16. What is an aquifer, and how does water reach it?

17. Explain why bedrock or soil must be both porous and permeable to be an aquifer.

18. Compare the movement of ground water in an aquifer with that of water in a stream.

19. How does an artesian aquifer differ from a normal one? Why does water from an artesian well rise without being pumped?

20. What is karst topography? How can it be recognized? How does it form?

21. Describe three types of heat source for a hot spring.

22. Describe wetlands. What types of environments are included in the wetlands category?

23. What proportion of wetlands that the early European settlers found in the United States no longer exists?

24. Describe how wetlands mitigate flooding.

For Discussion

1. Describe the ways in which (a) a rise and (b) a fall in the average global temperature could affect Earth's hydrologic cycle.

2. Describe ways in which a change in the hydrologic cycle could raise or lower the average global temperature.

3. A stream is 50 meters wide at a certain point. A bridge is built across the stream, and the abutments extend into the channel, narrowing it to 40 meters.

   Discuss the changes that might occur as a result of this constriction.

4. Describe and discuss effects of flood control structures, such as artificial levees, on the damage caused by the 1993 Mississippi River floods.

5. The National Flood Insurance Program (NFIP) is a federally sponsored insurance for people who live on flood plains. In exchange for low-cost insurance policies, the NFIP establishes rigorous building codes for new construction in flood plains. However, older buildings are not subject to these standards.

   As long as damage from one flood is less than 50 percent of the value of the structure, it is eligible for subsidized insurance against the next flood. Outline potential benefits and drawbacks of this program.

6. Most substances become denser when they freeze. Water is anomalous in that ice is less dense than water. Outline the seasonal temperature profile of a lake if ice were denser than water. How would aquatic ecosystems be affected?

7. Imagine that you live on a hill 25 meters above a nearby stream. You drill a well 40 meters deep and do not reach water. Explain.

8. Why are caverns, sinkholes, and karst topography most commonly found in limestone terrain?

9. Discuss why wetlands support such a large population of wildlife.

10. Discuss the mechanisms by which wetlands ecosystems purify contaminated water.

11. Discuss how changing public perceptions may have played a role in reversing governmental policy toward wetlands.

ThomsonNOW

Assess your understanding of this chapter's topics with additional comprehensive interactivities at **http://www.thomsonedu.com/login**, which also has up-to-date web links, additional readings, and exercises.

CHAPTER

12

Water Resources

Courtesy of Graham R. Thompson/Jonathan Turk

Glen Canyon Dam on the Colorado River in Arizona is one of the most controversial dams in America.

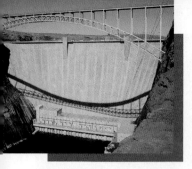

M ore than two-thirds of Earth's surface is covered with water. But most of this water is salty, and most of the freshwater is frozen into the Antarctic and Greenland ice caps. Humans divert half of all the flowing water from rain and snow that falls on the continents, but even that is not enough for our burgeoning population and per-capita consumption. More than two centuries ago, when freshwater seemed as inexhaustible as the buffalo that roamed the High Plains, Benjamin Franklin warned, "When the well's dry, we know the worth of water." Today, as we reach, and in some cases exceed, the limits of our water resources, the well is running dry. In March 2003, the United Nations published the "World Water Development Report" stating that within the next 50 years, up to 7 billion people—more than the entire current population of the world—will face water scarcity. Part of the problem is that much of the Earth's rain falls in the wrong places or at the wrong times to be of much use to humans. One prominent ecologist lamented, "The wet places of the world don't need the runoff because they are wet. The dry places of the world need the runoff to irrigate the land, but they don't get much."

In addition to water scarcity, much of the surface water and ground water in industrial areas is heavily polluted, destroying natural habitats and making the resource less useful for humans. ■

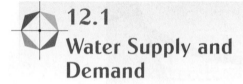

12.1
Water Supply and Demand

Rain and snow continuously replenish freshwater on land, so that the amount of available freshwater is about the same as it was two centuries ago. But the demand for water has risen dramatically until it has approached or even exceeded supply in many parts of the world (◆ Figure 12.1). In the early 1800s, there were 1 billion people on Earth, now there are 6.5 billion. Thus as the human population has grown, the amount of available freshwater per person has diminished. In addition, our technological society uses water at rates that would have been inconceivable to Ben Franklin.

Water use falls into two categories. Any process that uses water and then returns it to Earth locally is called **withdrawal.** Most of the water used by industry and homes returns to streams or ground water reservoirs near the place from which it was taken. For example, river water pumped through an electric generating station to cool the exhaust is returned almost immediately to the river. Water used to flush a toilet in a city is pumped to a sewage treatment plant, purified, and discharged into a nearby stream.

In contrast, a process that uses water and then returns it to Earth far from its source is called **consumption.** Most irrigation water evaporates, disperses with the wind, and returns to Earth as precipitation

hundreds or thousands of kilometers from its source (◆ Figure 12.2). Although industry accounts for about half of all water withdrawn in the United States, agriculture accounts for most of the water that is consumed and not returned to its place of origin (◆ Figure 12.3). Globally, about two-thirds of the total water that is used is consumed.

Water withdrawal and consumption accounts for about one-sixth (or about 17 percent) of the total runoff. Humans divert an additional 33 percent of the total runoff by damming rivers. Although this water is neither withdrawn nor consumed, dams create environmental problems, which are discussed in Section 12.2, "Dams and Diversion."

Water use (withdrawal and consumption) is subdivided into domestic, industrial (including power-plant cooling), and agricultural categories. As you can see from ◆ Figure 12.4, in the United States, agriculture accounts for 41 percent of our water use.

Domestic Water Use

A person in an industrial urban society can live a healthy, clean, hygienic life on 50 liters of water per day. The breakdown is 5 liters for drinking, 10 liters for cooking and dishwashing, 15 liters for bathing, and 20 liters for clothes washing and toilet flushing. These last two categories are lumped together because water used to wash clothes can be recycled and used to flush the toilet. In contrast, the average American uses 800 liters per day (◆ Figure 12.5). Of this total, 200 liters

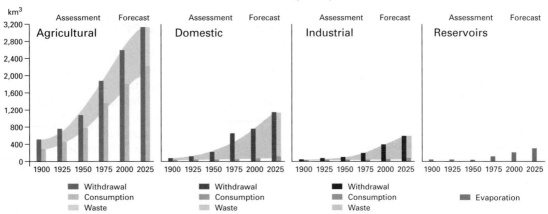

Evolution of Global Water Use
Withdrawal and Consumption by Sector

km³

Agricultural	Domestic	Industrial	Reservoirs
Assessment Forecast	Assessment Forecast	Assessment Forecast	Assessment Forecast

■ Withdrawal
■ Consumption
■ Waste

■ Evaporation

Note: Domestic water consumption in developed countries (500–800 litres per person per day) is about six times greater than in developing countries (60–150 litres per person per day).

◆ **FIGURE 12.1** Global water use, 1900 to 2000, with projections to 2025. Between 2000 and 2054, the world's population is expected to increase by 3 billion people, so water demand will continue to rise. (United Nations Environmental Programme.)

is used for watering lawns. However, domestic use accounts for only 10 percent of the water used in the United States.

In contrast, the World Health Organization defines reasonable water access in less-developed, agricultural societies as 20 liters of sanitary water per person per day, assuming that the source is less than 1 kilometer from the home. Yet 1.1 billion people, about one-sixth of the global population, lack even this small amount of clean water. Water access can fall short for three reasons: either there isn't enough water, the water is unfit to drink, or the source is so far away that people must carry their water long distances from public wells or streams to their homes. The World Health Organization estimates

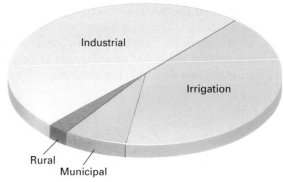

Withdrawal—1,800 billion liters per day
(excluding hydropower)

A

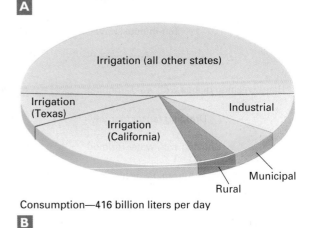

Consumption—416 billion liters per day

B

Courtesy of Graham R. Thompson/Jonathan Turk

◆ **FIGURE 12.2** Most irrigation water evaporates and is carried away by wind.

◆ **FIGURE 12.3** **(A)** Although industry accounts for about half of all water withdrawn in the United States, **(B)** agriculture accounts for more than three-fourths of the water that is consumed. Irrigation in two dry agricultural states, California and Texas, accounts for about 30 percent of all water consumed in the United States.

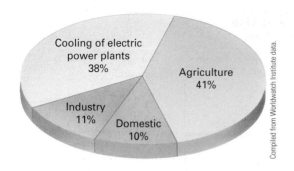

◆ **FIGURE 12.4** Industrial, agricultural, and domestic water use in the United States.

that diarrheal illnesses caused by unsafe water account for 4.1 percent of global disease and are responsible for the deaths of 1.8 million people each year, most of them children under five.

Industrial Water Use

The cooling systems in electric-power generating plants (run by fossil fuels or nuclear power) account for 38 percent of all water used in the United States. Although much of this water is returned to the stream from which it was taken, it is considerably warmer, which, as we shall see, affects aquatic ecosystems. All other industrial needs add an additional 11 percent to the national water demand. The quantities of water

required to produce some common industrial products are shown in ◆ Figure 12.6.

Agricultural Water Use

Agriculture accounts for 41 percent of water use in the United States. ◆ Figure 12.7 shows that different agricultural products use vastly different amounts of water. California's Central Valley was once a desert, but irrigation has transformed it into an immensely productive region for growing fruits and vegetables. The productivity of the Great Plains in the western United States would decline by one-third to one-half if irrigation were to cease.

Globally, many nations rely on irrigation for more than half of their food production. In dry regions an even higher proportion of farmland must be irrigated, and, as a result, surface water resources are stressed (◆ Figure 12.8).

12.2 Dams and Diversion

As we stated in the introduction to this chapter, much of Earth's rain falls in the wrong places or at the wrong times to be of much use to humans. For thousands of years, people realized that the problem could be

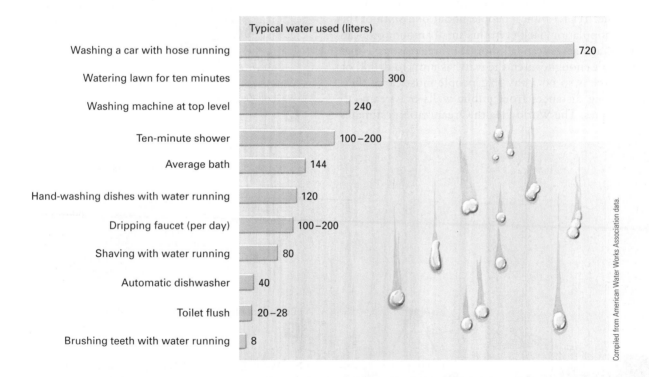

◆ **FIGURE 12.5** Domestic water for common activities use in the United States.

INTERACTIVE QUESTION: *Using the data displayed in this figure, list specific steps that you could take to conserve water this week. Order your list from most effective (greatest water savings) to least effective steps.*

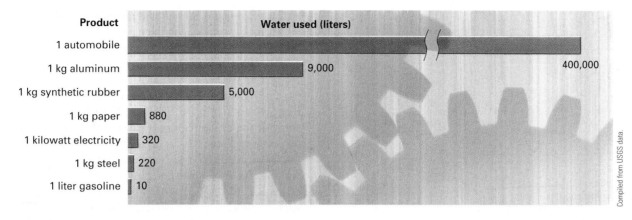

FIGURE 12.6 Large quantities of water are used to produce common industrial products in the United States.

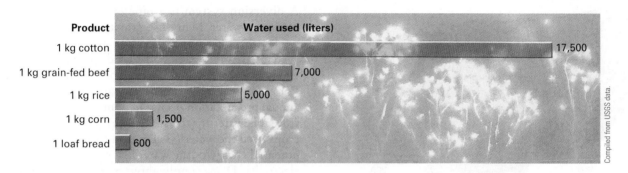

FIGURE 12.7 Different agricultural products require vastly different amounts of irrigation water in the United States.

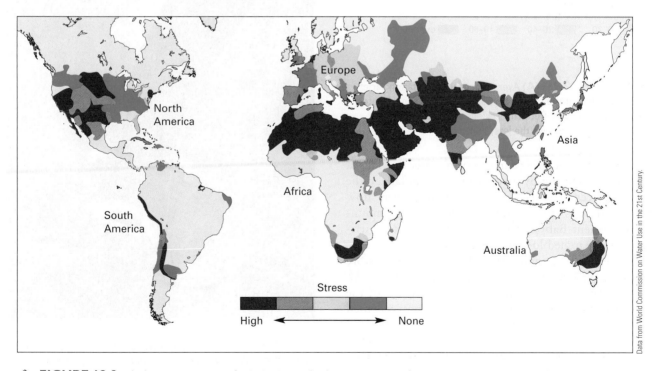

FIGURE 12.8 As humans use water for irrigation and other purposes, surface water resources become stressed. (Stress evaluation is based on the amount of water available compared with the amount used by people.)

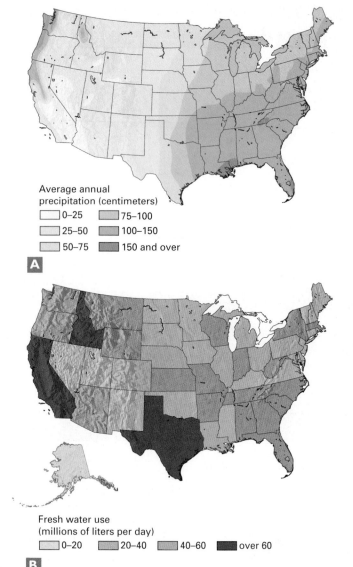

Average annual
precipitation (centimeters)
- 0–25
- 25–50
- 50–75
- 75–100
- 100–150
- 150 and over

A

Fresh water use
(millions of liters per day)
- 0–20
- 20–40
- 40–60
- over 60

B

◆ **FIGURE 12.9** **(A)** Average annual precipitation varies greatly throughout the United States. **(B)** Average annual water use also varies greatly. But many regions that use the greatest amounts of water receive the least rain.

against significant human and environmental costs.

In many regions, surface water is scarce, but huge reservoirs of ground water are accessible beneath the surface. This ground water can be pumped to the surface for human use.

The United States receives about three times more water from precipitation than it uses. However, some of the driest regions are those that use the greatest amounts of water (◆ Figure 12.9). For example, some of the most productive agricultural regions are located in the deserts of California, Texas, and Arizona, where crops are entirely supported by irrigation. Desert cities such as Phoenix, Los Angeles, Albuquerque, and Tucson support large populations that use hundreds of times more water than is locally available and renewable.

Surface Water Diversion

A **diversion system** is a pipe or canal constructed to transport water. Many use gravity to move the water, while others use pumps to lift water uphill. Diversion systems are often augmented by dams that store water. Dams are especially useful in regions of seasonal rainfall. For example, in the arid and semiarid west, much of the annual precipitation occurs in winter and spring. Dams store this water for the summer irrigation season, when crops need the water (◆ Figure 12.10). In addition, the potential energy of the water in a reservoir can generate electricity. Dams supply about 5 percent of the energy

ameliorated by building dams to store water that falls at the wrong time and by diverting water that falls in the wrong places. The Tigris and Euphrates rivers flow through deserts of the Middle East from distant mountains. Ancient Babylonians farmed the desert by irrigating the parched lowlands with diverted river water. In modern times, engineers build huge dams to store water. In 1950 there were 5,700 large dams in the world; by 2005 there were about 50,000.[1] Almost half of the new dams were built in China. This unprecedented building boom has increased water availability in many regions and produced large amounts of hydroelectric energy, but as we will see below, these benefits are offset

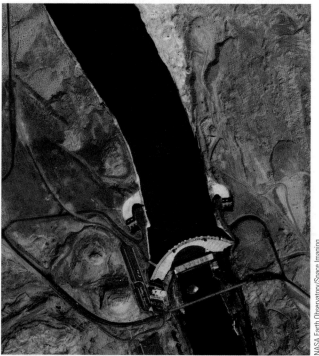

◆ **FIGURE 12.10** Construction of the Glen Canyon Dam on the Colorado River in the late 1950s created Lake Powell, shown in the upper half of this photo, between Utah and Arizona. The photo was taken from the Ikonos satellite on September 21, 2004.

1. A large dam is defined as rising 15 meters or more from base to top and storing more than 3 million cubic meters of water.

Courtesy of Graham R. Thompson/Jonathan Turk

◆ **FIGURE 12.11** White salts cover a vast expanse of Nevada desert. Salinization has reduced the productivity of millions of hectares of agricultural land globally.

terms, it is a brief moment of geological time and human civilization. In a few instances, engineers have made expensive miscalculations of sedimentation rates. The Tarbela Dam in Pakistan was completed in 1978 after nine years of construction, and by 2005 it was already more than 25 percent filled with silt. The reservoir behind the Sanmenxia Dam on the Yellow River in China filled four years after the dam was finished, making both the dam and reservoir useless.

Erosion

A stream erodes the beaches that line its banks during normal stream flow, but it deposits sediment to build the beaches back up during a flood. Recently, the beaches along the Colorado River in Grand Canyon were disappearing rapidly because management practices at the Glen Canyon Dam upstream prevented flooding in Grand Canyon. In a similar way, beaches are vanishing from many other dammed streams.

In the spring of 1996, scientists released enough water from Glen Canyon Dam to create a flood in Grand Canyon. As predicted, the floodwaters churned up sediment from the river bed and deposited it on the banks, thereby rebuilding old beaches and creating new ones (◆ Figure 12.12).

Because a reservoir accumulates silt and sand, it also interrupts the supply of sediment to the seacoast. As a result, coastal beaches erode, but natural processes do not replenish their sand. This problem has led to expensive rejuvenation projects on some popular beaches in the eastern United States.

Risk of Disaster

A dam can break, creating a disaster in the downstream flood plain. One of the greatest floods since the last ice age roared down Idaho's Teton River canyon when the Teton Dam broke on June 3, 1976. Floodwaters swept away 154 of the 155 buildings in the town of Wilford six miles below the dam; the town no longer exists. The flood inundated the lower half of Rexburg, a larger town farther downstream, damaging or destroying 4,000 homes and 350 businesses (◆ Figure 12.13). Damages totaled about $2 billion, and the raging waters stripped the topsoil from tens of thousands of hectares of farmland. Only 11 people died because civil authorities warned and evacuated most of the people. But if the flood had occurred in the middle of the night when warnings and evacuations would have been more difficult, thousands might have been killed.

Other dam failures have killed many more people. In 1889, a broken dam flooded Johnstown, Pennsylvania, killing 2,200 people; the St. Francis Dam in Saugus, California, collapsed in 1928 and drowned 450 people; in 1963 a rockslide created a huge wave in the lake behind a

used in the world today. Thus, dams are beneficial, but they can create numerous undesirable effects.

Loss of Water

The reservoir formed by a dam provides more surface area for evaporation and more bottom area for seepage into bedrock than did the stream that preceded it. For example, about 270,000 cubic meters of water per year evaporate from Lake Powell, above the Glen Canyon Dam on the Colorado River. Thus, less total water flows downstream. In addition, many canals built to carry water from a reservoir to farmland are simply ditches excavated in soil or bedrock, and they leak profusely.

Salinization

All rivers contain small concentrations of dissolved salts. When farmers use river water to irrigate cropland, the water evaporates and the salt remains in the soil. If desert or semi-desert soils are irrigated for long periods of time, salt accumulates and lowers the soil fertility. In the United States, **salinization** lowers crop yields on 25 to 30 percent of irrigated farmland, more than 5 million hectares (◆ Figure 12.11). Globally, the problem affects 10 percent of irrigated land, some 25 million hectares, and it is increasing at a rate of 1 to 1.5 million hectares per year.

Silting

A stream deposits its sediment in a reservoir, where the current slows. Rates of sediment accumulation in reservoirs vary with the sediment load of the dammed river. Lake Mead, behind Hoover Dam in Arizona and Nevada, lost 6 percent of its capacity in its first 35 years as a result of sediment accumulation. At a rate such as this, a reservoir lasts for hundreds of years. Although 100 years may be a long time in political or economic

◆ **FIGURE 12.12** Photos of the same Grand Canyon beach, taken **(A)** before and **(B)** after the spring 1996 flood, show that the flood more than doubled the size of the beach.

Disputes often arise among various groups that use water in dammed reservoirs. To prevent flooding, a reservoir should be nearly empty by spring so that it can store water and fill during spring runoff. Thus, a reservoir should be drawn down slowly during summer and fall. Agricultural users also prefer to have water stored during spring runoff and supplied for irrigation during summer months. However, recreational users object to a lowering of reservoir levels during summer, when they visit reservoirs most frequently. Managers of hydroelectric dams prefer to run water through their turbines during times of peak demands for electricity, which rarely correspond to flood management, irrigation, or recreational schedules.

Ecological Disruptions

A river is an integral part of the ecosystem that it flows through, and when the river is altered, the ecosystem changes. Dams interrupt water flow during portions of the year, prevent flooding, change the temperature of the downstream water, and often create unnatural daily fluctuations in stream flow. All of these changes alter relative abundances of aquatic species. In addition, dams may create specific ecological problems in individual rivers.

For example, before Egypt's Aswan Dam was built, the Nile River flooded every spring, depositing nutrient-rich sediment over the flood plain and delta. The delta was in balance between erosion by the sea and addition of new silt from the yearly floods. When the dam and reservoir cut off the silt supply, erosion then dominated, and the Nile Delta is now shrinking.

In addition, the loss of an annual supply of nutrients eliminated a source of free fertilizer for flood plain farms. Although the energy produced by the dam is used to manufacture commercial fertilizers, many of the poorer farmers on the flood plain cannot afford to purchase what they once received free from the river.

In another example, half a world away, salmon populations of the Pacific Northwest have decreased from 100 million fish in 1850 to 15 million today. In 2002, the number of coho salmon spawning in Pacific Northwest streams was less than 1 percent of their historic numbers. The Snake River sockeye and chinook have been declared endangered, and Snake River coho became extinct in 1985. Although overfishing and destruction of breeding habitats by logging, road build-

dam in Vaiont, Italy. The dam withstood the force of the wave, but the wave washed over the top of the dam, creating a flood that killed 1,800 people.

Recreational and Aesthetic Losses

Dams are often built across narrow canyons to minimize engineering and construction costs. But when the canyons are flooded, unique scenery and ecosystems are destroyed. Glen Canyon Dam on the Colorado River created Lake Powell by flooding one of the most spectacular desert canyons in the American West. Today, Glen Canyon is submerged for more than 300 kilometers above the dam and can only be seen in old photos. The flooding of canyons, however, provides lakes that can be used for fishing and other water sports.

U.S. Bureau of Reclamation

◆ **FIGURE 12.13** **(A)** The Teton Dam was brand new in the spring of 1976. **(B)** Rapidly rising water broke the dam on June 3, 1976, sending lake water pouring through the breach. **(C)** Flood waters from the failed Teton Dam inundated the city of Rexburg, Idaho.

ing, and construction contributed to the decline, wildlife biologists agree that the 18 giant dams on the Columbia River are the primary cause. The dams impede both upstream and downstream migration of the fish to and from their spawning grounds in the hundreds of small, gravel-bed tributaries of the Columbia. Young fish swimming toward the ocean slow their migration when they enter the lakes above dams, and

these fish are vulnerable to predators. In addition, the dam turbines stress young fish passing through them so they become easy prey for predators lurking below the dams.

Human Costs

Thayer Scudder, an anthropologist at the California Institute of Technology, was for years a proponent of dams, and he served on the board of the UN World Commission on Dams. Then, abruptly, he changed and became a critic of dam construction. In a recent book, *The Future of Large Dams: Dealing with Social, Environmental and Political Costs*, Professor Scudder argues that large dams impose unacceptable environmental and social costs, most of which burden "poor, uneducated, relatively powerless rural residents." In the past 50 years, 80 million people globally, approximately equal to 27 percent of the U.S. population, have been forcibly removed from their homes and farms to make way for the reservoirs behind the dams. When these people moved into new areas, they impacted the lives of the people already living in the relocation areas. In many regions, waterborne diseases are greater surrounding stagnant water than they were near free-flowing streams. In a study of 1.5 million people affected by 50 dams across the world, Scudder found that only 7 percent of the affected people realized an improvement in living standards and 70 percent were burdened by lower living standards. Twenty-three percent were unaffected.

 The Three Gorges Dam—China

The world's most populous nation, China, is building the world's largest dam, the Three Gorges Dam on the upper Yangtze River. This massive wall of earth and concrete stretches 1.5 kilometers across the world's third-longest river and rises 175 meters above the riverbed. Its reservoir will extend 550 kilometers upstream and force the displacement of 1.9 million people. Construction began in 1994 and is scheduled to take 20 years and over $24 billion to complete (◆ Figure 12.14).

Many people support the dam for the following reasons:

- The completed hydroelectric generators will supply 18.2 million kilowatts of electricity, enough to supply 150 million people with domestic and industrial needs. If this quantity of electricity were produced by conventional means, 40 million tons of coal would be burned annually, polluting the air and adding greenhouse gases.

- Yangtze floods have killed 500,000 people during the past 100 years and destroyed homes and farmlands. Engineers argue that the Three Gorges Dam will control floods.

- Irrigation water from the reservoir will increase crop productivities in nearby farms.

is maintained at a high level, as planned, to maximize power production, floodwaters could overflow the dam.

- Although irrigation water will increase food supply in some regions, the dam will flood an already-fertile valley and displace 1.9 million people from their homes and livelihoods.

- The reservoir will flood many historical sites and ruin the legendary scenery of the gorges and the local tourism industry.

- The concentration of water pollutants will increase below the dam because the quantity of pollutants will remain unchanged but the river flow will decrease.

- Heavy siltation will clog ports along the shores of the reservoir within a few years and negate improvements to navigation.

- Valuable and unique ecosystems will be destroyed.

Eddie Gerald/Alamy

◆ **FIGURE 12.14** The Three Gorges Dam on the upper Yangtze River will be the largest hydroelectric dam in the world and will require one of the largest construction projects ever undertaken. It will stretch 1.5 kilometers across, and tower 175 meters above, the world's third-longest river. Its reservoir will extend more than 550 kilometers upstream and displace 1.9 million people. Construction began in 1994 and is scheduled to take 20 years and $24 billion to complete. It will provide China with tremendous power generation and flood control services.

Conversely, critics argue that the Three Gorges Dam is an economic, ecological, and social disaster. These people argue that:

- There is already an oversupply of electricity in the area and the dam is so expensive that electricity from its generators will cost more than small, local, technologically sophisticated alternatives.

- In recent years, floods have been augmented by rapid runoff in areas that have been heavily logged. If nothing is done to control the upstream runoff and erosion, silting will seriously diminish the reservoir capacity. Furthermore, if the reservoir

Ground Water Diversion

Ground water provides drinking water for more than half of the population of North America and is a major source of water for irrigation and industry. It is a valuable resource because:

1. It is abundant. As described in Chapter 11, 60 times more freshwater exists underground than in streams and lakes combined.
2. Ground water moves very slowly; thus it is stored below Earth's surface and remains available during dry periods.
3. In some regions, ground water flows from wet environments to arid ones, making water available in dry areas.

Ground Water Depletion

If ground water is pumped to the surface faster than it can flow through the aquifer to the well, a **cone of depression** forms near the well (◆ Figure 12.15). When the pump

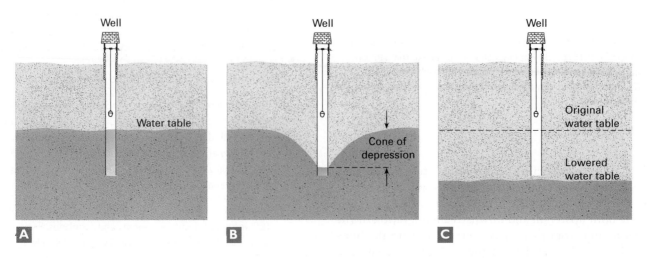

◆ **FIGURE 12.15** **(A)** A well is drilled into an aquifer. **(B)** A cone of depression forms because a pump draws water faster than the aquifer can recharge the well. **(C)** If the pump continues to extract water at the same rate, the water table falls.

is turned off, ground water flows back toward the well in a matter of days or weeks if the aquifer has good permeability, and the cone of depression disappears. Conversely, the water table drops if water is continuously removed more rapidly than it can flow to the well through the aquifer.

Before the development of advanced drilling and pumping technologies, human impact on ground water was minimal. Today, however, deep wells and high-speed pumps can extract ground water more rapidly than the hydrologic cycle recharges it. Where such excessive pumping is practiced, the water table falls as the ground water reservoir becomes depleted. In some cases the aquifer is no longer able to supply enough water to support the farms or cities that have overexploited it. This situation is common in the central, western, and southwestern United States (◆ Figure 12.16).

Most of the High Plains in western and midwestern North America receive scant rainfall. Early settlers prospered in rainy years and suffered during drought. In the 1930s, two events combined to change agriculture in this region. One was a great drought that destroyed crops and exposed the unprotected, plowed soil to erosion. Dry winds blew across the land, eroding the parched soil and carrying it for hundreds, and even thousands, of kilometers. Thousands of families lost their farms, and the region was dubbed the Dust Bowl. The second event was the arrival of inexpensive technology. Electric lines were built to service rural regions, and relatively cheap pumps and irrigation systems were developed. With the specter of drought fresh in people's memories and the tools available to avert future calamities, the age of modern irrigation began.

The Ogallala aquifer extends almost 900 kilometers from the Rocky Mountains eastward across the prairie, and from Texas to South Dakota (◆ Figure 12.17). It consists of water-saturated porous sandstone and conglomerate within 350 meters of the surface. The aquifer averages about 65 meters in thickness and is the world's largest known aquifer. Between 1930 and 2000, about 200,000 wells were drilled into the Ogallala aquifer, and extensive irrigation systems were installed throughout the region.

Most of the water in the Ogallala aquifer accumulated when the last Pleistocene ice sheet melted, about 15,000 years ago. But because the High Plains receive little rain, the aquifer is now mostly recharged by rain and snowmelt in the Rocky Mountains, hundreds of kilometers to the west. The problem is that water moves very slowly through the aquifer, at an average rate of about 15 meters per year. At this rate, ground water takes 60,000 years to travel from the mountains to the eastern edge of the aquifer. Presently, approximately 9.6 billion liters are returned to the aquifer every year through rainfall and ground water flow. However, farmers remove 90 billion liters annually. Under such conditions, the deep ground water is, for all practical purposes, nonrenewable.

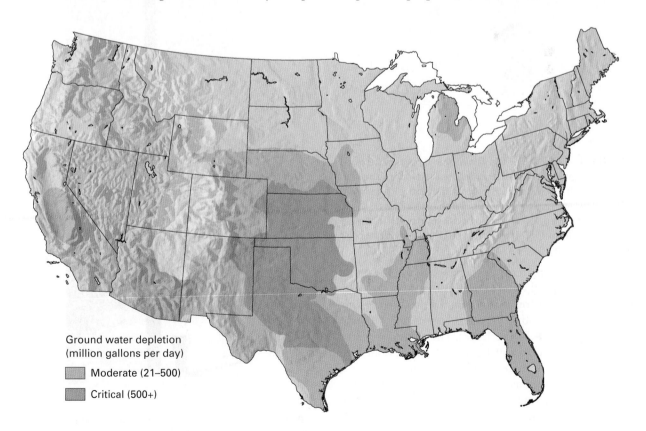

Ground water depletion
(million gallons per day)

███ Moderate (21–500)

███ Critical (500+)

◆ **FIGURE 12.16** Aquifer depletion is a common problem in the United States. The numbers refer to the amount of water used in excess of the recharge rate.

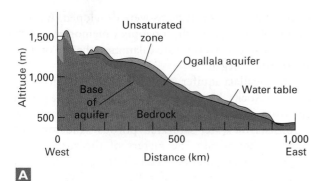

A

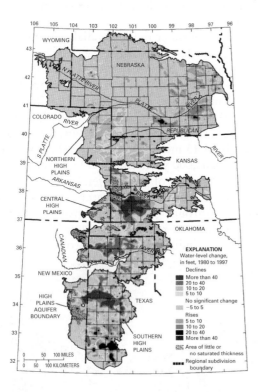

B

◆ **FIGURE 12.17** The Ogallala aquifer supplies water to much of the High Plains. **(A)** A cross-sectional view of the aquifer shows that much of its water originates in the Rocky Mountains and flows slowly as ground water beneath the High Plains. **(B)** This map shows the extent and water level changes in the Ogallala aquifer from 1980 and 1997. (www.choicesmagazine.org/ 2003-1/2003-1-04.htm)

If the present pattern of water use continues, one quarter of the original water in the Ogallala aquifer will be consumed by 2020. However, in some of the drier regions of northwest Texas and west central Kansas, the wells are already depleted (◆ Figure 12.17). The aquifer irrigates about 5 million hectares of farms and ranches (an area about the size of the states of Massachusetts, Vermont, and Connecticut combined). About 20 percent of the total U.S. agricultural output, including 40 percent of the feedlot beef, is produced in this region, with a total value of $32 billion. If the aquifer is depleted and another source of irrigation water is not found, productivity in the central High Plains is expected to decline by 80 percent.

Farmers will go bankrupt, and food prices will rise throughout the nation.

Subsidence

Excessive removal of ground water can cause **subsidence**, the sinking or settling of Earth's surface. When water is withdrawn from an aquifer, rock or soil particles shift to fill the space left by the lost water. As a result, the volume of the aquifer decreases and the overlying ground subsides.

Subsidence rates can reach 5 to 10 centimeters per year, depending on the rate of pumping and the nature of the aquifer. Some areas in the San Joaquin Valley of California have sunk nearly 10 meters (◆ Figure 12.18). The land surface has subsided by as much as 3 meters in the Houston–Galveston area of Texas. The problem is particularly severe in cities. For example, Mexico City is built on an old marsh. Over the years, as the weight of buildings and roadways has increased and much of the ground water has been removed, parts of the city have settled as much as 8.5 meters. Many millions of dollars have been spent to maintain the city on its unstable base. Similar problems are affecting Phoenix, Arizona, and other U.S. cities.

◆ **FIGURE 12.18** The land surface at this point in San Joaquin, California, was at the height of the 1925 marker in the year 1925. Subsidence resulting from ground water extraction for irrigation lowered the surface by about 10 meters in 52 years.

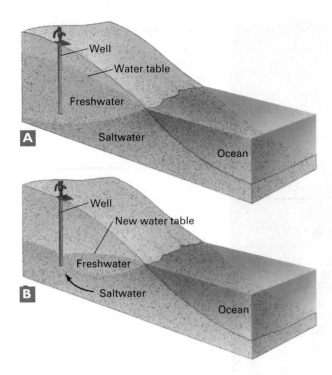

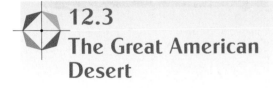

◆ FIGURE 12.19 Saltwater intrusion can pollute coastal aquifers. **(A)** Freshwater lies above salt water, and the water in the well is fit to drink. **(B)** If too much freshwater is removed, the water table falls. The level of salt water rises and contaminates the well.

Unfortunately, subsidence is irreversible. When rock and soil contract, their porosity is permanently reduced so that ground water reserves cannot be completely recharged, even if water becomes abundant again.

Saltwater Intrusion

Two types of ground water occur in coastal areas: fresh ground water and salty ground water that seeps in from the sea. Freshwater floats on top of salty water because it is less dense. If too much freshwater is removed from the aquifer, salty ground water rises to the level of wells (◆ Figure 12.19). It is unfit for drinking, irrigation, or industrial use. **Saltwater intrusion** has affected much of south Florida's coastal ground water reservoirs.

12.3
The Great American Desert

In the years following the Civil War, an American geologist, John Wesley Powell, explored the area between the western mountains (the Sierra Nevada–Cascade crest) and the 100° meridian (the line of longitude running through the Dakotas; Nebraska; Kansas; and Abilene, Texas). He recognized that most of this region is arid or semiarid,

and he called it the Great American Desert. Despite Powell's warning about inadequate water, Americans flocked to settle the Great American Desert. As described earlier, several parts of this region now use huge amounts of water for cities, industry, and agriculture, although they receive little precipitation. For example, arid California, Nevada, Arizona, Texas, and Idaho use the greatest quantities of water in the United States. Some desert cities, including Phoenix, El Paso, Reno, and Las Vegas, rely on surface-water diversion and/or overpumping of ground water for nearly all of their water (◆ Figure 12.20).

Billions of dollars have been spent on heroic projects to divert rivers and pump ground water to serve this area of the country. For example, 1,200 dams and the two largest irrigation projects in the world have been built in California alone.

The Colorado River

The Colorado River runs through the southwestern portion of the Great American Desert. Starting from the snowy mountains of Colorado, Wyoming, Utah, and New Mexico, it flows across the arid Colorado Plateau and southward into Mexico, where it empties into the Gulf of California (◆ Figure 12.21).

Because the river flows through a desert, farmers, ranchers, cities, and industrial users along the entire length of the river compete for rights to use the water. In the 1920s, the Colorado discharged 18 billion cubic meters of water per year into the Gulf of California. In 1922 the U.S. government apportioned 9 billion cubic meters for the Upper Basin (Colorado, Utah, Wyoming, and New Mexico) and the remaining 9 billion cubic meters for Lower Basin users in Arizona, California, and Nevada. Thus, all the water was allocated and none of it was set aside for maintenance of the river ecosystem or for Mexican users south of the border. No water flowed into the sea. Twenty years later, an international treaty awarded Mexico 1.8 billion cubic meters to be taken equally from Upper Basin and Lower Basin users.

The Colorado and many of its tributaries flow across sedimentary rocks, many of which contain soluble salt deposits. As a result, the Colorado is naturally a salty river. In addition, the U.S. government built 10 large dams on the Colorado, and evaporation from the reservoirs has further concentrated the salty water. In 1961, the water flowing into Mexico contained 27,000 ppm (parts per million) salts, compared with an average salinity of about 100 ppm for large rivers. For comparison, 27,000 ppm is 77 percent as salty as seawater.

Mexican farmers used this water for irrigation, and their crops died. In 1973, the U.S. government built a desalinization plant to reduce the salinity of Mexico's share of the water.

There are also significant questions about the possibility of a long-term drought in the Southwest. Tree ring studies have shown that more extreme and extended droughts have occurred in the past several centuries than have been experienced in modern times. With current demands on the Colorado River increasing, even a minor drought can have severe consequences. The end of 2004 marked five consecutive years of lower-than-average rainfall and

FIGURE 12.20 Desert cities, such as Las Vegas, use more water than is locally available in surface and ground water reservoirs. Consequently they must import much of their water through expensive, publicly funded, water-diversion systems.

D. C. Lowe/Getty Images

FIGURE 12.21 **(A)** The Colorado River drains much of the American Southwest. **(B)** Water is abundant where the river flows through the Grand Canyon. **(C)** Use of river water for irrigation often dries up the river completely near the Mexican border.

CHAPTER 12 • Water Resources

© Jon Bower/Alamy

© Ron Niebrugge/Alamy

A   **B**

◀▶ **FIGURE 12.22** **(A)** Lake Powell, Glen Canyon National Recreation Area, in Arizona and Utah, at full level. **(B)** The old concrete boat launch in the foreground of this photo is stranded above the current lake level because of drought.

subsequent decline in reservoir levels. In January 2005, Lake Powell was at 35 percent of its capacity, the lowest level since 1969 when the reservoir was being filled. Maintaining a reliable supply of water for the growing population of the Colorado River Basin states is one of the most challenging public policy issues of the Southwest (◀▶ Figure 12.22).

 The Los Angeles Water Project

In the mid-1800s Los Angeles was a tiny, neglected settlement on the California coast. Too far from the California gold fields to attract the miners or their money, it sat in its arid coastal basin with neither a seaport nor a railroad. The average annual precipitation of the Los Angeles area is 65 centimeters, most of which falls during a few winter weeks. The Los Angeles River frequently flooded in winter and diminished to a trickle in summer.

Among the early settlers were Mormons, who had become experts at irrigating dry farmland from their experience in the Utah desert. By the late 1800s, irrigated farms in the Los Angeles basin were producing a wealth of fruits and vegetables. It seemed that almost anything grew bigger and better with the combination of sun, warmth, rich soil, and irrigation. Suddenly, Los Angeles was an attractive, growing town. In 1848 the town had a population of 1,600. The population passed 100,000 by 1900 and 200,000 by 1904. Then the city ran out of water. Wells dried up and the river was inadequate to meet demand. To supply human needs, the city government prohibited lawn watering and shut off irrigation wells in the nearby San Fernando Valley. Los Angeles was surrounded on three sides by deserts and on the fourth by the Pacific Ocean. There was no nearby source of freshwater.

However, 250 miles to the northeast, the Owens River flowed out of the Sierra Nevada. Although the Owens Valley is arid, the abundant water from the high Sierra turned the region green. But the city of Los Angeles bought water rights, bit by bit, from farmers and ranchers in the Owens Valley and at the headwaters of the river. The city then spent millions of dollars building the Los Angeles Aqueduct across some of the most difficult, earthquake-prone terrain in North America. When finished, the aqueduct was 357 kilometers long, including 85 kilometers of tunnel. Siphons and pumps carried the water over hills and mountains too treacherous for tunneling (◀▶ Figure 12.23). On November 5, 1913, the first water poured from the aqueduct into the San Fernando Valley.

The following 10 to 15 years were unusually rainy in the Los Angeles area and in the Owens Valley. The rain recharged the ground water under the Los Angeles basin, and the Los Angeles River flowed freely again. As a result, little or none of the Owens River water coming through the aqueduct was needed by the city. Instead, most of the water was diverted to farms in the San Fernando Valley. Irrigated farmland in that valley increased from 1,200 hectares in 1913, when the aqueduct opened, to more than 30,000 hectares in 1918. In addition, some of the excess water now available to Los Angeles was used to water lawns again. Almost every house had a green lawn, and Santa Monica Boulevard changed from a dusty rut to a palm-lined oasis, creating the impression that Los Angeles wasn't really so dry after all.

In the 1920s, normal dryness and drought returned. Los Angeles and the farms of the San Fernando Valley demanded more water from the Owens River, and the Owens Valley dried up. The mood of the ranchers and farmers in the Owens Valley grew steadily worse, and in 1924 they threatened the aqueduct with sabotage. William Mulholland, superintendent of the Los Angeles Department of Water and Power, remarked that it was too bad that so many of the orchards in the Owens Valley had died of thirst because now there weren't enough trees to hang all the Owens Valley troublemakers.

◆ **FIGURE 12.23** **(A)** In this region upstream from the town of Bishop, California, most of the Owens River flows through a pipe **(B)** and only a small trickle remains in its old streambed.

By the mid-1930s, Los Angeles owned 95 percent of the farmland and 85 percent of the residential and commercial property in the Owens Valley. Although the city leased some of the farmland back to the farmers and ranchers, the water supply became so unpredictable that agriculture slowly dried up with the land. As Los Angeles continued to grow, the water from the Owens River was not sufficient, so the Department of Water and Power drilled wells into the Owens Valley and began pumping its ground water aquifer dry.

Today, the Owens Valley is parched. Its remaining citizens pump gas, sell beer, and make up motel beds for tourists driving through on their way to somewhere else (◆ Figure 12.24). Ironically, the water that once irrigated farms and ranches in the naturally arid Owens Valley was diverted through the Los Angeles Aqueduct, at great cost to taxpayers, to irrigate farms in the naturally arid San Fernando Valley. The aqueduct converted a farmer's paradise to a desert and a desert to a farmer's paradise.

More than 80 percent of the water diverted to Southern California is used to irrigate desert and near-desert cropland. The water

◆ **FIGURE 12.24** As most of the Owens River is diverted to Los Angeles, farms and ranches have dried up, and so have local businesses.

supplied to these farms commonly irrigates crops such as cotton, rice, and alfalfa. Cotton requires 17,000 liters per kilogram; rice and alfalfa also use great amounts of water. The U.S. government pays large subsidies to farmers in naturally wet southeastern and south central states not to grow those same crops. The irrigation water used to support cattle ranching and other livestock in California is enough to supply the needs of the entire human population of the state.

Controversy over water diversion projects is not limited to the American West. New York City obtains much of its water from reservoirs in upstate New York, and the New York City Board of Water Supply controls those reservoirs and watersheds hundreds of miles from the city. Opposing interests of local land-use groups and city water users have provoked heated conflicts in the state of New York.

12.4
Water and International Politics

The conflict between water users in Los Angeles and the Owens Valley involves two groups of people within the same nation. The conflict between Mexico and the United States over the Colorado River water involves two friendly allies. But now imagine what happens in the Middle East, a desert region where life-giving rivers flow between hostile nations.

The Jordan River flows along the border between Israel and its enemies, Jordan and Syria. In the north, it forms the boundary between Israel and the disputed Golan Heights and downstream it forms the boundary between Israel and the disputed West Bank. Population in this water-parched desert is expected to skyrocket from 32 million at the end of 2001 to 52 million by 2025.

- Syria has planned to build dams that will divert one of the major tributaries of the Jordan; Israel has threatened to bomb the dam before it is completed.
- Jordan's King Hussein said in 1990 that water was the only issue that could take him to war with Israel.
- While the Palestinians struggle for an independent homeland, the Israelis are pumping 35 percent of their sustainable freshwater supply from an aquifer that lies beneath the disputed West Bank. At the present time, the Israelis are overpumping the aquifer and at the same time rationing water use by the West Bank Arabs. If they cede the land to their enemies, they lose the water.

Similar conflicts are occurring over the water in the Nile River. The Nile flows from the relatively moist African highlands through the dry Egyptian desert. Tradi-tionally, Egypt has been the main user of this water, but in recent years, Ethiopia and Sudan have been taking an increased share. Because Ethiopia and Sudan are upstream, they claim that they can take as much water as they want, but Egypt claims to have historic rights to the Nile.

Current international water law offers little help in resolving conflicts over water. In general, downstream nations argue that a river or aquifer is a communal resource to be shared equitably among all adjacent nations. Predictably, many upstream nations maintain that they have absolute control over the fate of water within their borders and have no responsibility to downstream neighbors.

Any international code of water use must be based on three fundamental principles:

1. Water users in one country must not cause major harm to water users in other countries downstream.
2. Water users in one country must inform neighbors of actions that may affect them before the actions are taken. (For example, if a dam is built, engineers must inform downstream users that they plan to stop or reduce river flow to fill a reservoir.)
3. People must distribute water equitably from a shared river basin.

Unfortunately, these principles, especially the last one, are open to such subjective and self-interested interpretations that their intent is easily and often subverted.

12.5
Water Pollution

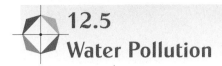

While consumption reduces the *quantity* of water in a stream, lake, or ground water reserve, **pollution** is the reduction of the *quality* of the water by the introduction of impurities (◁▶ Figure 12.25). A polluted stream may run full to the banks, yet it may be toxic to aquatic wildlife and unfit for human use.

In the early days of the Industrial Revolution, factories and sewage lines dumped untreated wastes into rivers. As the population grew and industry expanded, many of our waterways became fetid, foul-smelling, and unhealthy. The first sewage treatment plant in the United States was built in Washington, D.C., in 1889, more than 100 years after the Revolutionary War. Soon other cities followed suit, but few laws regulated industrial waste discharge.

In 1952 and again in 1969, the Cuyahoga River in Cleveland, Ohio, was so heavily polluted with oil and industrial chemicals that it caught fire, spreading flame and smoke across the water (◁▶ Figure 12.26). While the highly publicized photographs of the flaming river flowing through downtown Cleveland had a powerful visual impact, other water pollution disasters were equally insidious.

FIGURE 12.25 Pollution is the reduction of the quality of a resource by the introduction of impurities.

◆ FIGURE 12.26 Fireboats fight flames on the burning Cuyahoga River in November 1952.

 Love Canal

Love Canal in Niagara Falls, New York, was excavated to provide water to an industrial park that was never built. In the 1940s, the Hooker Chemical Company purchased part of the old canal as a site to dispose of toxic manufacturing wastes. The engineers at Hooker considered the relatively impermeable ground surrounding the canal to be a reasonably safe disposal area. During the following years, the company disposed of approximately 19,000 tons of chemical wastes by loading them into 55-gallon steel drums and dumping the drums in the canal. In 1953, the company covered one of the sites with dirt and sold the land to the Board of Education of Niagara Falls for one dollar, after warning the city of the buried toxic wastes. The city then built a school and playground on the site.

In the process of installing underground water and sewer systems to serve the school and growing neighborhood, the relatively impermeable soil surrounding the old canal and waste burial site was breached. Gravel used to backfill the water and sewer trenches then provided a permeable path connecting the toxic waste dump to surrounding parts of the city.

During the following decades, the buried drums rusted through and the chemical wastes seeped into the ground water. In the spring of 1977, heavy rains raised the water table to the surface, and the area around Love Canal became a muddy swamp. But it was no ordinary swamp; the leaking drums had contaminated the ground water with toxic and carcinogenic compounds. The poisonous fluids soaked the playground, seeped into basements of nearby homes, and saturated gardens and lawns. Children who attended the school and adults who lived nearby developed epilepsy, liver malfunctions, skin sores, rectal bleeding, and severe headaches. In the years that followed, an abnormal number of pregnant women suffered miscarriages, and large numbers of babies were born with birth defects. Thus a human tragedy at Love Canal was caused by deliberate, but legal, burial of industrial poisons, coupled with flawed hydrogeology, and complete disregard of public health by both the Hooker Chemical Company and local officials. These and similar incidents made it clear that water pollution was endangering health and reducing the quality of life for people throughout the United States. As a result, Congress passed the **Clean Water Act** in 1972. The legislation states that:

1. It is the national goal that the discharge of pollutants into the navigable waters be eliminated by 1985.
2. It is the national goal that, wherever attainable, an interim goal of water quality which provides for the protection and propagation of fish, shellfish, and wildlife and provides for recreation in and on the water be achieved by July 1, 1983.
3. It is the national policy that the discharge of toxic pollutants in toxic amounts be prohibited.

Now, 34 years later, we ask: What has been done? What hasn't been done? Before we answer these questions, we must understand the nature of water pollution.

Types of Pollutants

Biodegradable Pollutants

A biodegradable material is one that is consumed and destroyed in a reasonable amount of time by organisms that live naturally in soil and water. Several types of contaminants are **biodegradable pollutants**:

1. Sewage is wastewater from toilets and other household drains. It includes biodegradable organic material such as human and food wastes, soaps, and detergents. Sewage also includes some nonbiodegradable chemicals because people flush paints, solvents, pesticides, and other chemicals down the drains.
2. Disease organisms, such as typhoid and cholera, are carried into waterways in the sewage of infected people.
3. Phosphates and nitrate fertilizers flow into surface water and ground water mainly from agricultural runoff. Phosphate detergents and phosphates and nitrates from feedlots also fall into this category.

Nonbiodegradable pollutants

Many materials are not decomposed by environmental chemicals or consumed by decay organisms and therefore they are **nonbiodegradable**. Some of these are poisons, while others are not. In modern parlance, nonbiodegradable poisons are called **persistent, bioaccumulative toxic chemicals** (PBT). The Environmental Protection Agency (EPA) recognizes tens of thousands of PBTs. Yet, even nontoxic materials can adversely affect ecosystems. Nonbiodegradable pollutants can be roughly subdivided into three broad categories:

1. Many organic industrial compounds are particularly troublesome because they are toxic in high doses, suspected carcinogens in low doses, and they survive for decades in aquatic systems. Examples include some pesticides, dioxin, and PCBs.
2. Toxic inorganic compounds include mine wastes, road salt, and mineral ions such as those of cadmium, arsenic, and lead.
3. When sediment enters surface waters, it neither fertilizes nor poisons the aquatic system. However, the sediment muddies streams and buries aquatic habitats, thus degrading the quality of an ecosystem.

Radioactive materials

Radioactive materials include wastes from the mining of radioactive ores, nuclear power plants, nuclear weapons, and medical and scientific applications.

Heat

Heat can also pollute water. In a fossil-fuel or nuclear electric generator, an energy source (coal, oil, gas, or nuclear fuel) boils water to form steam. The steam runs a generator to produce electricity. The exhaust steam must then be cooled to maintain efficient operation.

The cheapest cooling agent is often river or ocean water. A 1,000-megawatt power plant heats 10 million liters of water by 35°C every hour. The warm water can kill fish directly, affect their reproductive cycles, and change the aquatic ecosystems to eliminate some native species and favor the introduction of new species. For example, if water is heated, indigenous cold-water species like trout will die out and be replaced by carp or other warm-water fish.

Any of these categories of pollutants can be released in either of two ways. **Point source pollution** arises from a specific site such as a septic tank, a gasoline spill, or a factory. In contrast, **nonpoint source pollution** is generated over a broad area. Fertilizer and pesticide runoff from lawns and farms fall into this latter category.

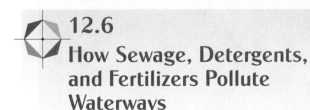

12.6
How Sewage, Detergents, and Fertilizers Pollute Waterways

Recall from Chapter 11 that an oligotrophic stream or lake contains few nutrients and clear, nearly pure water (◆ Figure 12.27A). Sewage, detergents, and agricultural fertilizers are plant nutrients. They do not poison aquatic systems; they nourish them. If humans dump one or more of these nutrients into an oligotrophic waterway, aquatic organisms multiply so quickly that they consume most of the dissolved oxygen. Many fish, such as trout and salmon, die. Other organisms, such as algae, carp, and water worms, proliferate. Thus an increased supply of nutrients transforms a clear, sparkling stream or lake into a slime-covered, eutrophic waterway (◆ Figure 12.27B).

If humans release even more sewage, detergent, or fertilizer, decay organisms multiply so rapidly that they consume all the oxygen. As a result, most aquatic life dies. Then, **anaerobic** bacteria (bacteria that live without oxygen) take over, releasing noxious hydrogen sulfide gas as they feast on the remaining organic matter (◆ Figure 12.27C). Even carp and worms cannot survive in such an environment. As they die and rot, the hydrogen sulfide that bubbles to the surface smells like rotten eggs.

When a discharge of raw sewage flows into a stream, decay organisms immediately begin feasting, and at the same time, the current carries the pollutants and the decay organisms downstream. Over time, the decay organisms consume most of the pollutants. When this food supply is consumed, the decay organisms die off. Natural turbulence replenishes oxygen in the water and aerobic organisms become reestablished. Thus, the river cleanses

◈ **FIGURE 12.27** **(A)** An oligotrophic waterway contains few nutrients and clear, pure water. **(B)** A eutrophic waterway is rich in nutrients, low in dissolved oxygen, and choked with plant life. **(C)** This stream flowing through La Paz, Bolivia, is an open sewer. Decay organisms have consumed all of the oxygen in the water, and anaerobic bacteria have taken over.

itself. However, if the original source of pollution is too large, or if every town along the river dumps pollutants into the same waterway, then these natural cleaning mechanisms become overwhelmed, and the river remains polluted all the way to the ocean.

12.7
Toxic Pollutants, Risk Assessment, and Cost–Benefit Analysis

Although natural decay organisms decompose and decontaminate human feces over a period of days or weeks, many organic compounds, such as the banned pesticide DDT, persist for decades, while heavy metal contami-

nants such as arsenic or lead never decompose. (These heavy metals may change chemical form and therefore become benign, but the atoms never decompose.)

Environmental laws in developed nations regulate the discharge of pollutants into waterways. Yet pollutants continue to enter streams and rivers. Even in situations where waterways are usually contaminated with only small or even minuscule amounts of these toxic nonbiodegradable compounds, a great many different pollutants may be found in a single river and small concentrations may cause cancer or birth defects.

Scientists do not know the doses at which many compounds become harmful to human health. The uncertainty results, in part, from delayed effects of some chemicals. For example, if a contaminant increases the risk of a certain type of cancer, the cancer may not develop until 10 to 20 years after exposure. Because scientists cannot perform direct toxicity experiments on humans, and because they cannot wait decades to observe results, they frequently attempt to assess the carcinogenic properties of a contaminant by feeding it to laboratory rats. If rats are fed very high doses, sometimes hundreds of thousands of times more concentrated than environmental pollutants, they may then contract cancer in weeks or months—rather than in years or decades. Suppose that a rat gets cancer after drinking the equivalent of 100,000 glasses of polluted well water each day for a few weeks. Can we say that the same contaminant will cause cancer in humans who drink 10 glasses a day for 20 years? No one knows. It may or may not be legitimate to extrapolate from high doses to low doses and from one species to another.

Scientists also use epidemiological studies to assess the risk of a pollutant. For example, if the drinking water in a city is contaminated with a pesticide and a high proportion of people in the city develop an otherwise rare disease, then the scientists may infer that the pesticide caused the disease and that its presence in drinking water constitutes a high level of risk to human health.

Because neither laboratory nor epidemiological studies can prove that low doses of a pollutant are harmful to humans, scientists are faced with a question: Should businesses and governments spend money to clean up the pollutant? Some argue that such expenditure is unnecessary until we can prove that the contaminant is harmful. Others invoke the precautionary principle, It's better to be safe than sorry. Proponents of the latter approach argue that people commonly act on the basis of incomplete proof. For example, if a mechanic told you that your brakes were likely to fail within the next 1,000 miles, you would recognize this as an opinion, not a fact. Yet would you wait for the brakes to fail or replace them now?

Pollution control is expensive. However, pollution is also expensive. If a contaminant causes people to sicken, the cost to society can be measured in medical bills and loss of income resulting from missed work. Many contaminants damage structures, crops, and livestock. People in polluted areas also bear expense because tourism diminishes and land values are reduced when people no longer want to visit or live in a contaminated area. All of these costs are called **externalities**.

Cost–benefit analysis compares the cost of pollution control with the cost of externalities. ◁▷ Figure 12.28 shows that the cost of externalities is very high when pollution levels are high and uncontrolled. As more controls are used to produce a cleaner environment, the cost of externalities decreases, but the cost of the controls rises. At a point in the middle of the blue curve, the total cost to the consumer, combining costs of externalities and pollution controls, reaches a minimum. Some people suggest that we should minimize the total cost, even though this approach accepts significant pollution, with its possible sickness and even death. Others argue that cost–benefit analysis is flawed because it ignores both the quality of life and the value of human life. How, they ask, can you place a dollar value on loss of recreation if people can no longer swim and fish in our waterways?

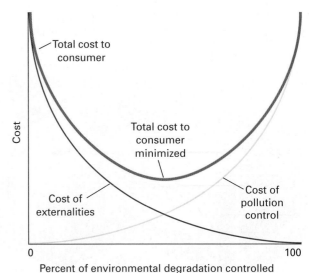

◁▷ **FIGURE 12.28** Pollution control becomes more expensive as pollution is reduced (yellow curve). Externalities (costs of living in a polluted environment) become less expensive as pollution is reduced (red curve). The blue curve is the sum of the other two curves and represents the total costs to the consumer.

Or even more poignant, how can you measure the economic value of good health or of a life cut short by cancer? If noneconomic costs of pollution are considered, then an increased level of pollution control becomes desirable despite its higher dollar cost.

12.8
Ground Water Pollution

Water in a sponge saturates tiny pores and passages. To clean a contaminant from a sponge, it would be necessary to clean every pore of the sponge that came into contact with contaminated water. Because this is nearly impossible to achieve completely, no one would wash dishes with a sponge that had been used to clean a toilet.

Many different types of sources contaminate ground water (◁▷ Figure 12.29). Because these contaminants permeate an aquifer in a manner analogous to the way in which contaminated water saturates a sponge, the removal of a contaminant from an aquifer is extremely difficult and expensive. Once a pollutant enters an aquifer, the natural flow of ground water disperses it as a growing **plume** of contamination. Because ground water flows slowly, usually at a few centimeters per day, the plume also spreads slowly (◁▷ Figure 12.30).

Most contaminants persist in a ground water aquifer for much longer than in a stream or lake. The rapid flow of water through streams and lakes replenishes their water quickly, but ground water flushes much more slowly (Table 12.1). In addition, oxygen, which decomposes

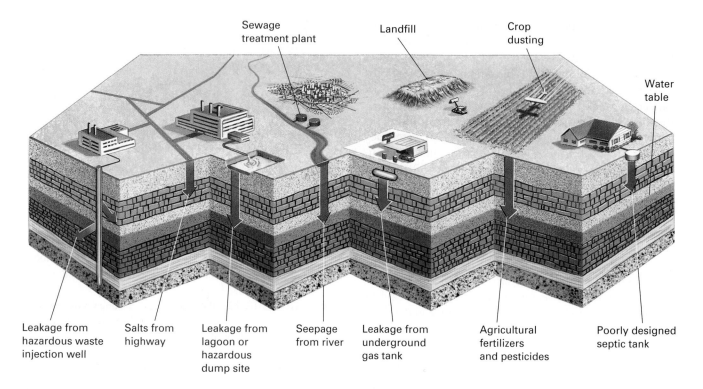

Sewage treatment plant

Landfill

Crop dusting

Water table

Leakage from hazardous waste injection well

Salts from highway

Leakage from lagoon or hazardous dump site

Seepage from river

Leakage from underground gas tank

Agricultural fertilizers and pesticides

Poorly designed septic tank

◁▷ **FIGURE 12.29** Many different sources contaminate ground water.

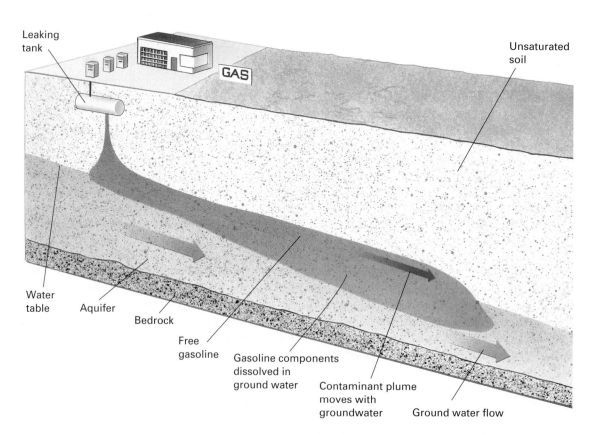

Leaking tank

Unsaturated soil

GAS

Water table

Aquifer

Bedrock

Free gasoline

Gasoline components dissolved in ground water

Contaminant plume moves with groundwater

Ground water flow

◁▷ **FIGURE 12.30** Pollutants disperse as a plume of contamination. Gasoline and many other contaminants are lighter than water. As a result, they float and spread on top of the water table. Soluble components may dissolve and migrate with ground water.
INTERACTIVE QUESTION: *Sketch a similar figure showing dispersion of a trichlorethylene plume in this same environment.*
(Hint: trichloroethylene is insoluble and heavier than water.)

Table 12.1 Residence* Times for Water in Various Reservoirs

Atmosphere	8 days
Rivers	16 days
Soil moisture	75 days
Seasonal snow cover	145 days
Glaciers	40 years
Large lakes	100 years
Shallow ground water	200 years
Deep ground water	10,000 years

*Residence time is the average time water remains in a reservoir.

◁▷ **FIGURE 12.31** An oil refinery in New Jersey. New Jersey suffers from some of the worst ground water pollution in North America as a result of heavy industry.

many contaminants, is less abundant in ground water than in surface water.

As a result, despite 34 years of government concern and action:

- Up to 25 percent of the usable ground water in the United States is contaminated. About 45 percent of municipal ground water supplies in the United States are contaminated with organic chemicals.
- The gasoline additive MTBE is a frequent and widespread contaminant of underground drinking water. According to the EPA, 5–10 percent of drinking water in areas using reformulated gasoline shows MTBE contamination.
- Wells in 38 states contain pesticide levels high enough to threaten human health. Every major aquifer in New Jersey is contaminated.
- In Florida, where 92 percent of the population drinks ground water, more than 1,000 wells have been closed because of contamination, and over 90 percent of the remaining wells have detectable levels of industrial or agricultural chemicals.
- In 2003 the EPA reported 436,494 confirmed releases of dangerous volatile organic compounds leaking from underground fuel storage tanks in the United States.

Treating a Contaminated Aquifer

The treatment, or **remediation,** of a contaminated aquifer commonly occurs in a series of steps.

Elimination of the Source

The first step in treating an aquifer is to eliminate the pollution source so additional contaminants do not escape. If an underground tank is leaking, the remaining liquid in the tank can be pumped out and the tank dug from the ground. If a factory is discharging toxic chemicals into the ground water, courts may issue an injunction ordering the factory to stop the discharge (◁▷ Figure 12.31).

Elimination of the source prevents additional pollutants from entering the ground water, but it does not solve the problem posed by the pollutants that have already escaped. For example, if a buried gasoline tank has leaked slowly for years, many thousands of gallons of gas may have contaminated the underlying aquifer. Once the tank has been dug up and the source eliminated, people must deal with the gasoline in the aquifer.

Monitoring

When aquifer contamination is discovered, a hydrogeologist monitors the contaminants to determine how far, in what direction, and how rapidly the plume is moving and whether the contaminant is becoming diluted. The hydrogeologist may take samples from domestic wells. If too few wells surround a pollution source, the hydrogeologist may drill wells to monitor the plume. He or she analyzes the water samples for the contaminant and can thereby monitor the movement of the plume through the aquifer.

Modeling

After measuring the rate at which the contaminant plume is spreading, the hydrogeologist develops a computer model to predict future dispersion of the contaminant through the aquifer. The model considers the permeability of the aquifer, directions of ground water flow, and mixing rates of ground water to predict dilution effects.

Remediation

Several processes are currently used to clean up a contaminated aquifer. Contaminated ground water can be contained by building an underground barrier to isolate it from other parts of the aquifer. If the trapped contaminant does not decompose by natural processes, hydrogeologists drill wells into the contaminant plume and pump the water to the surface, where it is treated to destroy the pollutant.

Courtesy of Graham R. Thompson/Jonathan Turk

Bioremediation uses microorganisms to decompose a contaminant. Specialized microorganisms can be fine-tuned by genetic engineers to destroy a particular contaminant without damaging the ecosystem. Once a specialized microorganism is developed, it is relatively inexpensive to breed it in large quantities. The microorganisms are then pumped into the contaminant plume, where they attack the pollutant. When the contaminant is destroyed, the microorganisms run out of food and die, leaving a clean aquifer. Bioremediation can be among the cheapest of all cleanup procedures.

Chemical remediation is similar to bioremediation. If a chemical compound reacts with a pollutant to produce harmless products, the compound can be injected into an aquifer to destroy contaminants. Common reagents used in chemical remediation include oxygen and dilute acids and bases. Oxygen may react with a pollutant directly or provide an environment favorable for microorganisms, which then degrade the pollutant. Thus, we can sometimes reduce contamination simply by pumping air into the ground. Acids or bases neutralize certain contaminants or precipitate dissolved pollutants.

Reclamation teams can also dig up the entire contaminated portion of an aquifer. The contaminated soil is treated by incineration or with chemical processes to destroy the pollutant. The treated soil is then returned to fill the hole. This process is prohibitively expensive and is used only in extreme cases.

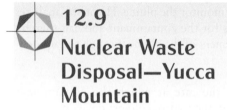

ThomsonNOW

CLICK ThompsonNOW Interactive to work through an activity on Contamination Through Water.

12.9
Nuclear Waste Disposal—Yucca Mountain

In a nuclear reactor, radioactive uranium nuclei split into smaller nuclei, many of which are also radioactive. Most of these radioactive waste products are useless and must be disposed of without exposing people to the radioactivity. In the United States, military processing plants, 104 commercial nuclear reactors, and numerous laboratories and hospitals generate more than 3,000 tons of high-level radioactive wastes every year.

Chemical reactions cannot destroy radioactive waste because radioactivity is a nuclear process, and atomic nuclei are unaffected by chemical reactions. Therefore, the only feasible method for disposing of radioactive wastes is to store them in a place safe from geologic hazards and human intervention and to allow them to decay

naturally. The U.S. Department of Energy defines a permanent repository as one that will isolate radioactive wastes for 10,000 years.[1] In addition, the EPA has set a separate limit on radiation that will extend for a million years. For a repository to keep radioactive waste safely isolated for such a long time, it must meet at least three geologic criteria:

1. It must be safe from disruption by earthquakes and volcanic eruptions.
2. It must be safe from landslides, soil creep, and other forms of mass wasting.
3. It must be free from floods and seeping ground water that might corrode containers and carry wastes into aquifers.

In December 1987, the U.S. Congress chose a site near Yucca Mountain, Nevada, about 175 kilometers from Las Vegas, as the national burial ground for all spent reactor fuel unless sound environmental objections were found. Since that time, scientists have spent $4 billion testing and analyzing the area, making Yucca Mountain probably the most closely studied piece of real estate in the history of geology.

The Yucca Mountain site is located in the Basin and Range province, a region noted for faulting and volcanism related to ongoing tectonic extension of that part of the western United States. Bedrock at the Yucca Mountain site is welded tuff, a hard volcanic rock. The tuffs erupted from several volcanoes that were active from 16 to 6 million years ago. The last eruption near Yucca Mountain occurred 15,000 to 25,000 years ago. In addition, geologists have mapped 32 faults that have moved during the past 2 million years adjacent to the Yucca Mountain site. The site itself is located within a structural block bounded by parallel faults (◆ Figure 12.32). On July 29, 1992, a magnitude 5.6 earthquake occurred about 20 kilometers from the Yucca Mountain site, and damaged the Department of Energy (DOE) facilities on the site. Critics of the Yucca Mountain site argue that recent earthquakes and volcanoes prove that the area is geologically active.

The site's environment is desert dry, and the water table lies 550 meters beneath the surface. The repository will consist of a series of tunnels and caverns dug into the tuff 300 meters beneath the surface and 250 meters above the water table. Thus, it is designed to isolate the waste from both surface water and ground water. However, it is impossible to predict the climatological changes that may occur over the next 10,000 years. If the climate becomes appreciably wetter, the water table may rise or surface water could percolate downward. In addition, if an earthquake fractures rocks beneath the site, then contaminated

---

1. This number is derived from human and political considerations more than scientific ones. The National Academy of Sciences issued a report stating that radioactive wastes will remain harmful for one million years because of the long half-lives of some waste isotopes.

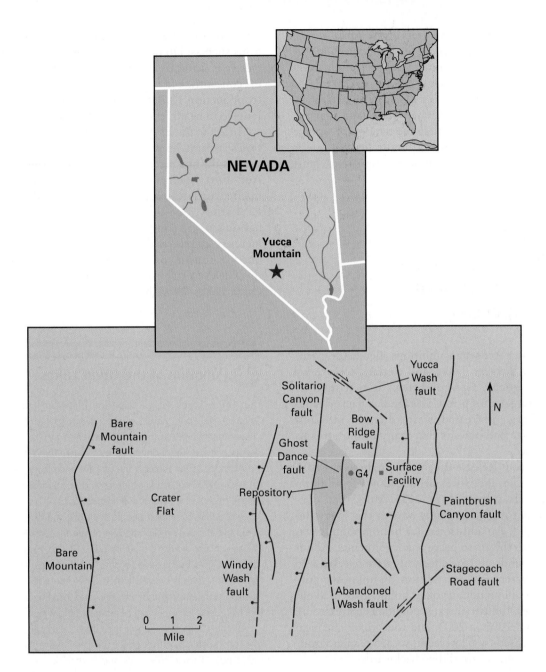

◀▶ **FIGURE 12.32** A map of the proposed Yucca Mountain repository shows numerous faults where rock has fractured and moved during the past 2 million years.

INTERACTIVE QUESTION: *In some endeavors there is a high probability that a minor accident will occur. For example, in the transportation industry, trucks are involved in accidents every day. In other endeavors, there is a low probability that a catastrophic disaster will occur. Discuss risk assessment and evaluation in this scenario and relate your discussion to the Yucca Mountain repository.*

ground water might disperse more rapidly than predicted. Furthermore, critics point out that construction of the repository will involve blasting and drilling, and these activities could fracture underlying rock, opening conduits for flowing water. In any case, water could lead to disaster in the repository. Because radioactive decay produces heat, the wastes may be hot enough to convert seeping ground water to steam. Steam trapped underground could build up enough pressure to explode containment vessels and cavern walls.

After decades of political wrangling, Congress approved the Yucca Mountain site in the summer of 2002. In July 2004 the EPA was forced to change its standard for the time period necessary to safeguard the nuclear material from leaking into the environment from 10,000 to one million years. The site could open as early as 2010.

But the debate continues. Supporters of the Yucca Mountain repository argue that we need nuclear power and that therefore as a society we must accept a certain level of risk. They argue further that at present, 45,000 tons of high-level radioactive waste lie in 126 unstable temporary storage facilities across the United States and that the Yucca Mountain site is safer than the temporary storage sites now being used. Opponents argue that the site is geologically unsound and that new alternatives must be found.

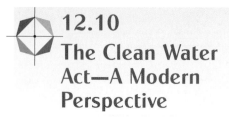

12.10
The Clean Water Act—A Modern Perspective

The Clean Water Act set an ambitious agenda for cleaning the nation's rivers, lakes, and wetlands. Have we achieved our goals? The good news is that municipalities no longer dump raw sewage and factories no longer discharge untreated waste directly into our waterways. The bad news is that many of our streams remain polluted.

In their Toxics Release Inventory database, the EPA reported that U.S. industries produced nearly 4 billion pounds of toxic substances in 2003. More than one-fourth of this total was mining wastes that were deposited on land. An additional 1.5 billion pounds was discharged into the air to become air pollutants. About 221 million pounds of toxic chemicals were released into the nation's waterways. In addition, an unknown portion of the 1.5 billion pounds of air pollutants settled out into waterways. Industrial water pollutants increased by 2 percent from 1995 to 2003, and total toxic chemical emission increased by 62 percent, so, from an environmental standpoint, we are sliding backwards.

Two hundred million pounds of poison sounds like a big number. Yet, when this material is diluted throughout all the nation's waterways and spread out over 365 days, the concentration of any given chemical will be quite low, measured as parts per million or parts per billion (one part pollutant per million or billion parts water). Should we be concerned?

Between 1999 and 2000, scientists from the United States Geological Society (USGS) sampled 139 streams in 30 states. Then they tested these samples for 95 different pharmaceuticals and household chemicals, such as steroids, insect repellents, birth-control drugs, caffeine, antibiotics, and fire retardants. They found at least one chemical on their list in 80 percent of the streams, with an average of seven and a maximum of 38. Birth-control drugs and steroids have severe long-term effects on fish and other aquatic organisms. We do not know how persistent small doses of these compounds affect humans.

Each watershed is unique and scientists must ask questions such as: Do people drink the water? If so, how effective is water treatment before drinking? Do the toxins concentrate in fish, and do people eat the fish? Recall from Section 12.7 that it is very difficult to assess the long-term health effects of small doses of pollutants.

In 2000, the EPA reported that 40 percent of the nation's streams, 45 percent of lakes, and 50 percent of estuaries do not meet the standards of the Clean Water Act. Forty percent of the country's freshwater remained unsafe for human use. According to a 2004 EPA study, one-third of lakes and one-fourth of the nation's rivers contain fish that are contaminated with mercury, dioxin, and PCBs at levels considered dangerous for human consumption. To look more closely at the Clean Water Act, in modern times, consider the Great Lakes Watershed.

Pollution of the Great Lakes

The Great Lakes—lakes Superior, Michigan, Huron, Erie, and Ontario—lie along the U.S.-Canadian border, filling ancient stream valleys that were scoured and deepened by glaciers during the Pleistocene Ice Age. Connected by streams that ultimately flow to the Atlantic Ocean through the St. Lawrence River, the lakes cover about 244,000 square kilometers and contain about one-fifth of Earth's fresh surface water (◆ Figure 12.33).

More than 33 million people live along the shores of the Great Lakes and about 38 million people obtain their drinking water from the lakes. Major industrial cities, including Duluth, Milwaukee, Chicago, Cleveland, Erie, Buffalo, and Toronto line the lakes, and much of the rural countryside is heavily farmed. Since the mid-1800s, the cities have poured industrial and human wastes into the lakes, and agriculture has added fertilizers and pesticides to the waters.

By the 1960s, pollution of the Great Lakes had become a conspicuous problem. Although all the lakes were contaminated by thousands of toxic chemicals and human wastes, lakes Erie and Ontario and shallow parts of lakes Huron and Michigan were the most strongly polluted. Lake Erie was the worst of all because it has the most strongly concentrated human population and industrial activity along its shores, and it is the shallowest of the Great Lakes. Algae, fertilized by sewage, consumed so much oxygen that millions of fish died. Floating garbage and rotting fish littered once-pristine beaches. The fish that survived were so badly contaminated with persistent toxins that they were unfit to eat. Waterborne bacteria became a health hazard to swimmers and to those people drinking inadequately treated lake water. Cormorants living near the lake were born with horrible birth defects and local populations of falcons and eagles were driven to extinction by pesticide poisoning.

In 1972, Canada and the United States began a joint cleanup effort in the Great Lakes watershed. After 30 years and a $20 billion effort, contamination by human wastes, detergents, industrial pollutants, and agricultural pesticides decreased dramatically. DDT levels in women's breast milk and PCB levels in fish and water birds have

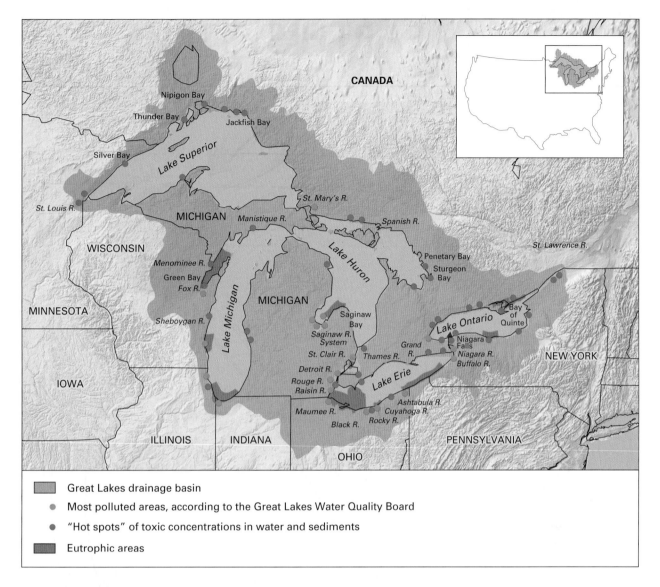

FIGURE 12.33 The Great Lakes drainage basin and the location of some of its water pollution problems. Lake Superior is the largest freshwater lake in the world; it is also the least polluted of the Great Lakes. Because the lakes drain eastward, Lakes Erie and Ontario contain pollutants flowing from the three upstream lakes plus pollutants that are locally generated.

declined. Populations of bald eagles and other birds that depend on the lakes for food and shelter have increased. Turbid water has become clear and the algae have died off in many places.

Yet, the Great Lakes watershed does not have a clean bill of health. While factories have cleaned up most of their discharges, coal-fired electric plants on the shores of the Great Lakes are still exempt from adhering to mercury emission restrictions mandated by the Federal Clean Air Act of 1970. Concentrations of PCBs in Great Lakes fish is 40 percent above levels deemed acceptable by the EPA, and in 2004 beach closures due to unsafe levels of bacteria were twice as frequent as in 2003. Pesticide and fertilizer runoff from farms and urban lawns, leaking septic tanks, waste from pets and livestock, and sewage overflows are some of the sources of pollution that still need to be addressed. Air currents from as far away

as Mexico transport pollutants—such as DDT and PCBs, which are banned in the United States—to the lake waters. Enough mercury air pollutants from coal-fired power plants fall into the lake to make many fish inedible. Many persistent toxic compounds remain in the lake waters and bottom sediment decades after they were introduced. From this example, we see that analyzing the results of the Clean Water Act today is like trying to decide whether a glass is half empty or half full. The Cuyahoga River does not catch fire any more, and no large chemical company would dare bury drums of untreated toxins in an abandoned canal. But water pollution remains a serious problem that threatens the health of people and aquatic organisms and reduces the quality of life for all.

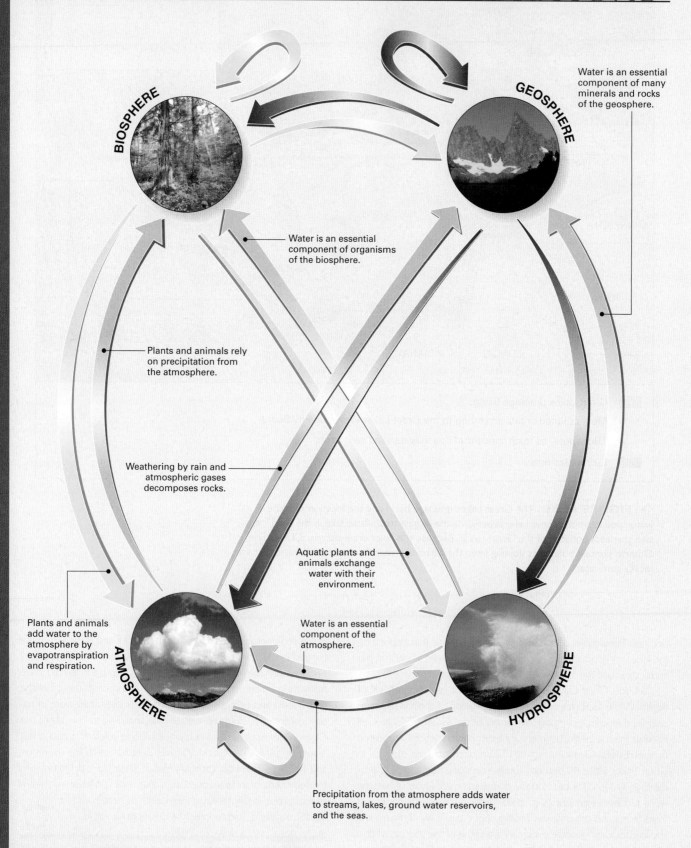

BIOSPHERE

GEOSPHERE

Water is an essential component of many minerals and rocks of the geosphere.

Water is an essential component of organisms of the biosphere.

Plants and animals rely on precipitation from the atmosphere.

Weathering by rain and atmospheric gases decomposes rocks.

Aquatic plants and animals exchange water with their environment.

Plants and animals add water to the atmosphere by evapotranspiration and respiration.

ATMOSPHERE

Water is an essential component of the atmosphere.

HYDROSPHERE

Precipitation from the atmosphere adds water to streams, lakes, ground water reservoirs, and the seas.

ALTHOUGH WATER IS THE material of Earth's hydrosphere, it is an essential component of organisms of the biosphere, it is contained in many minerals and rocks of the geosphere, and is a component of the atmosphere as well. For example, a human baby is about 78 percent water by weight, and an average adult male's body contains 60 percent water. Water, both as H_2O and the hydroxyl ion $(OH)^-$, are essential constituents of minerals such as gypsum $(CaSO_4 \cdot 2H_2O)$ and biotite $K(Mg, Fe)_3(Si,Al)_4(OH)_2$. Earth's atmosphere contains about 13,000 cubic kilometers of water, mostly in the form of water vapor and tiny droplets that make up clouds. Furthermore, the hydrologic cycle, described in Chapter 11, shows that water cycles continuously among all four of these spheres. Thus, water is a common thread connecting all four Earth systems. ■

SUMMARY

Growing populations have created demands for water that stretch, and in some cases exceed, the amount of water that is available both globally and in the United States. The United States receives about three times as much water as it uses, but some of the driest regions use the greatest amounts of water.

Most of the water used by homes and industry is **withdrawn** and then returned to streams or ground water reservoirs near the site of withdrawal. But most of the water used by agriculture is **consumed** because it evaporates. Water use falls into three categories. Domestic use accounts for 10 percent of U.S. water consumption, industrial use accounts for 49 percent, and agricultural use accounts for 41 percent.

Water **diversion** projects collect and transport surface water and ground water from places where water is available to places where it is needed. Dams are especially useful in regions that receive seasonal rainfall, and some dams also generate hydroelectric energy. But they can create undesirable effects, including water loss, salinization, silting, erosion, destruction when a dam fails, recreational and aesthetic losses, and ecological disruptions. The case study on the Three Gorges Dam project illustrates these problems.

Ground water projects pump water to Earth's surface for human use, but because ground water flows so slowly, the extraction often creates a **cone of depression** that may lead to a drop in the water table. This chapter's discussion of the Ogallala aquifer illustrates ground water depletion. Ground water withdrawal may also cause **subsidence** and **saltwater intrusion**.

The Great American Desert is a mostly arid region of the western United States. Americans have built great cities and extensive farms and ranches in the region, all supplied by irrigation systems. Diminishing water reserves and increasing costs of water diversion projects suggest that their future is uncertain. Case studies outline the Colorado and Owens River diversion projects. Water conflict leads to serious international conflict between hostile nations.

Water pollution is the reduction in the quality of water by the introduction of impurities. In the United States, the Clean Water Act was passed in response to serious water pollution disasters such as the Cuyahoga River fires and Love Canal. Water pollutants include **biodegradable materials**, such as sewage, disease organisms, and fertilizers; **non-biodegradable materials** (also called **persistent bioaccumulative toxic chemicals**, or PBTs), such as industrial organic compounds, toxic inorganic compounds, and sediment; **radioactive materials**; and heat. Pollution can originate from both **point sources** and **nonpoint** sources.

Biodegradable pollutants nourish an ecosystem and lead to oxygen depletion and growth of anaerobic bacteria, with resultant degradation of the ecosystem.

It is difficult to determine the health effects of low doses of water pollutants. **Cost–benefit analysis** compares the cost of pollution control with the cost of **externalities**. Ground water pollutants normally spread slowly into an aquifer as a contaminant **plume**. Because contaminants permeate all the tiny pores' spaces in an aquifer, cleanup, or **remediation,** is expensive and difficult. Techniques include: **elimination of the source, monitoring, modeling, bioremediation, chemical remediation**, and removal of contaminated rock and soil.

Radioactive wastes must be isolated from water resources because they are impossible to destroy and persist for long times. In the United States, Yucca Mountain is the designated site for spent fuels from nuclear power plants.

Over the past 34 years, waterways in the United States have been cleaned considerably, but they still remain polluted.

Key Terms

withdrawal   296
consumption   296
diversion system   300
salinization   301
cone of depression   304
subsidence   306
saltwater intrusion   307
pollution   311

Clean Water Act   312
biodegradable pollutants   313
nonbiogradable   313
persistent bioaccumulative toxic
    chemicals (PBTs)   313
point source pollution   313
nonpoint source pollution   313
anaerobic   313

externalities   315
cost–benefit analysis   315
plume   315
remediation   317
bioremediation   318
chemical remediation   318

For Review

1. The United States receives three times more water in the form of precipitation than it uses. Why do water shortages exist in many parts of the country?

2. Describe the three main categories of water use. What proportion of total U.S. water use falls into each category?

3. Explain the differences among water use, water withdrawal, and water consumption.

4. Why is agriculture responsible for the greatest proportion of water consumption?

5. What are the two main sources of water exploited by water-diversion projects?

6. Describe the beneficial effects of dams and their associated water-delivery systems.

7. Describe the negative effects of dams and their associated water-delivery systems.

8. Why does salinization commonly result from desert irrigation?

9. Describe the factors that make ground water a valuable resource.

10. Describe problems caused by excessive pumping of ground water.

11. Why does ground water depletion cause longer-term problems than might result from the draining of a surface reservoir?

12. If land subsides when an underlying aquifer is depleted, will it rise to its original level when pumping stops and the aquifer is recharged? Explain your answer.

13. Why does saltwater intrusion affect only coastal areas?

14. Describe the geographic region that John Wesley Powell called the Great American Desert.

15. List the three major goals of the Clean Water Act.

16. List eight categories of water pollutants. What is/are the source(s) of each, and what are the harmful effects?

17. How does a nontoxic substance such as cannery waste become a water pollutant?

18. Discuss the difficulties in assessing the health risks of low doses of toxic water pollutants.

19. What is cost–benefit analysis and how does it work?

20. Explain why pollutants persist in ground water longer than they do in surface water.

21. Outline possible procedures for cleaning polluted ground water.

22. Discuss the controversy over the Yucca Mountain nuclear waste repository.

23. Discuss the accomplishments and failings of the Clean Water Act.

For Discussion

1. Estimate how much water you use every day. For example, run your shower into a bucket for 15 seconds and measure the amount of water you collect. Then time an average shower. Repeat this process for all forms of water use. How does your consumption compare with that of an average Californian?

2. Consider how (a) a rise or (b) a fall in average global temperature might affect surface-water and ground water reserves in the United States.

3. Discuss how water conservation in each of the three main categories of water use in the United States can affect the total amount of water available for all types of uses.

4. Discuss the relative merits and disadvantages of surface-water and ground water diversion projects.

5. No major dam has been built in the United States without federal support, and the price charged to irrigators and other users for the water from a major dam has never paid for the cost of building the dam. Discuss why we build dams.

6. Develop scenarios for the future of farming and ranching in regions currently obtaining water from the Ogallala aquifer.

7. Discuss environmental problems that arise when cotton and rice are grown in California at the same time that farmers in naturally wet southeastern and south central states are paid federal subsidies to not grow the same crops.

8. In the book *Cadillac Desert*, Marc Reisner describes the transformation of the American West from its natural desert and semidesert environment to a region with great cities and a vast agricultural economy. He argues that the West cannot sustain the cities or the agriculture. Discuss how this transformation occurred and the validity of his hypothesis that the current system cannot continue indefinitely.

9. Consider reasons why desert and semidesert cities such as Phoenix, Las Vegas, and Los Angeles have become so heavily populated despite the lack of adequate water resources.

10. Abundant water exists in the Columbia River basin in the northwestern United States and in the Fraser River basin in Canada, whereas the Colorado River basin to the south is water deficient. Argue for or against a proposal to divert water from the Pacific Northwest to the arid southwest.

11. Discuss the role that water plays in the relations between the United States and its neighboring countries.

12. Of the eight categories of pollutants listed at the beginning of Section 12.5, which are most likely to originate from point sources, and which are most likely to originate from nonpoint sources?

13. Imagine that you live in a major industrial city and your drinking water has a slight "off" taste. Bottled water is expensive and the city health officials assure you that the tap water is safe. What facts would you search for in trying to make your own personal assessment of the safety of the water? Discuss the factors that you would weigh in making a personal decision of whether to drink city water or bottled water.

14. The EPA authorized U.S. industry to dispose of 160 million kilograms of toxic waste into deep injection wells. An injection well is approved if it is drilled into a porous rock layer protected from ground water by an impermeable rock layer. Discuss the pros and cons of this method of disposal.

CHAPTER
13

Glaciers and Ice Ages

Courtesy of Graham R. Thompson/Jonathan Turk

Weasel Glacier,
Bafflin Island,
Canada.

We often think of glaciers as features of high mountains and the frozen polar regions, yet anyone living in the northern third of the United States is familiar with landscapes created by a vast ice sheet that covered this region 18,000 years ago. The low, rounded hills of upper New York State, Wisconsin, and Minnesota are composed of gravel deposited by the ice. In addition, people in this region swim and fish in lakes created by those glaciers.

Numerous times during Earth's history, glaciers grew to cover large parts of Earth and then melted away. Before the most recent major glacial advance, beginning about 100,000 years ago, the world was free of ice except for high mountains and the polar ice caps of Antarctica and Greenland. Then, in a relatively short time—perhaps only a few thousand years—Earth's climate cooled. As winter snow failed to melt completely in summer, the polar ice caps spread into lower latitudes. At the same time, glaciers formed near mountain summits, even near the equator. They flowed down mountain valleys into nearby lowlands. When the glaciers reached their maximum size 18,000 years ago, they covered one-third of Earth's continents.

Then about 15,000 years ago, Earth's climate warmed again and the glaciers began to melt rapidly. Although 18,000 years is a long time when compared with a single human lifetime, it is but an eyeblink in geologic time. In fact, humans, who have been on this planet for about 100,000 years, lived through the most recent glaciation. In southwest France and northern Spain, humans developed sophisticated spearheads and carved body ornaments between 40,000 and 30,000 years ago. People first began experimenting with agriculture about 12,000 to 10,000 years ago. ■

13.1 Formation of Glaciers

In most temperate regions, winter snow melts completely in spring and summer. However, in certain cold, wet environments, some of the winter snow remains unmelted during the summer and accumulates year after year. During summer, the snow crystals become rounded and denser as the snowpack is compressed and alternately warmed during daytime and cooled at night. If snow survives through one summer, it converts to rounded ice grains called **firn**. Mountaineers like firn because the sharp points of their ice axes and crampons sink into it easily and hold firmly. If firn is buried deeper in the snowpack, it converts to glacial ice, which consists of closely packed ice crystals (◆ Figure 13.1).

A **glacier** is a massive, long-lasting, moving mass of compacted snow and ice. Glaciers form only on land, wherever the amount of snow that falls in winter exceeds the amount that melts in summer. Because ice under pressure is plastic, mountain glaciers flow downhill. Glaciers on level land flow outward under their own weight.

Glaciers form in two environments. Alpine glaciers form at all latitudes on high, snowy mountains. Continental ice sheets form at all elevations in the cold polar regions.

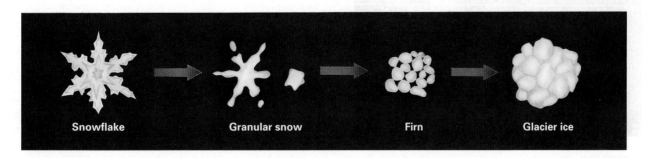

Snowflake   **Granular snow**   **Firn**   **Glacier ice**

◆ **FIGURE 13.1** Newly fallen snow changes through several stages to form glacier ice.

Alpine Glaciers

Mountains are generally colder and wetter than adjacent lowlands. Near the summits, winter snowfall is deep and summers are short and cool. These conditions create **alpine glaciers** (◆ Figure 13.2). Alpine glaciers exist on every continent—in the Arctic and Antarctica, in temperate regions, and in the tropics. Glaciers cover the summits of Mount Kenya in Africa and Mount Cayambe in South America, even though both peaks are near the equator.

Some alpine glaciers flow great distances from the peaks into lowland valleys. For example, the Kahiltna Glacier, which flows down the southwest side of Denali (Mount McKinley) in Alaska, is about 65 kilometers long, 12 kilometers across at its widest point, and about 700 meters thick. Although most alpine glaciers are smaller than the Kahiltna, some are larger.

The growth of an alpine glacier depends on both temperature and precipitation. The average annual temperature in the state of Washington is warmer than that in Montana, yet alpine glaciers in Washington are larger and flow to lower elevations than those in Montana.

Winter storms buffet Washington from the moisture-laden Pacific. Consequently, Washington's mountains receive such heavy winter snowfall that even though summer melting is rapid, snow accumulates every year. In much drier Montana, snowfall is light enough that most of it melts in the summer, and thus Montana's mountains have only a few small glaciers.

◆ **FIGURE 13.2** This alpine glacier flows around granite peaks in British Columbia, Canada.

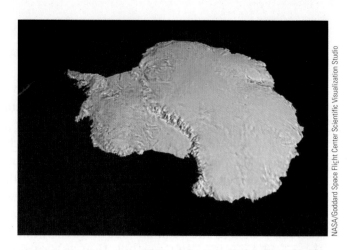

NASA/Goddard Space Flight Center Scientific Visualization Studio

◆ **FIGURE 13.3** The Antarctic ice sheet covers about 13 million square kilometers, an area almost 1.5 times the size of the United States. It blankets entire mountain ranges, and the mountains that rise above its surface are islands of rock in a sea of ice.

Continental Glaciers

Winters are so long and cold and summers so short and cool in polar regions that glaciers cover most of the land regardless of its elevation. An **ice sheet**, or **continental glacier**, covers an area of 50,000 square kilometers or more (◆ Figure 13.3). The ice spreads outward in all directions under its own weight.

Today, Earth has only two ice sheets, one in Greenland and the other in Antarctica. These two ice sheets contain 99 percent of the world's ice and about three-fourths of Earth's freshwater. The Greenland sheet is more than 2.7 kilometers thick in places and covers 1.8 million square kilometers. Yet it is small compared with the Antarctic ice sheet, which blankets about 13 million square kilometers, almost 1.5 times the size of the United States. The Antarctic ice sheet covers entire mountain ranges, and the mountains that rise above its surface are islands of rock in a sea of ice.

Whereas the South Pole lies in the interior of the Antarctic continent, the North Pole is situated in the Arctic Ocean. At the North Pole, only a few meters of ice freeze on the relatively warm sea surface, and the ice fractures and drifts with the currents. As a result, no ice sheet exists there.

⬡ 13.2 Glacial Movement

As an experiment, geologists have set two poles in dry ground on opposite sides of an alpine glacier, and a third pole in the ice to form a straight line with the other two. After a few months, the center pole moved downslope, and the three poles formed a triangle. The center pole moved because the glacier flowed downhill.

The rate of glacial movement varies with slope steepness, precipitation, and air temperature. In the coastal ranges of southeast Alaska, where annual precipitation is high and average temperature is relatively warm (for glaciers), some glaciers move 15 centimeters to a meter per day. In contrast, in the interior of Alaska, where conditions are colder and drier, glaciers move only a few centimeters per day. At these rates, ice can flow the length of an alpine glacier in a few hundred to a few thousand years. In some instances, a glacier may surge at a speed of 10 to 100 meters per day.

Glaciers move by two mechanisms: **basal slip** and **plastic flow**. In basal slip, the entire glacier slides over bedrock in the same way that a bar of soap slides down a tilted board. Just as wet soap slides more easily than dry soap, water between bedrock and the base of a glacier accelerates basal slip.

Several factors cause water to accumulate near the base of a glacier. Earth's heat melts ice near bedrock. Friction from glacial movement also generates heat. Water occupies less volume than an equal amount of ice. As a result, pressure from the weight of overlying ice favors melting. Finally, during the summer, water melted from the surface of a glacier may seep downward to its base.

A glacier also moves by plastic flow, in which the ice flows as a viscous fluid. Plastic flow is demonstrated by two experiments. In one, scientists set a line of poles in the ice (◆ Figure 13.4). After a few years, the ice moved downslope so that the poles formed a U-shaped array. This experiment shows that the center of the glacier moves faster than the edges. Frictional resistance with the valley walls slows movement along the edges and glacial ice flows plastically, allowing the center to move faster than the sides.

In another experiment, scientists drove a straight, flexible pipe downward into a glacier to study the flow of ice at depth (◆ Figure 13.5). At a later date, they discovered that the entire pipe moved downslope and also became bent. At the surface of a glacier, the ice is brittle, like an ice cube or the ice found on the surface of a lake. In contrast, at depths greater than about 40 meters, the pressure is sufficient to allow ice to deform plastically. The curvature in the pipe shows that the ice moved plastically and that middle levels of the glacier moved faster

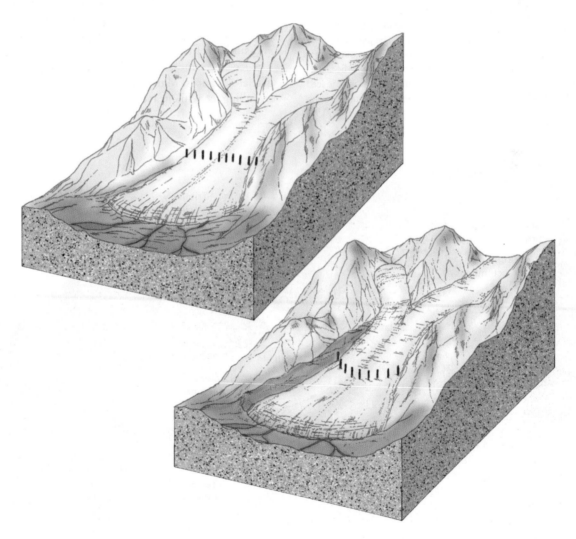

◆ **FIGURE 13.4** If a line of poles is set in a glacier, the poles near the center of the ice move downslope faster than those near the margin.

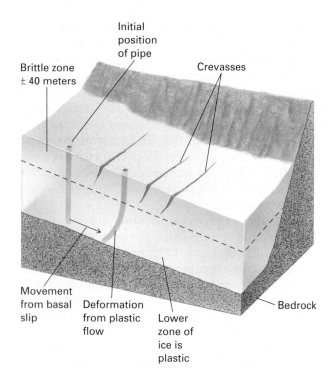

◆ **FIGURE 13.5** In this experiment, a pipe was driven through a glacier until it reached bedrock. The entire pipe moved downslope with the ice and also became curved. The pipe became curved because friction with bedrock slowed movement of the bottom of the glacier. Middle layers of ice flowed more rapidly because there the ice is plastic. At a depth shallower than 40 meters, the ice does not flow plastically, and the pipe in that zone remained straight.

than the lower part. The base of the glacier is slowed by friction against bedrock, so it moved more slowly than the plastic portion above it.

The relative rates of basal slip and plastic flow depend on the steepness of the bedrock underlying the glacier and the thickness of the ice. A small alpine glacier on steep terrain moves mostly by basal slip. In contrast, the bedrock beneath portions of the Antarctica and Greenland ice sheets is relatively level, so the ice has no slope to slide down. Thus these continental glaciers are huge plastic masses of ice (with a thin, brittle cap) that ooze outward, mainly under the forces created by their own weight.

When a glacier flows over uneven bedrock, the deeper plastic ice bends and flows over bumps, while the brittle, upper layer stretches and cracks, forming **crevasses** (◆ Figure 13.6). Crevasses form only in the brittle, upper 40 meters of a glacier, not in the lower plastic zone (◆ Figure 13.7). Crevasses open and close slowly as a glacier moves. An **ice fall** is a section of a glacier consisting of crevasses and towering ice pinnacles. The pinnacles form where ice blocks break away from the crevasse walls and rotate as the glacier moves. With crampons, ropes, and ice axes, a skilled mountaineer might climb into a crevasse. The walls are a pastel blue, and sunlight filters through the narrow opening above. The ice shifts and cracks, making creaking sounds as the glacier advances. Many mountaineers have been crushed by falling ice while traveling through icefalls.

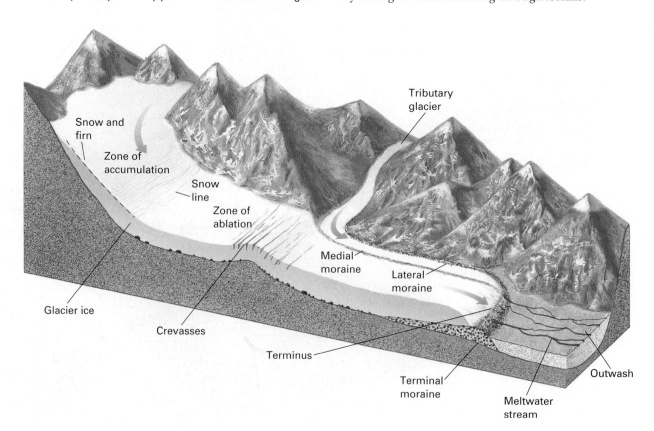

◆ **FIGURE 13.6** A schematic view of an alpine glacier. Crevasses form in the upper, brittle zone of a glacier where the ice flows over uneven bedrock.

◆ **FIGURE 13.7** Crevasses in the Bugaboo Mountains of British Columbia.

The Mass Balance of a Glacier

Consider an alpine glacier flowing from the mountains into a valley (◆ Figure 13.6). At the upper end of the glacier, snowfall is heavy, temperatures are below freezing for much of the year, and avalanches carry large quantities of snow from the surrounding peaks onto the ice. Thus, more snow falls in winter than melts in summer, and snow accumulates from year to year.

This higher-elevation part of the glacier is called the **zone of accumulation**. There the glacier's surface is covered by snow year-round.

Lower in the valley, the temperature is higher throughout the year, and less snow falls. This lower part of a glacier, where more snow melts in summer than accumulates in winter, is called the **zone of ablation**. When the snow melts, a surface of old, hard, glacial ice is left behind. The **snow line** is the boundary between permanent snow and seasonal snow. The snow line shifts up and down the glacier from year to year, depending on weather. Ice exists in the zone of ablation because the glacier flows downward from the accumulation area. Even farther down the valley, the rate of glacial flow cannot keep pace with melting, so the glacier ends at its **terminus**.

Glaciers grow and shrink. If annual snowfall increases or average temperature drops, more snow accumulates; then the snow line of an alpine glacier descends to a lower elevation, and the glacier grows thicker. At first the terminus may remain stable, but eventually it advances farther down the valley. The lag time between a change in climate and a glacial advance may range from a few years to several decades, depending on the size of the glacier, its rate of motion, and the magnitude of the climate change. If annual snowfall decreases or the climate warms, the zone of accumulation shrinks and the glacier retreats.

When a glacier retreats, its ice continues to flow downhill, but the terminus melts back faster than the glacier flows downslope. In Glacier Bay, Alaska, glaciers have retreated 60 kilometers in the past 125 years. Near the terminus, newly exposed rock is bare and lifeless. A few kilometers from the glacier, where rock has been exposed for a few decades, scattered lichens grow on otherwise bare rock. Even farther away, seabird droppings have mixed with windblown silt and weathered rock to form thin soil that supports mosses in sheltered cracks. Near the head of the bay, 60 kilometers from the terminus, tidal currents and ocean storms have washed enough sediment over the land to create soil and support stunted trees.

In equatorial and temperate regions, glaciers commonly terminate at an elevation of 3,000 meters or higher. However, in a cold, wet climate, a glacier may extend into the sea (◆ Figure 13.8). Giant chunks of ice break off, forming **icebergs**.

The largest icebergs in the world are those that break away from the Antarctic ice shelf. Between the years 2000 and 2002, two plates of ice the size of Connecticut and at least one the size of Rhode Island broke free from the West Antarctic Ice Sheet and floated into the Antarctic Ocean (see Chapter 21). The tallest icebergs in the world break away from tidewater glaciers in Greenland and are about 300 to 400 meters thick.

◆ **FIGURE 13.8** A kayaker paddles among small icebergs that calved from the Le Conte Glacier, Alaska.

Rock at the base and sides of a glacier may have been fractured by tectonic forces, frost wedging, or pressure-release fracturing. The moving ice dislodges the loosened rock (◆ Figure 13.9). Ice is viscous enough to pick up and carry particles of all sizes, from silt-sized grains to house-sized boulders. Thus glaciers erode and transport huge quantities of rock and sediment.

Ice itself is not abrasive to bedrock because it is too soft. However, rocks embedded in the ice scrape across bedrock, cutting deep, parallel grooves and scratches called **glacial striations** (◆ Figure 13.10). When glaciers melt and striated bedrock is exposed, the markings show the direction of ice movement. Glacial striations are used to map the flow directions of glaciers.

Erosional Landforms Created by Alpine Glaciers

Let's take an imaginary journey through a mountain range that was glaciated in the past but is now mostly ice free (◆ Figure 13.11). We start with a helicopter ride to the summit of a high, rocky peak. Our first view from the helicopter is of sharp, jagged mountains rising steeply above smooth, rounded valleys.

A mountain stream commonly erodes downward into its bed, cutting a steep-sided, V-shaped valley. A glacier, however, is not confined to a narrow streambed but in-

◆ **FIGURE 13.10** Stones embedded in the base of a glacier gouged these striations in bedrock in British Columbia.

stead fills its entire valley. As a result, it scours the sides of the valley as well as the bottom, carving a broad, rounded, **U-shaped valley** (◆ Figure 13.12).

We land on one of the peaks and step out of the helicopter. Beneath us, a steep cliff drops off into a horseshoe-shaped depression in the mountainside called a **cirque**. A small glacier at the head of the cirque reminds us of the larger mass of ice that existed in a colder, wetter time (◆ Figure 13.13A).

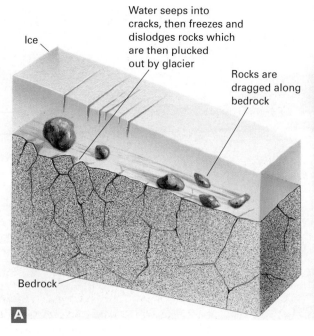

◆ **FIGURE 13.9** **(A)** A glacier plucks rocks from bedrock and then drags them along, abrading both the loose rocks and the bedrock. **(B)** These crescent-shaped depressions in granite at Le Conte Bay, Alaska, were formed by glacial plucking.

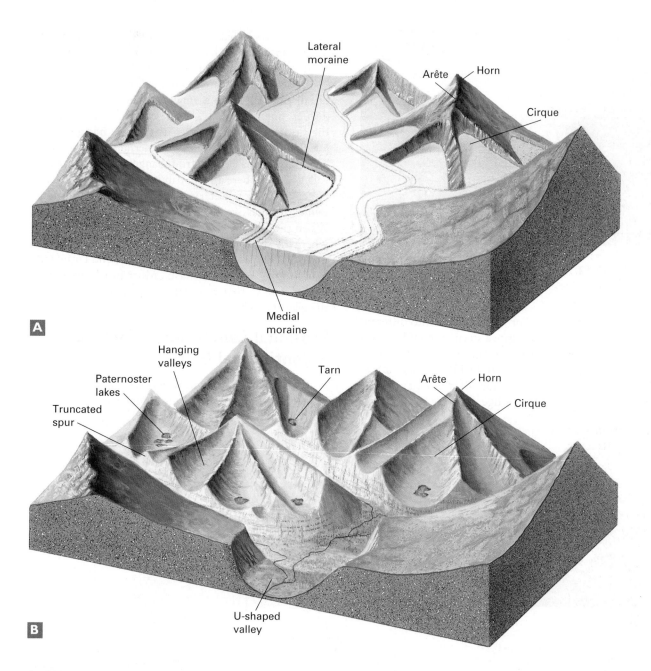

A

B

◁▷ **FIGURE 13.11** Two views of the same glacial landscape. **(A)** The landscape as it appeared when it was mostly covered by glaciers. **(B)** The same landscape as it appears now, after the glaciers have melted.

INTERACTIVE QUESTION: *Draw a third diagram showing the landscape in the future. Assume that there is no additional tectonic uplift or glacial advances.*

To understand how a glacier creates a cirque, imagine a gently rounded mountain. As snow accumulates and a glacier forms, the ice flows down the mountainside (◁▷ Figure 13.13B). The ice erodes a small depression that grows slowly as the glacier flows (◁▷ Figure 13.13C). With time, the cirque walls become steeper and higher. The glacier carries the eroded rock from the cirque to lower parts of the valley (◁▷ Figure 13.13D). When the glacier finally melts, it leaves a steep-walled, rounded cirque.

Streams and lakes are common in glaciated mountain valleys. As a cirque forms, the glacier may erode a depression into the bedrock beneath it. When the glacier melts, this depression fills with water, forming a small lake, or **tarn,** nestled at the base of the cirque. If we hike down the valley below the high cirques, we may encounter a series of lakes called **paternoster lakes,** which are commonly connected by rapids and waterfalls (◁▷ Figure 13.14). Paternoster lakes are a sequence of small basins plucked out by a glacier. The term *paternoster* refers to a string of rosary beads and evokes an image of a string of lakes in a glacial valley. When the glacier recedes, the basins fill with water.

◆ **FIGURE 13.12** A U-shaped valley in the Purcell Mountains of British Columbia.

If glaciers erode three or more cirques into different sides of a peak, they may create a steep, pyramid-shaped rock summit called a **horn**. The Matterhorn in the Swiss Alps is a famous horn (◆ Figure 13.15). Two alpine glaciers flowing along opposite sides of a mountain ridge may erode both sides of the ridge, forming a sharp, narrow rib of rock that lies approximately perpendicular to the main ridge. This feature, called an **arête** often forms a border between adjacent valleys.

Looking downward from our peak, we may see a waterfall pouring from a small, high valley into a larger, deeper one. A small glacial valley lying high above the floor of the main valley is called a **hanging valley** (◆ Figure 13.16). The famous waterfalls of Yosemite Valley in California cascade from hanging valleys. A hanging valley forms where a small tributary glacier joined a much larger one. The tributary glacier eroded a shallow valley while the massive main glacier gouged a deeper one. When the glaciers melted, they exposed an abrupt drop where the small valley joins the main valley.

Deep, narrow inlets called **fjords** extend far inland on many high-latitude seacoasts. Most fjords are glacially carved valleys that were later flooded by rising seas as the glaciers melted (◆ Figure 13.17).

Erosional Landforms Created by a Continental Glacier

A continental glacier erodes the landscape just as an alpine glacier does. However, a continental glacier is consider-

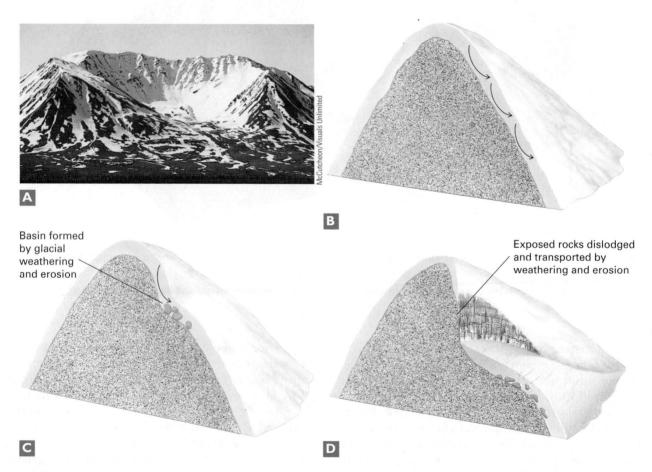

Basin formed by glacial weathering and erosion

Exposed rocks dislodged and transported by weathering and erosion

◆ **FIGURE 13.13** **(A)** A glacier eroded this concave cirque into a mountainside in the Alaska Range. **(B)** To form a cirque, snow accumulates, and a glacier begins to flow from the summit of a peak. **(C)** Glacial weathering and erosion form a small depression in the mountainside. **(D)** Continued glacial movement enlarges the depression. When the glacier melts, it leaves a cirque carved in the side of the peak, as in the photograph.

Courtesy of Graham R. Thompson/Jonathan Turk

◆ **FIGURE 13.14** Glaciers eroded bedrock to form this string of paternoster lakes in the Sierra Nevada.

Courtesy of Graham R. Thompson/Jonathan Turk

◆ **FIGURE 13.16** Yosemite Falls cascades from a hanging valley in Yosemite National Park.

Swiss Tourist Board

◆ **FIGURE 13.15** The Matterhorn in Switzerland formed as three alpine glaciers eroded cirques into the peak from three different sides. The ridge between two cirques is called an arête.

◆ **FIGURE 13.17** A steep-sided fjord bounded by 1,000-meter-high cliffs in Baffin Island, Canada.

ably larger and thicker and is not confined to a valley. As a result, it covers vast regions, including entire mountain ranges. The recent ice sheets have scoured deep basins that have filled with water to form the Great Lakes and the Finger Lakes in New York (◆ Figure 13.18).

In the 1800s, geologists recognized that the large deposits of sand and gravel found in some places had been transported from distant sources. A popular hypothesis at the time explained that this material had drifted in on icebergs during catastrophic floods. The deposits were called *drift* after this inferred mode of transport.

Today we know that continental glaciers covered vast parts of the land only 10,000 to 20,000 years ago and that these glaciers carried and deposited drift. Although the term *drift* is a misnomer, it remains in common use. Now geologists define **drift** as all rock or sediment transported and deposited by a glacier. Glacial drift averages 6 meters thick over the rocky hills and pastures of New England and 30 meters thick over the plains of Illinois.

Drift is divided into two categories. **Till** was deposited directly by glacial ice. **Stratified drift** was first carried by a glacier and then transported and deposited by a stream.

Landforms Composed of Till

Ice is so much more viscous than water that it carries a wide range of particle sizes. When a glacier melts, it deposits particles of all sizes—from fine clay to huge boulders—in an unsorted, unstratified mass

◆ **FIGURE 13.18** This satellite view shows the Finger Lakes (bottom right), which formed when Pleistocene glaciers scoured deep valleys and the valleys filled with water. Pleistocene glaciers also formed Lake Ontario (top).

◆ **FIGURE 13.19** Unsorted glacial till. Note that large cobbles are mixed with smaller sediment. The cobbles were rounded by stream action before they were transported and deposited by the glacier.

(◆ Figure 13.19). Within a glacier, each rock or grain of sediment is protected by the ice that surrounds it. Therefore, the pieces do not rub against one another, and glacial transport does not round sediment as a stream does. If you find rounded gravel in till, it became rounded by a stream before the glacier picked it up.

Occasionally, large boulders lie on the surface in country that was once glaciated. In many cases the boulders are of a rock type different from the bedrock in the immediate vicinity. Boulders of this type are called **erratics** and were transported to their present locations by a glacier. The origins of erratics can be determined by exploring the terrain in the direction the glacier came from until the parent rock is found. Some erratics were carried 500 or even 1,000 kilometers from their points of origin and provide clues to the movement of glaciers.

Moraines

A **moraine** is a mound or a ridge of till. Think of a glacier as a giant conveyor belt. An old-fashioned airport conveyor belt simply carries suitcases to the end of the belt and dumps them in a pile. Similarly, a glacier carries sediment and deposits it at its terminus. If a glacier is neither advancing nor retreating, its terminus may remain in the same place for years. During that time, sediment accumulates at the terminus to form a ridge called an **end moraine** (◆ Figure 13.20). An end moraine that forms when a glacier is at its greatest advance, before beginning to retreat, is called a **terminal moraine**.

If warmer conditions prevail, the glacier recedes. If the glacier then stabilizes again during its retreat and the terminus remains in the same place for a year or more, a new end moraine, called a **recessional moraine**, forms.

When a glacier recedes steadily, till is deposited in a relatively thin layer over a broad area, forming a **ground moraine**. Ground moraines fill old stream channels and other low spots. Often this leveling process disrupts drainage patterns. Many of the swamps in the northern Great Lakes region and northern New England lie on ground moraines formed when the most recent continental glaciers receded.

End moraines and ground moraines are characteristic of both alpine and continental glaciers. An end moraine deposited by a large alpine glacier may extend for several kilometers and be so high that even a person in good physical condition would have to climb for an hour to reach the top. Moraines may be dangerous to hike over if their sides are steep and the till is loose. Large boulders are mixed randomly with rocks, cobbles, sand, and clay. A careless hiker can dislodge boulders and send them tumbling to the base.

The most recent Pleistocene continental glaciers reached their maximum extent about 18,000 years ago. Their terminal moraines record the southernmost extent

Steve Sheriff

◆ **FIGURE 13.20** The end moraine of an alpine glacier on Baffin Island, Canada, in midsummer. Dirty, old ice forms the lower part of the glacier below the snow line, and clean snow lies higher up on the ice in the zone of accumulation.

of those glaciers. In North America, the terminal moraines lie in a broad, undulating front extending across the northern United States. Enough time has passed since the glaciers retreated that soil and vegetation have stabilized the till and most of the hills are now covered by vegetation (◆ Figure 13.21).

When an alpine glacier moves downslope, it erodes the valley walls as well as the valley floor. Therefore, the edges of the glacier carry large loads of sediment. Additional debris falls from the valley walls and accumulates on and near the sides of mountain glaciers. Sediment near the glacial margins forms a **lateral moraine** (◆ Figure 13.22).

If two alpine glaciers converge, their lateral moraines merge into the middle of the larger glacier. This till forms a visible dark stripe on the surface of the ice called a **medial moraine** (◆ Figure 13.23).

Drumlins

Elongate hills, called **drumlins**, cover parts of the northern United States (◆ Figure 13.24) and are best exposed across the rolling farmland in upstate New

◆ **FIGURE 13.21** This wooded terminal moraine in New York State marks the southernmost extent of glaciers in that region.

York. Drumlins usually occur in clusters. Each one looks like a whale swimming through the ground with its back in the air. An individual drumlin is typically about 1 to 2 kilometers long and about 15 to 50 meters high. Most are made of till, while others consist partly of till and partly of bedrock. In either case, when a glacier forms a drumlin, it erodes and deposits sediment to create the elongate, streamlined shape. The glacier generally erodes a steep-sided face as it advances. It then deposits some sediment on the downslope side to form a long, pointed slope. Thus a geologist can determine the direction of motion of an ancient glacier by studying drumlins.

Landforms Composed of Stratified Drift

Because a glacier erodes great amounts of sediment, streams flowing from a glacier are commonly laden with silt, sand, and gravel. The stream deposits this sediment beyond the glacier terminus as **outwash** (◆ Figure 13.25). Glacial streams carry such a heavy load of sediment that they often become braided, flowing in multiple channels. Outwash deposited in a narrow valley is called a **valley train**. If the sediment spreads out from the confines of the valley into a larger valley or plain, it forms an **outwash plain**

◆ **FIGURE 13.22** A lateral moraine lies against the valley wall in the Bugaboo Mountains, British Columbia.

Courtesy of Graham R. Thompson/Jonathan Turk

◆ **FIGURE 13.23** Merging lateral moraines from coalescing glaciers formed three separate medial moraines on Baffin Island, Canada.

(◆ Figure 13.26). Outwash plains are also characteristic of continental glaciers.

During summer, when snow and ice melt rapidly, streams form on the surface of a glacier. Many are too wide to jump across. Some of these streams flow off the front or sides of the glacier. Others plunge into crevasses and run beneath the glacier over bedrock or drift. These streams commonly deposit small mounds of sediment, called **kames**, at the margin of a receding glacier or where sediment collects in a crevasse or other depression in the ice. An **esker** is a long, sinuous ridge that forms as the channel deposit of a stream that flowed within or beneath a melting glacier (◆ Figure 13.27).

Because kames, eskers, and other forms of stratified drift are stream deposits and were not deposited directly by ice, they show sorting and sedimentary bedding, which distinguishes them from unsorted and unstratified till. In addition, the individual cobbles or grains are usually rounded.

Large blocks of ice may be left behind in a moraine or an outwash plain as a glacier recedes. When such an ice block melts, it leaves a depression called a **kettle**. Kettles fill with water, forming kettle lakes. A kettle lake is as large as the ice chunks that melted to form the hole. The lakes vary from a few tens of meters to a kilometer or so in diameter, with a typical depth of 10 meters or less.

Kevin Horan/Getty Images

◆ **FIGURE 13.24** Crop patterns emphasize glacially streamlined drumlins in Wisconsin.

13.4 Glacial Deposits

◆ **FIGURE 13.25** Streams flowing from the terminus of a glacier filled this valley with outwash on Baffin Island.

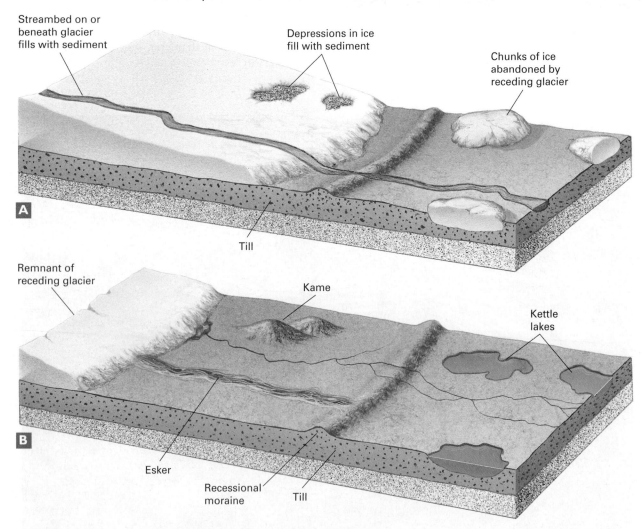

◆ **FIGURE 13.26** **(A)** A melting glacier exposes several different glacial landforms that were created beneath the ice. **(B)** Sediment flowing from the melting ice creates an outwash plain beyond the glacial terminus.

◆▶ **FIGURE 13.27** This esker in Manitoba, Canada, is a sinuous ridge of sand and gravel deposited in the bed of a stream that flowed beneath a continental glacier about 20,000 years ago.

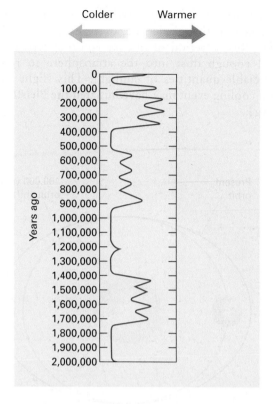

◆▶ **FIGURE 13.28** Glacial cycles during the Pleistocene Ice Age. Cold temperatures coincided with Pleistocene glacial advances, and warm intervals coincided with glacial melting. We are probably living within a warm period of the Pleistocene Ice Age, and continental ice sheets may advance once again.

13.5
The Pleistocene Ice Age

Geologists have found terminal moraines extending across all high-latitude continents. By studying those moraines, as well as lakes, eskers, outwash plains, and other glacial landforms, geologists are certain that massive glaciers once covered large portions of the continents, altering Earth systems. A time when alpine glaciers descend into lowland valleys and continental glaciers spread over land in high latitudes is called an **ice age**. During an ice age, glaciers several kilometers thick spread across the landscape. Beneath the burden of ice, the continents sink deeper into the asthenosphere. The ice weathers rock and erodes soil, altering the landscape.

Geologic evidence shows that Earth has been warm and relatively ice free for at least 90 percent of the past 1 billion years. However, at least six major ice ages occurred during that time. Each one lasted from 2 to 10 million years (see Chapter 21).

The most recent ice age took place mainly during the Pleistocene Epoch and is called the **Pleistocene Ice Age**. It began about 2 million years ago in the Northern Hemisphere, although evidence of an earlier beginning has been found in the Southern Hemisphere. However, Earth has not been glaciated continuously during the Pleistocene Ice Age; instead, climate has fluctuated and continental glaciers grew and then melted away several times (◆▶ Figure 13.28). During the most recent interglacial period, the average temperature was about the same as it is today, or perhaps a little warmer. Then, high-latitude temperature dropped

at least 15°C, causing the ice to advance.[1] Most climate models indicate that we are still in the Pleistocene Ice Age and the continental ice sheets may advance again.

Causes of the Pleistocene Ice Age and Glacial Cycles

For reasons that are poorly understood, prior to 2 million years ago, Earth's climate had been cooling for tens of millions of years, and then something happened to push the planet over a climate threshold and plunge it into an ice age. David Rea, a geologist from the University of Michigan, recently analyzed sediment cores of the mud on the sea floor in a region of the North Pacific.

He found that an increase in volcanic ash 2 million years ago coincided closely with a dramatic increase in sediment grains that showed glacial markings. Thus

---

1. Mark Chandler, "Glacial Cycles; Trees Retreat and Ice Advances," *Nature* 381 (June 6, 1996): 477–478.

an increase in volcanic activity occurred at about the same time that glaciers began to spread. According to Dr. Rea, a period of intense volcanic activity injected enough dust into the atmosphere to reflect appreciable quantities of sunlight. This slight additional cooling event may have initiated the Pleistocene Ice Age.

Although volcanic dust may have triggered the *onset* of the Pleistocene glacial epoch, no such events seem to be associated with the repeated growth and melting of glaciers that characterize the Pleistocene Epoch. Instead, scientists have found that slight, periodic variations in Earth's orbit and orientation coincided with Pleistocene glacial expansion and shrinking.

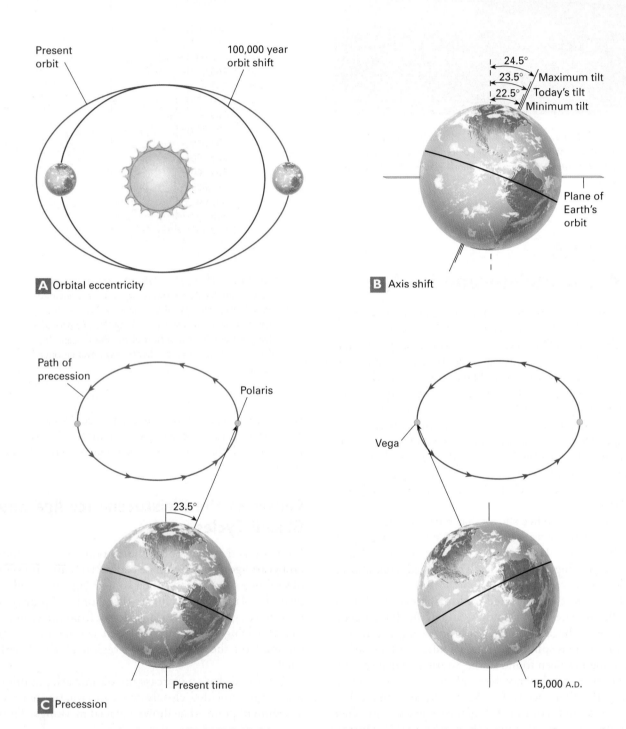

ThomsonNOW ◆ **ACTIVE FIGURE 13.29** Earth orbital variations may explain the temperature oscillations and glacial advances and retreats during the Pleistocene Epoch. Earth's orbit varies in three ways: **(A)** the elliptical shape of the orbit changes over a cycle of about 100,000 years; **(B)** the tilt of Earth's axis of rotation oscillates by about 2° over a cycle of about 41,000 years; **(C)** Earth's axis completes a full cycle of precession about every 26,000 years.

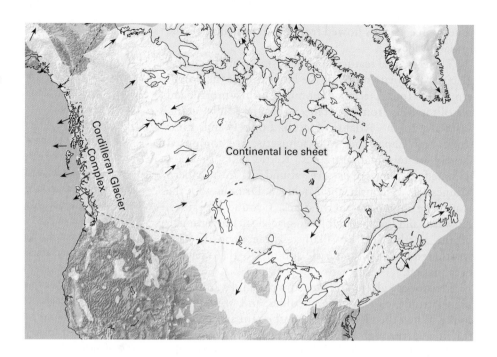

◆ **FIGURE 13.30** Maximum extent of the continental glaciers in North America during the latest glacial advance, approximately 18,000 years ago. The arrows show directions of ice flow.

Labels on figure: Cordilleran Glacier Complex · Continental ice sheet

In the nineteenth century, astronomers detected three periodic variations in Earth's orbit and spin axis (◆ Figure 13.29):

1. Earth's orbit around the Sun is elliptical rather than circular. The shape of the ellipse is called **eccentricity**. The eccentricity varies in a regular cycle lasting about 100,000 years.
2. Earth's axis is currently tilted at about 23.5° with respect to a line perpendicular to the plane of its orbit around the Sun. The **tilt** oscillates by 2.5° on about a 41,000-year cycle.
3. Earth's axis, which now points directly toward the North Star, circles like that of a wobbling top. This circling, called **precession**, completes a full cycle every 26,000 years.

These changes affect both the total solar radiation received by Earth and the distribution of solar energy with respect to latitude and season. Seasonal changes in sunlight reaching higher latitudes can reduce summer temperature. If summers are cool and short, winter snow and ice persist, leading to growth of glaciers.

Early in the twentieth century, a Yugoslavian astronomer, Milutin Milankovitch, calculated that the orbital variations generate alternating cool and warm climates in the midlatitudes and higher latitudes. Moreover, the timing of the calculated cooling coincided with that of Pleistocene glacial advances. Therefore, he concluded that orbital variations caused Pleistocene glacial cycles.

Modern calculations indicate that orbital cycles by themselves are not sufficient to cause glaciers to advance and retreat. Instead, orbital cycles disturb other Earth systems, which cause additional cooling. Thus a relatively small initial disturbance is amplified to cause a major climate change. In one recent study, researchers calculated that orbital variations probably caused high-latitude climate to cool enough to kill vast regions of Pleistocene northern forest. Forests control climate by absorbing solar energy and warming the atmosphere. When the forests died, more solar energy reflected back out to space. This loss of solar energy caused Earth to cool even more, and the glaciers advanced.[2] But ice reflects more solar radiation back into space, so the feedback process then became further amplified as the growth of glaciers amplified global cooling. Thus variations in Earth's orbit altered the biosphere and the altered biosphere triggered additional cooling that led to a major glacial advance.

Effects of Pleistocene Continental Glaciers

At its maximum extent about 18,000 years ago, the most recent North American ice sheet covered 10 million square kilometers—most of Alaska, Canada, and parts of the northern United States (◆ Figure 13.30). At the same

---

2. R. G. Gallimore and J. E. Kutzbach, "Role of Orbitally Induced Changes in the Tundra Area in the onset of Glaciation," *Nature* 381 (June 6, 1996): 503–505.

time, alpine glaciers flowed from the mountains into the lowland valleys.

The erosional features and deposits created by these glaciers dominate much of the landscape of the northern states. Today, terminal moraines form a broad band of rolling hills from Montana across the Midwest and eastward to the Atlantic Ocean. Long Island, New York, and Cape Cod, Massachusetts, are composed largely of terminal moraines. Kettle lakes or lakes dammed by moraines are abundant in northern Minnesota, Wisconsin, and Michigan. Drumlins dot the landscape in the northern states. Ground moraines, outwash, and loess (windblown glacial silt) cover much of the northern Great Plains. These deposits have weathered to form the fertile soil of North America's breadbasket.

Pleistocene glaciers advanced when midlatitude and high-latitude climates were colder and wetter than today. When the glaciers melted, the rain and meltwater flowed through streams and collected in numerous lakes. Later, as the ice sheets retreated and the climate became drier, many of these streams and lakes dried up.

The basin that is now Death Valley was once filled with water to a depth of 100 meters or more. Most of western Utah was covered by Lake Bonneville. As drier conditions returned, Lake Bonneville shrank to become Great Salt Lake, west of Salt Lake City.

When glaciers grow, they accumulate water that would otherwise be in the oceans, and sea level falls. When glaciers melt, sea level rises again. When the Pleistocene glaciers reached their maximum extent 18,000 years ago, global sea level fell to about 130 meters below its present elevation. As submerged continental shelves became exposed, the global land area increased by 8 percent (although about one-third of the land was ice covered).

When the ice sheets melted, most of the water returned to the oceans, raising sea level again. At the same time, portions of continents rebounded isostatically as the weight of the ice was removed. The effect along any specific coast depends upon the relative amounts of sea-level rise and isostatic rebound. The rising seas submerged some coastlines. Others rebounded more than sea level rose. Today, beaches in the Canadian Arctic lie tens to a few hundred meters above the sea. Portions of the shoreline of Hudson's Bay have risen isostatically 300 meters.

Pleistocene Glaciers and the Great Lakes

Before the onset of Pleistocene glaciation, several major rivers flowed through what is now the Great Lakes Basin and followed the path of the modern St. Lawrence River to the Atlantic Ocean. Then during Pleistocene time, glaciers spread southward across the St. Lawrence River valley, damming the rivers and forcing them to change direction and flow away from the ice, southward into the Mississippi drainage (◄► Figure 13.31A). The ice scoured and deepened the valleys and deposited moraines that blocked the St. Lawrence outlet.

At the same time, the great accumulation of ice on land had lowered sea level, thereby lowering the base level of the Mississippi River. This effect increased the gradient of the river, causing it to flow more rapidly. The fast-flowing Mississippi cut rapidly into its bed, forming a V-shaped valley with a narrow flood plain.

Starting about 18,000 years ago, the glaciers began to melt and retreat. The abundant meltwater carried more silt, sand, and gravel into the Mississippi than the current could transport, so the river spread out into many small channels, forming a braided stream and filling the narrow valley with sediment.

As the ice sheet melted, lakes formed behind moraine dams all along the receding ice front. Eventually the rising water level breached a dam near modern Lake Ontario, and much of the water from the Great Lakes region flowed into the Hudson River valley (◄► Figure 13.31B).

Over the next 5,000 years, the ice receded, sea level rose, and the continent rebounded isostatically. These changes altered the drainage patterns several times (◄► Figures 13.31C and D). About 9,000 years ago, the rivers abandoned the Hudson River valley outlet, and the Great Lakes drained eastward and southward through the St. Lawrence and Mississippi Rivers. A few thousand years later, the Mississippi River link was abandoned and all the water flowed through the St. Lawrence, as it does today.

When its supply of water was reduced, the Mississippi slowed and carried much less sediment. Gradually, the braided channels coalesced, forming the modern, meandering river, with its broad flood plain.

Glacial Lake Missoula and the Greatest Flood in North America

About 18,000 years ago, a lobe of the great continental ice sheet flowed southward from Canada into the United States along the Idaho–Montana border. The ice dammed the Clark Fork River, forming Glacial Lake Missoula, which was 600 meters deep and contained 2,000 cubic kilometers of water—as much as modern Lake Ontario. Ice is a disastrously poor material for a dam because it floats on water. When the lake became deep enough, the ice dam began to float and as a result, it no longer retained the lake water. Water poured across northern Idaho and Washington. But this was orders of magnitude more catastrophic than any modern disasters, even more catastrophic than large dam failures that have occurred in modern times.

Geologists estimate that a wall of water 600 meters high raced down valley. The water spread out as the valley opened up, and it continued down the Columbia River valley. The raging torrent transported rocks from western Montana and deposited some of them 300 meters above the level of the modern river near the Columbia River

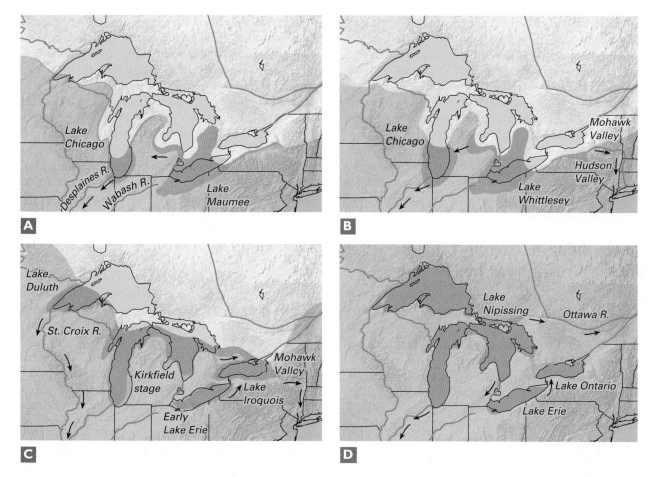

◈ FIGURE 13.31 Continental glaciers scoured the Great Lakes Basin and altered its drainage pattern several times.

Gorge, 450 kilometers from the failed dam. The rushing water eroded deep stream valleys in eastern and central Washington, which now contain tiny streams or no water at all (◈ Figure 13.32).

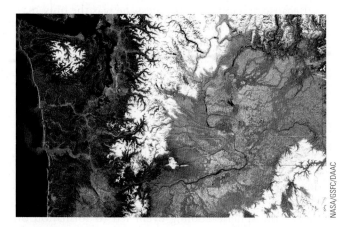

◈ FIGURE 13.32 Evidence of the immense floods from Lake Missoula is clearly visible in the region known as the Channeled Scablands of eastern and central Washington. The rushing water eroded deep stream valleys, which now contain tiny streams or no water at all.

13.6

Snowball Earth: The Greatest Ice Age in Earth History

Recent research has suggested that at least twice, and perhaps as many as five times, in late-Precambrian time between 800 and 550 million years ago, massive ice sheets completely covered all continents and the world's oceans froze over—even at the equator—entombing the entire globe in a 1-kilometer-thick shell of ice. This ice age, called Snowball Earth by the researchers who discovered it, contrasts sharply with the Pleistocene Ice Age, when ice covered only a third of the continents and only the polar seas froze over.[3]

The main evidence for these Precambrian global glaciations is based on a unique rock called **tillite**. Recall that till consists of an unsorted mixture of boulders, silt, and clay that was deposited by a glacier. Pleistocene tills are loose gravel; you can dig them up with a shovel. Tillite, however, is hard, solid rock that in every other

3. Hoffman et al., "A NeoProterozoic Snowball Earth," *Science* 281 (August 28, 1998): 1342–1346.

respect resembles the Pleistocene tills. It is till that was deposited by glaciers so long ago that it has become cemented into hard rock. Researchers have found at least two thick layers of tillite between 750 and 580 million years old on almost every continent. Recall that continents have moved around Earth through geologic time. Other types of evidence show that some of the continents lay at the equator when the tillites formed. Other continents were nearer to the poles at the same time, showing that the glaciations were global in scope. In some localities, the glacial deposits lie on top of thick limestone layers. The meaning of these limestone deposits is explained later.

Snowball Earth, Carbon Dioxide, and the Greenhouse Effect

Researchers have postulated that these ancient glacial episodes resulted from fluctuations in atmospheric carbon dioxide concentrations, caused by variations of weathering rates of rocks on continents, and by carbon exchanges among seawater, the atmosphere, and living organisms. Recall from discussions in earlier chapters that carbon dioxide in the atmosphere absorbs heat and warms Earth's atmosphere, in a process called the greenhouse effect. Conversely, if carbon dioxide is removed from the atmosphere, the atmosphere cools. What happened during the late Proterozoic that reduced the atmospheric carbon-dioxide concentration sufficiently to freeze Earth's surface? And then, why did the carbon dioxide concentration increase again to melt the global icecap?

Two papers, one published in 2003 and the other in 2004, offer different explanations, and either or both scenarios could have contributed to the overall effect. [4,5]

Donnadieu et al. start their explanation with the observation that weathering of continental silicate rocks removes carbon dioxide from the atmosphere and precipitates it in the ocean as limestone, according to the generalized reaction:

$$CaSiO_3 + CO_2 \xrightarrow{\text{Weathering}} CaCO_3 + SiO_2$$

calcium silicate + carbon dioxide → calcite + quartz

The decreases in the carbon-dioxide level in the atmosphere then result in global cooling.

The researchers then observe that the supercontinent Rodinia broke up between 800 and 700 million years ago, coinciding with the onset of Snowball Earth. When a supercontinent breaks apart, more coastline is exposed to the ocean. Rainfall is generally higher in coastal areas than in continental interiors. When precipitation and runoff increases, weathering also increases, atmospheric carbon dioxide decreases, and the planet cools.

This cooling could have caused glaciers to expand at high latitudes. But ice and snow reflect much more sunlight and solar heat back into space than soil and water do. Consequently, as the glaciers spread, less and less solar heat warmed Earth's surface and the atmosphere grew colder. Thus, a runaway feedback mechanism caused the glaciers to grow even larger, until they covered all continents and the seas froze over to create Snowball Earth.

Ridgwell and his coresearchers offer a different explanation for Snowball Earth. They point out that carbonate ions (CO_3^{2-}) dissolved in seawater react with atmospheric carbon dioxide to form water plus the bicarbonate ion (HCO_3^-) in the reaction:

$$CO_2 + CO_3^{2-} + H_2O \rightarrow 2HCO_3^-$$

Thus, an increased concentration of carbonate ions in seawater will cause CO_2 to move from the atmosphere to seawater, where it exists in dissolved form. When this occurs, the CO_2 concentration in the atmosphere decreases and the atmosphere cools.

Ridgwell and coworkers suggest that, during late-Proterozoic and earlier times, primitive organisms called cyanobacteria combined carbonate ions dissolved in seawater with dissolved calcium to precipitate calcite, the mineral that forms limestone. They further suggest that the cyanobacteria lived only in shallow parts of the oceans that could be reached by sunlight. Thus, limestone formed on continental margins in the shallow water of continental shelves during late Proterozoic and earlier times.

A continental shelf is a broad, shallow, submarine surface where the ocean depth is 200 meters or less and that extends tens or hundreds of kilometers offshore from the edge of land. At the edge of the continental shelf, ocean depths drop abruptly to five kilometers below sea level. Thus, if sea level lowers by 100 to 200 meters, most of the continental shelf becomes exposed above water and can no longer accumulate limestone.

Because precipitation of limestone was the only major process that removed carbonate ions from seawater during late-Proterozoic time, the amount of carbonate ion that could be transferred from seawater to solid limestone depended on the global surface area of continental shelves. Thus, if sea level should drop by 100 to 200 meters for some reason, the continental shelves would shrink by a large amount, less limestone would form, and more carbonate ion would be retained in seawater. Higher dissolved carbonate levels extract carbon dioxide from the atmosphere and lead to global atmospheric cooling.

Ridgwell and coworkers suggest that a fairly minor, global, atmospheric cooling event led to formation of glaciers during late-Proterozoic time that were similar in size to modern ice caps, lowering sea level by about 100 meters. The resulting shrinkage of the continental shelves then reduced the amount of limestone forming on the shelves, causing atmospheric cooling. Consequently, the

4. Yannick Donnadieu et al.: *Nature,* Vol 428, March 18, 2004, pg 303.
5. Andy J. Ridgwell et al.: *Science,* Vol. 302, October 31, 2003, pg 859.

glaciers grew larger. Growth of the ice caps further lowered sea level, causing even more CO_2 to migrate from the atmosphere to the oceans. In addition, the sparkling glaciers reflected large amounts of sunlight, causing further cooling. These feedback processes led to runaway global atmospheric cooling and Snowball Earth.

However, if some combination of supercontinent breakup, increased weathering rates, decreasing area of continental shelves, and global albedo can trigger a major, global, atmospheric cooling episode, then we might expect that the breakup of Pangea approximately 200 million years ago would have initiated another Snowball Earth. In fact, Pangea broke apart without a second Snowball Earth event. Why did Earth's climate system behave differently when Pangea broke apart than when Rodinia broke apart? Ridgwell and his coworkers suggest that by 200 million years ago, new types of organisms had emerged—tiny floating organisms called calcareous plankton that live in the shallow waters of the open oceans. Like the cyanobacteria, the plankton extract dissolved calcium and carbonate ions from seawater, and precipitate calcite to form their shells and skeleta. However, the plankton live in the surface waters throughout the oceans and are not restricted to continental shelves. When the plankton die, their shells and skeletons sink to the sea floor to form limestone, thus removing carbonate ions from seawater. Because these organisms are not restricted to the shallow water of continental shelves, their role in extracting carbonate from seawater is not sensitive to sea-level fluctuations. Thus, they provide a strong, stabilizing mechanism that modulates oceanic carbonate ion concentration, and consequently dampens atmospheric carbon dioxide response to sea-level change.

Thus, according to Ridgwell and his coresearchers, wide-ranging calcareous plankton buffered the global climate system during the breakup of Pangea to maintain relatively constant atmospheric temperatures. Yet when Rodinia broke apart in late pre-Cambrian time, cyanobacteria were restricted to the shallow water of continental shelves and were vulnerable to sea-level fluctuations, and a runaway feedback mechanism led to the dramatic cooling.

But if ice and snow reflect so much solar heat back into space without warming the atmosphere, why didn't Earth remain frozen forever? What eventually caused melting of the Precambrian ice crusts that enveloped Earth? Under normal conditions, some atmospheric carbon dioxide dissolves into the oceans, maintaining a moderate amount in the atmosphere. But when the seas froze over in late-Precambrian time, the ice covering the sea surface prevented atmospheric carbon dioxide from dissolving in seawater.

Volcanoes emit large quantities of carbon dioxide. Normally, this injection of carbon dioxide into the atmosphere is quickly absorbed into seawater, but when the seas froze over, carbon dioxide emitted by normal volcanic activity simply accumulated in the atmosphere in ever-increasing concentrations. Eventually the green-house effect warmed the atmosphere to the point that the global ice crust began to melt away.

Thick beds of limestone form only in warm seas. The limestone layers that overlie the tillites in some locations indicate that global climate switched from icy to tropical in a remarkably short time. The explanation is that the high levels of carbon dioxide required to melt the global ice sheets—as much as 350 times greater than present-day levels—must have created a greenhouse effect that raised global atmospheric temperatures to 40°C to 50°C in only a few thousand years. However, most of the carbon dioxide would then have dissolved into the seas, restoring temperatures to a more moderate level.

The last thaw occurred about 580 million years ago, about the same time that multicellular life bloomed on Earth. At that time both the number and the diversity of living organisms rapidly increased. Most scientists attribute this rapid proliferation of life to a gradual increase in atmospheric oxygen levels that finally reached a threshold at which organisms could thrive, multiply, and evolve at greatly accelerated rates. Some researchers, however, have suggested that the late-Precambrian explosion of life resulted from environmental conditions at the end of Snowball Earth. The freezing temperatures and icy crust would have killed off most living things, leaving only tiny isolated populations of survivors huddled around volcanic hot springs or submarine volcanic vents. Small populations and geographic isolation are two of the main factors responsible for rapid evolution and emergence of new species.

When the ice melted for the last time, these populations found a multitude of ecologic habitats ready to be colonized in the newly warm Earth, and abundant sources of food. Plants and animals rapidly multiplied and evolved into new species as they took advantage of the wealth of unused resources.

In summary: A generation ago, most scientists thought that planetary environments were relatively constant over time. Now we know that climates and conditions change dramatically. The Snowball Earth hypothesis invokes feedback and threshold mechanisms involving all of Earth's four spheres. It involves weathering of continental rocks of the geosphere, glaciation of the hydrosphere, exchanges of carbon between the atmosphere and the oceans, and exchanges of carbon between living organisms, seawater, and the atmosphere. This example reminds us of the fragility of our climate and the complex interactions that control it.

13.7
The Earth's Disappearing Glaciers

Glaciers are now shrinking in more places and at more rapid rates than at any other time since scientists began

Table 13.1   Selected Examples of Ice Melt Around the World

Name	Location	Measured Loss
Arctic Sea Ice	Arctic Ocean	Has shrunk by 6 percent since 1978, with a 14 percent loss of thicker, year-round ice. Has thinned by 40 percent in less than 30 years.
Greenland Ice Sheet	Greenland	Has thinned by more than a meter a year on its southern and eastern edges since 1993.
Columbia Glacier	Alaska	Has retreated nearly 13 kilometers since 1982. In 1999, retreat rate increased from 25 meters per day to 35 meters per day.
Glacier National Park	Rocky Mtns., U.S.	Since 1850, the number of glaciers has dropped from 150 to fewer than 50. Remaining glaciers could disappear completely in 30 years.
Antarctic Sea Ice	Southern Ocean	Ice to the west of the Antarctic Peninsula decreased by some 20 percent between 1973 and 1993, and continues to decline.
Pine Island Glacier	West Antarctica	Grounding line (where glacier hits ocean and floats) retreated 1.2 kilometers a year between 1992 and 1996. Ice thinned at a rate of 3.5 meters per year.
Larsen B Ice Shelf	Antarctic Peninsula	Calved a 200 km2 iceberg in early 1998. Lost an additional 1,714 km2 during the 1998–1999 season, and 300 km2 so far during the 1999–2000 season.
Tasman Glacier	New Zealand	Terminus has retreated 3 kilometers since 1971, and main front has retreated 1.5 kilometers since 1982. Has thinned by up to 200 meters on average since the 1971–1982 period. Icebergs began to break off in 1991, accelerating the collapse.
Meren, Carstenz, and Northwall Firn Glaciers	Irian Jaya, Indonesia	Rate of retreat increased to 45 meters a year in 1995, up from only 30 meters a year in 1936. Glacial area shrank by some 84 percent between 1936 and 1995. Meren Glacier is now close to disappearing altogether.
Dokriani Bamak Glacier	Himalayas, India	Retreated by 20 meters in 1998, compared with an average retreat of 16.5 meters over the previous 5 years. Has retreated a total of 805 meters since 1990.
Duosuogang Peak	Ulan Ula Mtns., China	Glaciers have shrunk by some 60 percent since the early 1970s.
Tien Shan Mountains	Central Asia	Twenty-two percent of glacial ice volume has disappeared in the past 40 years.
Caucasus Mountains	Russia	Glacial volume has declined by 50 percent in the past century.
Alps	Western Europe	Glacial area has shrunk by 35 to 40 percent and volume has declined by more than 50 percent since 1850. Glaciers could be reduced to only a small fraction of their present mass within decades.
Mt. Kenya	Kenya	Largest glacier has lost 92 percent of its mass since the late 1800s.
Speka Glacier	Uganda	Retreated by more than 150 meters between 1977 and 1990, compared with only 35–45 meters between 1958 and 1977.
Upsala Glacier	Argentina	Has retreated 60 meters a year on average over the last 60 years, and rate is accelerating.
Quelccaya Glacier	Andes, Peru	Rate of retreat increased to 30 meters a year in the 1990s, up from only 3 meters a year between the 1970s and 1990.

Worldwatch Institute, www.worldwatch.org/alerts/000306.html

keeping records (Table 13.1). During a single, hot summer in Europe in 2003, 10 percent of the glacial ice in the Alps melted. Scientists point out that this accelerated loss of glacial ice coincides with an increase in atmospheric carbon dioxide levels of about 50 percent in the past century, and they suggest that the melting may be one of the first symptoms of human-caused global warming.

Alpine glaciers reflect these changes in a highly visible way. The larger glaciers in Glacier National Park, Mon-

◆ **FIGURE 13.33** Boulder Glacier, Glacier National Park, Montana, in July 1932 (left) and July 1988 (right), 56 years later photographed from the same point. The glacier had disappeared completely by 1988.

tana, have shrunk to a third of their size since 1850, and they continue to melt away today. Many of the smaller glaciers have disappeared completely (◆ Figure 13.33). Since 1850, the number of glaciers in the park has decreased from 150 to fewer than 50. One computer model predicts that all of the glaciers will be gone from the park by 2030 if global temperatures continue to rise as predicted. Even if temperatures remain constant, the glaciers will disappear by 2100.

On a global average, "small" (relative to the Greenland and Antarctic ice sheets) glaciers similar to those of Glacier National Park lost about 7 meters in thickness between 1961 and 1998. In some instances, disintegration has been even more dramatic. For example, half of Alaska's Columbia glacier has melted in the past 20 years and the glacier currently releases five cubic kilometers of meltwater into Prince William Sound every year. Scientists predict that up to a quarter of the total global mass of alpine ice could disappear by 2050, and as much as half by 2100.

Global climate changes do not occur uniformly around the globe. The North Polar regions are warming faster than the average rate for all of Earth. Arctic sea ice covers an area about the size of the United States. It diminished in size by an average of 34,000 square kilometers per year between 1978 and 1996, losing a total of 6 percent of its area during that interval. The sea ice has

also thinned from an average thickness of 3.1 meters to an average of 1.8 meters since the 1960s. The Greenland Continental Ice Sheet has thinned by more than a meter per year on its southern and eastern edges since 1993. These rates are accelerating. For example, in 2005, the Kangerdlugssuaq Glacier was retreating at a rate of 14 kilometers a year, compared with 4.8 kilometers a year in 2001.

The Antarctic ice cap, which contains 91 percent of Earth's ice and averages 2.3 kilometers thick, has been shrinking for the past 10,000 years as the most recent of the Pleistocene glaciers melted away.

Ice shelves are thick masses of ice that are floating in the ocean but are connected to glaciers on land. They gain ice by flow from the land glaciers and lose ice by melting and when large chunks break off and float into the ocean. Most of Earth's ice shelves surround Antarctica. Ice shelves respond to rising temperature more sensitively than glaciers do. Since 1974, seven Antarctic ice shelves have shrunk by a total of 13,500 square kilometers. These ice shelves are described further in Chapter 21.

The melting of the Arctic and Antarctic ice has profound effects for the planet. This melting contributes to the rise in sea level, and the freshwater flowing into the ocean may ultimately alter ocean currents and global climate.

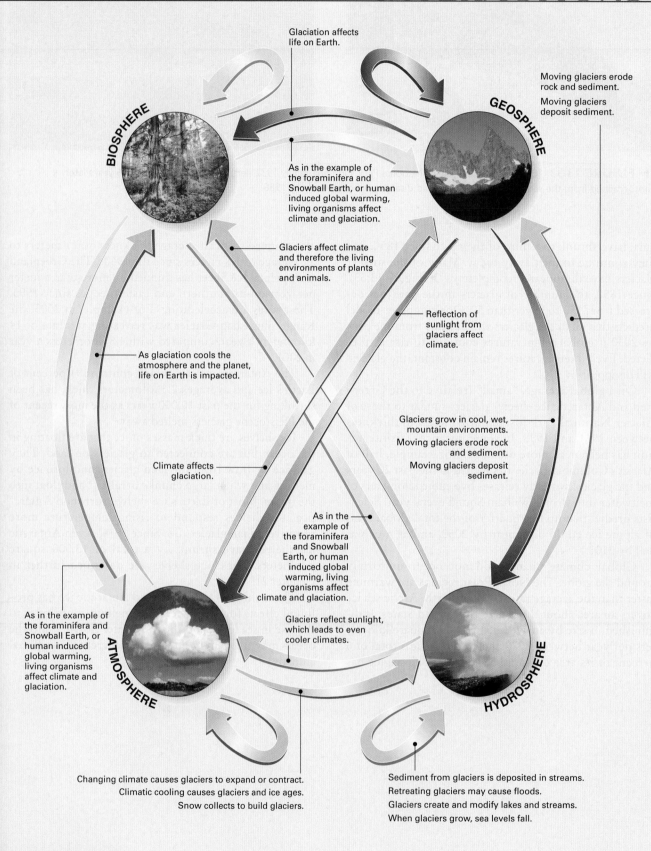

Glaciation affects life on Earth.

Moving glaciers erode rock and sediment.

Moving glaciers deposit sediment.

BIOSPHERE

GEOSPHERE

As in the example of the foraminifera and Snowball Earth, or human induced global warming, living organisms affect climate and glaciation.

Glaciers affect climate and therefore the living environments of plants and animals.

Reflection of sunlight from glaciers affect climate.

As glaciation cools the atmosphere and the planet, life on Earth is impacted.

Glaciers grow in cool, wet, mountain environments.

Moving glaciers erode rock and sediment.

Moving glaciers deposit sediment.

Climate affects glaciation.

As in the example of the foraminifera and Snowball Earth, or human induced global warming, living organisms affect climate and glaciation.

As in the example of the foraminifera and Snowball Earth, or human induced global warming, living organisms affect climate and glaciation.

Glaciers reflect sunlight, which leads to even cooler climates.

ATMOSPHERE

HYDROSPHERE

Changing climate causes glaciers to expand or contract.

Climatic cooling causes glaciers and ice ages.

Snow collects to build glaciers.

Sediment from glaciers is deposited in streams.

Retreating glaciers may cause floods.

Glaciers create and modify lakes and streams.

When glaciers grow, sea levels fall.

GLACIERS WEATHER AND ERODE ROCK and then they transport the eroded rock and sediment to deposit it downslope. Thus, water of the hydrosphere shapes landforms of the geosphere. These interactions are relatively straightforward. However, when we ask why Earth cools enough to form glaciers, or why Earth warms to melt the glaciers, then we enter the much more complex realm of global climate, which involves interactions among all four spheres. The example of Snowball Earth, given in the text, illustrated feedback and threshold mechanisms and interactions among all of Earth's spheres. In that example, we saw how microscopic foraminifera have a profound effect on climate. In the modern world, humans are another component of the biosphere that exerts a profound effect on climate. These interactions will be discussed further in the chapters to follow. ■

SUMMARY

If snow survives through one summer, it becomes a relatively hard, dense material called **firn**. A **glacier** is a massive, long-lasting accumulation of compacted snow and ice that forms on land and creeps downslope or outward under the influence of its own weight. **Alpine glaciers** form in mountainous regions; **continental glaciers** cover vast regions. Glaciers move both by **basal slip** and by **plastic flow**. The upper 40 meters of a glacier is too brittle to flow, and large cracks called **crevasses** develop in this layer.

In the **zone of accumulation**, the annual rate of snow accumulation is greater than the rate of melting, whereas in the **zone of ablation**, melting exceeds accumulation. The **snow line** is the boundary between permanent snow and seasonal snow. The end of the glacier is called the **terminus**.

Glaciers erode bedrock, forming **U-shaped valleys, cirques**, and other landforms. **Drift** is any rock or sediment transported and deposited by a glacier. The unsorted drift deposited directly by a glacier is **till**. Most glacial terrain is characterized by large mounds of till known as **moraines. Terminal moraines, ground moraines, recessional moraines, lateral moraines, medial moraines**, and **drumlins** are all depositional features formed by glaciers. **Stratified drift** consists of sediment first carried by a glacier and then transported, sorted, and deposited by streams. **Valley trains, outwash plains, kames**, and **eskers** are composed of stratified drift. A **kettle** is a depression created by melting of a large block of ice abandoned by a retreating glacier.

During the past 1 billion years, at least six major ice ages have occurred. The most recent was the **Pleistocene Ice Age**. One hypothesis contends that Pleistocene advances and retreats were caused by climate change induced by variations in Earth's orbit and the orientation of its rotational axis. Ice sheets isostatically depress continents, which later rebound when the ice melts. Sea level falls when continental ice sheets form and rises again when the ice melts.

The greatest ice age in Earth's history occurred during late-Precambrian time, when glaciers covered all continents and the seas froze over completely. Glaciers and sea ice are shrinking in more places and at more rapid rates than they did at any other time since scientists began keeping records.

Key Terms

firn   327

glacier   327

alpine glacier   328

ice sheet   328

continental glacier   328

basal slip   329

plastic flow   329

crevasse   330

ice fall   330

zone of accumulation   331

zone of ablation   331

snow line   331

terminus   331

iceberg   331

glacial striation   332

For Review

1. Outline the major steps in the metamorphism of newly fallen snow to glacial ice.

2. Differentiate between alpine glaciers and continental glaciers. Where are alpine glaciers found today? Where are continental glaciers found today?

3. Distinguish between basal slip and plastic flow.

4. Why are crevasses only about 40 meters deep, even though many glaciers are much thicker?

5. Describe the surface of a glacier in the summer and in the winter in (a) the zone of accumulation and (b) the zone of ablation.

6. Describe how glacial erosion can create (a) a cirque, (b) a paternoster lake, and (c) striated bedrock.

7. Describe the formation of arêtes, horns, and hanging valleys.

8. Distinguish among ground, recessional, terminal, lateral, and medial moraines.

9. Why are kames and eskers features of receding glaciers? How do they form?

10. What topographic features were left behind by the continental ice sheets? Where can they be found in North America today?

11. Briefly outline the changes in drainage patterns in the Great Lakes over the past 18,000 years.

12. Discuss possible mechanisms for the growth and collapse of Snowball Earth.

For Discussion

1. Compare and contrast the movement of glaciers with stream flow.

2. Outline the changes that would occur in a glacier if (a) the average annual temperature rose and the precipitation decreased, (b) the temperature remained constant but the precipitation increased, and (c) the temperature decreased and the precipitation remained constant.

3. Explain why plastic flow is a minor mechanism of movement for thin glaciers but is likely to be more important for a thick glacier.

4. In some low-elevation regions of northern Canada, both summer and winter temperatures are cool enough for glaciers to form, but there are no glaciers. Speculate on why continental glaciers are not forming in these regions.

5. If you found a large boulder lying in a field, how would you determine whether or not it was an erratic?

6. A bulldozer can only build a pile of dirt when it is moving forward. Yet a glacier can build a terminal moraine when it is retreating. Explain.

7. Imagine you encountered some gravelly sediment. How would you determine whether it was a stream deposit or a ground moraine?

8. Explain how medial moraines prove that glaciers move.

9. If you were hiking along a wooded hill in Michigan, how would you determine whether or not it was a moraine?

10. Outline various systems interactions involved in the formation and melting of Snowball Earth.

11. Many of the fertile farmlands of central Asia are irrigated from river water that flows from high glaciers in the Tien Shan and Pamir mountains. Discuss how this water resource might be affected by glacial retreat. In this case, would surface water be a nonrenewable resource, like the ground water of the Ogallalla aquifer? Why or why not?

ThomsonNOW

Assess your understanding of this chapter's topics with additional comprehensive interactivities at **http://www.thomsonedu.com/login**, which also has up-to-date web links, additional readings, and exercises.

CHAPTER
14

Deserts and Wind

Widely-spaced shrubs bloom among erosion gullies in the Painted Desert near Cameron, Arizona.

© Tom Bean/CORBIS

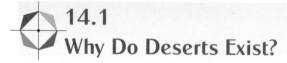

Deserts evoke an image of thirsty travelers crawling across lifeless sand dunes. Although this image accurately depicts some deserts, others are rocky and even mountainous, with colorful cliffs or peaks towering over plateaus and narrow canyons. Rain may punctuate the long, hot summers, and in winter a thin layer of snow may cover the ground. Although plant life in deserts is not abundant, it is diverse. Cactus, sage, grasses, and other plants may dot the landscape. After a rainstorm, millions of flowers bloom.

Earth's continents are categorized into climate zones based primarily on precipitation and temperature. In turn, climate determines the communities of plants and animals that live in a region. Thus the nature of Earth's surface is determined by complex interactions among the four spheres. A **desert** is any region that receives less than 25 centimeters (10 inches) of rain per year and consequently supports little or no vegetation.[1] Most deserts are surrounded by semiarid zones that receive 25 to 50 centimeters of annual rainfall, more moisture than a true desert but less than adjacent regions.

Deserts cover 25 percent of Earth's land surface outside of the polar regions and make up a significant part of every continent. If you were to visit the great deserts of Earth, you might be surprised by their geologic and topographic variety. You would see coastal deserts along the beaches of Chile; shifting dunes in the Sahara; deep, red sandstone canyons in southern Utah; stark granite mountains in Arizona; and bitter-cold polar deserts, with a few lichens hanging tenaciously to the otherwise-barren rock. The world's deserts are similar only in that they all receive scant rainfall.

Throughout human history, cultures have adapted to the low water and sparse vegetation of desert ecosystems. Many desert people were nomadic, exploiting resources where and when they were available. Others developed irrigation systems to water crops close to rivers and wells. Modern irrigation systems have improved human adaptation to dry environments and enabled 13 percent of the world's population to live in deserts. Two-thirds of the world's oil reserves lie beneath the deserts of the Middle East, transforming some of the poorest nations of the world into the richest. In the future, vast arrays of solar cells may convert desert sunlight to electricity. ■

14.1
Why Do Deserts Exist?

Rain and snow are unevenly distributed over Earth's surface. The wettest place on Earth, Mount Waialeale, Hawaii, receives an average of 1,168 centimeters (38 feet) of rain annually. In contrast, 10 years or more may pass between rains or snowfalls in the Atacama Desert of Peru and Chile. Several factors control rainfall patterns and therefore the global distribution of deserts and semi-arid lands.

Latitude

The Sun shines most directly near the equator, warming air near Earth's surface. The air absorbs moisture from the equatorial oceans and rises because it is warmer, and therefore less dense, than surrounding air. Rising air cools as pressure decreases. But cool air cannot hold as much water as warm air, so the water vapor condenses and falls as rain (◆ Figure 14.1). For this reason, vast tropical rainforests grow near the equator.

This rising equatorial air, which is now drier because of the loss of moisture, flows northward and southward at high altitudes. The air cools, becomes denser, and sinks back toward Earth's surface at about 30° north and south latitudes. As the air falls, it is compressed and becomes warmer, which enables it to hold more water vapor. As a result, water evaporates from the land surface into the air. Because the sinking air absorbs water, the ground surface is dry and rainfall is infrequent. Thus,

---

1. The definition of *desert* is linked to soil moisture and depends on temperature and amount of sunlight in addition to rainfall. Therefore, the 25-centimeter criterion is approximate.

ThomsonNOW◈ **ACTIVE FIGURE 14.1**
Falling air creates deserts at 30° north and south
latitudes. The red arrows inside the globe indicate
surface winds. The blue arrows on the right show
airflow on the surface and at higher elevations.

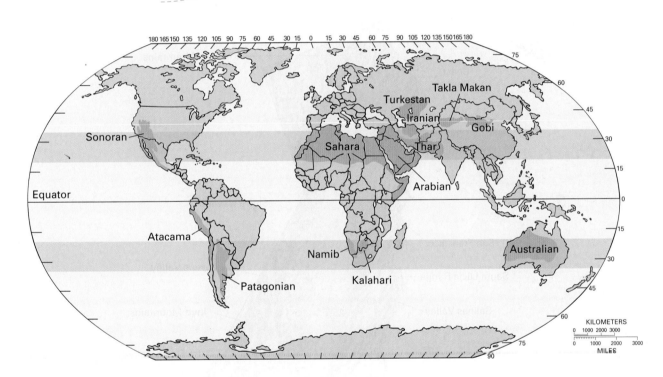

Falling air,
high pressure,
deserts

Rising air,
low pressure,
rain

Falling air,
high pressure,
deserts

◈ **FIGURE 14.2** The major deserts of the world are concentrated at approximately 30° north and south latitudes.

many of the world's largest deserts lie at about 30° north and south latitudes (◈ Figure 14.2).

Mountains: Rain-Shadow Deserts

When moisture-laden air flows over a mountain range, it rises. As the air rises, it cools and its ability to hold water decreases. As a result, the water vapor condenses into rain or snow, which falls as precipitation on the windward side and on the crest of the range (◈ Figure 14.3).

This cool air flows down the leeward (or downwind) side and sinks. As in the case of sinking air at 30° latitude, the air is compressed and warmed as it falls and it has already lost much of its moisture. This warm, dry air creates an arid zone called a **rain-shadow desert** on the leeward side of the range. ◈ Figure 14.4 shows the rainfall distribution in California. Note that the leeward valleys are much drier than the mountains to the west.

In this way, tectonic forces, which create mountains, affect rainfall patterns. In turn, as we learned in Chapter

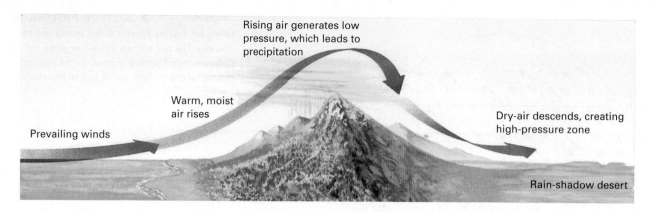

◇ **FIGURE 14.3** A rain-shadow desert forms where warm, moist air from the ocean rises as it flows over mountains. As it rises, it cools and water vapor condenses to form rain. The dry, descending air on the lee side absorbs moisture, forming a desert.

◇ **FIGURE 14.4** Rainfall patterns in the state of California, where prevailing winds carry moist Pacific air eastward over the mountains. Note that rain-shadow deserts lie east of the mountain ranges. Rainfall is given in centimeters per year.

CHAPTER 14 • Deserts and Wind

10, falling and flowing water weather and erode rocks and define living conditions for plants and animals. Thus we encounter yet another example of Earth systems interactions as the building of a mountain range (a tectonic process) alters rainfall (an atmospheric process) and ultimately defines the types of ecosystems that exist in a region (the biosphere).

Coastal and Interior Deserts

Because most evaporation occurs over the oceans, one might expect that coastal areas would be moist and climates would become drier with increasing distance from the sea. This is generally true, but a few notable exceptions exist.

The Atacama Desert along the west coast of South America is so dry that portions of Peru and Chile often receive no rainfall for a decade or more. Cool ocean currents flow along the west coast of South America. When the cool marine air encounters warm land, the air is heated. The warm, expanding air absorbs moisture from the ground, creating a coastal desert.

The Gobi Desert is a broad, arid region in central Asia. The center of the Gobi lies at about 40°N latitude, and its eastern edge is a little more than 400 kilometers from the Yellow Sea. As a comparison, Pittsburgh, Pennsylvania, lies at about the same latitude and is 400 kilometers from the Atlantic Ocean. If latitude and distance from the ocean were the only factors, both regions would have similar climates. However, the Gobi is a barren desert and western Pennsylvania receives enough rainfall to support forests and rich farmland. The Gobi is bounded by the Himalayas to the south and the Altai and Tien Shan mountain ranges to the west, which shadow it from the prevailing winds. In contrast, winds carry abundant moisture from the Gulf of Mexico, the Great Lakes, and the Atlantic Ocean to western Pennsylvania.

Thus, in some regions, deserts extend to the seashore and in other regions the interior of a continent is humid. The climate at any particular place on Earth results from a combination of many factors. Latitude and proximity to the ocean are important, but complex interactions involving the direction of prevailing winds, the direction and temperature of ocean currents, and the positions of mountain ranges also control climate.

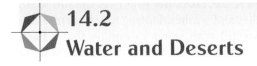

14.2
Water and Deserts

Although rain and snow rarely fall in deserts, water plays an important role in these dry environments. Thus the hydrosphere affects the geosphere in even the driest places on Earth. Water can reach a desert from three sources. Streams flow from adjacent mountains or other

Courtesy of Graham R. Thompson/Jonathan Turk

◆ **FIGURE 14.5** The Colorado River flows from the Rocky Mountains through the arid southwestern United States.

wetter regions, bringing surface water to some desert areas. Ground water may also flow from a wetter source to an aquifer beneath a desert. Finally, rain and snow fall occasionally on deserts.

Vegetation is sparse in most deserts because of the limited water supply. Thus much bare soil is exposed, unprotected from erosion. As a result, rain easily erodes desert soils and flowing water is an important factor in the evolution of desert landscapes.

Desert Streams

Large rivers flow through some deserts. For example, the Colorado River crosses the arid southwestern United States, and the Nile River flows through North African deserts (◆ Figure 14.5). Desert rivers receive most of their water from wetter, mountainous regions bordering the arid lands.

In a desert, the water table is commonly deep below a streambed, so that water seeps downward from the stream into the ground. As a result, many of the smaller desert streams flow for only a short time after a rainstorm or during the spring, when winter snows are melting. A streambed that is dry for most of the year is called a wash (◆ Figure 14.6).

Desert Lakes

While most lakes in wetter environments lie at the level of the water table and are fed, in part, by ground water, many desert lakes lie above the water table. During the wet season, rain and streams fill a desert lake. Some desert lakes are drained by outflowing streams, while many lose water only by evaporation and seepage. During the dry season, inflowing streams may dry up and evaporation and seepage may be so great that the lake dries up completely. An intermittent desert lake is called a **playa lake**, and the dry lake bed is called a **playa** (◆ Figure 14.7).

A

B

Courtesy of Graham R. Thompson/Jonathan Turk

◆ **FIGURE 14.6** Courthouse Wash, Utah **(A)** in the spring, when rain and melting snow fill the channel with water, and **(B)** in midsummer, when the creek bed is a dry wash.

Courtesy of Graham R. Thompson/Jonathan Turk

◆ **FIGURE 14.7** Mud cracks pattern the floor of a playa in Utah.

Recall from Chapter 11 that streams and ground water contain dissolved salts. When this slightly salty water fills a desert lake and then evaporates, the ions precipitate to deposit the salts on the playa. Over many years, economically valuable mineral deposits, such as those of Death Valley, may accumulate (◆ Figure 14.8).

Flash Floods

Bedrock or tightly compacted soil covers the surface of many deserts, and little vegetation is present to absorb moisture. As a result, rainwater runs over the surface to collect in gullies and washes. During a rainstorm, a dry streambed may fill with water so rapidly that a **flash flood** occurs. Occasionally, novices to desert camping pitch their tents in a wash, where they find soft, flat sand to sleep on and shelter from the wind. However, if a thunderstorm occurs upstream during the night, a flash flood may fill the wash with a wall of water mixed with rocks and boulders, creating disaster for the campers. By

◆ **FIGURE 14.8** Borax and other valuable minerals are abundant in the evaporite deposits of Death Valley. Mule teams hauled the ore from the valley in the 1800s.

◆ **FIGURE 14.9** In August 1997, 11 hikers perished when a flash flood filled a slot canyon similar to this one in the Utah desert. The steep walls made escape impossible.

midmorning of the next day, the wash may contain only a tiny trickle, and within 24 hours it may be completely dry again.

In August 1997, 11 hikers perished near Lake Powell in Utah when a flash flood filled a narrow, steep-walled canyon with a torrent of water, rocks, and mud. The flood was created by a thunderstorm that occurred 25 miles upstream and out of sight of the group (◆ Figure 14.9).

When rainfall is unusually heavy and prolonged, the desert soil itself may become saturated enough to create a mudflow that carries boulders and anything else in its path downslope. Some of the most expensive homes in Phoenix, Arizona, and other desert cities are built on alluvial fans and steep mountainsides, where they afford good views but are prone to mudflows during wet years.

Pediments and Bajadas

When a steep, flooding mountain stream empties into a flat valley, the water slows abruptly and deposits most of its sediment at the mountain front, forming an alluvial fan. Although fans form in all climates, they are particularly conspicuous in deserts (◆ Figure 14.10). A large fan may be several kilometers across and rise a few hundred meters above the surrounding valley floor.

If the mouths of several canyons are spaced only a few kilometers apart, the alluvial fans extending from each canyon may merge. A **bajada** is a broad, gently sloping depositional surface formed by merging alluvial fans and extending into the center of a desert valley. Typically, the fans merge incompletely, forming an undulating surface that may follow the mountain front for tens of kilometers. The sediment that forms the bajada may fill the valley to a depth of several thousand meters.

A **pediment** is a broad, gently sloping surface eroded into bedrock. Pediments commonly form along the front of desert mountains. The bedrock surface of a pediment is covered with a thin veneer of gravel that is in the

◆ **FIGURE 14.10** An alluvial fan forms where a steep mountain stream deposits sediment as it enters a valley. This photograph shows a fan in Death Valley.

INTERACTIVE QUESTION: *Would you expect the alluvial fan to grow continuously, month by month, or intermittently? Explain.*

◀▶ **FIGURE 14.11** The bajada in the foreground merges with a gently sloping pediment to form a continuous surface in front of mountains in Mongolia. This basin is filling with sediment from the surrounding mountains because it has no external drainage.

◀▶ **FIGURE 14.12** The Colorado Plateau and the Great Basin together make up a large desert and semiarid region of the western United States. The Colorado River flows through the Colorado Plateau to the Gulf of California, but no streams flow out of the Great Basin.

process of being transported from the mountains, across the pediment, to the bajada.

Together, a pediment and bajada form a smooth surface from the mountain front to the valley center (◇ Figure 14.11). The surface steepens slightly near the mountains, so it is concave. To distinguish a pediment from a bajada, you would have to dig or drill a hole. If you were on a pediment, you would strike bedrock after only a few meters, but on a bajada, bedrock may be buried beneath hundreds or even thousands of meters of gravel.

14.3
Two American Deserts

The Colorado Plateau

The Colorado Plateau covers a broad region encompassing portions of Utah, Colorado, Arizona, and New Mexico (◇ Figure 14.12). During the past one billion years of Earth's history, this region has been alternately covered by shallow seas, lakes, and deserts. Sediment accumulated, sedimentary rocks formed, and tectonic forces later uplifted the land to form the plateau. The Colorado River cut through the bedrock as the plateau rose, to form the 1.6-kilometer-deep Grand Canyon and its tributary canyons. The modern Colorado River receives most of its water from snowmelt and rains in the high Rocky Mountains east and north of the plateau, and the river then flows through the heart of the great desert to empty into the Gulf of California, in northern Mexico.

A stream forms a canyon by eroding downward into bedrock. If the downcutting stream reaches a resistant rock layer, it may erode laterally, widening the canyon. In some places on the Colorado Plateau, downcutting predominates and streams erode deep, narrow canyons. In other regions, lateral erosion undercuts canyon walls, and the rock collapses along vertical joints to form flat-topped mesas and buttes that rise above a relatively flat plain. The river eventually carries the sediment from the plateau to the Gulf of California.

The flat tops of mesas and buttes usually form on a bed of sandstone or other sedimentary rock that is relatively resistant to erosion. In many cases, extensive lateral erosion leaves spectacular pinnacles, isolated remnants of once-continuous rock layers that the streams have all but completely eroded away. A **plateau** is a large elevated area of fairly flat land. The term *plateau* is used for regions as large as the Colorado Plateau as well as for smaller, elevated flat surfaces. A **mesa** is smaller than a plateau and is a flat-topped mountain shaped like a table. A **butte** is also a flat-topped mountain characterized by steep cliff faces and is smaller and more tower-like than a mesa (◇ Figure 14.13). These landforms are common features of the Colorado Plateau.

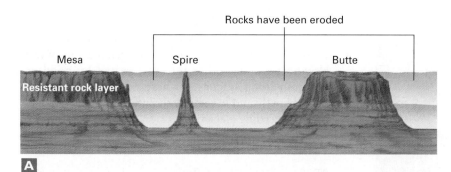

Rocks have been eroded

Mesa   Spire   Butte

Resistant rock layer

A

◇ **FIGURE 14.13** **(A)** Spires and buttes form when streams reach a temporary base level and erode laterally. The streams transport the eroded sediment away from the region. **(B)** Spires and buttes in Monument Valley, Arizona.

B

Courtesy of Graham R. Thompson/Jonathan Turk

Death Valley and the Great Basin

Death Valley lies in the rain shadow of the High Sierras in California. The deepest part of the valley is 82 meters below sea level. It is a classic rain-shadow desert, receiving a scant 5 centimeters of rainfall per year. The mountains to the west receive abundant moisture, and during the winter rainy season and spring snowmelt, streams flow from the mountains into the valley.

In the Colorado Plateau, the Colorado River and its tributaries carry sediment away from the desert to the Gulf of California, leaving deep canyons. In contrast, streams flow into Death Valley from the surrounding mountains, but no streams flow out. Because Death Valley has no external drainage, the valley is filling with sediment eroded from the surrounding mountains. The sediment collects to form vast alluvial fans and bajadas (◆ Figure 14.14). Stream water collects in broad playa lakes that dry up under the hot summer sun.

Death Valley is just one small part of a vast desert region in the American West that has no external drainage (◆ Figure 14.12). The Great Basin—which includes most of Nevada; the western half of Utah; and parts of California, Oregon, and Idaho—is a large desert region between the Sierra Nevada and the Rocky Mountains.

Although streams flow into the region from surrounding mountain ranges and from ranges within the basin, no streams flow out of the Great Basin. Consequently, sediment is not flushed out and has accumulated to become thousands of meters thick.

Frank M. Hanna/Visuals Unlimited

◆ **FIGURE 14.14** Sediment eroded from surrounding mountains is slowly filling Death Valley.

Bajadas and pediments are common features of the Great Basin. Both features result from a combination of tectonic, erosional, and depositional processes. The mountains and valleys initially formed by block faulting. The valleys are downdropped blocks of rock called grabens, and the mountain ranges are uplifted blocks called horsts (◆ Figure 14.15A). Grabens and horsts are discussed in Chapter 9. Desert streams partially eroded the mountains and deposited sediment to form pediments, alluvial fans, and bajadas (◆ Figure 14.15B).

Faulting continued to deepen the valleys, and at the same time, streams filled them with sediment. As a result, the mountains are slowly drowning in their own sediment (◆ Figure 14.15C).

14.4 Wind

Not only does water from the hydrosphere affect deserts, but moving air, of the atmosphere, also shapes and sculpts the landscape. When wind blows through a forest or across a prairie, the trees or grasses protect the soil from wind erosion. In addition, rain accompanies most windstorms in wet climates; the water dampens the soil and binds particles together. Therefore little wind erosion occurs. In contrast, a desert commonly has little or no vegetation and rainfall, so wind erodes bare, unprotected desert soil. Wind erosion is not limited to deserts, however. Wind is an important agent of erosion wherever the wind blows over unvegetated soil. Wind-blown dunes and other features created by wind are common along seacoasts, where salty sea spray limits plant growth, and in regions recently abandoned and left bare by receding glaciers.

Wind Erosion

Wind erosion, called **deflation**, is a selective process. Because air is much less dense than water, wind moves only small particles, mainly silt and sand. (Clay particles usually stick together, and, consequently, wind does not erode clay effectively.) Imagine bare soil containing silt, sand, pebbles, and cobbles. When wind blows, it removes only the silt and sand, leaving the pebbles and cobbles as a continuous cover of stones called **desert pavement** (◆ Figure 14.16). Desert pavement prevents the wind from eroding additional sand and silt, even though this finer sediment may be abundant beneath the layer of stones. As a result of this process, approximately 80 percent of the world's desert area is rocky and only 20 percent is covered by sand (◆ Figure 14.17).

Transport and Abrasion

Because sand grains are relatively heavy, wind rarely lifts sand more than 1 meter above the ground and carries it

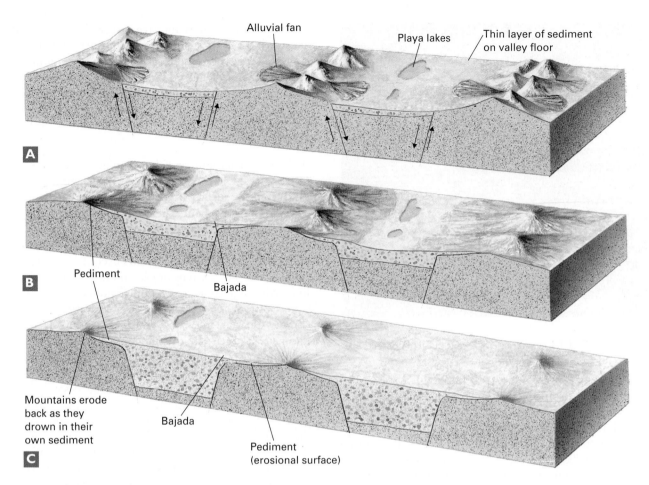

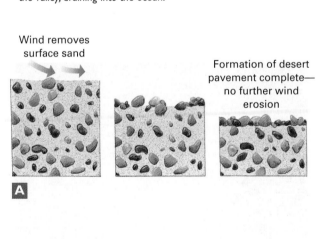

◆ FIGURE 14.15 A scenario for the formation of bajadas and pediments. **(A)** The mountains and valleys formed by block faulting. Desert streams deposited sediment to form alluvial fans. **(B)** As the streams eroded the mountains and deposited sediment in the valley, they formed both the erosional surface called a pediment, and a depositional surface called a bajada. **(C)** Today the mountains are drowning in their own sediment.

INTERACTIVE QUESTION: *Explain how the landscape would evolve if a stream flowed through the valley, draining into the ocean.*

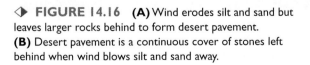

◆ FIGURE 14.16 **(A)** Wind erodes silt and sand but leaves larger rocks behind to form desert pavement. **(B)** Desert pavement is a continuous cover of stones left behind when wind blows silt and sand away.

Courtesy of Graham R. Thompson/Jonathan Turk

◆ **FIGURE 14.17** About 80 percent of Earth's deserts are covered by stony desert pavement.

only a short distance. In a windstorm, the sand grains bounce over the ground in a process called **saltation**. (Recall from Chapter 11 that sand also moves by saltation in a stream.) In contrast, wind carries fine silt in suspension. Skiers in the Alps commonly encounter a silty surface on the snow, blown from the Sahara Desert across the Mediterranean Sea.

◆ **FIGURE 14.18** Wind abrasion near Grand Canyon, Arizona, selectively eroded the base of this rock because wind-blown sand moves mostly near the surface.

Windblown sand is abrasive and erodes bedrock. Because wind carries sand close to the surface, wind erosion occurs near ground level. If the base of a pinnacle is sculpted as in ◆ Figure 14.18, wind may be the responsible agent, although salt cracking at ground level can also erode the base of a desert pinnacle.

Dunes

A **dune** is a mound or ridge of wind-deposited sand (◆ Figure 14.19). As explained earlier, wind removes sand from the surface in many deserts, leaving behind a rocky, desert pavement. The wind then deposits the sand in a topographic depression or other place where the wind slows down. Dunes commonly grow to heights of 30 to 100 meters, and some giants exceed 500 meters. In some places they are tens or even hundreds of kilometers long. Although some desert dune fields cover only a few square kilometers, the largest is the Rub Al Khali (Empty Quarter) in Arabia, which covers 560,000 square kilometers, larger than the state of California.

Dunes also form where glaciers have recently melted and along sandy coastlines. A glacier deposits large quantities of bare, unvegetated sediment. A sandy beach is commonly unvegetated because sea salt prevents plant growth. As a result, both of these environments contain the essentials for dune formation: an abundant supply of sand and a windy environment with sparse vegetation.

Dunes form when wind erodes sand from one location and deposits it nearby. A saucer or trough-shaped hollow formed by wind erosion is called a **blowout**. In the 1930s, intense, dry winds eroded large areas of the Great Plains and created the Dust Bowl. Deflation formed tens of thousands of blowouts, many of which remain today. Some are small, measuring only 1 meter deep and 2 or 3 meters across, but others are much larger (◆ Figure 14.20). One of the deepest blowouts in the world is the Qattara Depression in western Egypt. It is

◆ **FIGURE 14.19** Windblown sand formed these dunes near Lago Poopo, Bolivia.

FIGURE 14.20 Wind eroded sandy soil to create this blowout near Highway 2 in the Sand Hills of western Nebraska.

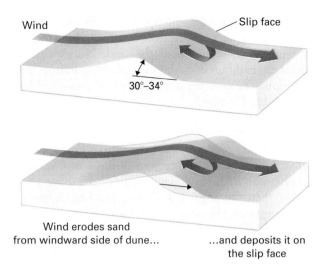

FIGURE 14.21 Sand dunes migrate in a downwind direction.

more than 100 meters deep and 10 kilometers in diameter. Ultimately, the lower limit for a blowout is the water table. If the bottom of the depression reaches moist soil near the water table, where water binds the sand grains, wind erosion is no longer effective.

Most dunes are asymmetrical. Wind erodes sand from the windward side of a dune, carries it up to the dune crest, and then the sand slides down the sheltered leeward side. In this way, dunes migrate in the downwind direction (◆ Figure 14.21). The leeward face of a dune is called the **slip face**. Typically, the slip face dips about 35 degrees from the horizontal, which is the angle of repose for sand as described in Chapter 10. That is about twice as steep as the windward face. In addition, the sand on a slip face is usually very loose, whereas the sand on the windward face is packed by the wind.

Migrating dunes overrun buildings and highways. For example, near the town of Winnemucca, Nevada, dunes advance across U.S. Highway 95 several times a year. Highway crews must remove as much as 4,000 cubic meters of sand to reopen the road. Engineers often attempt to stabilize dunes in inhabited areas. One method is to plant vegetation to reduce deflation and stop dune migration. The main problem with this approach is that desert dunes commonly form in regions that are too dry to support

vegetation. Another solution is to build artificial windbreaks to create dunes in places where they do the least harm. For example, a fence traps blowing sand and forms a dune, thereby protecting areas downwind. Fencing is a temporary solution, however, because eventually the dune covers the fence and resumes its migration. In Saudi Arabia, dunes are sometimes stabilized by covering them with tarry wastes from petroleum refining.

Fossil Dunes

When dunes are buried by younger sediment and lithified over geologic time, the resulting sandstone retains the original sedimentary structures of the dunes. ◆ Figure 14.22 shows a rock face in Zion National Park in Utah. The steep layering is not evidence of tectonic tilting but is

FIGURE 14.22 Cross-bedded sandstone in Zion National Park preserves the sedimentary bedding of ancient sand dunes.

the original, steeply dipping layering of the dune slip face. The beds dip in the direction in which the wind was blowing when it deposited the sand. Notice that the planes dip in different directions, indicating shifting wind direction. The layering is an example of **cross-bedding**, described in Chapter 3.

Types of Sand Dunes

Wind speed and sand supply control the shapes and orientation of dunes. **Barchan dunes** form in rocky deserts with little sand. The center of the dune grows higher than the edges (◆ Figure 14.23A). When the dune migrates, the edges move faster because there is less sand to transport (◆ Figure 14.23B). The resulting barchan dune is crescent shaped with its tips pointing downwind (◆ Figure 14.23C). Barchan dunes are not connected to one another but instead migrate independently. In a rocky desert, barchan dunes cover only a small portion of the land; the remainder is bedrock or desert pavement.

If sand is plentiful and evenly dispersed, it accumulates in long ridges called **transverse dunes** aligned perpendicular to the prevailing wind (◆ Figures 14.24A and B). If sparse desert vegetation is present, the wind may form a blowout in a bare area among the desert plants. As sand is carried out of the blowout, it accumulates in a parabolic dune, the tips of which are anchored by plants on each side of the blowout (◆ Figures 14.25A and B). A **parabolic dune** is similar in shape to a barchan dune, except that the tips of the parabolic dune point into the

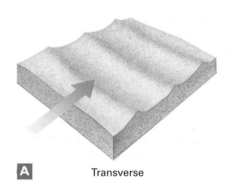

A   Transverse

Wind

Courtesy of Graham R. Thompson/Jonathan Turk

Martin G. Miller/Visuals Unlimited

◆ **FIGURE 14.23** (**A** and **B**) When sand supply is limited, the tips of a barchan dune travel faster than the center and point downwind. (**C**) A barchan dune in Coral Pinks, Utah.

◆ **FIGURE 14.24** (**A**) Transverse dunes form perpendicular to the prevailing wind direction in regions with abundant sand. (**B**) These transverse dunes formed in Death Valley, California.

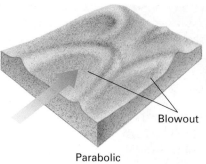

Parabolic

A

B

◆ **FIGURE 14.25** **(A)** A parabolic dune is crescent shaped with its tips upwind. It forms where wind blows sand from a blowout, and grass or shrubs anchor the dune tips. **(B)** Grass and shrubs anchor the tip of this parabolic dune in the Southern California desert.

Longitudinal

A

B

◆ **FIGURE 14.26** **(A)** Longitudinal dunes are long, straight dunes that form where the wind is erratic and sand supply is limited. **(B)** These longitudinal dunes formed on the Oregon coast.

wind. Parabolic dunes are common in moist semidesert regions and along seacoasts, where sparse vegetation grows in the sand.

If the wind direction is erratic but prevails from the same general quadrant of the compass and the supply of sand is limited, then long, straight **longitudinal dunes** form parallel to the prevailing wind direction (◆ Figures 14.26A and B). In portions of the Sahara Desert, longitudinal dunes reach 100 to 200 meters in height and are as much as 100 kilometers long.

Loess

Wind can carry silt for hundreds, or even thousands, of kilometers and then deposit it as **loess** (pronounced *luss*). Loess is porous, uniform, and typically lacks layering. Often the angular silt particles interlock. As a result, even though the loess is not cemented, it typically forms vertical cliffs and bluffs (◆ Figure 14.27).

The largest loess deposits in the world, found in central China, cover 800,000 square kilometers and are more than 300 meters thick. The silt was blown from

the Gobi and the Takla Makan deserts of central Asia. The particles interlock so effectively that people have dug caves into the loess cliffs to make their homes. However, in 1920 a great earthquake caused the cave system to collapse, burying and killing an estimated 100,000 people.

Large loess deposits accumulated in North America during the Pleistocene Ice Age, when continental ice sheets ground bedrock into silt. Streams carried this fine sediment from the melting glaciers and deposited it in vast plains. These zones were cold, windy, and devoid of vegetation, and wind easily picked up and transported the silt, depositing thick layers of loess as far south as Vicksburg, Mississippi. Residents of Vicksburg took shelter from federal bombardment in nearby loess caves during the Civil War's Battle of Vicksburg.

Loess deposits in the United States range from about 1.5 meters to 30 meters thick (◆ Figure 14.28). Soils formed on loess are generally fertile and make good farmland. Much of the rich soil of the central plains of the United States and eastern Washington State formed on loess.

◈ **FIGURE 14.27** Villagers in Askole, Pakistan, have dug caves in these vertical loess cliffs.

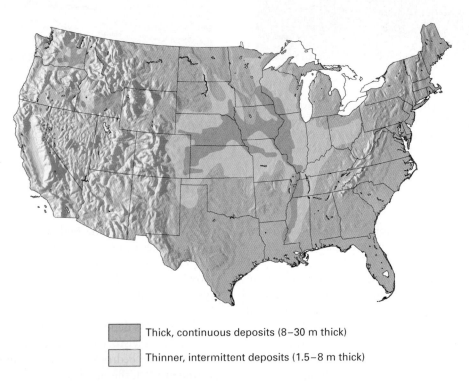

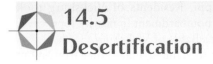 Thick, continuous deposits (8–30 m thick)

Thinner, intermittent deposits (1.5–8 m thick)

◈ **FIGURE 14.28** Loess deposits cover large areas of the United States.

◈ 14.5
Desertification

The Sahara is the largest desert on the planet. South of the Sahara lies the semiarid Sahel (◈ Figure 14.29). During the 1960s, unusually heavy rains caused the Sahel to bloom. People expanded their flocks to take advantage of the additional forage. Rich countries contributed foreign aid. As a result, medical attention and sanitation improved, and the human population grew dramatically. Many people predicted a new era of prosperity for the Sahel, but the favorable rains were an anomaly. In the late 1960s and early 1970s, drought destroyed the range. During this period, governments in North Africa began to enforce national borders more strictly, curtailing nomadism. When people settled in specific regions, their flocks grazed the same area throughout the year. Plants did not have time to regenerate, and the hungry animals chewed

◆ **FIGURE 14.29** The Sahara Desert and the semi-arid Sahel region dominate the ecosystems of Northern Africa.

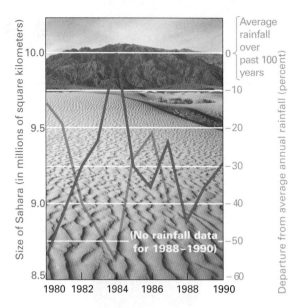

◆ **FIGURE 14.30** The Sahara Desert expanded and contracted between 1980 and 1990. Note that the Sahara expanded (blue line) when rainfall decreased (red line).

the grasses down to the roots. Civil and international war brought instability to the region, and famine struck.

Reports issued in the 1970s and 1980s claimed that the Sahara was expanding southward into the Sahel at a rate of 5 kilometers per year. Scientists argued that overgrazing, farming, and firewood gathering had caused the desert expansion. This growth of the desert caused by human mismanagement has been called **desertification**.

More recent research has shown that overgrazing a semiarid region causes land degradation but does not cause a desert to expand. The Sahel–Sahara desert boundary is clearly visible on satellite photographs as a boundary separating a region of sparse vegetation from one with almost no plants. Researchers plotted changes in the size of the desert by studying satellite photographs taken between 1980 and 1990. As shown in ◆ Figure 14.30, the desert expands (blue line) when rainfall declines (red line).[2] Thus, decreasing rainfall, not overgrazing, may have been responsible for expansion of the Sahara.

In another project, researchers studied natural and overgrazed range in semiarid grasslands in southern New Mexico.[3] The natural range consisted mainly of a homogeneous blanket of grass. In regions where cattle had overgrazed the land, woody creosote and mesquite had replaced the grass. The net primary productivities (the total plant growth) in the natural and overgrazed systems were nearly identical; thus, overgrazing did not reduce plant growth. It did, however, change the types of plants and their spatial distribution. The shrubs that invaded after overgrazing provided poorer forage for cattle and therefore reduced the economic value of the range.

Grasses have shallow roots that absorb falling rain. Excessive grazing destroys the grasses as the cattle eat the grass down to the roots. In addition, the heavy animals pack the soil with their hooves, blocking the natural seepage of air and water. Where soil is devoid of vegetation and baked in the sunlight, it becomes so impermeable that water evaporates or runs off before it soaks in. Increased runoff erodes the soil and carries off nutrients. The water, soil, and nutrients collect in shallow depressions and desert washes. In some regions, shrubs accumulate in these moist, nutrient-rich low points. Thus, the original range had a homogeneous distribution of plants, water, and nutrients, whereas the overgrazed range was heterogeneous, alternating thick shrubbery with bare ground.

With this background, let us return to the Sahel and consider changes that have occurred between 1990 and 2002. Increased rainfall has moisturized the soil and caused the desert to retreat again. But, in addition, massive agricultural aid to the region has led to improved farming and herding practices in some regions. For example, advisors from the Free University of Amsterdam have worked with villagers in the country of Burkina Faso, in the western Sahel. Farmers have been digging trenches between their fields and filling the trenches with manure to absorb water and attract dung-digesting termites that fertilize the soil. In addition, foreign aid has enabled people to construct stone barriers to slow surface-water runoff. As a result, thousands of acres of land have been rehabilitated and millet and sorghum yields have increased by 50 to 75 percent. This story reminds us once again that while a desert ecosystem is created by low rainfall, human management can significantly alter the productivity of the land.

2. William H. Schlesinger et al., "Biological Feedbacks in Global Desertification," *Science* 247 (March 1990): 1043.

3. Compton J. Tucker, Harold E. Dregne, and Wilbur W. Newcomb, "Expansion and Contraction of the Sahara Desert from 1980 to 1990," *Science* 253 (July 19, 1991): 299ff.

EARTH SYSTEMS INTERACTIONS

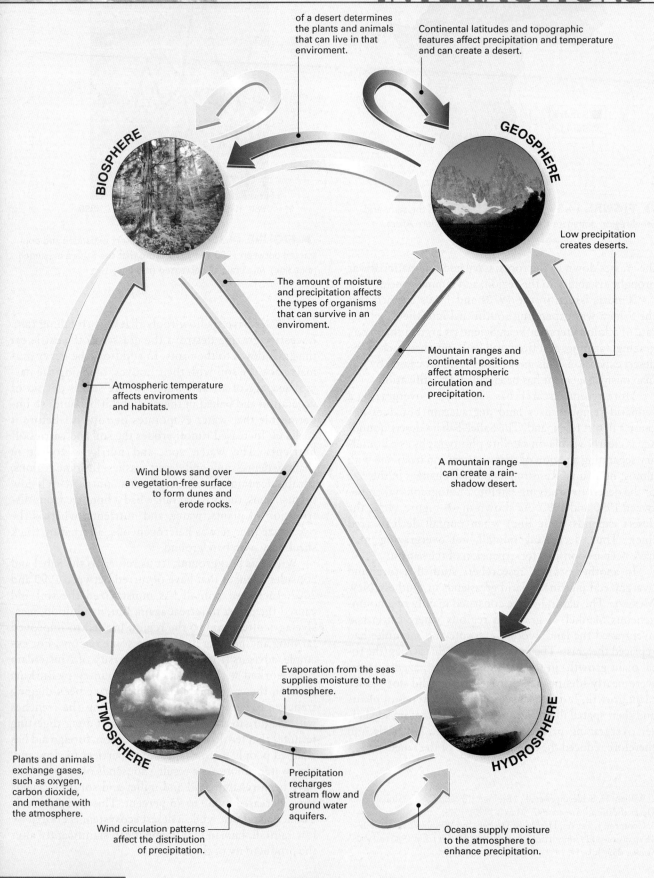

of a desert determines the plants and animals that can live in that enviroment.

Continental latitudes and topographic features affect precipitation and temperature and can create a desert.

BIOSPHERE

GEOSPHERE

Low precipitation creates deserts.

The amount of moisture and precipitation affects the types of organisms that can survive in an enviroment.

Mountain ranges and continental positions affect atmospheric circulation and precipitation.

Atmospheric temperature affects enviroments and habitats.

A mountain range can create a rain-shadow desert.

Wind blows sand over a vegetation-free surface to form dunes and erode rocks.

ATMOSPHERE

HYDROSPHERE

Evaporation from the seas supplies moisture to the atmosphere.

Plants and animals exchange gases, such as oxygen, carbon dioxide, and methane with the atmosphere.

Precipitation recharges stream flow and ground water aquifers.

Wind circulation patterns affect the distribution of precipitation.

Oceans supply moisture to the atmosphere to enhance precipitation.

DESERTS AND OTHER CLIMATE zones around the world are characterized principally by precipitation and temperature. Those two factors control environments and habitats of the biosphere. Precipitation and temperature, in turn, are products of solar radiation, latitude, proximity to the seas, atmospheric and oceanic circulation patterns, and regional topography. Thus, deserts and other regional climates are products of multiple interactions among all four of the Earth's systems: tectonic activity of the geosphere, atmospheric conditions and processes, oceanic circulation, and organisms of the biosphere. ■

EARTH SYSTEMS INTERACTIONS

SUMMARY

Deserts have an annual precipitation of less than 25 centimeters. The world's largest deserts occur near 30° north and south latitudes, where warm, dry, descending air absorbs moisture from the land. Deserts also occur in rain shadows of mountains, continental interiors, and coastal regions adjacent to cold ocean currents.

Desert streams are often dry for much of the year but may develop **flash floods** when rainfall occurs. **Playa lakes** are desert lakes that dry up periodically, leaving abandoned lake beds called **playas**. Alluvial fans are common in desert environments. A **bajada** is a broad depositional surface formed by merging alluvial fans. A **pediment** is a planar erosional surface that may lie at the base of a mountain front in arid and semiarid regions and that merges imperceptibly with a bajada.

The through-flowing Colorado River drains the Colorado Plateau desert. Thus, streams carry sediment away from the region, forming canyons and eroding the **plateaus** to form **mesas** and **buttes**. Death Valley and the Great Basin have no external drainage, and, as a result, the valleys are filling with sediment eroded from the surrounding mountains.

Deflation is erosion by wind. Silt and sand are removed selectively, leaving larger stones on the surface and creating **desert pavement**. Sand grains are relatively large and heavy and are carried only short distances by **saltation**, seldom rising more than a meter above the ground. Silt can be transported great distances at higher elevations. Wind erosion forms **blowouts**. Windblown particles are abrasive, but because the heaviest grains travel close to the surface, abrasion occurs mainly near ground level.

A **dune** is a mound or ridge of wind-deposited sand. Most dunes are asymmetrical, with gently sloping, windward sides and steeper **slip faces** on the lee sides. Dunes migrate. The various types of dunes include **barchan dunes**, **transverse dunes**, **longitudinal dunes**, and **parabolic dunes**. Wind-deposited silt is called **loess**.

Key Terms

For Review

1. Why are many deserts concentrated along zones at 30° latitude in both the Northern Hemisphere and the Southern Hemisphere?

2. List three conditions that produce deserts.

3. Why do flash floods and debris flows occur in deserts?

4. Why are alluvial fans more prominent in deserts than in humid environments?

5. Compare and contrast floods in deserts with those in more humid environments.

6. Compare and contrast pediments and bajadas.

7. Why is wind erosion more prominent in desert environments than it is in humid regions?

8. Describe the formation of desert pavement.

9. Describe the evolution and shape of a dune.

10. Describe the differences among barchan dunes, transverse dunes, parabolic dunes, and longitudinal dunes. Under what conditions does each type of dune form?

11. Compare and contrast desert plateaus, mesas, and buttes. Describe the formation of each.

12. Compare the effects of stream erosion and deposition in the Colorado Plateau and Death Valley.

For Discussion

1. Coastal regions boast some of the wettest and some of the driest environments on Earth. Briefly outline the climatological conditions that produce coastal rainforests versus coastal deserts.

2. Explain why soil moisture content might be more useful than total rainfall in defining a desert. How could one region have a higher soil moisture content and lower rainfall than another region?

3. Discuss two types of tectonic change that could produce deserts in previously humid environments.

4. Imagine that you live on a planet in a distant solar system. You have no prior information on the topography or climate of Earth and are designing an unmanned spacecraft to land on Earth. The spacecraft has arms that can reach out a few meters from the landing site to collect material for chemical analysis. It also has instruments to measure the immediate meteorological conditions and cameras that can focus on anything within a range of 100 meters.

The batteries on your radio transmitter have a life expectancy of 2 weeks. The spacecraft lands and you begin to receive data. What information will convince you that the spacecraft has landed in a desert?

5. Deserts are defined as areas with low rainfall, yet water is an active agent of erosion in desert landscapes. Explain this apparent contradiction.

6. Explain how and why many desert and semiarid parts of North America have become rich agricultural areas.

UNIT 4

THE OCEANS

CHAPTER
15
Ocean Basins

NOAA/National Geophysical Data Center

This image shows the Earth's ocean basins mapped with satellite altimetry. Sea floor features larger than 10 km are detected by gravitational distortion of sea surface.

15.1
The Origin of Oceans

The primordial Earth, heated by the impacts of colliding planetesimals and the decay of radioactive isotopes, was molten, or near molten. The sky, without an atmosphere, was black. There were no oceans—no life. Today, we consider Earth in terms of four spheres: the geosphere, hydrosphere, atmosphere, and biosphere. Each sphere is as different from the others as a rock is different from a flowing stream, a breath of air, or a butterfly.

But to understand Earth at the very beginning, we need another perspective. Rock and metal, which compose the geosphere, are nonvolatile, that is, they do not boil and become gases readily. In contrast, air, water, and living organisms are all composed of light, volatile compounds that boil or vaporize at relatively low temperatures. Air is a gas. Water readily becomes a gas by evaporating or boiling. At flame temperatures, the complex molecules in most living organisms break apart and the components evaporate as gases.

For the moment, let's abandon our view of Earth's four spheres and think of only two kinds of Earth materials: volatile substances and nonvolatile ones. Most scientists agree that the surface of primordial Earth contained few volatiles. How, then, did enough volatile compounds collect to form a thick atmosphere, vast oceans, and a global biosphere of living organisms?

For many years, geologists hypothesized that abundant volatiles, including water and carbon dioxide, were trapped within early Earth's interior. This reasoning was based on three observations and inferences. First, our cosmogenic models show that volatiles were evenly dispersed in the cloud of dust, gas, and planetesimals that coalesced to form the planets. It seemed likely that some of those volatiles would have become trapped within Earth as it formed. Second, scientists have detected volatiles in modern comets, meteoroids, and asteroids. If volatiles were trapped within the small objects that passed through our neighborhood in space, it seemed logical to infer that they also accumulated in Earth's interior as the original cloud of dust and gas coalesced. Finally, modern volcanic eruptions eject gases and water vapor into the air. Geologists inferred that these gases originate in the mantle and are remnants of the original volatiles trapped during Earth's formation. Geologists concluded that some of these volatiles escaped during volcanic eruptions early in Earth's history and that they formed the atmosphere, the oceans, and living organisms.

Today, many scientists question this conclusion. To understand their questions, let's return to the cloud of dust and gas that coalesced to form the planets. Recall that our region of space heated up as dust, gas, and planetesimals collided to become planets. At the same time, hydrogen fusion began within the Sun and solar energy radiated outward to heat the inner Solar System. The newly born Sun also emitted a stream of ions and electrons called the solar wind that swept across the inner planets, blowing their volatile compounds into outer regions of the Solar System. As a result, most of Earth's volatile compounds boiled off and were swept into the cold outer regions of the Solar System (◆ Figure 15.1).

According to a currently popular hypothesis, shortly after our planet formed and lost its volatiles, a Mars-sized object smashed into Earth. The cataclysmic impact blasted through the crust and deep into the mantle, ejecting huge quantities of pulverized rock into orbit. The fragments eventually coalesced to form the Moon. The impact also ejected most of Earth's remaining volatiles with enough velocity that they escaped Earth's gravity and disappeared into space. According to this hypothesis, Earth's surface then was left barren and rocky, with few volatiles either on the surface or in the deep mantle. Thus it had neither water nor an atmosphere. The hot mantle churned and volcanic eruptions repaved the surface with lava, but these events added few volatiles to Earth's surface. According to one estimate, outgassing of the deep mantle accounted for no more than 10 percent of Earth's hydrosphere, atmosphere, and biosphere.[1]

If this scenario is correct, why do modern volcanoes emit volatiles? According to one hypothesis, most of the gases given off by modern volcanoes are recycled from the surface. Water, carbon (in the form of carbonate rocks such as limestone), and other light compounds are carried into shallow parts of the mantle by subducting slabs. These volatiles return to the surface during volcanic eruptions. Therefore, modern volcanic eruptions, like their primordial ancestors, do not outgas appreciable quantities of volatiles from the deep mantle.

Now let's return to the volatiles that streamed away from the hot, inner Solar System. As they flew away from the Sun, they entered a cooler region beyond Mars. Most of the volatiles were captured by the outer planets—Jupiter, Saturn, Uranus, and Neptune—but some continued their journey toward the outer fringe of the Solar System. Here, beyond the orbits of the known planets, volatiles from the inner Solar System combined with residual dust and gas to form comets. A comet's nucleus has been compared to a dirty snowball because it is composed mainly of ice and rock. Other compounds not

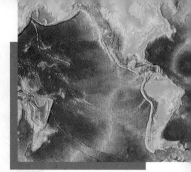

ThomsonNOW™ Throughout this chapter, the ThomsonNOW logo indicates an opportunity for online self-study, which:
- Assesses your understanding of important concepts and provides a personalized study plan
- Links you to animations and simulations to help you study
- Helps to test your knowledge of material and prepare for exams

Visit ThomsonNOW at http://www.thomsonedu.com/login to access these resources.

---

1. Paul Thomas, Christopher Chyba, and Christopher McKay, *Comets and the Origin of Life* (New York: Springer-Verlag, 1997).

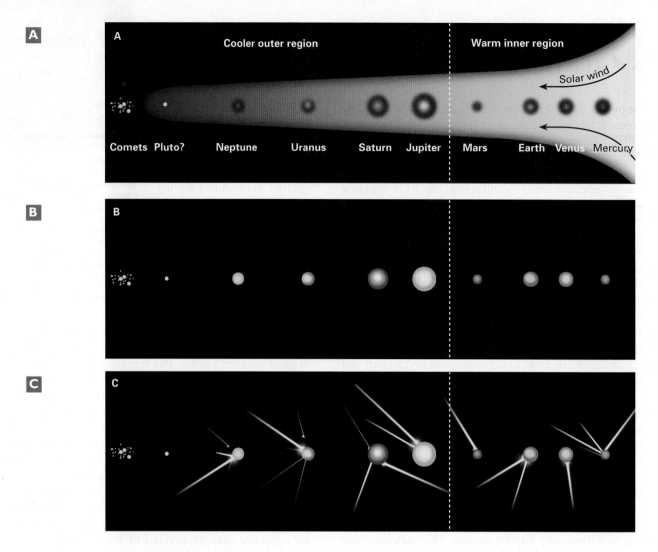

◇ **FIGURE 15.1** **(A)** As the Solar System formed, volatiles boiled away from the inner planets. The solar wind then blew them into the cool region beyond Mars. The outer planets captured some of the volatiles, and some condensed to form comets in the frigid zone beyond Neptune. **(B)** As a result, the inner four planets were left with no atmospheres or oceans. The outer planets grew to become giants. (Pluto is an anomalous planet, and its origin is discussed in Chapter 23.) **(C)** Comets and meteorites crashed into Earth and other planets, returning some of the volatiles to their surfaces. The volatiles accumulated to form the atmosphere, oceans, and the foundation of life on Earth. (Sizes and distances in this drawing are not to scale.)

common in snowballs exist in comets as well. These include frozen carbon dioxide, ammonia, and simple organic molecules. Volatiles are also abundant in certain types of meteoroids and asteroids in the region between Mars and Jupiter.

Astronomers calculate that the early Solar System was crowded with comets, meteoroids, and asteroids—space debris left over from planetary formation (◇ Figure 15.2). Many contained volatile compounds. When a large piece of space debris crashes into a planet, it is called a **bolide**. A large number of bolides crashed into Earth, nearby planets, and moons (◇ Figure 15.1C). While falling space debris added only one thousandth of a percent to Earth's total mass, it imported 90 percent of its modern reservoir of volatiles.

Upon entry and impact, the frozen volatiles in the bolides vaporized, releasing water vapor, carbon dioxide, ammonia, simple organic molecules, and other volatiles.

As the planet cooled and atmospheric pressure increased, the water vapor condensed to liquid, forming the first oceans.

The light molecules transported to Earth in bolides also provided gases that formed the atmosphere and the raw materials for life. (The formation and evolution of the atmosphere is discussed in Chapter 17.)

Thus, at least some, and probably most, of the compounds needed to produce the hydrosphere, the atmosphere, and the biosphere traveled to Earth from outer regions of the Solar System. The water that fills Earth's oceans came from interplanetary space.

Later in Earth's history, impacts from outer space blasted rock and dust into the sky, causing mass extinctions and killing large portions of life on Earth. Thus extraterrestrial impacts may have provided the raw materials for the oceans, the atmosphere, and for life, and later caused mass extinctions.

◆ FIGURE 15.2 Halley's Comet. The early Solar System was crowded with comets, meteoroids, and asteroids.

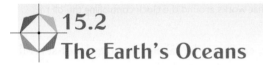

15.2
The Earth's Oceans

If you were to ask most people to describe the difference between a continent and an ocean, they would almost certainly reply, "Why, obviously, a continent is land and an ocean is water!" This observation is true, of course, but to a geologist another distinction is more important. The geologist would explain that rocks beneath the oceans are different from those of a continent. The accumulation of seawater in the world's ocean basins is a *result* of that difference.

Modern oceanic crust is dense basalt and varies from 4 to 7 kilometers thick. Continental crust is made of lower-density granite and averages 20 to 40 kilometers thick. In addition, the entire continental lithosphere is both thicker and less dense than oceanic lithosphere. As a result of these differences, the thick, lower-density continental lithosphere floats isostatically at high elevations, whereas oceanic lithosphere sinks to low elevations. Most of Earth's water flows downhill to collect in the depressions formed by oceanic lithosphere. Even if no water existed on Earth's surface, oceanic crust would form deep basins and continental crust would rise to higher elevations.

Oceans cover about 71 percent of Earth's surface. The sea floor is about 5 kilometers deep in the central parts of the ocean basins, although it is only 2 to 3 kilometers deep above the Mid-Oceanic Ridge and plunges to 11 kilometers in the Mariana trench (◆ Figure 15.3).

The ocean basins contain 1.4 billion cubic kilometers of water—18 times more than the volume of all land above sea level. So much water exists at Earth's surface

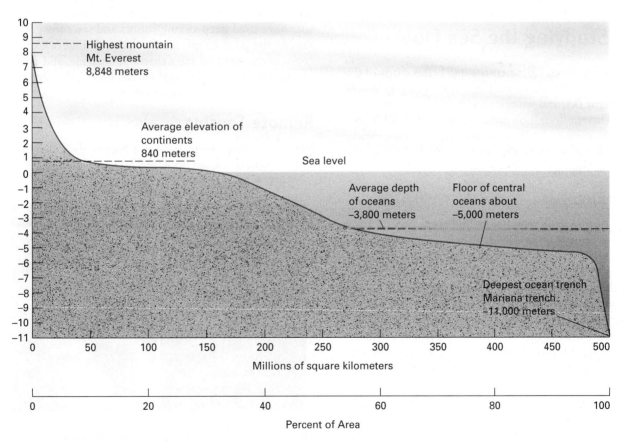

◆ FIGURE 15.3 A schematic cross-section of the continents and ocean basins. The vertical axis shows elevations relative to sea level. The horizontal axis shows the relative areas of the types of topography. Thus, 30 percent, or about 150 million square kilometers, of Earth's surface lies above sea level.

that if Earth were a perfectly smooth sphere, it would be covered by a global ocean 2,000 meters deep.

The size and shape of Earth's ocean basins change over geologic time. At present, the Atlantic Ocean is growing wider at a rate of a few centimeters each year as the sea floor spreads apart at the Mid-Atlantic Ridge and as the Americas move away from Europe and Africa. At the same time, the Pacific is shrinking at a similar rate, as oceanic crust sinks into subduction zones around its edges. In short, the Atlantic Ocean basin is now expanding at the expense of the Pacific.

The oceans (hydrosphere) affect global climate (atmosphere) and the biosphere in many ways. The seas absorb and store solar heat more efficiently than do rocks and soil. As a result, oceans are generally warmer in winter and cooler in summer than adjacent land is. Most of the water that falls as rain or snow is water that evaporated from the seas. In addition, ocean currents transport heat from the equator toward the poles, cooling equatorial climates and warming polar environments. Because plate tectonic activities alter the sizes and shapes of ocean basins, they also alter oceanic currents and profoundly affect regional climates over geologic time. In these and other ways, the oceans play a large role in Earth systems interactions.

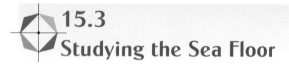

15.3
Studying the Sea Floor

Seventy-five years ago, scientists had better maps of the Moon than of the sea floor. The Moon is clearly visible in the night sky, and we can view its surface with a telescope. The sea floor, however, is deep, dark, and inhospitable to humans. Modern oceanographers use a variety of techniques to study the sea floor, including several types of sampling and remote sensing.

Sampling

Several devices collect sediment and rock directly from the ocean floor. A **rock dredge** is an open-mouthed steel net dragged along the sea floor behind a research ship. The dredge breaks rocks from submarine outcrops and hauls them to the surface. Oceanographers sample sea-floor mud by lowering a weighted, hollow steel pipe from a research vessel. The weight drives the pipe into the soft sediment, which is forced into the pipe. The sediment **core** is retrieved from the pipe after it is winched back to the surface. If the core is removed from the pipe carefully, even the most delicate sedimentary layering is preserved.

Sea-floor drilling methods developed for oil exploration also take core samples from oceanic crust. Large drill rigs are mounted on offshore platforms and on research vessels. The drill cuts cylindrical cores from both sediment and rock, which are then brought to the surface for study (◈ Figure 15.4). Although this type of sampling

◈ **FIGURE 15.4** The brown pipe in the center of the photo is the drill stem turning on the Discoverer Deep Seas drillship as Chevron drills for oil off the coast of Louisiana in the Gulf of Mexico in 2006. Nearly three football fields long, the ship holds a 200-person crew that works around the clock controlling the oil rig.

is expensive, cores can be taken from depths of several kilometers into oceanic crust.

A number of countries, including France, Japan, Russia, and the United States, have built small research submarines to carry oceanographers to the sea floor, where they view, photograph, and sample sea-floor rocks, sediment, and deep sea life. More recently, scientists have used deep-diving robots and laser imagers to sample and photograph the sea floor (◈ Figure 15.5). A robot is cheaper and safer than a submarine, and a laser imager penetrates up to eight times farther through water than a conventional camera does.

Remote Sensing

Remote sensing methods do not require direct physical contact with the ocean floor, and for some studies this ap-

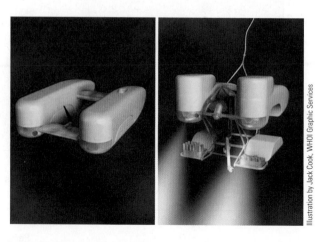

◈ **FIGURE 15.5** The deep sea exploration robot HROV can operate as a free-swimming vehicle(left) and as a tethered vehicle. In each of the two modes the HROV can spend up to 36 hours working in the deepest parts of the oceans.

proach is both effective and economical (◆ Figure 15.6). The **echo sounder** is commonly used to map sea-floor topography. It emits a sound signal from a research ship and then records the signal after it bounces off the sea floor and travels back up to the ship. The water depth is calculated from the time required for the sound to make the round trip. A topographic map of the sea floor is constructed as the ship steers a carefully navigated course with the echo sounder operating continuously. Modern echo sounders, called SONAR, transmit 1,000 signals at a time to create more complete and accurate maps.

The **seismic profiler** works in the same way but uses a higher-energy signal that penetrates and reflects from layers in the sediment and rock. This gives a picture of the layering and structure of oceanic crust, as well as the sea-floor topography (◆ Figure 15.7).

A **magnetometer** is an instrument that measures a magnetic field. Magnetometers towed behind research ships measure the magnetism of sea-floor rocks. Data collected by ship-borne magnetometers resulted in the now-famous discovery of symmetric magnetic stripes on the sea floor. That discovery rapidly led to the development of the

sea-floor-spreading hypothesis and of the theory of plate tectonics shortly thereafter, as described in Chapter 6.

Satellite-based **microwave radar** instruments have recently been used to measure subtle swells and depressions on the sea surface. These features reflect sea-floor topography. For example, the mass of a sea-floor mountain 4,000 meters high creates sufficient gravitational attraction to create a gentle, 6-meter-high swell on the sea surface directly above it. This technique is used to make modern sea-floor maps.

Scientists are currently working on sensors and robots that will be able to monitor the oceans in real time. One project, called Argo, is a set of 3,000 floating sensors distributed across the oceans to measure daily patterns of salinity and temperature as part of an effort to determine the nature and pace of global climate change. In 2007 the Hybrid Remotely Operated Vehicle (HROV) will begin collecting images and samples from the deepest parts of the sea floor. A joint project between the United States and Canada, called NEPTUNE (North East Pacific Time Series Undersea Networked Experiments), is in the process of installing a 200,000-square-mile network of fiber optic cables that will monitor seismic activity, seabed geology and chemistry, ocean climate change, and deep-sea ecosystems.

◆ 15.4 Features of the Sea Floor

The Mid-Oceanic Ridge System

In the early-sixteenth century, a few explorers, lowering hand lines to measure the depth of the oceans, found that the seas vary considerably in depth. However, despite limited data to the contrary, even into the early-nineteenth century, most people thought that the ocean floor was flat and featureless and had been unchanged since Earth formed.

Systematic ocean floor surveys began in the nineteenth century, when navigators routinely made line soundings in the Atlantic and Caribbean. In 1855, a map of the central Atlantic sea floor published by the U.S. Navy depicted a submarine mountain chain called Middle Ground. The existence of Middle Ground was later confirmed by survey ships laying trans-Atlantic telegraph cables between the late 1850s and 1900.

Following World War I (1914–18), oceanographers began using early versions of echo-sounding devices to measure ocean depths. Those surveys showed that the sea floor was much more rugged than previously thought, and the surveys further identified the continuity and size of Middle Ground, which is now called the Mid-Atlantic Ridge.

During World War II, naval commanders needed topographic maps of the sea floor to support submarine warfare. Those detailed maps, made with early versions

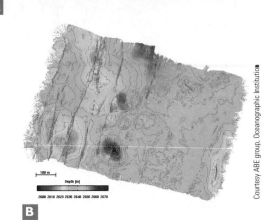

◆ **FIGURE 15.6** **(A)** Crew members recover the deep sea remote sensor ABE aboard R/V Atlantis after a dive to detect hydrothermal vents on the Galápagos Rift in 2002. **(B)** ABE can image the seafloor in several ways on a variety of scales. This bathymetric map from the Lau Basin near Fiji covers an area of 1 by 0.6 kilometers, showing mounds, fissures, and hydrothermal vent spires with a resolution of about 1 meter.

A

Marie Tharp

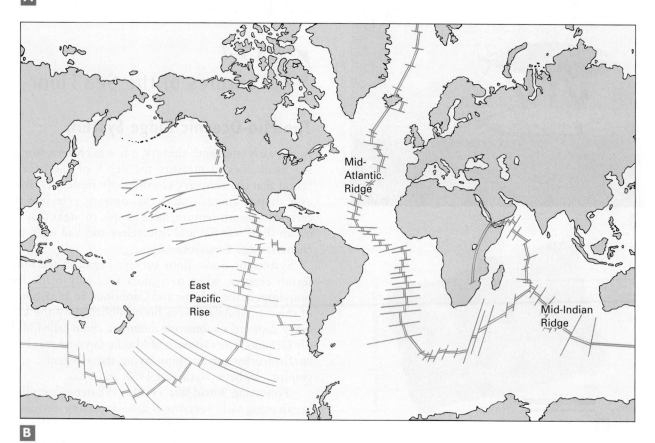

B

◇ **FIGURE 15.7** **(A)** An artist's rendition of sea-floor topography based on sonar and seismic profiling. **(B)** This map shows that divergent plate boundaries, or spreading centers, coincide exactly with the Mid-Oceanic Ridge system in the world's oceans. The spreading centers are shown in double red lines; the single red lines are transform faults.

of the echo sounder, were kept secret by the military. When they became available to the public after peace was restored, scientists were surprised to learn that the ocean floor has at least as much topographic diversity and relief as the continents (◆ Figure 15.7). Broad plains, high peaks, and deep valleys form a varied and fascinating submarine landscape. In the 1950s, oceanographic surveys conducted by several nations led to the discovery that Middle Ground, or the Mid-Atlantic Ridge, is just part of a great submarine mountain range, now called the **Mid-Oceanic Ridge** system.

The Mid-Oceanic Ridge system is a continuous submarine mountain chain that encircles the globe. Its total length exceeds 80,000 kilometers, and it is more than 1,500 kilometers wide in places. The ridge rises an average of 2 to 3 kilometers above the surrounding deep sea floor. Although it lies almost exclusively beneath the seas, it is Earth's largest mountain chain, covering more than 20 percent of Earth's surface, about two-thirds as much as all continents combined. Even the Himalayas, Earth's largest continental mountain chain, occupy only a small fraction of that area.

A **rift valley** 1 to 2 kilometers deep and several kilometers wide splits many segments of the ridge crest. Oceanographers use small research submarines to dive into the rift valley. They see gaping vertical cracks up to 3 meters wide on the floor of the valley. Recall that the Mid-Oceanic Ridge system is a spreading center, where two lithospheric plates are spreading apart from each other. The cracks form as brittle oceanic crust separates at the ridge axis. Basaltic magma then rises through the cracks and flows onto the floor of the rift valley. This basalt becomes new oceanic crust as two lithospheric plates spread outward from the ridge axis.

The new crust (and the underlying lithosphere) at the ridge axis is warmer, and therefore of relatively lower density, than older crust and lithosphere located farther from the ridge axis. Its buoyancy causes it to float high above the surrounding sea floor, elevating the Mid-Oceanic Ridge system 2 to 3 kilometers above the deep sea floor. The new lithosphere cools as it spreads away from the ridge. As a result of cooling, it becomes thicker and denser and sinks to lower elevations, forming the deeper sea floor on both sides of the ridge (◆ Figure 15.8).

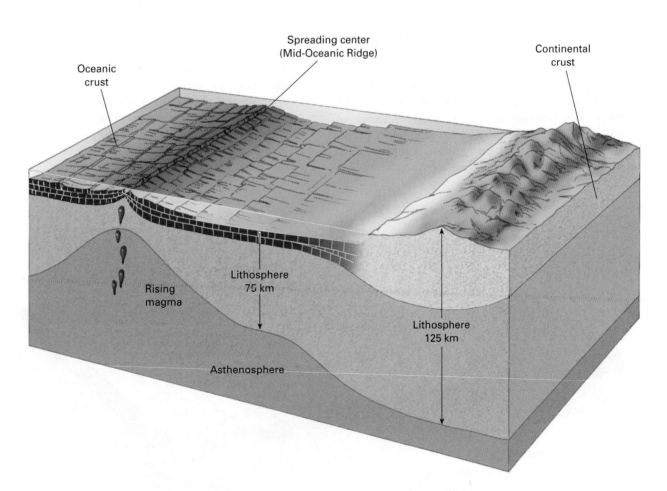

◆ **FIGURE 15.8** The sea floor sinks as it grows older. At the Mid-Oceanic Ridge, new lithosphere is buoyant because it is hot and of low density. It ages, cools, thickens, and becomes denser as it moves away from the ridge and consequently sinks. The central portion of the sea floor lies at a depth of about 5 kilometers.

INTERACTIVE QUESTION: *Why does the lithosphere thicken as it becomes older and cooler and migrates from the Mid-Oceanic Ridge?*

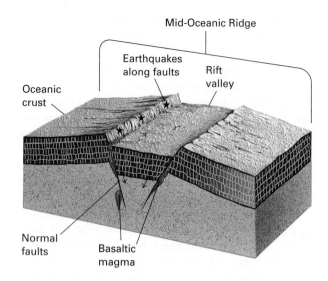

◈ FIGURE 15.9 A cross-sectional view of the central rift valley in the Mid-Oceanic Ridge. As the plates separate, blocks of rock drop down along the fractures to form the rift valley. The moving blocks cause earthquakes.

Normal faults and shallow earthquakes are common along the Mid-Oceanic Ridge system because oceanic crust fractures as the two plates separate (◈ Figure 15.9). Blocks of crust drop downward along the sea-floor cracks, forming the rift valley.

Hundreds of fractures, called **transform faults,** cut across the rift valley and the ridge (◈ Figure 15.10). These fractures extend through the entire thickness of the lithosphere. They develop because the Mid-Oceanic Ridge system consists of many short segments. Each segment is slightly offset from adjacent segments by a

transform fault. Transform faults are original features of the Mid-Oceanic Ridge; they form when lithospheric spreading begins.

Some transform faults displace the ridge by less than a kilometer, but others offset the ridge by hundreds of kilometers. In some cases, a transform fault can grow so large that it forms a transform plate boundary. The San Andreas Fault in California is a transform plate boundary.

Global Sea-Level Changes and the Mid-Oceanic Ridge System

A thin layer of marine sedimentary rocks blankets large areas of Earth's continents. These rocks tell us that those places must have been below sea level when the sediment accumulated.

Tectonic activity can cause a continent to sink, allowing the sea to flood a large area. However, at particular times in the past (most notably during the Cambrian, Carboniferous, and Cretaceous Periods), marine sediments accumulated on low-lying portions of all continents simultaneously, indicating simultaneous global flooding of low parts of all continents.

Although our plate tectonics model explains the sinking of individual continents, or parts of continents, it does not explain why all continents should sink at the same time. Therefore, we need to explain how sea level could rise globally by hundreds of meters to flood all continents simultaneously.

Continental glaciers have advanced and melted numerous times in Earth's history. During the growth of continental glaciers, seawater evaporates and is frozen into the ice that rests on land. As a result, sea level drops. When glaciers melt, the water runs back into the

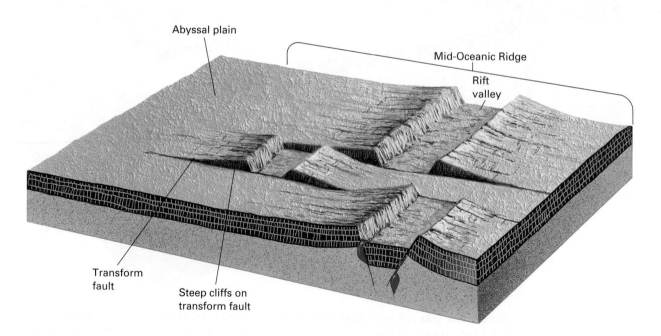

◈ FIGURE 15.10 Transform faults offset segments of the Mid-Oceanic Ridge. Adjacent segments of the ridge may be separated by steep cliffs 3 kilometers high. Note the flat abyssal plain far from the ridge.

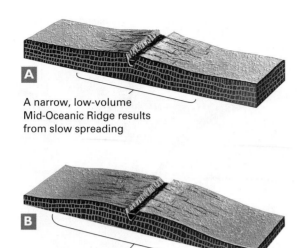

A narrow, low-volume
Mid-Oceanic Ridge results
from slow spreading

A broad, high-volume
Mid-Oceanic Ridge results
from rapid spreading

◆ **FIGURE 15.11** **(A)** Slow sea-floor spreading creates a narrow, low-volume Mid-Oceanic Ridge that displaces less seawater and lowers sea level. **(B)** Rapid sea-floor spreading creates a wide, high-volume ridge that displaces more seawater and raises sea level.

INTERACTIVE QUESTION: *What processes, other than variations in sea-floor spreading rates and growth and melting of glaciers, might cause global sea-level fluctuations? What evidence would indicate that other processes caused the fluctuations?*

oceans and sea level rises. The alternating growth and melting of glaciers during the Pleistocene Epoch caused sea level to fluctuate by as much as 200 meters. However, the ages of most marine sedimentary rocks on continents do not coincide with times of glacial melting. Therefore, we must look for a different cause to explain continental flooding.

Recall that the new, hot lithosphere at a spreading center is buoyant, causing the Mid-Oceanic Ridge system to rise above the surrounding sea floor. This submarine mountain chain displaces a huge volume of seawater. If the Mid-Oceanic Ridge system were smaller, it would displace less seawater and sea level would fall. If it were larger, sea level would rise.

The Mid-Oceanic Ridge rises highest at the spreading center, where new lithosphere rock is hottest and has the lowest density. The elevation of the ridge decreases on both sides of the spreading center because the lithosphere cools and shrinks as it moves outward.

Now consider a spreading center where spreading is very slow (perhaps 1 to 2 centimeters per year). At such a slow rate, the newly formed lithosphere would cool before it migrated far from the spreading center. As a result, the ridge would be narrow and of low volume, as shown in ◆ Figure 15.11A. In contrast, rapid sea-floor spreading of 10 to 20 centimeters per year would create a high-volume ridge because the newly formed, hot lithosphere would be carried a considerable distance away from the spreading center before it cooled and

shrank (◆ Figure 15.11B). This high-volume ridge would displace considerably more seawater than a low-volume ridge would displace, and the high-volume ridge would cause a global sea-level rise. If the modern Mid-Oceanic Ridge system were to disappear completely, sea level would fall by about 400 meters.

Sea-floor age data indicate that the rate of sea-floor spreading has varied from about 2 to 16 centimeters per year since Jurassic time, about 200 million years ago. Sea-floor spreading was unusually rapid during Late Cretaceous time, between 110 and 85 million years ago. That rapid spreading should have formed an unusually high-volumed Mid-Oceanic Ridge and resulted in flooding of low-lying portions of continents. Geologists have found marine sedimentary rocks of Late Cretaceous age on nearly all continents, indicating that Late Cretaceous time was, in fact, a time of abnormally high global sea level. Thus, a process initiated by heat transfer deep within the geosphere profoundly affected sea level of the hydrosphere and life throughout the biosphere. Unfortunately, because no oceanic crust is older than about 200 million years, the hypothesis cannot be tested for earlier times when extensive marine sedimentary rocks accumulated on continents.

Life on the Mid-Oceanic Ridge System

Oceanographers had long thought that little life could exist on the deep sea floor because no sunlight penetrates to those depths to support photosynthesis. However, scientists using modern diving and sampling techniques have discovered thriving communities and a unique food chain in isolated parts of the deep sea floor.

On the volcanically active Mid-Oceanic Ridge system, the hot rocks heat seawater as it circulates through fractures in oceanic crust. The hot water dissolves metals and sulfur from the rocks. Eventually, the hot, metal-and-sulfur-laden water rises back to the sea-floor surface, spouting from fractures as a jet of black water called a **black smoker**. The black color is caused by precipitation of fine-grained metal sulfide minerals as the solutions cool on contact with seawater (◆ Figure 15.12).

These scalding, sulfurous waters are as hot as 400°C and would be toxic to life on land. Yet the deep-sea floor around a black smoker teems with life (◆ Figure 15.13). At the vents, bacteria produce energy from hydrogen sulfide in a process called **chemosynthesis**. Thus, the bacteria release energy from chemicals and are not dependent on photosynthesis. The chemosynthetic bacteria are the foundation of a deep-sea food chain—either larger vent organisms eat them or the larger organisms live symbiotically with the bacteria. For example, instead of a digestive tract, the red-tipped tube worm has a special organ that hosts the chemosynthetic bacteria. It provides a home for the bacteria and in return receives nutrition from the bacteria's wastes. Other vent organisms in this unique food chain include giant clams and mussels, eyeless shrimp, crabs, and fish.

◆ FIGURE 15.12 A black smoker spouts from the East Pacific rise. Seawater is heated as it circulates through hot sea-floor rocks, and it dissolves metals and sulfur from the rocks. The ions precipitate as "smoke," consisting of tiny mineral grains, when the hot solution spouts into cold ocean water.

Oceanic Trenches and Island Arcs

In many parts of the Pacific Ocean and in some other ocean basins, two oceanic plates converge. One dives beneath the other, forming a subduction zone. The sinking plate drags the sea floor downward, forming a long, narrow depression called an **oceanic trench**. The deepest place on Earth is in the Mariana trench, north of New Guinea in the southwestern Pacific, where the ocean floor sinks to nearly 11 kilometers below sea level. Depths of 8 to 10 kilometers are common in other trenches.

◆ FIGURE 15.13 These red tubeworms are part of a thriving plant and animal community living near a black smoker in the Guaymas Basin in the Gulf of California.

INTERACTIVE QUESTION: *Discuss similarities and differences between chemosynthesis and photosynthesis.*

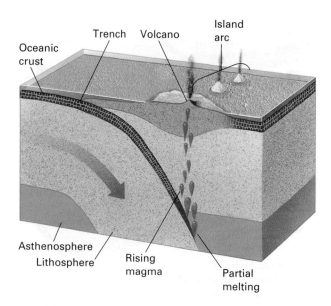

◆ FIGURE 15.14 An oceanic trench forms at a convergent boundary between two oceanic plates. One of the plates sinks, generating magma that rises to form a chain of volcanic islands called an island arc.

INTERACTIVE QUESTION: *Why is the rock that comprises an island arc generally different from (and less dense than) the surrounding oceanic crust?*

Huge amounts of magma are generated in the subduction zone. The magma rises and erupts on the sea floor to form submarine volcanoes next to the trench. The volcanoes eventually grow to become a chain of islands, called an **island arc** (◆ Figure 15.14). The western Aleutian Islands are an example of an island arc. Many others occur at the numerous convergent plate boundaries in the western Pacific (◆ Figure 15.15).

If subduction stops after an island arc forms, volcanic activity also ends. The island arc may then ride

◆ FIGURE 15.15 The Lesser Sunda Islands are part of the Indonesian Island arc.

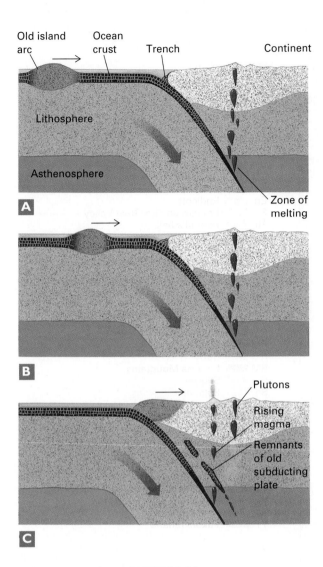

ThomsonNOW ◁▶ **FIGURE 15.16** **(A)** An island arc is part of a lithospheric plate that is sinking into a subduction zone beneath a continent. **(B)** The island arc reaches the subduction zone but cannot sink into the mantle because of its low density. **(C)** The island arc is jammed onto the continental margin and becomes part of the continent. The subduction zone and trench step back to the seaward side of the island arc.

quietly on a tectonic plate until it arrives at another subduction zone at a continental margin. However, the density of island arc rocks is relatively low, making them too buoyant to sink into the mantle. Instead, the island arc collides with the continent (◁▶ Figure 15.16A and B). When this happens, the subducting plate commonly fractures on the seaward side of the island arc to form a new subduction zone. In this way, the island arc breaks away from the ocean plate and becomes part of the continent (◁▶ Figure 15.16C). Much of western California, Oregon, Washington, and western British Columbia were added to North America in this way from 180 million to about 50 million years ago. These late additions to our continent, called **accreted terranes,** are shown in ◁▶ Figure 15.17.

Note the following points:

1. An island arc forms as magma rises from the mantle at an oceanic subduction zone.
2. The island arc eventually migrates toward the edge of a continent and becomes part of it.
3. The continent with the added island arc cannot sink into the mantle at a subduction zone because of its buoyancy.
4. Thus, material is transferred from the mantle to a continent. This aspect of the plate tectonics model suggests that the amount of continental crust has increased throughout geologic time. However, some geologists feel that small amounts of continental crust can also return to the mantle in subduction zones, and that the total amount of continental crust has been approximately constant for the past 2.5 billion years.

Seamounts, Oceanic Islands, and Atolls

A **seamount** is a submarine mountain that rises 1 kilometer or more above the surrounding sea floor. An **oceanic island** is a seamount that rises above sea level. Both are common in all ocean basins but are particularly abundant in the southwestern Pacific Ocean. Seamounts and oceanic islands sometimes occur as isolated peaks on the sea floor, but they are more commonly found in chains. Dredge samples show that seamounts, oceanic islands, and the ocean floor itself are all made of basalt.

Most seamounts and oceanic islands are volcanoes that formed at a hot spot above a mantle plume, and most form within a tectonic plate rather than at a plate boundary. An isolated seamount or short chain of small seamounts probably formed over a plume that lasted for only a short time. In contrast, a long chain of large islands, such as the Hawaiian Island–Emperor Seamount chain, formed over a long-lasting plume. In this case the lithospheric plate migrated over the plume as the magma continued to rise from a source beneath the lithosphere. Each volcano formed directly over the plume and then became extinct as the moving plate carried it away from the plume. As a result, the seamounts and oceanic islands become progressively younger toward the end of the chain that is volcanically active today (◁▶ Figure 15.18).

After a volcanic island forms, it begins to sink. Three factors contribute to the sinking:

1. If the mantle plume stops rising, it stops producing magma. Then the lithosphere beneath the island cools and becomes denser, and the island sinks. Alternatively, a moving plate may carry the island away from the hot spot. This also results in cooling, contraction, and sinking of the island.
2. The weight of the newly formed volcano causes isostatic sinking.
3. Erosion lowers the top of the volcano.

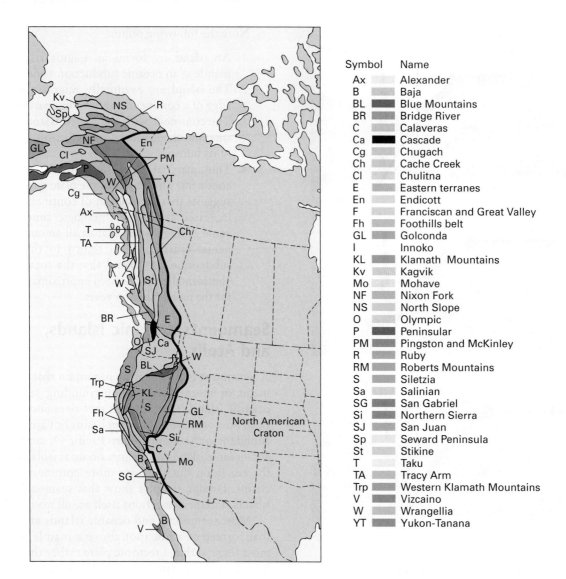

Symbol		Name
Ax		Alexander
B		Baja
BL		Blue Mountains
BR		Bridge River
C		Calaveras
Ca		Cascade
Cg		Chugach
Ch		Cache Creek
Cl		Chulitna
E		Eastern terranes
En		Endicott
F		Franciscan and Great Valley
Fh		Foothills belt
GL		Golconda
I		Innoko
KL		Klamath Mountains
Kv		Kagvik
Mo		Mohave
NF		Nixon Fork
NS		North Slope
O		Olympic
P		Peninsular
PM		Pingston and McKinley
R		Ruby
RM		Roberts Mountains
S		Siletzia
Sa		Salinian
SG		San Gabriel
Si		Northern Sierra
SJ		San Juan
Sp		Seward Peninsula
St		Stikine
T		Taku
TA		Tracy Arm
Trp		Western Klamath Mountains
V		Vizcaino
W		Wrangellia
YT		Yukon-Tanana

◆ **FIGURE 15.17** The accreted terranes of western North America are microcontinents and island arcs from the Pacific Ocean that were added to the continent.

These three factors gradually transform a volcanic island to a seamount (◆ Figure 15.19). If the Pacific Ocean plate continues to move at its present rate, the island of Hawaii may sink beneath the sea within 10 to 15 million years. Sea waves may erode a flat top on a sinking island, forming a flat-topped seamount called a **guyot** (◆ Figure 15.20).

The South Pacific and portions of the Indian Ocean are dotted with numerous islands called **atolls** (◆ Figure 15.21). An atoll is a circular coral reef that forms a ring of islands around a central lagoon. Atolls vary from 1 to 130 kilometers in diameter and are surrounded by deep water of the open sea. If corals live only in shallow water, how did atolls form in the deep sea? Charles Darwin studied this question during his famous voyage on the *Beagle* from 1831 to 1836. He reasoned that a coral reef must have formed in shallow water on the flanks of

a volcanic island. Eventually the island sank, but the reef continued to grow upward, so that the living portion always remained in shallow water (◆ Figure 15.22). This proposal was not accepted at first because scientists could not explain how a volcanic island could sink. However, when scientists drilled into a Pacific atoll shortly after World War II and found volcanic rock hundreds of meters beneath the reef, Darwin's hypothesis was revived. It is considered accurate today, in light of our ability to explain why volcanic islands sink.

ThomsonNOW

CLICK ThompsonNOW Interactive to work through an activity on Triple Junctions and Sea Floor Studies Through Plate Tectonics.

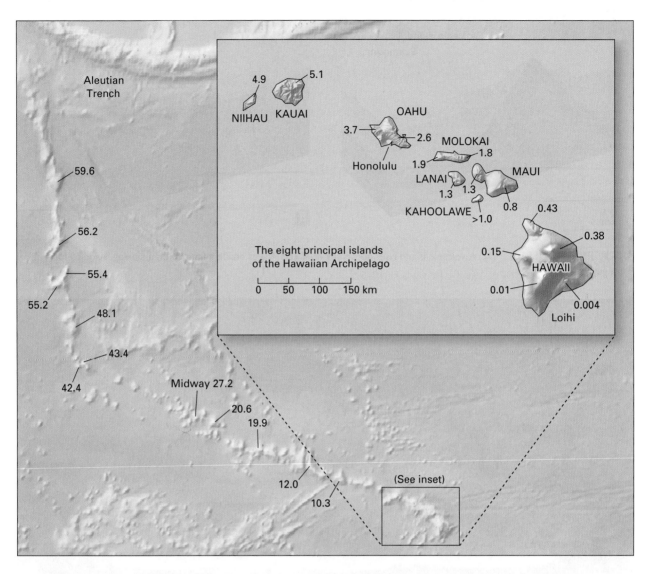

The eight principal islands
of the Hawaiian Archipelago

0   50   100   150 km

◆ **FIGURE 15.18**   The Hawaiian Island—Emperor Seamount chain becomes older in a direction going away from the island of Hawaii. The ages, in millions of years, are for the oldest volcanic rocks of each island or seamount.

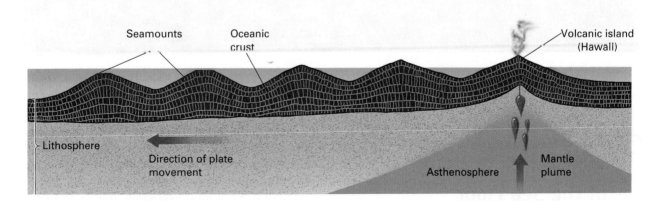

ThomsonNOW™   ◆ **FIGURE 15.19**   The Hawaiian Islands and Emperor Seamounts sink as they move away from the mantle plume.

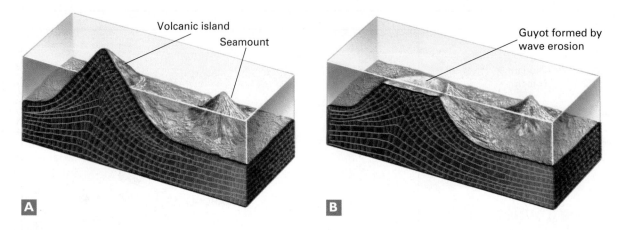

◇ **FIGURE 15.20** **(A)** A volcanic island rises above sea level. **(B)** Waves erode a flat top on a sinking island to form a guyot.

Tahiti Tourist Board

◇ **FIGURE 15.21** The Tetiaroa Atoll in French Polynesia formed by the process described in Figure 15.22. Over time, storm waves wash coral sands on top of the reef and vegetation grows on the sand, forming the individual islands in the atoll.

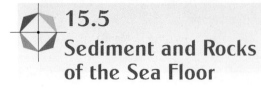

15.5
Sediment and Rocks of the Sea Floor

Early oceanographers had believed that the oceans are 4 billion years old, so mud on the ancient sea floor should have been very thick, having had so much time to accumulate. In 1947, however, scientists on the U.S. research ship *Atlantis* discovered that the mud layer on the bottom of the Atlantic Ocean is much thinner than they expected. Why is there so little mud on the sea floor? The answer to this question would be a crucial piece of evidence in development of the theory of plate tectonics.

Earth is 4.6 billion years old, and rocks as old as 3.96 billion years have been found on continents. Once formed, most continental crust remains near Earth's surface because of its buoyancy. In contrast, no parts of the sea floor are older than about 200 million years because oceanic crust forms continuously at the Mid-Oceanic Ridge and then recycles into the mantle at subduction

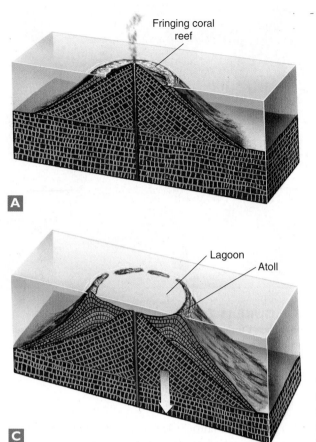

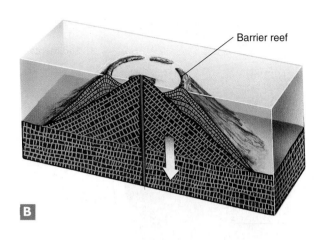

FIGURE 15.22 **(A)** A fringing reef grows along the shore of a young volcanic island. **(B)** As the island sinks, the reef continues to grow upward to form a barrier reef that encircles the island. **(C)** Finally the island sinks below sea level and the reef forms a circular atoll.

zones. As the theory of plate tectonics was emerging in the early 1960s, it became apparent that the fact that only a relatively thin layer of mud covers the sea floor could be easily explained if the oldest sea floor is less than 200 million years old, not 4 billion years old as oceanographers had once believed. This evidence supported the fledgling plate tectonics theory in its early days.

Seismic profiling and sea-floor drilling show that oceanic crust consists of three layers. The uppermost layer consists of sediment, and the lower two are basalt (◆ Figure 15.23).

Ocean-Floor Sediment

The uppermost layer of oceanic crust, called **layer 1**, consists of two types of sediment. **Terrigenous sediment** is sand, silt, and clay eroded from the continents and carried to the deep sea floor by gravity and submarine currents. Most of this sediment is found close to the continents. **Pelagic sediment**, however, collects even on the deep sea floor far from continents. It is a gray and red-brown mixture of clay that was mostly carried from continents by wind, and of the remains of tiny plants and animals that live in the surface waters of the oceans (◆ Figure 15.24). When these organisms die, their remains slowly settle to the ocean floor.

Pelagic sediment accumulates at a rate of about 2 to 10 millimeters per 1,000 years. Near the Mid-Oceanic Ridge system there is virtually no sediment because the

sea floor is so young. The sediment thickness increases with distance from the ridge because the sea floor becomes older as it spreads away from the ridge. Close to shore, pelagic sediment gradually merges with the much-thicker layers of terrigenous sediment, which can be 3 kilometers or more thick. The observation of increasing thickness of sea-floor mud away from the ridge also supported the plate tectonics theory in its early days.

Parts of the ocean floor beyond the Mid-Oceanic Ridge system are flat, level, featureless submarine surfaces called the **abyssal plains**. They are the flattest surfaces on Earth. Seismic profiling shows that the basaltic crust is rough and jagged throughout the ocean. On the abyssal plains, however, pelagic sediment buries this rugged profile, forming the smooth abyssal plains. If you were to remove all of the sediment, you would see rugged topography similar to that of the Mid-Oceanic Ridge.

Basaltic Oceanic Crust

Layer 2 lies below layer 1 and is about 1 to 2 kilometers thick. It consists mostly of **pillow basalt**, which forms as hot magma oozes onto the sea floor. Contact with cold sea water causes the molten lava to contract into pillow-shaped spheroids (◆ Figure 15.25).

Layer 3, 3 to 5 kilometers thick, is the deepest and thickest layer of oceanic crust. It directly overlies the mantle. The upper part consists of vertical basalt dikes, which formed as the magma oozing toward the surface froze in

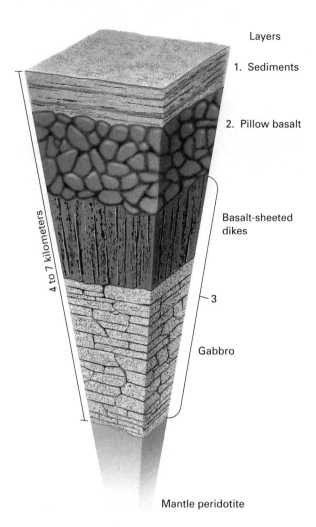

Layers

1. Sediments

2. Pillow basalt

Basalt-sheeted dikes

3

Gabbro

4 to 7 kilometers

Mantle peridotite

◆ **FIGURE 15.23** The three layers of oceanic crust. Layer 1 consists of sediment. Layer 2 is pillow basalt. Layer 3 consists of vertical dikes overlying gabbro. Below layer 3 is the upper mantle.

the cracks of the rift valley. The lower portion of layer 3 consists of gabbro, the coarse-grained equivalent of basalt. The gabbro forms as pools of magma cool slowly (because they are insulated by the basalt dikes above them).

The basaltic crust of layers 2 and 3 forms at the Mid-Oceanic Ridge. However, these rocks make up the foundation of all oceanic crust because all oceanic crust forms at the ridge axis and then spreads outward. In some places, chemical reactions with sea water have altered the basalt of layers 2 and 3 to a soft, green rock that contains up to 13 percent water.

15.6
Continental Margins

A continental margin is a place where continental crust meets oceanic crust. Two types of continental margins exist. A **passive continental margin** occurs where continental and oceanic crust are firmly joined together. Because it is not a plate boundary, little tectonic activity occurs at a passive margin. Continental margins on both

Texas A&M University Ocean Drilling Program

◆ **FIGURE 15.24** This scanning electron microscope photo shows foraminifera, tiny organisms that float near the surface of the seas. When these organisms die, their remains sink to the sea floor to become part of the pelagic mud layer. Each of the fossils is the size of a fine sand grain.

sides of the Atlantic Ocean are passive margins. In contrast, an **active continental margin** occurs at a convergent plate boundary, where oceanic lithosphere sinks beneath the continent in a subduction zone. The west coast of South America is an active continental margin.

Passive Continental Margins

Recall from Chapter 6 that, about 250 million years ago, all of Earth's continents were joined into the supercontinent called Pangea. Shortly thereafter, Pangea began to rift apart into the continents as we know them today. The Atlantic Ocean opened as the east coast of North

OAR/National Undersea Research Program (NURP)

◆ **FIGURE 15.25** Underwater photo of pillow lavas off the island of Hawaii.

America separated from Europe and Africa. As Pangea broke up, the continental crust fractured and thinned near the fractures (◆ Figure 15.26A). Basaltic magma rose at the new spreading center, forming oceanic crust between North America and Africa (◆ Figure 15.26B). All tectonic activity then focused on the spreading Mid-Atlantic Ridge, and no further tectonic activity occurred at the continental margins; hence the term *passive continental margin* (◆ Figure 15.26C).

The Continental Shelf

On all continents, streams and rivers deposit sediment on coastal deltas, like the Mississippi River Delta. Then, ocean currents redistribute the sediment along the coast, depositing it both on the thin margin of continental crust and on oceanic crust close to the continent. The sediment forms a shallow, gently sloping, submarine surface called a **continental shelf** on the edge of the continent (◆ Figure 15.27). As sediment accumulates on a continental shelf, the edge of the continent sinks isostatically because of the added weight. This effect keeps the shelf slightly below sea level.

Over millions of years, thick layers of sediment accumulated on the passive east coast of North America, forming a broad continental shelf along the entire coast. The depth of the shelf increases gradually from the shore to about 200 meters at the outer shelf edge. The average inclination of the continental shelf is about 0.1°. A continental shelf on a passive margin can be a large feature. The shelf off the coast of southeastern Canada is about 500 kilometers wide, and parts of the shelves of Siberia and northwestern Europe are even wider.

In some places, a supply of sediment may be lacking, either because no rivers bring sand, silt, or clay to the shelf or because ocean currents bypass that area. In warm regions where sediment does not muddy the water, reef-building organisms thrive. As a result, thick beds of limestone accumulate in tropical and subtropical latitudes where clastic sediment is lacking. Limestone accumulations of this type may be hundreds of meters thick and hundreds of kilometers across and are called **carbonate platforms**. The Florida Keys and the Bahamas are modern day examples of carbonate platforms on continental shelves (◆ Figure 15.28).

Some of the world's richest petroleum reserves occur on the continental shelves of the North Sea between England and Scandinavia, in the Gulf of Mexico, and in the Beaufort Sea on the northern coast of Alaska and western Canada. In recent years, oil companies have explored and developed these offshore reserves. Deep drilling has revealed that granitic continental crust lies beneath the sedimentary rocks, confirming that the continental shelves are truly parts of the continents, despite the fact that they are covered by seawater.

The Continental Slope and Rise

At the outer edge of a shelf, the sea floor suddenly steepens to an average slope of about 4° to 5° as it falls away from 200 meters to about 5 kilometers in depth. This steep region of the sea floor averages about 50 kilometers wide and is called the **continental slope**. It is a surface formed by sediment accumulation, much like the shelf. Its steeper angle is due primarily to thinning of continental crust where it nears the junction with oceanic crust. Seismic profiler exploration shows that the sedimentary layering is commonly disrupted where sediment has slumped and slid down the steep incline.

A continental slope becomes less steep as it gradually merges with the deep ocean floor. This region, called the **continental rise**, consists of an apron of terrigenous sediment that was transported across the continental shelf and deposited on the deep ocean floor at the foot of the slope. The continental rise averages a few hundred kilometers wide. Typically, it joins the deep sea floor at a depth of about 5 kilometers.

In essence, then, the shelf–slope–rise complex is a smoothly sloping, submarine surface on the edge of a continent, formed by accumulation of sediment eroded from the continent.

Submarine Canyons and Abyssal Fans

In many places, sea-floor maps show deep valleys, called **submarine canyons**, eroded into the continental shelf and slope. They look like submarine stream valleys. A canyon typically starts on the outer edge of a continental shelf and continues across the slope to the rise. At its lower end, a submarine canyon commonly leads into an **abyssal fan** (or **submarine fan**), a large, fan-shaped pile of sediment lying on the continental rise.

Most submarine canyons occur where large rivers enter the sea. When they were first discovered, geologists thought the canyons had been eroded by rivers during the Pleistocene Epoch, when accumulation of glacial ice on land lowered sea level by as much as 150 meters. However, this explanation cannot account for the deeper portions of submarine canyons cut into the lower continental slopes at depths of a kilometer or more. Therefore, the deeper parts of the submarine canyons must have formed underwater, and a submarine mechanism must be found to explain them.

Geologists subsequently discovered that **turbidity currents** erode the continental shelf and slope to create the submarine canyons. A turbidity current develops when loose, wet sediment tumbles down the slope in a submarine landslide. The movement may be triggered by an earthquake or simply by oversteepening of the slope as sediment accumulates. When the sediment starts to move, it mixes with water. Because the mixture of sediment and water is denser than water alone, it flows down the shelf and slope as a turbulent, chaotic avalanche. A turbidity current can travel at speeds greater than 100 kilometers per hour and for distances up to 700 kilometers.

Sediment-laden water traveling at such speed has tremendous erosive power. Once a turbidity current cuts a small channel into the shelf and slope, subsequent currents follow the same channel, just as a stream uses the same

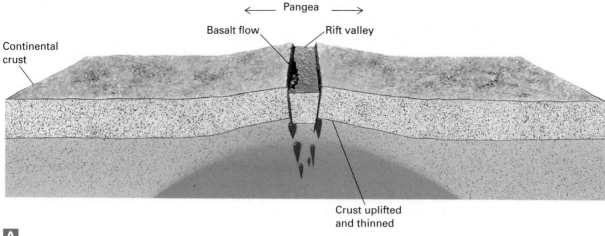

Pangea

Basalt flow   Rift valley

Continental crust

Crust uplifted and thinned

A

Rift valley   Fault blocks

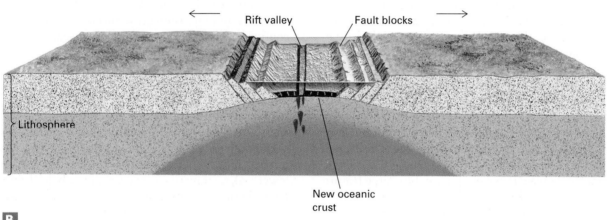

Lithosphere

New oceanic crust

B

Abyssal plains

Mid-Atlantic Ridge

East coast of North America

Continental shelf
Continental slope
Continental rise

Rift valley   Atlantic Ocean

West coast of Africa

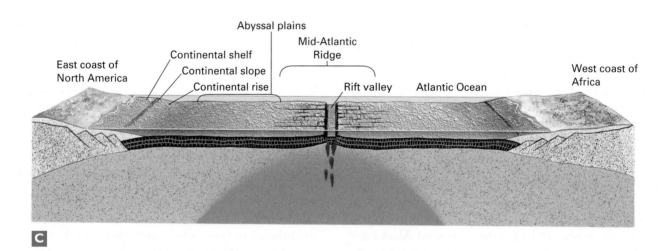

C

ThomsonNOW™ ◆ **ACTIVE FIGURE 15.26** **(A)** Continental crust fractured as Pangea began to rift. **(B)** Faulting and erosion thinned the crust as it separated. Rising basaltic magma formed new oceanic crust in the rift zone. **(C)** Sediment eroded from the continents formed broad continental shelves on the passive margins of North America and Africa.

INTERACTIVE QUESTION: *Draw a new diagram, (D), showing a cross-sectional view of North America reversing direction and migrating eastward.*

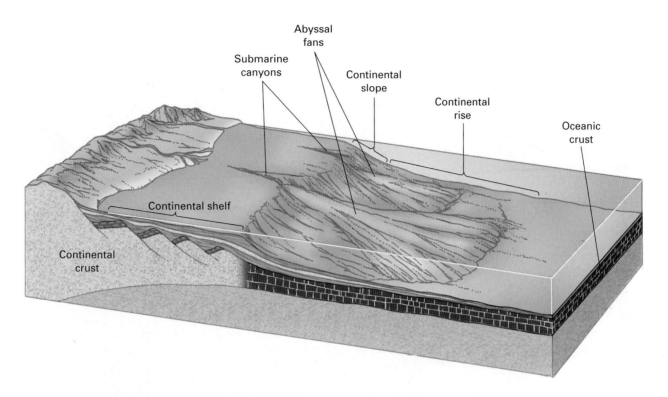

◆ ACTIVE FIGURE 15.27   A passive continental margin consists of a broad continental shelf, slope, and rise formed by accumulation of sediment eroded from the continent.

channel year after year. Over time, the currents erode a deep submarine canyon into the shelf and slope. Turbidity currents slow down when they reach the deep sea floor. The sediment accumulates there to form an abyssal fan. Most submarine canyons and fans form near the mouths

◆ FIGURE 15.28   Clear skies and calm waters allowed cameras on a satellite to capture this stunning image of southern Florida, the Bahamas and Cuba. In the image's center, Andros Island is surrounded by the bright blue halo of Great Bahama Bank, a carbonate platform that was inundated by rising sea level between 10,000 and 2,500 years ago, as the last ice age glaciers were melting. In most places the water above the platform doesn't exceed 6 m.

of large rivers because the rivers supply the great amount of sediment needed to create turbidity currents.

Large abyssal fans form only on passive continental margins. They are uncommon at active margins because in that environment, the trench swallows the sediment. Furthermore, most of the world's largest rivers drain toward passive margins. The largest known fan is the Bengal fan, which covers about 4 million square kilometers beyond the mouth of the Ganges River in the Indian Ocean east of India. More than half of the sediment eroded from the rapidly rising Himalayas ends up in this fan. Interestingly, the Bengal fan has no associated submarine canyon, perhaps because the sediment supply is so great that the rapid accumulation of sediment prevents erosion of a canyon.

Active Continental Margins

An active continental margin forms at a subduction zone, where an oceanic plate converges with a continent. The oceanic plate sinks into the mantle, forming an oceanic trench similar to the trenches associated with island arcs described in Section 15.4 (◆ Figure 15.29). A trench can form wherever subduction occurs—where oceanic crust sinks beneath the edge of a continent, or where it sinks beneath another oceanic plate.

Recall that at a passive margin, the continental crust merges gradually with oceanic crust and a wide continental shelf–slope–rise system develops as thick deposits of sediment accumulate on the passive margin. At an active margin, however, the thick continental crust meets the

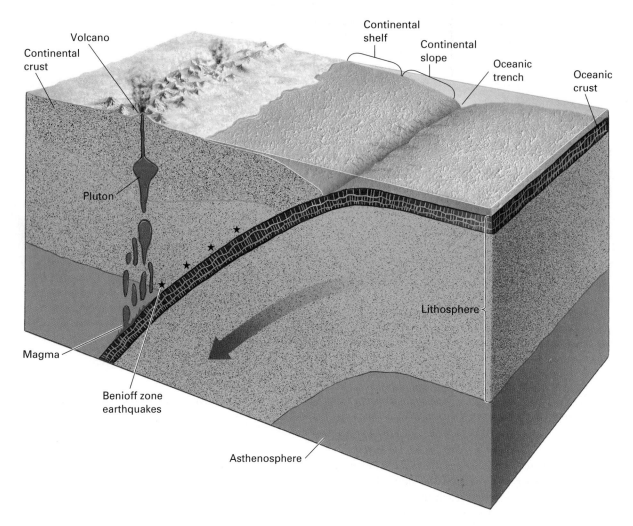

ThomsonNOW ◈ **ACTIVE FIGURE 15.29** At an active continental margin, an oceanic plate sinks beneath a continent, forming an oceanic trench. The continental shelf is narrow, the slope is steep, and the continental rise is nonexistent.

thin oceanic crust at a subduction zone. Most of the sediment transported from a continent to an active margin is swallowed up in the trench. As a result, an active margin commonly has a narrow continental shelf or none at all. The landward wall (the side toward the continent) of the trench is the continental slope of an active margin. It typically inclines at 4° or 5° in its upper part and steepens to 15° or more near the bottom of the trench. The continental rise is absent because sediment flows into the trench instead of accumulating on the ocean floor.

EARTH SYSTEMS INTERACTIONS

COMETS, ASTEROIDS, AND METEORITES crashing into early Earth imported the volatile elements and compounds that now comprise seawater, the atmosphere, and living organisms. Thus, the hydrosphere, the atmosphere, and the biosphere have a common source. Ocean basins have formed as a consequence of plate tectonic processes that create two different kinds of lithosphere and crust: those of the oceans and those of the continents. Oceans (the hydrosphere) affect global climate (atmosphere) and the biosphere in many ways. Seawater absorbs, reflects, and stores solar heat more efficiently than do rocks and soil. As a result, oceans are generally warmer in winter and cooler in summer than nearby land. Nearly all precipitation consists of water that evaporated from the seas. In addition, ocean currents transport heat from the equator toward the poles, cooling equatorial climates and warming polar environments. Because the movement of lithospheric plates alters the sizes and shapes of ocean basins, changing global geography alters oceanic currents and profoundly affects regional climates over geologic time. Thus, the oceans play a large role in Earth systems interactions. ■

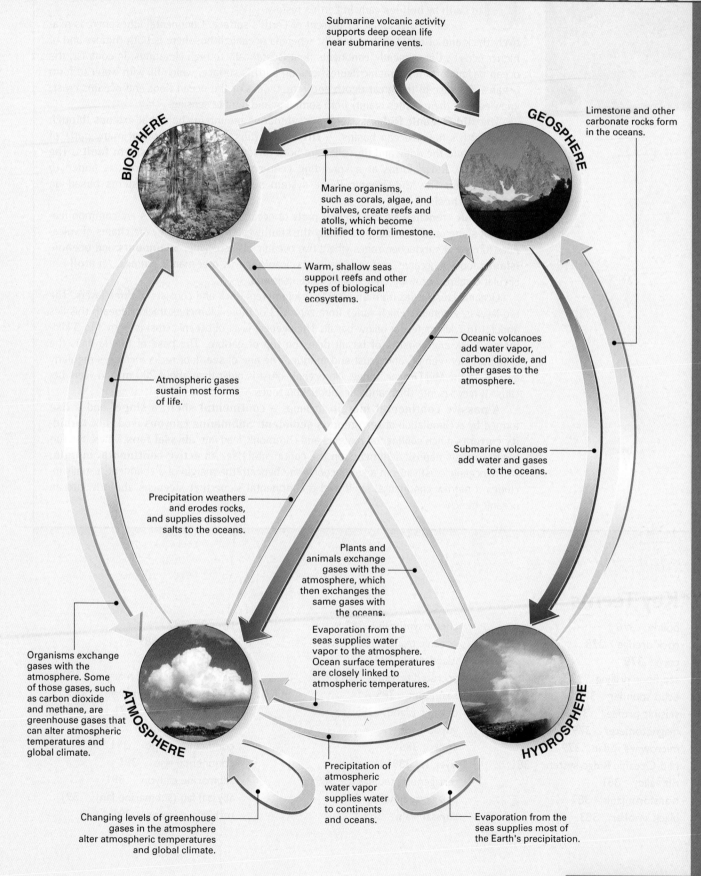

Submarine volcanic activity supports deep ocean life near submarine vents.

BIOSPHERE

GEOSPHERE

Limestone and other carbonate rocks form in the oceans.

Marine organisms, such as corals, algae, and bivalves, create reefs and atolls, which become lithified to form limestone.

Warm, shallow seas support reefs and other types of biological ecosystems.

Oceanic volcanoes add water vapor, carbon dioxide, and other gases to the atmosphere.

Atmospheric gases sustain most forms of life.

Submarine volcanoes add water and gases to the oceans.

Precipitation weathers and erodes rocks, and supplies dissolved salts to the oceans.

Plants and animals exchange gases with the atmosphere, which then exchanges the same gases with the oceans.

Evaporation from the seas supplies water vapor to the atmosphere. Ocean surface temperatures are closely linked to atmospheric temperatures.

Organisms exchange gases with the atmosphere. Some of those gases, such as carbon dioxide and methane, are greenhouse gases that can alter atmospheric temperatures and global climate.

ATMOSPHERE

HYDROSPHERE

Precipitation of atmospheric water vapor supplies water to continents and oceans.

Changing levels of greenhouse gases in the atmosphere alter atmospheric temperatures and global climate.

Evaporation from the seas supplies most of the Earth's precipitation.

SUMMARY

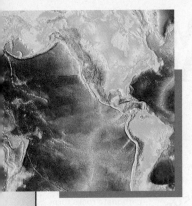

Ocean water, atmospheric gases, and the molecular building blocks of life were carried to Earth by **bolides** early in Earth's history.

Oceans cover about 71 percent of Earth's surface. Continental lithosphere is relatively thick and of relatively low density, whereas oceanic lithosphere is both thinner and of higher density. Consequently, continents float isostatically to high elevations. In contrast, the ocean basins form topographic depressions on Earth's surface, which fill with water to form oceans. Because of the great depth and remoteness of the ocean floor and oceanic crust, knowledge of them comes mainly from sampling and remote sensing.

The **Mid-Oceanic Ridge** system is a submarine mountain chain that extends through all of Earth's major ocean basins. A **rift valley** runs down the center of many parts of the ridge, and the ridge and rift valley are both offset by numerous **transform faults**. The Mid-Oceanic Ridge forms at a spreading center where new oceanic crust is added to the sea floor. The Mid-Oceanic Ridge system supports thriving ecosystems based on **chemosynthesis**.

Oceanic trenches are the deepest parts of ocean basins. **Island arcs** are common features of some ocean basins, particularly the southwestern Pacific. They are chains of volcanoes formed at subduction zones where two oceanic plates collide. **Seamounts** and **oceanic islands** form in oceanic crust as a result of volcanic activity over mantle plumes. An **atoll** is a circular coral reef growing on a sinking volcanic island.

Oceanic crust varies from about 4 to 7 kilometers thick and consists of three layers. The top layer is sediment, which varies from zero to 3 or more kilometers thick. Beneath this lies about 1 to 2 kilometers of pillow basalt. The deepest layer of oceanic crust is from 3 to 5 kilometers thick and consists of basalt dikes on top of gabbro. The base of this layer is the boundary between oceanic crust and mantle. The age of sea-floor rocks increases regularly away from the Mid-Oceanic Ridge. No oceanic crust is older than about 200 million years because it recycles into the mantle at subduction zones.

A **passive continental margin** includes a **continental shelf**, a **slope**, and a **rise** formed by accumulation of **terrigenous sediment**. **Submarine canyons** eroded by **turbidity currents** notch continental margins and commonly lead into **abyssal fans**, where the turbidity currents deposit sediments on the continental rise. An **active continental margin**, where oceanic crust sinks in a subduction zone beneath the margin of a continent, usually includes a narrow continental shelf and a continental slope that steepens abruptly into an oceanic trench.

Key Terms

For Review

1. Describe the main differences between oceans and continents.

2. Sketch a cross section of the Mid-Oceanic Ridge, including the rift valley.

3. Describe the dimensions of the Mid-Oceanic Ridge.

4. Explain why the Mid-Oceanic Ridge is topographically elevated above the surrounding ocean floor. Why does its elevation gradually decrease away from the ridge axis?

5. Explain the origin of the rift valley in the center of the Mid-Oceanic Ridge.

6. Why are the abyssal plains characterized by such low relief?

7. Sketch a cross section of oceanic crust from a deep sea basin. Label, describe, and indicate the approximate thickness of each layer.

8. Describe the two main types of sea-floor sediment. What is the origin of each type?

9. Compare the ages of oceanic crust with the ages of continental rocks. Why are they so different?

10. Sketch a cross section of both an active continental margin and a passive continental margin. Label the features of each. Give approximate depths below sea level of each of the features.

11. Explain how a continental shelf–slope–rise complex forms on a continental margin.

12. Why does an active continental margin typically have a steeper continental slope than a passive margin?

13. Why does an active margin typically have no continental rise?

14. Explain the relationships among submarine canyons, abyssal fans, and turbidity currents.

15. Why are turbidity currents often associated with earthquakes or with large floods in major rivers?

16. Explain the role played by an island arc in the growth of a continent.

17. Explain the origins of, and differences between, seamounts and island arcs.

18. Compare the ocean depths adjacent to an island arc and a seamount.

19. Why do oceanic islands sink after they form?

20. Explain how guyots and atolls form.

For Discussion

1. The east coast of South America has a wide continental shelf, whereas the west coast has a very narrow shelf. Discuss and explain this contrast.

2. Seismic data indicate that continental crust thins where it joins oceanic crust at a passive continental margin, such as on the east coast of North America. Other than that, we know relatively little about the nature of the junction between the two types of crust. Speculate on the nature of that junction. Consider rock types, geologic structures, ages of rocks, and other features of the junction.

3. Discuss the topography of Earth in an imaginary scenario in which all conditions are identical to present ones except that there is no water. In contrast, what would be the effect if there were enough water to cover all of Earth's surface?

4. If you look at maps of Earth, you will see that most of the world's largest rivers drain toward passive continental margins. Confirm this for yourself and explain this observation.

CHAPTER 16

Oceans and Coastlines

Courtesy of Graham R. Thompson/Jonathan Turk

A rare clear day and calm seas grace the rocky headland of Cape Horn at the Southern tip of South America. More commonly, storms, strong winds, and large waves make the cape notorious as a sailor's graveyard.

n December 2005, the United Nations reported that rising seas had forced 100 people on the Pacific island of Tegua in Vanuatu to move to higher ground in what may be the first historical example of a village displaced by rising sea level. With coconut palms standing in water, inhabitants in the Lateu village on Tegua began moving their wooden homes about 600 meters inland in August of that year. The U.N. Environment Programme stated that the Lateu settlement "has become one of, if not the first, to be formally moved out of harm's way as a result of climate change."

Near Papua New Guinea, about 2,000 people on the Cantaret Islands are planning to take a four-hour boat ride southwest to nearby Bougainville Island to escape coastal flooding. Two uninhabited islands, Tebua Tarawa and Abanuea, in the Pacific nation of Kiribati, disappeared beneath the waves in 1999.

The Pacific Ocean is also slowly inundating Tuvalu, the world's fourth-smallest country, consisting of six coral atolls and three limestone reefs in the South Pacific, 600 miles north of Fiji. Every year, the sea creeps further inland contaminating groundwater and rising toward the surface, killing crops. During hurricanes, waves wash across the entire nation, leaving only a few tall buildings standing above the angry sea, like apparitions in a Melanesian legend.

The coastal city of New Orleans experienced severe flooding and damage during Hurricane Katrina in September 2005, as described in Chapter 11 and Chapter 19. Much of the adjacent Gulf Coast suffered similar damage in the same storm.

About 60 percent of the world's human population lives within 100 kilometers of the coast. One million people live on low coral islands, and many millions more live on low-lying coastal land vulnerable to coastal flooding. At risk are not just individuals and villages but unique human cultures. Faced with changing coastal environments, these people are threatened with forced abandonment of their nations.

Just as the UN report attributes the drowning of Lateu village on Tegua to rising sea level caused by global warming, the residents of Tuvalu claim that rising sea level caused by global warming is drowning their islands. They are preparing to sue the industrial nations of the world for emitting the carbon dioxide that they claim is responsible for their national disaster. Similarly, popular media coverage has linked damage along the American Gulf Coast to sea-level rise induced by global warming. However, many geologists and oceanographers argue that other factors such as crustal subsidence, described in Chapter 15, may be responsible for the submergence of these Pacific Island nations and coastal regions of the United States.

Regardless of the cause, the image of an entire country disappearing beneath the sea, or a great seaport city being ravaged by a hurricane, is receiving considerable media attention. For many people, it is disturbing that solid land should sink beneath the waves or that a modern city was defenseless against a sea storm. Yet geologists know that continents and islands have risen from the sea and sunk back beneath the waves throughout Earth's history. In this chapter, we will continue studying Earth's oceans and the processes that form, shape, or destroy islands and coastlines.

The seashore is an attractive place to live or visit. Because the ocean moderates temperature, coastal regions are cooler in summer and warmer in winter than continental interiors are. Vacationers and residents sail, swim, surf, and fish along the shore. In addition, the sea provides both food and transportation. For all of these reasons, coastlines have become heavily urbanized and industrialized. However, coastlines are among the most geologically active environments on Earth. Sea level rises and falls, flooding shallow parts of continents and stranding beaches high above sea level. Regions of continental crust also rise and sink, with similar results to those caused by fluctuating sea level. Rivers deposit great quantities of sand and mud on coastal deltas. Waves and currents erode beaches and transport sand along hundreds or even thousands of kilometers of shoreline. Converging tectonic plates buckle coastal regions, creating mountain ranges, earthquakes, and volcanic eruptions. ■

16.1 Geography of the Oceans

All of Earth's oceans are connected, and water flows from one to another, so in one sense Earth has just one global ocean. However, several distinct ocean basins exist within the global ocean (◆ Figure 16.1). The largest and deepest is the Pacific. It covers one-third of Earth's surface, more than all land combined, and contains more than half of the world's water. The Atlantic Ocean has about half the surface area of the Pacific. The Indian Ocean is slightly smaller than the Atlantic. The Arctic Ocean surrounds the North Pole and extends southward to the shores of North America, Europe, and Asia. Therefore, it is bounded by land, with only a few straits and channels connecting it to the Atlantic and Pacific Oceans. The surface of the Arctic Ocean freezes in winter, and parts of it melt for a few months during summer and early fall (◆ Figure 16.2).

The Antarctic Ocean, feared by sailors for its cold and ferocious winds, has no sharp northern boundary. The northernmost limit of the Antarctic Ocean is the zone where warm currents from the north converge with cold Antarctic water.

16.2 Seawater

Salinity

The **salinity** of seawater is the total quantity of dissolved salts, expressed as a percentage. In this case, the term *salts* refers to all dissolved ions, not only sodium and chloride.

Dissolved ions make up about 3.5 percent of the weight of ocean water. The six ions listed in ◆ Figure 16.3

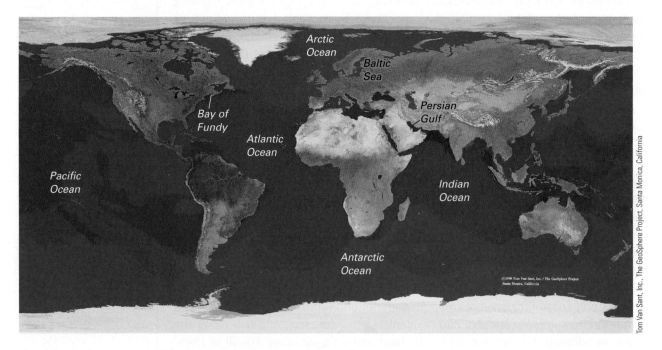

◆ **FIGURE 16.1**   The oceans of the world. The "seven seas" are the North Atlantic, South Atlantic, North Pacific, South Pacific, Indian, Arctic and Antarctic. However, these designations are related more closely to commerce than to the geology and oceanography of the ocean basins. Geologists and oceanographers recognize four major ocean basins: the Atlantic, Pacific, Indian, and Arctic. The map also designates a few smaller seas and bays that are mentioned in the text.

◀▶ **FIGURE 16.2** Ice floes cover large parts of the Arctic Ocean, even in summer.

make up 99 percent of the ocean's dissolved material. However, almost every other element found on land is also found dissolved in seawater, albeit in trace amounts. For example, seawater contains about 0.000000004 (4×10^{-9}) percent gold. Although the concentration is small, the oceans are large and therefore contain a lot of gold. About 4.4 kilograms of gold are dissolved in each cubic kilometer of seawater. Because the oceans contain about 1.3 billion cubic kilometers of water, about 5.7 billion kilograms of gold exist in the oceans. Unfortunately, it would be hopelessly expensive to extract even a small portion of this amount.

In addition to trace elements and salts, seawater also contains dissolved gases, especially carbon dioxide and oxygen. These dissolved gases exchange freely and continuously with the atmosphere. If the atmospheric concentration of carbon dioxide or oxygen rises, much of the gas quickly dissolves into seawater; if the atmospheric concentration of a gas falls, that gas exsolves (comes out of solution) from the sea and enters the atmosphere. Thus, the seas buffer atmospheric concentrations of gases. The capacity of seawater to absorb carbon dioxide from the atmosphere has a profound effect on the concentration of this greenhouse gas in the atmosphere, and therefore is an important factor in global climate and climate change. These processes are discussed in detail in Chapter 21. Oxygen dissolved in seawater is critical to most forms of marine life, which require oxygen for survival.

Although the average salinity of the oceans is 3.5 percent, it varies geographically. High rainfall near the equator dilutes salinity to 3.45 percent. Conversely, in dry subtropical regions, where evaporation is high and precipitation low, salinity can be as high as 3.6 percent (◀▶ Figure 16.4). Salinity varies more dramatically along coastlines. The Baltic Sea (◀▶ Figure 16.1) is a shallow ocean basin fed by many large, freshwater rivers and diluted further by rain and snow. As a result, its salinity is as low as 2.0 percent. In contrast, the Persian Gulf has low rainfall; high evaporation; and few large, inflowing rivers; its salinity exceeds 4.2 percent.

The world's rivers carry more than 2.5 billion tons of dissolved salts to the oceans every year. Underwater volcanoes contribute additional dissolved ions. However, the salinity of the oceans has been relatively constant throughout much of geologic time because salt has been removed from seawater at the same rate at which it has been added. When a portion of a marine basin becomes cut off from the open oceans, the water evaporates, precipitating thick sedimentary beds of salt. Additionally, large amounts of salt become incorporated into shale and other sedimentary rocks.

Temperature

Recall from Chapter 11 that lakes are comfortably warm for swimming during the summer because warm water floats on the surface and does not readily mix with the deep, cooler water. Oceans develop a similar temperature layering, but because sea waves and currents stir the surface water, the warm layer in an ocean extends from the surface to a depth as great as 450 meters (◀▶ Figure 16.5). Below this layer is the **thermocline**, a zone in which the temperature drops rapidly with depth. The thermocline extends to a depth of 2 kilometers. Beneath the thermocline, the temperature of ocean water varies from about 1°C to 2.5°C. The cold, dense water in the ocean depths mixes very little with the surface. Thus, there are three distinct temperature zones in the ocean, as shown in ◀▶ Figure 16.5. This layered structure does not exist in the polar seas because cold surface water sinks, causing vertical mixing.

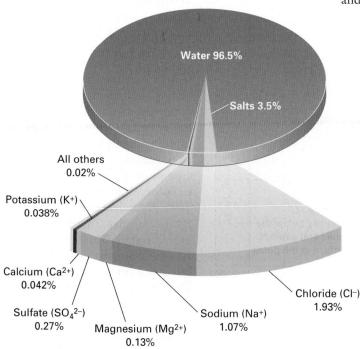

Water 96.5%

Salts 3.5%

All others
0.02%

Potassium (K+)
0.038%

Calcium (Ca²⁺)
0.042%

Sulfate (SO₄²⁻)
0.27%

Magnesium (Mg²⁺)
0.13%

Sodium (Na⁺)
1.07%

Chloride (Cl⁻)
1.93%

◀▶ **FIGURE 16.3** Six common ions form most of the salts in seawater.

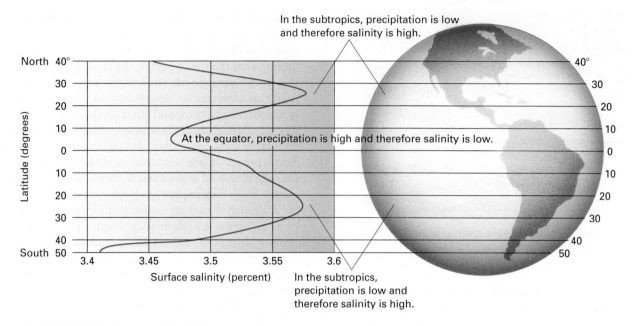

In the subtropics, precipitation is low and therefore salinity is high.

At the equator, precipitation is high and therefore salinity is low.

In the subtropics, precipitation is low and therefore salinity is high.

◆ **FIGURE 16.4** The salinity of the surface of the central oceans changes with latitude.

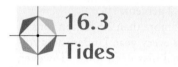

16.3 Tides

Even the most casual observer will notice that on any beach the level of the ocean rises and falls on a cyclical basis. If the water level is low at noon, it will reach its maximum height at about 6:13 p.m. and be low again at about 12:26 a.m. These vertical displacements are called **tides**.

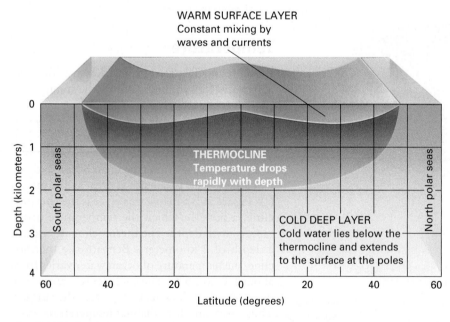

WARM SURFACE LAYER
Constant mixing by waves and currents

THERMOCLINE
Temperature drops rapidly with depth

COLD DEEP LAYER
Cold water lies below the thermocline and extends to the surface at the poles

◆ **FIGURE 16.5** There are three temperature layers in the ocean. The surface depth to about 450 meters is warm; temperature cools rapidly with depth in the thermocline, and the ocean depths are cold.

INTERACTIVE QUESTION: *Explain why the thermocline does not exist at high latitudes.*

Most coastlines experience two high tides and two low tides during an interval of about 24 hours and 53 minutes.

Tides are caused by the gravitational pull of the Moon and Sun on the sea surface. Although the Moon is much smaller than the Sun, it is so much closer to Earth that its influence predominates. At any time, one region of Earth (marked *A* in ◆ Figure 16.6) lies directly under the Moon. Because gravitational force is greater for objects that are closer together, the part of the ocean nearest to the Moon is attracted with the strongest force. The water rises, resulting in a high tide in that region.

But so far our explanation is incomplete. As Earth spins on its axis, a given point on Earth passes directly under the Moon approximately once every 24 hours and 53 minutes, but the period between successive high tides is only 12 hours and 26 minutes. Why are there ordinarily two high tides in a day? The tide is high not only when a point on Earth is directly under the Moon but also when it is 180° away. To understand this effect, we must consider the Earth–Moon orbital system. Most people visualize the Moon orbiting around Earth, but it is more accurate to say that Earth and the Moon orbit around a common center of gravity. The two celestial partners are locked together like dancers spinning around in each other's arms. Just as the back of a dancer's dress flies out-

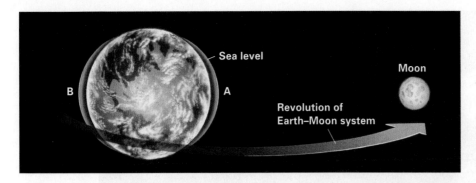

FIGURE 16.6 The Moon's gravity causes a high tide at point A, directly under the Moon. The motion of the Earth–Moon system causes a high tide at point B, opposite point A.

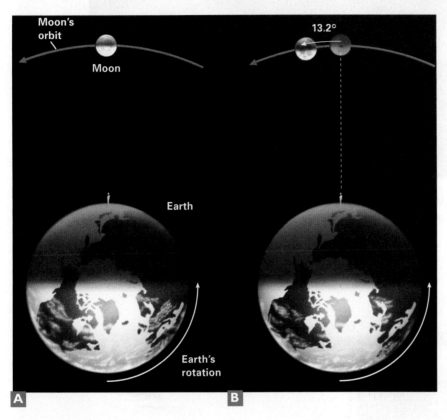

◆ **FIGURE 16.7** The Moon moves 13.2° every day. **(A)** The Moon is directly above an observer on Earth. **(B)** One day later, Earth has completed one complete rotation, but the Moon has traveled 13.2°. Earth must now travel for another 53 minutes before the observer is directly under it.

INTERACTIVE QUESTION: *Predict how the timing of the tides would be affected if the Moon orbited Earth at twice its present speed.*

ward as she twirls, the oceans on the opposite side of Earth from the Moon bulge outward. This bulge is the high tide 180° away from the Moon (point B in ◆ Figure 16.6). Thus, the tides rise and fall twice daily.

High and low tides do not occur at the same time each day but are delayed by approximately 53 minutes every 24 hours. Earth makes one complete rotation on its axis in 24 hours, but at the same time, the Moon is orbiting Earth in the same direction. After a point on Earth makes one complete rotation in 24 hours, that point must spin for an additional 53 minutes to catch

up with the orbiting Moon. This is why the Moon rises approximately 53 minutes later each day. In the same manner, the tides are approximately 53 minutes later each day (◆ Figure 16.7).

Although the Sun's gravitational pull on the oceans is smaller than the Moon's, it does affect tides. When the Sun and Moon are directly in line with Earth, their gravitational fields combine to create a strong tidal bulge. During these times, the variation between high and low tides is large, producing **spring tides** (◆ Figure 16.8A). When the Moon is 90° out of alignment with the Sun and Earth, each partially offsets the effect of the other and the differences between the levels of high and low tide are smaller. These relatively small tides are called **neap tides** (◆ Figure 16.8B).

Tidal variations differ from place to place. For example, the Bay of Fundy is shaped like a giant funnel (◆ Figure 16.1). The shape of the bay concentrates the rising and falling tides, and as a result the tidal variation is as much as 15 meters during a spring tide (◆ Figure 16.9). In contrast, tides vary by less than 2 meters along a straight stretch of coast such as that in Santa Barbara, California. In the central oceans, the tides average about 1.0 meter. Mariners consult tide tables that give the time and height of the tides in any area on any day.

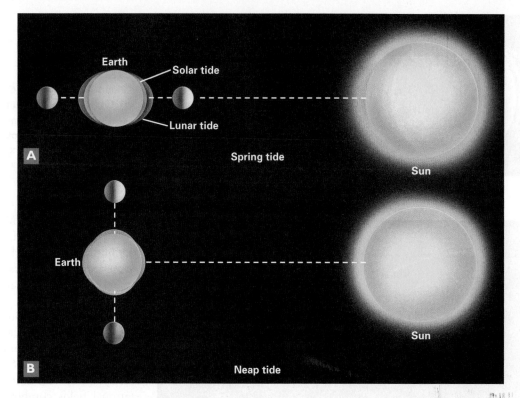

A

B

Spring tide

Neap tide

Earth
Solar tide
Lunar tide
Sun
Sun

ThomsonNOW ◆ **ACTIVE FIGURE 16.8** **(A)** Spring tides occur when Earth, Sun, and Moon are lined up. **(B)** Neap tides occur when the moon lies at right angles to a line drawn between the Sun and Earth.

A

B

◆ **FIGURE 16.9** **(A)** Low tide and **(B)** high tide vary by as much as 15 meters in the Bay of Fundy, Nova Scotia.

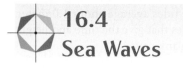

16.4
Sea Waves

Most ocean waves develop when wind blows across water. Waves vary from gentle ripples to destructive giants that can erode coastlines, topple beach houses, and sink ships. In deep water, the size of a wave depends on (1) the wind speed, (2) the length of time that the wind has blown, and (3) the distance that the wind has traveled (sailors call this last factor *fetch*). A 25-kilometer-per-hour

wind blowing for 2 to 3 hours across a 15-kilometer-wide bay generates waves about 0.5 meter high. But if a storm blows at 90 kilometers per hour for several days over a fetch of 3,500 kilometers, it can generate 30-meter-high waves, as tall as a ship's mast.

The highest part of a wave is called the **crest**; the lowest is the **trough** (◆ Figure 16.10). The **wavelength** is the distance between successive crests. The **wave height** is the vertical distance from the crest to the trough.

Recall from our discussion of earthquakes that when a wave travels through rock, the energy of the wave is

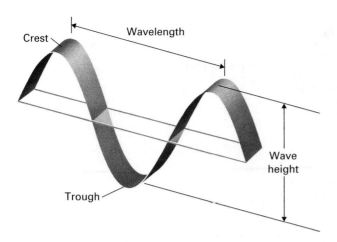

◆ FIGURE 16.10 Terminology used to describe a wave.

transmitted rapidly over large distances, but the rock itself moves only slightly. In a similar manner, a single water molecule in a water wave does not travel with the wave. While the wave moves horizontally, the water molecules move in small circles, as shown in ◆ Figure 16.11. That is why a ball on the ocean bobs up and down and sways back and forth as the waves pass, but it does not travel along with the waves. In addition, the circles of water movement become smaller with depth. At a depth equal to about one half the wavelength, the disturbance becomes negligible. Thus if you dive deep enough, you escape wave motion. No one gets seasick in a submarine.

ThomsonNOW™

CLICK Thomson Interactive to work through an activity on World Currents and Dissect a Current Through Waves, Tides, and Currents.

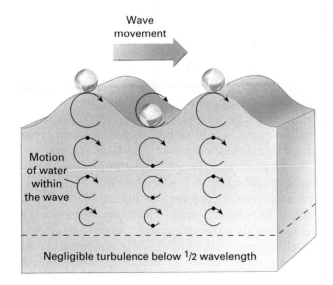

ThomsonNOW™ **◆ ACTIVE FIGURE 16.11** While a wave moves across the sea surface, the water itself moves only in small circles.

16.5
Storm Surge

Storm surge is responsible for about 90 percent of all human fatalities during hurricanes and tropical cyclones. The hurricane that struck Galveston, Texas, on September 8, 1900, created a storm surge that broke through the Galveston sea wall and killed between 6,000 and 12,000 people in the deadliest natural disaster in U.S. history. The second deadliest occurred in 1928 when a hurricane-driven surge from Lake Okeechobee flooded parts of southern Florida, causing approximately 2,500 human deaths.

A storm surge is an onshore flood of water created by a low-pressure weather system such as a hurricane or tropical cyclone. The surge is caused primarily by strong winds of the hurricane or cyclone blowing over the sea surface and causing seawater to pile up above sea level. Low pressure at the center of the storm also sucks the sea surface upward to enhance the height of the water mound. In the deep ocean, this dome of water eventually sinks and flows away harmlessly. But as a storm nears land, the shallowing sea floor prevents the piled-up water from escaping, and it flows inland as a deadly storm surge. Storm surges are highest and most damaging when they coincide with high tide, combining the effects of the surge and the tide.

The highest storm surge on record was a 13-meter-high flood that occurred in Bathurst Bay, Australia, in 1899. The greatest surge in U.S. history was a 9-meter-high flood that swept through parts of Bay St. Louis, Mississippi, during the 2005 Hurricane Katrina. The U.S. storm surge fatalities, tragic as they were, are overshadowed by 142 surge events that occurred along the shores of the Bay of Bengal on the coastlines of India and Southeast Asia from 1582 to 1991, which killed hundreds of thousands of people. This region is known as the storm surge capital of the world.

16.6
Ocean Currents

The water in an ocean wave oscillates in circles, ending up where it started. In contrast, a **current** is a continuous flow of water in a particular direction. Although river currents are familiar and easily observed, ocean currents were not recognized by early mariners. **Surface currents,** caused by wind blowing over the sea surface, flow in the upper 400 meters of the seas and involve about 10 percent of the water in the world's oceans. In contrast, **deep-sea currents** transport seawater both vertically and horizontally below a depth of 400 meters and are driven by gravity as denser water sinks and less-dense water rises. Furthermore, in certain places several

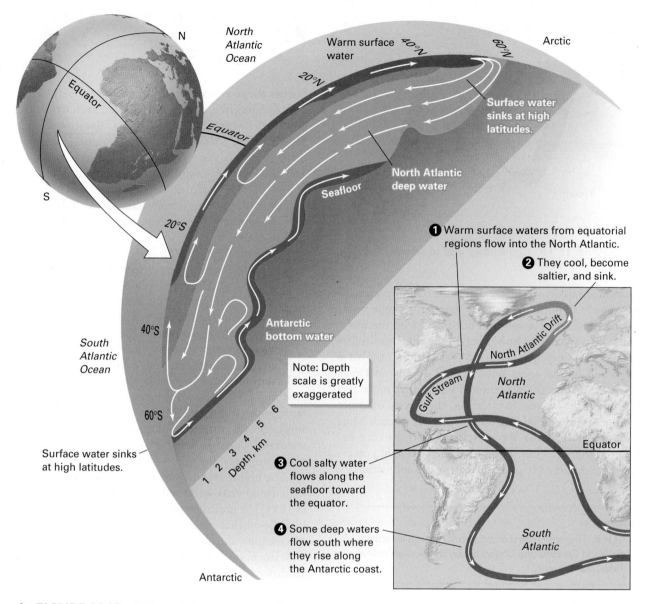

FIGURE 16.12 Surface and deep water currents of the Atlantic Ocean. Both types of currents profoundly affect climate and other components of Earth systems. This artist's rendition has simplified the currents to emphasize the major flow patterns.

processes (discussed later in the chapter) cause surface water to sink to great depths, and in other places deep water rises to the sea surface. Thus, the entire global ocean circulates continuously. ◀▶ Figure 16.12 shows both major surface currents and deep-sea *currents* in the Atlantic Ocean.

Surface Currents

In the mid-1700s, Benjamin Franklin lived in London, where he was deputy postmaster general for the American colonies. He noticed that mail ships took 2 weeks longer to sail from England to North America than did merchant ships. Franklin learned that the captains of the merchant ships had discovered a current flowing northward along the east coast of North America and then across the Atlantic to England. When sailing from Europe to North America, the merchant ships saved time by avoiding the current. The captains of the mail ships were unaware of this current and lost time sailing against it on their westward journeys. In 1769, Franklin and his cousin, Timothy Folger, a merchant captain, charted the current and named it the **Gulf Stream**.

As ship traffic increased and navigators searched for the quickest routes around the globe, they discovered other ocean currents (◀▶ Figure 16.13). Ocean currents have been described as rivers in the sea. The analogy is only partially correct; ocean currents have no well-defined banks, and they carry much more water than even the largest river. The Gulf Stream is 80 kilometers wide and 650 meters deep near the east coast of Florida and moves at approximately 5 kilometers per hour, a moderate walking speed. As it moves northward and eastward, the current widens and slows; east of

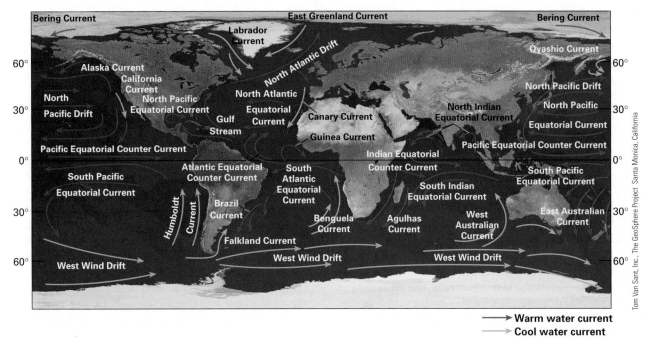

→ Warm water current
→ Cool water current

ThomsonNOW ◆ **FIGURE 16.13** Major oceanic surface currents of the world.

New York it is more than 500 kilometers wide and travels at less than 8 kilometers per day.

Another important difference between rivers and surface currents is that rivers flow in response to gravity, whereas ocean surface currents are driven primarily by wind. When wind blows across water in a constant direction for a long time, it drags surface water along with it, forming a current. In many regions, the wind blows in the same direction throughout the year, forming currents that vary little from season to season. However, in other places, changing winds cause currents to change direction. For example, in the Indian Ocean prevailing winds shift on a seasonal basis. When the winds shift, the ocean currents follow.

Ocean currents profoundly affect Earth's climate. The Gulf Stream transports one million cubic meters of warm water northward past any point every second, warming North America and Europe. For example, Churchill, Manitoba, is a village in the interior of Canada. It is frigid and icebound for much of the year. Polar bears regularly migrate through town. Yet it is at about the same latitude as Glasgow, Scotland, which is warmed by the Gulf Stream and therefore experiences relatively mild winters. In general, the climate in western Europe is warmer than that at similar latitudes in other regions not heated by tropical ocean currents.

Why Surface Currents Flow in the Oceans

Surface currents are driven primarily by friction between wind blowing over the sea surface and surface water. The winds simply drag the sea surface along in the same direction that the winds are blowing. In ◆ Figure 16.14,

the green arrows show the major prevailing winds over the North and South Atlantic Oceans. The orange arrows show the elliptical surface currents, called **gyres**, in the same regions, simplified from Figure 16.13. The North Atlantic gyre circulates in a clockwise direction, and the southern gyre circulates counterclockwise. But notice that the currents do not flow in exactly the same directions as the prevailing winds. Instead, the east–west surface currents in the Northern Hemisphere are deflected to the right of the winds. Where their flow is blocked by a continent, the currents veer clockwise to

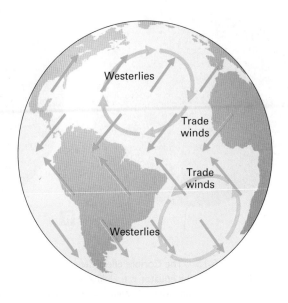

◆ **FIGURE 16.14** The green arrows show major prevailing winds over the North and South Atlantic Oceans—the Trade Winds (easterlies) and the westerlies. The orange arrows show the surface oceanic currents in the same regions.

continue their circuit. In the Southern Hemisphere the currents are deflected to the left of the winds and turn counterclockwise when they encounter land. These differences between prevailing wind directions and the flow directions of the surface currents suggest that forces other than wind also affect the great gyres.

One of those forces that deflect the gyres away from the prevailing winds is the **Coriolis effect.** To understand this effect, consider the rotating Earth: The circumference of Earth is greatest at the equator and decreases to zero at the poles. But all parts of the planet make one complete rotation every day. Therefore, a point on the equator must travel farther and faster than any point closer to the poles. At the equator, all objects move eastward with a velocity of about 1,600 kilometers per hour; at the poles there is no eastward movement at all and the velocity is 0 kilometers per hour.

Now imagine a rocket fired from the equator toward the North Pole. Before it was launched it was traveling eastward at 1,600 kilometers per hour with the rotating Earth. As it takes off, it is moving eastward at 1,600 kilometers per hour and northward at its launch speed. As it moves north from the equator, it is still traveling eastward at 1,600 kilometers per hour, but points on Earth beneath it move eastward at a slower and slower speed as the rocket approaches the pole. As a result, the rocket curves toward the east, or the right. In a similar manner, a mass of water or air deflects in an easterly direction as it moves poleward from the equator, as shown in ◀▶ Figure 16.15A.

Conversely, consider an ocean current flowing southward from the Arctic Ocean toward the equator. Since it started near the North Pole, this water moves more slowly than Earth's surface near the equator, and therefore the current veers toward the west, or to the right, as shown in ◀▶ Figure 16.15B. Thus, north–south currents always veer to the right in the Northern Hemisphere. In the Southern Hemisphere currents turn toward the left for the same reason.

At the sea surface, current directions are affected by both the prevailing winds and the Coriolis effect. However, below the surface, the water doesn't "feel" the wind; it only "feels" the movement of water directly above. But the Coriolis effect is as strong at depth as it is at the sea surface. As a result, each successively deeper layer is less affected by the wind and more affected by the Coriolis effect. Consequently, deeper layers of water in the surface currents are deflected even more to the right in the Northern Hemisphere, and to the left in the Southern Hemisphere, than is the shallowest water. The net effect of this process on the flow directions of the gyres, down to a depth of about 100 meters, is called **Ekman transport,** after the Swedish scientist who developed the mathematics that describe the process.

Yet another factor deflects surface currents. The combination of wind and Ekman transport cause the sea surface to pile up and form a hill of water in the middle of the North Atlantic Ocean. It isn't a big hill as hills on land go; it is only 2 meters high and slopes gradually from the middle of the ocean near the Sargasso Sea to land on both sides of the North Atlantic. It is so subtle that it can be detected only by satellite imaging of the sea surface. However, as surface currents flow toward the mound of water, they veer away from the hill rather than flowing uphill. This process deflects surface currents from directions controlled by prevailing winds and Ekman transport. In theory, Ekman transport should deflect surface currents 90 degrees to the right of the prevailing wind direction in the Northern Hemisphere, and 90 degrees left of prevailing winds in the Southern Hemisphere. However, the actual deflection of water flow from the prevailing wind direction due to Ekman transport is 45 degrees or less because of the topography of the sea surface.

Surface current gyres controlled by the combination of wind, the Coriolis effect modified by Ekman transport, and

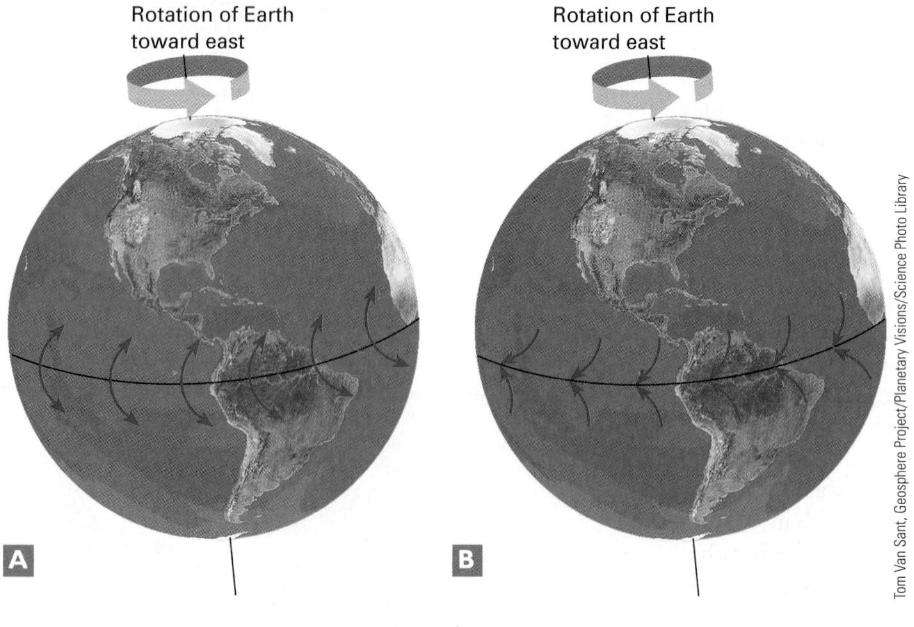

Rotation of Earth toward east

Rotation of Earth toward east

A   **B**

Tom Van Sant, Geosphere Project/Planetary Visions/Science Photo Library

◀▶ **FIGURE 16.15** The Coriolis effect deflects water and wind currents. **(A)** Water or air moving poleward from the equator is traveling east faster than the land beneath it and veers to the east (turns right in the Northern Hemisphere and left in the Southern Hemisphere). **(B)** Water or air moving toward the equator is traveling east slower than the land beneath it and veers to the west (turns right in the Northern Hemisphere and left in the Southern Hemisphere).

INTERACTIVE QUESTION: *Predict how a wind would be affected if it were moving due east. Due west?*

sea surface topography are called **geostrophic gyres** (*geos* for Earth, and *strophe* for turning). Of the six great surface currents in the world oceans shown in Figure 16.13—two in the Northern Hemisphere and four in the Southern Hemisphere—five are geostrophic gyres. The sixth, the West Wind Drift, or Antarctic Circumpolar Current, is driven by strong, ceaseless westerly winds in that inhospitable part of the world.

Deep-Sea Currents

Wind does not affect the ocean depths, and oceanographers once thought that deep ocean water was almost motionless. In her book *The Sea Around Us,* Rachael Carson wrote that the ocean depths are "a place where change comes slowly, if at all." However, in 1962 ripples and small dunes were photographed on the floor of the North Atlantic. Because flowing water forms these features, the photographs suggested that water was moving in the ocean depths. More recently, oceanographers have measured deep-sea currents directly with flow meters, and photographed moving sand and mud with underwater television cameras.

Wind drives surface currents, but deep-sea currents are driven by differences in water density. Dense water sinks and flows horizontally along the sea floor to form a deep-sea current. Two factors cause water to become dense and sink: cooling temperature and rising salinity. The global deep-sea circulation shown in ◆ Figures 16.12 and 16.16 is caused by these two factors and is called **thermohaline circulation** (*thermo* for temperature, and *haline* for salinity).

Recall that water is densest when it is cold, close to freezing. Therefore, as tropical surface water moves poleward and cools, it becomes denser and sinks. In addition, water density increases as salinity increases.

Therefore, water sinks when it becomes saltier. Seawater can become saltier if surface water evaporates. Polar seas also become saltier when the surface freezes, because salt does not become incorporated in the ice. Arctic and Antarctic water is dense because the water is both cold and salty. In contrast, addition of freshwater makes seawater less salty and less dense. This effect is pronounced in enclosed bays with abundant, inflowing rivers and also in the polar regions when icebergs float into the ocean and melt. As we will see in Chapter 21, recent rapid melting of Greenland glaciers introduces enough freshwater into the North Atlantic to reduce the density of surface water and alter its buoyancy.

◆ Figure 16.13 shows that the Gulf Stream originates in the subtropics. When this water reaches the northern part of the Atlantic Ocean near the tip of Greenland, it cools and sinks. When the sinking water reaches the sea floor, it is deflected southward to form the North Atlantic Deep Water, which flows along the sea floor all the way to Antarctica (◆ Figure 16.16). An individual water molecule that sinks near Greenland may travel for 500 to 2,000 years before resurfacing half a world away in the south polar sea.

Upwelling

If water sinks in some places, it must rise in others to maintain mass balance. This upward flow of water is called **upwelling**. Upwelling carries cold water from the ocean depths to the surface. Upwelling also brings nutrients from the deep ocean to the surface, creating rich fisheries along the coasts of California and Peru. Several processes can cause upwelling, both in the open oceans and along coastlines.

In ◆ Figure 16.13, note that the California Current flows southward along the coast of California. In the

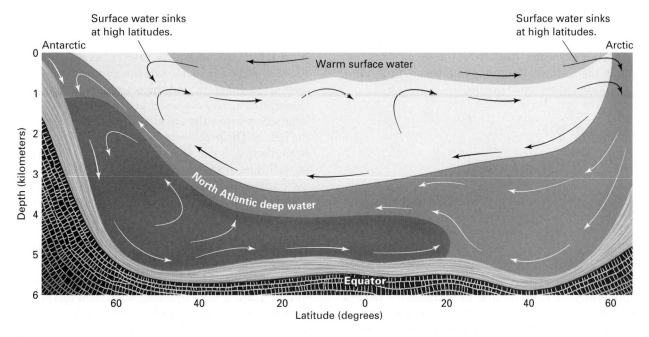

◆ **FIGURE 16.16**   A profile of the Atlantic Ocean shows surface and subsurface currents.

Southern Hemisphere, the Humboldt Current moves northward along the west coast of South America. Both currents are deflected westward—away from shore—by the Coriolis effect modified by Ekman transport. As these surface currents veer away from shore, water from the ocean depths rises toward the surface along the California coast and the west coast of South America. In August, water on the East Coast of the United States is warmed by the Gulf Stream and may be a comfortable 21°C. However, on the central California coast, the cool California Current combines with the upwelling deep water to produce water that is only 15°C, and surfers and swimmers must wear wetsuits to stay in the water for long.

The frictional drag of a prevailing offshore wind can pull surface water away from a coast. Deep water then upwells along the continental shelf to replace the surface water flowing away from shore. Winds blowing parallel to shore can also create upwelling. The wind causes water to flow parallel to shore, but the Coriolis effect and Ekman transport deflect the current, driving it away from the coast. Deep water then rises to replace the surface water.

In most years, an offshore wind drives surface water away from the coast of Peru and adjacent portions of western South America, creating a strong, nutrient-rich upwelling current that produces rich fisheries along that coast. In El Niño years, however, the offshore wind weakens and the upwelling does not occur. As a result unusually warm, nutrient-poor water accumulates along the west coast of South America, displacing the cold Humboldt Current. This occurs about every 3 to 7 years, and its effects last for about a year before conditions return to normal. Many meteorologists now think that El Niño affects weather patterns for nearly three quarters of Earth. The causes and effects of El Nino are described in Chapter 19.

Figure 16.13 shows that the south equatorial currents of both the Atlantic and Pacific flow near the equator. Although the Coriolis effect is weak near the equator, water in those currents is deflected poleward, and deep, cold water rises from the ocean depths to replace the surface water in **equatorial upwelling**, which occurs in the open oceans.

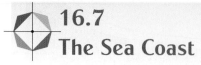

16.7
The Sea Coast

As we mentioned in the introduction to this chapter, coastlines are among the most geologically active zones on Earth, where atmosphere, geosphere, hydrosphere, and biosphere all affect the local environments. Subduction occurs along many continental coasts. Waves and currents weather, erode, transport, and deposit sediment continuously on all coastlines. In addition, the shallow waters of continental shelves are among the most productive biological ecosystems on Earth. Many organisms that live in these regions build hard shells or skeletons of cal-

cium carbonate. Eventually, the remains of these shells and skeletons become lithified to form limestone. Thus living organisms become part of the rock cycle.

Weathering and Erosion on the Sea Coast

Waves batter most coastlines. If you walk down to the shore, you can watch turbulent water carry sand grains or even small cobbles along the beach. Even on a calm day, waves steepen as they approach shore and then crash against the beach. When a wave enters shallow water, the bottom of the wave drags against the sea floor. This drag compresses the circular motion of the wave into ellipses. This deformation slows the lower part of the wave, so that the upper part moves more rapidly than the lower. As the front of the wave rises over the base, the wave steepens until it collapses forward, or **breaks** (◀ Figure 16.17). Chaotic, turbulent waves breaking along a shore are called **surf**.

Most coastal erosion occurs during intense storms because storm waves, and particularly storm surges, which were described in Section 16.5, are much larger and more energetic than normal waves (◀ Figure 16.18). A 6-meter-high wave strikes shore with 40 times the force of a 1.5-meter-high wave. A giant, 10-meter-high storm wave strikes a 10-meter-wide sea wall with four times the thrust energy of the three main orbiter engines of a space shuttle.

Seawater weathers and erodes coastlines by hydraulic action, abrasion, solution, and salt cracking, processes that we are familiar with from our earlier discussions of weathering and streams.

A wave striking a rocky cliff drives water into cracks or crevices in the rock, compressing air in the cracks. The air and water combine to create hydraulic forces strong enough to dislodge rock fragments or even huge boulders. Storm waves create forces as great as 25 to 30 tons per square meter. Engineers built a breakwater in Wick Bay, Scotland, of car-sized rocks weighing 80 to 100 tons each. The rocks were bound together with steel rods set in concrete, and the sea wall was topped by a steel-and-concrete cap weighing more than 800 tons. A large storm broke the cap and scattered the rocks about the beach. The breakwater was rebuilt, reinforced, and strengthened, but a second storm destroyed this wall as well. On the Oregon coast, the impact of a storm wave tossed a 60-kilogram rock over a 25-meter-high lighthouse. After sailing over the lighthouse, it crashed through the roof of the keeper's cottage, startling the inhabitants.

While images of flying boulders are spectacular, most wave erosion occurs gradually, by abrasion. Water is too soft to abrade rock, but waves carry large quantities of silt, sand, and gravel. Breaking waves roll this sediment back and forth over bedrock, acting like liquid sandpaper, eroding the rock. At the same time, smaller cobbles are abraded as they roll back and forth in the surf zone.

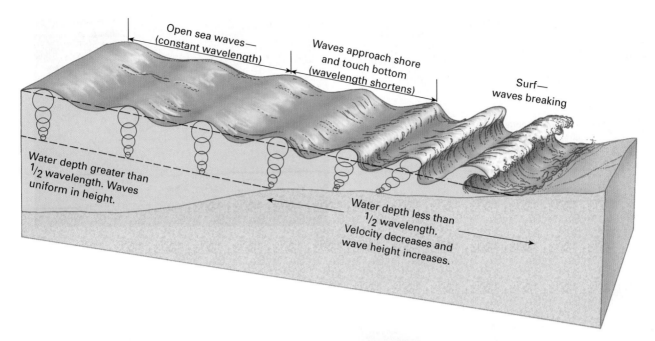

When a wave approaches the shore, the circular motion at the bottom of the wave flattens out and becomes elliptical. The bottom of the wave drags against the seafloor. As a result, the wavelength shortens and the wave steepens until it finally breaks, creating surf.

Susan Blanchet/Dembinsky Photo Assoc.

◀▶ **FIGURE 16.18** Expensive seashore houses are often threatened by coastal erosion that occurs during intense storms; Georgetown in Grand Cayman Island.

Seawater slowly dissolves rock and carries ions in solution. Saltwater also soaks into bedrock; when the water evaporates, the growing salt crystals pry the rock apart.

Sediment Transport Along Coastlines

Most waves approach the shore at an angle rather than head on. When this happens, one end of the wave encounters shallow water and slows down, while the rest of the wave is still in deeper water and continues to advance at a relatively faster speed. As a result, the wave bends (◀▶ Figures 16.19A and B). This effect is called **refraction**. Consider the analogy of a sled gliding down a snowy hill onto a cleared road. If the sled hits the road at an angle, one runner will reach it before the other. The runner that hits the pavement first slows down, while the other, which is still on the snow, continues to travel rapidly (◀▶ Figure 16.19C). As a result, the sled turns abruptly.

As waves approach an irregular coast, they reach the headlands first, breaking against the point and eroding it. The waves then refract around the headland and travel parallel to its sides. Surfers seek such refracted waves because these waves move nearly parallel to the coast and therefore travel for a long distance before breaking. For example, the classic big-wave mecca at Waimea Bay of Oahu, Hawaii, forms along a point of land that juts into the sea. Waves build when they strike an offshore coral reef, then refract along the point and charge into the bay (◀▶ Figure 16.20).

Refracted waves transport sand and other sediment toward the interior of a bay. As the headlands erode and the interiors of bays fill with sand, an irregular coastline eventually straightens (◀▶ Figure 16.21).

When waves strike shore at an angle, they form a **longshore current** that flows parallel to the shore. Longshore currents flow in the surf zone and a little farther out to sea and may travel for tens or even hundreds of kilometers. They transport sand for great distances along coastlines.

Sediment transport also occurs by **beach drift**. If a wave strikes the beach obliquely, it pushes sand up and along the beach in the direction that the wave is traveling. When water recedes, the sand flows straight down the

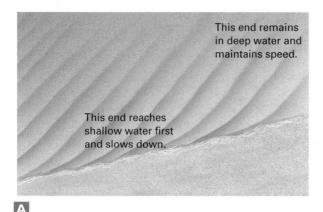

This end remains in deep water and maintains speed.

This end reaches shallow water first and slows down.

A

Courtesy of Graham R. Thompson/Jonathan Turk

B

Sled analogy

Snowy hill

Road

C

ThomsonNOW ◆ **ACTIVE FIGURE 16.19**
(A) When a water wave strikes the shore at an angle, one end slows down, causing the wave to bend, or refract. **(B)** Wave refraction on a lake shore. **(C)** A sled turns upon striking a paved roadway at an angle because one runner hits the roadway and slows down before the other does.

beach as shown in ◆ Figure 16.22. Thus, at the end of one complete wave cycle, the sand has moved a short distance parallel to the coast. The next wave transports the sand a little further, until, over time, sediment moves long distances.

Longshore currents and beach drift work together to transport and deposit huge amounts of sand along a coast. Much of the sand found at Cape Hatteras, North Carolina, originated hundreds of kilometers away, from the mouth of the Hudson River in New York and from glacial deposits on Long Island and southern New Eng-

Warren Bolster/Getty Images

◆ **FIGURE 16.20** Surfers in Waimea Bay on Oahu, Hawaii, cherish the long-lasting surf that forms when ocean waves reflect around a headland and break as they enter the bay.

land. Midway along this coast, at Sandy Hook, New Jersey, an average of 2,000 tons a day move past any point on the beach. As a result of this process, beaches have been called rivers of sand.

Tidal Currents

When tides rise and fall along an open coastline, water moves in and out from the shore as a broad sheet. If the flow is channeled by a bay with a narrow entrance or by islands, the moving water funnels into a **tidal current**, which is a flow of ocean water caused by tides. Tidal currents can be intense where large differences exist between high and low tides and narrow constrictions occur in the shoreline. On parts of the west coast of British Columbia, a diesel-powered fishing boat cannot make headway against tidal currents flowing between closely spaced islands. Fishermen must wait until the tide, and hence the tidal currents, reverse direction before proceeding.

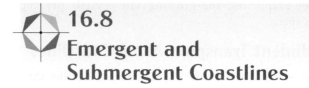

16.8
Emergent and Submergent Coastlines

Geologists have found drowned river valleys and fossils of land animals on continental shelves beneath the sea. They have also found sedimentary rocks containing

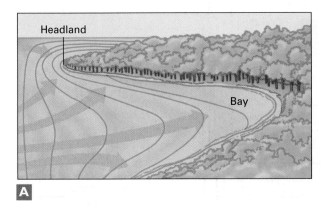

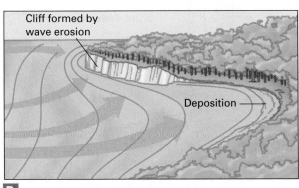

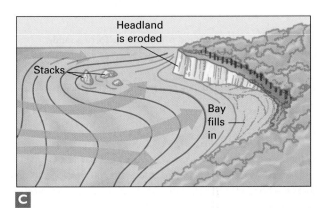

FIGURE 16.21 **(A and B)** When a wave strikes a headland, the shallow water causes that portion of the wave to slow down. Part of the wave breaks against the headland, weathering the rock. **(C)** A portion of the wave refracts, transporting sediment and depositing it on the beach inside the bay. Eventually, this selective weathering, erosion, and deposition will straighten an irregular coastline.

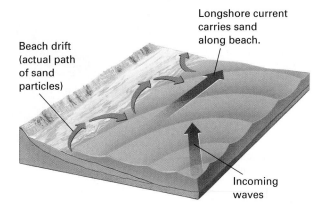

FIGURE 16.22 Longshore currents and beach drift transport sediment along a coast.

occurs when sea level rises or coastal land sinks. A submergent coast is commonly irregular, with many bays and headlands. The coast of Maine, with its numerous inlets and rocky bluffs, is a submergent coastline (◆ Figure 16.24). Small, sandy beaches form in protected coves, but most of the headlands are rocky and steep.

Factors That Cause Coastal Emergence and Submergence

Tectonic processes can cause a coastline to rise or sink. Isostatic adjustment can also depress or elevate a portion of a coastline. About 18,000 years ago, a huge continental glacier covered most of Scandinavia, causing it to sink isostatically. As the lithosphere settled, the displaced asthenosphere flowed southward, causing the Netherlands to rise. When the ice melted, the process reversed as the asthenosphere flowed back from below the Netherlands to Scandinavia. Today, Scandinavia is rebounding and the Netherlands is sinking. During the Pleistocene Ice Age, Canada was depressed by the ice, and asthenosphere rock flowed southward. Today, the asthenosphere is flowing back north, much of Canada is rebounding, and much of the United States is sinking.

Sea level can also change globally. A global sea-level change, called **eustatic sea-level change**, occurs by three mechanisms: the growth or melting of glaciers, changes in water temperature, and changes in the volume of the Mid-Oceanic Ridge.

During an ice age, vast amounts of water move from the sea to form continental glaciers, and sea level falls, resulting in global emergence. Similarly, when glaciers melt, sea level rises globally, causing submergence.

Seawater expands when it is heated and contracts when it is cooled. Although this change is not noticeable in a glass of water, the volume of the oceans is so great that a small temperature change can alter sea level measurably. As a result, global warming causes sea-level rise, and cooling leads to falling sea level.

Temperature changes and glaciation are linked. When global temperature rises, seawater expands and glaciers melt. Thus, even minor global warming can lead

fossils of fish and other marine organisms in continental interiors. As a result, we infer that sea level has changed, sometimes dramatically, throughout geologic time. An **emergent coastline** forms when a portion of a continent that was previously under water becomes exposed as dry land. Falling sea level or rising land can cause emergence. As explained in the following section, many emergent coastlines are sandy. In contrast, a **submergent coastline** develops when the sea floods low-lying land and the shoreline moves inland (◆ Figure 16.23). Submergence

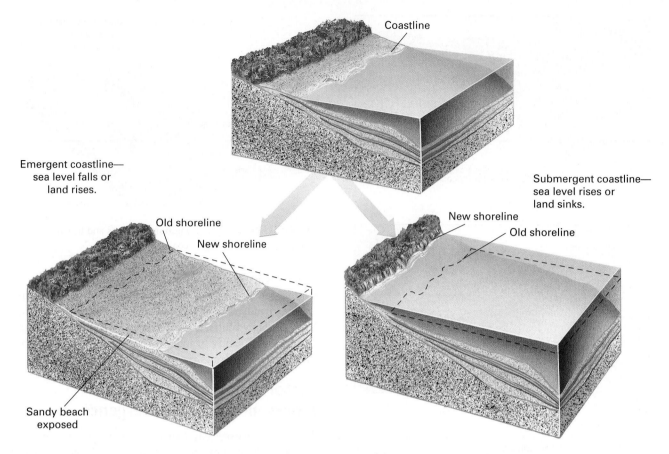

Coastline

Emergent coastline—
sea level falls or
land rises.

Submergent coastline—
sea level rises or
land sinks.

Old shoreline

New shoreline

New shoreline

Old shoreline

Sandy beach
exposed

◀▶ **FIGURE 16.23**   If sea level falls or if the land rises, the coastline is emergent. Offshore sand is exposed to form a sandy beach. If coastal land sinks or sea level rises, the coastline is submergent. Areas that were once land are flooded. Irregular shorelines develop and beaches are commonly rocky.

INTERACTIVE QUESTION: *Draw a sequence of diagrams to show how barrier islands could form during coastal emergence, or how small, more circular, islands could form during coastal submergence.*

to a large sea-level rise. The opposite effect is also true. When temperature falls, seawater contracts, glaciers grow, and sea level falls.

As explained in Chapter 15, changes in the volume of the Mid-Oceanic Ridge can also affect sea level. The Mid-Oceanic Ridge displaces seawater. When lithospheric plates spread slowly from the Mid-Oceanic Ridge, they create a narrow ridge that displaces relatively little seawater, resulting in low sea level. In contrast, rapidly spreading plates produce a high-volume ridge that displaces more water, causing a global sea-level rise. At times in Earth's history, spreading has been relatively rapid, and as a result, global sea level has been high.

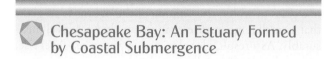

Chesapeake Bay: An Estuary Formed by Coastal Submergence

An **estuary** forms where rising sea level or a sinking coastline submerges a broad river valley or other basin. Estuaries are ordinarily shallow and have gently sloping beaches. Streams transport nutrients

to the estuary, and the shallow water provides habitats for marine organisms. Therefore, estuaries are among the most biologically productive ecosystems on Earth. However, estuaries also make excellent harbors and therefore are prime sites for industrial activity. As a result, many estuaries have become seriously polluted in recent years.

Chesapeake Bay is an estuary formed by submergence of the Susquehanna River valley (◀▶ Figure 16.25). Geologists have recently discovered a series of concentric faults in the bedrock beneath the bay. These faults provide evidence that a meteorite impact 35 million years ago may have formed a depression in Earth's crust. Later, this depression filled with water as sea level rose following the latest glacial retreat.

Chesapeake Bay is the nation's largest and most productive estuary. It is approximately 100 kilometers long, averages 10 to 15 kilometers wide, and contains numerous bays and inlets. Despite its great size, it is only 7 to 10 meters deep near its mouth, and its greatest depth is 50 meters. Three major cities—Washington, Baltimore, and Harrisburg—lie along Chesapeake Bay or its tributaries. Between 1950 and 2003, the population in the watershed doubled to nearly 16 million people, the urban land area tripled, and the bay became polluted.

Water runoff from farms and suburban developments carries silt, fertilizers, and pesticides into Chesapeake Bay. In addition,

◁▷ **FIGURE 16.24**   The Maine coast is a rocky, irregular, submergent coastline.

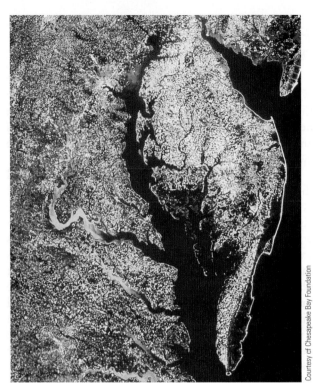

◁▷ **FIGURE 16.25**   Satellite view of Chesapeake Bay.

sewage treatment plants release nutrients into the bay, and numerous factories discharge toxic compounds. Silt destroys habitats by clogging spaces between rocks and covering plant roots, thereby reducing the flow of oxygen between the bay's waters and the bottom sediments, where many shellfish, plants, and other aquatic organisms live and feed. Fertilizers and sewage supply large amounts of nitrogen and phosphorous to the bay, which nourish algal blooms that deprive the bay grasses and other plants of sunlight, and deplete the waters of oxygen. The lowered oxygen levels kill fish and animals living in Chesapeake Bay and thereby degrade the recreational and commercial industries of the bay.

Pesticides are often toxic to marine organisms. Pollution and overfishing have devastated animal and plant populations in Chesapeake Bay. In 1960, fishermen harvested 2.7 million kilograms of striped bass from the bay; by 1985, the catch had dwindled by 90 percent to 270,000 kilograms. The bay once supported the richest oyster beds in the world; today the oyster population has dropped by 99 percent. The oyster harvest in Maryland alone decreased from 1.6 million bushels in 1986 to 26,500 bushels in 2004.

By the early 1980s, people had become alarmed by the pollution and ecosystem destruction. In 1984, the Chesapeake Bay Foundation and the Natural Resources Defense Council filed a lawsuit against the Bethlehem Steel Corporation for illegally dumping wastes into the estuary. The environmental groups won the suit: Bethlehem Steel was fined more than $1 million and was required to reduce the quantity

of pollutants emitted. Other industrial polluters began cleanups to avoid similar suits. In the same year, the governments of Maryland, Virginia, Pennsylvania, and the District of Columbia tightened regulations on municipal sewage disposal. State-run agricultural education programs showed farmers that they can maintain high yields with fewer fertilizers and pesticides than they had been using.

All of these efforts have improved the water quality of Chesapeake Bay, although there is still a long way to go. For example, water sampling between 1985 and 1993 show a decrease in phosphorous levels in waters of the estuary. At the same time, nitrogen levels have increased, indicating continued influx of certain pollutants. Fish populations are recovering slowly. In 1985, the Atlantic States Marine Fisheries Commission (AFMFC) declared a moratorium on fishing for striped bass on the Atlantic coast in an effort to restore the population. Following evidence of increasing striped-bass populations, the AFMFC declared the Atlantic coast striped bass fully recovered in 1995. Management efforts continue today, and the striped-bass population continues to increase in the bay. The goal of the Chesapeake Bay Oyster Recovery Program, described in the Chesapeake Bay 2000 agreement, is to increase oyster habitat tenfold by 2010.

But as long as cities line its coasts, the bay can never return to a pristine state. Developers have built concrete wharves over saltwater marshes. Runoff from roadways, parking lots, and storm drains carries oil and other contaminants into the estuary. Air pollutants drift into the water; runoff from farmers' fields and suburban lawns can be reduced but not eliminated.

Chemicals that seeped into nearby aquifers decades ago slowly percolate into the estuary's waters.

When most people think about going to the beach, they think of gently sloping expanses of sand. However, a **beach** is any strip of shoreline that is washed by waves and tides. Although many beaches are sandy, others are swampy, or rocky (◆ Figure 16.26).

A beach is divided into two zones, the **foreshore** and the **backshore**. The foreshore, also called the **intertidal zone**, lies between the high- and low-tide lines and is alternately exposed to the air at low tide and covered by water at high tide. The backshore is usually dry but is washed by waves during storms. Many terrestrial plants cannot survive in saltwater, so specialized, salt-resistant plants live in the backshore. The backshore can be wide or narrow, depending on its topography, the local tidal difference, and the frequency and intensity of storms. In a region where the land rises steeply, the backshore may be a narrow strip. In contrast, if the coast consists of low-lying plains and if coastal storms occur regularly, the backshore may extend several kilometers inland.

If weathering and erosion occur along all coastlines, why are some beaches sandy and others rocky? The answer lies partly in the fact that most sand is not formed by weathering and erosion at the beach itself. Instead, several processes transport sand to a seacoast. Rivers carry large quantities of sand, silt, and clay to the sea and deposit it on deltas that may cover thousands of square kilometers. In some coastal regions, glaciers deposited large quantities of sandy till along coastlines during the Pleistocene Ice Age. In tropical and subtropical latitudes, eroding reefs supply carbonate sand to nearby beaches. A sandy coastline is one with abundant sediment from any of these sources.

Longshore currents transport and deposit the sand along the coast. Much of the sand carried by these currents accumulates on underwater offshore bars. Thus, a great deal of sand may be stored offshore from a beach. If such a coastline emerges, this vast supply of sand becomes exposed as dry land. Thus, sandy beaches are abundant on emergent coastlines.

In contrast, rocky coastlines occur where sediment from any of these sources is scarce. With no abundant sources of sand, small sandy beaches may form in protected bays but most of the coast is rocky. On a submergent coastline, rising sea level puts the stored offshore sand even farther out to sea and below the depth of waves. As a result, submergent coastlines commonly have rocky beaches.

Sandy Coastlines

A **spit** is a small, finger-like ridge of sand or gravel that extends outward from a beach (◆ Figure 16.27). As sediment migrates along a coast, the spit may continue to grow. A well-developed spit may rise several meters above high-tide level and may be tens of kilometers long. A spit may block the entrance to a bay, forming a **baymouth bar**. A spit may also extend outward into the sea, creating a trap for other moving sediment.

A **barrier island** is a long, low-lying island that extends parallel to the shoreline. It looks like a beach or spit and is separated from the mainland by a sheltered body of water called a **lagoon** (◆ Figure 16.28). Barrier islands extend along the east coast of the United States from New York to Florida. They are so nearly continuous that a sailor in a small boat can navigate the entire coast inside the barrier island system and remain protected from the open ocean most of the time. Barrier islands also line the Texas Gulf Coast.

Barrier islands form in several ways. The two essential ingredients are a large supply of sand and waves or currents to transport it. If a coast is shallow for several kilometers outward from shore, breaking storm waves may carry sand toward shore and deposit it just offshore as a barrier island. Alternatively, if a longshore current

Courtesy of Graham R. Thompson/Jonathan Turk

A

B

◆ **FIGURE 16.26** **(A)** Sea lions doze on a sandy California beach. **(B)** Rocky beaches are found along the Oregon coast to the north.

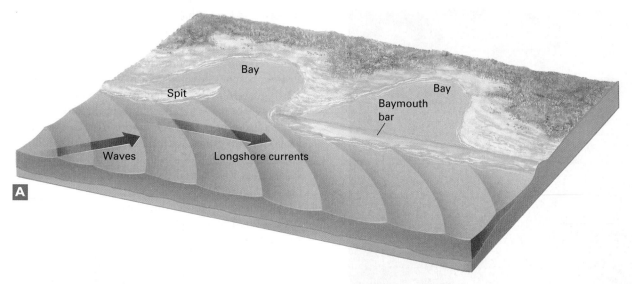

◆ **FIGURE 16.27** **(A)** Spits and baymouth bars are common features of sandy emergent coastlines. **(B)** Aerial photograph of a spit that formed along a low-lying coast in northern Siberia.

veers out to sea, it slows down and deposits sand where it reaches deeper water. Waves may then pile up the sand to form a barrier island.

Other mechanisms involve sea-level change. Underwater sand bars may be exposed as a coastline emerges. Alternatively, sand dunes or beaches may form barrier islands if a coastline sinks.

Development on Sandy Coastlines

The Atlantic coast of the United States is fringed with the longest chain of barrier islands in the world. Many seaside resorts are built on these islands, and developers often ignore the fact that they are transient and changing landforms (◆ Figure 16.29). If the rate of erosion exceeds that of deposition for a few years in a row, a barrier island can shrink or disappear completely, leading to destruction of beach homes and resorts. In addition, barrier islands are especially vulnerable to hurricanes, which

can wash over low-lying islands and move enormous amounts of sediment in a very brief time. In September of 1996, Hurricane Fran flattened much of Topsail Island, a low-lying barrier island in North Carolina. Geologists were not surprised because the homes were not only built on sand—they were built on sand that was virtually guaranteed to move.

As a second example, Long Island extends eastward from New York City and is separated from Connecticut by Long Island Sound. Longshore currents flow westward, eroding sand from glacial deposits at the eastern end of the island and depositing it to form beaches and barrier islands on the south side of the island (◆ Figure 16.30). At any point along the beach, the currents erode and deposit sand at approximately the same rates (◆ Figure 16.31A). Geologists calculate that supply of sand at the eastern end of the island is large enough to last for a few hundred years. When the glacial deposits at the eastern end

◆ **FIGURE 16.28** This system of barrier islands near Houston, Texas, forms protected waters inside the bay.

of the island become exhausted, the flow of sand will cease. Then the entire coastline will erode and the barrier islands and beaches will disappear.

However, if we narrow our time perspective and look at a Long Island beach over a season or during a single storm, the rates of erosion and deposition are not equal. Thus beaches shrink and expand with the seasons or the passage of violent gales. In the winter, violent waves and currents erode beaches, whereas sand accumulates on the beaches during the calmer summer months. In an effort to prevent these seasonal fluctuations and to protect their personal beaches, Long Island property owners have built stone barriers called **groins** from shore out into the water. The groin intercepts the steady flow of sand moving from

the east and keeps that particular part of the beach from eroding. But the groin impedes the overall flow of sand. West of the groin the beach erodes as usual, but the sand is not replenished because the upstream groin traps it. As a result, beaches downcurrent from the groin erode away (◆ Figure 16.31B). The landowner living downcurrent from a groin may then decide to build another groin to protect his or her beach (◆ Figure 16.31C). The situation has a domino effect, with the net result that millions of dollars are spent in ultimately futile attempts to stabilize a system that was naturally stable in its own dynamic manner (◆ Figure 16.32).

Storms pose another dilemma. Hurricanes commonly strike Long Island in the late summer and fall, generating storm waves that completely overrun the barrier islands, flattening dunes and eroding beaches. When the storms are over, gentler waves and longshore currents carry sediment back to the beaches and rebuild them. As the sand accumulates again, salt marshes rejuvenate and the dune grasses grow back within a few months.

These short-term fluctuations are incompatible with human ambitions. People build houses, resorts, and hotels on or near the shifting sands. The owner of a home or resort hotel cannot allow the buildings to be flooded or washed away. Therefore, property owners construct large sea walls along the beach. When a storm wave rolls across an undeveloped low-lying beach, it dissipates its energy gradually as it flows over the dunes and transports sand. The beach is like a judo master who defeats an opponent by yielding with the attack, not countering it head on. A sea wall interrupts this gradual absorption of wave energy. The waves crash violently against the barrier and erode sediment at its base until the wall collapses. It may seem surprising that a reinforced concrete sea wall is more likely to be permanently destroyed than a beach of grasses and sand dunes, yet this is often the case (◆ Figure 16.33).

◆ **FIGURE 16.29** Many seaside resorts, such as the Fountainebleau Hilton Hotel at South Beach, Miami, are built on transient and changing barrier islands.

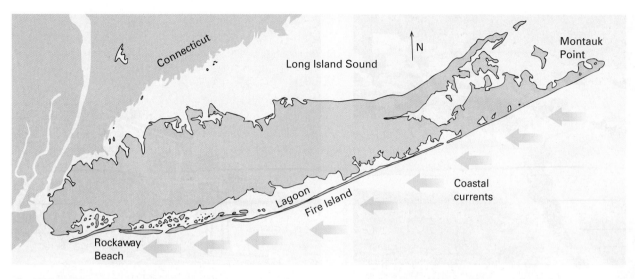

◆ **FIGURE 16.30** Longshore currents carry sand westward along the south shore of Long Island to create a series of barrier islands.

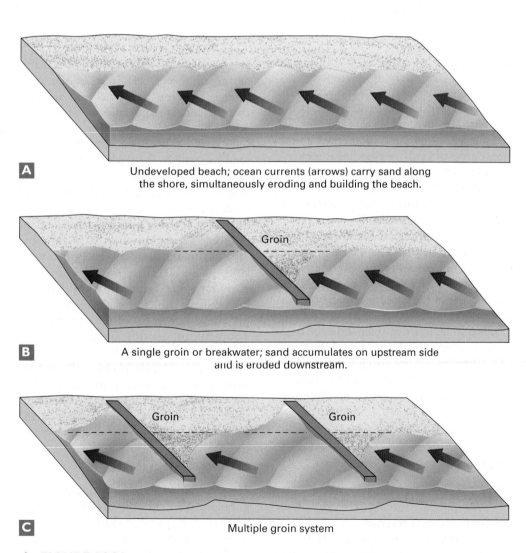

A Undeveloped beach; ocean currents (arrows) carry sand along the shore, simultaneously eroding and building the beach.

B A single groin or breakwater; sand accumulates on upstream side and is eroded downstream.

C Multiple groin system

◆ **FIGURE 16.31** **(A)** Longshore currents simultaneously erode and deposit sand along an undeveloped beach. **(B)** A single groin or breakwater traps sand on the upstream side, resulting in erosion on the downstream side. **(C)** A multiple groin system propagates the uneven distribution of sand along the entire beach.

◆ **FIGURE 16.32** **(A)** This aerial photograph of a Long Island beach shows sand accumulating on the upstream side of a groin and erosion on the downstream side. **(B)** A close-up of the house in A shows waves lapping against the foundation.

Rocky Coastlines

A rocky coastline is one without any of the abundant sediment sources described previously. In many areas on land, bedrock is exposed or covered by a thin layer of soil. If this type of sediment-poor terrain is submerged, and if there are no other sources of sand, the coastline is rocky.

A **wave-cut cliff** forms when waves erode the headland into a steep profile. As the cliff erodes, it leaves a flat or gently sloping **wave-cut platform** (◆ Figure 16.34). If waves cut a cave into a narrow headland, the cave may eventually erode all the way through the headland, forming a scenic **sea arch**. When an arch collapses or when the inshore part of a headland erodes faster than the tip, a pillar of rock called a **sea stack** forms (◆ Figure 16.35). As waves continue to batter the rock, eventually the sea stack crumbles.

If the sea floods a long, narrow, steep-sided coastal valley, a sinuous bay called a **fjord** is formed. Fjords are common at high latitudes, where rising sea level has flooded coastal valleys scoured by Pleistocene glaciers. Fjords may be hundreds of meters deep, and often the cliffs drop straight into the sea.

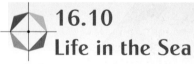

16.10
Life in the Sea

On land, most photosynthesis is conducted by multicellular plants such as mosses, ferns, grasses, and trees. Large animals such as cows, deer, elephants, and bison consume the plants. In contrast, most of the photosynthesis and consumption in the ocean is carried out by small organisms called **plankton**. Many plankton are single-celled and microscopic; others are more complex and are up to a few centimeters long. Another major difference between terrestrial and aquatic ecosystems is that on land, soil nutrients are abundant on the surface, where light is also abundant. However, in the oceans, light is available only on the surface, whereas nutrients tend to settle to the dark depths. Plankton live mostly within a few meters of the sea surface, where light is available. However, the growth and productivity of these organisms is limited by the fact that most of the nutrients such as nitrates, iron, and phosphates, are pulled downward toward the sea floor by gravity.

Phytoplankton conduct photosynthesis like land-based plants do. Therefore, they are the base of the food chain for aquatic animals. Although phytoplankton are not readily visible, they are so abundant that they supply about 50 percent of the oxygen in our atmosphere. **Zooplankton** are tiny animals that feed on the phytoplankton (◆ Figure 16.36). The larger and more familiar marine plants (such as seaweed) and animals (such as fish, sharks, and whales) play a relatively small role in oceanic photosynthesis and consumption. However these organisms are important to humans, because people depend on fish for a vital source of protein.

World Fisheries

The shallow water of a continental shelf supports large populations of marine organisms. In addition, many deep-sea fish spawn in shallow water within a kilometer or two of shore. Shallow zones in bays, lagoons, and estuaries are especially hospitable to life because they have (1) easy access to the deep sea, (2) lower salinity than the open ocean, (3) a high concentration of nutrients originating from land and sea, (4) shelter, and (5) abundant

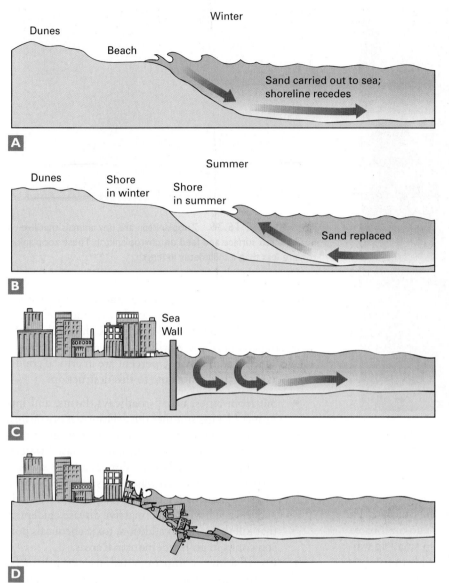

A

B

Sea
Wall

C

D

◈ **FIGURE 16.33** **(A)** In a natural beach, the violent winter waves often move sand out to sea. **(B)** The gentler summer waves push sand toward shore and rebuild the beach. **(C)** Wave energy concentrates against a sea wall and **(D)** may eventually destroy it.

◈ **FIGURE 16.34** Waves hurl sand and gravel against solid rock to erode cliffs and create a wave-cut platform along the Oregon coast.

Courtesy of Graham R. Thompson/Jonathan Turk

plant life rooted to the sea floor in addition to the phytoplankton floating on the surface. As a result, about 99 percent of the marine fish caught every year are harvested from the shallow waters adjacent to shore.

World fish harvests increased steadily from 1950 to 1989 and then leveled off during the 1990s at about 90 million tons annually (◈ Figure 16.37). However, during the past 20 years, harvests of some species have declined while others have risen. For example, Atlantic cod is a valuable fish that thrived off the northeast coast of North America. The first cod fishermen fished this dangerous but fertile coast 12 years after Columbus had landed in the New World. After 300 years of intensive fishing, the cod stocks began to decline. When cod populations declined, their prey—herring, mackerel, and squid—thrived. In 1816, Atlantic fishermen began harvesting mackerel, which had previously been used only as bait. By the 1850s, boats set out to net herring.

In 1970, fishermen harvested 3 million tons of cod. But this catch was biologically unsustainable and declined to 1 million tons by 1993. In 1995, biologists suspended cod fishing in many areas to allow the fish populations to recover. In 1978, the herring and mackerel stocks began to decline, so boats switched nets to trawl for squid. In the 1980s the squid population began to decline. Thus fishing pressure has worked through the food chain, disrupting the entire ecosystem.

In a recent study, oceanographers documented that industrial fishing fleets have caused a 90 percent reduction of large predatory fish such as tuna, marlin, swordfish, cod, halibut, and flounder. At the advent of intensive commercial fishing 100 years ago, most fleets caught 6 to 12 fish for every 100 baited hooks. Today, despite sophisticated electronic and aerial fish-finding techniques, the catch rate has plummeted to 1 fish per 100 hooks. The Worldwatch Institute has pointed out that as industrial fishing depletes "large, long-lived predatory species . . . that occupy the highest levels of the food chain, they move down to the next level—to species that tend to be smaller, shorter-lived, and less valuable. As a result, fishers worldwide now fill their nets with plankton-eating species such as squid, jacks, mackerel, sardines, and invertebrates including oysters, mussels, and shrimp. Commercial fishermen now work harder, spend more time, and

FIGURE 16.35 Massive waves of the Antarctic Ocean eroded cliffs to form these sea stacks near Cape Horn.

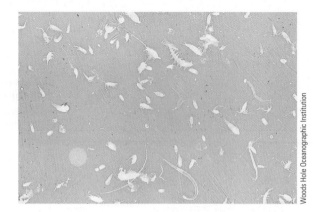

FIGURE 16.36 Zooplankton are tiny animals that live near the sea surface and feed on phytoplankton. These zooplankton are less than a millimeter in length.

consume more fuel to capture smaller quantities of less valuable species—they are fishing down the marine food chain." The authors of this study continue: "But the cycle of fishing down the marine food web can't go on forever. (At lower trophic levels, the species are so small and diluted that it is no longer economically feasible to fish.) At the current rate of descent, it will take only 30 to 40 years to fish down to the level of plankton."[1]

The United Nations Food and Agricultural Organization predicts that world fish harvests will remain fairly constant at around 90 million tons per year until the year 2010. After 2010, ecosystem destruction and overfishing will lead to a sharp decline in populations and harvests.

Reefs

A **reef** is a wave-resistant ridge or mound built by corals, oysters, algae, or other marine organisms. Because corals need sunlight and warm, clear water to thrive, coral reefs develop in shallow, tropical seas where little suspended clay or silt muddies the water (◆ Figure 16.38). As the corals die, their offspring grow on their remains. Oyster reefs form in temperate estuaries and can grow in more turbid water.

Globally, coral reefs cover about 600,000 square kilometers, about the area of France, but they spread

out in long, thin lines. They are extraordinarily productive ecosystems because the corals provide shelter for many fish and other marine species. Within the past 50 years, 10 percent of the world's coral reefs have been destroyed and an additional 30 percent are in critical condition. Several factors contribute to this destruction:

- Silt from cities, urban roadways, farms, and improper logging smother the delicate reef organisms.
- Fertilizer runoff from farms and sewage runoff from cities have added nutrients to coastal waters, feeding coral predators. For example, starfish thrive in nutrient-rich water and starfish eat corals. Microorganisms fertilized by agricultural runoff and sewage have caused massive disease epidemics that kill the coral. In addition, toxic chemicals poison corals in polluted, industrial areas.
- Overfishing or improper fishing can kill reefs. Parrot fish and sea urchins eat algae that smother

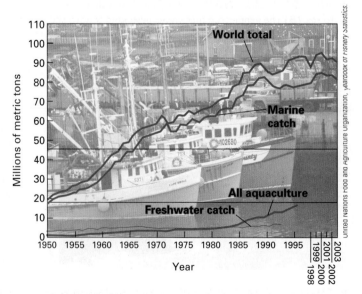

FIGURE 16.37 World fish harvest from 1950 to 1998. The harvest peaked twice, in 1989 and again in 1996, and declined after 1996.

---

1. http://www.worldwatch.org/mag/1998/98-3b.html.

A

B

◆ **FIGURE 16.38** **(A)** Coral reefs form abundantly in clear, shallow, tropical water. This small island on the east coast of Australia is surrounded by reefs that are part of the Great Barrier Reef. **(B)** Reefs grow in the clear, shallow water near Vanuatu and many other South Pacific Islands.

corals. If fishermen harvest too many parrot fish and sea urchins, the algae grow unchecked and kill the reefs. Also, in many parts of the world, fishermen dynamite coral reefs so their nets do not get tangled. Unfortunately, when the reefs are destroyed, fish populations decline. Therefore, while dynamiting reefs improves short-term gain, the practice diminishes long-term, sustainable harvests.

• Corals thrive best in a narrow temperature range. In recent years, oceanographers have compiled considerable evidence that sea surface temperatures have become warmer in recent decades and the warm water is leading to massive deaths of the corals.

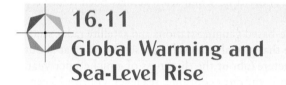

16.11
Global Warming and Sea-Level Rise

Sea level has risen and fallen repeatedly in the geologic past, and coastlines have emerged and submerged throughout Earth's history. During the past 40,000 years, sea level has fluctuated by 150 meters, primarily in response to growth and melting of glaciers (◆ Figure 16.39). The rapid sea-level rise that started about 18,000 years ago began to level off about 7,000 years ago. By coincidence, humans began to build cities about 7,000 years ago. Thus, civilization has developed

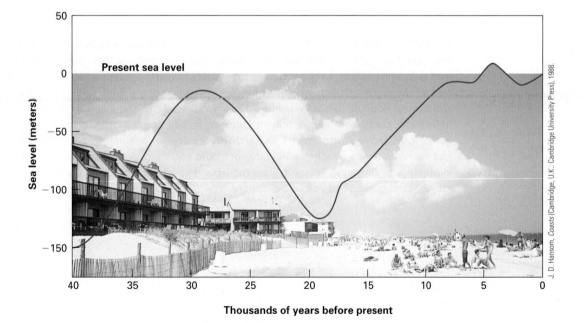

◆ **FIGURE 16.39** Sea level has fluctuated more than 150 meters during the past 40,000 years.

◆ FIGURE 16.40 A 1-meter sea-level rise would flood 17 percent of Bangladesh and displace 38 million people.

cause glaciers to melt, which adds freshwater to the oceans. The expansion of seawater with increasing temperature is gradual because a small temperature increase causes a small sea-level rise. However, melting of glaciers can be caused by threshold mechanisms and therefore can occur rapidly. This topic was introduced in Chapter 13 and will be discussed further in Chapter 21.

Consequences of rising sea level vary with location and economics. As mentioned in the introduction, some villages on South Pacific islands have already been impacted. The wealthy, developed nations could build massive barriers to protect cities and harbors from a small sea-level rise. In regions where global sea-level rise is compounded by local tectonic sinking, dikes are already in place or planned. Portions of Holland lie below sea level, and the land is protected by a massive system of dikes. In London, where the high-tide level has risen by 1 meter in the past century, multimillion-dollar storm gates have been built on the Thames River. Venice, Italy, which is built over sea-level canals, has flooded frequently in recent years, and here, too, expensive engineering projects are underway. However, it is unlikely that people could protect against a dramatic sea-level rise. Coastal cities worldwide would be inundated.

Many poor countries cannot afford coastal protection even for a small sea-level rise. A 1-meter rise in sea level would flood 17 percent of the land area of Bangladesh, displacing 38 million inhabitants (◆ Figure 16.40).

during a short time when sea level has been relatively constant.

Shore-based gauging stations and satellite radar studies agree that global sea level is presently rising at about 3 millimeters (about the thickness of a nickel) per year. Thus if present rates continue, sea level will rise 20 centimeters in 100 years. Records from the last century indicate an average rise in sea level of 1 to 2 millimeters per year. Such a rise would be significant along very low-lying areas such as the Netherlands, Bangladesh, and islands such as those described in the introduction to this chapter. As explained previously, global warming causes sea level rise by two mechanisms. First, water expands when it is heated. Second, warm air temperatures

EARTH SYSTEMS INTERACTIONS

COASTAL REGION PROVIDE ATTRACTIVE places to live and visit. In addition, the seas provide both food and transportation. For these reasons, coastlines have become heavily urbanized and industrialized. Sixty percent of the world's human population lives within 100 kilometers of the coast, and many millions live on low-lying coastal land vulnerable to hurricanes and coastal flooding. Global warming, rising global sea level, and intensifying storms threaten individuals, cities, and unique human cultures. In addition, coastlines are among the most geologically active environments on Earth. Rivers deposit sediment on coastal deltas. Waves and currents erode beaches and transport sand and mud along hundreds or thousands of kilometers of shoreline. Tectonic plate movements buckle coastal regions, creating mountain ranges, earthquakes, and volcanic eruptions.

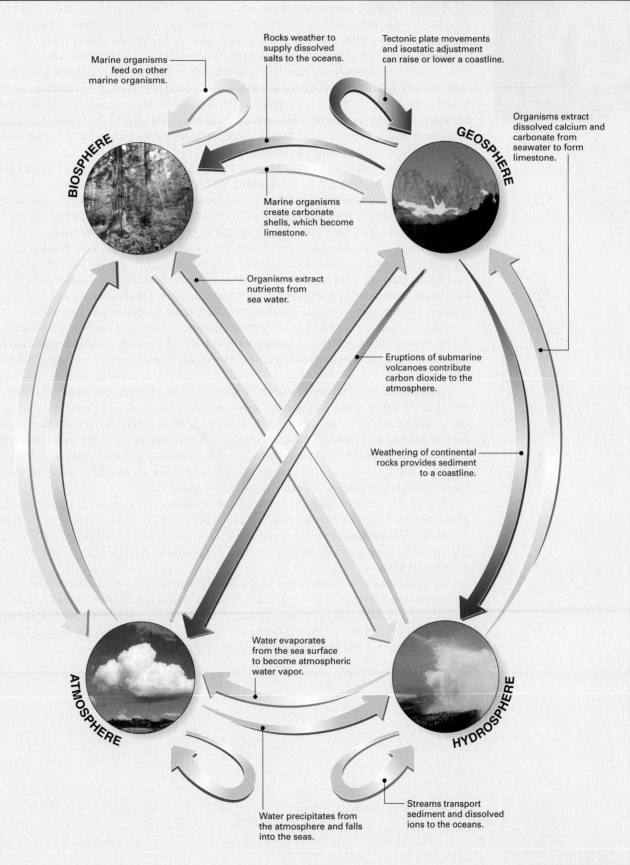

Marine organisms feed on other marine organisms.

Rocks weather to supply dissolved salts to the oceans.

Tectonic plate movements and isostatic adjustment can raise or lower a coastline.

Organisms extract dissolved calcium and carbonate from seawater to form limestone.

BIOSPHERE

GEOSPHERE

Marine organisms create carbonate shells, which become limestone.

Organisms extract nutrients from sea water.

Eruptions of submarine volcanoes contribute carbon dioxide to the atmosphere.

Weathering of continental rocks provides sediment to a coastline.

Water evaporates from the sea surface to become atmospheric water vapor.

ATMOSPHERE

HYDROSPHERE

Streams transport sediment and dissolved ions to the oceans.

Water precipitates from the atmosphere and falls into the seas.

SUMMARY

All of Earth's oceans are connected, so Earth has a single global ocean. Seawater contains about 3.5 percent dissolved salts. The upper layer of the ocean, about 450 meters thick, is relatively warm. In the **thermocline,** below the warm surface layer, temperature drops rapidly with depth. Deep ocean water is consistently around 1°C to 2.5°C.

Tides are caused by gravitational pull of the Moon and Sun. Two high tides and two low tides occur approximately every day. Wave size depends on (1) wind speed, (2) the amount of time the wind has blown, and (3) the distance that the wind has traveled. The highest part of a wave is the **crest;** the lowest is the **trough.** The distance between successive crests is called the **wavelength. Wave height** is the vertical distance from the crest to the trough. The water in a wave moves in a circular path. **Storm surge** is responsible for 90 percent of human fatalities during hurricanes and tropical cyclones.

A **current** is a continuous flow of water in a particular direction. Surface currents are driven by wind and deflected by the **Coriolis effect, Ekman transport,** and sea-surface topography. Deep-sea currents are driven by differences in seawater density. Cold, salty water is dense and therefore sinks and flows along the sea floor. When a coastal surface current is deflected offshore, cold deep water rises to replace it in a process called **upwelling.**

When a wave nears the shore, the bottom of the wave slows and the wave **breaks,** creating **surf.** Ocean waves weather and erode rock by hydraulic action, abrasion, solution, and salt cracking. **Refraction** is the bending of a wave as it strikes shore. Refracted waves often form **longshore currents** that transport sediment along a shore. **Tidal currents** also transport sediment in some areas. Irregular coastlines are straightened by erosion and deposition.

If land rises or sea level falls, the coastline migrates seaward and old beaches are abandoned above sea level, forming an **emergent coastline.** In contrast, a **submergent coastline** forms when land sinks or sea level rises. An **estuary** is a submerged riverbed and flood plain.

A **beach** is a strip of shoreline washed by waves and tides. Most coastal sediment is transported to the sea by rivers. Glacial drift, reefs, and local erosion also add sand in certain areas. Coastal emergence may expose large amounts of sand. **Spits, baymouth bars,** and **barrier islands** are common on sandy coastlines. Human intervention such as the building of **groins** may upset the natural movement of coastal sediment and alter patterns of erosion and deposition on beaches. A rocky coast is dominated by **wave-cut cliffs, wave-cut platforms, arches,** and **stacks.** A **fjord** is a submerged glacial valley.

Most of the photosynthesis and nutrient consumption in the ocean is carried out by **plankton,** tiny organisms that live mostly within a few meters of the sea surface where light is available. **Phytoplankton** conduct photosynthesis, just as land-based plants do, and are the base of the food chain for aquatic animals. **Zooplankton** are tiny animals that feed on the phytoplankton.

The shallow water of a continental shelf supports large populations of marine organisms. Commercial overfishing and habitat destruction threaten world fish harvests. A **reef** is a wave-resistant ridge or mound built by corals and other organisms. Sea level rises when ocean water warms or when glaciers melt and freshwater flows into the sea. Global sea level has risen over the past century and may continue to rise into the next.

Key Terms

For Review

1. Name the major ocean basins and give their locations.

2. Using Figure 15.3 in the previous chapter, what percentage of Earth's area (a) lies above an elevation of 840 meters, (b) lies 4 kilometers or more below sea level, (c) lies 5 kilometers or more below sea level?

3. What factors affect the salinity of seawater near shore?

4. Describe the temperature profile of the open oceans.

5. Explain why two high tides occur every day, even though the Moon lies directly above any portion of Earth only once a day.

6. Explain the difference between spring and neap tides.

7. List the three factors that determine the size of a wave.

8. Draw a picture of a wave and label the crest, the trough, the wavelength, and the wave height.

9. Describe the motion of both the surface water and the deeper layers of water as a wave passes.

10. What is a storm surge and how does it form?

11. Explain the Coriolis effect. What is a gyre?

12. What drives surface currents, deep-sea currents, tidal currents, and longshore currents?

13. What is Ekman transport? How does it interact with surface winds and the Coriolis effect to produce the currents that we observe in the deep ocean.

14. What is upwelling? How and where does it form?

15. Explain how surf forms.

16. What is refraction? How does it affect coastal erosion?

17. How is a tidal current different from a longshore current?

18. Draw a picture of a beach and name the major zones along the interface between land and sea.

19. What is an emergent coastline, and how does it form? What is a submergent coastline, and how does it form?

20. List the major mechanisms of weathering and erosion along coastlines.

21. List three sources of coastal sediment.

22. Compare and contrast a beach, a barrier island, and a spit.

23. Explain how coastal processes straighten an irregular coastline.

24. Describe some dominant features of sediment-rich and sediment-poor coastlines.

25. What is a groin? How does it affect the beach in its immediate vicinity? How does it affect the entire shoreline?

26. Explain how warming of the atmosphere could lead to a rise in sea level.

For Discussion

1. Earthquake waves were discussed in Chapter 7. Compare and contrast earthquake waves with water waves.

2. How can a ship survive 10-meter-high storm waves, whereas a beach house can be smashed by waves of the same size?

3. During World War II, few maps of the underwater profile of shorelines existed. When planning amphibious attacks on the Pacific islands, the Allied commanders needed to know near-shore water depths. Explain how this information could be deduced from aerial photographs of breaking waves and surf.

4. Imagine that an oil spill occurs from a tanker accident. Discuss the effects of mid-ocean currents, deep-sea currents, longshore currents, storm waves, and tides on the dispersal of the oil.

5. Compare and contrast coastal erosion with stream erosion.

6. In Section 16.7, we explained how erosion and deposition tend to smooth out an irregular coastline by eroding headlands and depositing sediment in bays. If coastlines are affected in this manner, why haven't they all been smoothed out in the 4.6- billion-year history of Earth?

7. The text states that many submergent coastlines are sediment poor. The sandy East Coast of the United States, between Long Island and Georgia, is a predominantly submergent coastline. Where did the sand come from?

8. Prepare a three-way debate. Have one side argue that the government should support the construction of groins. Have the second side argue that the government should prohibit the construction of groins. Have the third position defend the argument that groins should be permitted but not supported.

9. Prepare another debate on whether government funding should be used to repair storm damage to property on barrier islands.

10. Solutions to environmental problems can be divided into two general categories: social solutions and technical solutions. Social solutions involve changes in attitudes and lifestyles but do not generally require expensive industrial or technological processes. Technical solutions do not mandate social adjustments but require advanced, and often expensive, engineering. Describe a social solution and a technical solution to the problem of coastline changes in Long Island. Which approach do you feel would be more effective?

11. Imagine that sea level rises by 25 centimeters in the next century. How far inland would the shoreline advance if the beach (a) consisted of a vertical cliff, (b) sloped steeply at a 45° angle, (c) sloped gently at a 5° angle, (d) were almost flat and rose only 1°?

12. Explain why sea level does not necessarily rise in a linear relationship to air temperature.

13. Explain how changes in the atmosphere can alter the hydrosphere in the form of changing ocean currents.

14. Explain how a feedback mechanism can develop where a change in ocean currents can then change the atmosphere even more, thereby affecting the biosphere.

UNIT 5

CIMSS/University of Wisconsin, Madison

THE ATMOSPHERE: EVOLUTION AND COMPOSITION

CHAPTER 17

The Atmosphere

© Eric Nguyen/Jim Reed Photography/Corbis

A tornado backlit by the setting sun creates a dirt-filled debris cloud as the twister churns across rural Kansas farmland. May 29, 2004

ThomsonNOW™ Throughout this chapter, the ThomsonNOW logo indicates an opportunity for online self-study, which:
• Assesses your understanding of important concepts and provides a personalized study plan
• Links you to animations and simulations to help you study
• Helps to test your knowledge of material and prepare for exams
Visit ThomsonNOW at
http://www.thomsonedu.com/login to access these resources.

I f you decided to climb Mount Everest, you might start walking in the Nepali lowlands, then hike through mountain foothills where hearty farmers cultivate barley and peas in the thin mountain soil. When you reach an elevation around 4,000 meters, your breathing would be labored and even simple activities might seem difficult. Yet, the local people, acclimatized to living at high altitudes, would be scurrying up steep trails to tend their terraced fields. Near Everest base camp, at about 5,000 meters, there is half as much oxygen in the air as there is at sea level. Most climbers stop at base camp for a week to acclimatize. Above this point, even the strongest climbers struggle for breath in the frigid air as they ascend the glaciers and snowfields of Everest's lower flanks. At 7,000 meters, a lungful of air contains only 44 percent of the oxygen at sea level and the nighttime temperature frequently plunges to −20° or even to −30°C. Above this level, you enter the "death zone," where people cannot survive for long periods of time and where even the fittest athletes may perish.

Every multicellular organism needs oxygen to survive. If the oxygen abundance in the atmosphere were to drop below 44 percent of its current value, life on Earth as we know it would perish. If oxygen is essential to life, would we be better off if we had an even greater supply? The answer is yes, to a limit. Even at sea level, athletes can enhance their performance by breathing a small amount of bottled oxygen. But, paradoxically, too much oxygen is poisonous. If you breathe air that has 55 percent or more oxygen than is found at sea level, your body metabolism is so rapid that essential molecules and enzymes decompose. In addition, fires burn more rapidly with increased oxygen concentration. If the oxygen level in the atmosphere were to rise significantly, fires would burn uncontrollably across the planet, altering ecosystems as we know them.

Temperatures in the Solar System range from the frigid −270°C in the void of interplanetary space to the torrid 15 million°C plasma in the Sun's core. Yet life thrives within a precariously narrow temperature range, from about 0°C to 40°C. Although many organisms, including people, can survive for long periods in arctic cold, almost all photosynthesis occurs at above-freezing temperatures. While other organisms live in boiling hot-springs or hot deep-sea vents, they are the exceptions, not the rule.

Earth is the near-perfect size and distance from a stable and medium-temperature star to permit such conditions. But in addition, atmospheric composition and temperature are not determined solely by planetary size and distance from the Sun. Our planet's environment is finely regulated by Earth systems interactions. Over the past 4 billion years, the sun's output has slowly increased, although there have been numerous fluctuations during this period. However, Earth's temperature has remained remarkably constant. In this and subsequent chapters, we will study the interactions that have maintained this life-supporting climate. During this study, we will ask whether humans could disrupt Earth's climate and diminish conditions for life. ■

Scientists have a nearly continuous record of rocks of different ages, from 3.96 billion years ago to the present. Therefore when they study the history of the crust, they can analyze the chemical composition of ancient rocks. But there are no samples of very old atmospheres. So how can we determine atmospheric composition millions to billions of years ago? There are many gaps in our understanding, but the history chronicled next comes from two sources: modeling and the study of rocks. Modeling involves calculations about how atmospheric gases would have behaved under the presumed environment of early Earth. To test these models, scientists study the geochemistry of ancient rocks. Rocks react with water, rock, and air. By studying the rocks that existed at a specific time period, scientists deduce the other components of the environment that would have produced those reactions. For example, as we will discuss, iron reacts with oxygen to produce iron oxides. Thus, if we find iron oxides in certain types of sedimentary rocks that formed 2.6 billion years ago, we deduce that oxygen must have been present in the air and water.

The First Atmospheres: 4.6 to 4.0 Billion Years Ago

Our Solar System formed from a cold, diffuse cloud of interstellar gas and dust. About 99.8 percent of this cloud was composed of the two lightest elements, hydrogen and helium. Consequently, when Earth formed, its primordial atmosphere was composed almost entirely of these two light elements. But because Earth is relatively close to the

Sun and its gravitational force is relatively weak, its primordial hydrogen and helium atmosphere rapidly boiled off into space and escaped (◆ Table 17.1).

In Chapter 15, we learned that most of the volatile compounds that form Earth's hydrosphere, atmosphere, and biosphere originated from outer parts of the Solar System. Recall that, in its infancy, the Solar System was crowded with bits of rock, comets, ice chunks, and other debris left over from the initial coalescence of the planets. These bolides crashed into the planet in a near-continuous rain that lasted almost 800 million years. Carbonate compounds and carbon-rich rocks reacted under the heat and pressure of impact to form carbon dioxide. Ice quickly melted into water. Ammonia, common in the icy tail of comets, reacted to form nitrogen (◆ Figure 17.1).

When Life Began: 4.0 to 2.6 Billion Years Ago

This carbon dioxide, water, and nitrogen atmosphere changed as volatiles escaped from Earth's mantle to the surface in volcanic eruptions, in a process called **outgassing.** In 1952, Stanley Miller and Harold Urey hypothesized—on the basis of little direct modeling—that by 4 billion years ago, Earth's atmosphere consisted primarily of methane (CH_4), ammonia (NH_3), hydrogen (H_2), and water (H_2O). They mixed these gases in a glass tube and fired sparks across the tube to simulate Earth's early atmosphere, beset by lightning storms. Amino acids, the building blocks of proteins, formed in the tube. The Miller–Urey model immediately became popular because scientists speculated that the first living organisms formed by accretion of these abiotic (nonliving) amino acids.

In the early 1970s, the idea of a methane–ammonia–hydrogen atmosphere was largely discredited, and most scientists postulated that the Hadean atmosphere was

Table 17.1 The Earth's Atmosphere Through Time

Events that Formed the Atmosphere	Age of Atmosphere	Composition of Atmosphere
Primordial atmosphere: From initial accretion of planets.	4.6 billion years ago	Hydrogen (H_2) and helium (He).
Secondary atmosphere: Bolide impact from outer space.	4.5 billion years ago	Carbon dioxide (CO_2), water (H_2O), nitrogen (N_2).
Atmosphere formed by outgassing and modified by reactions of gases with geosphere.	4.5 to 2.7 billion years ago	Hydrogen (H_2) and carbon dioxide (CO_2) are predominant; water (H_2O), nitrogen (N_2) remain.
Cyanobacteria evolve and begin producing oxygen.	2.7 billion years ago	Reactions with the environment remove oxygen (O_2) produced by cyanobacteria as quickly as it is produced.
First Great Oxidation Event	2.4 billion years ago	Oxygen (O_2) begins to accumulate in the atmosphere; hydrogen (H_2) becomes a trace gas.
Second Great Oxidation Event	600 million years ago	Biological and geological processes slow decay so the oxygen concentration in the atmosphere builds.
Modern atmosphere: High oxygen concentration maintained by biological photosynthesis.	Today	Primarily nitrogen (N_2) and oxygen (O_2), with smaller concentrations of other gases.

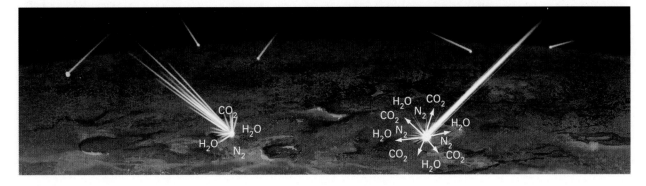

◆ **FIGURE 17.1** Comets, meteoroids, and asteroids imported Earth's volatiles from outer parts of the Solar System.

composed mainly of carbon dioxide (CO_2), with smaller amounts of nitrogen (N_2), water (H_2O), and other gases. However, in 2005, scientists used models of Earth's Hadean geosphere to calculate how gases trapped in the interior would react with rocks and minerals in the planet's interior and surface. One pivotal question is: When did Earth's core form? Earth's core is the deepest layer of the geosphere. The atmosphere is a thin veneer surrounding the crust. How could one affect the other? Recall that the core is composed primarily of iron and nickel. Before Earth's interior dissociated into a layered core and mantle, the mantle contained more iron than it does today. Then, when iron settled into the core, the mantle became relatively iron-poor.

Mantle rocks rise to the surface through volcanic eruptions. In turn, gases in the atmosphere react with surface rocks. When gases react with rocks, the reactions not only change the chemical composition of the rocks, they also alter the composition of the atmosphere. Thus, the formation of the core affects the evolution of the atmosphere. As we have said many times before, Earth is a system. Rocks affect the air. Air affects rocks.

Modern hypotheses state that Earth was hot at the time of its formation and most of the planet's iron was sequestered in its core shortly after the planet evolved. By modeling the reactions of volatiles in this iron-poor mantle, Tian and fellow researchers conclude that Earth's early atmosphere contained large amounts of hydrogen, as Miller and Urey postulated, but also high concentrations of carbon dioxide, as later models had proposed. The argument is ongoing, and we can expect further discussion and modifications.

Our models of Earth and its evolution are constantly changing. Scientists propose hypotheses and theories, consider new ideas, and often other scientists disagree. This is the nature and the joy of science. In a recent review article about Earth's early atmospheres, professor Christopher Chyba of SETI Institute and Stanford University wrote that the argument about the composition of the Hadean atmosphere "makes it a great time for young scientists to enter the field, but it also reminds us that some humility regarding our favorite models is in order."

17.2 Life, Iron, and the Evolution of the Modern Atmosphere

As explained previously, the first living organisms may have formed by accretion of complex abiotic (nonliving) organic molecules. But these complex organic molecules are oxidized and destroyed in an oxygen-rich environment. (This oxidation is analogous to slow burning.) If large amounts of oxygen were present in Earth's early atmosphere, the abiotic precursors to living organisms could not have formed.

By studying the minerals in a rock, a geochemist can determine whether it formed in an oxygen-rich or an oxygen-poor environment. Recent studies of Earth's oldest rocks indicate that the atmospheric oxygen concentration in the Hadean atmosphere was extremely low. Thus, the molecules necessary for the emergence of living organisms would have been preserved in the primordial atmosphere.

Although life could not have emerged in an oxygen-rich environment, complex multicellular life requires an oxygen-rich atmosphere to survive. How did oxygen become abundant in our atmosphere?

The world's earliest organisms probably obtained their energy from reactions with minerals such as iron and sulfur in an extremely inefficient process. Later, organisms subsisted, in part, by eating each other. But these food chains were limited because there were only a few organisms on Earth. A crucial step in evolution occurred when primitive bacteria evolved the ability to harness the energy in sunlight and produce organic tissue. This process, known as **photosynthesis**, is the foundation for virtually all modern life. During photosynthesis, organisms convert carbon dioxide and water to organic sugars. They release oxygen as a by-product. In 1972, an English chemist named James Lovelock hypothesized that the oxygen produced by primitive organisms gradually accumulated to create the modern atmosphere. When

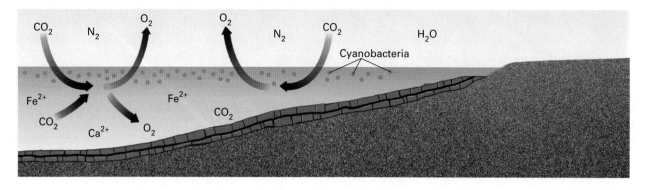

◆ **FIGURE 17.2** The oxygen content of the atmosphere gradually increased as cyanobacteria plants released oxygen.

the oxygen concentration reached a critical level to sustain efficient metabolism in late-Precambrian time, multicellular organisms evolved and the biosphere as we know it was born. Lovelock was so overwhelmed by the intimate connection between living and nonliving components of Earth's systems that he likened our planet to a living creature, which he called **Gaia** (Greek for Earth).

The Lovelock hypothesis, now 35 years old, remains generally accepted, but scientists are still investigating many of the details. For example, blue-green algae called **cyanobacteria** began producing oxygen 2.7 billion years ago, but appreciable quantities of oxygen didn't appear in the atmosphere until 2.4 billion years ago (◆ Figure 17.2). The production of oxygen by cyanobacteria occurred slowly and steadily. But for 300 million years, the concentration of oxygen in the atmosphere didn't rise. Then, in a process called **The First Great Oxidation Event,** the concentration of atmospheric oxygen rose abruptly.

To understand the mechanism of this threshold reaction, we must study atmosphere–geosphere and atmosphere–biosphere systems interactions.

Systems Interactions that Affected Oxygen Concentration in the Hadean Atmosphere

Geosphere–Atmosphere Interactions

Even though large quantities of iron coalesced into Earth's core, appreciable quantities remained in the mantle and crust. Today, about 5 percent of the weight of the crust is iron.

Oxygen dissolves in water. As a result, when early cyanobacteria released oxygen, some of this gas dissolved in seawater. In an oxygen-poor environment, iron also dissolves in water. However, when oxygen is abundant, the dissolved oxygen reacts with dissolved iron and causes the iron to precipitate rapidly. Thus, if air contains little or no oxygen, there is also little oxygen in seawater, and large amounts of iron dissolve in the seas. If the oxygen concentration of the atmosphere and the ocean rises to a

threshold level, iron precipitates rapidly, forming a layer of iron oxide minerals on the sea floor. Thus, when enough oxygen had accumulated in seawater to react with the dissolved iron, vast quantities of iron minerals precipitated onto the sea floor, producing layers of iron-rich minerals (◆ Figure 17.3). This process also removed oxygen, explaining why the oxygen concentration in the atmosphere didn't rise, even though cyanobacteria were releasing this gas. Iron, the main ingredient in steel, is the world's most commonly used metal. About 1 billion tons of iron are mined every year, 90 percent from **banded iron formations,** the sedimentary layers of iron-rich minerals sandwiched between beds of clay and other silicate minerals. The alternating layers are a few centimeters thick and give the rocks their banded appearance (◆ Figure 17.4). A single iron formation of this type may be hundreds of meters thick and cover tens of square kilometers (◆ Figure 17.5). Most of Earth's banded iron deposits formed from 2.6 to 1.9 billion years ago, although a few are older and some are younger.

The alternating layers of iron minerals and other minerals may have developed because, for a long time, the oxygen level in the seas hovered near the threshold at which soluble iron converts to the insoluble variety. When the dissolved oxygen concentration increased, the oxygen reacted with the dissolved iron to precipitate iron-oxide minerals on the sea floor. But the formation of those minerals extracted oxygen as well as iron from the seawater, and lowered its oxygen concentration below the threshold. Then, dissolved iron accumulated again in the seas while the oxygen was slowly replenished. During that time, clay and other minerals washed from the continents and accumulated on the sea floor as they do today, forming the thin layers of silicate minerals that lie between the iron-rich layers. When the oxygen concentration rose above the threshold again, another layer of iron minerals formed.

Banded iron formations contain thousands of alternating layers of iron minerals and silicates. The great thickness of the iron formations, coupled with the fact that they continued to form from 2.6 to 1.9 billion years ago, suggests that these reactions must have kept the levels of

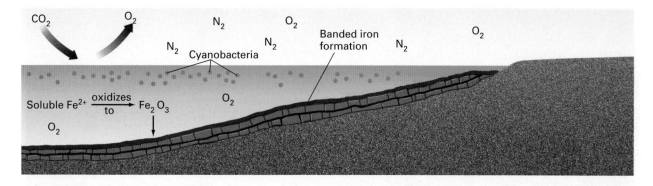

◆ **FIGURE 17.3**   The oxygen concentration in the atmosphere and seawater reached a threshold at which the oxygen combined with dissolved iron to form iron minerals. The iron minerals precipitated to the seafloor to form the first layer of a banded iron formation.

dissolved oxygen close to the threshold for 700 million years. Thus the iron-rich rocks that support our industrial society were formed by interactions among early photosynthetic organisms, sunlight, air, and the oceans.

Biosphere–Atmosphere Interactions

In the primordial atmosphere, free oxygen also reacted with hydrogen to form water. This process helped keep the oxygen concentration in the atmosphere low. However, after life evolved, bacteria in the oceans removed atmospheric hydrogen in a process that produced methane. When the hydrogen concentration decreased sufficiently, oxygen became chemically stable in the atmosphere and the oxygen concentration could rise.

Whatever the exact combination of biological and geochemical processes, evidence in the rocks indicates that the oxygen concentration in the atmosphere remained low and then jumped suddenly from trace to appreciable quantities approximately 2.4 billion years ago.

Evolution of the Modern Atmosphere

Several layers of banded iron formed after The First Great Oxidation Event, and this process continued to remove oxygen that was released during photosynthesis. The last, major, banded iron layer was deposited about 1.9 billion years ago, but the oxygen concentration in the atmosphere did not increase dramatically. At least two more critical steps were required before efficient multicellular organisms could evolve. The Sun emits energy largely in the form of high-energy ultraviolet light. These rays are energetic enough to break complex molecules apart and kill evolving multicellular organisms. But high-altitude oxygen absorbs ultraviolet radiation in a process that forms ozone (see Section 17.7). Thus the oxygen concentration couldn't increase in the lower atmosphere until appreciable concentrations accumulated in the upper atmosphere. To summarize: oxygen, largely produced by the earliest photosynthetic organisms, was not only necessary for life as we know it today, it also protected multicellular life by filtering out harmful solar rays.

Multicellular plants and animals emerged in late-Precambrian time, between 1 billion and 543 million years ago. About 600 million years ago, the oxygen level in the atmosphere increased rapidly a second time, in a process called **The Second Great Oxidation Event**. What changed abruptly 1.3 billion years after the last banded iron layers were deposited to allow oxygen to accumulate? Scientists propose that prior to 600 million years ago, biological decay was almost as rapid as photosynthesis. Therefore, the oxygen

Barbara Gerlach/Visuals Unlimited

◆ **FIGURE 17.4**   In this banded iron formation from Michigan, the red bands are iron oxide minerals and the dark layers are chert (silica).

◆ **FIGURE 17.5** The Mesabi Range in northern Minnesota is a banded iron formation that is one of the chief iron-producing regions in the world.

that was released into the atmosphere was immediately consumed during respiration, according to the following reactions:

During photosynthesis: Carbon dioxide + Water ⟶ Sugars + Oxygen

During respiration and decay: Sugars + Oxygen ⟶ Carbon dioxide + Water

Then abruptly, at 600 million years ago, several processes locked up organic matter in sediments before it could decay. When decay and respiration slowed down, excess oxygen accumulated in the atmosphere. All of the proposed processes involve complex reactions among Earth's four spheres:

- Geochemical processes produced an abundance of clays 600 million years ago. These clays buried and preserved organic matter on the sea floor.
- Zooplankton evolved in the seas. These organisms produced dense, organic-laden feces that fell to the sea floor and accumulated in the clays mentioned previously.
- Simple lichens evolved on land. The lichens accelerated weathering, and the weathered ions washed into the sea and provided nutrients for phytoplankton. In turn, the phytoplankton fed the zooplankton, which sequestered nutrients as described previously.

Thus numerous complex chemical, physical, and biological processes combined to set the stage for the Second Great Oxidation Event. In 1972, when James Lovelock first proposed the Gaia hypothesis, he assumed a relatively simple exchange between living organisms and the atmosphere. Now we know that although Lovelock's basic premise is reasonable, the mechanisms are much more complicated than he envisioned. Living organisms release the oxygen needed for living organisms to survive. If all life on Earth were to cease, the atmosphere would revert to an oxygen-poor composition and become poisonous to modern plants and animals. But at the same time the geochemical mechanisms also contribute to the atmosphere that sustains us.

Fires burn rapidly if oxygen is abundant; if its concentration were to increase even by a few percent, fires would burn uncontrollably across the planet. If atmospheric oxygen levels were to decrease appreciably, most modern plants and animals would not survive. If the carbon dioxide concentration were to increase by a small amount, atmospheric temperature would rise as a result of greenhouse warming. Earth's atmosphere not only sustains us, it insulates Earth's surface and winds distribute the Sun's heat around the globe, so the surface is neither too hot nor too cold for life. Clouds form from water vapor in the atmosphere and rain falls from clouds. In addition, the atmosphere filters out much of the Sun's ultraviolet radiation, which can destroy living tissue and cause cancer. The atmosphere carries sound; without air we would live in silence. Without an atmosphere, airplanes and birds could not fly, wind would not transport pollen and seeds, the sky would be black rather than blue, and no reds, purples, and pinks would color the sunset. When we understand this complex web of interacting processes, and realize that we are the only planet in the Solar System to be so fortunate, we can only wonder at the fragility of Earth's atmosphere.

17.3 The Modern Atmosphere

The modern atmosphere is mostly gas, with small quantities of water droplets and dust. The gaseous composition of dry air is roughly 78 percent nitrogen, 21 percent oxygen, and 1 percent other gases (◆ Figure 17.6 and ◆ Table 17.1). Nitrogen, the most abundant gas, does not react readily with other substances. Oxygen, though, reacts chemically as fires burn, iron rusts, and plants and animals respire. Notice that carbon dioxide, which formed 80 percent of the secondary Hadean atmosphere, is a trace gas with a concentration of only 0.035 percent. Carbon dioxide and the greenhouse effect are discussed further in Chapter 21.

In addition to the gases listed in Table 17.1, air contains water vapor, water droplets, and dust. The types and quantities of these components vary with both location and altitude. In a hot, steamy jungle, air may contain 5 percent water vapor by weight, whereas in a desert or cold polar region, only a small fraction of a percent may be present.

If you sit in a house on a sunny day, you may see a sunbeam passing through a window. The visible beam is light reflected from tiny specks of suspended dust. Clay, salt, pollen, bacteria, and viruses, bits of cloth, hair, and skin are all components of dust. People travel to the sea-

side to enjoy the "salt air." Visitors to the Great Smoky Mountains in Tennessee view the bluish, hazy air formed by sunlight reflecting from pollen and other dust particles.

Within the past century, humans have altered the chemical composition of the atmosphere in many different ways. We have increased the carbon dioxide concentration by burning fuels and igniting wildfires. Factories release chemicals into the air—some are benign, others are poisonous. Smoke and soot change the clarity of the atmosphere. These changes are discussed in Section 17.6.

17.4 Atmospheric Pressure

The molecules in a gas zoom about in a random manner. For example, at 20°C an average oxygen molecule is traveling at 425 meters/second (950 miles per hour). In the absence of gravity, temperature differences, or other perturbations, a gas will fill a space homogeneously.

Thus, if you floated a cylinder of gas in space, the gas would disperse until there was an equal density of molecules and an equal pressure throughout the cylinder. But gases that surround Earth are perturbed by many influences, which ultimately create the complex and turbulent atmosphere that helps shape the world we live in.

Within our atmosphere, gas molecules zoom about, as in the imaginary cylinder, but in addition, gravity pulls them downward. As a result of this downward force, more molecules concentrate near the surface of Earth than at higher elevations. Therefore, the atmosphere is denser at sea level than it is at higher elevations—and the pressure is higher. Density and pressure then decrease exponentially with elevation (◆ Figure 17.7). As explained in the introduction, at an elevation of about 5,000 meters, the atmosphere contains about half as much oxygen as it does at sea level. If you ascended in a balloon to 16,000 meters (16 kilometers) above sea level, you would be above 90 percent of the atmosphere and would need an oxygen mask to survive. At an elevation of 100 kilometers, pressure is only 0.00003 that of sea level, approaching the vacuum of outer space. There is no absolute upper boundary to the atmosphere.

Atmospheric pressure is measured with a **barometer** and is often called **barometric pressure**. A simple but accurate barometer is constructed from a glass tube that is sealed at one end. The tube is evacuated and the open end is placed in a dish of a liquid such as mercury.

The mercury rises in the tube because atmospheric pressure depresses the level of mercury in the dish but there is no air in the tube (◆ Figure 17.8). At sea level mercury rises approximately 76 centimeters, or 760 millimeters (about 30 inches), into an evacuated tube.

Meteorologists express pressure in inches or millimeters of mercury, referring to the height of a column of mercury in a barometer. They also express pressure in

Composition of the Modern Atmosphere

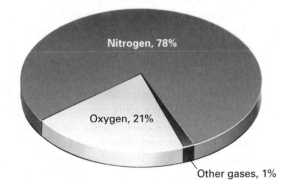

Nitrogen, 78%

Oxygen, 21%

Other gases, 1%

◆ **FIGURE 17.6** Composition of the modern atmosphere.

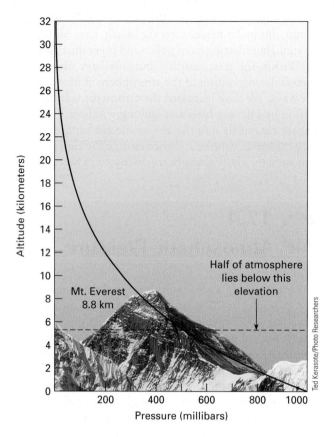

◆ **FIGURE 17.7** Atmospheric pressure decreases with altitude. One-half of the atmosphere lies below an altitude of 5,600 meters.

bars and millibars. A **bar** is approximately equal to sea-level atmospheric pressure. A millibar is 0.001 of a bar.

A mercury barometer is a cumbersome device nearly a meter tall, and mercury vapor is poisonous. A safer and more portable instrument for measuring pressure, called an aneroid barometer, consists of a partially evacuated metal chamber connected to a pointer. When atmospheric pressure increases, it compresses the chamber and the pointer moves in one direction. When pressure decreases, the chamber expands, directing the pointer the other way (◆ Figure 17.9).

Changing weather can also affect barometric pressure. On a stormy day at sea level, pressure may be 980 millibars (28.94 inches), although it has been known to drop to 900 millibars (26.58 inches) or less during a hurricane. In contrast, during a period of clear, dry weather, a typical high-pressure reading may be 1,025 millibars (30.27 inches). These changes are discussed in Chapter 19.

17.5 Atmospheric Temperature

The temperature of the atmosphere changes with altitude (◆ Figure 17.10). The layer of air closest to Earth,

the layer we live in, is the **troposphere**. Virtually all of the water vapor and clouds exist in this layer, and almost all weather occurs here. Earth's surface absorbs solar energy, and thus the surface of the planet is warm. But, as explained earlier, continents and oceans also radiate heat, and some of this energy is absorbed in the troposphere. At higher elevations in the troposphere, the atmosphere is thinner and absorbs less energy; in addition, lower parts of the troposphere have absorbed much of the heat radiating from Earth's surface. Consequently, temperature decreases at higher levels in the troposphere. Thus, mountaintops are generally colder than valley floors, and pilots flying at high altitudes must heat their cabins.

The top of the troposphere is the **tropopause**, which lies at an altitude of about 17 kilometers at the equator, although it is lower at the poles. At the tropopause the steady decline in temperature with altitude ceases abruptly. Cold air from the upper troposphere is too dense to rise above the tropopause. As a result, little mixing occurs between the troposphere and the layer above it, called the **stratosphere**.

In the stratosphere, temperature remains constant to 35 kilometers and then increases with altitude until, at about 50 kilometers, it is as warm as that at Earth's surface. This reversal in the temperature profile occurs because the troposphere and stratosphere are heated by different mechanisms. As already explained, the troposphere is heated primarily from below, by Earth. The stratosphere, however, is heated primarily from above, by solar radiation.

Oxygen molecules (O_2) in the stratosphere absorb energetic ultraviolet rays from the Sun. The radiant energy breaks the oxygen molecules apart, releasing free oxygen atoms. The oxygen atoms then recombine to form ozone (O_3). Ozone absorbs ultraviolet energy more efficiently than oxygen does, warming the upper stratosphere. Ultraviolet radiation is energetic enough to affect organisms. Small quantities give us a suntan, but large doses cause skin cancer and cataracts of the eye, inhibit the growth of many plants, and otherwise harm living tissue. The ozone in the upper atmosphere protects life on Earth by absorbing much of this high-energy radiation before it reaches Earth's surface.

Ozone concentration declines in the upper portion of the stratosphere, and therefore at about 55 kilometers above Earth, temperature once more begins to decline rapidly with elevation. This boundary between rising and falling temperature is the **stratopause**, the ceiling of the stratosphere. The second zone of declining temperature is the **mesosphere**. Little radiation is absorbed in the mesosphere, and the thin air is extremely cold. Starting at about 80 kilometers above Earth, the temperature again remains constant and then rises rapidly in the **thermosphere**. Here the atmosphere absorbs high-energy X-rays and ultraviolet radiation from the Sun. High-energy reactions strip electrons from atoms and molecules to produce ions. The temperature in the upper

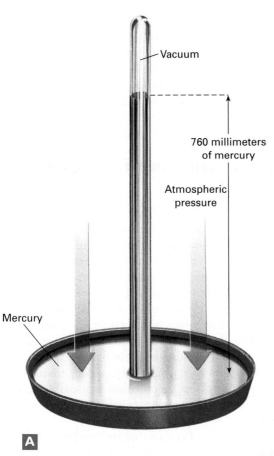

Vacuum

760 millimeters
of mercury

Atmospheric
pressure

Mercury

A

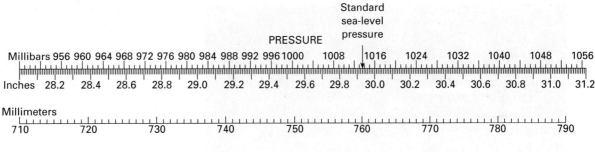

PRESSURE

Standard
sea-level
pressure

Millibars 956 960 964 968 972 976 980 984 988 992 996 1000   1008 |1016   1024   1032   1040   1048   1056

Inches 28.2   28.4   28.6   28.8   29.0   29.2   29.4   29.6   29.8   30.0   30.2   30.4   30.6   30.8   31.0   31.2

Millimeters
710   720   730   740   750   760   770   780   790

B

◆ **FIGURE 17.8** **(A)** Atmospheric pressure forces mercury upward in an evacuated glass tube. The height of the mercury in the tube is a measure of air pressure. **(B)** Three common scales for reporting atmospheric pressure and the conversion among them.

INTERACTIVE QUESTION: *Express 750 mm in inches and millibars.*

portion of the thermosphere is just below freezing, not extremely cold by surface standards.

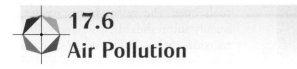

17.6
Air Pollution

Ever since the first cave dwellers huddled around a smoky fire, people have introduced impurities into the air. The total quantity of these impurities is minuscule compared with the great mass of our atmosphere and with the monumental changes that occurred during the evolution of the planet. Yet air pollution remains a significant health, ecological, and climatological problem for modern industrial society.

In 1948, Donora was an industrial town of about 14,000 located 50 kilometers south of Pittsburgh, Pennsylvania. One large factory in town manufactured

structural steel and wire and another produced zinc and sulfuric acid. During the last week of October 1948, dense fog settled over the town. But it was no ordinary fog; the moisture contained pollutants from the two factories. After four days, visibility became so poor that people could not see well enough to drive, even at noon with their headlights on. Gradually at first, and then in increasing numbers, residents sought medical attention for nausea, shortness of breath, and constrictions in the throat and chest. Within a week, 20 people had died and about half of the town was seriously ill.

Other incidents similar to that in Donora occurred worldwide. In response to the growing problem, the United States enacted the Clean Air Act in 1963. As a result of the Clean Air Act and its amendments, total emissions of air pollutants have decreased and air quality across the country has improved (◆ Figure 17.11). It is even more encouraging to note that this decrease in emissions has occurred at a time when population, energy

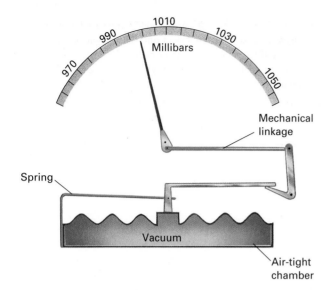

◆ **FIGURE 17.9** In an aneroid barometer, increasing air pressure compresses a chamber and causes a connected pointer to move in one direction. When the pressure decreases, the chamber expands, deflecting the pointer the other way.

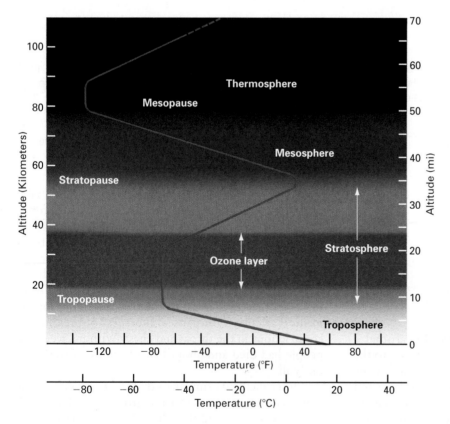

◆ **FIGURE 17.10** Atmospheric temperature varies with altitude. The atmospheric layers are zones in which different factors control the temperature.

INTERACTIVE QUESTION: *If you were ascending from Earth in a rocket, what layer would you be in at 5 km, 20 km, and 80 km? What would the outside temperature be at each of these elevations?*

consumption, vehicle miles traveled, and gross domestic product have increased dramatically (◆ Figure 17.12). Donora-type incidents have not been repeated. Smog has decreased, and rain has become less acidic. Yet some people believe that we have not gone far enough and that air pollution regulations should be strengthened further.

Sources and types of air pollution are listed in ◆ Figure 17.13 and discussed in the following section.

Gases Released When Fossil Fuels Are Burned

Coal is largely carbon, which, when burned completely, produces carbon dioxide. Petroleum is a mixture of hydrocarbons, compounds composed of carbon and hydrogen. When hydrocarbons burn completely, they produce carbon dioxide and water. Neither is poisonous, but both are greenhouse gases. If fuels were composed purely of compounds of carbon and hydrogen, and if they always burned completely, air pollution from burning of fossil fuels would pose little direct threat to our health (although combustion of fossil fuels would still contribute to global warming). However, fossil fuels contain impurities, and combustion is usually incomplete. As a result, other products form—most of which are harmful.

Products of incomplete combustion include hydrocarbons such as benzene and methane. Benzene is a carcinogen (a compound that causes cancer), and methane is another greenhouse gas. Incomplete combustion of fossil fuels releases many other pollutants, including carbon monoxide (CO), which is colorless and odorless yet very toxic.

Additional problems arise because coal and petroleum contain impurities that generate other kinds of pollution when they are burned. Small amounts of sulfur are present in coal and, to a lesser extent, in petroleum. When these fuels burn, the sulfur forms oxides, mainly sulfur dioxide, SO_2, and sulfur trioxide, SO_3. High sulfur-dioxide concentrations have been associated with major air pollution disasters of the type that occurred in Donora. Today the primary global source of sulfur-dioxide pollution is coal-fired electric generators.

Nitrogen, like sulfur, is common in living tissue and therefore is found in all fossil fuels. This nitrogen, together with a small amount of atmospheric nitrogen, reacts when coal or petroleum is burned. The products are mostly nitrogen oxide, NO, and

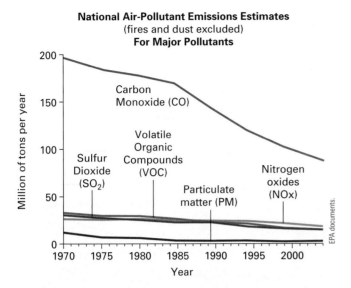

National Air-Pollutant Emissions Estimates
(fires and dust excluded)
For Major Pollutants

◇ **FIGURE 17.11** Emission of five air pollutants in the United States from 1970 to 2004. The Clean Air Act was first enacted in 1963. (This graph does not show local concentrations in heavily congested areas such as Los Angeles.)

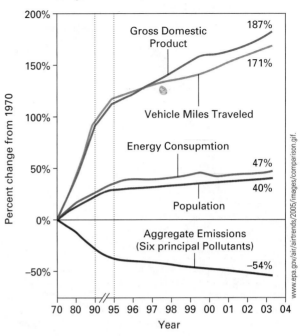

Comparison of Growth Areas and Emissions

◇ **FIGURE 17.12** Air pollution emissions declined by 54 percent between 1970 and 2004, despite the fact that gross domestic product, vehicle miles traveled, energy consumption, and population have all increased.

nitrogen dioxide, NO_2. Nitrogen dioxide is a reddish-brown gas with a strong odor. It therefore contributes to the "browning" and odor of some polluted urban atmospheres. Automobile exhaust is the primary source of nitrogen-oxide pollution.

Acid Rain

As explained earlier, sulfur and nitrogen oxides are released when coal and petroleum burn. These oxides are also released when metal ores are refined. In moist air, sulfur dioxide reacts to produce sulfuric acid and nitrogen oxides react to form nitric and nitrous acid. All of these are strong acids. Atmospheric acids dissolve in water droplets and fall as **acid precipitation**, also called **acid rain** (◇ Figure 17.14).

Acidity is expressed on the **pH scale**. A solution with a pH of 7 is neutral, neither acidic nor basic. On a pH scale, numbers lower than 7 represent acidic solutions, and numbers higher than 7 represent basic ones. For example, soapy water is basic and has a pH of about 10, whereas vinegar is an acid with a pH of 2.4.

Rain reacts with carbon dioxide in the atmosphere to produce a weak acid. As a result, natural rainfall has a pH of about 5.7. However, in the "bad old days" before the Clean Air Act was properly enforced, rain was much

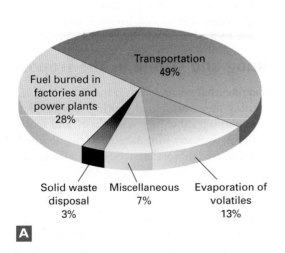

A

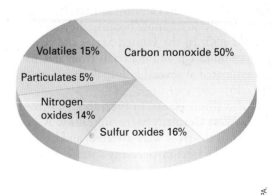

B

◇ **FIGURE 17.13** **(A)** Sources of air pollution in the United States. **(B)** Types of air pollutants in the United States. (Although carbon dioxide is a greenhouse gas, it is not listed as a pollutant because it is not toxic.)

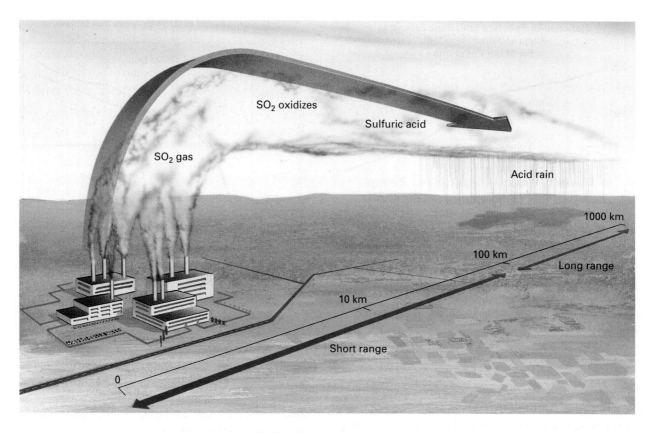

◆ FIGURE 17.14 Acid rain develops from the addition of sulfur compounds to the atmosphere by industrial smokestacks.

more acidic. A fog in Southern California in 1986 reached a pH of 1.7, which approaches the acidity of toilet bowl cleaners.

Consequences of Acid Rain

Sulfur and nitrogen oxides impair lung function, aggravating diseases such as asthma and emphysema. They also affect the heart and liver and have been shown to increase vulnerability to viral infections such as influenza.

Acid rain corrodes metal and rock. Limestone and marble are especially susceptible because they dissolve rapidly in mild acid. In the United States the cost of deterioration of buildings and materials from acid precipitation is estimated at several billion dollars per year (◆ Figure 17.15).

Acid rain also affects plants. In 1982, about 8 percent of the trees in West Germany were unhealthy. A year later, 34 percent of the trees were affected, and by 1995 more than half of the trees in Germany's western forests were sick or dying.

Air pollution control laws in Europe and North America have drastically reduced acid emissions. According to the summary statement from the 2005 International Acid Rain Conference (http://www.acidrain2005.cz/conclusion.pdf),

"In Europe and North America, the acid rain saga has really been a success story. From the bleak outlook of the 1970s, with extensive damage to fish and risk to forests, the picture now in 2005 is rosy indeed. . . . Acidification of soil and surface water has ceased in large areas and ecosystems are beginning to recover. . . . There is still a long way to go, however, as recovery so far has been slow and erratic. . . .

But the news is not all good. . . . Large new regions such as China, Japan, and the Indian sub-continent are now at risk for damage to human health and to freshwater and terrestrial ecosystems."

Smog and Ozone in the Troposphere

Imagine that your great-grandfather had entered the exciting new business of making moving pictures. Old-time photographic film was "slow" and required lots of sunlight, so many filmmakers left the polluted and overcast, industrial Northeast. Southern California, with its warm, sunny climate and little need for coal, was preferable. Thus, a district of Los Angeles called Hollywood became the center of the movie industry. Its population boomed, and after World War II automobiles became about as numerous as people. Then the quality of the air deteriorated in a strange way. People noted four different kinds of changes: (1) A brownish haze called

◆ **FIGURE 17.16** A brownish haze of smog settles over Los Angeles. Note that the smog is thick beneath a distinct line caused by an atmospheric inversion. Above the inversion level, the air is much cleaner.

◆ **FIGURE 17.15** Acid rain has destroyed this 1817 marble tombstone in England.

INTERACTIVE QUESTION: *In the United States, many large, coal-fired, electric-generating facilities are located in the desert in the Southwest. Argue for or against the statement: Because there is little rain in the desert, sulfur dioxides emitted by these generators are less likely to react to form acid precipitation. Therefore the air pollutants are less likely to cause harm.*

smog settled over the city (◆ Figure 17.16); (2) people felt irritation in their eyes and throats; (3) vegetable crops became damaged; and (4) the sidewalls of rubber tires developed cracks.

In the 1950s, air pollution experts worked mostly in the industrialized cities of the East Coast and the Midwest. When they were called to diagnose the problem in Southern California, they looked for the sources of air pollution they knew well, especially sulfur dioxide. But the smog was nothing like the pollution they were familiar with. These researchers eventually learned that incompletely burned gasoline in automobile exhaust reacts with nitrogen oxides and atmospheric oxygen in the presence of sunlight to form ozone, O_3. The ozone then reacts further with automobile exhaust to form smog (◆ Figure 17.17).

Earlier in this chapter, we read about the harmful effects of excessive ozone in the air over cities such as Los Angeles. In this chapter, we will learn that ozone in the stratosphere absorbs ultraviolet radiation and pro-

tects life on Earth. Is ozone a pollutant to be eliminated, or a beneficial component of the atmosphere that we want to preserve? The answer is that it is both, depending on *where* it is found. Ozone in the troposphere reacts with automobile exhaust to produce smog and therefore it is a pollutant. Ozone in the stratosphere is beneficial and the destruction of the ozone layer creates serious problems.

Ozone irritates the respiratory system, causing loss of lung function and aggravating asthma in susceptible individuals. Ozone also increases susceptibility to heart disease and is a suspected carcinogen. High ozone concentrations slow the growth of plants, which is a particularly serious problem in the rich, agricultural areas of California.

Nationwide, ozone levels have decreased by about 25 percent between 1980 and 2005. However, local problems remain. According to a recent EPA report, 49 counties, where 42 million people live, had poor ozone air quality in 2004, and scientists recorded unhealthy ozone levels in 7 counties, where 18 million people live, primarily in California and Texas.

Toxic Volatiles

A volatile compound is one that evaporates readily and therefore easily escapes into the atmosphere. Whenever chemicals are manufactured or petroleum is refined, some volatile by-products escape into the atmosphere. When metals are extracted from ores, gases such as sulfur dioxide are released. When pesticides are sprayed onto fields and orchards, some of the spray is carried off by wind. When you paint your house, the volatile parts of the paint evaporate into the air. As a result of all these processes, tens of thousands of different volatile compounds are present in polluted air: some are harmless, others are poisonous, and many have not been studied. Consider the case of dioxin.

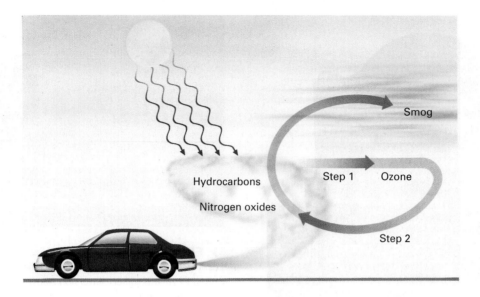

◆ **FIGURE 17.17** Smog forms in a sequential process. Step 1: Automobile exhaust reacts with air in the presence of sunlight to form ozone. Step 2: Ozone reacts with automobile exhaust to form smog.

Very little dioxin is intentionally manufactured. It is not an ingredient in any herbicide, pesticide, or other industrial formulation. You cannot buy dioxin at your local hardware store or pharmacy. Dioxin forms as an unwanted by-product in the production of certain chemicals and when specific chemicals are burned. For example, in the United States today, garbage incineration is the most common source of dioxin. When a compound containing chlorine, such as the plastic polyvinyl chloride, is burned, some of the chlorine reacts with organic compounds to form dioxin. The dioxin then goes up the smokestack of the incinerator, diffuses into the air, and eventually falls to Earth. Cattle eat grass lightly dusted with dioxin and store the dioxin in their fat. Humans ingest the compound mostly in meat and dairy products. The EPA estimates that the average U.S. citizen ingests about 0.0000000001 grams (100 picograms) of dioxin in food every day. Although this is a minuscule amount, the EPA has argued that dioxin is the most-toxic known chemical and that even these low background levels may cause adverse effects such as cancer, disruption of regulatory hormones, reproductive and immune system disorders, and birth defects. Others disagree. The Chemical Manufacturers Association wrote, "There is no direct evidence to show that any of the effects of dioxins occur in humans in everyday levels."

No one knows whether very small doses of potent poisons are harmful. Environmentalists argue that it is "better to be safe than sorry," and that therefore we should reduce ambient concentrations of volatiles like dioxin. Others counter that because the harmful effects are unproven, we should not burden our economy with the costs of control.

Particulates and Aerosols

A **particle** or **particulate** is any small piece of solid matter, such as dust or soot. An **aerosol** is any small particle that is larger than a molecule and suspended in air. These three terms are used interchangeably to discuss air pollution. Many natural processes release aerosols. Windblown silt, pollen, volcanic ash, salt spray from the oceans, and smoke and soot from wildfires are all aerosols. Industrial emissions add to these natural sources.

Smoke and soot are carcinogenic aerosols formed whenever fuels are burned. Coal always contains clay and other noncombustible minerals that accumulated when the coal formed in the muddy bottoms of ancient swamps. When the coal burns, some of these minerals escape from the chimney as **fly ash**, which settles as gritty dust. When metals are mined, the drilling, blasting, and digging raise dust, and this, too, adds to the total load of aerosols.

In 1988, EPA epidemiologists noted that whenever atmospheric aerosol levels rose above a critical level in Steubenville, Ohio, the number of fatalities from all causes—car accidents to heart attacks—rose. After several studies substantiated the Steubenville report, the EPA proposed additional reductions of the ambient aerosol levels in the United States. Opponents argued that it is unfair to target all aerosols because the term covers a wide range of substances from a benign grain of salt to a deadly mist of toxic volatiles.

17.7
Depletion of the Ozone Layer

Solar energy breaks oxygen molecules (O_2) apart in the stratosphere, releasing free oxygen atoms (O). The free oxygen atoms combine with oxygen molecules to form ozone (O_3). Ozone absorbs high-energy ultraviolet light. This absorption protects life on Earth because ultraviolet light causes skin cancer, inhibits plant growth, and otherwise harms living tissue.

In the 1970s, scientists learned that organic compounds containing chlorine and fluorine, called **chlorofluorocarbons (CFCs),** or compounds containing bromine and chlorine, called **halons,** rise into the upper atmosphere and destroy ozone (◆ Figure 17.18). At that time, CFCs were used as cooling agents in almost all refrigerators and air conditioners, as propellants in some aerosol cans, as cleaning solvents during the manufacture of weapons, and in plastic foam in coffee cups and some building insulation.

In 1985, scientists observed an unusually low ozone concentration in the stratosphere over Antarctica, called the **ozone hole.** The ozone concentration over Antarctica continued to decline between 1985 and 1993, until it was 65 percent below normal over 23 million square kilometers, an area almost the size of North America. Research groups also reported significant increases in ultraviolet radiation from the Sun at ground level in the region. In addition, scientists recorded ozone depletion in the Northern Hemisphere. In March 1995, ozone concentration above the United States was 15 to 20 percent lower than during March 1979.

Data on global ozone depletion persuaded the industrial nations of the world to limit the use of CFCs and other ozone-destroying compounds. In a series of international agreements signed between 1978 and 1992, many nations of the world agreed to reduce or curtail production of compounds that destroy atmospheric ozone. Most industrialized countries stopped production of CFCs on January 1, 1996.

The international bans have had positive results. The concentration of ozone-destroying chemicals peaked in the troposphere (lower atmosphere) in 1994 and has been declining ever since. As a result, fewer CFCs and halons

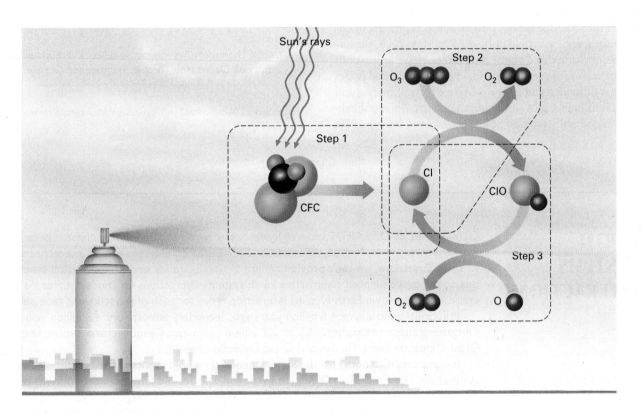

◆ **ACTIVE FIGURE 17.18** CFCs destroy the ozone layer in a three-step reaction. Step 1: CFCs rise into the stratosphere. Ultraviolet radiation breaks the CFC molecules apart, releasing chlorine atoms. Step 2: Chlorine atoms react with ozone, O_3, to destroy the ozone molecule and release oxygen, O_2. The extra oxygen atom combines with chlorine to produce ClO. Step 3: The ClO sheds its oxygen to produce another free chlorine atom. Thus, chlorine is not used up in the reaction, and one chlorine atom reacts over and over again to destroy many ozone molecules.

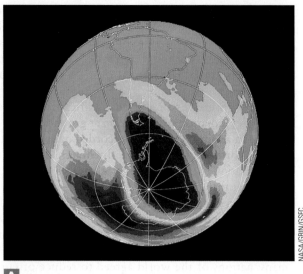

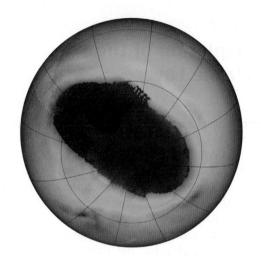

ThomsonNOW ◊ **FIGURE 17.19** **(A)** A satellite image of stratospheric ozone over Antarctica in 1994. **(B)** Similar data for 2004. In both images, dark purple shows the lowest ozone concentration.

have been drifting into the stratosphere. The CFCs and halons that are already in the stratosphere break down slowly, but the concentration of ozone-destroying chemicals in the stratosphere has peaked and is beginning to decline. In a review article published in May, 2006, the authors conclude that "ozone abundances have at least not decreased for most of the world." However, natural cycles make it difficult to judge the relative effects of human and natural influences. In summary, they write, "it is unlikely that ozone will stabilize at levels observed before 1980,

when a decline in ozone concentrations was first observed.[1] (◊ Figure 17.19).

ThomsonNOW

CLICK ThomsonNOW Interactive to work through an activity on Ozone Hole Through Weather and Climate.

---

[1] Elizabeth C. Weatherhead and Signe Bech Andersen, *Nature*, 441, May 4, 2006, 39.

EARTH SYSTEMS INTERACTIONS

OUR ANALYSIS OF EARTH systems interactions is complicated in this chapter because as Earth's environments evolved, different conditions created different sets of interactions. Due to space limitations, we can't provide an interaction figure for each of the critical time periods, so the figure supplied summarizes Earth systems interactions for modern times. As a exercise, draw your own Earth systems interaction figure for each of the following time periods: Primordial atmosphere: 4.6 billion years ago; Secondary atmosphere: 4.5 billion years ago; Outgassing alters atmosphere: 4.5 to 2.7 billion years ago; Cyanobacteria evolve; The First Great Oxidation Event; The Second Great Oxidation Event.

It is important to note also the importance of threshold events in the evolution of Earth's atmosphere. The First Great Oxidation Event and The Second Great Oxidation Event remind us that while component processes, such as the production of oxygen by cyanobacteria, may continue at a relatively constant rate, the atmosphere reacts readily with the other three Earth spheres, and the result of all these interactions proceeding simultaneously has produced massive threshold events in the past. ∎

EARTH SYSTEMS INTERACTIONS

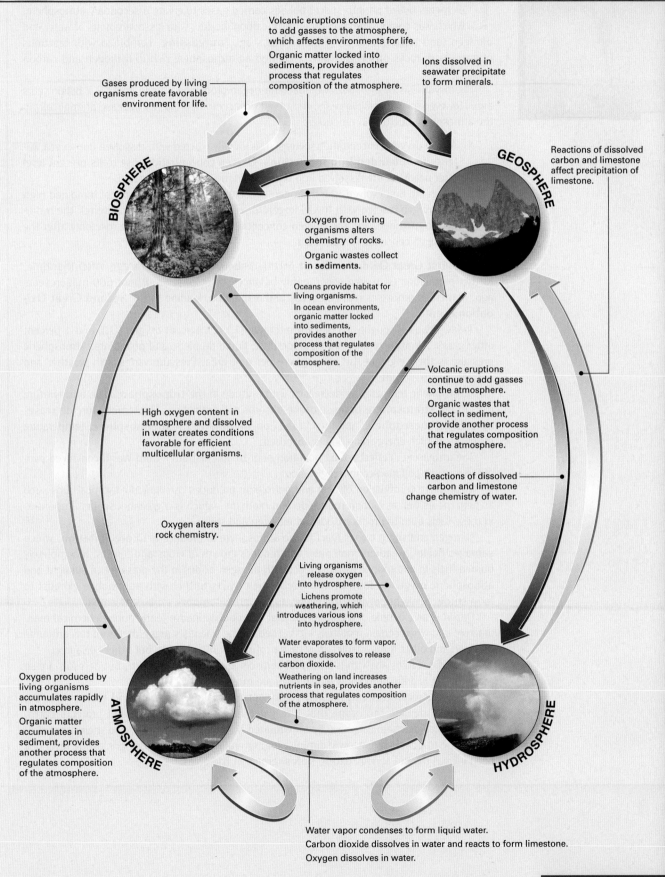

Volcanic eruptions continue to add gasses to the atmosphere, which affects environments for life.

Organic matter locked into sediments, provides another process that regulates composition of the atmosphere.

Ions dissolved in seawater precipitate to form minerals.

Gases produced by living organisms create favorable environment for life.

Reactions of dissolved carbon and limestone affect precipitation of limestone.

BIOSPHERE

GEOSPHERE

Oxygen from living organisms alters chemistry of rocks.

Organic wastes collect in sediments.

Oceans provide habitat for living organisms.

In ocean environments, organic matter locked into sediments, provides another process that regulates composition of the atmosphere.

Volcanic eruptions continue to add gasses to the atmosphere.

Organic wastes that collect in sediment, provide another process that regulates composition of the atmosphere.

High oxygen content in atmosphere and dissolved in water creates conditions favorable for efficient multicellular organisms.

Reactions of dissolved carbon and limestone change chemistry of water.

Oxygen alters rock chemistry.

Living organisms release oxygen into hydrosphere.

Lichens promote weathering, which introduces various ions into hydrosphere.

Water evaporates to form vapor.

Limestone dissolves to release carbon dioxide.

Weathering on land increases nutrients in sea, provides another process that regulates composition of the atmosphere.

Oxygen produced by living organisms accumulates rapidly in atmosphere.

Organic matter accumulates in sediment, provides another process that regulates composition of the atmosphere.

ATMOSPHERE

HYDROSPHERE

Water vapor condenses to form liquid water.

Carbon dioxide dissolves in water and reacts to form limestone.

Oxygen dissolves in water.

17.7   Depletion of the Ozone Layer

447

SUMMARY

Earth's primary atmosphere was predominantly hydrogen and helium. After these light elements boiled off into space, bolides carried gases to create a secondary atmosphere, which was composed predominantly of carbon dioxide, with lesser amounts of water and nitrogen, and trace gases. By 4 billion years ago, **outgassing,** combined with reactions of gases with rocks of the geosphere, created an atmosphere rich in hydrogen and carbon dioxide.

Photosynthesis by primitive cyanobacteria began producing oxygen about 2.7 billion years ago. However, for 300 million years, the oxygen concentration did not increase dramatically in the atmosphere due to two processes:

1. As the oxygen concentration increased, it initially reacted with dissolved iron in sea water to form **banded iron formations**. Oxygen concentration rose to its present level after nearly all the dissolved iron had been removed by precipitation.
2. When life first evolved, the hydrogen concentration of the atmosphere remained high and oxygen reacted with the hydrogen to produce water. Bacteria converted the hydrogen to methane, and the oxygen concentration in the atmosphere rose only after the hydrogen concentration declined.

The **First Great Oxidation Event** occurred about 2.4 billion years ago. After that time, further production of oxygen was roughly balanced by respiration. About 600 million years ago, numerous processes reduced decay and respiration, leading to the **Second Great Oxidation Event.**

Today, dry air is roughly 78 percent nitrogen (N_2), 21 percent oxygen (O_2), and 1 percent other gases. Air also contains water vapor, dust, liquid droplets, and pollutants. Atmospheric pressure is the weight of the atmosphere per unit area. Pressure varies with weather and decreases with altitude.

Atmospheric temperature decreases with altitude in the **troposphere**. The temperature rises in the **stratosphere** because ozone absorbs solar radiation. The temperature decreases again in the **mesosphere**, and then in the uppermost layer, the **thermosphere**, temperature increases as high-energy radiation is absorbed.

The increasing ill effects of air pollution prior to and just after Word War II convinced people to pass air-pollution-control legislation.

Incomplete combustion of coal and petroleum produces carcinogenic hydrocarbons such as benzene as well as carbon monoxide and methane, which is a greenhouse gas. Impurities in these fuels burn to produce oxides of nitrogen and sulfur.

Nitrogen and sulfur oxides react in the atmosphere to produce **acid precipitation**, which damages health, weathers materials, and reduces growth of crops and forests. Incompletely burned fuels in automobile exhaust react with nitrogen oxides in the presence of sunlight and atmospheric oxygen to form ozone. Ozone then reacts further with automobile exhaust to form **smog**. Sunlight provides the energy to convert automobile exhaust to smog.

Dioxin is an example of a compound that is produced inadvertently during chemical manufacture and when certain materials are burned. Some scientists argue that even tiny amounts of dioxin and other toxic volatiles may be harmful to human health, but others disagree.

Scientific studies show that industrial aerosols are harmful to health, but aerosols are so varied that it is difficult to know which ones are most harmful.

Chlorofluorocarbons and halons, compounds containing chlorine and bromine, in the stratosphere deplete the ozone that filters out harmful UV radiation and protects Earth. Ozone-destroying chemicals have been regulated by international treaty, and their concentration in the troposphere is slowly diminishing. Overall, Earth's stratospheric ozone level has been increasing, but serious **ozone holes** remain at the poles.

Key Terms

For Review

1. Discuss the formation and composition of Earth's earliest atmosphere.

2. How and when did Earth's secondary atmosphere form? Compare the composition of this atmosphere with that of the modern one.

3. How did outgassing affect the composition of the Hadean atmosphere?

4. What process first began producing oxygen about 2.7 billion years ago? Why did the oxygen concentration remain low for 300 million years?

5. Briefly describe the First Great Oxidation Event.

6. How did banded iron formations form? How did their formations affect the atmosphere?

7. Briefly describe the Second Great Oxidation Event.

8. List the two most abundant gases in the atmosphere. List three other, less abundant gases. List three nongaseous components of natural air.

9. Draw a graph of the change in pressure with altitude. Explain why the pressure changes as you have shown.

10. What is a barometer and how does it work?

11. Draw a figure showing temperature changes with altitude. Label all the significant layers in Earth's atmosphere.

12. Discuss the air pollution disaster in Donora. What pollutants were involved? Where did they come from? How did they become concentrated?

13. Briefly list the major sources of air pollution.

14. What air pollutants are generated when coal and gasoline burn?

15. What is acid rain, how does it form, and how does it affect people, crops, and materials?

16. What is smog, how does it form, and how does it differ from automobile exhaust?

17. What is a volatile? Why are volatiles hard to control?

18. Discuss the production, dispersal, and health effects of dioxin.

19. What is an aerosol? Briefly discuss the health effects of aerosols.

20. Explain how CFCs deplete the ozone layer.

21. Why is ozone in the troposphere harmful to humans, while ozone in the stratosphere is beneficial?

22. Discuss the effects of the international ban on ozone-destroying chemicals.

For Discussion

1. Explain how Earth systems interactions among the atmosphere, hydrosphere, and geosphere caused the First Great Oxidation Event and the Second Great Oxidation Event.

2. Explain how Earth systems interactions among the atmosphere, hydrosphere, and geosphere led to the formation of the banded iron formations.

3. Discuss the statement: Life could not have formed in the modern atmosphere and living organisms could not survive in the primordial one.

4. Given your knowledge of the evolution of Earth's atmosphere, do you think that it is likely that there is life on other planets? Defend your position.

5. Explain the importance of threshold events in the formation of Earth's primordial atmospheres.

6. Explain how feedback mechanisms affected Earth's primordial atmospheres.

7. Imagine that enough matter vanished from Earth's core so that Earth's mass decreased by half. In what ways would the atmosphere change? Would normal pressure at sea level be affected? Would the thickness of the atmosphere change? Explain.

8. If gasoline produces only carbon dioxide and water when it burns completely, why is automobile exhaust a source of air pollution?

9. State which of the following processes are potential sources of gaseous air pollutants, particulate air pollutants, both, or neither: (a) Gravel is screened to separate sand, small stones, and large stones into different piles. (b) A factory stores drums of liquid chemicals outdoors. Some of the drums are not tightly closed, and others have rusted and are leaking. The exposed liquids evaporate. (c) A waterfall drives a turbine, which makes electricity. (d) Automobile bodies in an assembly plant are sprayed with paint. The automobile bodies then move through an oven that dries the paint. (e) A garbage dump catches fire.

10. How can sulfur in coal contribute to the acidity of rainwater? What happens in a furnace when the coal is burned? What happens in the outdoor atmosphere?

11. Gasoline vapor plus ultraviolet lamps do not produce the same smog symptoms as do automobile exhaust plus ultraviolet lamps. What is missing from gasoline vapor that helps to produce smog?

12. Imagine that someone planned to build a municipal garbage incinerator in your neighborhood. The facility would burn domestic trash and use the energy to generate electricity. Moreover, stringent air pollution controls would keep toxic emissions to very low levels. Discuss the environmental benefits and drawbacks of this proposed incinerator.

CHAPTER 18

Energy Balance in the Atmosphere

Courtesy of Graham R. Thompson/Jonathan Turk

Glaciers and snowfields reflect light and cool the earth. The Jeffries Glacier, St Elias Range, Alaska

As we learned in Chapter 6, heat from the planet's interior causes mantle rock to circulate in great convection cells that drag the continents and ocean floors across the surface. In turn, moving plates generate earthquakes and volcanic eruptions. In contrast, almost all surface events are driven by solar energy. We feel that energy directly as we bask in sunshine on the first warm spring day after a hard winter. Without solar energy, there would be no life on our planet. But the Sun not only provides life-nurturing warmth and light, it generates wind, rain, and ocean currents. Sunlight heats the ocean, evaporates water, and thus, indirectly, causes rain to fall. A gentle zephyr that wafts two honeymooners across a lake in a small sailboat, a tornado that lifts the roof off a house, and a hurricane that drives storm waves over a seaside resort are all powered by the Sun.

When you wake up in the morning and look out the window, you may remark, "It is a nice day," or "What a nasty, gray day." Your remarks refer to the **weather**, the state of the atmosphere at a given place and time. Temperature, wind, cloudiness, humidity, and precipitation are all components of weather. The weather changes frequently, from day to day or even from hour to hour.

Climate is the characteristic weather of a region, particularly the temperature and precipitation, averaged over several decades. Miami and Los Angeles have warm climates; summers are hot and even the winters are warm. In contrast, New York and Chicago experience much greater temperature extremes. Even though winters are cool and often snowy, summers can be almost as hot as those in Miami. Seattle experiences moderate temperatures with foggy, cloudy winters. In this and the following two chapters, we discuss the atmosphere, weather, and climate. In Chapter 21, we will consider global climate change.

Many of the processes that drive energy balance in the atmosphere vary in gradual and predictable manners. But the weather and climate don't always vary gradually and predictably because they are affected by many complex, overlapping system interactions, and by threshold and feedback mechanisms. As a simple example: Incoming solar radiation is most intense at the equator and decreases predictably toward the poles. But some high-latitude locations are warmer than regions closer to the equator, and on any given day, it can be warmer in Montreal than in Houston. In this and subsequent chapters, we will explain the fundamental processes that drive weather and climate, and how they interact. ■

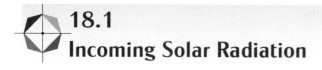

18.1
Incoming Solar Radiation

Solar energy streams from the Sun in all directions, and Earth receives only 1/2,000,000,000 (one two-billionth) of the total solar output. However, even this tiny fraction warms Earth's surface and makes it habitable.

The space between Earth and the Sun is nearly empty. How does sunlight travel through a vacuum? In the late 1600s and early 1700s, light was poorly understood. Isaac Newton postulated that light consists of streams of particles that he called "packets" of light. Two other physicists,

Robert Hooke and Christian Huygens, argued that light travels in waves. Today we know that Hook and Huygens were correct—light behaves as a wave, but Newton was also right—light acts as if it is composed of particles. But how can light be both a wave and a particle at the same time? In a sense this is an unfair question because light is fundamentally different from familiar objects. Light is unique; it behaves as a wave *and* a particle simultaneously.

Particles of light are called **photons**. In a vacuum, photons travel only at one speed, the speed of light, never faster and never slower. The speed of light is 3×10^8 meters/second. At that rate a photon covers the 150 million kilometers between Sun and Earth in about 8 minutes. Photons are unlike ordinary matter in that

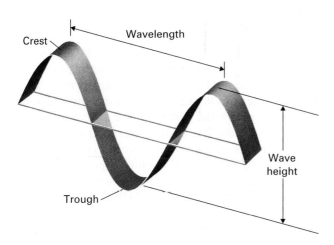

◈ **FIGURE 18.1** The terms used to describe a light wave are identical to those used for water, sound, and other types of waves.

they appear when they are emitted and disappear when they are absorbed.

Light also behaves as an electrical and magnetic wave, called **electromagnetic radiation**. Its **wavelength** is the distance between successive wave crests (◈ Figure 18.1). The **frequency** of a wave is the number of complete wave cycles, from crest to crest, that pass by any point in a second. (Think of how *frequently* the waves pass by.) Electromagnetic radiation occurs in a wide range of wavelengths and frequencies. The **electromagnetic spectrum** is the continuum of radiation of different wavelengths and frequencies (◈ Figure 18.2). At one end of the spectrum, radiation given off by ordinary household current has a long wavelength (5,000 kilometers) and low frequency (60 cycles/second). At the other end, cosmic rays

from outer space have a short wavelength (about one-trillionth of a centimeter, or 10^{-14} meter) and very high frequency (10^{-2} cycles/second). Visible light is a tiny portion (about one-millionth of one percent) of the electromagnetic spectrum.

Absorption and Emission

If you go outside on a cold, snowy, February day, you wear a heavy jacket, hat, and gloves. Yet the sunlight warms your face. If the sky is clear, even though the temperature may be well below freezing, you must wear sunglasses to prevent snow blindness and if you have unprotected fair skin, you will suffer sunburn.

Each photon is a tiny packet of concentrated energy. The energy of a single photon is related only to the frequency of the light, and *not* to the temperature. When a photon strikes your face, it may be absorbed. During **absorption of radiation**, the energy of the photon may initiate chemical and physical reactions in the molecules of your skin. One possible reaction is roughly analogous to cooking, and this reaction causes sunburn or tan. A photon may also cook a molecule in the retina of your eye, and if this reaction occurs frequently enough, you will become snow blind.

Some absorbed photons do not cause chemical reactions; instead they cause molecules to vibrate or rotate more rapidly. This rapid motion makes your skin feel warm.

Alternatively, the energy of a photon may be transferred to an electron in a molecule. In this case, the electron jumps to what scientists call an **excited state**. But electrons do not remain in their excited states forever. Frequently they fall back to a lower energy state, initiating **emission of radiation** in the process. All objects emit radiant energy at some wavelength (except at a temperature of absolute zero).

Let's shift our focus from your skin to an iron bar. An iron bar at room temperature emits infrared radiation. This radiation has low energy and long wavelengths. It is invisible because the energy is too low to activate the sensors in our eyes. Infrared radiation is sometimes called radiant heat, or heat rays. Thus even at room temperature an iron bar radiates heat. If you place the bar in a hot flame, it begins to glow with a dull, red color when it becomes hot enough. The heat has excited electrons in the iron bar, and the excited electrons then emit visible, red, electromagnetic radiation (◈ Figure 18.3). If you heat the bar further, it gradually

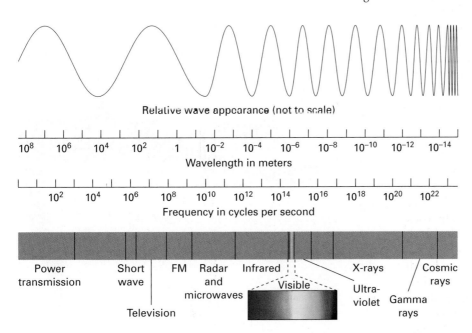

◈ **FIGURE 18.2** The electromagnetic spectrum. The wave shown is not to scale. In reality the wavelength varies by a factor of 10^{22}, and this huge difference cannot be shown.

◀▶ **FIGURE 18.3** Iron glows red when heated in a forge. If it is heated further, it emits white light.

Reflection

Radiation reflects from many surfaces. We are familiar with the images reflected by a mirror or the surface of a still lake. Some surfaces are better reflectors than others. **Albedo** is the proportional reflectance of a surface. A mirror reflects nearly 100 percent of the light that strikes it and has an albedo close to 100 percent. Even some dull-looking objects are efficient reflectors. Light is bouncing back to your eye from the white paper of this page, although very little is reflected from the black letters.

Snowfields and glaciers have high albedos and reflect 80 to 90 percent of sunlight. Clouds have the second-highest albedo and reflect 50 to 55 percent of sunlight. On the other hand, city buildings and dark pavement have albedos of only 10 to 15 percent (◀▶ Figure 18.4). Forests, with many independent surfaces of dark leaves, have an even lower albedo of about 5 percent. The oceans, which cover about two-thirds of Earth's surface, also have a low albedo. As a result, they absorb considerable solar energy and thus strongly affect Earth's radiation balance. Thus the temperature balance of the atmosphere is profoundly affected by the albedos of the hydrosphere, geosphere, and biosphere. If Earth's albedo were to rise by growth of glaciers or cloud cover, the surface of our planet would cool. Alternatively, a decrease in albedo (caused by the melting of glaciers) would cause warming. Thus the growth and shrinking of Earth's snow and ice is a classic example of a feedback mechanism. If Earth were to cool by a small amount, snowfields and glaciers would grow. But snow and ice reflect solar radiation back into space and lead to additional cooling—which causes further expansion of the snow and ice—and so on.

changes color until it becomes white. This demonstration shows another property of emitted electromagnetic radiation: The wavelength and color of the radiation are determined by the temperature of the source. With increasing temperature, the energy level of the radiation increases, the wavelength decreases, and the color changes progressively.

The Sun's surface temperature is about 6,000°C. Because of its high temperature, the Sun emits relatively high-energy (short-wavelength) radiation, primarily in the ultraviolet and visible portions of the spectrum. When this radiation strikes Earth, it is absorbed by rock and soil. After the radiant energy is absorbed, the rock and soil reemit it. But Earth's surface is much cooler than the Sun's. Therefore Earth emits low-energy, infrared heat radiation, which has relatively long wavelength and low frequency. Thus Earth absorbs high-energy, visible light and emits low-energy, invisible, infrared heat radiation.

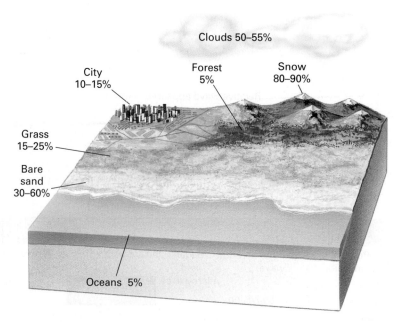

◀▶ **FIGURE 18.4** The albedos of common Earth surfaces vary greatly.

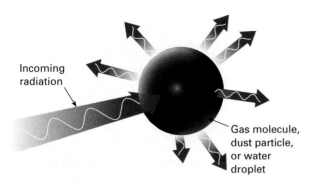

FIGURE 18.5 Atmospheric gases, water droplets, and dust scatter incoming solar radiation. When radiation scatters, its direction changes but the wavelength remains constant.

Scattering

On a clear day the Sun shines directly through windows on the south side of a building, but if you look through a north-facing window, you cannot see the Sun. Even so, light enters through the window, and the sky outside is blue. If sunlight were only transmitted directly, a room with north-facing windows would be dark and the sky outside the window would be black. Atmospheric gases, water droplets, and dust particles scatter sunlight in all directions, as shown in ◆ Figure 18.5.

It is this scattered light that illuminates a room with north-facing windows and turns the sky blue.

The amount of scattering is inversely proportional to the wavelength of light. Short-wavelength blue light, therefore, scatters more than longer-wavelength red light. The Sun emits light of all wavelengths, which combine to make up white light. Consequently, in space the Sun appears white. The sky appears blue from Earth's surface because the blue component of sunlight scatters more than other frequencies and colors the atmosphere. The Sun appears yellow from Earth because yellow is the color of white light with most of the blue light removed.

18.2 The Radiation Balance

With this background, let us examine the fate of sunlight as it reaches Earth (◆ Figure 18.6). Of all the sunlight that reaches Earth, 50 percent is absorbed, scattered, or reflected by clouds and atmosphere, 3 percent is reflected by Earth's surface, and 47 percent is absorbed by Earth's surface. The absorbed radiation warms rocks, soil, and water.

50% absorbed, reflected, and scattered by the atmosphere

50% reaches ground

19% absorbed by clouds and atmosphere

23% reflected by clouds

3% reflected by Earth's surface

8% scattered

47% absorbed by Earth's surface

ThomsonNOW ◆ **ACTIVE FIGURE 18.6** One half of the incoming solar radiation reaches Earth's surface. The atmosphere scatters, reflects, and absorbs the other half. All of the radiation absorbed by Earth's surface is reradiated as long-wavelength heat radiation.

INTERACTIVE QUESTION: *What effect would the doubling of the average global albedo of Earth's surface have on the average global atmospheric temperature?*

If Earth absorbs radiant energy from the Sun, why doesn't Earth's surface get hotter and hotter until the oceans boil and the rocks melt? The answer is that rocks, soil, and water reemit all the energy they absorb. As explained previously, most solar energy that reaches Earth is short-wavelength, visible and ultraviolet radiation. Earth's surface absorbs this radiation and then reemits the energy mostly as long-wavelength, invisible, infrared (heat) radiation. Some of this infrared heat escapes directly into space, but some is absorbed by the atmosphere. The atmosphere traps this heat radiating from Earth and acts as an insulating blanket.

If Earth had no atmosphere, radiant heat loss would be so rapid that Earth's surface would cool drastically at night. Earth remains warm at night because the atmosphere absorbs and retains much of the radiation emitted by the ground. If the atmosphere were to absorb even more of the long-wavelength radiant heat from Earth, the atmosphere and Earth's surface would become warmer. This warming process is called the **greenhouse effect**[1] (◀▶ Figure 18.7).

Some gases in the atmosphere absorb infrared radiation and others do not. Oxygen and nitrogen, which together make up almost 99 percent of dry air at ground level, do not absorb infrared radiation; water, carbon dioxide, methane, and a few other gases do. Thus, they are called greenhouse gases. Water is the most abundant greenhouse gas in Earth's atmosphere.

Carbon dioxide is also important because its abundance in the atmosphere can vary as a result of several natural and industrial processes. Methane has become important because large quantities are released by industry and agriculture. The greenhouse effect and global climate change are discussed in Chapter 21.

Dust, cloud cover, aerosols, and other particulate air pollutants also affect Earth's atmospheric temperature by altering the amount of sunlight that is absorbed or reflected. Large volcanic eruptions can inject dust into the upper atmosphere and reflect enough sunlight to cool Earth. The effect of man-made pollution is harder to interpret. Dust and aerosols close to the surface of Earth reflect incoming solar radiation but also absorb infrared radiation emitted by the ground. The net result depends on a complex balance of factors, including particle size, particle composition, and natural cloudiness. Whatever the outcome, it is certain that whenever people change the composition of air, they run the risk of altering weather and climate.

---

1. The comparison between the atmosphere and a greenhouse is only partially correct. Both the glass in a greenhouse and Earth's atmosphere are transparent to incoming, short-wavelength radiation and partially opaque to emitted, long-wavelength, heat radiation. However, the glass in a greenhouse is also a physical barrier that prevents heat loss through air movement, whereas the atmosphere is not. Yet in the Earth, the absorption of radiation is sufficient to warm the atmosphere and cause significant changes in Earth's climate. We use the term *greenhouse effect* because it has become common, both in atmospheric science and in everyday use.

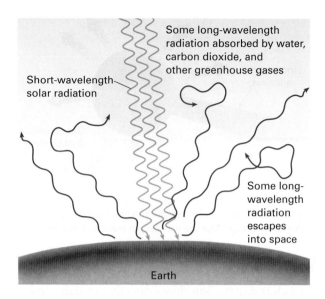

◀▶ **FIGURE 18.7**   The greenhouse effect can be viewed as a three-step process. Step 1: Rocks, soil, and water absorb short-wavelength solar radiation, and become warmer (orange lines). Step 2: The Earth reradiates the energy as long-wavelength infrared heat rays (red lines). Step 3: Molecules in the atmosphere absorb some of the heat, and the atmosphere becomes warmer.

INTERACTIVE QUESTION: *What effect would a doubling of the atmospheric concentration of carbon dioxide have on the average global atmospheric temperature?*

18.3
Energy Storage and Transfer—The Driving Mechanisms for Weather and Climate

Heat and Temperature

All matter consists of atoms and molecules that are in constant motion. They fly through space, they rotate, and they vibrate. **Temperature** is proportional to the average speed of the atoms or molecules in a sample.[2] In a teacup full of boiling water, the water molecules are racing around rapidly, smashing into each other, spinning like so many boomerangs, and vibrating like spheres connected by pulsating springs. In a bathtub full of ice water, the water molecules are also moving, but much more slowly. Molecules in the hot water are moving faster than those in the cold water, so the hot water has a higher temperature.

In contrast, **heat** is a measure of the total energy in a sample. It is related to the average energy of every

---

2. More precisely, temperature is proportional to the kinetic energy of the atoms and molecules. In turn, the kinetic energy equals (mass) × (velocity)^2.

molecule multiplied by the total number of molecules. It may seem counterintuitive, but there is more heat in a bathtub full of ice water than in a teacup full of boiling water. The average molecule in the ice water is moving slower and therefore has less energy than the average molecule in the boiling water. But there are so many more molecules in the bathtub than in the teacup that the total heat energy is greater.

Heat Transport by Conduction and Convection

If you place a metal frying pan on the stove, the handle gets hot, even though it is not in contact with the burner, because the metal conducts heat from the bottom of the pan to the handle. **Conduction** is the transport of heat by direct collisions among atoms or molecules. When the frying pan is heated from below, the metal atoms on the bottom of the pan move more rapidly. They then collide with their neighbors and transfer energy to them. Like a falling row of dominoes, energy is passed from one atom to another throughout the pan until the handle becomes hot.

Metals conduct heat rapidly and efficiently, but air is a poor conductor. To understand how air transports heat, imagine that a heater is placed in one corner of a cold room. The heated air in the corner expands, becoming less dense. This light, hot air rises to the ceiling. It flows along the ceiling, cools, falls, and returns to the stove, where it is reheated (◀▶ Figure 18.8A). **Convection** is the transport of heat by the movement of currents. Convection occurs readily in liquids and gases. Recall from Chapter 16 that ocean currents transport large quantities of heat northward or southward, thereby altering climate. For example, the Gulf Stream carries tropical water northward and warms the west coast of Europe.

Similar currents occur in the atmosphere (◀▶ Figure 18.8B). If air in one region is heated above the temperature of surrounding air, this warm air becomes less dense and rises, just as a log floats to the surface of a pond. As the warm air rises, cooler, denser air in another portion of the atmosphere sinks. Air then flows along the surface to complete the cycle. In common speech, this horizontal airflow is called wind. In meteorology, horizontal airflow is called **advection**, whereas *convection* is reserved for vertical airflow. The steady winds that blow across the tropical oceans, a tornado that ravages a city in the Midwest, and a thunderstorm that drops rain and hail on your Sunday picnic are all caused by convective and advective processes in the atmosphere.

Changes of State

Given the proper temperature and pressure, most substances can exist in three states: solid, liquid, and gas. However, at Earth's surface, many substances commonly exist in only one state. In our experience, rock is almost always solid and molecular oxygen is almost always a gas. Water commonly exists in all three states—as solid ice, as liquid, and as gaseous water vapor.

Latent heat (stored heat) is the energy released or absorbed when a substance changes from one state to another. About 80 calories are required to melt a gram of ice at a constant temperature of 0°C. As a comparison, 100 calories are needed to heat the same amount of water from freezing to boiling (0°C to 100°C). Another 540

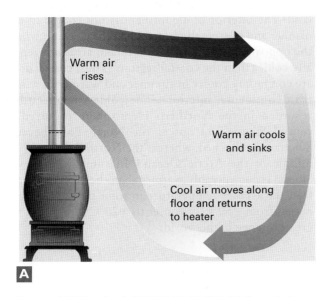

Warm air rises

Warm air cools and sinks

Cool air moves along floor and returns to heater

A

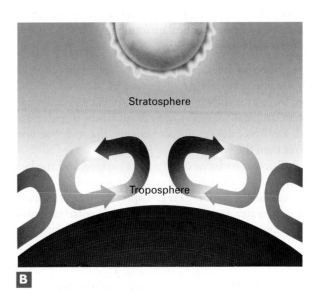

Stratosphere

Troposphere

B

ThomsonNOW ◀▶ **ACTIVE FIGURE 18.8** **(A)** Convection currents distribute heat throughout a room. **(B)** Convection also distributes heat through the atmosphere when the Sun heats Earth's surface. In this case the ceiling is the boundary between the troposphere and the stratosphere.

to 600 calories are needed to evaporate a gram of water at constant temperature and pressure.[3]

The energy transfers also work in reverse. When one gram of water vapor condenses to liquid, 540 to 600 calories are released. When one gram of water freezes, 80 calories are released (◆ Figure 18.9).

If you walk onto the beach after swimming, you feel cool, even on a hot day. Your skin temperature drops because water on your body is evaporating, and evaporation absorbs heat, cooling your skin. Similarly, evaporation from any body of water cools the water and the air around it. Conversely, condensation releases heat. The energy released when water condenses into rain during a single hurricane can be as great as the energy released by several atomic bombs.

The energy absorbed and released during freezing, melting, evaporation, and condensation of water is important in the atmospheric energy balance. For example, in the northern latitudes March is usually colder than September, even though equal amounts of sunlight are received in both months. However, in March much of the solar energy is absorbed by melting snow. Snow also has a high albedo and reflects sunlight efficiently. Evaporation cools seacoasts, and the energy of a hurricane comes, in part, from the condensation of massive amounts of water vapor.

Heat Storage

If you place a pan of water and a rock outside on a hot summer day, the rock becomes hotter than the water. Both have received identical quantities of solar radiation.[4]

Why is the rock hotter?

1. **Specific heat** is the amount of energy needed to raise the temperature of 1 gram of material by 1°C. Specific heat is different for every substance, and water has an unusually high specific heat. Thus, if water and rock absorb equal amounts of energy, the rock becomes hotter than the water.

2. Rock absorbs heat only at its surface, and the heat travels slowly through the rock. As a result, heat concentrates at the surface. Heat disperses more effectively through water for two reasons. First, solar radiation penetrates several meters below the surface of the water,

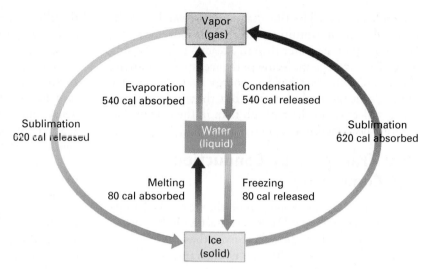

◆ **FIGURE 18.9** Water releases or absorbs latent heat as it changes among its liquid, solid, and vapor states. (Calories are given per gram at 0°C and 100°C. The values vary with temperature. Red arrows show processes that absorb heat; blue arrows show those that release heat.)

warming it to this depth. Second, water is a fluid and transports heat by convection.

3. Evaporation is a cooling process. Water loses heat and cools by evaporation, but rock does not.

Think of the consequences of the temperature difference between rock and water. On a hot summer day you may burn your feet walking across dry sand or rock, but the surface of a lake or ocean is never burning hot. Suppose that both the ocean and the adjacent coastline are at the same temperature in spring. As summer approaches, both land and sea receive equal amounts of solar energy. But the land becomes hotter, just as rock becomes hotter than water. Along the seacoast the cool sea moderates the temperature of the land. The interior of a continent is not cooled in this manner and is generally hotter than the coast. In winter the opposite effect occurs, and inland areas are generally colder than the coastal regions. Thus, coastal areas are commonly cooler in summer and warmer in winter than continental interiors. The coldest temperatures recorded in the Northern Hemisphere occurred in central Siberia and not at the North Pole, because Siberia is landlocked, whereas the North Pole lies in the middle of the Arctic Ocean.[5] In summer, however, Siberia is considerably warmer than the North Pole. In fact, the average temperature in some places in Siberia ranges from −50°C in winter to +20°C in summer, the greatest range in the world.

3. The heat of vaporization varies with temperature and pressure; at 100°C and one atmosphere pressure the value is 539.5 cal/g.
4. Water actually absorbs a bit more solar energy than rock because water has a lower albedo than rock, but this difference is overwhelmed by the other factors.

5. Even though the Arctic Ocean is covered with ice, water lies only a few meters below the surface and it still influences climate.

FOCUS ON

Latitude and Longitude

If someone handed you a perfectly smooth ball with a dot on it and asked you to describe the location of the dot, you would be at a loss to do so because all positions on the surface of a sphere are equal. How, then, can locations on a spherical Earth be described? Even if we ignore irregularities, continents, and oceans, Earth has points of reference because it rotates on its axis and has a magnetic field that nearly (but not exactly) coincides with the axis of rotation.

The North Pole and South Pole lie on the rotational axis, and lines of latitude form imaginary horizontal rings around the axis. Mathematicians measure distance on a sphere in degrees. Using this system, the equator is defined as 0° latitude, the North Pole is 90° north latitude, and the South Pole is 90° south latitude (Figure 1).

No natural east–west reference exists, so a line running through Greenwich, England, was arbitrarily chosen as the 0° line. The planet was then divided by lines of longitude, also measured in degrees. On a globe with the rotational axis vertically oriented, lines of longitude also run vertically. To a navigator they measure east–west angular distance from Greenwich, England.

The system is easy to use. Minneapolis–St. Paul lies at 45° north latitude and 93° west longitude. The latitude tells us that the city lies halfway between the equator (0°) and the North Pole (90°). Because a circle has 360°, the longitude tells us that Minneapolis is about one quarter of the way around the world in a westward direction from Greenwich, England.

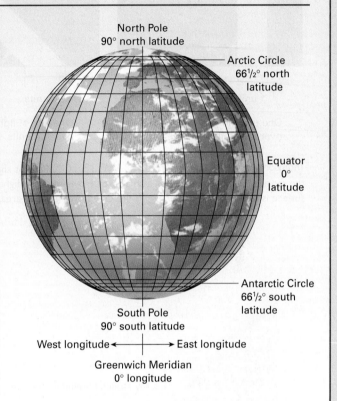

◇ **FIGURE 1** Latitude and longitude allow navigators to identify a location on a spherical Earth.

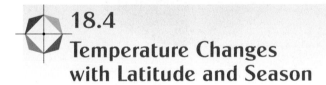

18.4
Temperature Changes with Latitude and Season

Temperature Changes with Latitude

The region near the equator is warm throughout the year, whereas polar regions are cold and ice-bound even in summer. To understand this temperature difference, consider first what happens if you hold a flashlight above a flat board. If the light is held directly overhead and the beam shines straight down, a small area is brightly lit. If the flashlight is held at an angle to the board, a larger area is illuminated. However, because the same amount of light is spread over a larger area, the intensity is reduced (◇ Figure 18.10).

Now consider what happens when the Sun shines directly over the equator. The equator, analogous to the flat board under a direct light, receives the most concentrated radiation. The Sun strikes the rest of the globe at an angle and thus radiation is less concentrated at higher latitudes (◇ Figure 18.11). Because the equator receives the most concentrated solar energy, it is generally warm throughout the year. Average atmospheric temperature becomes progressively cooler poleward (north and south of the equator). But, as

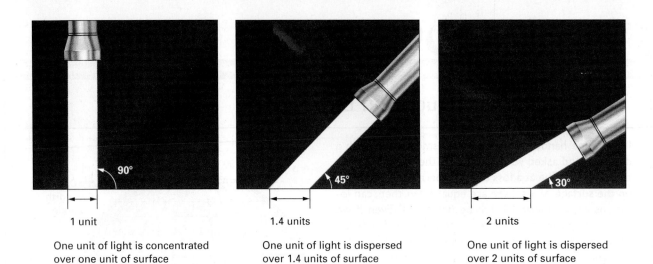

1 unit	1.4 units	2 units
One unit of light is concentrated over one unit of surface	One unit of light is dispersed over 1.4 units of surface	One unit of light is dispersed over 2 units of surface

ThomsonNOW ◈ **ACTIVE FIGURE 18.10** If a light shines from directly overhead, the radiation is concentrated on a small area. However, if the light shines at an angle, or if the surface is tilted, the radiant energy is dispersed over a larger area.

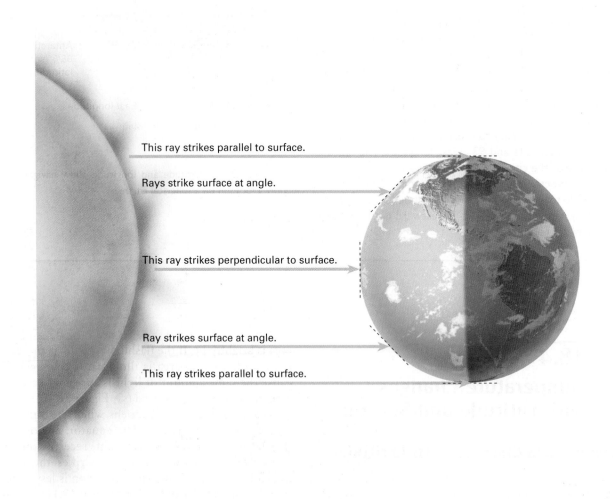

This ray strikes parallel to surface.

Rays strike surface at angle.

This ray strikes perpendicular to surface.

Ray strikes surface at angle.

This ray strikes parallel to surface.

◈ **FIGURE 18.11** When the Sun shines directly over the equator, the equator receives the most intense solar radiation, and the poles receive little.

mentioned in the introduction, the average temperature does not change steadily with latitude, because many other factors, such as winds, ocean currents, albedo, and proximity to the oceans also affect atmospheric temperature in any given region.

The Seasons

Earth circles the Sun in a planar orbit, while simultaneously spinning on its axis. This axis is tilted at 23.5° from a line drawn perpendicular to the orbital plane.

Earth revolves around the Sun once a year. As shown in ◄▶ Figure 18.12, the North Pole tilts toward the Sun in summer and away from it in winter. June 21 is the summer **solstice** in the Northern Hemisphere because at this time, the North Pole leans the full 23.5° toward the Sun. As a result, sunlight strikes Earth from directly overhead at a latitude 23.5° north of the equator. This latitude is the **tropic of Cancer**. If you stood on the tropic of Cancer at noon on June 21, you would cast no shadow. June is warm in the Northern Hemisphere for two reasons: (1) When the Sun is high in the sky, sunlight is more concentrated than it is in winter; (2) when the North Pole is

tilted toward the Sun, it receives 24 hours of daylight. Polar regions are called lands of the midnight Sun because the Sun never sets in the summertime (◄▶ Figure 18.13). Below the Arctic Circle the Sun sets in the summer, but the days are always longer than they are in winter (Table 18.1).

While it is summer in the Northern Hemisphere, the South Pole tilts away from the Sun and the Southern Hemisphere receives low-intensity sunlight and has short days. June 21 is the first day of winter in the Southern Hemisphere. Six months later, on December 21 or 22, the seasons are reversed. The North Pole tilts away from the Sun, giving rise to the winter solstice in the Northern Hemisphere, while it is summer in the Southern Hemisphere. On this day sunlight strikes Earth directly overhead at the tropic of Capricorn, latitude 23.5° south. At the North Pole, the Sun never rises and it is continuously dark, while the South Pole is bathed in continuous daylight.

On March 21 and September 21 or 22 Earth's axis lies at right angles to a line drawn between Earth and the Sun. As a result, the poles are not tilted toward or away from the Sun and the Sun shines directly overhead at the

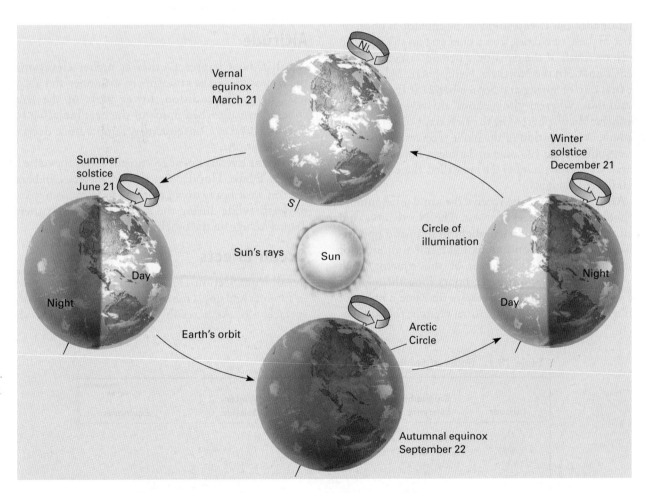

ThomsonNOW ◄▶ **ACTIVE FIGURE 18.12** Weather changes with the seasons because Earth's axis is tilted relative to the plane of its orbit around the Sun. As a result, the Northern Hemisphere receives more direct sunlight during summer but less during winter.

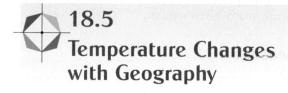

reaches the poles at a much lower angle and therefore delivers much less total energy per unit of surface area.

18.5
Temperature Changes with Geography

Even though all locations at a given latitude receive equal amounts of solar radiation, some places have cooler climates than others at the same latitude. ◆ Figure 18.14 shows temperatures around Earth in January and in July. Lines called **isotherms** connect areas of the same average temperature. Note that the isotherms loop and dip across lines of latitude. For example, the January 0°C line runs through Seattle, Washington, dips southward across the center of the United States, and then swings northward to northern Norway. Such variations occur with latitude because winds and ocean currents transport heat from one region of Earth to another. Other factors, discussed in Section 18.3, include the different heat storage properties of oceans and continents, evaporative cooling of the oceans, and the latent heat of snow and ice.

Altitude

Recall from Chapter 17 that at higher elevations in the troposphere, the atmosphere is thinner and absorbs less energy. In addition, lower parts of the troposphere have absorbed much of the heat radiating from Earth's surface. Consequently, temperature decreases with elevation.

Mount Everest lies at 28° north, about the same latitude as Tampa, Florida. Yet, at 8,000 meters, climbers on Everest wear heavy down clothing to keep from freezing to death, while during the same season, sunbathers in Tampa lounge on the beach in bikinis.

Ocean Effects

Recall from Section 18.3 that land heats more quickly in summer and cools more quickly in winter than ocean surfaces do. As a result, continental interiors show

◆ **FIGURE 18.13** The Sun shines at midnight in July at 70° north latitude in the Canadian Arctic. A small boat sits at anchor near Cape Parry in the Beaufort Sea, Northwest Territories.

equator at noon. If you stood at the equator at noon on either of these two dates, you would cast no shadow. But north or south of the equator, a person casts a shadow even at noon. In the Northern Hemisphere, March 21 is the first day of spring and September 21 is the first day of autumn, whereas the seasons are reversed in the Southern Hemisphere. On the first days of spring and autumn, every portion of the globe receives 12 hours of direct sunlight and 12 hours of darkness. For this reason, March 21 and September 21 are called the **equinoxes,** meaning equal nights.

All areas of the globe receive the same total number of hours of sunlight every year. The North Pole and South Pole receive direct sunlight in dramatic opposition, 6 months of continuous light and 6 months of continuous darkness, whereas at the equator, each day and night are close to 12 hours long throughout the year. Although the poles receive the same number of sunlight hours as do the equatorial regions, the sunlight

Latitude	Geographic Reference	Summer Solstice	Winter Solstice	Equinoxes
08	Equator	12 h	12 h	12 h
308	New Orleans	13 h 56 min	10 h 04 min	12 h
408	Denver	14 h 52 min	9 h 08 min	12 h
508	Vancouver	16 h 18 min	7 h 42 min	12 h
908	North Pole	24 hr	0 h 00 min	12 h

Table 18.1   Hours of Sunlight Per Day

January

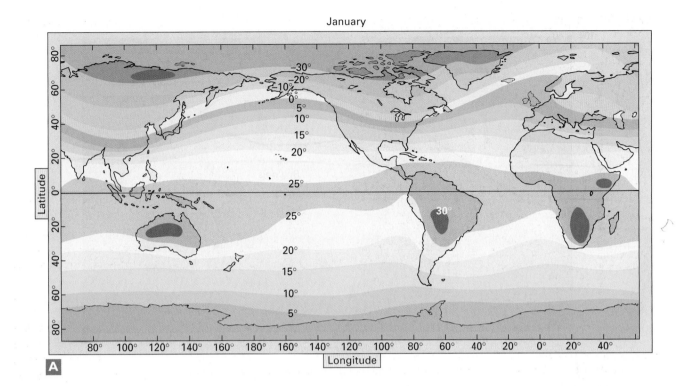

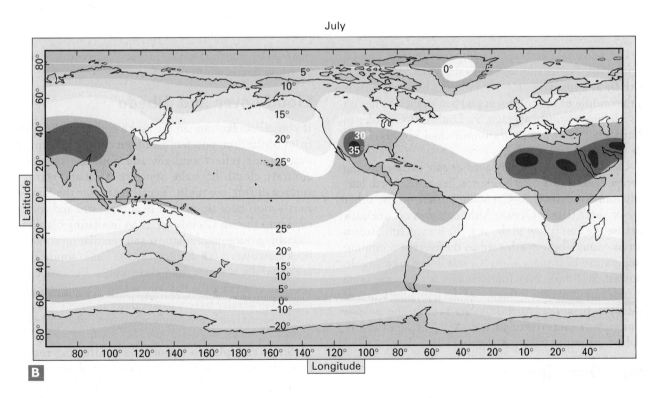

July

◈ **FIGURE 18.14** Global temperature distributions **(A)** in January, and **(B)** in July.
Isotherm lines connect places with the same average temperatures.

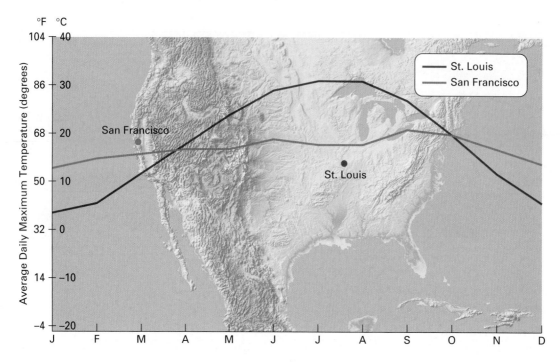

°F °C

Average Daily Maximum Temperature (degrees)

104 — 40
86 — 30
68 — 20
50 — 10
32 — 0
14 — −10
−4 — −20

J F M A M J J A S O N D

St. Louis
San Francisco

San Francisco

St. Louis

◆ **FIGURE 18.15** Continental St. Louis (red line) is colder in winter and warmer in summer than coastal San Francisco (blue line).

greater seasonal extremes of temperature than coastal regions do. For example, San Francisco and St. Louis both lie at approximately 38° latitude, but St. Louis is in the middle of the continent and San Francisco lies on the Pacific coast. On average, St. Louis is 9°C cooler in winter and 9°C warmer in summer than San Francisco (◆ Figure 18.15).

Ocean currents also play a major role in determining average temperature. Paris, France, lies at 48° north latitude, north of the U.S.–Canadian border on the other side of the Atlantic Ocean. Although winters are dark because the Sun is low in the sky, the warm Gulf Stream carries enough heat northward so the average minimum temperature in January is −1°C, just below freezing. In contrast, St. John's, Newfoundland, is at about the same latitude but under the influence of the Labrador Current that flows from the North Pole. As a result, the average minimum January temperature is −8°C, significantly colder than it is in Paris (◆ Figure 18.16).

Wind Direction

Winds carry heat from region to region just as ocean currents do. Vladivostok is a Russian city on the west coast of the Pacific Ocean at about 43° north latitude. It has a near-Arctic climate with frigid winters and heavy snow. Forests are stunted by the cold and trees are generally small.

In comparison, Portland, Oregon, lies at 45° north latitude on the east coast of the Pacific Ocean. Its temperate climate supports majestic cedar forests, and rain is more common than snow in winter (◆ Figure 18.17). The

main difference in this case is that Vladivostok is cooled by frigid Arctic winds from Siberia.

Cloud Cover and Albedo

If you are sunbathing on the beach and a cloud passes in front of the Sun, you feel a sudden cooling. During the day, clouds reflect sunlight and cool the surface. In contrast, clouds have the opposite effect and warm the surface during the night. Recall that after the Sun sets, Earth cools because radiant heat from soil and rock escape into the air. Clouds act as an insulating blanket by absorbing outgoing radiation and reradiating some of it back downward (◆ Figure 18.18). Thus cloudy nights are generally warmer than clear nights. Because clouds cool Earth during the day and warm it at night, cloudy regions generally have a lower daily temperature range than regions with predominantly blue sky and starry nights.

Snow also plays a major role in affecting regional temperature. Imagine that it is raining and the temperature is 0.5°C, or just above freezing. After the rain stops, the Sun comes out. The bare ground absorbs solar radiation and the surface heats up. Now imagine that the temperature dropped to −0.5°C during the storm. In this case, the precipitation would fall as snow. After the storm passed and the Sun came out, the surface would be covered with a sparkling white cover with 80 to 90 percent albedo. Most of the incoming solar radiation would reflect back into the atmosphere and the surface would remain much cooler. If the surface temperature did rise above freezing, the snow

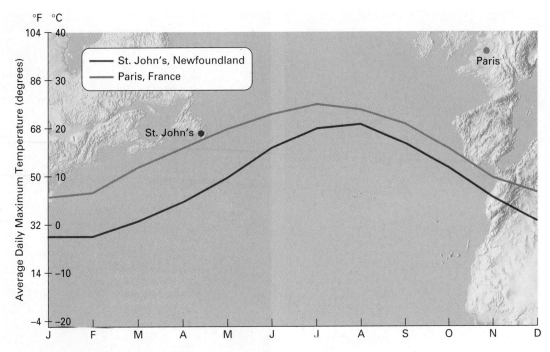

◈ **FIGURE 18.16** Paris is warmed by the Gulf Stream and the North Atlantic Drift. On the other hand, St. John's, Newfoundland, is alternately warmed by the Gulf Stream and cooled by the Labrador Current. The cooling effect of the Labrador Current depresses the temperature of St. John's year round.

would have to melt before the ground temperature would rise. But snow has a very high latent heat; it takes a lot of energy to melt the snow without raising the temperature at all. In this example, a seemingly insignificant, one-degree initial temperature difference during a storm would cause a much larger temperature difference when the Sun came out after the storm.

This type of threshold mechanism is very important in weather and climate and will be discussed in the following two chapters.

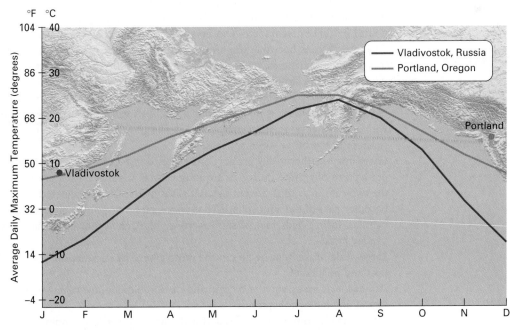

◈ **FIGURE 18.17** During the summer, temperatures in Vladivostok, Russia, and Portland, Oregon, are nearly identical. However, frigid Arctic wind from Siberia cools Vladivostok during the winter (red line) so that the temperature is significantly colder than that of Portland (blue line).

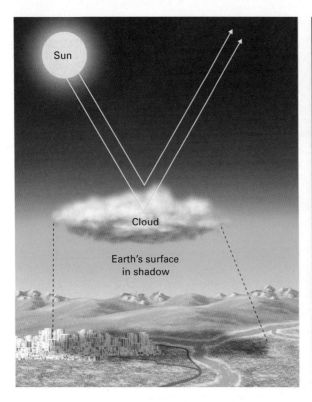
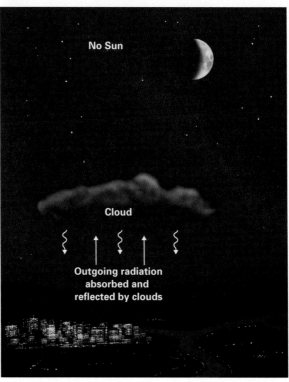

◆ **FIGURE 18.18** Clouds cool Earth's surface during the day but warm it during the night.

EARTH SYSTEMS INTERACTIONS

IN CHAPTER 17, WE learned that blue-green algae called cyanobacteria began producing oxygen 2.7 billion years ago, but appreciable quantities of oxygen didn't appear in the atmosphere until 2.4 billion years ago. The emergence of oxygen was a classic threshold effect caused by interactions between the atmosphere, biosphere, hydrosphere, and geosphere. In this and subsequent chapters, we see that many systems interactions, threshold, and feedback mechanisms affect the atmosphere and, hence, our climate and weather.

System interactions are ubiquitous in atmospheric physics and chemistry for several reasons:

- Molecular movement is unimpeded in a gas. Thus gas molecules are in constant contact with each other and with nearby surfaces, and they are able to react chemically with the surfaces of the other three spheres.
- Gases mix together in any proportions. Water doesn't mix readily with oil—or rock. If you put a piece of granite on top of a quartz crystal at surface temperatures and pressures, the two remain as separate entities. But gases mix, so any gas injected into the atmosphere becomes part of the atmosphere. Thus water vapor evaporates into air and later can condense to form clouds, rain, or snow. Aerosols, dust, and other particulates also remain suspended in air.
- Large-scale movement occurs in the atmosphere by convection. This mobility transfers heat and pollutants.
- Atmospheric gases, such as oxygen and carbon dioxide, are reactive in hydrological, geological, and biological systems.
- Gases readily compress and expand with changing physical conditions, such as temperature. We will explore changes in pressure and volume in more detail in Chapter 19.
- The atmosphere has relatively little mass compared with the rest of Earth. Therefore relatively small transfers of matter can have huge effects on the composition of the atmosphere. ■

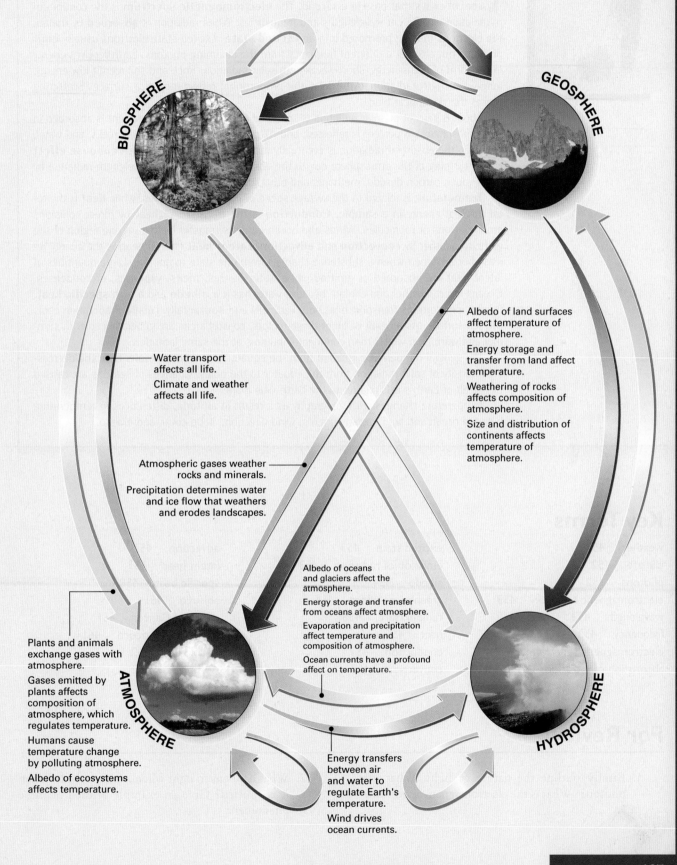

BIOSPHERE

GEOSPHERE

Water transport affects all life.

Climate and weather affects all life.

Atmospheric gases weather rocks and minerals.

Precipitation determines water and ice flow that weathers and erodes landscapes.

Albedo of land surfaces affect temperature of atmosphere.

Energy storage and transfer from land affect temperature.

Weathering of rocks affects composition of atmosphere.

Size and distribution of continents affects temperature of atmosphere.

Plants and animals exchange gases with atmosphere.

Gases emitted by plants affects composition of atmosphere, which regulates temperature.

Humans cause temperature change by polluting atmosphere.

Albedo of ecosystems affects temperature.

ATMOSPHERE

Albedo of oceans and glaciers affect the atmosphere.

Energy storage and transfer from oceans affect atmosphere.

Evaporation and precipitation affect temperature and composition of atmosphere.

Ocean currents have a profound affect on temperature.

HYDROSPHERE

Energy transfers between air and water to regulate Earth's temperature.

Wind drives ocean currents.

SUMMARY

Light is a form of **electromagnetic radiation** and exhibits properties of both waves and particles. **Wavelength** is the distance between wave crests, and **frequency** is the number of cycles that pass in a second. The **electromagnetic spectrum** is the continuum of radiation of different wavelengths and frequencies. When radiation is **absorbed** by matter, the electrons may be promoted into an **excited state**. Excited-state electrons usually **emit** photons, but often at a different frequency from the incoming photons. Earth receives energy in the form of high-energy, short-wavelength solar radiation. Rock and soil reemit low-energy, long-wavelength radiation. **Albedo** is the proportional reflectance of a surface. Scattering makes the sky appear blue.

About 8 percent of solar radiation is scattered back into space; 19 percent is absorbed in the atmosphere; 26 percent is reflected; and 47 percent is absorbed by soil, rocks, and water. However, the absorbed radiation is eventually reemitted into space. The **greenhouse effect** is the warming of the atmosphere due to the absorption of this long-wavelength radiation by water vapor, carbon dioxide, methane, and other greenhouse gases.

Temperature is related to the average speed of atoms or molecules, while **heat** is the total thermal energy in a sample. **Conduction** is the transport of heat by direct collisions among atoms or molecules. Winds and ocean currents transfer heat from one region of the globe to another by **convection** and **advection**. **Latent heat** (stored heat) is the energy released or absorbed when a substance changes from one state to another. Large quantities of latent heat are absorbed or emitted when water freezes, melts, vaporizes, or condenses. Oceans affect weather and climate because water has low **albedo** and a high **specific heat**. In addition, currents transport heat both vertically and horizontally. Finally, evaporation cools the sea surface. As a result of all of these factors, coastal areas are generally cooler in summer and warmer in winter than continental interiors at the same latitude.

The general temperature gradient from the equator to the poles results from the decreasing intensity of solar radiation from the equator to the poles. Changes of seasons are caused by the tilt of Earth's axis relative to the Earth–Sun plane.

Temperature changes with geography as a result of altitude, differences in temperature between ocean and land, ocean currents, wind direction, cloud cover, and albedo.

Key Terms

weather   452
climate   452
photons   452
electromagnetic radiation   453
wavelength   453
frequency   453
electromagnetic spectrum   453
absorption of radiation   453

excited state   453
emission of radiation   453
albedo   454
greenhouse effect   456
temperature   456
heat   456
conduction   457
convection   457

advection   457
latent heat   457
specific heat   458
solstice   461
tropic of Cancer   461
tropic of Capricorn   461
equinoxes   462
isotherms   462

For Review

1. Briefly outline the nature of light. What is a photon? What is the electromagnetic spectrum?

2. What happens to light when it is absorbed? When it is emitted? Give an example of each of these phenomena.

3. What is albedo? How does the albedo of a surface change when snow melts? When forests are converted to agricultural fields? When fields are paved into parking lots?

4. What is the fate of the solar radiation that reaches the Earth?

5. Explain the greenhouse effect.

6. Explain the difference between heat and temperature.

7. Explain the difference between conduction and convection.

8. What is latent heat? How does it affect Earth's surface temperature?

9. What is specific heat? How does it affect Earth's surface temperature?

10. Explain how the tilt of Earth's axis affects climate in the temperate and polar regions.

11. Discuss temperature and lengths of days at the poles, the midlatitudes, and the equator at the following times of year: June 21, December 21, and the equinoxes.

12. Explain how altitude, wind, the ocean, cloud cover, and albedo affect regional temperature.

For Discussion

1. An astronaut on a space walk must wear protective clothing as a shield against the Sun's rays, but the same person is likely to relax in a bathing suit in sunlight down on Earth. Explain.

2. Why must climbers wear dark glasses to protect their eyes while they are on high mountains? Why do they get sunburned even when the temperature is below freezing?

3. As the winter ends, the snow generally melts first around trees, twigs, and rocks. Explain why the line of melting radiates outward from these objects, and why the snow in open areas melts last.

4. Explain how the geosphere, hydrosphere, and biosphere can affect the temperature of the atmosphere by altering the albedo, by storing or releasing energy during changes of state, by heat storage, or by affecting heat transport by convection.

5. In central Alaska the sky is often red at noon in December. Explain why the sky is red, not blue, at this time.

6. Refer to Figure 18.6. (a) What percent of the solar energy is absorbed by Earth? (b) What percent is scattered by the atmosphere? (c) What percent is absorbed, reflected, or scattered by the atmosphere?

7. Refer to the figure of the hydrological cycle in Chapter 11. Assess the energy transfers for each of the processes listed.

8. If the North Pole receives the same number of hours of sunlight per year as do the equatorial regions, why is it so much colder than the equator?

9. Would a large inland lake be likely to affect the climate of the land surrounding it? Deep lakes seldom freeze completely in winter, whereas shallow ones do. Would a deep lake have a greater or a lesser effect on weather than a shallow one?

10. Neither oxygen nor nitrogen is an efficient absorber of visible or infrared radiation. Because these two gases together make up 99 percent of the atmosphere, the atmosphere is largely transparent to visible and infrared radiation. Predict what would happen if oxygen and nitrogen absorbed visible radiation. What would happen if these two gases absorbed infrared radiation?

11. From the information presented in this chapter, explain how the geosphere, hydrosphere, and atmosphere interact to produce climate.

12. How would Earth's climate be affected if: (a) Earth had no oceans, (b) Earth's spin axis were perpendicular to the plane of Earth's orbit, and (c) Earth's spin axis were tilted 50° to the plane of Earth's orbit?

13. What is the latitude of the city that you live in? Using the Internet, or any other source available, find the average temperature for each month in your city. Now find three other cities, at the same latitude, with different average monthly temperatures. Speculate on why these differences occur.

CHAPTER 19

Moisture, Clouds, and Weather

© AP/Wide World Photos

This lighthouse on the shore of Lake Pontchartrain, Louisiana, was heavily damaged by Hurricane Katrina. Shortly thereafter, in this photo, it is threatened by the rising tide and storm surge of Hurricane Rita, Friday, Sept. 23, 2005.

Today's weather may be sunny and warm, a sharp contrast to yesterday, when it was rainy and cold. In New York City, a winter wind from the northwest brings cool air, but when a breeze blows from the southeast, the temperature rises and a storm develops as warm, moist, maritime air flows into the city. Moisture, temperature, and wind combine to create the atmospheric conditions called weather.

The most severe weather, such as a winter blizzard or a hurricane, brings heavy precipitation, violent wind, and human misery. The energy that drives these storms ultimately is derived from the Sun. In this chapter, we will learn how the Sun's heat drives a blizzard that sweeps across the land with swirling snow and subzero temperatures, or a hurricane that blackens the sky and flattens houses. ∎

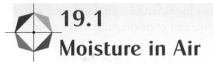

19.1
Moisture in Air

Precipitation occurs only when there is moisture in the air. Therefore to understand precipitation, we must first understand how moisture collects in the atmosphere and how it behaves.

Humidity

When water boils on a stove, a steamy mist rises above the pan and then disappears into the air. The water molecules have not been lost; they have simply become invisible. In the pan, water is liquid, and in the mist above, the water exists as tiny droplets. These droplets then evaporate, and the invisible water vapor mixes with air. Water also evaporates into air from the seas, streams, lakes, and soil. Winds then distribute this moisture throughout the atmosphere. Thus, all air contains some water vapor, even over the driest deserts.

Humidity is the amount of water vapor in air. **Absolute humidity** is the mass of water vapor in a given volume of air, expressed in grams per cubic meter (g/m^3).

Air can hold only a certain amount of water vapor, and warm air can hold more water vapor than cold air can. For example, air at 25°C can hold 23 g/m^3 of water vapor, but at 12°C, it can hold only half that quantity, 11.5 g/m^3 (◁▷ Figure 19.1). **Relative humidity** is the amount of water vapor in air relative to the maximum it can hold at a given temperature. It is expressed as a percentage:

$$\text{Relative Humidity (\%)} = \frac{\begin{array}{c}\text{actual quantity of water}\\\text{per unit of air}\end{array}}{\begin{array}{c}\text{maximum quantity}\\\text{at the same temperature}\end{array}} \times 100$$

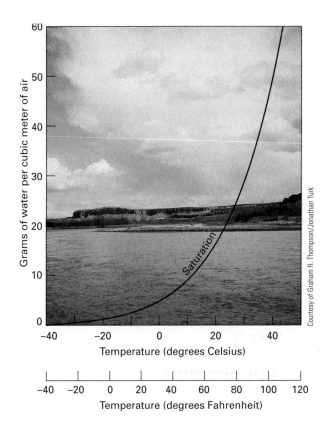

ThomsonNOW™ ◁▷ **ACTIVE FIGURE 19.1** Warm air can hold more water vapor than cold air can.

If air contains half as much water vapor as it can hold, its relative humidity is 50 percent. Suppose that air at 25°C contains 11.5 g/m^3 of water vapor. Since air at that temperature can hold 23 g/m^3, it is carrying half of its maximum, and the relative humidity is 11.5 g/23 g × 100 = 50 percent.

Now let us take some of this air and cool it without adding or removing any water vapor. Because cold air holds less water vapor than warm air holds, the relative humidity increases even though the *amount* of water vapor remains constant. If the air cools to 12°C, and it still contains 11.5 g/m³, the relative humidity reaches 100 percent because air at that temperature can hold only 11.5 g/m³.

When relative humidity reaches 100 percent, the air is **saturated**. The temperature at which saturation occurs, 12°C in this example, is the **dew point**. If saturated air cools below the dew point, some of the water vapor may condense into liquid droplets (although, as discussed next, under special conditions in the atmosphere, the relative humidity can rise above 100 percent).

◆ **FIGURE 19.2** Ice crystals condense on a window on a frosty morning.

Supersaturation and Supercooling

When the relative humidity reaches 100 percent (at the dew point), water vapor condenses quickly onto solid surfaces such as rocks, soil, and airborne particles. Airborne particles such as dust, smoke, and pollen are abundant in the lower atmosphere. Consequently, water vapor may condense easily at the dew point in the lower atmosphere, and there the relative humidity rarely exceeds 100 percent. However, in the clear, particulate-free air high in the troposphere, condensation occurs so slowly that for all practical purposes it does not happen. As a result, the air commonly cools below its dew point but water remains as vapor. In that case, the relative humidity rises above 100 percent, and the air becomes **supersaturated**.

Similarly, liquid water does not always freeze at its freezing point. Small droplets can remain liquid in a cloud even when the temperature is −40°C. Such water is **supercooled**.

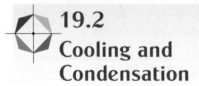

19.2
Cooling and Condensation

As you have just learned, moisture condenses to form water droplets or ice crystals when moist air cools below its dew point. Clouds and fog are visible concentrations of this airborne water and ice. Three atmospheric processes cool air to its dew point and cause condensation: (1) Air cools when it loses heat by radiation. (2) Air cools by contact with a cool surface such as water, ice, rock, soil, or vegetation. (3) Air cools when it rises.

Radiation Cooling

As described in Chapter 18, the atmosphere, rocks, soil, and water absorb the Sun's heat during the day and then radiate some of this heat back out toward space at night. As a result of heat lost by radiation, air, land, and water become cooler at night, and condensation may occur.

Contact Cooling—Dew and Frost

You can observe condensation on a cool surface with a simple demonstration. Heat water on a stove until it boils and then hold a cool drinking glass in the clear air just above the steam. Water droplets will condense on the surface of the glass because the glass cools the hot, moist air to its dew point. The same effect occurs in a house on a cold day. Water droplets or ice crystals appear on windows as warm, moist, indoor air cools on the glass (◆ Figure 19.2).

In some regions, the air on a typical summer evening is warm and humid. After the Sun sets, plants, houses, windows, and most other objects lose heat by radiation and therefore become cool. During the night, water vapor condenses on the cool objects. This condensation is called **dew**. If the dew point is below freezing, **frost** forms. Thus frost is not frozen dew, but ice crystals formed directly from vapor.

Cooling of Rising Air

Radiation and contact cooling close to Earth's surface form dew, frost, and some types of fog. However, clouds and precipitation normally form at higher elevations where the air is not cooled by direct contact with the ground. Almost all cloud formation and precipitation occur when air cools as it rises (◆ Figure 19.3).

Work and heat are both forms of energy. Work can be converted to heat or heat can be converted to work, but energy is never lost. If you pump up a bicycle tire, you are performing work to compress the air. This energy is not lost; much of it converts to heat. Therefore, both the pump and the newly filled tire feel warm. Conversely, if you puncture a tire, the air rushes out. It must perform

◆▶ **FIGURE 19.3** Most clouds form as rising air cools. The cooling causes invisible water vapor to condense as visible water droplets, or ice crystals, which we see as a cloud.

work to expand, so the air rushing from a punctured tire cools. Variations in temperature caused by compression and expansion of gas are called **adiabatic temperature changes**. Adiabatic means without gain or loss of heat. During adiabatic warming, air warms up because work is done on it, not because heat is added. During adiabatic cooling, air cools because it performs work, not because heat is removed.

As explained in Chapter 17, air pressure decreases with elevation. When dense surface air rises, it expands because the atmosphere around it is now of lower density, just as air expands when it rushes out of a punctured tire. Rising air performs work to expand, and therefore it cools adiabatically. Dry air cools by 10°C for every 1,000 meters it rises (5.5°F/1,000 ft). This cooling rate is called the **dry adiabatic lapse rate**. Thus, if dry air were to rise from sea level to 9,000 meters (about the height of Mount Everest), it would cool by 90°C (162°F).

Almost all air contains some water vapor. As moist air rises and cools adiabatically, its temperature may eventually decrease to the dew point. At the dew point, moisture may condense as droplets, and a cloud forms. But recall that condensing vapor releases latent heat. As the air rises

through the cloud, its temperature now is affected by two opposing processes. It cools adiabatically, but at the same time it is heated by the latent heat released by condensation. However, the warming caused by latent heat is generally less than the amount of adiabatic cooling. The net result is that the rising air continues to cool, but more slowly than at the dry adiabatic lapse rate. The **wet adiabatic lapse rate** is the cooling rate after condensation has begun. It varies from 5°C/1,000 m (2.7°F/1,000 ft) for air with a high moisture content, to 9°C/1,000 m (5°F/1,000 ft) for relatively dry air (◆▶ Figure 19.4). Thus, once clouds start to form, rising air no longer cools as rapidly as it did lower in the atmosphere. Rising air cools at the dry adiabatic lapse rate until it cools to its dew point and condensation begins. Then, as it continues to rise, it cools at the lesser, wet adiabatic lapse rate as condensation continues.

In contrast, sinking air becomes warmer because of adiabatic compression. Warm air can hold more water vapor than cool air can. Consequently, water does not condense from sinking, warming air, and the latent heat of condensation does not affect the rate of temperature rise. As a result, sinking air always becomes warmer at the dry adiabatic rate.

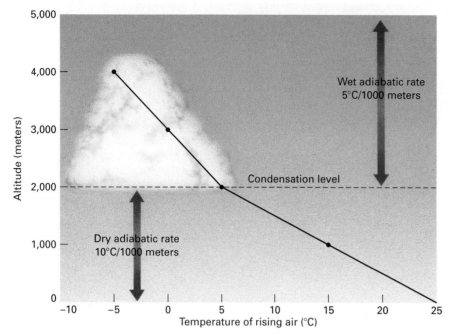

ThomsonNOW ◆ **ACTIVE**
FIGURE 19.4 A rising air mass initially cools rapidly at the dry adiabatic lapse rate. Then, after condensation begins, it cools more slowly at the wet adiabatic lapse rate.

INTERACTIVE QUESTION: *From the graph, estimate the air temperature at an elevation of 500 meters and 1,500 meters. Calculate the dry, adiabatic lapse rate from your estimates.*

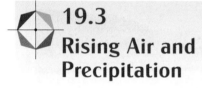

19.3
Rising Air and Precipitation

To summarize: When moist air rises, it cools and forms clouds. Three mechanisms cause air to rise (◆ Figure 19.5):

Orographic Lifting

When air flows over mountains, it is forced to rise. This rising air frequently causes rain or snow over the mountains, as will be explained in Section 19.7.

Frontal Wedging

A moving mass of cool, dense air may encounter a mass of warm, less-dense air. When this occurs, the cool, denser air slides under the warm air mass, forcing the warm air upward to create a weather front. We will discuss weather fronts in more detail later in this chapter.

Convection–Convergence

If one portion of the atmosphere becomes warmer than the surrounding air, the warm air expands, becomes less dense, and rises. Thus a hot-air balloon rises because it contains air that is warmer and less dense than the air around it. If the Sun heats one parcel of air near Earth's surface to a warmer temperature than that of surrounding air, the warm air will rise, just as the hot-air balloon rises.

Convective Processes and Clouds

On some days clouds hang low over the land and obscure nearby hills. At other times clouds float high in the

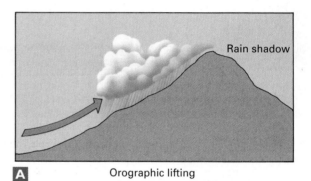

A   Orographic lifting

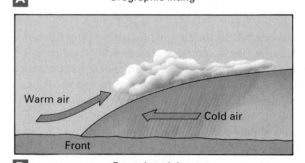

B   Frontal wedging

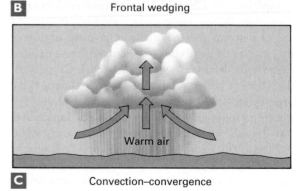

C   Convection–convergence

◆ **FIGURE 19.5** Three mechanisms cause air to rise and cool: **(A)** orographic lifting, **(B)** frontal wedging, and **(C)** Convection–convergence.

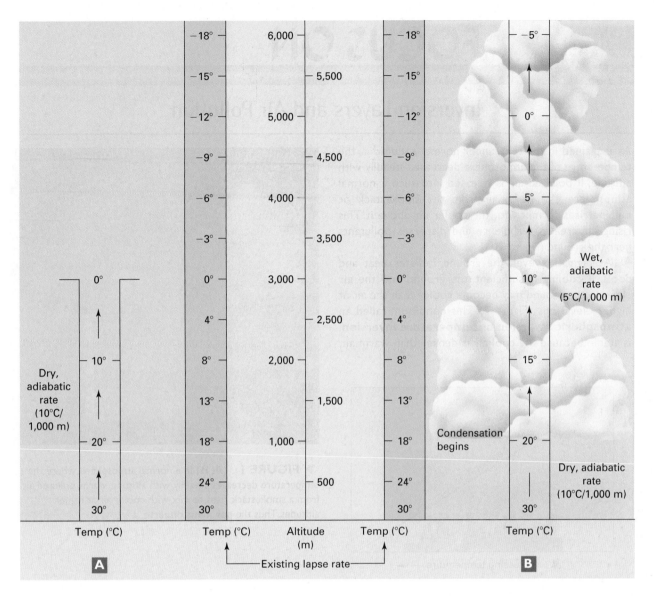

◆ **FIGURE 19.6** **(A)** As dry air rises, it expands and cools at the dry, adiabatic lapse rate. Thus, it soon cools to the temperature of the surrounding air, and it stops rising. **(B)** As moist air rises, initially it cools at the dry, adiabatic lapse rate. It soon cools to its dew point, and clouds form. Then, it cools more slowly at the wet, adiabatic lapse rate. As a result, it remains warmer than surrounding air and continues to rise for thousands of meters. It stops rising when all moisture has condensed, and the air again cools at its dry, adiabatic rate.

INTERACTIVE QUESTION: *How would cloud formation be affected if the wet, adiabatic lapse rate were 8°C /1,000 meters?*

sky, well above the mountain peaks. What factors determine the height and shape of a cloud?

Recall that air is generally warmest at Earth's surface and cools with elevation throughout the troposphere. The rate at which air that is neither rising nor falling cools with elevation is called the **normal lapse rate**. The average normal lapse rate is 6°C/1,000 m (3.3°F/1,000 ft) and thus is less than the dry adiabatic lapse rate. However, the normal lapse rate is variable. Typically, it is greatest near Earth's surface and decreases with altitude. The normal lapse rate also varies with latitude, the time of day, and the seasons. It is important to note that the normal lapse rate is simply the vertical temperature

structure of the atmosphere. In contrast, *rising* air cools because of adiabatic cooling.

◆ Figure 19.6 shows two rising, warm-air masses, one consisting of dry air and the other of moist air. The central part of the figure shows that the normal lapse rate is the same for both air masses: The temperature of the atmosphere decreases rapidly in the first few thousand meters and then more slowly with increasing elevation. However, the two air masses behave differently because of their different moisture contents.

The dry air mass (part A of the figure) rises and cools at the dry adiabatic lapse rate of 10°C/1,000 m. As a result, in this example, the mass's temperature and density

FOCUS ON

Inversion Layers and Air Pollution

As explained in the text, under normal conditions, the temperature of the atmosphere decreases steadily with altitude. If polluted air is released into such a normal atmosphere, the warm air from the smokestack or tailpipe rises to mix with the cooler air above it. This rising air creates turbulence and disperses pollutants near the ground (Figure 1).

At night, however, the ground radiates heat and cools. If cooling is sufficient, the ground and the air close to the ground may become cooler than the air at higher elevations (Figure 2). This condition, called an **atmospheric inversion** or **temperature inversion**, is stable because the cool air is denser than warm air.

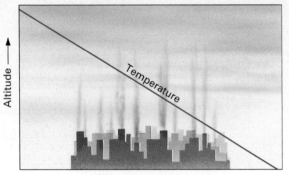

◈ **FIGURE 1** **(A, B)** In a normal atmosphere, where the temperature decreases steadily with altitude, warm, polluted air from a smokestack rises to mix with cooler air at higher altitudes. Thus the pollutants disperse.

become equal to that of surrounding air at an elevation of 3,000 meters. Because the density of the rising air is the same as that of the surrounding air, the rising air is no longer buoyant, and it stops rising. No clouds form because the air has not cooled to its dew point.

In part B of the figure, the rising, moist air initially cools at the dry adiabatic rate of 10°C/1,000 m, but only until the air cools to its dew point, at an elevation of 1,000 meters. At that point, moisture begins to condense, and clouds form. But the condensing moisture releases latent heat of condensation. This additional heat causes the rising air to cool more slowly, at the wet adiabatic rate of 5°C/1,000 m. As a result, the rising air remains warmer and more buoyant than surrounding air, and it continues to rise for thousands of meters, creating a towering, billowing cloud with the potential for heavy precipitation.

In simple terms, warm, moist air is **unstable** because it rises rapidly, forming towering clouds and heavy rainfall. Also, as shown in ◈ Figure 19.5C, air rushes along the ground to replace the rising air, thus generating surface winds. Most of us have experienced a violent thunderstorm on a hot, summer day. Puffy clouds seem to appear out of nowhere in a blue sky. These clouds grow vertically and darken as the afternoon progresses. Suddenly, gusts of wind race across the land, and shortly thereafter, heavy rain falls. These events, to be described in more detail in Section 19.7, are all caused by unstable, rising, moist air.

In contrast, warm, dry air doesn't rise rapidly, doesn't ascend to high elevations, and doesn't lead to cloud formation and precipitation. Thus warm, dry air is said to be **stable**. Yet, convection is only one of the three processes that leads to rising air. **Orographic lifting** and **frontal wedging** also lead to rising air, cloud formation, and rain.

Therefore, the air near the surface does not rise and mix with the air above it. During an inversion, pollutants concentrate in the stagnant layer of cool air next to the ground.

Usually, the morning Sun warms air near the ground and breaks the inversion. However, under some conditions, inversions last for days. For example, a large mass of warm air may move into a region at high altitude and float over the colder air near the ground, keeping the air over a city stagnant. Inversions are common along coastlines and large lakes, where the water cools surface air. In the Los Angeles basin, cool, maritime air is often trapped beneath a warm, subtropical, air mass. The cool air cannot move eastward because of the mountains, and it cannot rise because it is too dense. Pollutants from the city's automobiles and factories then concentrate until the stagnant air becomes unhealthy.

Two types of weather changes can break up an inversion.

If the Sun heats Earth's surface sufficiently, the cool air near the ground warms and rises, dispersing the pollutants. Alternatively, storm winds may dissipate an inversion layer.

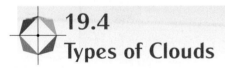

FIGURE 2 **(A, B)** During an inversion, when warm air lies on top of cooler air near the ground, warm, polluted air from a smokestack cannot rise above the inversion layer. In this photograph, taken in the Gdansk Shipyards in Poland, an early morning inversion layer concentrates clouds, steam, and pollutants close to the ground.

19.4
Types of Clouds

Even a casual observer of the daily weather will notice that clouds are quite different from day to day. Different meteorological conditions create the various cloud types and, in turn, a look at the clouds provides useful information about the daily weather.

Cirrus (Latin for "wisp of hair") clouds are wispy clouds that look like hair blowing in the wind or feathers floating across the sky. Cirrus clouds form at high altitudes, 6,000 to 15,000 meters (20,000 to 50,000 feet). The air is so cold at these elevations that cirrus clouds are composed of ice crystals rather than water droplets. High winds aloft blow them out into long, gently curved streamers (◆ Figure 19.7).

FIGURE 19.7 Cirrus clouds are high, wispy clouds composed of ice crystals.

Ralph F. Kresge/NOAA

◆ **FIGURE 19.8** Stratus clouds spread out across the sky in a low, flat layer.

NCAR

◆ **FIGURE 19.9** Cumulus clouds are fluffy white clouds with flat bottoms.

Stratus (Latin for "layer") clouds are horizontally layered, sheet-like clouds. They form when condensation occurs at the same elevation at which air stops rising and the clouds spread out into a broad sheet. Stratus clouds form the dark, dull-gray, overcast skies that may persist for days and bring steady rain (◆ Figure 19.8).

Cumulus (Latin for "heap" or "pile") clouds are fluffy, white clouds that typically display flat bottoms and billowy tops (◆ Figure 19.9). On a hot, summer day the top of a cumulus cloud may rise 10 kilometers or more above its base in cauliflower-like masses. The base of the cloud forms at the altitude at which the rising air cools to its dew point and condensation starts. However, in this situation the rising air remains warmer than the surrounding air and therefore continues to rise. As it rises, more vapor condenses, forming the billowing columns.

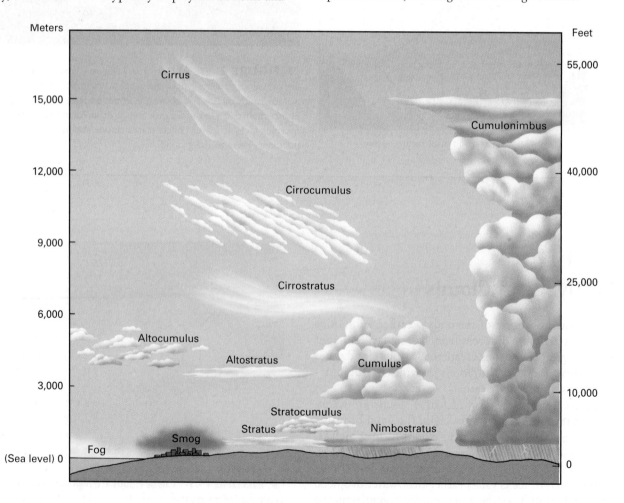

◆ **FIGURE 19.10** Cloud names are based on the shape and altitude of the clouds.

Other types of clouds are named by combining these three basic terms (◆ Figure 19.10). **Stratocumulus** clouds are low, sheet-like clouds with some vertical structure. The term *nimbo* refers to a cloud that precipitates. Thus a **cumulonimbus** cloud is a towering rain cloud. If you see one, you should seek shelter, because cumulonimbus clouds commonly produce intense rain, thunder, lightning, and sometimes hail. A **nimbostratus** cloud is a stratus cloud from which rain or snow falls. Other prefixes are also added to cloud names. For example, *Alti* is derived from the Latin root *altus,* meaning high. An **altostratus** cloud is simply a high stratus cloud.

Types of Precipitation

Rain

Why does rain fall from some clouds, whereas other clouds float across a blue sky on a sunny day and produce no rain? The droplets in a cloud are small, about 0.01 millimeter in diameter (about one-seventh the diameter of a human hair). In still air, such a droplet would require 48 hours to fall from a cloud 1,000 meters above Earth. But these tiny droplets never reach Earth because they evaporate faster than they fall.

If the air temperature in a cloud is above freezing, the tiny droplets may collide and coalesce. You can observe similar behavior in droplets sliding down a window pane on a rainy day. If two droplets collide, they merge to become one large drop. If the droplets in a cloud grow large enough, they fall as drizzle (0.1 to 0.5 millimeter in diameter) or light rain (0.5 to 2 millimeters in diameter). About one million cloud droplets must combine to form an average-size raindrop.

In many clouds, however, water vapor initially forms ice crystals rather than condensing as tiny droplets of supercooled water. Part of the reason for this is that the temperature in clouds is commonly below freezing, but another factor also favors ice formation. At near- or below-freezing temperatures, air that is slightly undersaturated with respect to water is slightly supersaturated with respect to ice. For example, if the relative humidity of air is 95 percent with respect to water, it is about 105 percent with respect to ice. Thus, as air cools toward its dew point, all the vapor forms ice crystals rather than supercooled water droplets. The tiny ice crystals then grow larger as more water vapor condenses on them, until they are large enough to fall. The ice then melts to form raindrops as it falls through warmer layers of air.

If you have ever been caught in a thunderstorm, you may remember raindrops large enough to be painful as they struck your face or hands. Recall that a cumulus cloud forms from rising air and that its top may be several kilometers above its base. The temperature in the upper part of the cloud is commonly below freezing. As a result, ice crystals form and begin to fall.

Condensation continues as the crystal falls through the towering cloud, and the crystal grows. If the lower

Courtesy of Graham R. Thompson/Jonathan Turk

◆ **FIGURE 19.11** Snow blankets the ground during the winter in temperate regions. Snow and ice cover the ground year-round in the high mountains and at the Poles.

atmosphere is warm enough, the ice melts before it reaches the surface. Raindrops formed in this manner may be 3 to 5 millimeters in diameter, large enough to hurt when they hit.

Snow, Sleet, and Glaze

As explained previously, when the temperature in a cloud is below freezing, the cloud is composed of ice crystals rather than water droplets. If the temperature near the ground is also below freezing, the crystals remain frozen and fall as snow (◆ Figure 19.11). In contrast, if raindrops form in a warm cloud and fall through a layer of cold air at lower elevation, the drops freeze and fall as small spheres of ice called **sleet**. Sometimes the freezing zone near the ground is so thin that raindrops do not have time to freeze before they reach Earth. However, when they land on subfreezing surfaces, they form a coating of ice called **glaze** (◆ Figure 19.12). Glaze can be heavy enough to break tree limbs and electrical transmission lines. It also coats highways with a dangerous icy veneer. In the winter of 1997–98,

◀◆ **FIGURE 19.12** Glaze forms when rain falls on a surface that is colder than the freezing temperature of water.

a sleet and glaze storm in eastern Canada and the northeastern United States caused billions of dollars in damage. The ice damaged so many electric lines and power poles that many people were without electricity for a few weeks.

Hail

Occasionally, precipitation takes the form of very large ice globules called **hail**. Hailstones vary from 5 millimeters in diameter to a record 14 centimeters in diameter; that record breaker weighed 765 grams (more than 1.5 pounds) and fell in Kansas. A 500-gram (1-pound) hailstone crashing to Earth at 160 kilometers (100 miles) per hour can shatter windows, dent car roofs, and kill people and livestock. Even small hailstones can damage crops. Hail falls only from cumulonimbus clouds. Because cumulonimbus clouds form in columns with distinct boundaries, hailstorms occur in local, well-defined areas. Thus, one farmer may lose an entire crop while a neighbor is unaffected.

A hailstone consists of ice in concentric shells, like the layers of an onion. Two mechanisms have been proposed for their formation. In one, turbulent winds blow falling ice crystals back upward in the cloud. New layers of ice accumulate as additional vapor condenses on the recirculating ice grain. An individual particle may rise and fall several times until it grows so large and heavy that it drops out of the cloud. In the second mechanism, hailstones form in a single pass through the cloud. During their descent, supercooled water freezes onto the ice crystals. The layering develops because different temperatures and amounts of supercooled water exist in different portions of the cloud, and each layer forms in a different part of the cloud.

✦ 19.5 Fog

Fog is a cloud that forms at or very close to ground level, although most fog forms by processes different from those that create higher-level clouds. **Advection fog** occurs when warm, moist air from the sea blows onto cooler land. The air cools to its dew point, and water vapor condenses at ground level. San Francisco; Seattle; and Vancouver, B.C.; all experience foggy winters as warm, moist air from the Pacific Ocean is cooled first by the cold California current and then by land.

The foggiest location in the United States is Cape Disappointment, Washington, where visibility is obscured by fog 29 percent of the time.

Radiation fog occurs when Earth's surface and air near the surface cool by radiation during the night (◆ Figure 19.13). Water vapor condenses as fog when the air cools below its dew point. Often the cool, dense, foggy air settles into valleys. If you are driving late at night in hilly terrain, beware, because a sudden dip in the roadway may lead you into a thick fog where visibility is low. A ground fog of this type typically "burns off" in the morning. The rising Sun warms the land or water surface which, in turn, warms the low-lying air. As the air becomes warmer, its capacity to hold water vapor increases, and the fog droplets evaporate. Radiation fog is particularly common in areas where the air is polluted because water vapor condenses readily on the tiny particles suspended in the air.

Recall that vaporization of water absorbs heat and, therefore, cools both the surface and the surrounding air. In addition, vaporization adds moisture to the air. The cooling and the addition of moisture combine to form conditions conducive to fog. **Evaporation fog** occurs when air is cooled by evaporation from a body of water, commonly a lake or river. Evaporation fogs are common in late fall and early winter, when the air has become cool but the water is still warm. The water evaporates, but the vapor cools and condenses to fog almost immediately upon contact with the cold air.

Upslope fog occurs when air cools as it rises along a land surface. Upslope fogs occur both on gradually

◆ **FIGURE 19.13**   Radiation fog is seen as a morning mist in this field in Idaho.

INTERACTIVE QUESTION: *Why does fog of this type commonly concentrate in low places?*

sloping plains and on steep mountains. For example, the Great Plains rise from sea level at the Mississippi Delta to 1,500 meters (5,000 feet) at the Rocky Mountain front. When humid air moves northwest from the Gulf of Mexico toward the Rockies, it rises and cools adiabatically to form upslope fog. The rapid rise at the mountain front also forms fog.

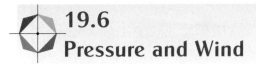

19.6
Pressure and Wind

Warm air is less dense than cold air. Thus warm air exerts a relatively low atmospheric pressure and cold air exerts a relatively high atmospheric pressure. Warm air rises because it is less dense than the surrounding cool air (◆ Figure 19.14A). Air rises slowly above a typical low-pressure region, at a rate of about 1 kilometer per day. In contrast, if air in the upper atmosphere cools, it becomes denser than the air beneath it and sinks (◆ Figure 19.14B).

Air must flow inward over Earth's surface toward a low-pressure region to replace a rising air mass. But a sinking air mass displaces surface air, pushing it outward from a high-pressure region. Thus vertical airflow in both high- and low-pressure regions is accompanied by horizontal airflow, called **wind**. Winds near Earth's surface always flow away from a region of high pressure and toward a low-pressure region. Ultimately, all wind

is caused by the pressure differences resulting from unequal heating of Earth's atmosphere.

Pressure Gradient

Wind blows in response to *differences* in pressure. Imagine that you are sitting in a room and the air is still. Now you open a can of vacuum-packed coffee and hear the hissing as air rushes into the can. Because the pressure in the room is higher than that inside the coffee can, wind blows from the room into the can. But if you blow up a balloon, the air inside the balloon is at higher pressure than the air in the room. When the balloon is punctured, wind blows from the high-pressure zone of the balloon into the lower-pressure zone of the room (◆ Figure 19.15).

Wind speed is determined by the magnitude of the pressure difference over distance, called the **pressure gradient**. Thus wind blows rapidly if a large pressure difference exists over a short distance. A steep pressure gradient is analogous to a steep hill. Just as a ball rolls quickly down a steep hill, wind flows rapidly across a steep pressure gradient. To create a pressure-gradient map, air pressure is measured at hundreds of different weather stations. Points of equal pressure are connected by map lines called **isobars**. A steep pressure gradient is shown by closely spaced isobars, whereas a weak pressure gradient is indicated by widely spaced isobars (◆ Figure 19.16). Pressure gradients change daily, or sometimes hourly, as high- and low-pressure zones move. Therefore, maps are updated frequently.

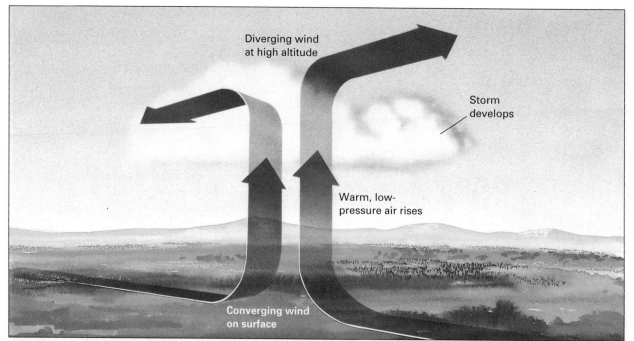

A

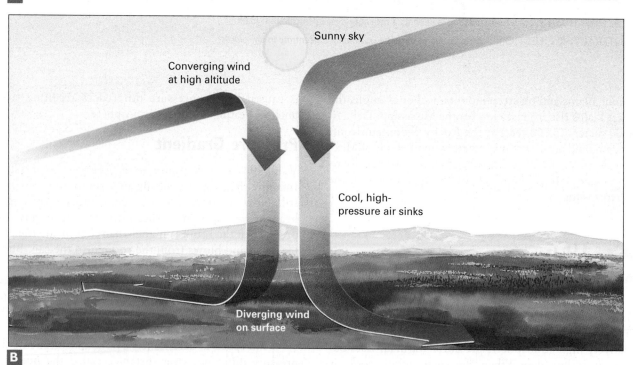

B

◈ **FIGURE 19.14** **(A)** Rising, low-pressure air creates clouds and precipitation. Air flows inward toward the low-pressure zone, creating surface winds. **(B)** Sinking, high-pressure air creates clear skies. Air flows outward from the high-pressure zone and also creates surface winds.

CHAPTER 19 • Moisture, Clouds, and Weather

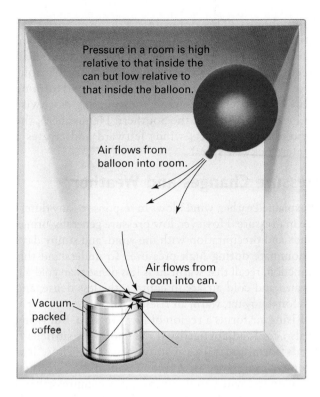

Pressure in a room is high relative to that inside the can but low relative to that inside the balloon.

Air flows from balloon into room.

Air flows from room into can.

Vacuum-packed coffee

◁▷ **FIGURE 19.15** Winds blow in response to differences in pressure.

Coriolis Effect

Recall from Chapter 16 that the Coriolis effect, caused by Earth's spin, deflects ocean currents. The Coriolis effect similarly deflects winds. In the Northern Hemisphere wind is deflected toward the right, and in the Southern Hemisphere, to the left (◁▷ Figure 19.17). The Coriolis effect alters wind direction but not its speed.

Friction

Rising and falling air generates wind both along Earth's surface and at higher elevations. Surface winds are affected by friction with Earth's surface, whereas high-altitude winds are not. As a result, wind speed normally increases with elevation. This effect was first noted during World War II. On November 24, 1944, U.S. bombers were approaching Tokyo for the first mass bombing of the Japanese capital. Flying between 8,000 and 10,000 meters (27,000 to 33,000 feet), the pilots suddenly found themselves roaring past landmarks 140 kilometers (90 miles) per hour faster than the theoretical top speed of their airplanes! Amid the confusion, most of the bombs missed their targets, and the mission was a military failure. However, this experience introduced meteorologists to **jet streams**, narrow bands of high-altitude wind. The jet stream in the

◁▷ **FIGURE 19.16** Pressure map and winds at 5,000 feet in North America on February 3, 1992. High-altitude data are shown because the winds are not affected by surface topography and thus the effect of pressure gradient is well illustrated. Note that in the Northeast and Northwest, steep pressure gradients, shown by closely spaced isobars, cause high winds that spiral counterclockwise into the low-pressure zones. Widely spaced isobars around high-pressure zones in the central United States cause weaker winds.

INTERACTIVE QUESTION: *Where on this map would you expect stormy conditions? Where would you expect fair weather?*

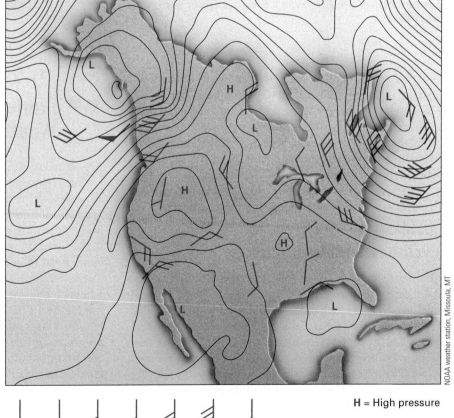

H = High pressure

L = Low pressure

5	10	15	20	30	40	50

Wind flags represent wind speed in knots. The ends of the flags point in the direction the wind is blowing.

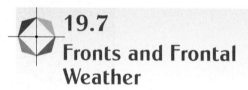

Winds are deflected
to the right in the
Northern Hemisphere

Rotation of Earth

Winds are deflected
to the left in the
Southern Hemisphere

◆ **FIGURE 19.17** The Coriolis effect deflects winds to the right in the Northern Hemisphere and to the left in the Southern Hemisphere. Only winds blowing due east or west are unaffected.

Northern Hemisphere flows from west to east at speeds between 120 and 240 kilometers per hour (75 and 150 mph). As a comparison, surface winds attain such velocities only in hurricanes and tornadoes. Airplane pilots traveling from Los Angeles to New York fly with the jet stream to gain speed and save fuel, whereas pilots moving from east to west try to avoid it.

Jet-stream influence on weather and climate will be discussed in more detail later in this chapter and in Chapter 20.

Cyclones and Anticyclones

◆ Figure 19.18A shows the movement of air in the Northern Hemisphere as it converges toward a low-pressure area. If Earth did not spin, wind would flow directly across the isobars, as shown by the black arrows. However, Earth does spin, and the Coriolis effect deflects wind to the right, as shown by the small, red arrows. This rightward deflection creates a counterclockwise vortex near the center of the low-pressure region, as shown by the large, magenta arrows.

A low-pressure region with its accompanying surface wind is called a **cyclone**. In this usage, *cyclone* means a system of rotating winds, not the violent storms that are sometimes called cyclones, hurricanes,

and typhoons. The opposite mechanism forms an **anticyclone** around a high-pressure region. When descending air reaches the surface, it spreads out in all directions. In the Northern Hemisphere, the Coriolis effect deflects the diverging winds to the right, forming a pinwheel pattern, with the wind spiraling clockwise (◆ Figure 19.18B). In the Southern Hemisphere, the Coriolis effect deflects winds leftward and creates a counterclockwise spiral.

Pressure Changes and Weather

As explained earlier, wind blows in response to any difference in pressure. However, low pressure generally brings clouds and precipitation with the wind, and sunny days predominate during high pressure. To understand this distinction, recall that warm air is less dense than cold air. If warm and cold air are in contact, the less dense, and therefore buoyant, warm air rises.

Rising air forms a region of low pressure. But rising air also cools adiabatically. If the cooling is sufficient, clouds form and rain or snow may fall. Thus, low barometric pressure is an indication of wet weather. Alternatively, when cool air sinks, it is compressed and the pressure rises. In addition, sinking air is heated adiabatically. Because warm air can hold more water vapor than cold air can, the sinking air absorbs moisture and clouds generally do not form over a high-pressure region. Thus, fair, dry weather generally accompanies high pressure.

19.7
Fronts and Frontal Weather

An **air mass** is a large body of air with approximately uniform temperature and humidity at any given altitude. Typically, an air mass is 1,500 kilometers or more across and several kilometers thick. Because air acquires both heat and moisture from Earth's surface, an air mass is classified by its place of origin. Temperature can be either polar (cold) or tropical (warm). Maritime air originates over water and has high moisture content, whereas continental air has low moisture content (◆ Figure 19.19, ◆ Table 19.1).

Air masses move and collide. The boundary between a warmer air mass and a cooler one is a **front**. The term was first used during World War I because weather systems were considered analogous to armies that advance and clash along battle lines. When two air masses collide, each may retain its integrity for days before the two mix. During a collision, one of the air masses is forced to rise, which often results in cloudiness and precipitation. Frontal weather patterns are determined by the types of air masses that collide and their relative speeds and

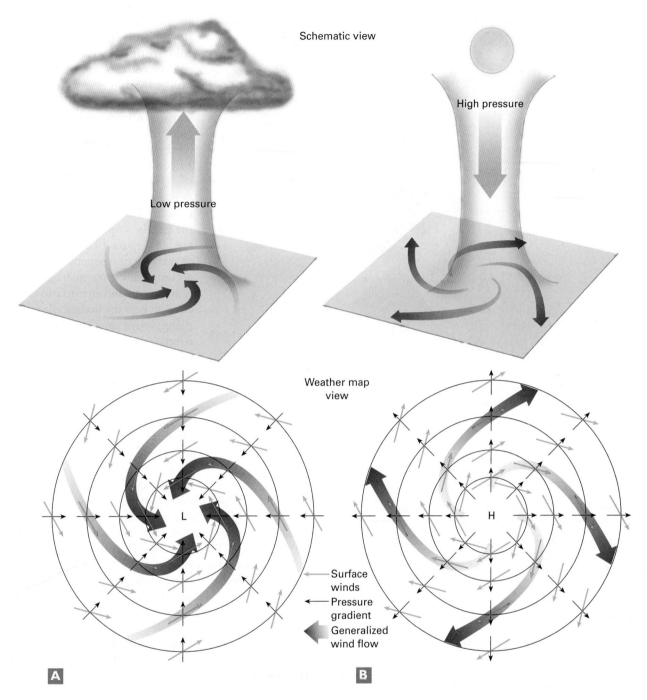

Schematic view

Low pressure

High pressure

Weather map view

Surface winds

Pressure gradient

Generalized wind flow

A

B

◇ **FIGURE 19.18** **(A)** In the Northern Hemisphere, a cyclone consists of winds spiraling counterclockwise into a low-pressure region. **(B)** An anticyclone consists of winds spiraling clockwise out from a high-pressure zone.

INTERACTIVE QUESTION: *Redraw this figure showing cyclones and anticyclones in the Southern Hemisphere.*

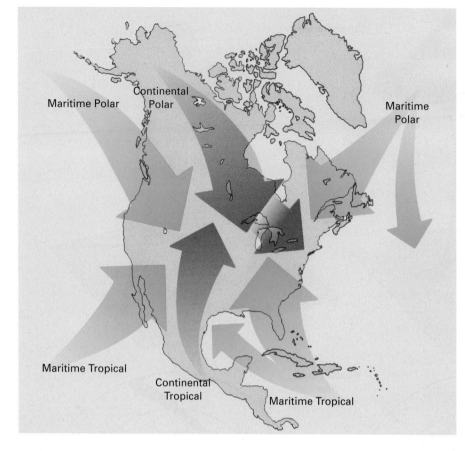

◀ **FIGURE 19.19** Air masses are classified by their source regions.

Table 19.1   Classification of Air Masses

Classification according to latitude (temperature):

Polar (P) air masses originate in high latitudes and are cold. Tropical (T) air masses originate in low latitudes and are warm.

Classification according to moisture content:

Continental (c) air masses originate over land and are dry. Maritime (m) air masses originate over water and are moist.

Symbol	Name	Characteristics
mP	Maritime polar	Moist and cold
cP	Continental polar	Dry and cold
mT	Maritime tropical	Moist and warm
cT	Continental tropical	Dry and warm

directions. The symbols commonly used on weather maps to describe fronts are shown in ◀ Figure 19.20.

Warm Fronts and Cold Fronts

Fronts are classified by whether a warm air mass moves toward a stationary (or more slowly moving) cold mass, or vice versa. A **warm front** forms when moving, warm

air collides with a stationary or slower-moving, cold air mass. A **cold front** forms when moving, cold air collides with stationary or slower-moving, warm air.

In a warm front, the moving, warm air rises over the denser, cold air as the two masses collide (◀ Figure 19.21). The rising, warm air cools adiabatically and the cooling generates clouds and precipitation. Precipitation is generally light because the air rises slowly along the gently sloping frontal boundary. ◀ Figure 19.21 shows that a characteristic sequence of clouds accompanies a warm front. High, wispy cirrus and cirrostratus clouds develop near the leading edge of the rising, warm air. These high clouds commonly precede a storm. They form as much as 1,000 kilometers ahead of an advancing band of precipitation that falls from thick, low-lying nimbostratus and stratus clouds near the trailing edge of the front. The cloudy weather may last for several days because of the gentle slope and broad extent of the frontal boundary.

A cold front forms when faster-moving, cold air overtakes and displaces warm air. The dense, cold air distorts into a blunt wedge and pushes under the warmer air (◀ Figure 19.22). Thus the leading edge of a cold front is much steeper than that of a warm front. The steep contact between the two air masses causes the warm air to rise rapidly, creating a narrow band of violent weather commonly accompanied by cumulus and cumulonimbus clouds. The storm system may be only 25 to 100 kilometers wide, but within this zone downpours, thunderstorms, and violent winds are common.

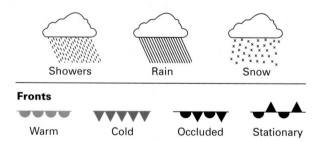

◀ **FIGURE 19.20**   Symbols commonly used in weather maps. *Warm* and *cold* are relative terms. Air over the central plains of Montana at a temperature of 0°C may be warm relative to polar air above northern Canada but cold relative to a 20°C air mass over the southeastern United States.

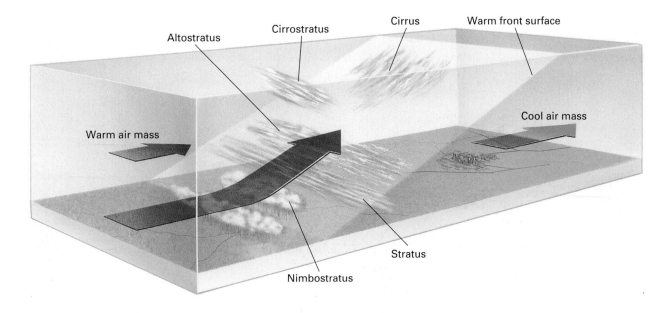

ThomsonNOW ◁▶ **ACTIVE FIGURE 19.21** In a warm front, moving warm air rises gradually over cold air.

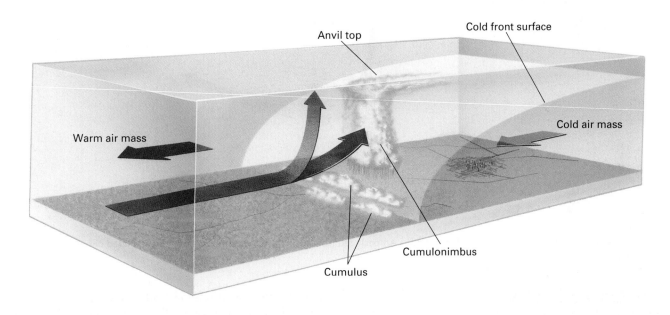

ThomsonNOW ◁▶ **ACTIVE FIGURE 19.22** In a cold front, moving cold air slides abruptly beneath warm air, forcing it steeply upward.

INTERACTIVE QUESTION: *Explain why the surface of the warm front rises gradually with elevation, while the cold front rises much more steeply.*

Occluded Front

An **occluded front** forms when a faster-moving, cold mass traps a warm air mass against a second mass of cold air. Thus the warm air mass becomes trapped between two colder air masses (◁▶ Figure 19.23). The faster-moving, cold air mass then slides beneath the warm air, lifting it completely off the ground. Precipitation occurs along both frontal boundaries, combining the narrow band of heavy precipitation of a cold front with the wider band of lighter precipitation of a warm front. The net result is a large zone of inclement weather. A storm of this type is commonly short-lived because the warm air mass is cut off from its supply of moisture evaporating from Earth's surface.

Stationary Front

A **stationary front** occurs along the boundary between two stationary air masses. Under these conditions, the front can remain over an area for several days. Warm air rises, forming conditions similar to those in a warm front. As a result, rain, drizzle, and fog may occur.

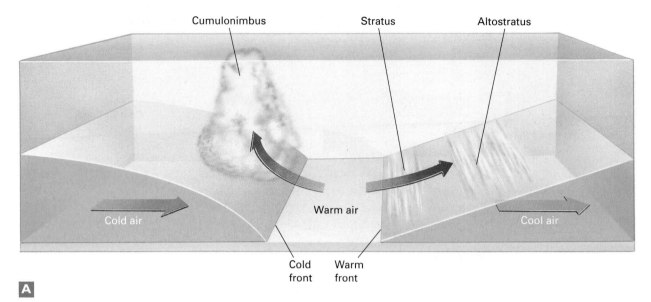

Cumulonimbus   Stratus   Altostratus

Cold air   Warm air   Cool air

Cold front   Warm front

A

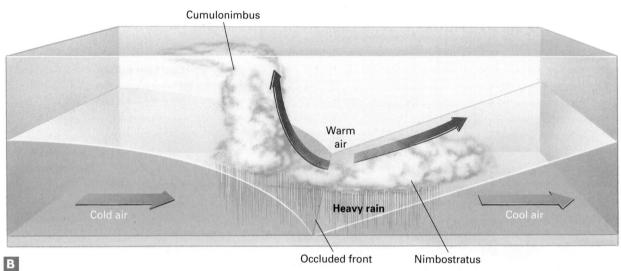

Cumulonimbus

Warm air

Cold air   Heavy rain   Cool air

Occluded front   Nimbostratus

B

ThomsonNOW ◆ **ACTIVE FIGURE 19.23** An occluded front forms where warm air is trapped and lifted between two, cold air masses.

The Life Cycle of a Middle-Latitude Cyclone

Most low-pressure cyclones in the middle latitudes of the Northern Hemisphere develop along a front between polar and tropical air masses. The storm often starts with winds blowing in opposite directions along a stationary front between the two air masses (◆ Figure 19.24A). In this figure, a warm air mass was moving northward and was deflected to the east by the Coriolis force. At the same time, a cold air mass traveling southward was deflected to the west.

In ◆ Figure 19.24B, the cold, polar air continues to push southward, creating a cold front and lifting the warm air off the ground. Then, some small disturbance deforms the straight, frontal boundary, forming a wavelike kink in the front. This disturbance may be a topographic feature such as a mountain range, airflow from a local storm, or a local temperature variation. Once the

kink forms, the winds on both sides are deflected to strike the front at an angle. Thus, a warm front forms to the east and a cold front forms to the west.

Rising warm air then forms a low-pressure region near the kink (◆ Figure 19.24C). In the Northern Hemisphere, the Coriolis effect causes the winds to circulate counterclockwise around the kink, as explained in Section 19.6. To the west, the cold front advances southward, and to the east, the warm front advances northward. At the same time, rain or snow falls from the rising, warm air (◆ Figure 19.24D). Over a period of one to three days, the air rushing into the low-pressure region equalizes pressure differences, and the storm dissipates. Many of the pinwheel-shaped storms seen on weather maps are cyclones of this type. In North America, the jet stream and other prevailing, upper-level, westerly winds generally move cyclones from west to east along the same paths, called **storm tracks** (◆ Figure 19.25).

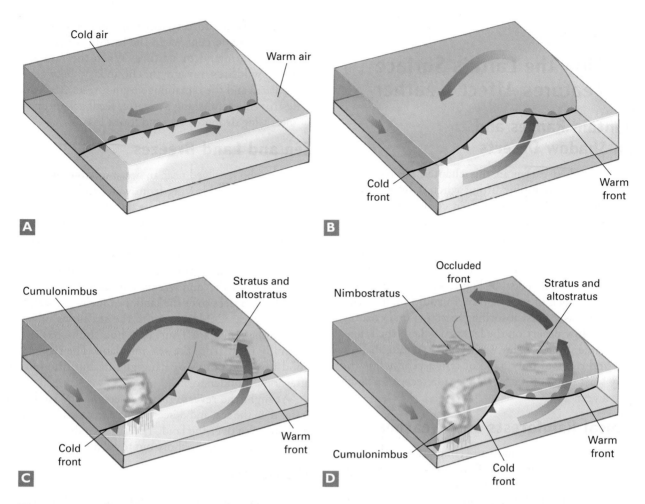

ThomsonNOW ◆ **ACTIVE FIGURE 19.24** A midlatitude cyclone develops along a front between polar air and a tropical air mass. **(A)** A front develops. **(B)** Some small disturbance creates a kink in the front. **(C)** A low-pressure region and cyclonic circulation develop. **(D)** An occluded front forms.

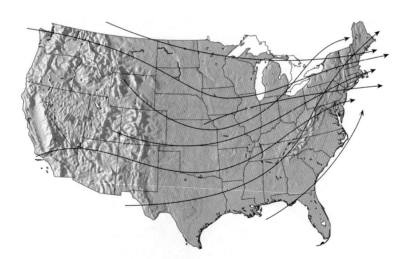

◆ **FIGURE 19.25** Most North American cyclones follow certain paths, called storm tracks, from west to east.

How the Earth's Surface Features Affect Weather

Mountain Ranges and Rain-Shadow Deserts

As we described earlier in this chapter, air rises in a process called orographic lifting when it flows over a mountain range. As the air rises, it cools adiabatically, and water vapor may condense into clouds that produce rain or snow. These conditions create abundant precipitation on the windward side and the crest of the range. When the air passes over the crest onto the leeward (downwind) side, it sinks (◄▶ Figure 19.26). This air has already lost much of its moisture. In addition, it warms adiabatically as it falls, absorbing moisture and creating a **rain-shadow desert** on the leeward side of the range. For example, Death Valley, California, is a rain-shadow desert and receives only 5 cm of rain a year, while the nearby west slope of the Sierra Nevada receives 178 cm of rain a year.

Tropical Rainforests and Weather

Recall that atmospheric moisture condenses when moist air cools below its dew point. Forests cool the air. Large quantities of water evaporate from leaf surfaces in a process called **transpiration,** and evaporation cools the surrounding air. In addition, forests shade the soil from the hot sun, and tree roots and litter retain moisture. In an open clearing, rainwater evaporates quickly after a storm or runs off the surface. But forest soils remain moist long after the rain dissipates. Evaporation from soil litter combines with transpiration cooling from leaf surfaces to maintain relatively cool temperatures during times when there is no rain. In yet another feedback mechanism: Forests cool the air. Cool air promotes rainfall. Rainfall supports forests.

In tropical rainforests, local rainfall has decreased by as much as 50 percent when the forests are cut and replaced by farmland or pasture. When the rainfall decreases, wildfires ravage the boundary between the logged area and the remaining virgin forests. More forest is destroyed, establishing a negative feedback mechanism of increasing drought, fire, and forest loss.

Sea and Land Breezes

Anyone who has lived near an ocean or large lake has encountered winds blowing from water to land and from land to water. Sea and land breezes are caused by uneven heating and cooling of land and water (◄▶ Figure 19.27). Recall that land surfaces heat up faster than adjacent bodies of water and cool more quickly. If land and sea are nearly the same temperature on a summer morning, during the day the land warms and heats the air above it. Hot air then rises over the land, producing a local, low-pressure area. Cooler air from the sea flows inland to replace the rising air. Thus, on a hot, sunny day, winds generally blow from the sea onto land. The rising air is good for flying kites or hang gliding but often brings afternoon thunderstorms.

At night the reverse process occurs. The land cools faster than the sea, and descending air creates a local, high-pressure area over the land. Then the winds reverse, and breezes blow from the shore out toward the sea.

Monsoons

A **monsoon** is a seasonal wind and weather system caused by uneven heating and cooling of continents and oceans. Just as sea and land breezes reverse direction with day and night, monsoons reverse direction with the seasons. In the summertime the continents become warmer than the sea. Warm air rises over land, creating a large, low-pressure area and drawing moisture-laden maritime air inland. When the moist air rises as it flows over the land, clouds form and heavy monsoon rains fall. In winter the process is reversed. The land cools below the sea temperature, and as a result, air descends over

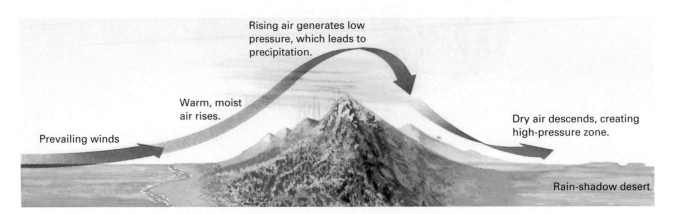

Rising air generates low pressure, which leads to precipitation.

Warm, moist air rises.

Prevailing winds

Dry air descends, creating high-pressure zone.

Rain-shadow desert

ThomsonNOW ◄▶ **ACTIVE FIGURE 19.26** A rain-shadow desert forms where moist air rises over a mountain range and precipitates most of its moisture on the windward side and crest of the range. The dry, descending air on the lee side absorbs moisture, forming a desert.

◀▶ **FIGURE 19.27** **(A)** Sea breezes blow inland during the day, and **(B)** land breezes blow out to sea at night.

land, producing dry, continental, high pressure. At the same time, air rises over the ocean and the prevailing winds blow from land to sea. More than half of the inhabitants of Earth depend on monsoons because the predictable, heavy, summer rains bring water to the fields of Africa and Asia. If the monsoons fail to arrive, crops cannot grow and people starve.

19.9
Thunderstorms

An estimated 16 million thunderstorms occur every year, and at any given moment about 2,000 thunderstorms are in progress over different parts of Earth. A single bolt of lightning can involve several hundred million volts of energy and for a few seconds produces as much power as a nuclear power plant. It heats the surrounding air to 25,000°C or more, much hotter than the surface of the Sun. The heated air expands instantaneously to create a shock wave that we hear as thunder.

Despite their violence, thunderstorms are local systems, often too small to be included on national weather maps. A typical thunderstorm forms and then dissipates in a few hours and covers from about ten to a few hundred square kilometers. It is not unusual to stand on a hilltop in the sunshine and watch rain squalls and lightning a few kilometers away. All thunderstorms develop when warm, moist air rises, forming cumulus clouds that develop into towering cumulonimbus clouds. Different conditions cause these local regions of rising air:

1. **Wind convergence.** Central Florida is the most-active thunderstorm region in the United States. As the subtropical Sun heats the Florida peninsula, rising air draws moist air from both the east and west coasts. Where the two air masses con-verge, the moist air rises rapidly to create a thunderstorm. Thunderstorms also occur in other environments where moist air masses converge.

2. **Convection.** Thunderstorms also form in continental interiors during the spring or summer, when afternoon sunshine heats the ground and generates cells of rising, moist air.

3. **Orographic Lifting.** Moist air rises as it flows over hills and mountain ranges, commonly generating mountain thunderstorms.

4. **Frontal thunderstorms.** Thunderstorms commonly occur along frontal boundaries, particularly at cold fronts.

A typical thunderstorm occurs in three stages. In the initial stage, moisture in rising air condenses, forming a cumulus cloud (◀▶ Figure 19.28A). As the cloud forms, the condensing vapor releases latent heat. In an average thunderstorm, 400,000 tons of water vapor condense within the cloud, and the energy released by the condensation is equivalent to the explosion of 12 atomic bombs the size of the one dropped on Hiroshima during World War II. This heat warms air in the cloud and fuels the violent convection characteristic of thunderstorms. Large droplets or ice crystals develop within the cloud at this stage, but the rising air keeps them in suspension, and no precipitation falls to the ground.

Eventually, water droplets or hailstones become so heavy that updrafts can no longer support them, and they fall as rain or hail. During this stage, warm air continues to rise and may attain velocities of more than 300 kilometers per hour in the central and upper portions of the cloud. The cloud may double its height in minutes. At the same time, ice falling through the cloud chills the lower regions and this cool air sinks, creating a downdraft. Thus, air currents rise and fall simultaneously within the same cloud. These conditions, known as **wind shear**, are dangerous for aircraft, and pilots avoid large thunderheads (◀▶ Figure 19.28B).

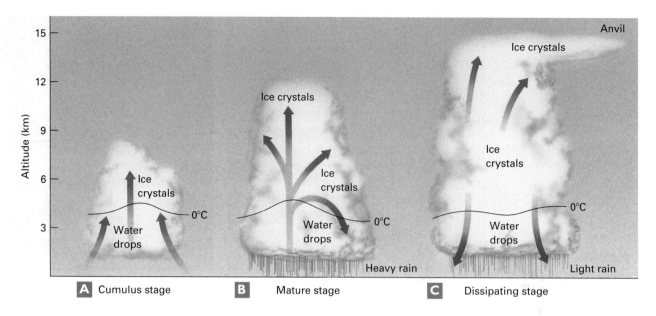

A Cumulus stage B Mature stage C Dissipating stage

ThomsonNOW ◆ **ACTIVE FIGURE 19.28** A typical thunderstorm develops in three stages. **(A)** Air rises, cools, and condenses, creating a cumulus cloud. **(B)** Latent heat of condensation energizes the storm, forming heavy rain and violent wind. **(C)** The cloud cools, convection weakens, and the storm wanes.

INTERACTIVE QUESTION: *Why does hail fall only from cumulonimbus clouds?*

As explained earlier, rainfall from a cumulonimbus cloud can be unusually heavy. In one extreme example in 1976, a sequence of thunderstorms dropped 25 centimeters of rain in about four hours over Big Thompson Canyon on the eastern edge of the Colorado Rockies. The river flooded the narrow canyon, killing 139 people.

The mature stage of a thunderstorm, with rain or hail and lightning, usually lasts for about 15 to 30 minutes and seldom longer than an hour. The cool downdraft reduces the temperature in the lower regions of the cloud. As the temperature drops, convection weakens and warm, moist air is no longer drawn into the cloud (◆ Figure 19.28C). Once the water supply is cut off, condensation ceases, and the storm loses it source of latent heat. Within minutes the rapid, vertical, air motion dies and the storm dissipates. Although a single thundercloud dissipates rapidly, new thunderheads can build in the same region, causing disasters such as that at Big Thompson Canyon.

In 1996, a group of climbers were slowly approaching the summit of Mount Everest. Below them, fluffy white clouds began to obscure the lower peaks. Most of the climbers thought that the clouds were benign and continued on toward the summit. But one of the climbers was an airplane pilot and realized that the view from 8,500 meters was analogous to a view from an airplane and not the perspective that we are normally accustomed to. To his trained eye, the clouds were the tops of rising thunderheads. Realizing that dangerous strong winds and heavy precipitation accompany a thunderstorm, he abandoned his summit attempt and retreated. The others pushed onward. A few hours later the intense storm engulfed the summit ridge and six climbers perished.

Lightning

Lightning is an intense discharge of electricity that occurs when the buildup of static electricity overwhelms the insulating properties of air (◆ Figure 19.29). If you walk across a carpet on a dry day, the friction between your feet and the rug shears electrons off the atoms on the rug. The electrons migrate into your body and concentrate there. If you then touch a metal doorknob, a spark consisting of many electrons jumps from your finger to the metal knob.

In 1752 Benjamin Franklin showed that lightning is an electrical spark. He suggested that charges separate within cumulonimbus clouds and build until a bolt of lightning jumps from the cloud. In the 250 years since Franklin, atmospheric physicists have been unable to agree upon the

◆ **FIGURE 19.29** Time-lapse photo captures multiple ground-to-ground lightning strikes during a nighttime thunderstorm in Norman, Oklahoma, March 1978.

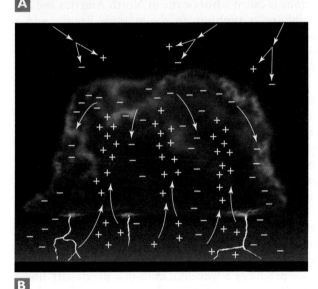

A

B

◁▷ **ACTIVE FIGURE 19.30** Two hypotheses for the origin of lightning. **(A)** Friction between intense winds and ice particles generates charge separation. **(B)** Charged particles are produced from above by cosmic rays and below by interactions with the ground. The particles are then distributed by convection currents.

exact mechanism of lightning. According to one hypothesis, friction between the intense winds and moving ice crystals in a cumulonimbus cloud generates both positive and negative electrical charges in the cloud, and the two types of charges become physically separated. The positive charges tend to accumulate in the upper portion of the cloud, and the negative charges build up in the lower reaches of the cloud. When enough charge accumulates, the electrical potential exceeds the insulating properties of air, and a spark jumps from the cloud to the ground, from the ground to the cloud, or from one cloud to another.

Another hypothesis suggests that cosmic rays bombarding the cloud from outer space produce ions at the top of the cloud. Other ions form on the ground as winds blow over Earth's surface. The electrical discharge

occurs when the potential difference between the two groups of electrical charges exceeds the insulating properties of air (◁▷ Figure 19.30). Perhaps neither hypothesis is entirely correct and some combination of the mechanisms causes lightning.

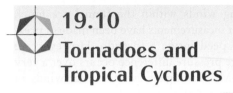

19.10
Tornadoes and Tropical Cyclones

Tornadoes and tropical cyclones are both intense, low-pressure centers. Strong winds follow the steep pressure gradients and spiral inward toward a central column of rising air.

Tornadoes

A **tornado** is a small, short-lived, funnel-shaped storm that protrudes from the base of a cumulonimbus cloud (◁▷ Figure 19.31 A). The base of the funnel can be from

A

B

◁▷ **FIGURE 19.31** **(A)** The dark funnel cloud of a tornado descends on Dimmit, Texas, on June 2, 1995. **(B)** Tornadoes that accompanied Hurricane Andrew added to the terror and confusion at La Place, Louisiana, where two people died and many homes were destroyed.

2 meters to 3 kilometers in diameter. Some tornadoes remain suspended in air while others touch the ground. After a tornado touches ground, it may travel for a few meters to a few hundred kilometers across the surface. The funnel travels from 40 to 65 kilometers per hour, and in some cases, as much as 110 kilometers per hour, but the spiraling winds within the funnel are much faster. Few direct measurements have been made of pressure and wind speed inside a tornado. However, we know that a large pressure difference occurs over a very short distance. Meteorologists estimate that winds in tornadoes may reach 500 kilometers per hour or greater. These winds rush into the narrow, low-pressure zone and then spiral upward. After a few seconds to a few hours, the tornado lifts off the ground and dissipates.

Tornadoes are the most violent of all storms. One tornado in 1910 lifted a team of horses and then deposited it, unhurt, several hundred meters away. They were lucky. In the past, an average of 120 Americans were killed every year by these storms, and property damage costs millions of dollars. The death toll has decreased in recent years because effective warning systems allow people to seek shelter, but the property damage has continued to increase (◁▷ Figure 19.31B). Tornado winds can lift the roof off a house and then flatten the walls. Flying debris kills people and livestock caught in the open. Even so, the total destruction from tornadoes is not as great as that from hurricanes because the path of a tornado is narrow and its duration short.

Although tornadoes can occur anywhere in the world, 75 percent of the world's twisters concentrate in the Great Plains, east of the Rocky Mountains. Approximately 700 to 1,000 tornadoes occur in the United States each year. They frequently form in the spring or early summer. At that time, continental, polar (dry, cold) air from Canada collides with maritime, tropical (warm, moist) air from the Gulf of Mexico. As explained previously, these conditions commonly create thunderstorms. Meteorologists cannot explain why most thunderstorms dissipate harmlessly but a few develop tornadoes. However, one fact is apparent: Tornadoes are most likely to occur when large differences in temperature and moisture exist between the two air masses and the boundary between them is sharp.

The probability that any particular place will be struck by a tornado is small. Nevertheless, Codell, Kansas, was struck three years in a row—in 1916, 1917, and 1918—and each time the disaster occurred on May 20! During a two-day period in 1974, 148 tornadoes occurred in 13 states.

Tropical Cyclones

A **tropical cyclone** or **tropical storm** is less intense than a tornado but much larger and longer-lived (Table 19.2). Tropical cyclones are circular disturbances that average 600 kilometers in diameter and persist for days or weeks. If the wind exceeds 120 kilometers per hour, a tropical cyclone is called a **hurricane** in North America and the Caribbean, a **typhoon** in the western Pacific, and a **cyclone** in the Indian Ocean (◁▷ Figure 19.32). Intense low pressure in the center of a hurricane can generate wind as strong as 300 kilometers per hour.

The low atmospheric pressure created by a tropical cyclone can raise the sea surface by several meters. Often, as a tropical cyclone strikes shore, strong onshore winds combine with the abnormally high water level created by low pressure to create a **storm surge** that floods coastal areas. In 2005 during Hurricane Katrina, sea level rose about 8.5 meters above normal on the Gulf Coast as a result of a storm surge.

Tropical cyclones form only over warm oceans, never over cold oceans or land. Thus, moist, warm air is crucial to the development of this type of storm. Recall that a midlatitude cyclone develops when a small disturbance produces a wavelike kink in a previously linear front. A similar mechanism initiates a tropical cyclone. In late summer, the Sun warms tropical air. The rising hot air creates a belt of low pressure that encircles the globe over the tropics. In addition, many local, low-pressure disturbances move across the tropical oceans at this time of year. If a local disturbance intersects the global tropical low, it creates a bulge in the isobars. Winds are deflected by the bulge and, directed by the

	Range	
Feature	**Tornado**	**Tropical Cyclone**
Diameter	2–3 km	400–800 km
Path length (distance traveled across terrain)	A few meters to hundreds of kilometers	A few hundred to a few thousand kilometers
Duration	A few seconds to a few hours	A few days to a week
Wind speed	300–800 km/hr	120–250 km/hr
Speed of motion	0–70 km/hr	20–30 km/hr
Pressure fall	20–200 mb	20–60 mb

Table 19.2 Comparison of Tornadoes and Tropical Cyclones

NASA-GSFC

ThomsonNOW ◆ **FIGURE 19.32** Hurricane Katrina making landfall near New Orleans on August 28, 2005.

The center of the storm is a region of vertical airflow, called the eye. In the outer, and larger part, of the eye, the air that has been rushing inward spirals upward. In the inner eye, air sinks. Thus, the horizontal wind speed in the eye is reduced to near zero (◆ Figure 19.33). Survivors who have been in the eye of a hurricane report an eerie calm. Rain stops, and the Sun may even shine weakly through scattered clouds. But this is only a momentary reprieve. A typical eye is only 20 kilometers in diameter, and after it passes the hurricane rages again in full intensity.

Thus, a hurricane is powered by a classic feedback mechanism: The low-pressure storm causes condensation; condensation releases heat; heat powers the continued low pressure. The entire storm is pushed by prevailing winds, and its path is deflected by the Coriolis effect. A hurricane usually dissipates only after it reaches land or passes over colder water because the supply of moist, warm air is cut off. Condensing water vapor in a single tropical cyclone releases as much latent heat energy as that produced by all the electric generators in the United States in a six-month period.

The numerical "category" of tropical cyclones is based on a rating scheme called the Saffir–Simpson scale, after its developers (Table 19.3). This scale, commonly mentioned in weather reports, rates the damage potential of a hurricane or other tropical storm, and typical values of atmospheric pressure, wind speed, and height of storm surge associated with storms of increasing intensity.

Coriolis effect, begin to spiral inward. Warm, moist air rises from the low. Water vapor condenses from the rising air, and the latent heat warms the air further, which causes even more air to rise. As the low pressure becomes more intense, strong surface winds blow inward to replace the rising air. This surface air also rises, and more condensation and precipitation occur. But the additional condensation releases more heat, which continues to add energy to the storm.

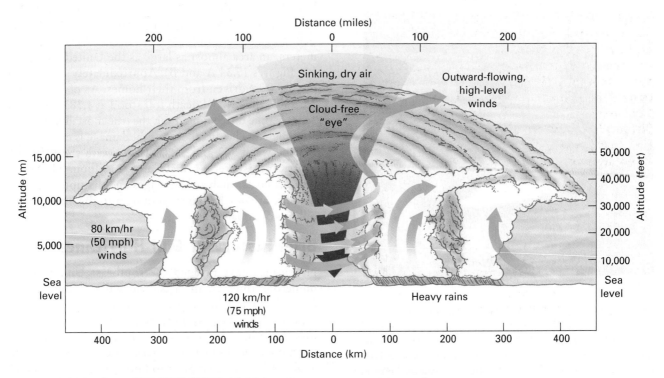

ThomsonNOW ◆ **ACTIVE FIGURE 19.33** Surface air spirals inward toward a hurricane, rises through the towering wall of clouds, and then flows outward above the storm. Falling air near the storm's center creates the eerie calm in the eye of the hurricane.

Table 19.3 The Saffir–Simpson Hurricane Damage Potential Scale

Type	Category	Damage	Pressure (millibar)	Winds (km/h)	Storm surge (m)
Depression				>56	
Tropical Storm				63–117	
Hurricane	1	minimal	980	119–152	1.2–1.5
Hurricane	2	moderate	965–979	154–179	1.8–2.4
Hurricane	3	extensive	945–964	179–209	2.7–3.7
Hurricane	4	extreme	920–944	211–249	4–5.5
Hurricane	5	catastrophic	<920	>249	<5.5

A

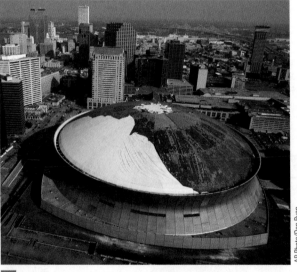

B

◆ **FIGURE 19.34** **(A)** A Hurricane Katrina flood victim lies on a road in New Orleans La., Sept. 8, 2005. Katrina was the most costly and one of the deadliest natural disasters in U.S. history, as an 8.5-meter storm surge inundated 233,000 square kilometers of the New Orleans and Mississippi coastlines. **(B)** The partly destroyed roof of the Lousiana Superdome in downtown New Orleans is a massive reminder of the tragedies suffered during Hurricane Katrina.

19.11
Hurricane Katrina

Scope of the Disaster

Hurricane Katrina formed in the South Atlantic on August 23, 2005, made landfall as a category 1 hurricane just north of Miami, then crossed the Florida peninsula and grew to a category 5 storm in the Gulf of Mexico with sustained winds of 280 km/hr. By the time it made a second landfall on the Mississippi delta, Hurricane Katrina had diminished to a category 3–4 storm, with winds hovering around 200 km/hr. Yet despite the reduced winds, Katrina quickly became the most costly and one of the deadliest natural disasters in U.S. history. Parts of New Orleans were flooded to the rooftops of houses. In addition, a record 8.5-meter storm surge inundated the New Orleans and Mississippi coastlines, flattening and flooding homes over 233,000 square kilometers, an area almost as large as the United Kingdom (◆ Figures 19.34 A and B). Approximately 1.3 million people were driven from their homes. As of December 20, 2005, the official death toll stood at 1,383, but with 4,000 people unaccounted for, the true mortality is probably higher. The direct structural damage is expected to cost between $40 and $120 billion, with the National Hurricane Center estimate at $75 billion. However, indirect costs, such as lost wages and disruption of human life are much higher, running into the hundreds of billions of dollars.

Brief History of Gulf Hurricanes

Hurricanes have ravaged the southeast United States numerous times in the past and will certainly do so again. But while hurricanes are not preventable, hurricane deaths and costs result from a combination of factors, including the severity of the hurricane, building practices, and human responses. For example, from Table 19.4 we see that prior to Hurricane Katrina, the 10 deadliest hurricanes all occurred before 1957. However, the 10 costliest storms all occurred after 1955. With the exception of Katrina, none of the 10 deadliest storms are

Table 19.4

Ten Deadliest Storms

Rank	Hurricane	Year	Category	Deaths*
1	Galveston, Texas	1900	4	8,000+
2	Florida	1928	4	2,500+
3	Katrina	2005	3/4	1,383+
4	Louisiana	1893	4	1,100+
5	South Carolina and Georgia	1893	3	1,000+
6	South Carolina and Georgia	1891	2	700
7	Florida Keys	1935	5	408
8	Louisiana	1856	4	400
9	Texas and Louisiana (Audrey)	1957	4	390
10	Florida, Alabama, and Mississippi	1926	4	372

*Does not include offshore deaths.

Ten Costliest Storms

Rank	Hurricane	Year	Category	Damage (in billions)*
1	Katrina	2005	3/4	$75
2	Andrew (FL, LA)	1992	5	$43.7
3	Charley (FL)	2004	4	$15.0
4	Ivan (FL, AL)	2004	3	$14.2
5	Hugo (SC)	1989	4	$12.3
6	Agnes (FL, NE U.S.)	1972	1	$11.3
7	Betsy (FL, LA)	1965	3	$10.8
8	Frances (FL)	2004	2	$8.9
9	Camille (MS, LA, VA)	1969	5	$8.9
10	Diane (NE U.S.)	1955	1	$7.0

*2004 dollars adjusted for inflation

on the list of costliest storms, and vice versa. Why is there so little correlation between structural damage (cost) and death toll?

In the last half century, structural damage has been high for three main reasons: (1) More Americans lived near the Atlantic and Gulf coasts in the late 1990s and early 2000s than earlier in the century; (2) homes and other structures have had an increased inflation-adjusted value; and (3) Americans now own more things than ever before and, consequently, the dollar value of their possessions is at an all-time high. But while structures are immobile, people can evacuate if given ample warning.

The death toll has been low in the past half century because accurate forecasting warns people of impending hurricanes. The deadliest hurricane in the United States struck Galveston, Texas, in September 1900. More than

8,000 people died because the population was caught unaware. A tropical cyclone has a sharp boundary, and even a few hundred kilometers outside that boundary, fluffy white clouds may be floating in a blue sky. Today, satellites track hurricanes and news reports give people ample time to evacuate.

Why then, with adequate warning systems, was Katrina so deadly? Most of the Katrina fatalities resulted from the unanticipated flooding of New Orleans. Therefore, to answer this question, we must ask, "Why did flood waters ravage the city?"

Human Factors that Augmented the Disaster

Much of New Orleans lies below sea level (◆ Figure 19.35). Over 500 kilometers of levees protect the city by holding back the Mississippi River and Lake Ponchartrain. But if hurricane-driven storm surges or floodwaters should breach even one of those levees (as they did) disaster occurs.

Wetland Loss

Coastal wetlands, barrier islands, and delta land blunt the force of incoming hurricane storm surges. Over the past century, these wetlands have eroded and disappeared rapidly. In 1900, there were approximately 15,000 square kilometers of low-lying wetlands in the Mississippi delta. By the time Hurricane Katrina struck in 2005, approximately one-third of those wetlands, an area larger than the state of Rhode Island, had disappeared (◆ Figure 19.36). With the ocean much closer to New Orleans than it had been previously, storm surges were more intense in urban areas than they would have been if the delta had remained intact. Three factors have led to wetland erosion in the Mississippi delta:

1. Recall from Chapter 11 that dams and river diversions in the Mississippi River basin have reduced the flow of sediment to the delta. As a result, for the past half century, erosion has exceeded delta buildup, and the delta has receded.
2. The Mississippi delta is heavy enough to depress Earth's crust. Under the huge weight of sediment, the delta subsides, cracking subterranean rock into thousands of subsurface faults that crisscross the region from east to west. Prior to 1900, subsidence was counterbalanced by an influx of additional sediment and the delta expanded. But after the sediment deposition was reduced, the edge of the delta began to slump into the sea. This subsidence has been augmented because thousands of wells remove petroleum from beneath the delta. As the liquid is removed, the surface settles, exacerbating the natural subsidence.

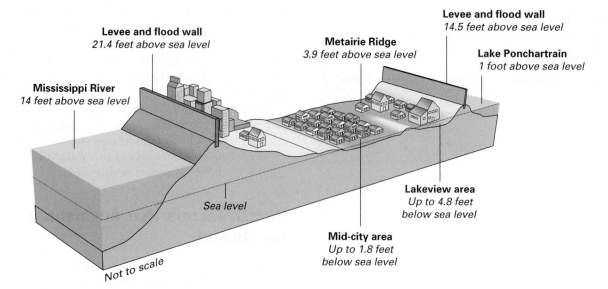

Levee and flood wall
21.4 feet above sea level

Mississippi River
14 feet above sea level

Metairie Ridge
3.9 feet above sea level

Levee and flood wall
14.5 feet above sea level

Lake Ponchartrain
1 foot above sea level

Sea level

Not to scale

Lakeview area
Up to 4.8 feet
below sea level

Mid-city area
Up to 1.8 feet
below sea level

◈ **FIGURE 19.35** Large parts of New Orleans are protected by levees because they lie below sea level, below Lake Pontchartrain, and below the level of the Mississippi River. When the levees fail, these portions of the city suffer flooding as they did during Hurricane Katrina.

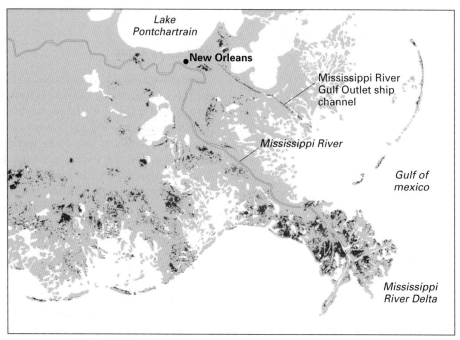

Lake
Pontchartrain

● New Orleans

Mississippi River
Gulf Outlet ship
channel

Mississippi River

*Gulf of
mexico*

*Mississippi
River Delta*

◈ **FIGURE 19.36** Between 1900 and 2005, about one-third of the wetlands in the Mississippi delta have been lost (lost wetlands are shown in blue) to erosion, subsidence, and dredging of channels.

3. Engineers have built more than 13,000 kilometers of shipping canals through the delta. Saltwater flowing inland through these canals has killed thousands of square kilometers of forests and marsh vegetation. When the vegetation is destroyed, erosion increases, and the delta washes away. In addition, storm surges move inland more readily through straight, artificial navigation channels than through sinuous, natural distributaries. In particular, the Mississippi River Gulf Outlet ship canal funneled storm surges directly into the city.

Levee Failure

Despite the fact that New Orleans lies below sea level and the delta has been disappearing at a rapid rate, the levees were designed to protect the city from category 4 hurricanes such as Katrina. In November 2005, engineers from the National Science Foundation, the American Society of Civil Engineers, and the state of Louisiana reported to the U.S. Senate that most of the flooding occurred because the levees were built on a weak soil layer of peat. When floodwaters rose, the levees failed catastrophically (◈ Figure 19.37).

Hurricane Frequency and Intensity: Greenhouse Effect and Global Warming

Recall that hurricanes form only over warm water and that their energy is driven by a cycle of evaporation and condensation. In 1987, Professor Kerry Emanuel of MIT calculated that if global atmospheric warming heated the ocean surface, it would increase evaporation and hence increase the intensity of tropical cyclones. Therefore, more tropical cyclones would attain the velocity to become hurricanes. In 2005, the average tropical cyclone released 50 to 80 percent more energy than the average over the past 30 years. Wind velocities had increased by 15 percent and storms were lasting longer. Thus the observed result matched the model.

In addition, global sea level is rising, so storm surges, like the ones that breached the levees in New Orleans, will inevitably become more destructive.

In 2005 there were a record number of tropical cyclones (26) and a record number of hurricanes (14). Pre-viously, 1969 had held the record with 12 hurricanes. Although it is possible that 2005 was simply anomalous, most scientists agree that hurricanes will be more frequent and more intense in a warmer world. Between 1985 and 1994, when sea-surface temperatures were relatively cool, hurricane frequency was low, and between 1995 and 2004, when sea-surface temperatures were higher, hurricane frequency was also higher (◆ Figure 19.38). This statistical correlation does not prove cause and effect—it may be a coincidence. But the statistics agree with the prediction.

19.12
El Niño

Hurricanes are common in the southeastern United States and on the Gulf Coast, but they rarely strike California. Consequently, Californians were taken by surprise in late September 1997 when Hurricane Nora ravaged Baja California and then, somewhat diminished by landfall, struck San Diego and Los Angeles. The storm brought the first rain to Los Angeles after a record 219 days of drought, then spread eastward to flood parts of Arizona, where it caused the evacuation of 1,000 people.

Other parts of the world also experienced unusual weather during the autumn of 1997. In Indonesia and Malaysia, fall monsoon rains normally douse fires intentionally set in late summer to clear the rainforest. The rains were delayed for two months in 1997, and, as a result, the fires raged out of control, filling cities with such dense smoke that visibility at times was no more than a few meters. Even an airliner crash was attributed to the smoke. Severe drought in nearby Australia caused ranchers to slaughter entire herds of cattle for lack of water and feed. At the same time, far fewer hurricanes than usual threatened Florida and the U.S. Gulf Coast. Floods soaked northern Chile's Atacama Desert, a region that commonly receives no rain at all for a decade at a time, record snowfalls blanketed the Andes, and heavy rains caused floods in Peru and Ecuador.

All of these weather anomalies have been attributed to **El Niño**, an ocean current that brings unusually warm water to the west coast of South America. But the current does not flow every year; instead, it occurs about every 3 to 7 years, and its effects last for about a year before conditions return to normal. Although meteorologists paid little attention to the phenomenon until the El Niño year of 1982–83, many now think that El Niño affects weather patterns for nearly three quarters of Earth.

Meteorologists first began recording El Niños in 1982–83, although they were known to Peruvian fishermen much earlier because they warm coastal waters and diminish fish harvests. The fishermen called the warming events *El Niño* because they commonly occur around Christmas, the birthday of El Niño—the Christ Child.

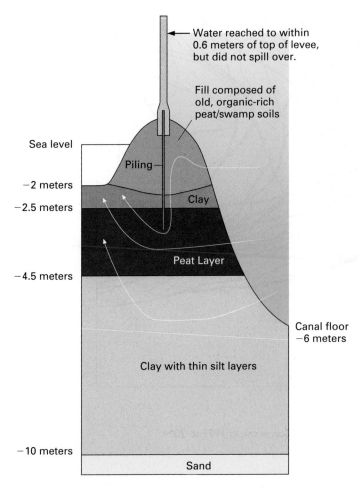

◆ **FIGURE 19.37** Many of the levees in New Orleans failed because foundations were faulty and anchored in weak layers of peat.

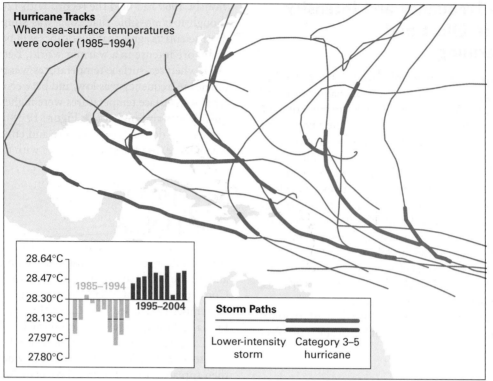

A

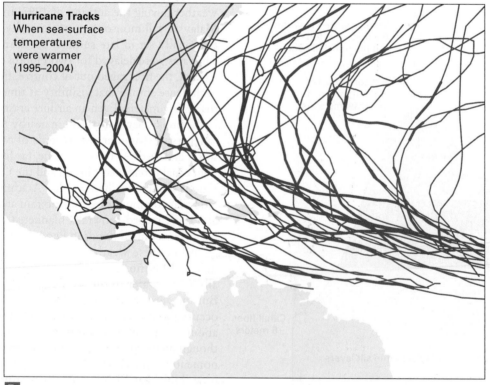

B

◈ **FIGURE 19.38** **(A)** Storm tracks 1985 to 1994 **(B)** Storm tracks 1995 to 2004.

To understand the El Niño effects, first consider interactions between southern Pacific sea currents and weather in a normal, non–El Niño year (◀▶ Figure 19.39). Normally in fall and winter, strong trade winds blow westward from South America across the Pacific Ocean. The winds drag the warm, tropical surface water away from Peru and Chile and pile it up in the western Pacific near Indonesia and Australia. In the western Pacific, the warm water forms a low mound thousands of kilometers across. The water is up to 10°C warmer and as much as 60 centimeters higher than the surface of the ocean near Peru and Chile.

As the wind-driven surface water flows away from the South American coast, cold, nutrient-rich water rises from the depths to replace the surface water. The nutrients support a thriving fishing industry along the coasts of Peru and Chile.

Abundant moisture evaporates from the surface of a warm ocean. In a normal year, much of this water condenses to bring rain to Australia, Indonesia, and other lands in the southwestern Pacific, which are adjacent to the mound of warm water. On the eastern side of the Pacific, the cold, upwelling ocean currents cool the air above the coasts of Peru and northern Chile.

This cool air becomes warmer as it flows over land. The warming lowers the relative humidity and creates the coastal Atacama Desert.

In an El Niño year, for reasons poorly understood by meteorologists, the trade winds slacken (◀▶ Figure 19.40). The mound of warm water near Indonesia and Australia

then flows downslope—eastward across the Pacific Ocean toward Peru and Chile. The anomalous accumulation of warm water off South America causes unusual rains in normally dry coastal regions, and heavy snowfall in the Andes. At the same time, the cooler water near Indonesia, Australia, and nearby regions causes drought. The mass of warm water that caused the 1997–98 El Niño was about the size of the United States, and caused one of the strongest El Niño weather disturbances in history.

El Niño has global effects that go far beyond regional rainfall patterns. For example, El Niño deflects the jet stream from its normal path as it flows over North America, directing one branch northward over Canada and the other across Southern California and Arizona. Consequently, those regions receive more winter precipitation and storms than usual, while fewer storms and warmer winter temperatures affect the Pacific Northwest, the northern plains, the Ohio River valley, the mid-Atlantic states, and New England. Southern Africa experiences drought, while Ecuador, Peru, Chile, southern Brazil, and Argentina receive more rain than usual.

Altered weather patterns created by El Niño wreak havoc. Globally, 2,000 deaths and more than $13 billion in damage are attributed to the 1982–83 El Niño effects. In the United States alone, more than 160 deaths and $2 billion in damage occurred, mostly from storm and flood damage. In southern Africa, economic losses of $1 billion and uncounted deaths due to disease and starvation have been attributed to the 1982–83 El Niño.

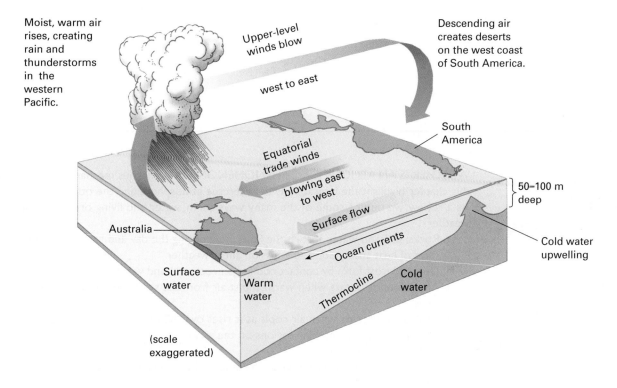

◀▶ **FIGURE 19.39**   In a normal year, trade winds drag warm surface water westward across the Pacific and pile it up in a low mound near Indonesia and Australia, where the warm water causes rain. The surface flow creates upwelling of cold, deep, nutrient-rich waters along the coast of South America.

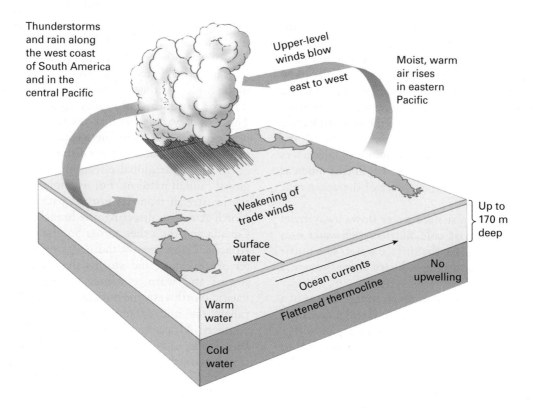

Thunderstorms and rain along the west coast of South America and in the central Pacific

Upper-level winds blow east to west

Moist, warm air rises in eastern Pacific

Weakening of trade winds

Up to 170 m deep

Surface water

Ocean currents

No upwelling

Warm water

Flattened thermocline

Cold water

◀▶ **FIGURE 19.40**   In an El Niño year, the trade winds slacken and the warm water flows eastward toward South America, causing the storms and rain to move over South America, and diminishing the upwelling currents.

EARTH SYSTEMS INTERACTIONS

WEATHER IS AN ATMOSPHERIC PHENOMENON, yet numerous mechanisms discussed in this chapter highlight the interactions between the atmosphere and the other three spheres. As a review, we outline some of the many ways earth, water, and living organisms affect the weather:

- Rocks, soil, and water absorb the Sun's heat during the day and release it at night. The rates of heat absorption and emission affect weather.
- Dew and frost form by contact cooling on rock, soil, and vegetation.
- Advection fog occurs when warm, moist air from the sea blows across cooler land surfaces.
- Upslope fog occurs when air cools as it rises over hilly terrain of the geosphere.
- Mountain ranges form rainy regions on the windward side and dry zones on the lee.
- Forests induce rainfall.
- Different temperatures between the land and the ocean cause sea breezes and monsoons.
- Tropical cyclones are powered by evaporation from warm oceans.
- Ocean currents such as El Niño affect local weather. ■

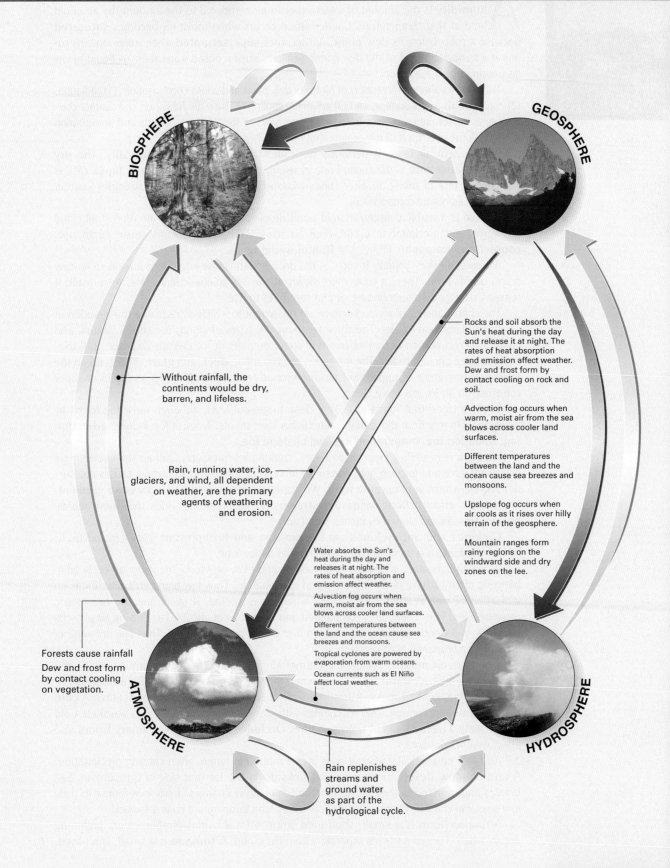

BIOSPHERE

GEOSPHERE

Rocks and soil absorb the Sun's heat during the day and release it at night. The rates of heat absorption and emission affect weather. Dew and frost form by contact cooling on rock and soil.

Advection fog occurs when warm, moist air from the sea blows across cooler land surfaces.

Different temperatures between the land and the ocean cause sea breezes and monsoons.

Upslope fog occurs when air cools as it rises over hilly terrain of the geosphere.

Mountain ranges form rainy regions on the windward side and dry zones on the lee.

Without rainfall, the continents would be dry, barren, and lifeless.

Rain, running water, ice, glaciers, and wind, all dependent on weather, are the primary agents of weathering and erosion.

Water absorbs the Sun's heat during the day and releases it at night. The rates of heat absorption and emission affect weather.

Advection fog occurs when warm, moist air from the sea blows across cooler land surfaces.

Different temperatures between the land and the ocean cause sea breezes and monsoons.

Tropical cyclones are powered by evaporation from warm oceans.

Ocean currents such as El Niño affect local weather.

Forests cause rainfall

Dew and frost form by contact cooling on vegetation.

ATMOSPHERE

HYDROSPHERE

Rain replenishes streams and ground water as part of the hydrological cycle.

SUMMARY

Absolute humidity is the mass of water vapor in a given volume of air. **Relative humidity** is the amount of water vapor in air compared to the amount the air could hold at that temperature. Condensation occurs when moist air becomes **saturated** because it cools below its **dew point**. Air becomes **supersaturated** when water remains vapor at a temperature below the dew point. Similarly, **supercooled** water remains liquid below its freezing point.

Three atmospheric processes cool air to its dew point and cause condensation: (1) radiation, (2) contact with a cool surface, and (3) adiabatic cooling of rising air. Radiation and contact cooling cause the formation of dew, frost, and some types of fog. However, clouds and precipitation normally form as a result of the cooling that occurs when air rises.

When warm air rises, it performs work and therefore cools **adiabatically**. The **dry adiabatic lapse rate** is the cooling rate of rising, dry air. The **wet adiabatic lapse rate** is the cooling rate of moist air after condensation has begun. Sinking air becomes warmer because of adiabatic compression.

A cloud is a visible concentration of water droplets or ice crystals in air. Almost all cloud formation and precipitation occur when air rises. Three mechanisms cause air to rise: **convection**, **orographic** lifting, and **frontal wedging**.

As moist air rises, initially it cools at the dry adiabatic lapse rate. Once it cools to its dew point, clouds form. Then, it cools more slowly at the wet adiabatic lapse rate. As a result, it remains warmer than surrounding air and continues to rise.

The characteristics of a cloud depend on the height to which air rises and the elevation at which condensation occurs. The three fundamental types of clouds are cirrus, stratus, and cumulus. Precipitation occurs when small water droplets or ice crystals coalesce until they become large enough to fall. The formation of rain, snow, **sleet**, and **glaze** all depend on the relative temperature of upper-level air and ground-level air. **Hail** forms as layers of vapor condense on an ice grain in a cloud.

Fog is a cloud that forms at or very close to ground level, although most fog forms by processes different from those that create higher-level clouds. Types of fog include: **advection fog**, **radiation fog**, **evaporation fog**, and **upslope fog**.

When air is heated, it expands and rises, creating low pressure. Cool air sinks, exerting a downward force and forming high pressure. Uneven heating of Earth's surface causes pressure differences, which, in turn, cause **wind**. Wind speed is determined by the **pressure gradient**. The Coriolis effect deflects winds. **Jet streams** are high-altitude winds that move rapidly because they are not slowed by friction with Earth.

Cyclones and **anticyclones** are low-pressure and high-pressure zones, respectively. Winds spiral into a cyclone, and outward from an anticyclone.

When air is heated, it expands and rises, creating low pressure. The rising air cools adiabatically, often causing cloud formation and precipitation. Thus low barometric pressure is an indication of wet weather. Alternatively, when air is cooled, it sinks and the barometric pressure rises. This falling air is compressed and heated adiabatically. The sinking air absorbs moisture, and clouds generally do not form over a high-pressure region. Thus fair, dry weather generally accompanies high pressure.

When two **air masses** collide, the warmer air rises along the **front**, forming clouds and often precipitation. A **warm front** forms when moving, warm air collides with a stationary or slower-moving, cold air mass. Warm fronts often bring several days of wet weather. A **cold front** forms when moving, cold air collides with stationary or slower-moving, warm air. Cold fronts create a narrow band of violent weather. **Occluded fronts** and **stationary fronts** also affect weather patterns.

Air cools adiabatically when it rises over a mountain range, often causing precipitation. A **rain-shadow desert** forms where air sinks down the leeward side of a mountain. Sea breezes and **monsoons** arise because ocean temperature changes slowly in response to daily and seasonal changes in solar radiation, whereas land temperature changes quickly.

A thunderstorm is a small, short-lived storm from a cumulonimbus cloud. Lightning occurs when charged particles separate within the cloud. A **tornado** is a small, short-lived, funnel-shaped storm that protrudes from the bottom of a cumulonimbus cloud and reaches the ground. A **tropical cyclone** is a larger, longer-lived storm that forms over warm oceans and is powered by the energy released when water vapor condenses to form clouds and rain. Death and property loss from Hurricane Katrina was caused by the intensity of the storm and by a variety of human factors, including: human-caused erosion of the delta, and levee failure. **El Niño** is a weather pattern caused by shifting motion of ocean currents.

Key Terms

humidity   471
absolute humidity   471
relative humidity   471
saturation   472
dew point   472
supersaturation   472
supercooling   472
dew   472
frost   472
adiabatic temperature changes   473
dry adiabatic lapse rate   473
wet adiabatic lapse rate   473
normal lapse rate   475
atmospheric inversion   476
temperature inversion   476
unstable air   476
stable air   476
orographic lifting   476
frontal wedging   476
cirrus cloud   477

stratus cloud   478
cumulus cloud   478
stratocumulus cloud   479
cumulonimbus cloud   479
nimbostratus cloud   479
altostratus cloud   479
sleet   479
glaze   479
hail   480
advection fog   480
radiation fog   480
evaporation fog   480
upslope fog   480
wind   481
pressure gradient   481
isobar   481
jet stream   483
cyclone   484
anticyclone   484
air mass   484

front   484
warm front   486
cold front   486
occluded front   487
stationary front   487
storm tracks   488
rain-shadow desert   490
transpiration   490
monsoon   490
wind shear   491
tornado   493
tropical cyclone   494
hurricane   494
typhoon   494
cyclone   494
storm surge   494
El Niño   499

For Review

1. Describe the difference between absolute humidity and relative humidity. How can the relative humidity change while the absolute humidity remains constant?

2. List three atmospheric processes that cool air.

3. What is the dew point? How does dew form?

4. List a set of atmospheric conditions that might produce supersaturation or supercooling.

5. What are the differences among dew, frost, and fog?

6. Explain why air cools as it rises.

7. What is an adiabatic temperature change? How does it differ from a nonadiabatic temperature change?

8. Compare and contrast the dry adiabatic lapse rate, the wet adiabatic lapse rate, and the normal lapse rate.

9. Discuss the factors that determine the height at which a cloud forms.

10. Describe cirrus, stratus, and cumulus clouds. Include their shapes and the type of precipitation to be expected from each.

11. How does rain form? How do droplets falling from a stratus cloud differ from those formed in a cumulus cloud?

12. Compare and contrast snow, sleet, glaze, and hail.

13. Describe four types of fog.

14. List three mechanisms that cause air to rise.

15. Why does low pressure often lead to rain? Why does high pressure often bring sunny skies?

16. If the wind is blowing southward in the Northern Hemisphere, will Earth's spin cause it to veer east or west? If the wind is moving southward in the Southern Hemisphere, which way will it veer? Explain your answer.

17. Compare and contrast sea and land breezes with monsoons.

18. What is a rain-shadow desert? How does it form?

19. How do warm and cold fronts form, and what types of weather does each cause?

20. Describe the three stages in the life of a thunderstorm.

21. Briefly describe two hypotheses of how lightning forms.

22. Compare and contrast tornadoes with tropical cyclones. How does each form, and how does each affect human settlements?

23. Discuss the interplay between human and natural factors in determining the destructiveness of Hurricane Katrina.

24. Explain why hurricanes are predicted to be more intense, on average, in a warm world than in a cooler world.

25. Explain why hurricane intensity correlates with sea-surface temperature.

26. Explain how ocean currents affect precipitation patterns during a normal and an El Niño year.

For Discussion

1. Using Figure 19.1, estimate the maximum absolute humidity at 0°C, 10°C, 20°C, and 40°C. Estimate the quantity of water in air, at 50 percent relative humidity and at each of these temperatures.

2. Explain why frost forms on the inside of a refrigerator (assuming it is an old-fashioned one and not a modern, frost-free unit). Would more frost tend to form in (a) summer or winter, and (b) in a dry, desert region or a humid one? Explain.

3. Which of the following conditions produces frost? Which produces dew? Explain. (a) A constant temperature throughout the day. (b) A warm, summer day followed by a cool night. (c) A cool, fall afternoon followed by freezing temperature at night.

4. Draw a chart showing temperatures at cloud level, between the cloud and the ground, and at ground level that will cause condensing water vapor to fall as rain, snow, sleet, and glaze.

5. What is the energy source that powers the wind?

6. Are sea breezes more likely to be strong on an overcast day or on a bright, sunny one? Explain.

7. What is an air mass? Describe what would likely happen if a polar air mass collided with a humid, subtropical air mass.

8. Study the weather map in today's newspaper, and predict the weather two days from now in your area, Salt Lake City, Chicago, and New York City. Defend your prediction. Check the paper in two days to see if you were right or wrong.

9. Compare the annual precipitation in the place where you live with precipitation patterns in adjacent regions at the same latitude. Explain any differences you observe.

10. Immediately after Hurricane Katrina, a few policy makers advocated relocating the city rather than rebuilding it in its traditional location. Argue for or against changing the location or rebuilding in the present position.

11. Give an example of how the geosphere, the hydrosphere, and the biosphere affect, and are affected by, weather.

ThomsonNOW

Assess your understanding of this chapter's topics with additional comprehensive interactivities at **www.thomsonedu.com/login**, which also has up-to-date web links, additional readings, and exercises.

CHAPTER
20 Climate

Courtesy of Graham R. Thompson/Jonathan Turk

Late Fall in the
Japanese Alps.

ThomsonNOW™ Throughout this chapter, the ThomsonNOW logo indicates an opportunity for online self-study, which:
• Assesses your understanding of important concepts and provides a personalized study plan
• Links you to animations and simulations to help you study
• Helps to test your knowledge of material and prepare for exams
Visit ThomsonNOW at www.thomsonedu.com/login to access these resources.

f you are planning a picnic for next Sunday, you hope for bright sunshine, but at the same time, you understand that your plans may be spoiled by rain. Next Sunday's temperature and precipitation are determined by daily fluctuations of weather. But weather only varies within a relatively narrow range. You are certain that it is not going to snow on the Fourth of July in Texas and that if you live in Montana, you are not going to bask outside in shorts and a T-shirt in January.

Over the time span of generations or centuries, climate is stable enough so that we plan our lives around it. Farmers in Kansas plant wheat and never attempt to raise bananas. In winter, hotel owners in Florida prepare for an influx of tourists escaping the northern winter. As a result, climate strongly influences many aspects of our lives: our outdoor recreation, the houses we live in, and the clothes we wear. Humans tend to migrate toward warm, sunny regions. In the United States, the populations of California and Texas have increased rapidly in the past generation, while North Dakota and Montana have seen little growth.

Yet such stability is only guaranteed over relatively short periods of geologic time. Dinosaur bones and coal deposits indicate that parts of Antarctica were once covered by warm, humid swamps. Desert dunes lie buried beneath the fertile wheat fields of Colorado. When geologists study the climate record in a specific region, they must separate the effects of tectonic movement from global climate change. For example, the coal deposits in Antarctica do not tell us that Earth was once warm enough for coal-producing swamps to grow near the South Pole. Instead, other evidence indicates that the continent of Antarctica was once close to the equator, as explained in our study of tectonics. Yet, even when the effects of tectonic plate movements are factored out, it is clear that global climate has changed dramatically and often abruptly over the long spans of Earth's history. In this chapter, we will examine today's global climate. In the following chapter, we will look at climate change throughout Earth's history and then project into the future. ■

20.1
Major Factors That Control Earth's Climate

In Chapter 1, we outlined feedback processes and loops among the geospheres, hydrospheres, and atmospheres of Venus, Earth, and Mars. On Venus, which is closer to the Sun than Earth and therefore a little hotter early in the history of the Solar System, feedback loops eventually produced torridly hot temperatures and an atmosphere hostile to life. Mars, farther from the Sun and initially a little cooler, lost most of its atmosphere and became frigid. On Earth, temperatures were moderate—in the range where water exists in liquid, vapor, and solid states. In Chapter 17, we learned that Precambrian organisms slowly added oxygen to Earth's early atmosphere, to permit efficient metabolism and the evolution of large, multicellular plants and animals by the beginning of Cambrian time. In this section, we discuss the factors that control

Earth's climate: latitude, wind, oceans, altitude, and albedo. While these are unquestionably important, it is critical to remember that the temperate climate of Earth depends on the composition of the atmosphere, which is strongly affected by living organisms.

In Chapters 17, 18, and 19, we introduced many factors that regulate regional temperature and rainfall. The same factors influence global climate. We will review these briefly.

Latitude

Ultimately, the Sun controls Earth's climate. Sunlight warms Earth and provides the energy to evaporate water and power winds and ocean currents. As we learned in Chapter 18, high latitudes receive less total sunlight than equatorial regions. High latitudes also experience great seasonal differences in sunlight, causing winter and summer seasons. Temperature and precipitation are not constant at a given latitude, however, because many other factors are important.

Wind

Wind transports heat and moisture across the globe. We discussed the relationship between wind and weather in Chapters 18 and 19; wind and climate are discussed in Section 20.2.

Oceans

Oceans affect climate in numerous ways. As explained in Chapter 16, currents transport heat. However, even in the absence of currents, the ocean stores and releases heat differently than land does. As a result, coastlines are generally warmer in winter and cooler in summer than continental interiors are.

Oceans also affect precipitation over coastal regions. In regions where warm oceans lie adjacent to cooler land, humid maritime air blows inland, leading to abundant precipitation (◀▶ Figure 20.1A). However, not all coastal areas are humid. Large portions of the west coast of South America, from southern Ecuador through northern Chile, contain some of the driest deserts in the world (◀▶ Figure 20.1B). These deserts exist because maritime air chilled by a cold Pacific Ocean current becomes warmer as it passes over the hot coasts of Ecuador, Peru, and Chile.

Altitude

Air is cooler at higher elevations, so mountainous regions and high plateaus are almost always colder than adjacent lowlands. Thus glaciers exist on Mount Kenya in Africa and Mount Cayambe in South America, even though both lie directly on the equator.

Mountains also alter the movement of air and generate local climate zones. In general, air rises on the windward side of a range, causing abundant precipitation. After it has passed over the mountains, it sinks and warms adiabatically, forming a rain-shadow desert on the downwind side of the mountains.

Albedo

Temperature not only depends on the amount of sunlight received but on the albedo, which is the amount reflected. If the albedo of a region is high, much of the solar energy is reflected out into space and the region will be cooler than a similar region with a lower albedo.

Climate As a Complex System

Many factors affect climate in interactive ways. Recall from Chapter 14 that the Gobi Desert and Pittsburgh are geographically similar in latitude and distance from the ocean, but the Gobi is a desert and central Pennsylvania has a moist climate (◀▶ Figure 20.2). Prevailing winds, mountain ranges, ocean currents, and many other factors combine to control the climate in any region.

◀▶ **FIGURE 20.1** **(A)** This Costa Rican rainforest lies along the Pacific coast of Central America at about 10° north latitude. **(B)** The Atacama Desert, Chile, at 20° south latitude. The contrasting climates and ecosystems result from different ocean currents and winds in the two regions.

As another example, high altitudes and latitudes are often glacier covered. In turn, glaciers have high albedo. But the albedo of white, sparkling, glacial ice reflects sunlight and leads to further cooling and the continued growth of glaciers. Feedback mechanisms such as the glacier–albedo interaction are important to climate and climate change and will be discussed further in Chapter 21.

A

B

Courtesy of Graham R. Thompson/Jonathan Turk

Martin Rogers/Getty Images

◆ **FIGURE 20.2** **(A)** The Gobi Desert and **(B)** central Pennsylvania lie at the same latitude and the same distance from the ocean. The two regions are each influenced by differences in prevailing winds and surrounding mountains.

INTERACTIVE QUESTION: *Locate three other regions with the same latitude and distance from the oceans as the Gobi Desert and central Pennsylvania. Record their climates using Figure 20.8. Explain what factors regulate each climate. (The map of ocean currents in Figure 16.12 may be helpful in answering this question.)*

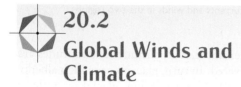

20.2
Global Winds and Climate

In 1735 a British meteorologist, George Hadley, reasoned that global winds are generated solely by the temperature difference between the equator and the poles. Hadley inferred that air is heated at the equator, rises, and then

flows poleward at high elevations. At the poles, it cools, sinks, and flows back toward the equator along Earth's surface (◆ Figure 20.3). Thus, Hadley hypothesized that the atmosphere circulates in two giant convection cells, one in the Northern Hemisphere and one in the Southern. According to this model, high-altitude winds in the Northern Hemisphere would flow from south to north and surface winds would flow predominately from north to south. However, when meteorologists mapped atmospheric circulation patterns, they did not find this pattern at all. For example, in the Northern Hemisphere the jet stream is an upper-level wind that blows the midlatitudes from west to east. Moreover the jet stream is not constant but writhes around and loops, like a writhing snake (◆ Figure 20.4). A Norwegian scientist, Carl Rossby, concluded that the Hadley model did not explain the jet stream because it did not account for Earth's rotation and the consequent Coriolis effect.

In the 1950s, global wind systems were modeled experimentally by climatologists at the University of Chicago. They mounted a circular pan on a variable-speed turntable and placed a heating element around the rim and a cooling coil at the center (◆ Figure 20.5). The pan represented Earth, with the rim analogous to the equator and the center analogous to the poles. They filled the pan with water and added dye to trace currents.

When the pan was stationary, water rose at the heated edge, traveled across the surface, sank at the cooled center, and returned to the edge along the bottom, thus forming a

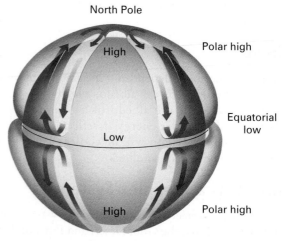

North Pole

Polar high

High

Equatorial low

Low

High

Polar high

South Pole

◆ **FIGURE 20.3** The Hadley model for global circulation patterns, proposed in 1735. It has been replaced by the three-cell model.

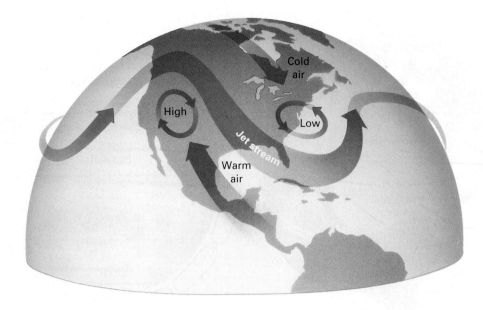

◆ **FIGURE 20.4** The Hadley model cannot explain upper-level winds that snake their way across the Northern Hemisphere.

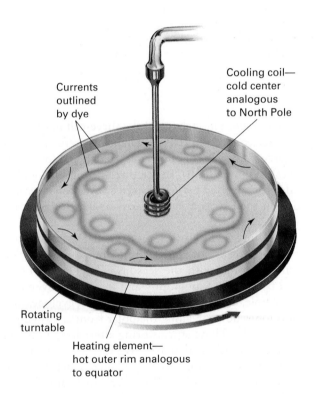

Currents outlined by dye

Cooling coil— cold center analogous to North Pole

Rotating turntable

Heating element— hot outer rim analogous to equator

◆ **FIGURE 20.5** An experimental model to demonstrate the Earth's wind patterns.

Hadley cell. When the pan rotated slowly, this current was deflected by the Coriolis effect, but the single-cell pattern was retained. However, when the scientists increased the rotational speed, the cell broke apart. Midway between the edge and the center—the area representing Earth's middle latitudes—the current diverged into whirls and eddies that looked very much like the storms observed in the middle latitudes. Today, scientists

repeat this experiment mathematically, on giant computers. They then compare the computer data with millions of direct measurements of natural wind patterns.

All climate models start with the observation that the Sun is the ultimate energy source for winds and evaporation. The Sun shines most directly at or near the equator and warms the air near Earth's surface. The warm air gathers moisture from the equatorial oceans. The warm, moist, rising air forms a vast region of low pressure near the equator, with little horizontal airflow. As the rising air cools adiabatically, the water vapor condenses and falls as rain. Therefore, local squalls and thunderstorms are common, but steady winds are rare. This hot, still region was a serious barrier in the age of sailing ships. Mariners called the equatorial region the **doldrums**, and the old sailing literature is filled with stories recounting the despair and hardship of being becalmed on the vast, windless seas. On land, the frequent rains near the equatorial low-pressure zone nurture lush tropical rainforests.

The air rising at the equator splits to flow north and south at high altitudes. However, these high-altitude winds do not continue to flow due north and south as Hadley predicted, because they are deflected by the Coriolis effect. Thus their poleward movement is interrupted.

In both the Northern and the Southern Hemispheres, this air veers until it flows due east at about 30° north and south latitudes (◆ Figure 20.6). The air then cools enough to sink to the surface, creating subtropical high-pressure zones at 30° north and south latitudes. The sinking air warms adiabatically, absorbing water and forming clear, blue skies. At the center of the high-pressure area, the air moves vertically and not horizontally, and therefore few steady surface winds blow. This calm, high-pressure belt circling the globe is called the **horse latitudes**.

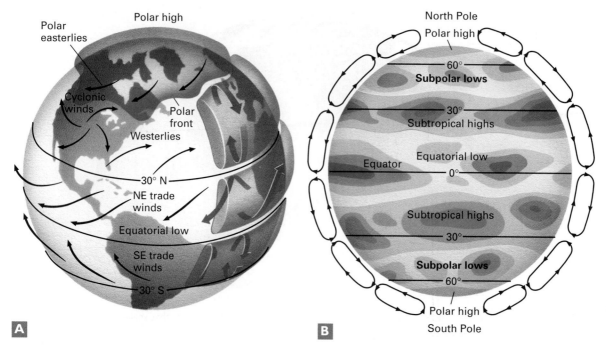

A   **B**

ThomsonNOW ◆ **FIGURE 20.6**   Global wind patterns predicted by the three-cell model. **(A)** Air rising at the equator moves poleward at high elevations, falls at about 30° north and south latitudes, and returns to the equator, forming trade winds. The orange arrows show both upper-level and surface wind patterns. The black arrows show only surface winds. **(B)** High- and low-pressure belts with surface and upper-level wind patterns shown on the edges of the sphere.

INTERACTIVE QUESTION: *Predict how the three-cell model would change if the Earth rotated twice as fast as it does now.*

The region was named because sailing ships were becalmed and horses transported as cargo often died of thirst and hunger. The warm, dry, descending air in this high-pressure zone forms many of the world's great deserts, including the Sahara in northern Africa, the Kalahari in southern Africa, and the Australian interior desert.

Descending air at the horse latitudes splits and flows over Earth's surface in two directions, toward the equator and toward the poles. The surface winds moving toward the equator are deflected by the Coriolis effect, so they blow from the northeast in the Northern Hemisphere and from the southeast in the Southern Hemisphere. Sailors depended on these reliable winds and called them the **trade winds**. The winds moving toward the poles are also deflected by the Coriolis effect, forming the **prevailing westerlies**. They flow from the southwest in the Northern Hemisphere and from the northwest in the Southern Hemisphere (◆ Figure 20.6A).

The poles are cold year-round. The cold polar air sinks, creating yet another band of high pressure. The sinking air flows over the surface toward lower latitudes. In the Northern Hemisphere these surface winds are deflected by the Coriolis effect to form the **polar easterlies**. The polar easterlies and prevailing westerlies converge at about 60° latitude. Warm air rises at the convergence, forming a low-pressure boundary zone called the **polar front**.

Recall that the Hadley model for global winds depicts one convection cell in each hemisphere. The more-accurate modern model, called the **three-cell model**, depicts three convection cells in each hemisphere. The cells are bordered by alternating bands of high and low pressure (◆ Figure 20.6B). In the three-cell model, global winds are generated by heat-driven convection currents, and then their direction is altered by Earth's rotation (Coriolis effect).

◆ Figure 20.6 shows stationary planar boundaries between the cells. In reality these boundaries migrate north and south with the seasons. They are also distorted by surface topography and local air movement. For example, in the Northern Hemisphere, cyclones and anticyclones develop along the polar front as explained in Chapter 19. These storms bring alternating rain and sunshine, conditions that are favorable for agriculture. Thus the great wheat belts of the United States, Canada, and Russia all lie between 30° and 60° north latitude.

Recall from Chapter 19 that a jet stream is a narrow band of fast-moving, high-altitude air. Jet streams form at boundaries between Earth's climate cells as high-altitude air is deflected by the Coriolis effect. The **subtropical jet stream** flows between the trade winds and the westerlies, and the **polar jet stream** forms along the polar front. When you watch a weather forecast on TV, the meteorologist may show the movement and direction of the polar jet stream as it snakes across North America. Storms commonly occur along this line because the jet stream marks the boundary between cold, polar air and the warm, moist, westerly flow that originates in the sub-

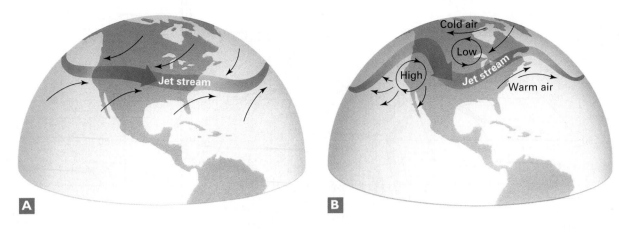

◈ **FIGURE 20.7** The polar front and the polar jet stream migrate with the seasons and with local conditions. Storms commonly occur along the jet stream.

tropics. The storms develop where the two contrasting air masses converge (◈ Figure 20.7).

20.3
Ocean Currents and Climate

Both London and central Newfoundland lie near the Atlantic Ocean at 51° north latitude, but Newfoundland has a polar climate whereas London is temperate. Although both regions receive the same amount of solar energy, the Gulf Stream and the North Atlantic Drift, related ocean currents that carry tropical water to northern Europe, warm London. Yet the Labrador Current flows from the North Pole southward to cool Newfoundland.

Winds and ocean currents transport heat, altering regional temperatures. In the absence of these heat transfers, the equatorial regions would be much warmer than they are today and the polar regions would be colder. Winds and ocean currents also affect rainfall. Recall that a cold ocean current creates the Atacama Desert along the west coast of South America. In contrast, warm currents bring frequent fog and rain to Scotland.

Vertical Mixing of Ocean Water

As the Gulf Stream pushes into the North Atlantic, both the temperature and salinity change so that the water becomes dense enough to sink. Much of this cold, dense water flows near the ocean floor all the way to Antarctica. The Gulf Stream water flowing northward is, on average, 8°C warmer than the deep water flowing southward and has a flow equal to 100 Amazon Rivers. Therefore, it transports immense amounts of heat.

Between 1100 and 1000 AD, Earth became immersed in a millennium-long cold snap called the Younger Dryas. Evidence from sea-floor sediment indicates that during this period, the North Atlantic deep current slowed down or stopped. When the deep current slowed, the circulation cell was interrupted and the Gulf Stream stopped, slowed, or veered southward.

As a result, heat transport to the North Atlantic region slowed. North America and Europe became frigid. At the same time, global temperatures dropped. Thus a change in ocean circulation caused a severe temperature change in one region. In turn, regional climate change altered winds or other marine currents to cause global change.

20.4
Climate Zones of Earth

Earth's major climate zones are classified primarily by temperature and precipitation. But an area with both wet and dry seasons has a different climate from one with moderate rainfall all year long, even though the two areas may have identical total annual precipitation.

Therefore, climatic zones are also classified on the basis of seasonal variations in temperature and precipitation. The **Koeppen Climate Classification**, used by climatologists throughout the world, defines five principal groups (Table 20.1).

Subclassifications of these five groups are shown in Table 20.2.

Although climate zones are defined by temperature and precipitation, you can often estimate climate types from a photograph of an area. Visual classification is possible because specific plant communities grow in specific climates. For instance, cactus grows in the desert, and trees grow where moisture is more abundant. A **biome** is a community of plants living in a large geographic area characterized by a particular climate. Following is a brief discussion of some of the major climate zones and their biomes (◈ Figure 20.8).

The climate in any location is summarized by a climograph that records annual and seasonal temperature and

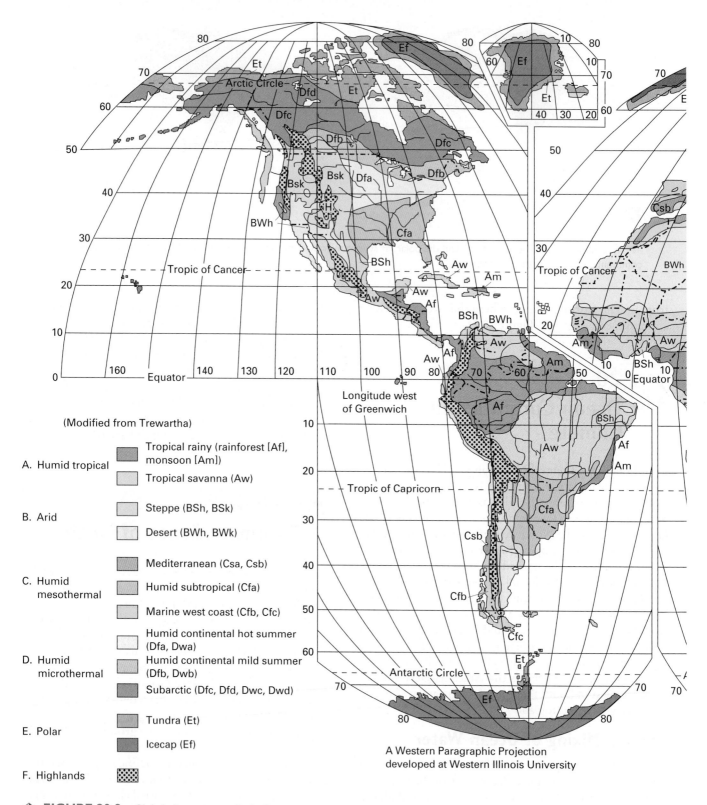

(Modified from Trewartha)

A. Humid tropical
- Tropical rainy (rainforest [Af], monsoon [Am])
- Tropical savanna (Aw)

B. Arid
- Steppe (BSh, BSk)
- Desert (BWh, BWk)

C. Humid mesothermal
- Mediterranean (Csa, Csb)
- Humid subtropical (Cfa)
- Marine west coast (Cfb, Cfc)

D. Humid microthermal
- Humid continental hot summer (Dfa, Dwa)
- Humid continental mild summer (Dfb, Dwb)
- Subarctic (Dfc, Dfd, Dwc, Dwd)

E. Polar
- Tundra (Et)
- Icecap (Ef)

F. Highlands

A Western Paragraphic Projection developed at Western Illinois University

◆ **FIGURE 20.8** Global climate zones. Each climate zone supports a unique biome.

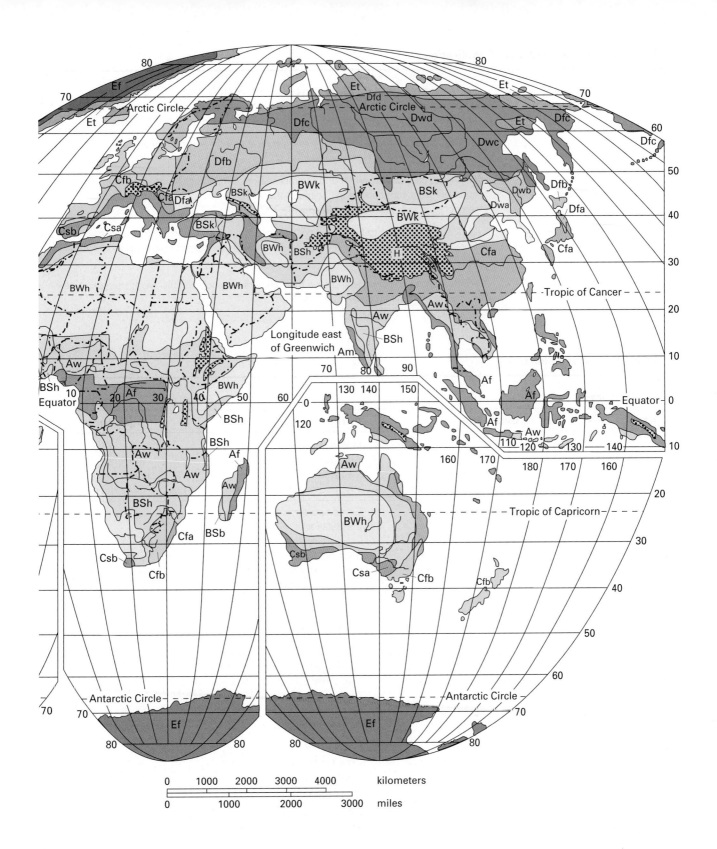

Table 20.1 Koeppen Climate Classification

A	Humid tropical	Every month is warm with a mean temperature over 18°C (64°F). The temperature difference between day and night is greater than the difference between December and June averages. There is enough moisture to support abundant plant communities.
B	Dry	B climates have a chronic water deficiency; in most months evaporation exceeds precipitation.
C	Humid midlatitudes with mild winters	C climates have distinct winter and summer seasons and enough moisture to support abundant plant communities. In C climates the winters are mild, with the average temperature in the coldest month above −3°C (27°F).
D	Humid midlatitudes with severe winters	D climates are similar to C climates, with distinct summer and winter seasons. However, D climates are colder, with the average temperature in the coldest month below −3°C (27°F).
E	Polar	Winters are extremely cold and even the summers are cool, with the average temperature in the warmest month below 10°C (50°F).

Table 20.2 World Climates

Symbol					
1st	2nd	3rd	Climate Type	Description	Vegetation
A				Humid tropical; no winter	
	f		Af = tropical rainforest	Abundant year-round rainfall	Rainforest
	m		Am = tropical monsoon	Wet with short dry season	Rainforest
	w		Aw = tropical savanna	Winter dry season	Grassland savanna
B				Dry; evaporation greater than precipitation	
	S		Steppe		
		h	BSh = tropical steppe	Semiarid	Steppe grassland
		k	BSk = midlatitude steppe	Cool and dry continental interior or rain-shadow location	Steppe grassland
	W			Desert	
		h	BWh = tropical desert	Warm	Desert
		k	BWk = midlatitude desert	Cool and dry; continental interior or rain-shadow location	Desert
C				Humid midlatitude; mild winter	
	f	a	Cfa = humid subtropical	Warm, wet summers with winter cyclonic storms	Forest
	s	a	Csa = Mediterranean	Dry summer and wet winter with cyclonic storms; subtropical	Chaparral; shrubby plants and oak savanna
	f	b	Cfb = marine west coast	Cool summer and mild winter; maritime, westerlies, and cyclonic storms	Forest to temperate rainforest
D				Humid midlatitude; severe winter	
	f	a	Dfa = humid continental; warm summer, cold winter	Midlatitude continental warm summer, cold winter	Prairie and forest
	f	b	Dfb = humid continental; cool summer, cold winter	High midlatitude continental	Prairie and forest
	f	c & d	Dfc/d = subarctic	High-latitude continental	Boreal forest taiga
E				Polar with little warmth, even in summer	Tundra

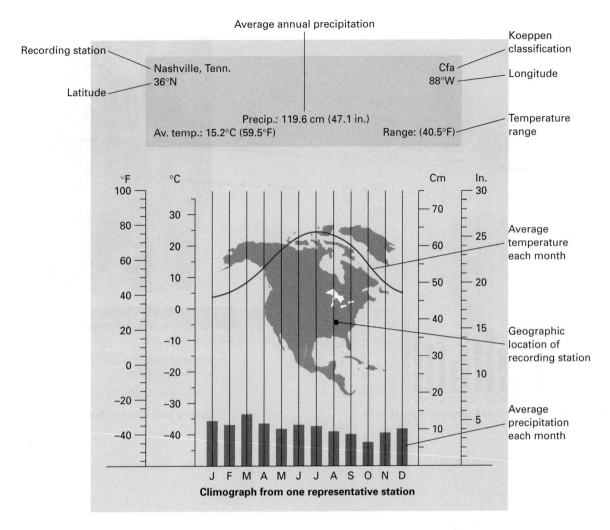

Recording station

Nashville, Tenn.
36°N

Latitude

Average annual precipitation

Precip.: 119.6 cm (47.1 in.)
Av. temp.: 15.2°C (59.5°F)

Koeppen classification

Cfa
88°W

Longitude

Range: (40.5°F)

Temperature range

Average temperature each month

Geographic location of recording station

Average precipitation each month

J F M A M J J A S O N D

Climograph from one representative station

◆ **FIGURE 20.9** A model climograph. This graph shows the average monthly temperature (curved line) and average monthly precipitation for Nashville, Tennessee. The labels highlight other features of the climograph.

precipitation. ◆ Figure 20.9 is a model climograph for Nashville, Tennessee.

A. Humid Tropical Climates; No Winter

Tropical Climates with Abundant Year-Round Rainfall

The large, low-pressure zone near the equator causes abundant rainfall, often exceeding 400 centimeters per year, which supports **tropical rainforests** (◆ Figure 20.10). The dominant plants in a tropical rainforest are tall trees with slender trunks. These trees branch only near the top, covering the forest with a dense canopy of leaves. The canopy blocks out most of the light, so as little as 0.1 percent of the sunlight reaches the forest floor, which consequently has relatively few plants. The ground in a tropical forest is soggy, the tree trunks are wet, and water drips everywhere.

Tropical Climates with Distinct Wet and Dry Seasons

Tropical monsoon (◆ Figure 20.11) and **tropical savanna** (◆ Figure 20.12) both have large, seasonal variations in rainfall, but a monsoon climate has greater total precipitation, greater monthly variation, and a shorter dry season. Precipitation is great enough in tropical-monsoon biomes to support rainforests. The seasonal precipitation is also ideal for agriculture. Some of the great rice-growing regions in India and Southeast Asia lie in tropical monsoon climates.

A tropical savanna is a grassland with scattered small trees and shrubs. Such grasslands extend over large areas, often in the interiors of continents, where rainfall is insufficient to support forests or where forest growth is prevented by recurrent fires. Savannas are most extensive in Africa, where they support a rich collection of grazing animals such as zebras, wildebeest, and gazelles. Grasses sprout and grow with the seasonal rains, and the great African herds migrate with the foliage.

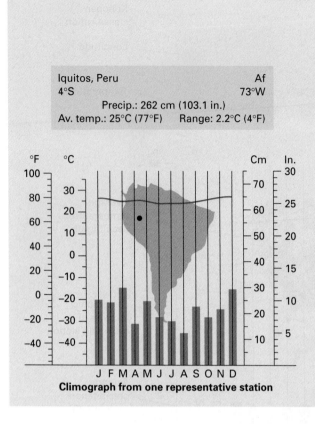

Iquitos, Peru   Af
4°S   73°W
Precip.: 262 cm (103.1 in.)
Av. temp.: 25°C (77°F)   Range: 2.2°C (4°F)

Climograph from one representative station

Tropical rainforest
Latitude: 0° to 15°
Av. temperature difference: 23°C to 28°C (winter to summer)
Av. annual precip.: 200 cm/yr. to 500 cm/yr.

General statistics for climate type

◈ **FIGURE 20.10**   Tropical rainforest.

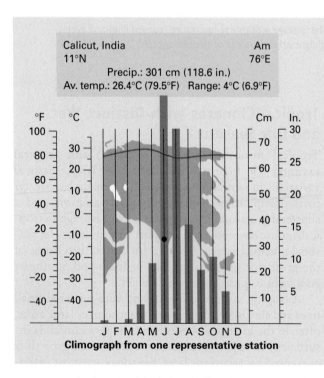

Calicut, India   Am
11°N   76°E
Precip.: 301 cm (118.6 in.)
Av. temp.: 26.4°C (79.5°F)   Range: 4°C (6.9°F)

Climograph from one representative station

Tropical monsoon
Latitude: 5° to 30°
Av. temperature difference: 20°C to 30°C (winter to summer)
Av. annual precip.: 150 cm/yr. to 400 cm/yr.

General statistics for climate type

◈ **FIGURE 20.11**   Tropical monsoon.

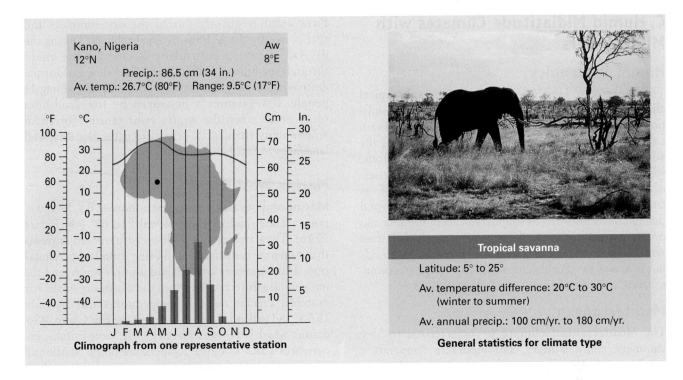

Kano, Nigeria   Aw
12°N   8°E
Precip.: 86.5 cm (34 in.)
Av. temp.: 26.7°C (80°F)   Range: 9.5°C (17°F)

Climograph from one representative station

Tropical savanna

Latitude: 5° to 25°

Av. temperature difference: 20°C to 30°C
(winter to summer)

Av. annual precip.: 100 cm/yr. to 180 cm/yr.

General statistics for climate type

◆ **FIGURE 20.12**   Tropical savanna.

B. Dry Climates; Evaporation Greater than Precipitation

In dry zones where the annual precipitation varies from 25 to 35 centimeters per year, the climate is semiarid and grassland **steppes** predominate. The great steppe grasslands of Central Asia fall into these categories.

If the rainfall is less than 25 centimeters per year, **deserts** form and support only sparse vegetation (◆ Figure 20.13). The world's largest deserts lie along the high-pressure zones of the 30° latitude, although rain-shadow and coastal deserts exist in other latitudes. Thus some deserts are torridly hot, while Arctic deserts are frigid for much of the year.

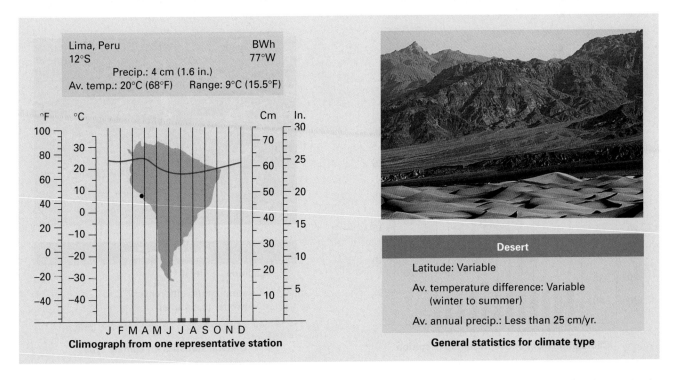

Lima, Peru   BWh
12°S   77°W
Precip.: 4 cm (1.6 in.)
Av. temp.: 20°C (68°F)   Range: 9°C (15.5°F)

Climograph from one representative station

Desert

Latitude: Variable

Av. temperature difference: Variable
(winter to summer)

Av. annual precip.: Less than 25 cm/yr.

General statistics for climate type

◆ **FIGURE 20.13**   Desert.

C. Humid Midlatitude Climates with Mild Winters

Humid Subtropics

The southeastern United States has a **humid subtropical climate** (◆ Figure 20.14). During the summer, conditions can be as hot and humid as in the tropics, and rain and thundershowers are common. However, during the winter, arctic air pushes southward, forming cyclonic storms. Although the average monthly temperature seldom falls below 7°C, cold fronts occasionally bring frost and snow. Precipitation is relatively constant year-round due to convection-driven thunderstorms in summer and cyclonic storms in winter. This zone supports both trees with needles (conifers) and trees with broad leaves (deciduous) as well as valuable crops such as vegetables, cotton, tobacco, and citrus fruits.

Mediterranean

The **Mediterranean climate** is characterized by dry summers, rainy winters, and moderate temperature (◆ Figure 20.15). These conditions occur on the west coasts of all continents between latitudes 30° and 40°. In summer the subtropical high migrates to higher latitudes, producing near-desert conditions with clear skies as much as 90 percent of the time. In winter the prevailing westerlies bring warm, moist air from the ocean, leading to fog and rain. Thus, 75 percent or more of the annual rainfall occurs in winter. Although redwoods, the largest trees on Earth, grow in specific environments in central California, the summer heat and drought of Mediterranean climates generally retard the growth of large trees. Instead, shrubs and scattered trees dominate.

Fires occur frequently during the dry summers and spread rapidly through the dense shrubbery. During the dry fall of 1991, a particularly devastating fire swept through the hills of Oakland and Berkeley, California, destroying 3,354 homes worth $1.5 billion and killing 24 people. If vegetation is destroyed by fire, landslides often occur when the winter rains return. Torrential winter rains frequently bring extensive flooding and landslides to Southern California.

Marine West Coast

Marine west coast climate zones border the Mediterranean zones and extend poleward to 65° (◆ Figure 20.16). They are severely influenced by ocean currents that moderate temperature and bring abundant precipitation. Thus summers are cool and winters warm, and the temperature difference between the seasons is small. For example, average monthly temperatures vary by only 15.5°C in Portland, Oregon. In contrast, Eau Claire, Wisconsin, which is a continental city at the same latitude, experiences a 31.5°C annual temperature range. Seattle and other northwestern coastal cities experience rain and drizzle for days at a time, especially during the winter, when the warm, moist, maritime air from the Pacific flows first over cool currents close to shore and then over cool land surfaces. The total rainfall varies from moderate, 50 centimeters per year (20 in./yr.), to wet, 250 centimeters per year (100 in./yr.). The wettest climates occur where mountains interrupt the maritime air. **Temperate rainforests** grow where rainfall is greater than 100 centimeters per year and is constant throughout the year. Temperate rainforests are common along the northwest coast of North America, from Oregon to Alaska.

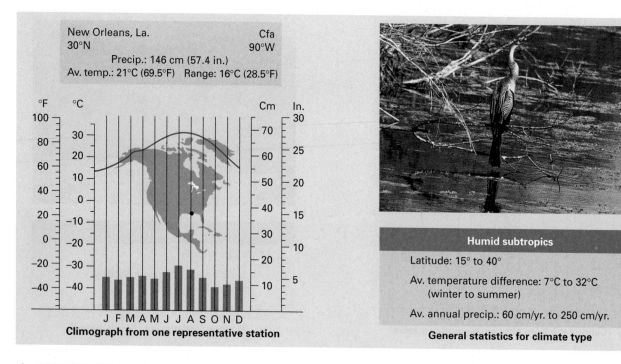

New Orleans, La.   Cfa
30°N   90°W
Precip.: 146 cm (57.4 in.)
Av. temp.: 21°C (69.5°F)   Range: 16°C (28.5°F)

Climograph from one representative station

Humid subtropics

Latitude: 15° to 40°

Av. temperature difference: 7°C to 32°C (winter to summer)

Av. annual precip.: 60 cm/yr. to 250 cm/yr.

General statistics for climate type

◆ **FIGURE 20.14**   Humid subtropics.

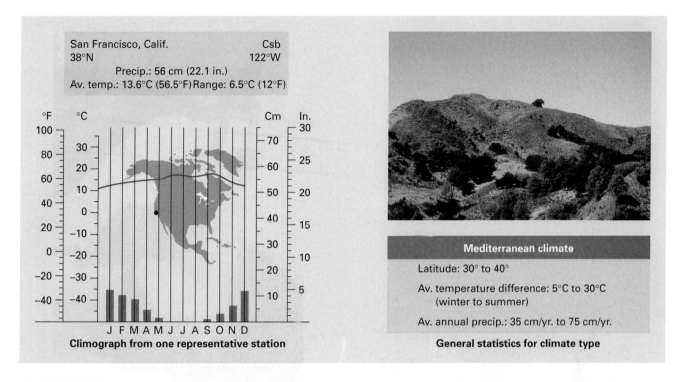

◈ FIGURE 20.15 Mediterranean climate.

San Francisco, Calif.   Csb
38°N   122°W
Precip.: 56 cm (22.1 in.)
Av. temp.: 13.6°C (56.5°F) Range: 6.5°C (12°F)

Climograph from one representative station

Mediterranean climate
Latitude: 30° to 40°
Av. temperature difference: 5°C to 30°C (winter to summer)
Av. annual precip.: 35 cm/yr. to 75 cm/yr.

General statistics for climate type

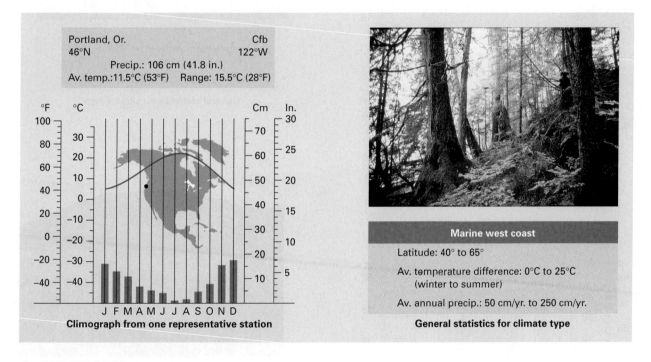

◈ FIGURE 20.16 Marine West Coast.

Portland, Or.   Cfb
46°N   122°W
Precip.: 106 cm (41.8 in.)
Av. temp.: 11.5°C (53°F)   Range: 15.5°C (28°F)

Climograph from one representative station

Marine west coast
Latitude: 40° to 65°
Av. temperature difference: 0°C to 25°C (winter to summer)
Av. annual precip.: 50 cm/yr. to 250 cm/yr.

General statistics for climate type

D. Humid Midlatitude Climate with Severe Winters

Continental interiors in the midlatitudes are characterized by hot summers and cold winters, giving rise to **humid continental climates** (◈ Figure 20.17). Thus in the northern Great Plains the temperature can drop to −40°C in winter and soar to 38°C in summer. Even in a given season, the temperature may vary greatly as the polar front moves northward or southward. For example, in winter the northern continental United States may experience arctic cold one day and rain a few days later. If rainfall is sufficient, this climate supports abundant coniferous forests, whereas grasslands dominate the drier regions. Millions of bison once roamed the continental grasslands of North America, and today wheat

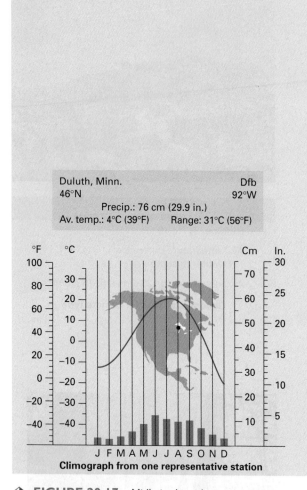

Duluth, Minn.   Dfb
46°N   92°W
          Precip.: 76 cm (29.9 in.)
Av. temp.: 4°C (39°F)   Range: 31°C (56°F)

Climograph from one representative station

Midlatitude with severe winters

Latitude: 40° to 65°

Av. temperature difference: –15°C to 25°C
(winter to summer)

Av. annual precip.: 50 cm/yr. to 150 cm/yr.

General statistics for climate type

◈ **FIGURE 20.17**   Midlatitude with severe winters.

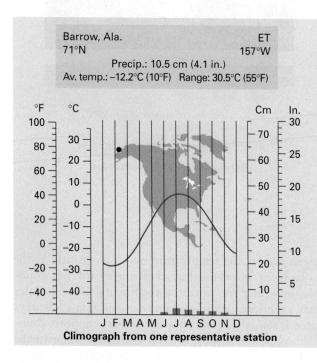

Barrow, Ala.   ET
71°N   157°W
          Precip.: 10.5 cm (4.1 in.)
Av. temp.: –12.2°C (10°F)   Range: 30.5°C (55°F)

Climograph from one representative station

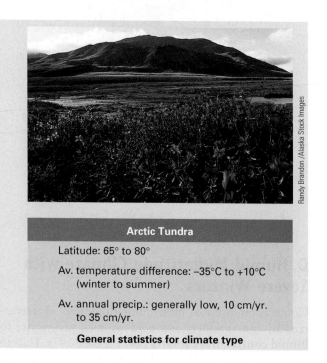

Arctic Tundra

Latitude: 65° to 80°

Av. temperature difference: –35°C to +10°C
(winter to summer)

Av. annual precip.: generally low, 10 cm/yr.
to 35 cm/yr.

General statistics for climate type

◈ **FIGURE 20.18**   Arctic tundra.

and other grains grow from horizon to horizon. The northernmost portion of this climate zone is the **subarctic**, which supports the taiga biome. **Taiga** consists of conifers that survive extremely cold winters.

E. Polar Climate

In the Arctic, winters are harsh and long, and the temperature remains above freezing only during a short summer. Trees cannot survive, and low-lying plants such as mosses, grasses, flowers, and a few small bushes cover the land. This biome is called **tundra** (◆ Figure 20.18).

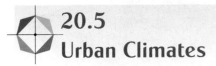

20.5
Urban Climates

If you ride a bicycle from the center of a city toward the countryside, you may notice that the air gradually feels cooler and more refreshing as you leave the city streets and enter the green fields or hills of the outlying area. This feeling is not entirely psychological; the climate of a city is measurably different from that of the surrounding rural regions (◆ Table 20.3).

As shown in ◆ Figure 20.19, the average winter minimum temperature in the center of Washington, D.C., is more than 3°C warmer than that in outlying

Table 20.3 Average Changes in Climatic Elements Caused by Urbanization

Element	Comparison with Rural Environment
Cloudiness:	
Cover	5–10% more
Fog—winter	100% more
Fog—summer	30% more
Precipitation, total	5–10% more
Relative humidity:	
Winter	2% lower
Summer	8% lower
Radiation:	
Total	15–20% less
Direct sunshine	5–15% less
Temperature:	
Annual mean	0.5–1.0°C higher
Winter minimum (average)	1.0–3.0°C higher
Wind speed:	
Annual mean	20–30% lower
Extreme gusts	10–20% lower
Calms	5–20% higher

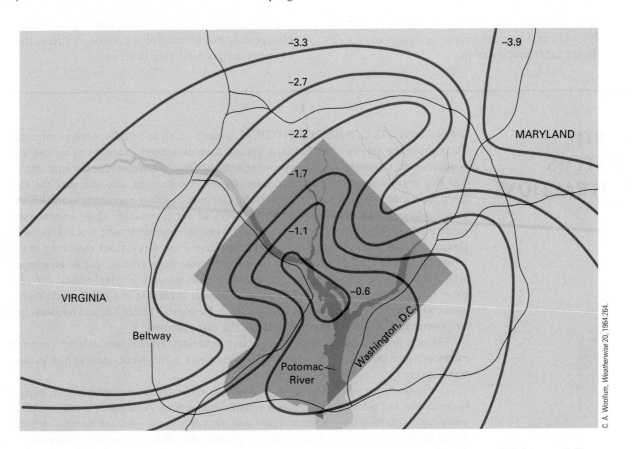

◆ **FIGURE 20.19** The urban heat island effect. The average minimum temperature in and around Washington, D.C., during the winter.

areas. This difference, called the **urban heat island effect**, is caused by numerous factors:

- Stone and concrete buildings and asphalt roadways absorb solar radiation and reradiate it as infrared heat.
- Cities are warmer because little surface water exists and, as a result, little evaporative cooling occurs. In contrast, in the countryside, water collects in the soil and evaporates for days after a storm. Roots draw water from deeper in the soil, and this water evaporates from leaf surfaces.
- Urban environments are warmed by the heat released when fuels are burned. In New York City in winter, the combined heat output of all the vehicles, buildings, factories, and electrical generators is 2.5 times the solar energy reaching the ground.
- Tall buildings block winds that might otherwise disperse the warm air.
- Air pollutants absorb long-wave radiation (heat rays) emitted from the ground and produce a local greenhouse effect.

As warm air rises over a city, a local low-pressure zone develops, and rainfall is generally greater over the city than in the surrounding areas (◈ Figure 20.20). Water condenses on dust particles, which are abundant in polluted, urban air. Weather systems collide with the city buildings and linger, much as they do on the windward side of mountains. Thus, a front that might pass quickly over rural farmland remains longer over a city and releases more precipitation.

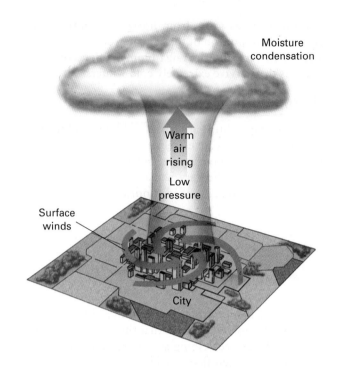

◈ **FIGURE 20.20**   Warm air rising over a city creates a low- pressure zone. As a result, precipitation is greater in the city than over the surrounding countryside.

In 1600, less than 1 percent of the global population lived in cities. By 1950, 30 percent of the world's population was urban, and by 2000 that ratio had grown to 55 percent. Therefore, although urban climate change may not affect global climate, it affects the lives of the many people living in cities.

EARTH SYSTEMS INTERACTIONS

ENVIRONMENTS CONSIST OF PHYSICAL factors—such as latitude, wind, oceans, altitude. Throughout the course of evolution, organisms have developed strategies to survive in those environments, or moved to other environments, or perished. However, scientists also recognize that organisms not only adapt to their physical environment, they also alter the environment to fit their needs. A recent book, *Niche Construction*, offers numerous examples of feedback loops in which organisms or communities of organisms alter their environments to construct favorable habitats for themselves. For example, cyanobacteria that live in the Negev Desert of Israel secrete sugary substances that form small depressions consisting of circular crusts of sand and soil, protecting the bacterial colonies from erosion. But when sparse rains arrive, water collects in these tiny pools, which then provide an ideal habitat for seeds of other plants to germinate. In turn, the plants provide shade and the roots retain even more water, which makes the soil even more hospitable for plants. Thus the desert becomes greener and wetter than it would have been without the cyanobacteria.

Humans are the penultimate niche constructionists. We irrigate deserts, turn tropical rainforests into banana plantations, build heated houses in the Arctic, and live high in the air in skyscrapers.

In this textbook, we describe niche construction as biosphere–atmosphere, biosphere–hydrosphere, and biosphere–geosphere systems interactions. Whatever the title, for billions of years organisms have been a factor in the evolution of planetary environments—including the atmosphere–climate system. In the following chapter, we will learn how humans are altering the composition of the modern atmosphere—and will try to understand the implications of that change for life on Earth. ∎

EARTH SYSTEMS INTERACTIONS

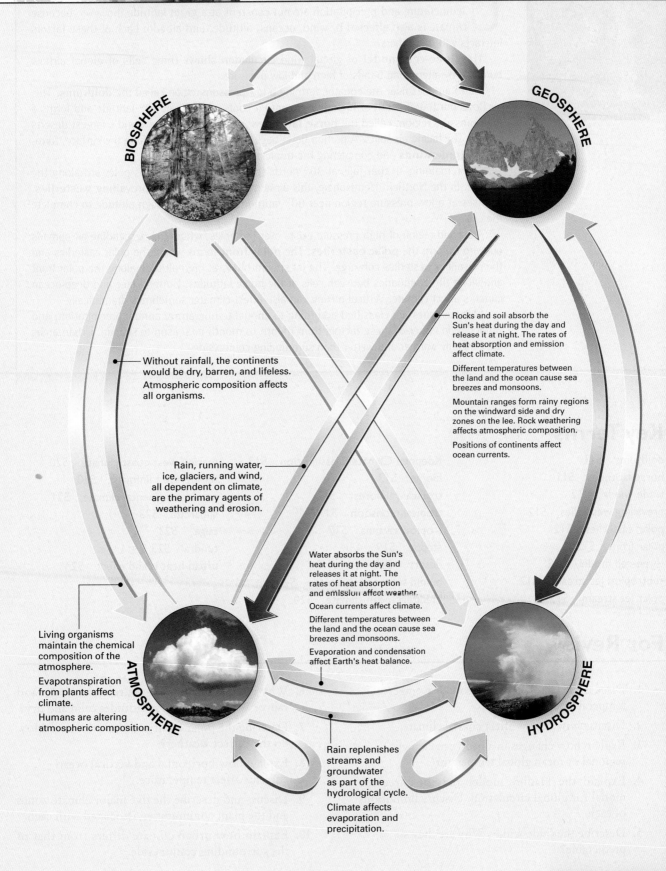

BIOSPHERE

GEOSPHERE

Without rainfall, the continents would be dry, barren, and lifeless. Atmospheric composition affects all organisms.

Rocks and soil absorb the Sun's heat during the day and release it at night. The rates of heat absorption and emission affect climate.

Different temperatures between the land and the ocean cause sea breezes and monsoons.

Mountain ranges form rainy regions on the windward side and dry zones on the lee. Rock weathering affects atmospheric composition.

Positions of continents affect ocean currents.

Rain, running water, ice, glaciers, and wind, all dependent on climate, are the primary agents of weathering and erosion.

Water absorbs the Sun's heat during the day and releases it at night. The rates of heat absorption and emission affect weather.

Ocean currents affect climate.

Different temperatures between the land and the ocean cause sea breezes and monsoons.

Evaporation and condensation affect Earth's heat balance.

Living organisms maintain the chemical composition of the atmosphere.

Evapotranspiration from plants affect climate.

Humans are altering atmospheric composition.

ATMOSPHERE

HYDROSPHERE

Rain replenishes streams and groundwater as part of the hydrological cycle.

Climate affects evaporation and precipitation.

SUMMARY

Ultimately, the Sun controls Earth's climate; seasonal sunlight is dependent on latitude. Temperature and precipitation are not constant at a given latitude, however, because climate is also affected by wind, oceans, altitude, and albedo. Each of these factors interacts with the others.

The **three-cell model** of global wind circulation shows three cells of global airflow bordered by alternating bands of high and low pressure.

Warm air rises near the equator, forming a low-pressure region called the **doldrums**. This air flows north and south at high altitude until it cools and falls at 30° latitude and forms a high-pressure region, called the **horse latitudes**, where most of the world's largest deserts prevail. The falling air splits. A portion flows back toward the equator along the surface, forming the **trade winds** and completing the tropical cell.

The remaining air that falls at 30° north and south latitudes flows poleward along the surface. In the Northern Hemisphere, this air is deflected to form the **prevailing westerlies**. Air rises at a low-pressure region near 60° latitude and returns at high altitude to complete the cell.

A second region of high pressure exists over the poles, where the descending air spreads outward to form the **polar easterlies**. The **polar front** forms where the polar easterlies and the prevailing westerlies converge. The jet stream blows at high altitude along the polar front and along the boundaries between cells at the horse latitudes. Both surface and deep-ocean currents affect climate. Vertical mixing can also affect climate, sometimes dramatically.

Climate zones are classified according to annual temperature, annual precipitation, and variability in either of these factors from month to month or season to season. Urban areas are generally warmer and wetter than surrounding countryside.

Key Terms

doldrums 511
horse latitudes 511
trade winds 512
prevailing westerlies 512
polar easterlies 512
polar front 512
three-cell model 512
subtropical jet stream 512
polar jet stream 512

Koeppen Climate Classification 513
biome 513
tropical rainforest 517
tropical monsoon 517
tropical savanna 517
steppe 519
desert 519
humid subtropical climate 520
Mediterranean climate 520

marine west coast climate 520
temperate rainforest 520
humid continental climate 521
subarctic 523
taiga 523
tundra 523
urban heat island effect 524

For Review

1. Discuss the factors that control the average temperature in any region.

2. Discuss how oceans affect coastal climate.

3. Explain how changes in albedo can change regional or mean global temperature.

4. Explain the Hadley model and the three-cell model for global circulation. Discuss limitations of each.

5. Describe the trade winds. Why are they so predictable?

6. Why is the doldrums region relatively calm and rainy? Why are the horse latitudes calm and dry?

7. Describe the polar front and the jet stream. How do they affect weather?

8. Explain how horizontal and vertical ocean currents affect temperature.

9. Discuss and describe the five major climate zones and the plant communities associated with each.

10. Explain how urban climate differs from that of the surrounding countryside.

For Discussion

1. Give an example of how winds and precipitation are affected by (a) altitude and (b) ocean currents.

2. Would the exact location of the low-pressure area in the doldrums be likely to change from month to month? From year to year? Explain.

3. Sailors traveling in the Northern Hemisphere expect to incur predictable winds from the northeast between latitudes about 5° north and 30° north. Should airplane pilots expect northeast trade winds while flying at high altitudes in the same region? Explain.

4. Why doesn't the air that is heated at the equator continue to rise indefinitely?

5. Explain how changes in the temperature in the surface water can lead to rapid vertical mixing in the ocean. Explain why this mechanism can operate as a threshold effect.

6. Predict the climate at the following locations: (a) at 45° north latitude, 200 kilometers from the ocean, on the leeward side of a mountain range; (b) at 45° north latitude, on a coastline influenced by warm currents; (c) on the equator in a continental interior; (d) at 30° north latitude in a continental interior; (e) at 30° north latitude, influenced by warm currents.

CHAPTER 21

Climate Change

© Blickwinkel/Alamy

A dead giraffe thorn acacia in Namib Naukluft Park, North Africa, stands as silent testimony to recent expansion of the desert environment.

In March 2000, an 11,000-square-kilometer iceberg as large as the state of Connecticut broke free from the Ross Ice Shelf in Antarctica and drifted north, where it melted in warmer water. Two months later, a massive chunk of ice cracked off the nearby Ronne Ice Shelf. In September, a second, Connecticut-sized chunk of the Ross Ice Shelf disintegrated. Then in January of 2002, a Rhode Island–sized section of the Larsen Ice Shelf splintered into millions of small fragments ($\triangleright$ Figure 21.1). Since the height of the last Ice Age, 18,000 years ago, the Ross Ice Shelf has receded 700 kilometers, shedding 5.3 million cubic kilometers of ice. And, as evidenced by the recent disintegration, the process continues to this day.

ThomsonNOW™ Throughout this chapter, the ThomsonNOW logo indicates an opportunity for online self-study, which:
• Assesses your understanding of important concepts and provides a personalized study plan
• Links you to animations and simulations to help you study
• Helps to test your knowledge of material and prepare for exams
Visit ThomsonNOW at www.thomsonedu.com/login to access these resources.

In 2003, Switzerland witnessed its hottest June in 250 years. Two months later, the temperature in France soared to 40°C (104°F) and remained at or above record-breaking levels for several weeks. Health workers estimate that the intense heat killed 30,000 people across central Europe.

During the past 100 years, the average temperature of Earth's atmosphere has risen by a little less than 0.8°C. It is tempting to make the seemingly obvious conclusions that warm air has melted sections of the Antarctic ice and has directly caused the torrid European summer of 2003. But associations between cause and effect must be made carefully, with conclusions backed by scientific data. Earth's climate system is complicated by many opposing processes, threshold effects, and feedback mechanisms.

In order to have a balanced perspective on what is happening today, we must first study historical climate change. The geological record in almost every locality on Earth provides evidence that past regional climates were different from modern climates. Geologists have discovered sand dunes beneath prairie grasslands near Denver, Colorado, indicating that this semiarid region was recently desert. Moraines on Long Island, New York, tell us that this temperate region was once glaciated. Fossil ferns in nearby Connecticut indicate that, before the glaciers, the northeastern United States was warm and wet.

Many of these regional climatic fluctuations resulted from global climate change. Thus, 18,000 years ago, Earth was cooler than it is today and glaciers descended to lower latitudes and altitudes. During Mississippian time, from 360 to 325 million years ago, Earth was warmer than it is today. Vegetation grew abundantly, and some of it collected in huge swamps to form coal.

During the past 10,000 years, global climate has been mild and stable as compared with the preceding 100,000 years. During this time, humans have developed from widely separated bands of hunter-gatherers, to agrarian farmers, and then moved into crowded communities in huge industrial megalopolises. Today, with a global population of more than 6 billion, people have stressed the food-producing capabilities of the planet. If temperature or rainfall patterns were to change, even slightly, crop failures could lead to famines. In addition, cities and farmlands on low-lying coasts could be flooded under rising sea level. As a result, climate change may be one of the most important issues that we face in the twenty-first century. ■

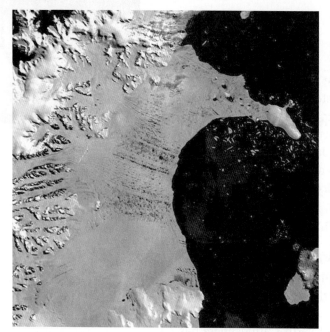

January 31, 2002

February 23, 2002

National Snow and Ice Data Center

March 5, 2002

◆ **FIGURE 21.1** The Larsen B ice shelf rims a small portion of the eastern coastline of the Antarctic Peninsula. Between January 31 and March 5 of 2002, 3,250 square kilometers of ice, about the size of the state of Rhode Island, broke off the ice shelf and floated northward.

21.1
Climate Change
in Earth's History

Recall from Chapter 17 that Earth's primordial atmosphere contained high concentrations of carbon dioxide (CO_2) and water vapor (H_2O). Both of these greenhouse gases absorb infrared radiation in the atmosphere. Astronomers have calculated that the Sun was 20 to 30 percent fainter early in Earth's history than it is today. Yet oceans did not freeze. The high concentrations of atmospheric carbon dioxide and water vapor retained enough of the Sun's lesser radiation to warm Earth's atmosphere and surface to temperatures that kept the oceans liquid. Luckily for us, the concentration of carbon dioxide and water in the atmosphere declined gradually as the Sun warmed.

◆ Figure 21.2 is a graph of mean global temperature and precipitation throughout Earth's history. Note, for example, that the planet plunged into a deep-freeze on at least five occasions. There is evidence that the cold

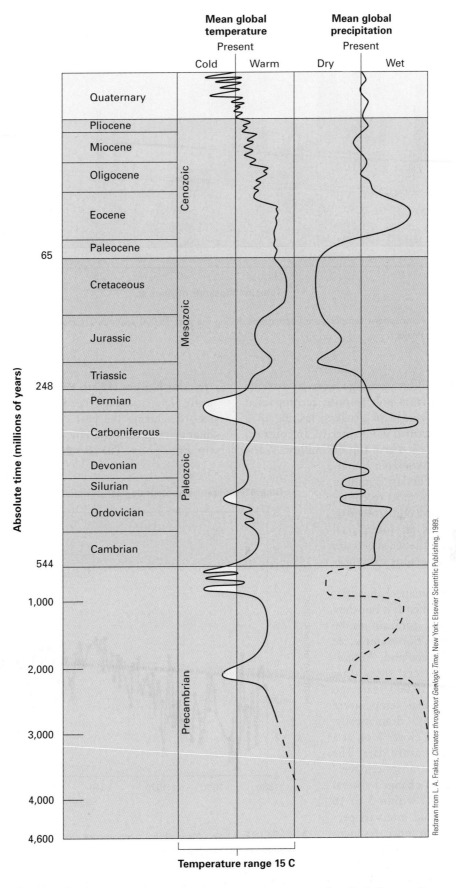

Mean global temperature

Present

Cold | Warm

Mean global precipitation

Present

Dry | Wet

Temperature range 15 C

Absolute time (millions of years)

65

248

544

1,000

2,000

3,000

4,000

4,600

Quaternary

Pliocene

Miocene

Oligocene

Eocene

Paleocene

Cenozoic

Cretaceous

Jurassic

Triassic

Mesozoic

Permian

Carboniferous

Devonian

Silurian

Ordovician

Cambrian

Paleozoic

Precambrian

Redrawn from L. A. Frakes, *Climates throughout Geologic Time*. New York: Elsevier Scientific Publishing, 1989.

◈ **FIGURE 21.2** Mean global temperature and precipitation have both fluctuated throughout Earth's history.

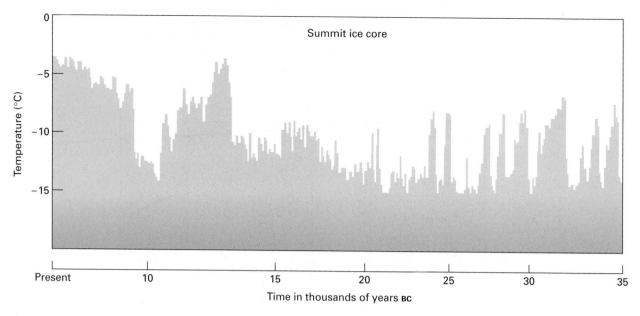

FIGURE 21.3 Atmospheric temperature fluctuations during the past 35,000 years in Greenland. The data were collected by the Greenland Ice Core Project.

periods 700 to 600 million years ago were so extreme that the oceans froze from pole to pole, accompanied by near-total land coverage of glaciers, leading to a "Snowball Earth," described in Chapter 13. In contrast, our planet was relatively warm for 248 million years, from the start of the Mesozoic Era almost to the present. The last 2 million years have witnessed the most recent ice age. According to many climatologists, the current period is an interglacial warming episode and the ice sheets are likely to return in the geologically near future.

Many additional climate changes occurred during Earth's history, but they were of too short a duration to be apparent on this graph. For example, ice core records from Greenland, Antarctica, and alpine glaciers show that between 110,000 and 10,000 years ago the mean annual global temperature changed frequently and dramatically. ◆ Figure 21.3 gives a closer look at a portion of this temperature data. The graph shows that the atmospheric temperature in Greenland changed several times by 5°C to 10°C within 5 to 10 years. Many of the cold intervals persisted for 1,000 years or more. As mentioned earlier, the past 10,000 years, during which civilization developed, witnessed anomalously stable climate.

◆ Figure 21.4 gives us an even more detailed look at a 125-year span from 1880 to 2005. Note that the temperature rose slowly from 1880 to 1970 and then increased dramatically. The last year recorded, 2005, was the hottest on record. During the past century, people burned large quantities of fuel, thereby injecting carbon dioxide into the atmosphere. The correlation between carbon dioxide

Global Temperature (meteorological stations)

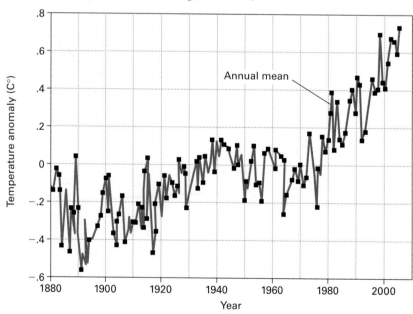

FIGURE 21.4 Mean global temperature changes from 1880 to 2005. The zero line represents the average from 1951 to 1980, and plus or minus values represent deviations from the average.

INTERACTIVE QUESTION: *Compare the magnitudes of some of the large temperature changes in Figure 21.3 with the temperature change shown in Figure 21.4. Discuss the comparative magnitudes of possible human-induced climate change during the past century with past examples of natural climate change.*

emission and global temperature implies—but does not prove—that human activities have caused the current global warming. The temperature rise has been a little less than 0.8°C. Temperature changes much larger than this have occurred repeatedly throughout Earth's history. This chapter is devoted to understanding and interpreting this 125-year warming trend.

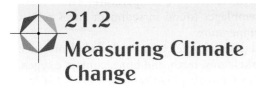

21.2 Measuring Climate Change

For the past 100 years, meteorologists have used instruments to measure temperature, precipitation, wind speed, and humidity. But how do we interpret prehistoric climates? ◆ Figure 21.5 reviews several techniques for determining past climate.

Historical Records

Historians search for written records or archaeological data that chronicle climate change. In 985 CE the Viking explorer Eric the Red sailed to southwest Greenland with a few hundred immigrants. They established two colonies, exported butter and cheese to Iceland and Europe, and the population flourished. Some Vikings sailed farther west, colonized the Labrador Coast of North America, and visited Ellesmere Island, near 80° north latitude. Then within 300 to 400 years, the colonies vanished. Sagas tell of heavy sea ice in summer, crop failures, starvation, and death.

During the same time, European glaciers descended into lowland valleys. This period of global cooling, called the Little Ice Age, lasted from about 1450 to 1850 CE and is documented by old landscape paintings and writings that depict a glacial advance between the fifteenth and nineteenth centuries. Other historical and archaeological evidence chronicles climate changes at different times and places.

Tree Rings

Growth rings in trees also record climatic variations. Each year, a tree's growth is recorded as a new layer of wood called a tree ring. Trees grow slowly during a cool, dry year and more quickly in a warm, wet year; therefore, tree rings grow wider during favorable years than during unfavorable ones. Paleoclimatologists date ancient logs preserved in ice, permafrost, or glacial till by carbon-14 techniques to determine when trees died. They then count and measure the rings to reconstruct the history of past climate recorded in the wood. Interpretations of climate change from tree-ring data coincide well with historical data. For example, growth rings are narrow in trees that lived during the Little Ice Age.

Plant Pollen

Plant pollen is widely distributed by wind and is coated with a hard, waxy cover that resists decomposition. As a result, pollen grains are abundant and well preserved in sediment in lake bottoms and bogs. For example, 11,000 years ago, spruce was the most abundant tree species in a Minnesota bog. In modern forests, spruce dominates in colder Canadian climates but is less abundant in Minnesota. Therefore, scientists deduce that the climate in Minnesota was colder 11,000 years ago than it is at present. Pollen in younger layers of sediment shows that about 10,500 years ago, pines displaced the spruce, indicating that the temperature became warmer.

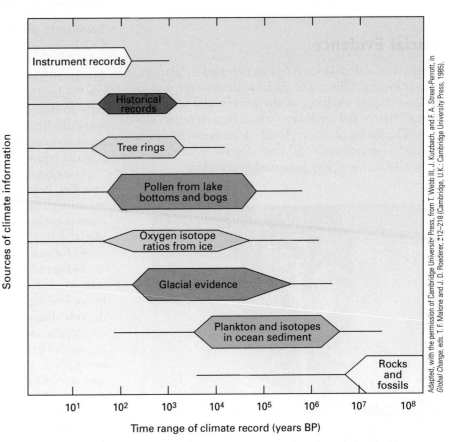

◆ **FIGURE 21.5** Several methods, each with its own useful time range, allow scientists to determine historical and ancient climates.

Adapted, with the permission of Cambridge University Press, from T. Webb III, J. Kutzbach, and F. A. Street-Perrott, in *Global Change*, eds. T. F. Malone and J. D. Roederer, 212–218 (Cambridge, U.K.: Cambridge University Press, 1985).

Oxygen Isotope Ratios in Glacial Ice

Oxygen consists mainly of two isotopes, abundant ^{16}O and rare ^{18}O. Both isotopes are incorporated into water, H_2O. Water molecules containing ^{16}O are lighter and evaporate more easily than those containing ^{18}O. At high temperature, however, evaporating water vapor contains a higher proportion of ^{18}O than it does at lower temperatures. Therefore, the ratio of $^{18}O/^{16}O$ in vapor from warm water is higher than that in vapor from cool water. Some of the water vapor condenses as snow, which accumulates in glaciers. Thus the $^{18}O/^{16}O$ ratios in glacial ice reflect water temperature at the time the water evaporated. Because most of the atmospheric water vapor that falls as snow originated from evaporation of ocean water, scientists then use the $^{18}O/^{16}O$ data from glacial ice to estimate mean ocean surface temperatures. Because the sea surface and the atmosphere are in close contact, mean ocean surface temperature reflects mean global atmospheric temperature.

Geologists have drilled deep into Greenland and Antarctic glaciers, where the ice is up to 110,000 years old, and have carefully removed ice cores (◆ Figure 21.6). The age of the ice at any depth is determined by counting annual ice deposition layers or by carbon-14 dating of windblown pollen within the glacier. The oxygen isotope ratios in each layer reflect the air temperature at the time the snow fell. The temperature data in ◆ Figure 21.3 were obtained from Greenland ice cores.

Glacial Evidence

Erosional and depositional features created by glaciers, such as the tills, tillites, and glacial striations described in Chapter 13, are evidence of the growth and retreat of alpine glaciers and ice sheets, which in turn reflect climate. The timing of recent glacial advances can be determined by several methods. One effective technique is carbon-14 dating of logs preserved in glacial till.

◆ **FIGURE 21.6** Scientists remove an ice core from a glacier in Greenland. Studies of ancient ice provide information about past climate.

R. Gaillarde/Getty Images

Plankton and Isotopes in Ocean Sediment

In a technique that parallels pollen studies, scientists estimate climate by studying fossils in deep-sea sediment. The dominant life forms in the ocean are microscopic plankton that float near the sea surface. Just as pollen ratios change with air temperature, plankton species ratios change with sea-surface temperature. Thus, fossil plankton assemblages found in sediment cores reflect sea-surface temperature.

Most hard tissues formed by animals and plants, such as shells, exoskeletons, teeth, and bone, contain oxygen. Many organisms absorb a high ratio of $^{18}O/^{16}O$ at low temperatures, but the ratio decreases with increasing temperature. For example, foraminifera are tiny marine organisms. During a time of Pleistocene cooling and glacial growth, their shells contain an average of 2 percent more ^{18}O than similar shells formed during a warm interglacial interval. Thus, just as scientists estimate paleoclimate by measuring oxygen isotope ratios in glacial ice, they can estimate ancient climate by measuring oxygen isotope ratios in fossil corals, plankton, teeth, and the remains of other organisms. Oxygen is also incorporated into soil minerals, so isotope ratios in soil and sea-floor sediment also reflect paleoclimate.

The Rock and Fossil Record

Fossils are abundant in many sedimentary rocks of Cambrian age and younger. Geologists can approximate climate in ancient ecosystems by comparing fossils with modern relatives of the ancient organisms (◆ Figure 21.7). For example, modern coral reefs grow only in tropical water. Therefore, we infer that fossil reefs also formed in the tropics. Coal deposits and ferns formed in moist tropical environments; cactus indicate that the region was once desert.

Looking backward even farther—into the Proterozoic Era, before life became abundant—it is difficult to measure climate with fossils. Thus, geologists search for clues in rocks. Tillite is a sedimentary rock formed from glacial debris and thus indicates a cold climate. Lithified dunes formed in deserts or along coasts.

Sedimentary rocks form in water, so their existence tells us that the temperature was above freezing and below boiling. Carbonate rocks precipitate from carbon dioxide dissolved in seawater. Geochemists know the chemical conditions under which carbon dioxide dissolves and precipitates, so they can calculate a range of atmospheric and oceanic compositions and temperatures that would have produced limestone and other carbonate rocks. Most thick limestones formed in warm, shallow seas. Some ancient mineral deposits, such as the banded iron deposits described in Chapter 17, also reflect the chemistry of the ancient atmosphere.

After scientists learned that climates have changed, they began to search further to understand how climates change. In the following four sections, we will discuss natural mechanisms of climate change. The final three sections will focus on climate change caused by humans and the possible consequences of that change.

Before proceeding, let us summarize the fundamental mechanisms of Earth's atmospheric heat balance. Keep these mechanisms in mind as you read the rest of this chapter.

Despite its immense complexity, Earth's atmospheric temperature is determined by three basic processes:

- The heat of the Sun. Solar output fluctuates, so in the absence of mitigating factors, when the Sun becomes hotter, Earth warms. When the Sun cools, so does our planet.
- Albedo. If the incoming solar radiation is absorbed at Earth's surface, the planet becomes warmer; if heat is reflected back into space, the planet becomes cooler. Therefore, the reflectivity, or albedo of Earth's surface is a critical factor in determining the planet's temperature.
- Heat retention by the atmosphere. Heat that is reflected back into space may be absorbed by gases in the atmosphere, in a process called the greenhouse effect. Since different gases retain heat differently, atmospheric composition is the third basic factor that regulates our planet's atmospheric temperature.

These three factors determine the amount of heat available to drive the planet's climate and weather engines. Multitudinous feedback loops and threshold effects—involving atmosphere, hydrosphere, geosphere, and biosphere—then perturb the basic system to produce the resultant climate.

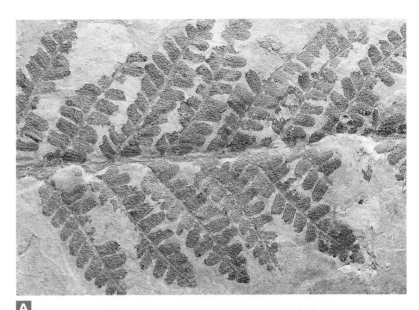

A

B

Courtesy of Graham R. Thompson/Jonathan Turk

◆ **FIGURE 21.7** **(A)** A fossil fern indicated that a region was wet and warm at the time the fern grew. **(B)** These fossil sand dune cross-beds indicate that this region was dry at the time the dunes formed.

21.3
Astronomical Causes of Climate Change

Recall from Chapter 13 that variations in Earth's orbit and rotational pattern may have caused the climate fluctuations responsible for the glacial advances and retreats of the Pleistocene Ice Age. Other astronomical factors may also cause climate change.

Changes in Solar Radiation

A star the size of our Sun produces energy by hydrogen fusion for about 10 billion years. During this time, its energy output increases slowly. As mentioned earlier, solar output has increased by 20 to 30 percent during Earth's history. This slow evolution has influenced climate over long expanses of geologic time. Fortunately for us, the atmospheric carbon dioxide concentration was high during Earth's early history, so temperature was relatively warm even though solar output was low.

Within the past few hundred million years, solar output has changed by only one fifty-millionth of 1 percent per century, and therefore the variation had no measurable influence on climate change over thousands, to even millions, of years. Over shorter periods of time, solar magnetic storms, sunspots, and anomalous variations cause fluctuations in solar output that may affect Earth's climate. Magnetic storms on the Sun create dark, cool regions on the solar surface called sunspots. Sunspot activity alternates dramatically on an 11-year cycle and on longer cycles spanning hundreds of years. During periods of high sunspot activity, the Sun is slightly cooler than during periods of low activity. Several studies show that

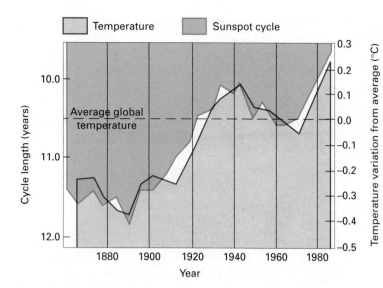

E. Friis-Christensen and K. Lassen, "Length of the Solar Cycle: An Indicator of Solar Activity Closely Associated with Climate," *Science* 254 (November 1, 1991): 698ff.

◆ **FIGURE 21.8** Times of maximum sunspot abundance (blue curve) correlate closely to mean global atmospheric temperature (brown curve) between 1860 and 1990. Global temperature is expressed as change from the average annual temperature in the years 1951 to 1980.

changes in global temperatures coincide with changes in sunspot activity (◆ Figure 21.8). Critics argue that the correlation must be a statistical coincidence because differences in solar output resulting from sunspot cycles are too small to alter Earth's temperature. Proponents counter that a feedback mechanism must be involved. Perhaps changes in solar radiation affect the stratospheric ozone concentration, which alters heat transfer mechanisms between the stratosphere and the troposphere. The issue remains unresolved.

Bolide Impacts

Recall that the evidence strongly suggests that a bolide crashed to Earth about 65 million years ago. The impact blasted enough rock and dust into the sky to block out sunlight and cool the planet. According to one current hypothesis, this cooling led to the extinction of the dinosaurs. Other bolide impacts may have caused rapid and catastrophic climate changes throughout Earth's history.

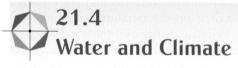

21.4
Water and Climate

Recall that water is abundant in all four of Earth's spheres. It occurs in rocks and soil of the geosphere, comprises over 90 percent of the living organisms of the biosphere, constitutes essentially the entire hydrosphere, and exists in the atmosphere as vapor, liquid droplets, and solid crystals of snow and ice. Moreover, water moves freely from one system to another. As a result, "water acts as the

Venetian blind of our planet, as its central heating system, and as its refrigerator, all at the same time."[1] We have already discussed many of the components of the complex, and often conflicting, relationship between water and climate. As a brief review:

- Water vapor is the most abundant greenhouse gas. Thus it warms the atmosphere and Earth's surface.
- Clouds reflect sunlight and therefore cool the atmosphere and Earth's surface. But clouds also absorb heat radiating from Earth's surface. Thus, water in the atmosphere causes both warming and cooling.
- Glaciers and snowfields have a high albedo (80 to 90 percent); they reflect sunlight and also cool Earth's climate. Conversely, surface water has a very low albedo, only about 5 percent. As a result, any change from surface water to glaciers, or vice versa, can have a dramatic effect on the planet's albedo and temperature.
- When water evaporates from the ocean surface, solar energy is stored as latent heat of the resultant vapor. The vapor moves vast distances, transporting this heat. The heat is then released when the water vapor condenses to form rain or snow.
- Similarly, heat is released when water freezes to form snow and ice. But once a glacier forms or a portion of Earth's surface is snow covered, a lot of heat is required to melt the ice.
- Flowing water weathers rocks and initiates chemical reactions that alter the carbon dioxide concentration in the atmosphere. Carbon dioxide is a greenhouse gas that warms Earth's surface.
- Ocean currents move heat to and from the polar regions. But because more currents flow from the equator toward the poles than the other way around, there is a net transport of heat toward the polar regions. This polar warming effect is counterbalanced by the fact that currents also transport moisture. When the moisture falls as snow, the increased albedo cools the polar regions.

Climate models attempt to quantify all of these factors, but clearly the balances are delicate to unravel. For this reason, it is difficult to predict climate changes.

---

1. Thomas Karl, Neville Nichols, and Jonathan Gregory, "The Coming Climate," *Scientific American* (May 1997): 79ff.

21.5
The Natural Carbon Cycle and Climate

Carbon circulates among the atmosphere, the hydrosphere, the biosphere, and the geosphere and is stored in each of these reservoirs (◆ Figure 21.9).

Carbon in the Atmosphere

As explained in Chapter 17, oxygen and nitrogen, the most abundant gases in the atmosphere, are transparent to infrared radiation and are not greenhouse gases. Carbon exists in the atmosphere mostly as carbon dioxide (CO_2), and in smaller amounts as methane (CH_4).

Although only 0.1 percent of the total carbon near Earth's surface is in the atmosphere, this reservoir plays an important role in controlling atmospheric temperature because carbon dioxide and methane are greenhouse gases; they absorb infrared radiation and heat the lower atmosphere. If either of these compounds is removed from the atmosphere, the atmosphere cools; if they are released into the atmosphere, the air becomes warmer.

Carbon in the Biosphere

Carbon is the fundamental building block for all organic tissue. Plants extract carbon dioxide from the atmosphere and build their body parts predominantly of carbon. This process occurs both on land and in the sea. Most of the aquatic fixation of carbon is conducted by microscopic photoplankton. Therefore, healthy terrestrial and aquatic ecosystems play a vital role in removing carbon from the atmosphere.

Most of the carbon is released back into the atmosphere by natural processes, such as respiration, fire, or rotting (◆ Figure 21.10). However, at certain times and places, organic material does not decompose completely and is stored as fossil fuels—coal, oil, and gas. Thus, plants transfer carbon from the biosphere to rocks of the upper crust.

Carbon in the Hydrosphere

Carbon dioxide dissolves in seawater. Most of it then reacts to form bicarbonate, HCO_3^- (commonly found in your kitchen as baking soda or bicarbonate of soda), and carbonate ($CO_3)^{2-}$.

The amount of carbon dioxide dissolved in the oceans depends in part on the temperature of the atmosphere and the oceans. When seawater warms, it releases dissolved carbon dioxide into the atmosphere, causing greenhouse warming. In turn, greenhouse warming further heats the oceans, causing more carbon dioxide to escape. Warmth evaporates seawater as well, and water vapor also absorbs infrared radiation. Clearly, such a feedback mechanism can escalate. A runaway greenhouse effect may be responsible for the high temperature on Venus.

Carbon in the Crust and Upper Mantle

As shown in ◆ Figure 21.9, the atmosphere contains about 750 billion tons of carbon. In contrast, the crust and upper mantle contain 1,000 times as much, or 750 trillion tons of carbon. The upper geosphere, combined

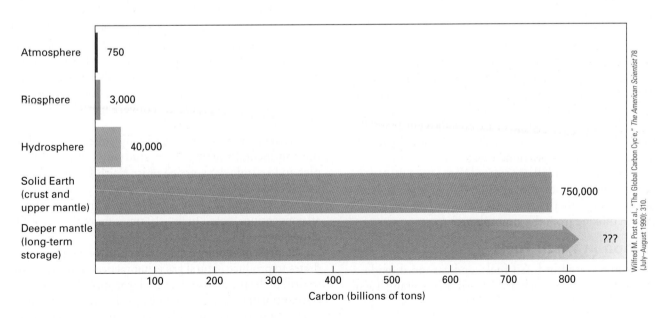

◆ **FIGURE 21.9** Carbon reservoirs in the atmosphere, biosphere, hydrosphere, and solid Earth. The numbers represent billions of tons of carbon.

INTERACTIVE QUESTION: *How would the amount of carbon in the atmosphere be affected if 1 percent of the carbon dissolved in seawater were released into the atmosphere?*

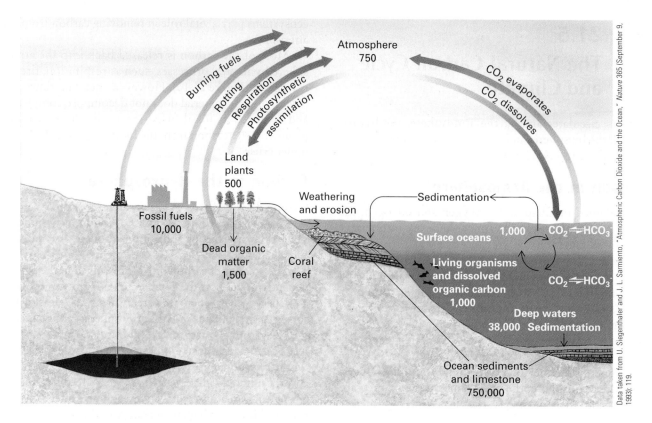

Data taken from U. Siegenthaler and J. L. Sarmiento, "Atmospheric Carbon Dioxide and the Ocean," *Nature* 365 (September 9, 1993): 119.

◆ **FIGURE 21.10**   The carbon cycle. The numbers show the size of the reservoirs and represent billions of tons of carbon.

with the hydrosphere and biosphere, contain almost 800 trillion tons of carbon. Thus, if only a minute portion of the carbon in the geosphere, hydrosphere, and biosphere is released, the atmospheric concentration of carbon dioxide can change dramatically, with a draconian effect on climate.

Carbonate Rocks

Marine organisms absorb calcium and carbonate ions from seawater and convert them into calcium carbonate ($CaCO_3$) in shells and other hard parts. This process removes carbon from seawater and causes more atmospheric carbon dioxide to dissolve into the seawater. Thus, formation of shells removes carbon dioxide from the atmosphere. The shells and skeletons of these organisms gradually collect to form limestone.

When sea level falls or tectonic processes raise portions of the sea floor above sea level, limestone and silicate rocks weather by processes that extract additional carbon dioxide from the atmosphere.[2]

Carbon in Fossil Fuels

Carbon is stored in fossil fuels, and carbon dioxide is released when these fuels are burned. Recoverable fossil

fuels contain about 4 trillion tons of carbon, five times the amount in the atmosphere today. For this reason, scientists are concerned that burning fossil fuels will raise atmospheric carbon dioxide levels. This topic is discussed in Section 21.5.

Methane in Sea-Floor Sediment

When organic material falls to the sea floor and is buried with mud, bacteria decompose it, releasing methane, commonly called natural gas. Between a depth of about 500 meters and 1 kilometer, the temperature of water-saturated mud on continental shelves is low enough and the pressure is favorable to convert methane gas to a frozen solid called methane hydrate.

Methane hydrate then gradually collects in mud on the continental shelves. After studying both drill samples and seismic data, geochemist Keith Kvenvolden of the United States Geological Survey estimates that methane hydrate deposits hold twice as much carbon as all conventional fossil fuels—10 times more than is in the atmosphere.

At present, commercial extraction of methane hydrates to produce natural gas is impractical. It is expensive to drill in deep water, and a thin layer spread throughout the continental shelves would be prohibitively expensive to exploit. However, scientists are studying links between methane hydrates and climate. Tectonic activity at subduction zones or landslides on continental slopes could release methane from hydrate deposits. Changes in bottom temperatures on continental shelves resulting from

---

2. The complete reaction is:
   $CaCO_3 + CO_2 + H_2O = Ca(HCO_3)_2$
   limestone plus carbon dioxide plus water equals calcium bicarbonate (soluble).

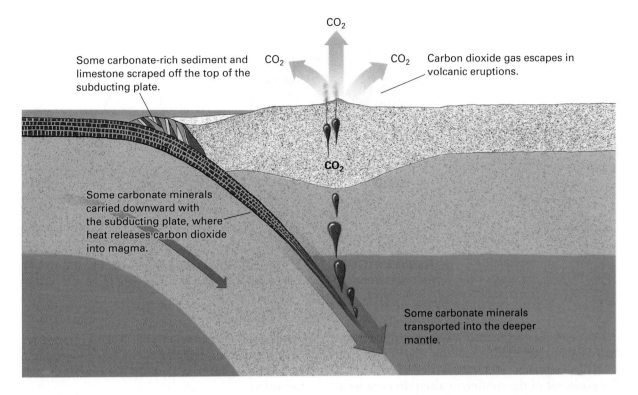

Some carbonate-rich sediment and limestone scraped off the top of the subducting plate.

CO_2

CO_2

CO_2 Carbon dioxide gas escapes in volcanic eruptions.

CO_2

Some carbonate minerals carried downward with the subducting plate, where heat releases carbon dioxide into magma.

Some carbonate minerals transported into the deeper mantle.

◆ **FIGURE 21.11** A subducting oceanic plate carries limestone and other carbonate-rich sediment into the mantle. Some of the carbonate minerals are heated to produce carbon dioxide, which escapes during volcanic eruptions. Some of the carbonate minerals may be stored in the mantle.

warming of seawater could also release methane from the frozen hydrates. In turn, increased atmospheric methane could trigger greenhouse warming. Ice core studies show that global atmospheric methane concentration has changed rapidly in the past, perhaps by sudden releases of oceanic methane hydrates.

About 55 million years ago, near the end of the Paleocene Epoch, climate suddenly warmed and many aquatic and terrestrial species became extinct. According to one model, a change in sea-surface circulation caused equatorial waters to remain in low latitudes. High equatorial temperatures evaporated enough water to increase the salinity of the sea surface. When the salinity reached a threshold value where the surface water was denser than the cold deep water, the warm, salty water sank. The warm, sinking water melted the methane hydrates and released the methane. Aquatic species were poisoned by the methane in the water, and many terrestrial species succumbed to the rapid greenhouse warming.[3]

Carbon in the Deeper Mantle

During subduction, oceanic crust sinks into the mantle (◆ Figure 21.11). The descending plate may carry carbonate rocks and sediment. As this material sinks to greater depths, the carbonate minerals become hot and

release carbon dioxide, which is carried back to the surface by volcanic eruptions. Some carbonate rock, although geologists are uncertain how much, may be carried into deeper regions of the mantle during subduction. Large quantities of carbon were trapped within Earth during its formation. Much of this carbon escaped early in Earth's history, but some remains in the deep mantle and may rise from the mantle to the surface during volcanic eruptions. Carbon exchanges between the deep mantle and the surface are an important topic of current research.

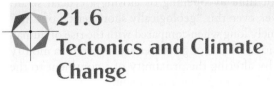

21.6
Tectonics and Climate Change

Positions of the Continents

A map of Pangea shows that 200 million years ago, Africa, South America, India, and Australia were all clustered near the South Pole (◆ Figure 21.12). Because climate is colder at high latitudes than near the equator, continental position alters continental climate.

In addition, continental interiors generally experience colder winters and hotter summers than coastal areas do. When all the continents were joined into a

---

3. Gerald Dickens et al., "A Blast of Gas in the Latest Paleocene; Simulating First-Order Effects of Massive Dissociation of Oceanic Methane Hydrate," *Geology* (March 1997): 259–262.

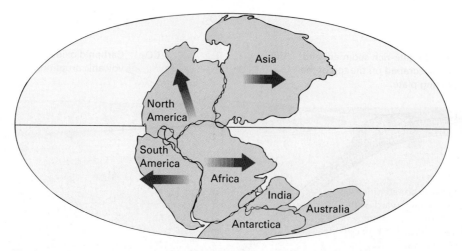

ThomsonNOW ◆ **ACTIVE FIGURE 21.12** Two hundred million years ago, when the Pangea supercontinent was assembled, Africa, South America, India, and Australia were all positioned close to the South Pole.

supercontinent, the continental interior was huge, and regional climates were different from the climates on many smaller continents with extensive coastlines.

The positions of the continents also influence wind and sea currents, which, in turn, affect climate. For example, today the Arctic Ocean is nearly landlocked, with three straits connecting it with the Atlantic and Pacific oceans (◆ Figure 21.13). The Bering Strait between Alaska and Siberia is 80 kilometers across, Kennedy Channel between Ellesmere and Greenland is only 40 kilometers across, and a third, wider seaway runs along the east coast of Greenland. Presently, cold currents run southward through Kennedy Channel and the Bering Strait, and the North Atlantic Drift carries warm water northward along the coast of Norway. If any of these straits were to widen or close, global heat transfer would be affected. Deep-sea currents also transport heat and are affected by continental positions.

Tectonic plates move from 1 to 16 centimeters per year. A plate that moves 5 centimeters per year travels 50 kilometers in a million years. Thus continental motion can change global climate within a geologically short period of time by opening or closing a crucial strait. (However, even this "geologically short period of time" is extremely long when compared with the rise of human civilization.) Much longer times are required to modify climate by altering the proximity of a continent to the poles or by creating a supercontinent.

Mountains and Climate

Global cooling during the past 40 million years coincided with the formation of the Himalayas and the North American Cordillera.[4] Mountains interrupt airflow, altering regional winds. Air cools as it rises and

passes over high, snow-covered peaks. However, it is unclear whether this regional cooling could account for the global cooling that accompanied this episode of mountain formation.

Large portions of the Himalayas and the North American Cordillera are composed of marine limestone. Recall from the previous section that when marine lime-

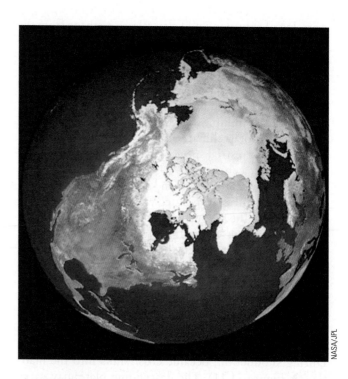

◆ **FIGURE 21.13** With the modern distribution of continents, the Arctic Ocean is nearly landlocked. Most of this seawater is covered with ice for most of the year, as indicated by the white zone in this photograph. When the continents move sufficiently to open or close the narrow straits connecting the Arctic Ocean to more southern waters, currents and global heat transfer will be affected.

4. William F. Ruddiman and John Kutzbach, "Plateau Uplift and Climatic Change," *Scientific American* (March 1991): 66ff.

stone weathers, carbon dioxide is removed from the atmosphere. When sea-floor rocks are thrust upward to form mountains, they become exposed to the air. Rapid weathering then may remove enough atmospheric carbon dioxide to cause global cooling.

Volcanoes and Climate

Recall from Chapter 8 that volcanoes emit ash and sulfur compounds that reflect sunlight and cool the atmosphere. For two years after Mt. Pinatubo erupted in 1991, Earth cooled by a few tenths of a degree Celsius. Temperature rose again in 1994 after the ash and sulfur settled out.

Volcanoes also emit carbon dioxide that warms the atmosphere by absorbing infrared radiation. The net result—warming or cooling—depends on the size of the eruption, its violence, and the proportion of solids and gases released. Some scientists believe that a great eruption in Siberia 250 million years ago cooled the atmosphere enough to cause or contribute to the Permian extinction. A huge sequence of eruptions 120 million years ago, called the mid-Cretaceous superplume, may have emitted enough carbon dioxide to warm the atmosphere by 7°C to 10°C. Dinosaurs flourished in huge swamps, and some of the abundant vegetation collected to form massive coal deposits.

How Tectonics, Sea Level, Volcanoes, and Weathering Interact to Regulate Climate

Tectonics, sea level, volcanoes, and weathering are all part of a tightly interconnected Earth system that affects both global and regional climate. When tectonic plates spread slowly, the Mid-Oceanic Ridge system is so narrow that it displaces relatively small amounts of seawater. As a result, sea level falls. When sea level falls, large marine limestone deposits on the continental shelves are exposed as dry land. The limestone weathers. Weathering of limestone removes carbon dioxide from the atmosphere, leading to global cooling. At the same time, when sea-floor spreading is slow, subduction is also slow. Volcanic activity at both the spreading centers and the subduction zones slows down, so relatively small amounts of carbon dioxide are emitted. With small additions of carbon dioxide from volcanic eruptions and removal of atmospheric carbon dioxide by weathering, the atmospheric carbon dioxide concentration decreases and the global temperature cools. In addition, dropping sea level decreases the surface area of the oceans and increases the surface area of the higher-albedo continents. This results in an increase of average global albedo, and, consequently, reinforces the global cooling (◁▶ Figure 21.14A). These conditions may have caused the cooling at the end of the Carboniferous Period shown in ◁▶ Figure 21.2.

In contrast, during periods of rapid sea-floor spreading, a high-volume Mid-Oceanic Ridge system

raises sea level. Marine limestone beds are submerged, weathering slows, and weathering removes less carbon dioxide from the atmosphere. Volcanic activity is high during periods of rapid plate movement, so large amounts of carbon dioxide are released into the atmosphere. Rising sea level decreases continental area and therefore decreases the average global albedo. All of these factors lead to global warming (◁▶ Figure 21.14B). But rapid spreading also coincides with rapid subduction and accelerated mountain-building, leading to accelerated weathering on the continents, which consumes carbon dioxide. Once again, climate systems are driven by so many opposing mechanisms that it is often difficult to determine which will prevail.

21.7
Greenhouse Effect: The Carbon Cycle and Global Warming

We have learned that the amount of carbon in the atmosphere is determined by many natural factors, including rates of plant growth, mixing of surface ocean water and deep ocean water, growth rates of marine organisms, weathering, the movement of tectonic plates, and volcanic activity. Within the past few hundred years, humans have become an important part of the carbon cycle. Modern industry releases four greenhouse gases—carbon dioxide, methane, chlorofluorocarbons (CFCs), and nitrogen oxides.

People release carbon dioxide whenever they burn fossil or bio fuels. This release is inherent in the chemistry of combustion. Carbon in the fuel reacts with oxygen in the air to produce carbon dioxide. Furthermore, once carbon dioxide is released, it is, for all practical purposes, impossible to remove this gas from the atmosphere. If you drive your car to town today, the carbon dioxide released will remain in the atmosphere for centuries. Logging also frees carbon dioxide because stems and leaves are frequently burned and forest litter rots more quickly when it is disturbed by heavy machinery. The recent rise in the concentration of atmospheric carbon dioxide has attracted considerable attention because it is the most abundant industrial greenhouse gas (◁▶ Figure 21.15).

In addition, several other greenhouse gases are released by modern agriculture and industry. Small amounts of methane are released during some industrial processes. Larger amounts are released from the guts of cows, other animals, and termites, and from rotting that occurs in rice paddies. Today, industry and agriculture combined add about 37×10^{12} grams of methane into the atmosphere every year. Chlorofluorocarbons were, until recently, used as refrigerants and as

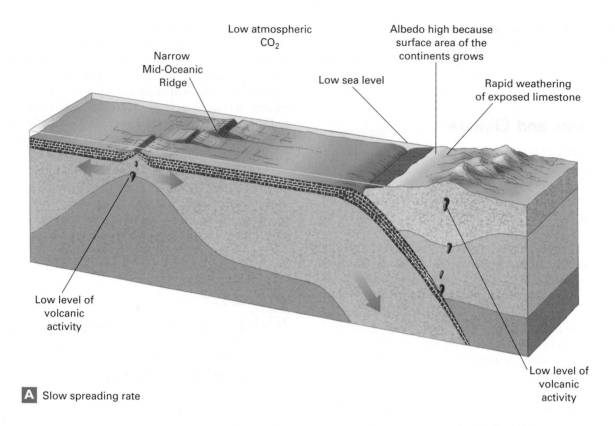

A Slow spreading rate

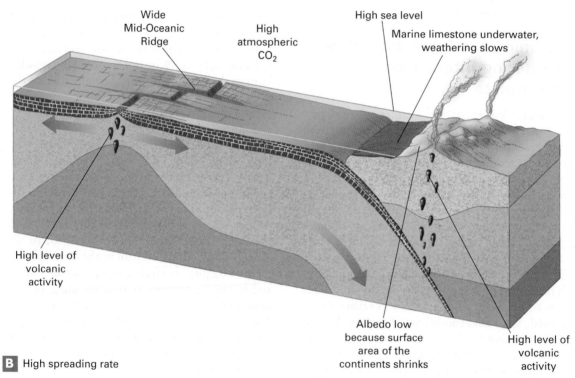

B High spreading rate

◇ **FIGURE 21.14** **(A)** When subduction is slow, the Mid-Oceanic Ridge is narrow. As a result, sea level is low, exposing marine limestone to weathering. Volcanic activity is also low. The atmospheric carbon dioxide concentration is low because of rapid weathering and low volcanic activity. Also, global albedo is high because the surface of the continents grows at the expense of shrinking oceans. All of these factors cool Earth. **(B)** When subduction is fast, the Mid-Oceanic Ridge is wide. As a result, sea level is high, flooding coastal regions. Volcanic activity is also high. The atmospheric carbon dioxide concentration is high because weathering is slow and volcanic activity is high. Also, global albedo is low because of the low surface area of the continents and greater surface area of the seas. All of these factors warm Earth.

Mauna Loa Observatory, Hawaii
Monthly Average Carbon Dioxide Concentration

(Y-axis: CO_2 Concentration, (PPM); values 310, 315, 320, 325, 330, 335, 340, 345, 350, 355, 360, 365, 370, 375, 380, 385, 390)

(X-axis: Year; 1958, 60, 62, 64, 66, 68, 70, 72, 74, 76, 78, 80, 82, 84, 86, 88, 90, 92, 94, 96, 98, 00, 02, 04)

19- May-05

◆ **FIGURE 21.15** Atmospheric carbon dioxide concentration has risen by about 20 percent since 1958. The short-term fluctuations are caused by seasonal changes in carbon dioxide absorption by plants. (www.epa.gov/air/airtrends/2005/econemissions.html)

propellants in aerosol cans. This source is rapidly diminishing because international treaties have banned CFC production. N_2O, yet another greenhouse gas, is released from the manufacture and use of nitrogen fertilizer, some industrial chemical syntheses, and from the exhaust of high-flying jet aircraft.

With this background, let us ask one simple question: Has the human release of greenhouse gases caused a warming of Earth over the past century?

Let us first summarize the data:

- Human activities release greenhouse gases.
- The concentration of these gases in the atmosphere has risen since the beginning of the Industrial Revolution.
- Greenhouse gases absorb infrared radiation and trap heat.
- The atmosphere has warmed by almost 0.8°C during the last century.

Of the twenty hottest years on record, nineteen have occurred after 1980 and 2005 was the hottest year in more than a century.

The obvious connection is that rising atmospheric concentrations of industrial greenhouse gases have caused the recent global temperature rise. However, many government and industry sources have disagreed, stating that the current warming trend may be an unrelated natural event that just happened to coincide with the atmospheric increase in carbon dioxide. Although this debate has raged for a few decades, today scientists are essentially unanimous in concluding that industrial emissions are causing global warming. Of 928 peer-reviewed papers appearing in scientific journals between 1993 and 2003, every author and every study concluded that anthropogenic greenhouse gas emissions are causing the observed warm temperatures.

Recall from the introduction to this chapter that we asked whether the hot European summer of 2003 was an anomaly or a consequence of global warming. A study of this question, released in 2004, concludes, "Anthropogenic warming trends in Europe imply an increased probability of very hot summers . . . . Past human influence has more than doubled the risk of European mean summer temperatures as hot as 2003".

Consequences of Greenhouse Warming

Many people ask, "What's the big deal? What difference will it make if the planet is a few degrees warmer than it is today?" ◆ Figure 21.16 summarizes the major predicted consequences of global warming.

Temperature Effects on Agriculture

A warmer global climate would mean a longer frost-free period in the high latitudes, which would benefit agriculture. In parts of North America, the growing season is now a week longer than it was a few decades ago. On the negative side, warmth affects plants in many ways that decrease crop yields. In one study, researchers found that rice yields decrease by 10 percent for every one degree Celsius increase in nighttime temperatures. The scientists concluded that warmer nighttime temperatures increased plant respiration and therefore decreased the energy available for storage in the rice kernels. In another study,

scientists estimated that heat stress, increased virulence of parasites and disease organisms, and loss of soil moisture led to a 30 percent reduction in forest and grassland growth during the 2003 European drought. The effect on crops was variable. Winter wheat yields were barely affected, because the maximum growth period occurred before the heat wave. On the other end of the spectrum, the Italian corn yields decreased by 36 percent, despite extensive irrigation.

Precipitation and Soil-Moisture Effects on Agriculture

In a warmer world, both precipitation and evaporation would increase. Computer models show that the effects would differ from region to region. Between 1900 and 2000, rainfall increased in Pakistan and in the great grain-growing regions in North America and Russia. However, parts of North Africa, India, and Southeast Asia received less rainfall over the same time. The resulting drought has led to famine during recent decades.

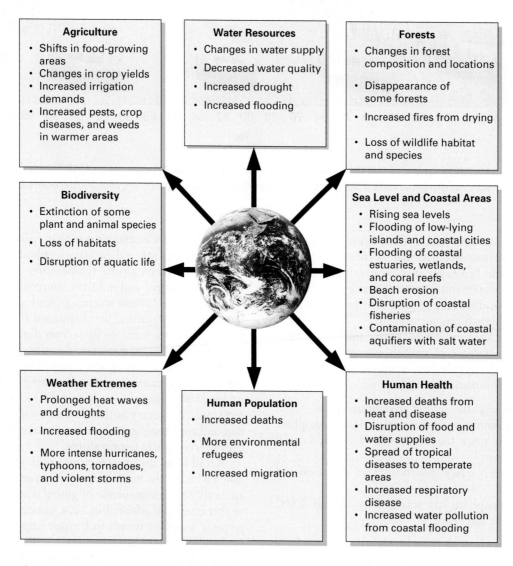

◆ **FIGURE 21.16** Predicted consequences of global warming.

Systems Perspective

Geologic and Human Components of the Carbon Cycle

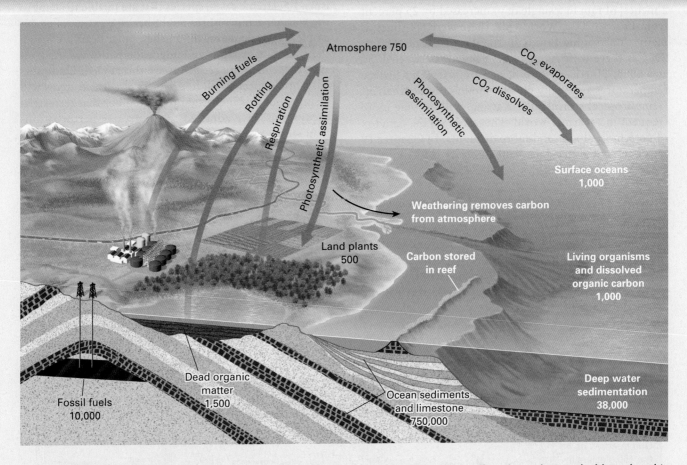

Natural geologic events, such as volcanic eruptions, combine with human influences to affect the carbon cycle. Note that this figure is an amalgam of Figure 21.10 and Figure 21.11. The figure is schematic because all of the pictured reservoirs and mechanisms do not usually appear in a single landscape.

In a hotter world, more soil moisture will evaporate. As mentioned earlier, in the Northern Hemisphere, scientists predict that global warming will lead to increased rainfall, adding moisture to the soil. At the same time, warming will increase evaporation that removes soil moisture. Most computer models forecast a net loss of soil moisture and depletion of groundwater, despite the prediction that rainfall will increase. Drought and soil moisture depletion would increase demands for irrigation. But, irrigation systems are already stressing global water resources, causing both shortages and political instability in many parts of the world. As a result of all these factors, most scientists predict that higher mean global temperature would decrease global food production, perhaps dramatically.

In many regions of the world, winter precipitation accumulates as snow in the high mountains. Snowmelt feeds rivers during the summer and this river water is used to irrigate crops. In a warmer world, more precipitation falls as rain, and the snow that does fall is more likely to melt in early spring rather than in late summer. Also, glaciers recede, thus diminishing the amount of water in long-term storage. As a result of all these factors, river levels during the midsummer to late summer are lower in a warmer world than in a cooler world. Thus, unless people build expensive dams, there is less water available for irrigation.

Extreme Weather Events

Computer models predict that weather extremes, such as intense rainstorms, flooding, heat waves, prolonged droughts, and violent storms such as hurricanes, typhoons, and tornados, will become more common in a warmer, wetter world. In the United States, the number of heavy downpours (defined as 5 centimeters, or 2 inches, of rain in a single day) increased by 25 percent from 1900 to 2000. Worldwide, there were ten times as many catastrophic floods in the decade from 1990 to 2000 as there were in an average decade between 1950 and 1985. Hurricanes are driven by warm sea surface temperatures and many climate models predict that extreme tropical storms, like Katrina, will become more frequent in a warmer world. Thus, we face the paradox of *both* more floods and more droughts. In Chapter 19 we documented that hurricane intensity is likely to increase in a warmer world.

Changes in Biodiversity

According to a study published by the World Wildlife Fund in 2000, global warming could alter one-third of the world's wildlife habitats by 2100. In some northern-latitude regions, 70 percent of habitats will be significantly affected. As soil moisture decreases, trees will die and wildfires will become more common, changing forests into savannah. Plants and animals that thrive in cold temperatures will die off. Several recent studies

have shown that many plant and animal pathogens thrive better in warmer temperatures than in a colder world. Thus, the pine beetle in northern North America has flourished in recent years, decimating huge swaths of forests in the Rocky Mountains of western United States and Canada. Populations of frogs have declined dramatically as they have succumbed to virulent disease organisms. In extreme cases, species that are unable to adapt to the warmer temperatures might become extinct. For example, as Arctic sea ice diminishes, polar bears are losing their hunting grounds and migration pathways. As a result, these majestic creatures are threatened. Undoubtedly, other species will flourish. Ecological systems will change, but it is important to remember that terrestrial ecosystems have adjusted to far greater perturbations than we are experiencing today.

Sea-Level Change

When water is warmed, it expands slightly. From 1900 to 2000, mean global sea level rose 10 to 25 centimeters, or roughly the thickness of a dime every year. Oceanographers suggest that even this modest rise may be affecting coastal estuaries, wetlands, and coral reefs. However, sea-level rise is accelerating dramatically because the polar ice sheets of Antarctica and Greenland are melting. In 2005, 150 cubic kilometers of Antarctic ice melted and the water flowed into the oceans. To exacerbate the problem, the flow of ice from Greenland has more than doubled over the past decade. (◆ Figure 21.17).

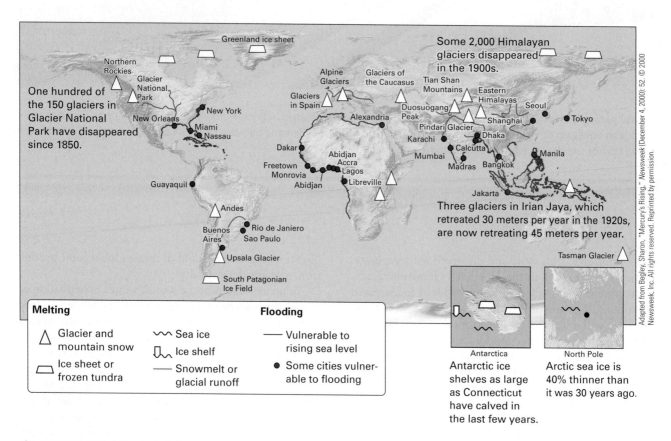

◆ **FIGURE 21.17** Global changes resulting from melting glaciers and polar ice caps.

Effects on People

Humans evolved during a period of rapid climate change in the savannahs and forests of Africa. In fact, many anthropologists argue that we flourished because we adapted to climate change faster and more efficiently than other species in the African ecosystem. A few million years later, we survived the Pleistocene Ice Age. We are a clever and resourceful species. Yet the consequences of global warming could be severe. In a warmer world, tropical diseases such as malaria have been spreading to higher latitudes. But the biggest problem arises because expanding human population is already stressing global systems. Due to high population density, land ownership, and political boundaries, people cannot migrate as freely as they once could. As a result, small shifts in food production or water availability could lead to mass starvation, water scarcity, and political instability. In addition, rising sea level could flood coastal cities and farmlands, causing trillions of dollars worth of damage throughout the world.

21.8
Feedback and Threshold Mechanisms in Climate Change

In a linear relationship, if you raise the concentration of greenhouse gases a small amount, the mean global temperature will rise by a small amount, and further small increases of greenhouse gas concentrations will cause additional small temperature increases, and so on. However, many scientists argue that climate change does not follow linear relationships.

In Chapter 1 we mentioned that the melting of ice is a threshold phenomenon. If the air above a glacier warms from $-1.5°C$ to $-0.5°C$, the ice does not melt and the warming may cause only minimal environmental change. However, if the air warms another degree to $+0.5°C$, the ice warms beyond the threshold defined by its melting point. Melting ice can cause sea level rise and a host of cascading effects.

Recall that a feedback mechanism occurs when a small initial perturbation affects another component of Earth's systems, which amplifies the original effect, which perturbs the system even more, which leads to a greater effect, and so on. Four important climate feedback mechanisms are discussed next:

Albedo Effects

Sparkling snowfields and icy glaciers have high albedos and cool Earth by reflecting sunlight. When Earth warms a little, snow and ice cover decrease, and the albedo decreases. But when the albedo decreases, the atmosphere gets warmer, which melts more snow, which warms the atmosphere even more. A recent study has shown that a second, interlocking feedback mechanism is exacerbating the first. In the Arctic, warmer temperatures have promoted the growth of bushes that often protrude above the spring snow and trap solar heat—thereby depressing the albedo. Thus snow melts faster, providing favorable conditions for bushes to grow, which decreases the albedo even further, which warms the atmosphere.

The accelerating nature of this double feedback loop has contributed to the fact that between 1960 and 1990, Arctic landscapes warmed by roughly 0.15°C per decade. However, between 1990 and 2004, this warming more than doubled to 0.3°C to 0.4°C per decade (◆ Figure 21.18).

Imbalances in Rates of Plant Respiration and Photosynthesis

Any process that increases the total photosynthesis in an ecosystem will remove carbon dioxide from the air and sequester the carbon in plant tissue. In contrast, any process that increases respiration over photosynthesis will introduce more carbon dioxide into the air and increase global warming. Numerous studies have shown that decay respiration in forest and grassland soils accelerates in a warmer world, thus warming the world more by introducing carbon dioxide into the atmosphere.

In another study, scientists showed that both respiration and photosynthesis decreased during the 2003 European drought, but photosynthesis declined more than respiration, leading to a net increase in carbon emissions into the atmosphere.

West Antarctic Ice Shelf

The Antarctic Ice Shelf collapse discussed in the introduction to this chapter could initiate a catastrophic feedback mechanism. Glacial ice in western Antarctica occupies three environments. Continental glaciers rest on solid ground that lies above sea level. Grounded ice rests on the continental shelf below sea level. Floating ice is attached to grounded ice but is floating on the ocean surface.

Within the past few decades, some process has caused the grounded ice to become thinner. As a result, some of this ice has lifted off the bottom and started to float. In turn, large chunks of the floating ice have broken up and drifted away. Scientists do not know why the grounded ice has thinned. Much of Antarctic has actually become *cooler* in recent decades. Even in the Antarctic Peninsula, where slight warming has occurred, the temperature is usually so far below freezing that a small warming would not melt the ice.

When scientists study sea-floor sediments, they find evidence for earlier oscillations in the Antarctic ice shelves. Thus, the changes occurring today are not unusual. Even though we do not know *why* the ice shelves have been disintegrating, it is important to study the effects of this breakup.

Glaciers flow slowly downslope and large ones spread outward under their own weight. As grounded

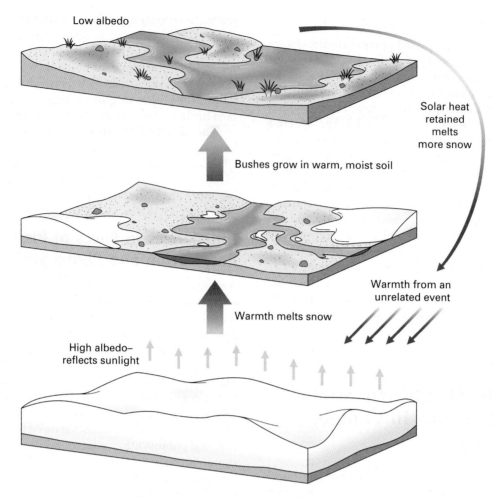

Low albedo

Solar heat retained melts more snow

Bushes grow in warm, moist soil

Warmth from an unrelated event

Warmth melts snow

High albedo–reflects sunlight

◆ **FIGURE 21.18** An arctic climate feedback mechanism. An initial warming melts ice and increases the growth of bushes. These changes lower the albedo, causing increased warming.

ice begins to float, seawater creeps beneath the glacier and saturates the bottom mud. Wet mud is an ideal lubricant, so the glacier flows faster. As flow increases, more ice pushes out into the sea, breaks off, and floats away. But the grounded ice had been acting as a barrier that reduced the flow of the part of the ice that rested on land. When the grounded ice barrier begins to float and flow faster, the land-based portion of the ice is no longer held back effectively and it too may speed up.

If this process accelerates, enough ice may spill into the ocean to cause a significant rise in sea level. Oceans have a lower albedo than land does. Therefore, if sea level rises and floods low-lying land, the oceans would increase in surface area. Average global albedo would decrease, and both the oceans and the atmosphere would become warmer. This would lead to further breakup of the West Antarctic Ice Sheet and more sea-level rise. Also, as the oceans warm, the warm surface water releases additional carbon dioxide and water vapor, leading to even greater warming. According to this scenario, climate would spiral out of control. Thus the breakup of the West Antarctic Ice Sheet could eventually lead to rising sea level and flooding of coastal cities and farmlands (◆ Figure 21.19).

Thermohaline Circulation: How Global Warming Could Cause Global Cooling

Paradoxically, some scientists have calculated that a small, initial global warming could lead to an eventual global cooling. The Gulf Stream and the North Atlantic Drift flow from the tropical Atlantic Ocean northward toward northern Europe. This surface water becomes saltier as it approaches the high latitudes because the warm water evaporates. As the water reaches the coast of Greenland, it cools. Cooling and increased salt content make the water denser, so it sinks. According to some models, this sinking is essential in maintaining the conveyor-belt thermohaline flow of the Gulf Stream (◆ Figure 21.20A).

Now, what would happen if the air and the water surface were to become warmer? Warm water is more buoyant than cold water. Warming would increase evaporation, but it would also melt portions of the Greenland Ice Cap. Abundant freshwater flowing onto the surface of the North Atlantic would overwhelm the evaporation effect and make the seawater less salty and also more buoyant. If the water became warm and buoyant enough, it would stop sinking. According to some models, this change in vertical motion could shut off the flow of the Gulf Stream

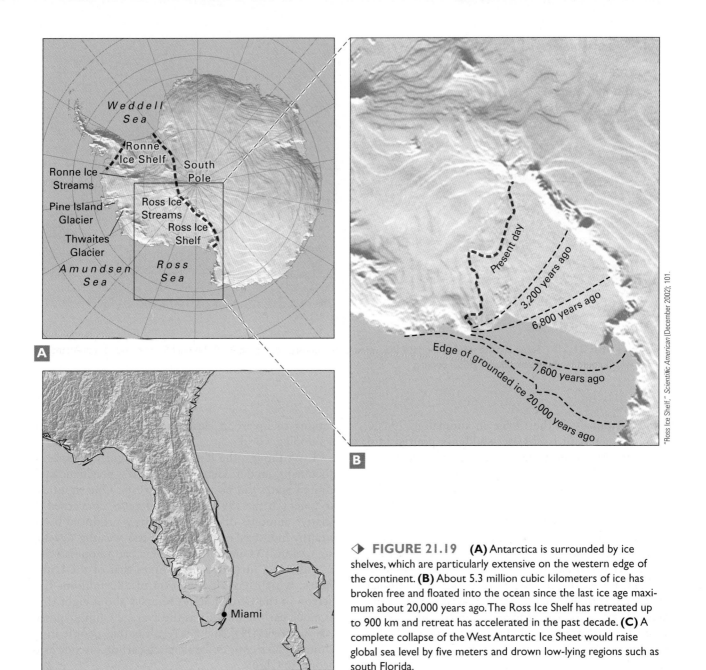

FIGURE 21.19 **(A)** Antarctica is surrounded by ice shelves, which are particularly extensive on the western edge of the continent. **(B)** About 5.3 million cubic kilometers of ice has broken free and floated into the ocean since the last ice age maximum about 20,000 years ago. The Ross Ice Shelf has retreated up to 900 km and retreat has accelerated in the past decade. **(C)** A complete collapse of the West Antarctic Ice Sheet would raise global sea level by five meters and drown low-lying regions such as south Florida.

"Ross Ice Shelf," *Scientific American* (December 2002): 101.

(◆ Figure 21.20B). In turn, if the Gulf Stream stopped or veered southward, global heat distribution would be disrupted and Earth would cool rapidly and dramatically.[5] Such a change could cool the Earth by 5°C to 10°C in a decade, causing a catastrophic drop in global food production. Thus, paradoxically, a small initial warming would cause a rapid cooling.

Studies of both paleoclimates and modern climates support this model. The last ice age maximum occurred approximately 18,000 years ago. Over time, Earth became warmer and the glaciers started to melt. During the early

stages of deglaciation, most of the meltwater from North America emptied into the Mississippi River basin. But recall from section 13.5 that drainage from central Canada eventually diverted into the Great Lakes, and hence to the North Atlantic, via the St. Lawrence River. According to one model, increased warming coupled with a huge influx of fresh meltwater into the North Atlantic disrupted the thermohaline conveyor in approximately 1100 to 1000 AD. The world abruptly cooled, leading to the Little Ice Age, also called the Younger Dryas Event.

So what is happening today? Increased winter snowfall over Greenland, coupled with warmer summer temperatures has led to a massive melting of the Greenland ice cap. Huge and erratic pulses of freshwater have entered the North Atlantic. For example, during the Great

5. Wallace C. Broecker, "Thermohaline Circulation, the Achilles Heel of our Climate System: Will Man-Made CO₂ Upset the Current Balance?" *Science* 278 (November 28, 1997): 1582.

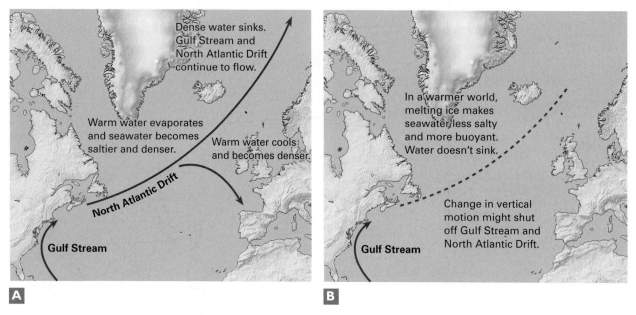

◆ FIGURE 21.20 Some climate models indicate that warmer temperatures may shut off the Gulf Stream and North Atlantic Drift, which could result in subsequent global cooling.

Salinity Anomaly in the 1990s, an anomalous 10,000 cubic kilometers of freshwater suddenly poured into the ocean. As a result, the surface salinity decreased. In December 2005, scientists reported that the thermohaline currents had decreased by 30 percent from 1988 to 2004.

No one knows what the future will bring, but we do know that climates have changed radically in the past, from the frigid cold of Snowball Earth to the warmth of the Cretaceous. Today, by altering the atmospheric composition, we are undergoing a great and unintentional experiment with the climate system that sustains us. Furthermore global climate change has already affected the Earth, its ecosystems, and the humans that call this planet home.

◆ 21.9
The Kyoto Treaty on Greenhouse Warming

In December 1997, representatives from 160 nations met in Kyoto, Japan, to discuss global climate change. The major issues were: How seriously are humans altering climate? How will climate change affect humans and global ecosystems? Can the nations of the world cooperate to reduce carbon emissions and defuse the problem?

After years of acrimonious debate, the Kyoto Treaty was ratified on February 16, 2005. Kyoto is the first international agreement to set numerical targets for reductions in carbon dioxide emissions. It also establishes a global market so nations can trade the right to emit carbon dioxide and to create incentives for industrialized

nations to invest in clean-energy technology for the developing world.

Even though the United States signed the original document, under George W. Bush's presidency, the United States did not ratify the treaty. One significant argument at the 1997 Kyoto conference concerned the relative amounts of greenhouse gases produced by the wealthy industrial nations and the less-wealthy, developing nations. As an example, China has a population of 1.2 billion, a little over four times that of the United States, which has a population of 280 million. Each person in the United States releases about eight times as much carbon dioxide as the average Chinese person (◆ Figure 21.21). The net result is that people in the United States emit twice as much carbon dioxide as do people in China.

At the Kyoto conference, the United States argued that all countries must decrease emission of greenhouse gases by the same proportion. However, representatives of developing nations, including China, argued that the United States' position is unfair because it condemns people in developing nations to continued poverty. They reasoned that the world's poor people should be allowed to raise their standards of living first, before they worry about global warming. The United States countered that because of the great number of people in China and the rest of the developing world, even a small per-capita increase in fossil fuel use would lead to an unacceptably large increase in total carbon dioxide emissions.

Furthermore, President Bush has argued that he will never sign the Kyoto Treaty because emissions reduction would harm the U.S. economy, whereas no one has proven that global warming would have adverse economic effects.

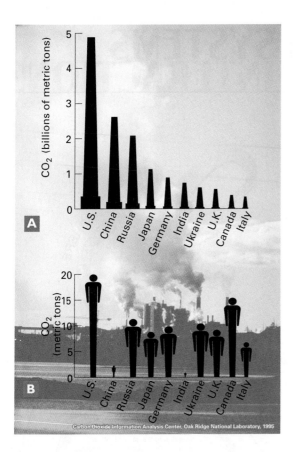

◇ FIGURE 21.21 **(A)** Carbon dioxide emissions from the top 10 emitting countries in 1994. **(B)** Per capita carbon dioxide emissions from the top 10 emitting countries in 1994.

In rebuttal, supporters of the Kyoto Treaty argue the following:

- Wealth is not necessarily dependent on profligate fuel consumption. For example, between 1975 and 2005, the demand for petroleum has been almost constant in Western Europe and Japan. In the United States, petroleum consumption has doubled during the same time period. Yet, the economies and Western Europe and Japan remain strong.

- Furthermore, reduction of fuel consumption and carbon emissions could actually help the economy. People save money when they use less fuel. Moreover, if individuals and businesses shifted to the more fuel-efficient technologies, the massive implementation of new infrastructure would be a boon to the economy. For example, in the early 1970s, California residents used 7,000 kilowatt-hours per person per year, about the same amount of electricity as the national average. Over the past 30 years, as a result of aggressive energy conservation programs, per-capita electricity consumption in California has *declined* to 6,800 kilowatt-hours, compared with an almost doubling of the national average to 12,800 kilowatt-hours. In addition, California has been aggressive about using non-fossil-fuel energy sources. In 2005, California produced 11 percent of its electricity from geothermal, wind, and solar sources, compared with a national average of 2 percent. In an ambitious program, the state plans to increase its share of renewable energy sources to 20 percent by 2010 and to 33 percent by 2020. Yet, in the past three decades, the California economy rose more rapidly than the national average.

- Computer models show that the longer we procrastinate and fail to decrease emissions, the longer it will take the system to recover once emissions are reduced.

- Finally, the argument continues, the alternative—climate change—could possibly initiate runaway feedback mechanisms that would spiral out of control. Global food production could be disrupted, leading to mass starvation and global political unrest. Therefore, on the grounds that it is better to be "safe than sorry," we should reduce emissions now.

The Kyoto Protocol expires in 2012. Representatives gathered in Montreal in December 2005 to write a sequel, but there was little motivation for mandatory caps on emission or for other dramatic steps toward reduction of atmospheric carbon dioxide.

EARTH SYSTEMS INTERACTIONS

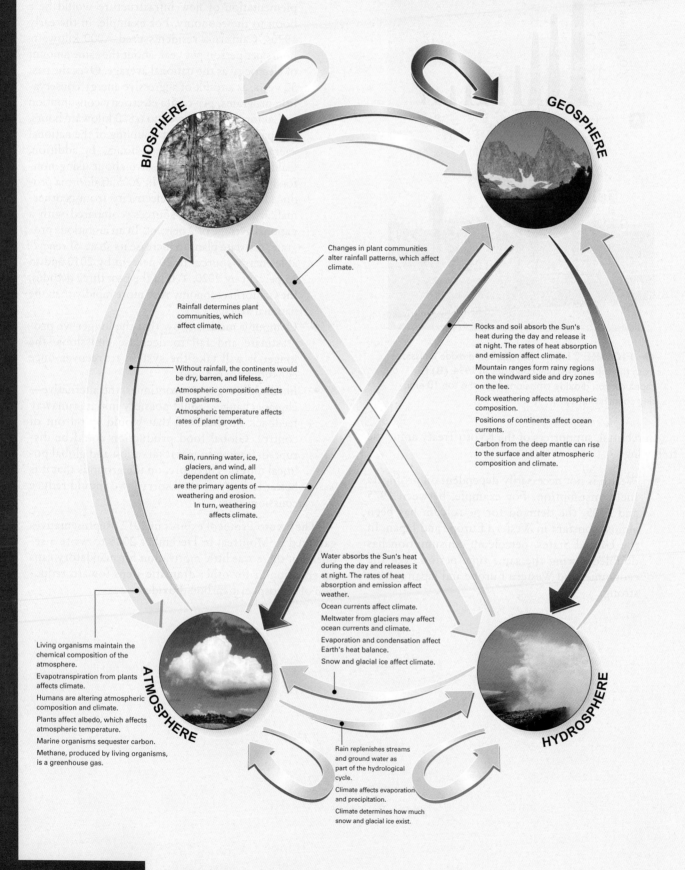

BIOSPHERE

GEOSPHERE

Changes in plant communities alter rainfall patterns, which affect climate.

Rainfall determines plant communities, which affect climate.

Without rainfall, the continents would be dry, barren, and lifeless.

Atmospheric composition affects all organisms.

Atmospheric temperature affects rates of plant growth.

Rain, running water, ice, glaciers, and wind, all dependent on climate, are the primary agents of weathering and erosion. In turn, weathering affects climate.

Rocks and soil absorb the Sun's heat during the day and release it at night. The rates of heat absorption and emission affect climate.

Mountain ranges form rainy regions on the windward side and dry zones on the lee.

Rock weathering affects atmospheric composition.

Positions of continents affect ocean currents.

Carbon from the deep mantle can rise to the surface and alter atmospheric composition and climate.

Water absorbs the Sun's heat during the day and releases it at night. The rates of heat absorption and emission affect weather.

Ocean currents affect climate.

Meltwater from glaciers may affect ocean currents and climate.

Evaporation and condensation affect Earth's heat balance.

Snow and glacial ice affect climate.

Living organisms maintain the chemical composition of the atmosphere.

Evapotranspiration from plants affects climate.

Humans are altering atmospheric composition and climate.

Plants affect albedo, which affects atmospheric temperature.

Marine organisms sequester carbon.

Methane, produced by living organisms, is a greenhouse gas.

ATMOSPHERE

HYDROSPHERE

Rain replenishes streams and ground water as part of the hydrological cycle.

Climate affects evaporation and precipitation.

Climate determines how much snow and glacial ice exist.

SEVERAL EXAMPLES IN THIS chapter illustrate the long-range connectivity of various components of the Earth's systems. People driving cars in North America or Europe inadvertently warm the atmosphere, which melts portions of the Antarctic Ice cap, which causes a rise in sea level, which could flood cities. Or as another example, melting glaciers, combined with changing drainage patterns led to a historical disruption of the North Atlantic thermohaline current systems, which caused the Little Ice Age. With the background you have now, go back and read the opening essay in Chapter 1, comparing the atmospheres of Venus, Earth, and Mars. This example, coupled with an understanding of the numerous feedback and threshold loops in climate systems, reminds us of the fragility of our planet's environment. ■

SUMMARY

Global climate has changed throughout Earth's history, from extreme cold with extensive glaciation to a 243-million-year warm period from the start of the Mesozoic Era almost to the present. During the last 100,000 years, the mean annual global temperature has changed frequently and dramatically, although the past 10,000 years, during which civilization developed, witnessed anomalously stable climate.

Past climates are measured by historical records, tree rings, pollen assemblages, oxygen isotope ratios in glacial ice, glacial evidence, plankton assemblages and isotope studies in ocean sediment, and fossils in sedimentary rocks.

Despite its immense complexity, Earth's atmospheric temperature is determined by the heat of the Sun, albedo, and heat retention by the atmosphere.

Astronomical causes of climate change include the variations in Earth's orbit and the tilt of its axis, changes in solar radiation, and bolide impacts.

Water in its various forms can cause warming or cooling. Water vapor warms the atmosphere, but clouds can cause warming or cooling. Glaciers and snowfields reflect sunlight and cause cooling. Evaporation is a cooling process, while condensation releases heat. The evaporation/condensation cycle transports heat from the equatorial regions to the poles. Weathering alters the carbon dioxide concentration in the atmosphere. Ocean currents transport heat and moisture.

Carbon circulates among all four of Earth's realms. Carbon dioxide is a greenhouse gas that warms the atmosphere. Carbon exists in the biosphere as the fundamental building block for organic tissue. Carbon dioxide gas dissolves in seawater to form bicarbonate and carbonate ions. Carbon exists in the crust and mantle in several forms: (1) Marine organisms absorb calcium and carbonate ions and convert them to solid calcium carbonate (shells and skeletons). As a result, large quantities of carbon exist in limestone and marine sediment. (2) Fossil fuels are largely composed of carbon. (3) Large quantities of carbon exist as methane hydrates on the sea floor. (4) Carbon also exists in the deeper mantle. Some of this is primordial carbon trapped during Earth's formation and some is carried into the mantle on subducting plates.

Natural climate change also results from changes in the positions of continents, the growth of mountains, and volcanic eruptions. Independent climate-changing factors are linked by Earth systems interactions.

Humans release carbon dioxide into the atmosphere when they burn fuel. Logging, some industrial processes, and some aspects of agriculture contribute greenhouse gases to the atmosphere. Most scientists agree that human introduction of greenhouse gases has altered climate in the past century.

On the positive side, global warming would lead to a longer growing season in the high latitudes. However, studies have shown that rice yields decline in a warmer world and the ratio of respiration/photosynthesis increases, leading to more carbon dioxide in the atmosphere.

Global warming will affect precipitation and snowmelt, extreme weather, biodiversity, and sea level.

Climate feedback loops include snowmelt and albedo and changes in rates of plant respiration and photosynthesis.

The collapse of the West Antarctic Ice Sheet could lead to rising sea level and further global warming. Alternatively, some scientists have calculated that slight global warming could disrupt the North Atlantic thermohaline circulation, leading to global cooling.

The United States has not signed the Kyoto Treaty. Proponents of the treaty argue that the economic downside would be minimal, whereas the possible effects of global warming could be catastrophic.

For Review

1. Explain why early Earth was warm, even though the Sun emitted 20 to 30 percent less energy at that time.

2. Briefly outline climate changes both from 100,000 to 10,000 years ago and from 10,000 years ago to the present.

3. Explain why oxygen isotope measurements provide paleotemperature data from cores in glacial ice.

4. Explain how pollen and plankton assemblages are used to measure past climates.

5. List several techniques for measuring climate 100 to 500 million years ago.

6. Discuss the role of solar output in determining climate.

7. Outline the complex, and often conflicting, relationship between water and climate.

8. In what forms is carbon present in the atmosphere, the biosphere, the hydrosphere, and the crust?

9. Draw a rough sketch of the carbon cycle and explain the chemical transformations shown in your cycle.

10. Explain how weathering can affect atmospheric composition and climate.

11. Explain how continental positions can affect climate. Discuss the time scale for various tectonic influences on climate.

12. Explain how volcanic eruptions can cause either a cooling or a warming of the atmosphere.

13. Explain why falling sea level contributes to global cooling.

14. List and explain four feedback loops that affect climate. For each one, outline the contributions of geosphere, atmosphere, biosphere, and hydrosphere.

15. Discuss the scientific arguments for our concern that humans are causing global warming.

16. Discuss the consequences of a warmer Earth. How would temperature changes affect precipitation? How would changes in temperature and precipitation affect agriculture, ecosystems, and human society?

17. Explain how an increase in greenhouse gases in the atmosphere could lead to global cooling.

18. Explain how break-up of the West Antarctic Ice Sheet could lead to global warming and flooding of coastal areas.

19. Discuss the current status of the Kyoto Treaty.

For Discussion

1. Give an example of a feedback and a threshold effect in a science other than earth science (such as psychology, political science, or any other science).

2. Based on Figure 21.5, list the useful time ranges for historical records, pollen, and plankton as indicators of paleoclimates. Explain why each of these techniques is not useful farther back in time.

3. Discuss the difficulties and costs involved in reducing carbon dioxide emissions.

4. Type the keywords "climate feedback loops" into an Internet search engine. Find an example of a feedback loop not mentioned in this text and discuss with your classmates.

5. Explain why a small change in average global temperatures can have environmental, economic, and political effects.

6. Do you feel that the United States should ratify the Kyoto Treaty? Defend your answer.

7. Discuss the argument that the U.S. economy would be damaged if the country set a priority to reduce greenhouse gas emissions.

8. Stephen Jay Gould wrote, "During most of the past 600 million years, the Earth has been sufficiently warm so that even the bottom of an orbital cycle produced no ice caps" (*Natural History*, May 1991, p. 18). Explain this statement in your own words.

9. According to David des Marais of Ames Research Center, about 2.2 billion years ago, rapid plate motion led to rapid rise of mountains throughout the world. As the mountains rose, large quantities of organic-rich sediment eroded and were transported into the sea, where they were buried. Predict how this burial might alter atmospheric composition.

ThomsonNOW

Assess your understanding of this chapter's topics with additional comprehensive interactivities at **www.thomsonedu. com/login**, as well as current and up-to-date web links, additional readings, and exercises.

UNIT 6

ASTRONOMY

CHAPTER 22

Motions in the Heavens

© NASA/Roger Ressmeyer/CORBIS

NASA astronauts replace the optics and gyroscopes on the Hubble Space Telescope.

Ancient astronomers used cycles of the Sun, Moon, and stars to tell time and to mark the seasons. However, these objects are so far away that, until recently, astronomers knew very little about them. The ancient Greeks believed that Earth is stationary and that the Sun rises and sets because it orbits our planet daily. Today we realize that Earth's rotation causes the cycle of night and day.

The Sun shines brightly in the sky and warms Earth, whereas the stars twinkle feebly in the darkness. Modern astronomers understand that the Sun and stars are similar and that the Sun is merely much closer. However, the ancients had little concept of astronomical distances or sizes, and, as a result, they did not realize that the Sun is just an average-sized star.

In this and the following two chapters, we will study the motions, compositions, and structures of the celestial bodies. ■

22.1 The Motions of the Heavenly Bodies

Even a casual observer notices that both the Sun and Moon rise in the east and set in the west. In the midlatitudes, summer days are long and the Sun rises high in the sky. Winter days are shorter, and at midlatitudes in the Northern Hemisphere, the Sun never rises very high above the southern horizon, even at noon. In contrast to the yearly cycle of seasons, the Moon completes its cycle once a month. It is full and round one night, then darkens slowly until it is a thin crescent. It disappears completely after two weeks, then reappears as a sliver that grows again, completing the entire cycle in about 29.5 days.

If you had been a cowboy in the 1800s, the foreman might have told you to guard the herd at night until the "dipper holds water." This phrase refers to two features of the night sky. First, stars remain in fixed positions relative to one another. This fact led the ancients to identify groups of stars, which they called **constellations** (◁ Figure 22.1). Second, in the Northern Hemisphere, the Pole Star, or North Star, is a motionless fixed point in the sky, and all other stars appear to revolve around it (◁ Figure 22.2). This motion provided the clock for the cowboy on night watch because the dipper alternately holds and spills water as it revolves around the Pole Star.

Constellations appear and disappear with the seasons. For example, the Egyptians noted that Sirius, the brightest star in the sky, became visible at dawn just before the Nile began to flood. Farmers would therefore plant crops when this star first appeared, with the assurance that the high water soon would irrigate their fields.

Ancient astronomers noted several objects that appeared to be stars but were different because they changed position with respect to the stars. The ancient Greeks called these objects **planets**, from the word meaning "wanderers." For most of the year, planets appear to drift eastward with respect to the stars, but sometimes they seem to reverse direction and drift westward. This apparent reverse movement is called **retrograde motion** (◁Figure 22.3).

22.2 Aristotle and the Earth-Centered Universe

The Greek philosopher and scientist Aristotle proposed a **geocentric**, or Earth-centered, Universe. In this model, Earth is stationary and positioned at the center of the Universe. A series of concentric **celestial spheres** made of transparent crystal surrounds Earth. The Sun, Moon, planets, and stars are imbedded in the spheres like jewels (◁ Figure 22.4). At any one time a person can see only a portion of each sphere, but as the sphere revolves around Earth, objects appear and disappear.

Aristotle based his conclusions on two observations. First, he reasoned that Earth must be stationary because people have no sensation of motion. People tend to fall off the back of a chariot when horses start to gallop, but they do not fall off Earth, so Aristotle reasoned that Earth must not be moving.

FOCUS ON

The Constellations

The oldest-known written record of the constellations appears in a 2000 BCE Sumerian manuscript. Other accounts occur in old manuscripts and legends from numerous cultures.

Ancient astronomers and religious leaders imagined that constellations represented animals such as Leo, the lion, or gods such as Aquarius, the Sumerian deity who pours the waters of immortality onto Earth. The constellations have no particular significance to modern astronomy other than as a convenience to naming parts of the sky map. For example, Orion, the hunter, is easily recognized by the three bright stars of his sword belt (Figure 1). The brightest star in Orion is Rigel, which forms his left kneecap. If you look at Rigel through a telescope or a good set of binoculars, you will note that it is bluish. In contrast,

Betelgeuse (pronounced "beetle juice"), which defines Orion's right shoulder, is red. As we will learn in Chapter 24, color is an indication of the temperature of a star, and the temperature is an indication of its size and age. Blue stars such as Rigel are quite different from red ones such as Betelgeuse. In addition, the two stars are not even close to one another: Rigel is almost twice as far from Earth as Betelgeuse is. The only relationship between the two is that they appear close together in the night sky.

Each star in a constellation is independent from the others and travels in its own path. These relative motions are not apparent in a single lifetime, but they become significant over geologic time. As a result, constellations change shape over geologic time.

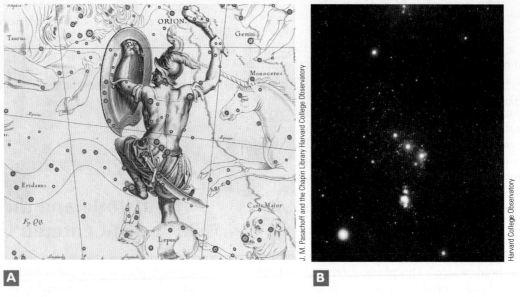

A

J. M. Pasachoff and the Chapin Library Harvard College Observatory

B

Harvard College Observatory

◆ **FIGURE 1** **(A)** A seventeenth-century star atlas depicts the constellation Orion. **(B)** As seen in the night sky, the three stars in Orion's belt form a line in the center of the photograph. The bright stars below it represent his sword.

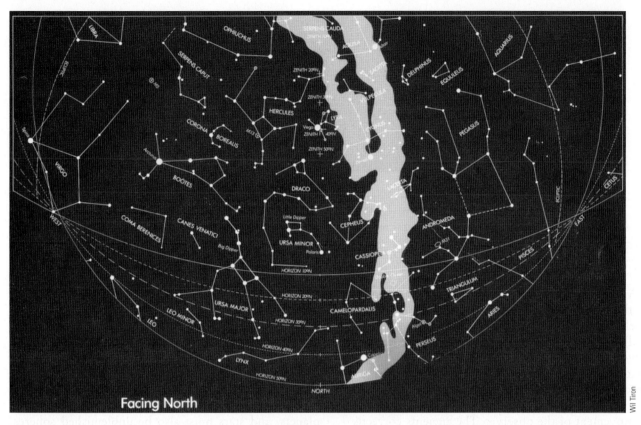

Facing North

◇▶ **FIGURE 22.1**   Different constellations appear at different seasons. This view shows the summer sky, as viewed by an observer facing north. The light shaded area is the Milky Way.

◇▶ **FIGURE 22.2**   A time exposure of the night sky shows the rotation of the stars around the Pole Star, which is nearly motionless.

◆ **FIGURE 22.3** A planetarium simulation of the movement of Mars from August 1, 1990, through April 1, 1991. Mars appears to reverse direction, forming a retrograde loop. This motion was difficult to explain in the geocentric model.

Aristotle's second observation was based on **parallax**, the apparent change in position of an object due to the change in position of the observer. To understand parallax, consider the fence posts in ◆ Figure 22.5. They appear to stand one in front of the other when the photographer is in position A and seem offset when the photographer steps to the side (B). Thus the relative positions of the stationary posts appear to change with the movement of the observer. The ancients correctly reasoned that if Earth moved around the Sun, the stars should change position relative to one another (◆ Figure 22.6). Because they observed no parallax shift, they concluded that Earth must be stationary. The mistake arose not out of faulty reasoning but because

stars are so far away that their parallax shift is too small to be detected with the naked eye.

Aristotle incorporated philosophical concepts, in addition to observation, into scientific theory. He argued that the gods would create only perfection in the heavens and that a sphere is a perfectly symmetrical shape. Therefore, the celestial spheres must be a natural expression of the will of the gods. In addition, the Sun, Moon, planets, and stars must also be unblemished spheres. This incorporation of philosophy into science retarded debate and allowed dogma to rule over logic.

Aristotle's theory, although incorrect, did explain the motions of the Sun, Moon, and stars (◆Figure 22.7). However, it failed to explain the retrograde motion

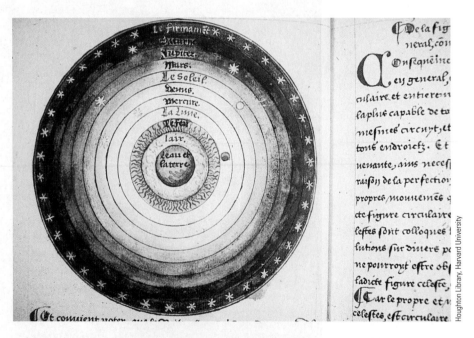

◆ **FIGURE 22.4** In Aristotle's cosmology, water and Earth lie in the center of the universe, surrounded by air and fire. Beyond these four basic elements, the Moon, Mercury, Venus, the Sun, Mars, Jupiter, Saturn, and the stars lie in concentric celestial spheres.

Courtesy of Graham R. Thompson/Jonathan Turk

◀▷ **FIGURE 22.5** Parallax is illustrated by two photographs of a fence on the Montana prairie. **(A)** The photographer is nearly in line with the fence, so the distant posts appear to be in line. **(B)** When the photographer moved, the posts appear to have shifted position. Now we can see spaces between the more distant posts. Of course, the posts have not moved, only the photographer has. The same effect has been observed in astronomical studies. As Earth revolves about the Sun, the stars appear to shift position relative to one another.

INTERACTIVE QUESTION: *Imagine that there are two poles in the ground set 10 meters apart. You are directly in line with the two poles, 100 meters from the closest one. Now you step one meter to the left. Draw a scale diagram of your position and the positions of the two poles. With a protractor, estimate the parallax angle between you and the poles. Discuss the magnitude of the parallax angle if the units in your diagram were changed from meters to light-years.*

◀▷ **FIGURE 22.6** A nearby star appears to change position with respect to the distant stars as Earth orbits around the Sun. This drawing is greatly exaggerated; in reality, the distance to the nearest stars is so much greater than the diameter of Earth's orbit that the parallax angle is only a small fraction of a degree.

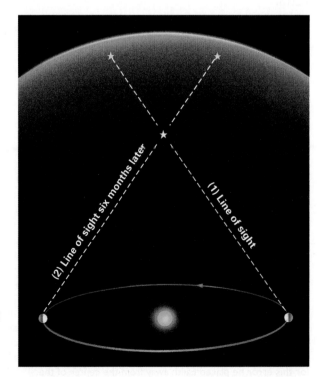

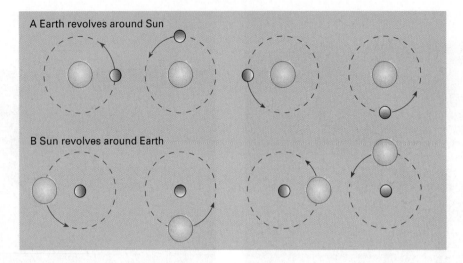

A Earth revolves around Sun

B Sun revolves around Earth

◆ **FIGURE 22.7** Series A shows Earth revolving around the Sun, and series B shows the Sun revolving around Earth. Now lay some thin paper over series A and trace the outlines of the Sun and Earth but not the arrows or orbits. Lay this tracing over series B and note that they match exactly, after you shift the paper for each sketch to make sure the Sun and Earth superimpose. Conclusion: There is no apparent difference between Earth revolving around the Sun and vice versa, provided you do not refer to anything else, such as the outline of these diagrams or another star.

of the planets. In about CE 150, Claudius Ptolemy modified the celestial sphere model to incorporate retrograde motion. In Ptolemy's model, each planet moves in small circles as it follows its larger orbit around Earth (◆ Figure 22.8). Ptolemy's sophisticated mathematics accurately described planetary motion and therefore his model was accepted. Still, he retained Aristotle's erroneous idea that the Sun and the planets orbit a stationary Earth.

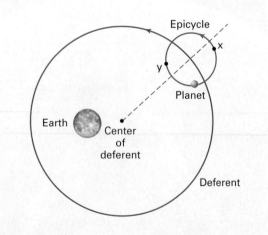

◆ **FIGURE 22.8** Ptolemy's explanation of retrograde motion. Each planet revolves in a small orbit (the epicycle) around the larger orbit (the deferent). When the planet is in position x, it appears to be moving eastward. When it is in position y, it appears to have reversed direction and moves westward. Ptolemy did not realize that the planets moved in elliptical orbits. To compensate for this error, he placed Earth away from the center of the deferent.

22.3
The Renaissance and the Heliocentric Solar System

Aristotle's and Ptolemy's ideas remained essentially unchallenged for 1,400 years. Then, in the 120 years from 1530 to 1650, several Renaissance scholars changed our understanding of motion in the Solar System and, in the process, revolutionized scientific thought.

Copernicus

In 1530 a Polish astronomer and cleric, Nicolaus Copernicus, proposed that the Sun, not Earth, is the center of the Solar System and that Earth is a planet, like the other "wanderers" in the sky (◆ Figure 22.9). Copernicus based his hypothesis on the philosophical premise that the Universe must operate by the simplest possible laws. Copernicus believed Ptolemy's model was too complex and that the motions of the heavenly bodies could be explained more concisely by a **heliocentric** model with the Sun at the center of the Solar System.

◆ Figure 22.10 shows how the heliocentric model explains retrograde motion. Assume that initially Earth is in position 1 and Mars is in position 1'. An observer on Earth looks past Mars (as shown by the white line) and records its position relative to more distant stars. Mars appears to be in position 1" in the night sky. After a few weeks Earth has moved to position 2, Mars has moved to 2', and Mars's position relative to the stars is indicated by 2". In the Copernican model, Earth moves

◆ **FIGURE 22.9** Nicolaus Copernicus was the first scientist to propose that the Sun, not Earth, is the center of the Solar System.

faster than Mars because Earth is closer to the Sun. Therefore, Earth catches up to and eventually passes Mars, as shown by positions 3, 3′ and 4, 4′. During this passage, Mars appears to turn around and move backward, although, of course, this appearance is merely an illusion against the backdrop of the stars. Mars appears to reverse direction again through positions 5 and 6, thus completing one cycle of retrograde motion. In reality, Mars never reverses direction, it just appears to behave in this way as Earth catches up to it and then passes it.

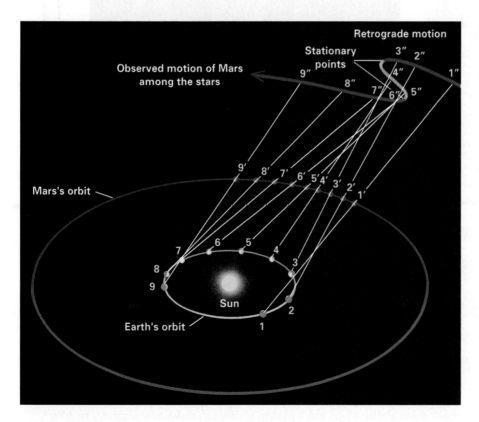

◆ **FIGURE 22.10** Retrograde motion is explained within the heliocentric model by changes in the relative positions of Earth and the planet we are observing.

Brahe and Kepler

In the late 1500s, Tycho Brahe, a Danish astronomer, accurately mapped the positions and motions of all known bodies in the Solar System. His maps enabled him to predict where any planet would be seen at any time in the near future, but he never explained their motions. Brahe drank himself to death at a party in 1601, and his student Johannes Kepler inherited the vast amount of data Brahe had collected. Kepler calculated that the planets moved in elliptical orbits, not circular ones, and he derived a set of mathematical formulas to describe their paths. However, Kepler never answered the important question, Why do the planets move in orbits around the Sun rather than flying off into space in straight lines?

Galileo

Galileo was an Italian mathematician, astronomer, and physicist who made so many contributions to science that he is often called the father of modern science (◆ Figure 22.11). Perhaps most significant was his realization that the laws of nature must be understood through observation, experimentation, and mathematical analysis. This concept freed scientists from the confines of Aristotelian dogma.

Recall that Aristotle's geocentric theory was based on two observations and one philosophical premise: (1) If Earth were moving, people should fall off, but they do not. (2) Aristotle was unable to observe parallax shift of the stars. (3) The gods would create only an unblemished, symmetrical Universe. Galileo's experiments and observations showed that Aristotle's model was incorrect and led him to support Copernicus's heliocentric model.

Galileo studied the motion of balls rolling across a smooth marble floor, and legend tells us he also dropped objects from the leaning Tower of Pisa. He organized the results of these experiments into laws of motion that were later expanded and quantified by Isaac Newton. One of these laws states that "an object at rest remains at rest and an object in uniform motion remains in uniform motion until forced to change." This corresponds to Newton's first law of motion, the law of **inertia**. Inertia is the tendency of an object to resist a change in motion.

According to the law of inertia, if Earth were in uniform motion, a person on its surface would

◀▶ **FIGURE 22.11** Galileo Galilei is often called the father of modern science.

be in uniform motion along with it. The person therefore would travel with Earth and would not fall off and be left behind, as Aristotle had assumed. In fact, the person could not even feel the motion.

Galileo built his first telescope in 1609 and turned it to the heavens soon afterwards. He learned that the hazy white line across the sky called the Milky Way was not a cloud of light as Aristotle had proposed but a vast collection of individual stars. Next, he trained his telescope at the Moon and saw hills, mountains, giant craters, and broad, flat regions on its surface, which he thought were seas and therefore named **maria** (Latin for "seas") (◀▶ Figure 22.12). Looking at the Sun, Galileo recorded dark regions, called sunspots, that appeared and then vanished.

Although these discoveries had no direct bearing on the controversy of a geocentric versus a heliocentric Universe, they were important because they led Galileo to question Aristotle's views. The prevailing scientific opinion at the time was that if the Milky Way were a collection of stars, Aristotle would have known about it. Furthermore, Galileo's observations of the Sun and the Moon did not agree with Aristotle's philosophical assumptions that the heavenly bodies were perfectly homogeneous and unblemished. Galileo reasoned that if Aristotle was wrong about the structures of the Milky Way, the Sun, and the Moon, perhaps he was also wrong about the motions of these celestial bodies.

When Galileo studied Jupiter, he saw four moons orbiting the giant planet. (Today we know that Jupiter has 60 moons, but only 4 are large enough to have been seen with Galileo's telescope.) According to the geocentric model, every celestial body orbits Earth.

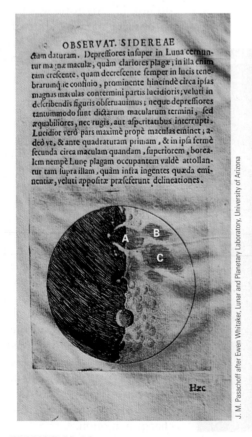

◀▶ **FIGURE 22.12** A comparison of Galileo's drawing of the Moon with a modern photograph shows how accurately he drew topographical features, such as the maria A, B, and C. Although Galileo thought that the maria were seas, today we know that they are flat lava flows.

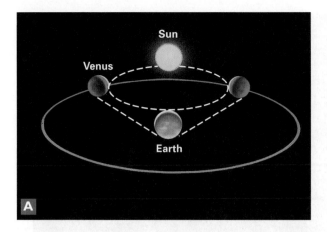
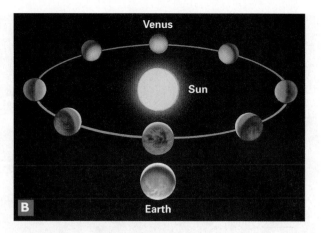

◆ FIGURE 22.13   (A) In Ptolemy's theory, Venus could never move farther from the Sun than is shown by the dotted lines. Therefore it would always appear as a crescent. (B) In the heliocentric theory, Venus passes through phases like the Moon. Galileo observed phases of Venus through a telescope and concluded that the planets must orbit the Sun.

However, Jupiter's moons clearly orbited Jupiter, not Earth. The contradiction increased Galileo's doubt about Aristotelian theories.

Finally, he observed that the planet Venus passes through phases as the Moon does. Such cyclical phases could not be readily explained in the geocentric model, even with Ptolemy's modifications (◆ Figure 22.13). As a result of his observations, Galileo proposed that the Sun is the center of the Solar System and that the planets orbit around it. The Roman Catholic Church rejected Galileo's model and ordered him to recant his conclusions. He bowed to the pressure rather than undergo imprisonment and torture. Despite his recantations, he died blind, poor, and under arrest.

Isaac Newton and the Glue of the Universe

Galileo successfully described the motions in the Solar System, but he never addressed the question, Why do the planets orbit the Sun rather than fly off in straight lines into space? Aristotle had observed that an arrow shot from a bow flies in a straight line, but celestial bodies move in curved paths. He reasoned that arrows have an essence that compels them to move in straight lines, and planets and stars have an essence that compels them to move in circles. In the Renaissance, however, this answer was no longer acceptable.

Isaac Newton was born in 1643, the year Galileo died (◆ Figure 22.14). During his lifetime Newton made important contributions to physics and developed calculus. A popular legend tells that Newton was sitting under an apple tree one day when an apple fell on his head and—presto—he discovered gravity. Of course, people knew that unsupported objects fall to the ground long before Newton was born. But he was the first to recognize that gravity is a universal force that governs all objects, including a falling apple, a flying arrow, and an orbiting planet.

According to the laws of motion introduced by Galileo and expanded by Newton, a moving body travels in a straight path unless it is acted on by an outside force. Just as a ball rolls in a straight line unless it is forced to change direction, a planet also moves in a straight line unless a force is exerted on it. The gravitational attraction between the Sun and a planet forces the planet to change direction and move in an elliptical orbit. Gravity is the glue of the Universe and affects the motions of all celestial bodies.

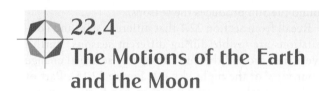

22.4
The Motions of the Earth and the Moon

By about 1700 astronomers knew that the Sun is the center of our Solar System and that planets revolve around it in elliptical orbits. In addition to revolving around the

National Portrait Gallery, London

◆ FIGURE 22.14   Sir Isaac Newton developed mathematical laws that describe the motions of all objects larger than atomic particles.

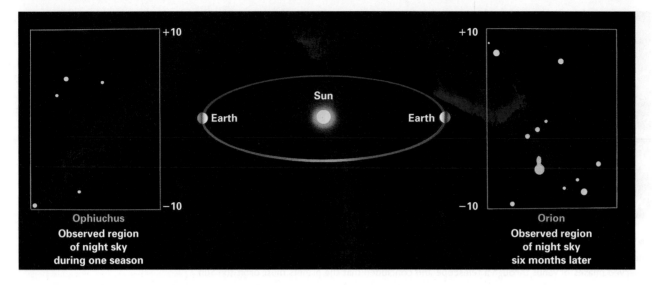

Sun, the planets simultaneously spin on their axes. Earth spins approximately 365 times for each complete orbit around the Sun. Each complete **rotation** of Earth represents one day. As Earth rotates about its axis, the Sun, Moon, and stars appear to move across the sky from east to west. We explained in Chapter 18 that Earth's axis is tilted and that this tilt combined with Earth's orbit around the Sun produces the seasons.

Recall from section 22.1 that different stars and constellations are visible during different seasons. Earth's **revolution** around the Sun causes this seasonal change in our view of the night sky (◀▶ Figure 22.15). Part of the sky is visible on a winter night when Earth is on one

side of the Sun, and a different part is visible on a summer night 6 months later.

More recent measurements have revealed several additional types of planetary motion. As Earth rotates, its axis wobbles like a spinning top. This motion is called **precession** (◀▶ Figure 22.16). At the present time, the axis points toward Polaris, the North Star. In 12,000 years, the axis will point toward Vega, and Vega will become the North Star. Because precession cycles on a 26,000-year period, Earth's axis will point toward Polaris again by the year 28000. Recall from Chapter 13 that precession is one factor that contributed to glacial advances and retreats within the Pleistocene Ice Age.

In addition, the Moon's gravity pulls Earth slightly out of its orbit. Although we normally draw Earth's orbit around the Sun as a line, the Moon's effect actually causes Earth to spiral slightly as it circles the Sun. At the same time, the entire Solar System is moving toward the star Vega. Our Sun also orbits the center of the Milky Way galaxy. Traveling at a speed of 220 kilometers per second, it completes its orbit in about 200 million years. The entire Milky Way galaxy speeds along, carrying our Sun and planets through intergalactic space.

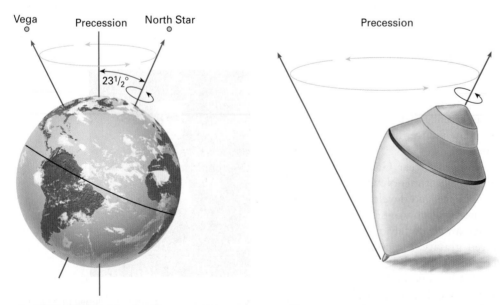

FIGURE 22.16 Earth's axis wobbles or precesses like a top, completing one precession cycle every 26,000 years.

Motion of the Moon

The gravitational attraction between the Moon and Earth not only holds the two in orbit around each other but it also affects the surfaces and interiors of both. We learned that the Moon's gravitation causes tides on Earth. In turn, Earth's gravitation pulls on the Moon sufficiently to cause it to bulge, despite the fact that it is solid rock. Earth's gravity attracts the bulge, so the side of the Moon with the bulge always faces Earth. As a result, the Moon rotates on its axis at the same rate at which it orbits Earth. Thus, we always see the same side of the Moon, and the other side was invisible to us until the space age.

Almost everyone has observed that the Moon appears to change shape over a month's time. When the moon is dark, it is called the **new moon**. Four days after the new moon, a thin **crescent moon** appears. The crescent grows and the Moon is **waxing** until it forms the first quarter 7 days after the new moon. Ten days after the new moon, we see the bright, waxing, **gibbous moon**, with only a sliver of dark. About 14 to 15 days after the new moon, the moon appears circular and is a **full moon** (◁▶ Figure 22.17). A few evenings later, part of the disk is darkened. As the days progress, the visible portion shrinks and the Moon is said to be **waning**.

Eventually, only a tiny, curved sliver, the waning crescent, is left. After a total cycle of about 29.5 Earth days, the Moon is dark because it has returned to the new-moon position.

When the Moon is full, it rises approximately at sunset. On each successive evening it rises about 53 minutes later, so that in 7 days it rises in the middle of the night. In about a month the cycle is complete, and the next full moon rises in the early evening again.

To understand the phases of the Moon, we must first realize that the Moon does not emit its own light but reflects light from the Sun. The half of the Moon facing the Sun is always bathed in sunlight, while the other half is always dark. The phases of the Moon depend on how much of the sunlit area is visible from Earth. In turn, this visible area depends on the relative positions of the Sun, Moon, and Earth. Earth's orbit around the Sun forms an elliptical plane, with the Sun in the same plane and near its center. In a similar way, the Moon's orbit around Earth describes another plane with Earth at its center. But the plane of the Moon's orbit is tilted 5.2° with respect to Earth's orbital plane. As a result, the Moon is not usually in the same plane as that of Earth and the Sun (◁▶ Figure 22.18). Therefore Earth's shadow does not normally fall on the Moon, and the Moon's shadow does not normally fall on Earth.

As the Moon orbits Earth, the side of the Moon facing the Sun is fully illuminated. When the Moon is on the opposite side of Earth from the Sun, the entire sunlit area is visible and the Moon appears full (◁▶ Figure 22.17E). However, if the Moon moves to a position between Earth and the Sun, the Moon's sunlit side faces away from us. In this position the Moon appears new because the surface that faces Earth is not illuminated. Midway between these two extremes, when the Moon is located 90° from the line between Earth and the Sun, half of the illuminated side is visible and the Moon is quartered. With each complete revolution around Earth, the Moon passes through one complete cycle, from new to first quarter, full, and third quarter, and back to new again.

But why does the Moon rise at a different time each day? It travels in a complete elliptical orbit about Earth approximately once a month. Imagine that today the Moon rises at 6:00 p.m. Twenty-four hours later Earth will have rotated once, but in the meantime the Moon will have moved a short distance across the sky; therefore, Earth must rotate an additional 13.2° in order to catch up with the Moon (◁▶ Figure 22.19). Thus, the timing of the moonrise depends on the Moon's position above a point on Earth, and this also changes in a cyclical, monthly pattern.

Eclipses of the Sun and the Moon

Recall that the plane of the Moon's orbit around Earth is tilted with respect to that of Earth's orbit about the Sun, so normally the Moon lies slightly out of the plane of the Earth–Sun orbit (◁▶ Figure 22.20A). As a result, during a new moon the Moon's shadow misses the Earth (◁▶ Figure 22.20B) and at a full moon Earth's shadow misses the Moon (◁▶ Figure 22.20C).

However, on rare occasions, the new Moon passes through the Earth–Sun orbital plane. At these times, the Moon passes directly between Earth and the Sun (◁▶ Figure 22.20D). When this happens, the Moon's shadow falls on Earth, producing a **solar eclipse** (◁▶ Figure 22.20E). As the Moon slides in front of the Sun, an eerie darkness descends, and Earth becomes still and quiet. Birds return to their nests and stop singing. While the eclipse is total, the Moon blocks out the entire surface of the Sun, but the outer solar atmosphere, or **corona**, normally invisible owing to the Sun's brilliance, appears as a halo around the black Moon (◁▶ Figure 22.21). Due to the relative distances between Sun, Moon, and Earth and their respective sizes, the Moon's shadow is only a narrow band on Earth (◁▶ Figure 22.22). The band where the Sun is totally eclipsed, called the **umbra**, is never wider than 275 kilometers. In the **penumbra**, a wider band outside of the umbra, only a portion of the Sun is eclipsed. During a partial eclipse of the Sun, the sky loses some of its brilliance but does not become dark. Viewed through a dark filter such as a welder's mask, a semicircular shadow cuts across the Sun (◁▶ Figure 22.23).

If the Moon passes through the Earth–Sun orbital plane when it is full, then Earth lies directly between the Sun and the Moon. At these times Earth's shadow falls on the Moon and the Moon temporarily darkens to produce a **lunar eclipse** (◁▶ Figure 22.20F). Lunar eclipses are more common and last longer than solar eclipses because Earth is larger than the Moon and therefore its shadow is

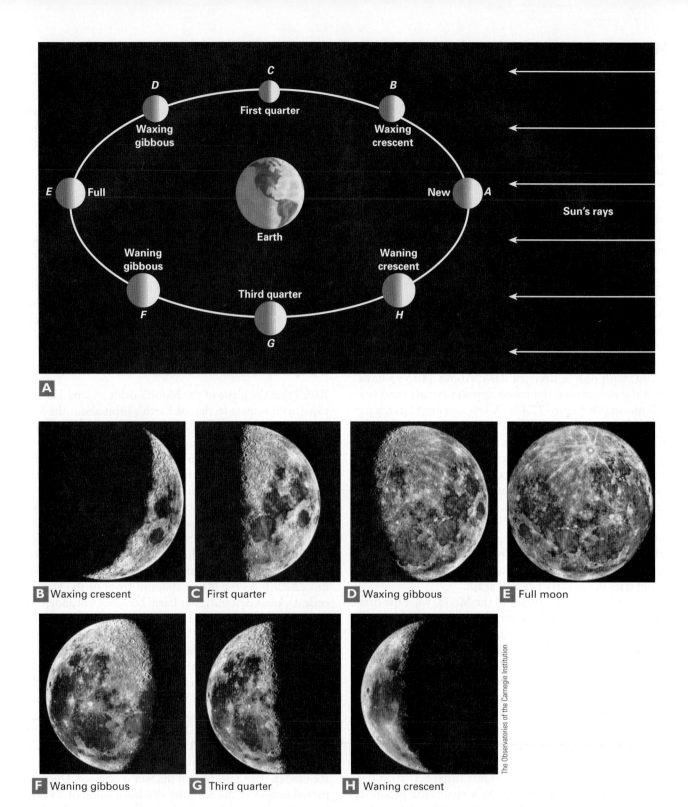

B Waxing crescent   **C** First quarter   **D** Waxing gibbous   **E** Full moon

F Waning gibbous   **G** Third quarter   **H** Waning crescent

The Observatories of the Carnegie Institution

◀▶ **FIGURE 22.17** Approximately every 29.5 Earth days, the Moon passes through a complete cycle of phases. The upper part of the drawing shows the Earth–Moon orbital system viewed over the course of a month. The lower portion shows what the Moon looks like from Earth at the various phases.

INTERACTIVE QUESTION: *How would the phases of the Moon differ if the Moon orbited Earth within the Earth–Sun orbital plane?*

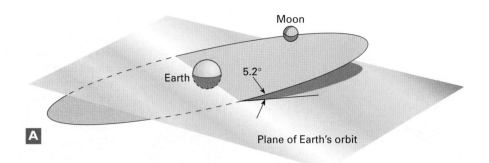

A

Plane of Earth's orbit

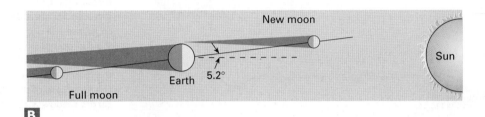

B

◇▶ **FIGURE 22.18** **(A)** The Sun and Earth lie in one plane, while the Moon's orbit around Earth lies in another. **(B)** A sideways look at the Sun, Moon, and Earth shows that most of the time the Moon's shadow misses Earth and Earth's shadow misses the Moon. Scales are exaggerated for emphasis.

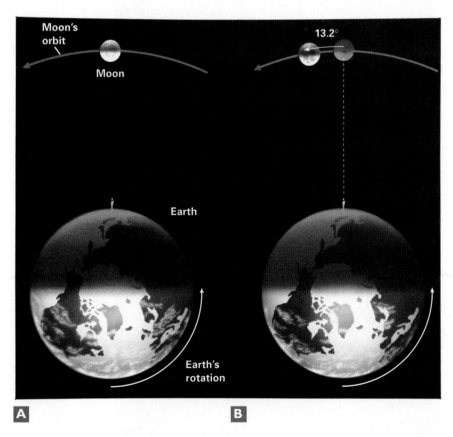

A **B**

◇▶ **FIGURE 22.19** The Moon moves 13.2° every day. **(A)** The Moon is directly above an observer on Earth. **(B)** One day later, Earth has completed one complete rotation, but the Moon has traveled 13.2°. Earth must now travel for another 53 minutes before the observer is directly under it again.

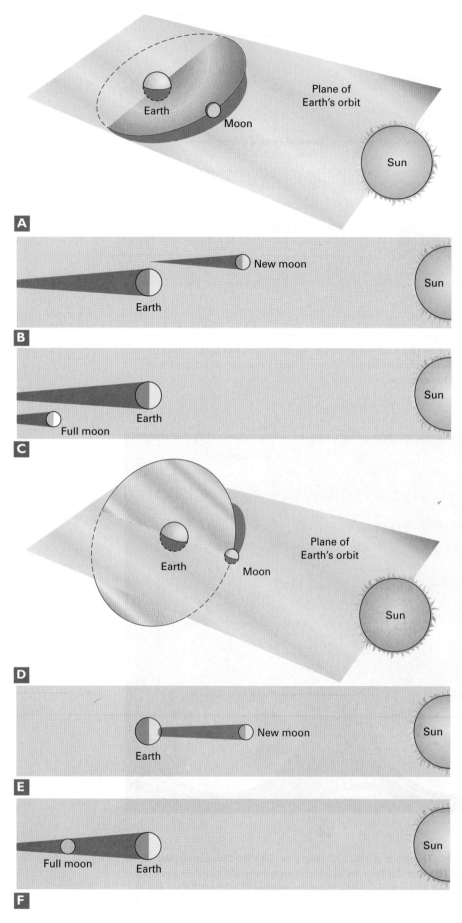

◈ **FIGURE 22.20** Eclipses of the Sun and Moon. The Sun and Earth lie in one plane, while the Moon's orbit around Earth lies in another. Scales are exaggerated for emphasis. **(A)** Normally the Moon lies out of the plane of the Earth–Sun orbit. **(B)** During the new moon, the Moon's shadow misses Earth. **(C)** During the full moon, Earth's shadow misses the Moon. **(D)** However, if the Moon passes through the Earth–Sun plane when the three bodies are aligned properly, then an eclipse will occur. **(E)** An eclipse of the Sun occurs when the Moon is directly between the Sun and Earth, and the Moon's shadow is cast on Earth. **(F)** An eclipse of the Moon occurs when Earth's shadow is cast on the Moon.

FIGURE 22.23 During a partial solar eclipse, a dark shadow obliterates a portion of the Sun.

FIGURE 22.21 The solar corona appears as a bright halo around the eclipsed Sun. This photograph was taken during the July 11, 1991, total eclipse of the Sun, in La Paz, Baja California.

more likely to cover the entire lunar surface. A lunar eclipse can last a few hours.

22.5 Modern Astronomy

Once astronomers understood the relative motions of the Sun, Moon, and planets, they began to ask questions about the nature and composition of these bodies and to probe more deeply into space to study stars and other objects in the Universe. How do we gather data about such distant objects?

If you stand outside at night and look at a speck of light in the sky, the information you receive is limited by several factors. For one, your eye detects only visible light, which is only one millionth of 1 percent of the electromagnetic spectrum. Thus, more than 99.99 percent of the spectrum is invisible to the naked eye. In addition, the naked eye collects little light, and you may not see faint or distant objects at all. Your eye also has poor **resolution**; it may see one dot when two exist (Figure 22.24). Finally, the light you see has been distorted by Earth's atmosphere. Modern astronomers attempt to overcome these difficulties with telescopes and other instruments.

Optical telescopes

A telescope is a device that collects light from a wide area and then focuses it where it can be detected. Detection devices may be simple, such as your eye or a photographic plate. In other instances, complex instruments analyze the images. In order to create a detectable image from a weak signal, a telescope may collect light for several hours.

Galileo and many other early astronomers used **refracting telescopes**, which employ two lenses. The first, called the **objective lens**, collects light from a distant object, and the **eyepiece** is a small magnifying lens. Light bends, or refracts, when it passes through the curved

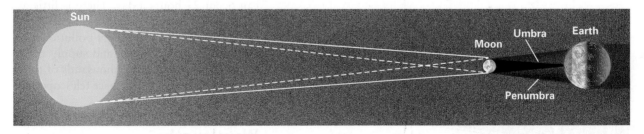

FIGURE 22.22 A total solar eclipse is viewed in the narrow band, called the umbra, formed by the projection of the Moon's shadow on Earth. The penumbra is the wider band where a partial eclipse is visible. (Drawing is not to scale.)

◀▶ **FIGURE 22.24** Resolution is the degree to which details are distinguishable in an image. The photograph on the left has poor resolution and appears to show a single object. With increasing resolution (center and right photos), we clearly see two distinct objects.

surface of the objective lens (◀▶ Figure 22.25). The bent light rays converge on the focus, forming an image of a distant object. The eyepiece then magnifies the image.

The problem with refracting telescopes is that different colors in the spectrum refract by different amounts. Therefore, if you focus the telescope to collect blue light sharply, red light will be fuzzy. As a result, most modern optical telescopes are **reflecting telescopes**. They collect light with a large curved mirror and reflect it to an eyepiece (◀▶ Figure 22.26). The reflecting mirror at the Palomar Observatory is 508 centimeters (200 inches) in diameter. It has 2.8 million times as much surface area as your pupil and thus collects 2.8 million times as much light. Modern telescopes are outfitted with both a camera and electronic detectors at the eyepiece.

The Palomar telescope took its first photo in 1949, over 57 years ago. Technology has advanced radically over the past generation, so while this observatory remains a giant of the past, astronomers have built many other newer and more powerful instruments. In addition, in 1949, Palomar Mountain, in North San Diego County, was quite rural, far away from city lights. As cities and suburban sprawl have encroached outward, city lights now glow in the night sky, competing with the lights from distant stars and reducing the vision of the telescope, in a process called **light pollution.**

In the past few decades, astronomers have built more powerful telescopes, both on land and in space. In 1990 the Hubble Space Telescope (HST) was launched into orbit around Earth (◀▶ Figure 22.27). In the black vacuum of space, the HST isn't adversely affected by either light pollution or atmospheric interference. The Hubble has been repaired four times while in space, and its high-resolution images have altered our understanding of many celestial bodies (◀▶ Figure 22.28). The Hubble telescope is outfitted with sensors to collect visible light and radiation in other portions of the spectrum. A fifth service mission is scheduled for 2006 or 2007, but, unless Congress allocates more money, the HST is scheduled for termination around 2010.

The HST is enormously expensive and repair or maintenance missions involve dangerous and costly flights into space. As a result, astronomers are also building powerful observatories on land. Because a mirror much larger than 600 centimeters sags under its own weight, the most recent telescopes use an array of smaller mirrors. The Keck telescopes on Mauna Kea in Hawaii contain 36 individual mirrors that are aligned and focused by computer (◀▶ Figure 22.29). The total mirror area is four times larger than the one at Palomar.

One of the largest proposed telescopes for the future may be sited in Antarctica. This frozen continent experiences several months of total darkness where observations can occur 24 hours a day. There is little atmospheric water vapor to distort light and no nearby cities to produce light pollution. Although winters are harsh and supply transport expensive, costs are thousands of times cheaper than they are for space telescopes.

Telescopes Using Other Wavelengths

Visible light is only a small portion of the electromagnetic spectrum. The wavelengths of electromagnetic radiation emitted by a

◀▶ **FIGURE 22.25** In a refracting telescope, light is collected and focused by a large objective lens. A second lens, called the eyepiece, magnifies the image produced by the objective lens.

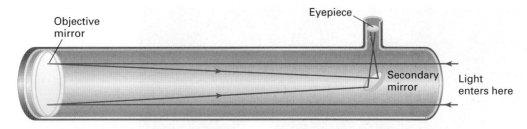

◆ FIGURE 22.26   A Newtonian reflecting telescope. Incoming light (right) is collected and focused by a curved objective mirror. The light is then reflected to the eyepiece by a secondary mirror.

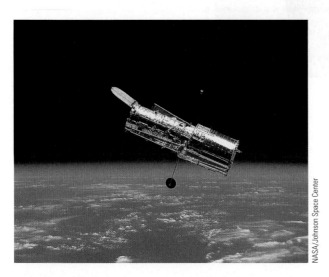

NASA/Johnson Space Center

◆ FIGURE 22.27   The orbiting Hubble Space Telescope has enabled astronomers to make many new discoveries of the Solar System, our galaxy, and intergalactic space.

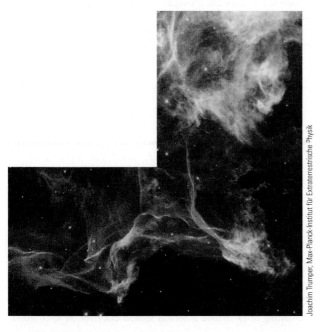

Joachim Trumper, Max-Planck-Institut für Extraterrestrische Physik

◆ FIGURE 22.28   When a massive star dies, it explodes, blasting filaments of gas into space. The Hubble Space Telescope has photographed these stellar shreds in much more detail than is possible from ground-based instruments.

star are determined by several factors, including the types of nuclear reactions that occur in the star, its chemical composition, and its temperature.

In recent years, astronomers have enhanced our knowledge of stars and other objects in space by studying many different wavelengths, from low-energy radio and infrared signals to high-energy gamma rays and X-rays. The telescopes used in these studies often do not look like conventional optical telescopes but rather are tailored to the characteristics of the wavelength being studied (◆ Figure 22.30). Each wavelength provides specific information not available from other wavelengths. In Chapter 23 and Chapter 24, we will discuss some of the measurements made at various wavelengths.

Emission and Absorption Spectra

If light passes through a prism, it separates into a **spectrum**, an ordered array of colors[1] (◆ Figure 22.31). A rainbow is such an array, with white sunlight separated into its individual colors. Each color is formed by a band of wavelengths.

As light passes from the hot interior of a star through the cooler, outer layers, some wavelengths are selectively absorbed by atoms in the star's outer atmosphere. Therefore, in a spectrum of starlight, dark lines cross the band of colors. This is called an **absorption spectrum** (◆ Figure 22.32). Each dark line represents a wavelength that is absorbed by atoms of a particular element. Thus an absorption spectrum enables us to determine the chemical composition of a star. This type of analysis is so effective that astronomers discovered helium in the Sun 27 years before chemists detected it on Earth. Because an atom's spectrum changes with temperature and pressure, spectra can also be used to determine surface temperatures and pressures of stars.

Whenever an atom absorbs radiation, it must eventually reemit it. **Emission spectra** are often hard to detect against the bright background of starlight, but they can be seen along the outer edges of some stars and in large clouds of dust and gas in space.

---

1. In modern instruments light is dispersed by a diffraction grating, not a prism, but the effect is the same.

A

B

◆ FIGURE 22.29 **(A)** The 10-meter Keck Telescope is housed in the white dome on the left, located on Mauna Kea in Hawaii. In this photograph, a twin telescope, named Keck II, is under construction on the right. **(B)** The 10-meter mirror of the Keck Telescope is composed of 36 separate hexagonal sections, each adjusted by computer. This view shows the mirror under construction, with 18 of the mirrors installed. Note how small the worker appears compared with the overall size of the mirror.

◆ FIGURE 22.30 The very large array (VLA) is a radio telescope consisting of 27 mobile antennas spread out over 36 kilometers.

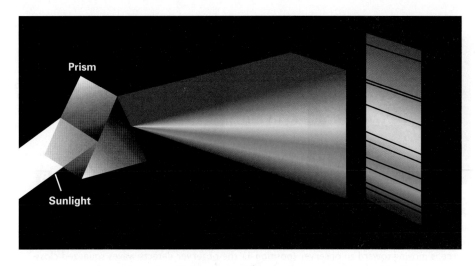

Prism

Sunlight

◆ **FIGURE 22.31**   A prism disperses a beam of sunlight into a spectrum of component colors. Each color represents a different band of wavelengths.

Doppler Measurements

Have you ever stood by a train track and listened to a train speed by, blowing its whistle? As it approaches, the pitch of the whistle sounds higher than usual, and after it passes, the pitch lowers. The same effect can be duplicated with an electric razor. If you turn on an electric razor, hold it still, and listen to it, it produces a constant sound. Now move it quickly past your ear and listen to the change in pitch. This change, called the **Doppler effect**, was first explained for both light and sound waves by the Austrian physicist Johann Christian Doppler in 1842. Three years later, a Dutch meteorologist, Christopher Heinrich Buys-Ballot, mounted an orchestra of trumpeters on an open railroad car and measured the frequency shifts as the musicians rode past him through the Dutch countryside.

A stationary object remains in the center of the circular waves it generates (◆ Figure 22.33A). The waves from a moving object crowd each other in the direction of the object's motion. The object, in effect, is catching up with its own waves (◆ Figure 22.33B). If the object is moving toward you, you receive more waves per second (higher frequency) than you would if it were stationary, and if it is moving away from you, you receive fewer waves per second (lower frequency).

In the same way, the frequency of light waves changes with relative motion. The Doppler effect causes light from an object moving away from Earth to reach us at a lower frequency than it had when it was emitted. Lower-frequency light is closer to the red end of the spectrum. Thus, a Doppler shift to lower frequency is called a **red shift**. Alternatively, light from an object traveling toward us reaches us at higher frequency and is **blue shifted**. Using these principles, astronomers measure the relative velocities of stars, galaxies, and other celestial objects millions or billions of light-years away.

Spacecraft

On October 4, 1957, the U.S.S.R. launched the first spacecraft, *Sputnik I*. A few months later the Americans launched their first space probe. In the generation that followed, hundreds of spacecraft were launched to study Earth, the Solar System, and distant galaxies. Some of the most notable missions include 6 manned lunar landings, 25 orbital or landing missions to Venus, four successful landings on Mars, and numerous probes to the outer planets, planetary moons, comets, and asteroids. As we will learn in Chapter 23, new planetary missions have increased our knowledge of the Solar System. In addition, astronauts in the Russian *Mir* space station have

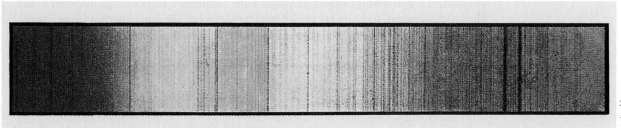

Deutsches Museum

◆ **FIGURE 22.32**   A copy of the first solar absorption spectrum made in 1811. The dark lines are the absorption lines.

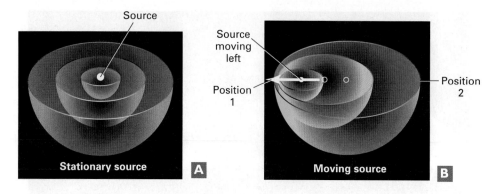

◆ FIGURE 22.33 Doppler effect. **(A)** When a stationary source emits a signal, the frequency of the signal is unaffected by the source, and observers in any direction detect the same frequency. **(B)** The Doppler effect is the change in the frequency of a signal when it moves. If the source is moving toward an observer (position 1), the observer detects waves squeezed close together and therefore with higher frequency than those detected from the same source when it is stationary. If the source is moving away from an observer (position 2), the observer detects waves stretched farther apart and therefore with lower frequency.

performed numerous experiments and demonstrated that people can live in space for long periods of time. Future missions will alter some of the information in this book by the time you read it. That is one reason why science is so exciting; we are constantly probing deeper, with better instruments and new experiments.

SUMMARY

Aristotle proposed a **geocentric**, or Earth-centered, Universe in which a stationary, central Earth is surrounded by **celestial spheres** that contain the Sun, the Moon, the planets, and the stars. Ptolemy modified the geocentric model to explain the **retrograde motion** of the planets. In Ptolemy's model, each planet moves in small circles within its larger orbit. Copernicus believed that the Universe should operate in the simplest manner possible and showed that a **heliocentric** Solar System best explains planetary motion. Kepler described the elliptical orbits of the planets mathematically. Galileo used observation and experimentation to discredit the geocentric model and show that the planets revolve around the Sun. Newton proved that gravity holds the planets in elliptical orbits.

The revolution of the Moon around Earth causes the phases of the Moon. A **lunar eclipse** occurs when Earth lies directly between the Sun and the Moon. A **solar eclipse** occurs when the Moon lies directly between the Sun and Earth.

Objects in space are studied with both optical telescopes and telescopes sensitive to invisible wavelengths. **Emission** and **absorption spectra** provide information about the chemical compositions and temperatures of stars and other objects. The **Doppler effect** causes light from an object moving away from Earth to reach us at a lower frequency than it had when it was emitted. Light from an object moving toward Earth has a higher frequency than it had when it was emitted. Instruments carried aloft by spacecraft eliminate interference by Earth's atmosphere and allow close inspection of objects in the Solar System.

Key Terms

For Review

1. Describe the apparent motion of the stars, as seen by an observer on Earth, over the course of a single night and over the course of a year.

2. List and explain Aristotle's observations and the reasoning he used to support his geocentric model of the Solar System.

3. What is retrograde motion? How is it explained in Ptolemy's and Copernicus's models?

4. Explain how Galileo's studies of physics contributed to his rejection of Aristotle's geocentric model.

5. Explain how Galileo's observations of the Milky Way, Sun, and Moon led him to question Aristotle's geocentric model.

6. What evidence convinced Galileo that Earth revolves around the Sun?

7. What was the astronomical significance of Newton's studies of gravity?

8. How is the Moon positioned with respect to Earth and the Sun when it is (a) full, (b) new, (c) gibbous, (d) crescent?

9. Why does the Moon rise approximately 53 minutes later each day than it did the previous day?

10. Draw a picture of the Sun, Earth, and Moon as they will be during an eclipse of the Sun. Draw a picture of the Sun, Earth, and Moon as they are aligned during an eclipse of the Moon. Explain how these positions produce eclipses.

11. Explain how optical telescopes work. How is a refracting telescope different from a reflecting telescope?

12. Discuss the relative advantages and disadvantages of space telescopes vs. ground-based observatories.

13. Describe the differences between absorption and emission spectra.

14. Explain how the frequency of radiation changes with the motion of an object that emits radiation.

For Discussion

1. Earth simultaneously rotates on its axis and revolves around the Sun. Which of these two motions is responsible for each of the following observations?

   (a) The Sun rises in the east and sets in the west.

   (b) Different constellations appear in the summer and winter.

   (c) On any given night the stars in the Northern Hemisphere appear to revolve around the Pole Star.

2. The Moon revolves around Earth while Earth rotates on its axis. Which of these two motions is responsible for each of the following observations?

   (a) The Moon rises in the east and sets in the west.

   (b) The Moon rises approximately 53 minutes later each day.

   (c) The Moon passes through monthly phases.

3. If you were the editor of a modern scientific journal, how would you respond to Aristotle's arguments for the geocentric model of the Universe? Defend your position.

4. Suppose that you are driving a car and the speedometer reads 80 km/hr. A person sitting next to you and looking at the speedometer at the same time may read a different value, perhaps 70 km/hr. Explain how two people can look at the same instrument at the same time and read different values.

5. With modern telescopes, astronomers can determine how far away some stars are by observing the parallactic shift as Earth travels around the Sun. Using this technique, would it be easier to estimate distances to nearby stars or to distant stars, or would it be equally difficult for all stars? Defend your answer.

6. Describe the lunar cycle if the Sun, Earth, and Moon lie in a single plane.

7. Many societies use a lunar calendar instead of a solar one. In a lunar calendar, each month represents one full cycle of the Moon. How many days are there in a lunar month? If there are 12 months in a lunar year, is a lunar year longer or shorter than a solar year?

CHAPTER
23

Planets and Their Moons

NASA/JPL/Space Science Institute

While cruising around Saturn at a distance of about 6.3 million kilometers in early October 2004, Cassini captured a series of images that have been composed into the most detailed natural color view of Saturn and its rings ever made.

Recall from Chapter 6 that the entire Solar System formed from a single, homogeneous cloud of dust and gas. Today, however, each of the planets is quite different from the others. Some differences are easy to explain. Pluto, the farthest planet from the Sun, is frigid; Mercury, the closest, is torridly hot. However, neighboring planets, such as Venus, Earth, and Mars, have greater differences from one another than can be explained solely by their different distances from the Sun. As explained in Chapter 1, small, initial differences in planetary environments have been amplified to create vastly dissimilar worlds. A runaway greenhouse effect on Venus created a hot, inhospitable environment. The climate on Mars was once temperate, and water flowed across the surface, but today this planet is cold and dry. Meanwhile, on nearby Earth, flowers bloom, forests and prairies thrive, and animals, including humans, scurry across the surface. ■

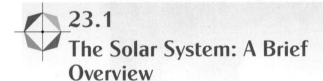

23.1
The Solar System: A Brief Overview

The Solar System formed about 4.6 billion years ago from a cold, diffuse cloud of dust and gas rotating slowly in space. The cloud was composed of about 92 percent hydrogen and 7.8 percent helium, about the same elemental composition as the Universe.[1] All of the other elements composed only 0.2 percent of the Solar System.

A portion of the cloud gravitated toward its center to form the Sun. Here the pressure became so intense that hydrogen fused, producing energy as some of the hydrogen converted to helium. Hydrogen fusion is still the source of the Sun's energy and will be discussed in Chapter 24. The remaining matter in the original cloud formed a disk-shaped, rotating nebula that eventually coalesced into separate spheres to produce the planets. The evolution of the planets is an example of a feedback mechanism occurring in space. Any object—including a rocket ship or a gas molecule, can escape from a planet's gravity when it reaches a speed known as the **escape velocity**. The escape velocity is proportional to the mass, and hence the gravitational force, of the planet.

Let us now compare the evolution of Mercury, the closest planet to the Sun, with that of Jupiter, one of the more-distant planets. In their primordial states, all planets were composed mainly of gases. Gases in a planetary atmosphere are in constant motion, and the higher the temperature, the higher the average speed. Mercury, being closer to the Sun, was originally hotter than Jupiter. Its gases were moving faster, so they were more likely to

escape the planet's gravity and fly off into space. As gases escaped, the planet lost mass, so the escape velocity decreased, making it easier for gases to escape. In addition, the **solar wind**, a stream of positive ions and electrons radiating outward from the sun at high speed, blew even more gases away from the primordial planets. Combining all processes, the inner planets, Mercury, Venus, Earth, and Mars, lost most of their gases, leaving behind spheres composed mostly of nonvolatile metals and silicate rocks. These four are now called the **terrestrial planets**. In contrast, the protoplanets in the outer reaches of the Solar System were so far from the Sun that they were initially cool. As a result, the outer **Jovian planets**, Jupiter, Saturn, Uranus, and Neptune, retained large amounts of hydrogen, helium, and other light elements. In fact, they actually grew larger as they captured gases that escaped from the terrestrial planets. As the mass increased, the escape velocity also increased, and gas escape became very slow. Today, the Jovian planets all have relatively small, rocky or metal cores surrounded by swirling liquid and gaseous atmospheres. The outermost planet, Pluto, is anomalous; it is relatively small and may be an escaped moon of Neptune.

Table 23.1 provides an overview of the nine major planets. Due to the differences in composition, the terrestrial planets are much denser than the Jovian planets. However, the Jovian planets are much larger and more massive than the Earth and its neighbors.

23.2
The Terrestrial Planets

As the planets coalesced, the primordial Solar System was crowded with asteroids, meteoroids, comets, and other

---

1. Composition is given here in percentage of number of atoms. Percentage by mass is significantly different.

TABLE 23.1   Comparison of the Nine Major Planets

Planet	Distance from Sun (Millions of Kilometers)	Radius (Compared to Radius of Earth = 1)	Mass (Compared to Mass of Earth = 1)	Density (Compared to Density of Water = 1)	Composition of Planet	Density of Atmosphere (Compared to Earth's Atmosphere = 1)	Number of Satellites
Terrestrial Planets							
Mercury	58	0.38	0.06	5.4		One billionth	0
Venus	108	0.95	0.82	5.2	Rocky with	90	0
Earth	150	1	1	5.5	metallic core	1	1
Mars	229	0.53	0.11	3.9		0.01	2
Jovian Planets							
Jupiter	778	11.2	318	1.3	Liquid hydrogen surface with liquid metallic mantle and solid core	Dense and turbulent	60
Saturn	1420	9.4	94	0.7			31
Uranus	2860	4.0	15	1.3 }	Hydrogen and helium outer layers with solid core	Similar to Jupiter except that some compounds that are gases on Jupiter are frozen on the outer planets	22
Neptune	4490	3.9	17	1.7 }			11
Most Distant Planet							
Pluto	5910	0.17	0.0025	2.0	Rock and ice	0.00001	1

chunks of rock, gases, and ices. This space debris crashed into the planets, pock-marking their surfaces with millions of craters. Over geologic time, craters can be obliterated by tectonic events, weathering, and erosion. However, if these processes do not occur, then the craters remain for billions of years. Thus, scientists learn a lot about a planet's history simply by observing crater density. With this background, let us compare the geology and the atmospheres of the four terrestrial planets.

Atmospheres and Climates of the Terrestrial Planets

Mercury

Mercury, which is the closest planet to the Sun and therefore initially the hottest, has lost essentially all of its atmosphere, and today it is a rocky sphere with a radius of 2,400 kilometers, less than 0.4 that of Earth's radius. Mercury makes a complete circuit around the Sun faster than any other planet; each Mercurial year is only about 88 Earth days long. Mercury rotates slowly on its axis, so there are only 3 Mercurial days every 2 Mercurial years. Because Mercury is so close to the Sun and its days are so long, the temperature on its sunny side

reaches 427°C, hot enough to melt lead. In contrast, the temperature on its dark side drops to −175°C, cold enough to freeze methane. The lack of an atmosphere is partly responsible for these extremes of temperature, because there is no wind to carry heat from the hot, sunlit regions to the dark, frigid shadows.

In 1991, radar images of Mercury revealed highly reflective regions at the planet's poles. Data indicated that these regions were composed of ice. Mercury's spin axis is almost perpendicular to its orbital plane around the Sun, so the Sun never rises or sets at the poles but remains low on the horizon throughout the year. Because the Sun is so low in the sky, regions inside meteorite craters are perpetually in the shade. With virtually no atmosphere to transport heat, the shaded regions have remained below the freezing point of water for billions of years.

Venus

Recall from Chapter 1 that **Venus** closely resembles Earth in size, density, and distance from the Sun. As a result, Venus and Earth probably had similar atmospheres early in their histories. However, Venus is closer to the Sun than Earth is, and therefore it was initially hotter. One hypothesis suggests that because of the higher temperature, water never condensed, or if it did, it quickly

Systems Perspective

The Solar System

About five billion years ago, exploding stars accelerated dust and gas into the great void of space.

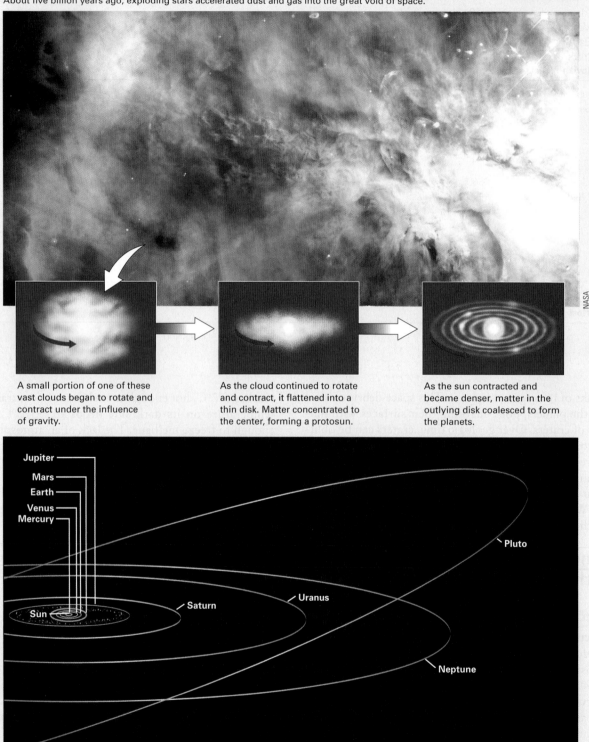

NASA

A small portion of one of these vast clouds began to rotate and contract under the influence of gravity.

As the cloud continued to rotate and contract, it flattened into a thin disk. Matter concentrated to the center, forming a protosun.

As the sun contracted and became denser, matter in the outlying disk coalesced to form the planets.

Jupiter
Mars
Earth
Venus
Mercury

Pluto

Sun

Saturn

Uranus

Neptune

A schematic view of the modern Solar System.

Systems Perspective

The Solar System

The planets drawn to scale. The terrestrial planets are so small on this scale that they are enlarged below. The Jovian planets (right) are predominantly composed of gases and liquids with small rocky and metallic cores, while the Terrestrial planets (below) are composed primarily of rock and metal.

Jupiter

Neptune

Saturn

Uranus

Terrestrial planets

Mercury

Moon

Venus (radar image)

Mars

Mercury

Sun

Venus

Earth

Moon

Mars

Jupiter

Saturn

Uranus

Neptune

Pluto

The planets and the Sun drawn to scale. Note that the Earth-Moon orbital system would easily fit into a portion of the Sun's surface. The Earth is the most massive of the terrestrial planets, but is only 1/300th the mass of Jupiter. Saturn's rings would reach from the Earth to the Moon.

evaporated again. Because there were no seas for carbon dioxide to dissolve into, most of the carbon dioxide also remained in the atmosphere. Water and carbon dioxide combined to produce a runaway greenhouse effect, and surface temperatures became torridly hot. The molecular weight of water is less than half that of carbon dioxide. Because water is so light, the hot surface temperature boiled the water into space and the solar wind swept most of the vapor toward the outer reaches of the Solar System. Today, the Venusian atmosphere is 90 times denser than that of Earth. Thus atmospheric pressure at the surface of Venus is equal to the pressure 1,000 meters beneath the sea on our planet. The Venusian atmosphere is more than 97 percent carbon dioxide, with small amounts of nitrogen, helium, neon, sulfur dioxide, and other gases. Corrosive sulfuric acid aerosols float in a dense cloud layer that perpetually obscures the surface (◆ Figure 23.1). Due to greenhouse warming, the Venusian surface is hotter than that of Mercury, hot enough to destroy the complex organic molecules necessary for life.

Earth

On Earth, outgassing from the mantle modified the primordial atmosphere. As life evolved, complex interactions among the geosphere, hydrosphere, and biosphere further altered the atmosphere, creating conditions favorable for life. This complex sequence of events is discussed in Chapter 17.

Mars

Today, the surface of **Mars** is frigid and dry. The surface temperature averages −56°C and never warms up enough to melt ice. At the poles, the temperature can dip to −120°C, freezing carbon dioxide to form a dry ice.

The atmosphere at the Martian surface is as thin as Earth's atmosphere 43 kilometers high, which for us is the outer edge of space. Although water ice exists in the Martian polar ice caps and in the soil, there are no oceans, rivers, or lakes, and the water vapor in the air, if condensed, would form a global layer only a fraction of a millimeter thick. In contrast, a totally smooth Earth would have enough water to covers its entire surface with an ocean three kilometers thick.

However, abundant evidence indicates that the Martian climate was once much warmer and that water flowed across the surface. Photographs from Mariner and Viking spacecraft show eroded crater walls and extinct streambeds and lake beds. One giant canyon, Valles Marineris, is approximately 10 times longer and 6 times wider than Grand Canyon in the American Southwest (◆ Figure 23.2). Massive alluvial fans at the mouths of Martian canyons indicate that floods probably raced across the land at speeds up to 270 kilometers per hour.

In January 2004, NASA landed two mobile robots, called **Spirit** and **Opportunity**, on the Martian surface. Each robot was equipped with six wheels and a drive system, communications equipment to receive instructions from Earth and transmit data, and solar panels for power (◆ Figure 23.3). Scientific research was conducted with a variety of sensitive instruments: a camera, a magnifying glass, a grinding wheel to burrow into a rock and expose fresh surfaces, a device to analyze the chemical elements in rock, two remote-sensing devices to identify minerals, and a magnet to collect magnetic dust particles. One of the main functions of these rovers was to look for evidence of the existence of water in the Martian past.

Spirit landed first, in the 160-kilometer-wide Gusev crater. Scientists chose this site because photographs from orbital spacecraft indicated that the crater floor was once the site of a huge lake. Instead of a lake bed,

◆ **FIGURE 23.1** The solid surface of Venus is obscured by a dense, corrosive atmosphere and a turbulent cloud cover.

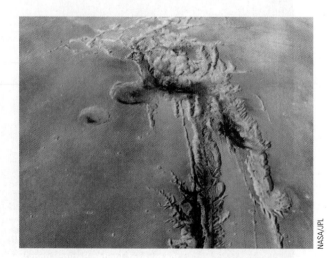

◆ **FIGURE 23.2** The giant canyon, Valles Marineris, was eroded by flowing water and is many times larger than the Grand Canyon in the American Southwest.

◀▶ **FIGURE 23.3** Mars Exploration Rover drives over the surface of Mars.

◀▶ **FIGURE 23.4** On March 8, 2006, Spirit photographed these cross-bedded sands that show sedimentary structures similar to those commonly found on Earth.

Spirit detected volcanic rocks with large concentrations of the minerals olivine and pyroxene. Both olivine and pyroxene react to form other minerals in the presence of water. Therefore these minerals can only persist in a dry environment. Spirit data indicates that the bottom of Gusev crater has been drier than any desert on Earth for three billion years.

Three weeks later, the second rover, Opportunity, landed on the opposite side of the planet on the plains of Meridiani. The first photographs showed layered sedimentary rocks (◀▶ Figure 23.4). Chemical and mineralogical analysis indicated the abundance of iron-rich minerals that usually form in the presence of water. Opportunity also found sulfate salts that had clearly precipitated from a water-rich solution. Several weeks later, Opportunity took spectacular images of ripple marks preserved on the floor of a long-dead lake or ocean.

Meanwhile, Spirit climbed off the lava-covered Gusev plain and found precipitated sulfur salts on the crater rim, indicating that some water had existed here, too, although this environment was never as wet at the Meridiani site.

As this chapter is being written, Spirit and Opportunity, which had been warrantied for 90 days, had been active for more than two Earth years. The Rover experiments, coupled with observations from orbital spacecraft, have given us the following picture of Martian climate history:

Although water was definitely present in the Martian past, the lakes and oceans were either cold, short-lived, or ice covered. This conclusion is derived mainly from the observation that carbonate rocks, such as limestone, precipitate from warm, carbon-dioxide-rich oceans. It seems probable that carbon dioxide was present in the Martian atmosphere, but there are no carbonates.

Scientists now believe that water existed in isolated regions on Mars for relatively brief periods of time. Thus the climate alternated from relative warmth to frigid cold, and back again. Earth's climate has also fluctuated from the frigid cold of Snowball Earth to the warmth of the Mesozoic Era when dinosaurs roamed steamy swamps, but the Martian temperature variations were more extreme. Today, Mars has abundant water frozen into soils and the polar ice caps. As evidenced so graphically by the photos, mentioned earlier, of canyons and alluvial fans, liquid water ran across the surface as recently as a few million years ago. These cycles remind us, once again, that planetary climates are delicate and subject to radical change. The Focus On box explores the mystery, Did life evolve during one of the warm, wet cycles on Mars?

Geology and Tectonics of the Terrestrial Planets

Mercury

Little was known about the surface of Mercury before the spring of 1974, when the spacecraft Mariner 10 passed within a few hundred kilometers of the planet. Images relayed to Earth revealed a cratered surface remarkably similar to that of our Moon (◀▶ Figure 23.5).

Recall that craters formed on all planets and their moons during intense meteorite bombardment early in

◆ **FIGURE 23.5** Mercury has a cratered surface. The craters are remarkably well preserved because there is no erosion or tectonic activity on Mercury. The photograph shows an area 580 km from side to side.

the history of the Solar System. However, tectonic activity and erosion have erased Earth's early meteorite craters. Yet Mercury is so close to the Sun that its water and atmosphere have boiled off into space, so no wind, rain, or rivers have eroded its surface. Today, 4-billion-year-old craters look as fresh as if they formed yesterday.

Flat plains on Mercury are lava flows that formed early in its history, when the planet's interior was hot enough to produce magma. However, Mercury is so small that its interior cooled quickly and igneous activity ceased when the planet was still young. Yet, curiously, Mercury has a magnetic field. On Earth, the magnetic field is generated by the flow of molten metals in the core. But the Mercurial surface shows no evidence of tectonic activity or of a hot interior. Therefore astronomers cannot explain how the Mercurial magnetic field is generated.

Venus

Astronomers use spacecraft-based radar to penetrate the Venusian atmosphere and produce photolike images of its surface. Gravity studies, also conducted by spacecraft, are used to infer the density of rocks near the Venusian surface. These density measurements provide information about the planet's mineralogy and internal structure. The most spectacular data were obtained by the orbiting Magellan spacecraft, which was launched in May 1989. In October 1994, its mission 99 percent accomplished and federal funding running out, scientists sent Magellan on one final suicide mission into the Venusian atmosphere to provide additional information about its density and composition. During the early history of the Solar System, thick swarms of meteorites bombarded all the planets. However, Magellan's detailed maps show few meteorite craters on Venus. This observation indicates that the Venusian surface was reshaped after the major meteorite bombardments.

More detailed analysis shows that most of the landforms on Venus are 300 to 500 million years old.

James Head, Magellan investigator, suggests that a catastrophic series of volcanic eruptions occurred 300 to 500 million years ago, creating volcanic mountains and covering much of the surface with basalt flows (◆ Figure 23.6). According to this model, rising mantle plumes generated magma that repaved the planet in a short time with a rapid series of cataclysmic volcanic eruptions. Head concludes, "The planet's entire crust and lithosphere turned itself over. It certainly makes a strong case that catastrophic events are in the geological records of planets and it ought

◆ **FIGURE 23.6** The volcano Maat Mons, on Venus, has produced large lava flows, shown in the foreground. According to one hypothesis, a catastrophic series of volcanic eruptions altered the surface of Venus 300 to 500 million years ago. This image was produced from radar data recorded by Magellan spacecraft. Simulated color is based on color images supplied by Soviet spacecraft.

to make us think about the possibilities of such catastrophic events in Earth's past." An alternative hypothesis contends that the resurfacing occurred more slowly.

Crater abundances indicate that most tectonic activity on Venus stopped after the intense volcanic activity 300 to 500 million years ago. Some evidence indicates that the volcanic activity has ceased permanently because the planet's interior cooled. This cooling may have occurred because radioactive elements floated toward the surface during the volcanic events, removing the mantle heat source. However, other data imply that Venus remains volcanically active and that another repaving event may occur in the future.

Most Earth volcanoes form at tectonic plate boundaries or over mantle plumes, so the discovery of volcanoes on Venus led planetary geologists to look for evidence of plate tectonic activity there. Radar images from spacecraft show that 60 percent of Venus's surface consists of a flat plain. Two large, and several smaller, mountain chains rise from the plain (◆ Figure 23.7). The tallest mountain is 11 kilometers high—2 kilometers higher than Mount Everest. The images also show large, crustal fractures and deep canyons. If Earth-like horizontal motion of tectonic plates caused these features, then spreading centers and subduction zones should exist on Venus. The images show no features like a mid-oceanic ridge system, transform faults, or other evidence of lithospheric spreading. However,

geophysicists have located 10,000 kilometers of trench-like structures that they believe to be subduction zones.

Despite the apparent existence of subduction zones, the most popular current model suggests that Venusian tectonics has been dominated by mantle plumes. In some regions the rock has melted and erupted from volcanoes by processes similar to those that formed the Hawaiian Islands. In other regions the hot, Venusian mantle plumes have lifted the crust to form nonvolcanic mountain ranges. Some geologists have suggested the term **blob tectonics** to describe Venusian tectonics because Venus is dominated by rising and sinking of the mantle and crust. In contrast, tectonic activity on Earth causes significant horizontal movement of its plates.

Mantle plumes on Earth may initiate rifting of the lithosphere and formation of a spreading center. Why have spreading centers not developed over mantle plumes on Venus? Perhaps surface temperature on Venus is so high that the surface rocks are more plastic than those on Earth. Therefore, rock flows plastically rather than fracturing into lithospheric plates. It is also possible that Venus has a thicker lithosphere, which can move vertically but does not fracture and slide horizontally.

Earth

Earth is large enough to have retained considerable internal heat. This heat drives convection within the mantle that produces tectonic motion. In turn Earth tectonics

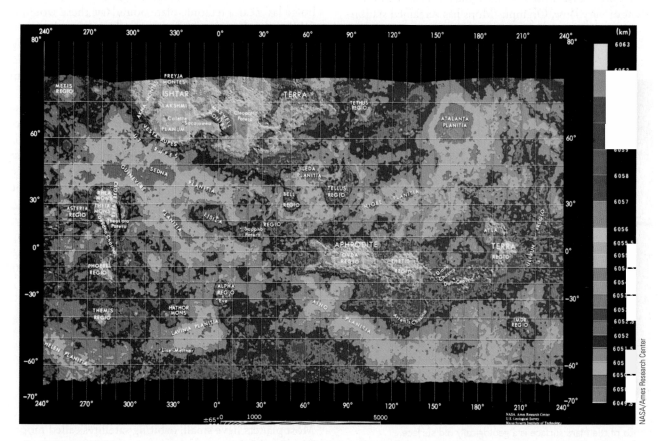

◆ **FIGURE 23.7** Scientists used radar images from the Pioneer Venus orbiter to produce this map of Venus. The lowland plains are shown in blue, and the highlands are shown in yellow and red-brown. The colors are assigned arbitrarily.

continuously reshapes the surface of our planet and is partially responsible for the atmosphere that sustains us. This topic was discussed in detail in Unit II.

Mars

The Martian surface consists of old, heavily cratered plains and younger regions that have been altered by tectonic activity (◆ Figure 23.8). Lava flows much like those on Venus and the Moon cover the plains. The Tharsis bulge is the largest plain, crowned by Olympus Mons, the largest volcano in the Solar System (◆ Figure 23.9). Olympus Mons is nearly three times higher than Mount Everest, with a height of 25 kilometers and a diameter of 500 kilometers. Its central crater is so big that Manhattan Island would easily fit inside.

One hypothesis suggests that the geology of Mars is similar to that of Venus and is dominated by blob tectonics. Supporters of this concept point out that the largest volcanic mountain on Earth, Mauna Kea in the Hawaiian Islands, is limited in size because the mountain is riding on a tectonic plate. Therefore, it will drift away from the underlying hot spot before it can grow much larger. In contrast, because horizontal movement on Mars is nonexistent or very slow, Olympus Mons has remained station-

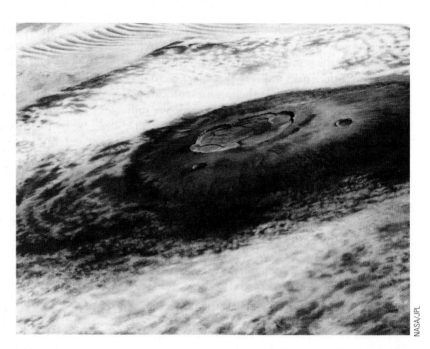

◆ **FIGURE 23.9**　Olympus Mons is the largest volcano on Mars and also the largest in the Solar System.

ary over its hot spot. As a result, it has grown bigger and bigger.

Another line of evidence for Venusian-type tectonics on Mars is that tremendous parallel cracks split the crust adjacent to the Tharsis bulge (◆ Figure 23.10). If this bulge lay near a tectonic plate boundary, there would be folding or offsetting of the cracks. However, the cracks are neither folded nor offset. Therefore, scientists suggest that a rising mantle plume formed the Tharsis bulge and its volcanoes. The parallel cracks may be the result of stretching as the crust uplifted.

An alternative hypothesis suggests that Earth-like, horizontal tectonics is occurring on Mars. As evidence, researchers have identified what they believe to be strike-slip faults. To support this hypothesis, scientists note that a linear distribution of volcanoes on and near the Tharsis bulge is similar to linear chains of volcanoes along terrestrial subduction zones.

Recent observations, reported in 2005, show very few meteor craters in some lava flows, indicating that eruptions have occurred as recently as a few million years ago (◆ Figure 23.11). Thus the planet has remained hot and active until relatively recent times.

23.3 The Moon: Our Nearest Neighbor

Most planets have small, orbiting satellites called moons. The Earth's Moon is close enough so that we can see some of its surface features with the naked eye. In the

◆ **FIGURE 23.8**　The cratered plain on the left in this photo of the Mars surface is a geologically old surface, whereas the mountains on the lower right are evidence of more recent tectonic activity.

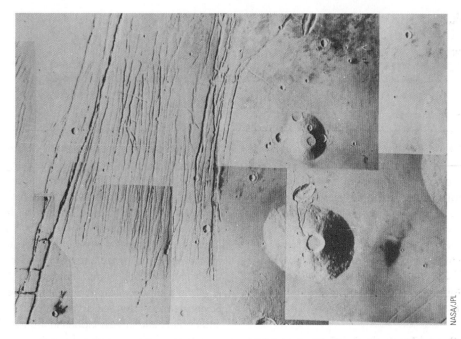

◆ **FIGURE 23.10** The huge parallel cracks near the Tharsis bulge on Mars are neither folded nor offset. Scientists speculate that the bulge and the cracks were formed by a rising mantle plume. However, the absence of folding or offset movement provides evidence that horizontal tectonics is not active in this region. Two large volcanoes appear at right.

early 1600s, Galileo studied the Moon with a telescope and mapped its mountain ranges, craters, and plains. Galileo thought that the plains were oceans and called them seas, or **maria**. The word *maria* is still used today, although we now know that these regions are dry, barren, flat expanses of volcanic rock (◆ Figure 23.12). Much of the lunar surface is heavily cratered, similar to that of Mercury (◆ Figure 23.13).

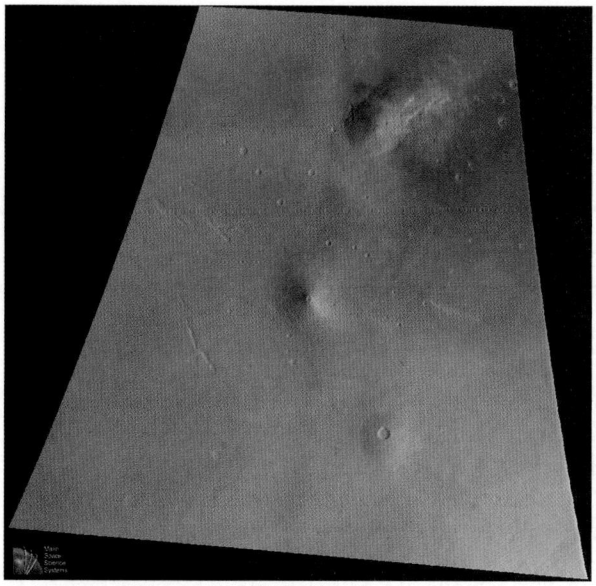

A Soviet orbiter took the first close-up photographs of the Moon in 1959. A decade later the United States landed the first of six manned Apollo spacecraft on the lunar surface (◆ Figure 23.14). The Apollo program was designed to answer several questions about the

NASA

◆ **FIGURE 23.12** The Moon as photographed from the Apollo spacecraft from a distance of 18,000 kilometers. Even from this distance, we see cratered regions and the maria, which are smooth lava flows.

INTERACTIVE QUESTION: *Use or modify the principles of relative age dating (Chapter 4) to deduce the relative ages of different regions of the lunar surface in this photograph.*

NASA/JPL/Malin Space Science Systems

◆ **FIGURE 23.11** Elysium Mons is one of three large Martian volcanoes that occur on the Elysium Rise. The volcano rises about 12.5 kilometers above the surrounding plain.

Extraterrestrial Life

Few scientific inquiries excite both scientists and nonscientists as much as the search for extraterrestrial life. Close scrutiny of our Solar System reveals no densely vegetated regions or advanced civilizations except for those on Earth. But even the presence of a few microscopic bacteria would be exciting proof that we are not alone in the Universe.

Temperatures are too hot to support life as we know it on Mercury and Venus. However, Mars once had a temperate climate, and life could have evolved there.

The Search for Martian Life

Although Mars is inhospitable today, some scientists speculate that life may have evolved during its more temperate past. Supporters of this hypothesis point out that the primordial Martian atmosphere contained water, carbon dioxide, carbon monoxide, oxygen, hydrogen, nitrogen, and sulfur. On Earth these gases combined to form organic molecules, the building blocks of life. Moreover, the record of surface water on Mars tells us that at one time the Martian temperature was neither too hot nor too cold to support life. This evidence does not prove that life evolved on Mars, just that the necessary components and conditions were once present.

Spacecraft that landed on Mars have searched for microorganisms in the soil and found none. However, they analyzed only a minuscule portion of the Martian surface.

Some biologists have also speculated that primitive life may have evolved in the atmosphere of Jupiter. The outermost region of the Jovian atmosphere is too cold for organisms to survive, and the interior is too hot. But in between, the temperature is just right. The atmosphere is composed primarily of hydrogen, helium, ammonia, water, methane, and hydrogen sulfide, the elements needed to build living tissue. In short, all the ingredients are present, and it would be thrilling, but not totally surprising, to find microorganisms floating about in the clouds of the giant planet. To many scientists, it seems more likely that life could exist in the dark, warm, subterranean oceans or lakes of Europa. Venus was once much cooler than it is today, so life could conceivably have evolved there as well. Titan has all the chemical ingredients for life, but the planet is so cold that it is unlikely that life evolved there.

One can only guess about life in other solar systems. Scientists have observed organic molecules on meteorites and comets and in dust clouds in interstellar space. Many scientists infer that, under the right conditions and with enough time, these molecules can combine to form living organisms. An average galaxy contains roughly 100 billion stars, and astronomers estimate that there are 10 to 50 billion galaxies in the Universe. In recent years, researchers found evidence for about 150 planets orbiting stars outside the Solar System. One is estimated to have a surface temperature of about 80°C, roughly the temperature of warm tea, and a perfect temperature for life to emerge. Thus evidence indicates that planets may be common, stellar companions, and there are many, many stars in the Universe. Almost certainly conditions are favorable for life on some of the planets. Therefore, one argument concludes that the emergence of living organisms—and even their evolution into intelligent beings—is so probable that it almost surely must have happened in many parts of the Universe. Still, we cannot rule out the possibility that life is unique to Earth and exists nowhere else.

> **FOCUS QUESTION**
> Discuss the difficulties in detecting life outside the Solar System, even if it does exist.

Moon. How did it form? What is its geologic history? Was it once hot and molten like Earth? If so, does it still have a molten core, and is it tectonically active?

Formation of the Moon

According to the most popular current hypothesis, the Moon was created when a huge object—the size of Mars or even larger—smashed into Earth shortly after our planet formed. This massive bolide plowed through Earth's mantle, and silica-rich rocks from the mantles of both bodies vaporized and created a cloud around Earth. The vaporized rock condensed and aggregated to form the Moon (Figure 23.15A).

Perhaps the single most significant discovery of the Apollo program was that much of the Moon's surface consists of igneous rocks. The maria are mainly basalt flows. The highland rocks are predominantly anorthosite, a feldspar-rich igneous rock not common on Earth. Additionally, the rock of both the maria and the highlands has been crushed by meteorite impacts and

◆ **FIGURE 23.13** Most of the lunar surface is heavily cratered. In places where smaller craters lie within the larger ones, scientists deduce that the larger craters formed first.

then welded together by lava. Since igneous rocks form from magma, it is clear that portions of the Moon were once hot and liquid.

How did the Moon become hot enough to melt? Earth was heated initially by energy released by collisions among particles as they collapsed under the influence of gravity. Later, radioactive decay and intense meteorite bombardment heated Earth further. But what about the Moon? Radiometric dating shows that the oldest lunar igneous rocks formed before there was enough time for radioactive decay to have melted a significant amount of lunar rock. Thus, gravitational coalescence and meteorite bombardment must have been the main causes of early melting of the Moon.

◆ **FIGURE 23.14** The six Apollo missions answered many scientific questions about the origin, structure, and history of the Moon.

History of the Moon

The Moon formed about 4.45 to 4.5 billion years ago, shortly after Earth. So much energy was released during the Moon's rapid accretion that, as the lunar sphere grew, it melted to a depth of a few hundred kilometers, forming a magma ocean. Meteorite bombardment kept the Moon's outermost layer molten. Eventually meteorite bombardment diminished enough for the Moon's surface to cool. The igneous rocks of the lunar highlands are about 4.4 billion years old, indicating that the highlands were solid by that time.

Swarms of meteorites, some as large as Rhode Island, bombarded the Moon again between 4.2 and 3.9 billion years ago (◆ Figure 23.15B). In the meantime, radioactive decay was also heating the lunar interior. As a result, by 3.8 billion years ago, most of the Moon's interior was molten and magma erupted onto the lunar surface. The maria formed when lava filled circular meteorite craters (◆ Figure 23.15C). This episode of volcanic activity lasted approximately 700 million years. The Moon and Earth shared a similar history until this time, but the Moon is so much smaller that it soon cooled and has remained geologically inactive for the past 3.1 billion years.

Apollo astronauts left seismographs on the lunar surface. Initial analysis of the data indicated that the energy released by moonquakes is only one billionth to one trillionth that released by earthquakes on our own planet. However, in 2005, 25 years after the first reports were released, scientists realized that the computers used in the study had only 64 kilobytes of memory, insufficient to properly record and analyze the data. Under re-examination, Yosio Nakamura from the University of Texas detected numerous deep moonquakes, indicating the moon has a hot, possibly molten, core.

Data from the lunar Prospector spacecraft, released in March 1998, indicate that water may be present in shadowed craters at the lunar poles. This discovery complements the earlier conclusion that ice exists on Mercury. The water on the Moon could be used to support human colonies or to synthesize hydrogen fuel from a Moon base for exploration of more distant planets. However, this evidence for water is indirect and uncertain.

23.4
The Jovian Planets: Size, Compositions, and Atmospheres

Jupiter

Jupiter is the largest planet in the Solar System, 71,000 kilometers in radius. It is composed mainly of hydrogen

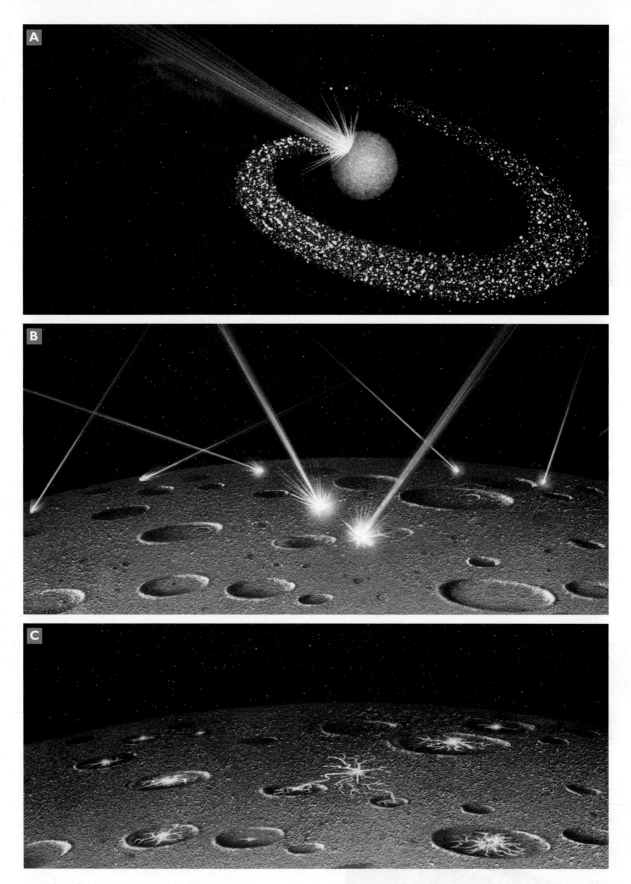

◈ **FIGURE 23.15** **(A)** According to the most widely held hypothesis, the Moon was formed about 4.5 billion years ago when a Mars-sized object struck Earth, blasting a cloud of vaporized rock into orbit. The vaporized rock rapidly coalesced to form the Moon. **(B)** During its first 0.6 billion years, intense meteorite bombardment cratered the Moon's surface. **(C)** The dark, flat maria formed about 3.8 billion years ago as lava flows spread across portions of the surface. Today, all volcanic activity has ceased.

and helium, similar to the composition of the Sun. However, Jupiter is not quite massive enough to generate fusion temperatures, so it never became a star.

Jupiter has no hard, solid, rocky crust on which an astronaut could land or walk. Instead, its surface is a vast sea of cold, liquid, molecular hydrogen, H_2, and atomic helium, He, 12,000 kilometers deep. Beneath the hydrogen/helium sea is a layer where temperatures are as high as 30,000°C and pressures are as great as 100 million times the Earth's atmospheric pressure at sea level ($\triangleleft\triangleright$ Figure 23.16). Under these extreme conditions, hydrogen molecules dissociate to form atoms. Pressure forces the atoms together so tightly that the electrons move freely throughout the packed nuclei, much as electrons travel freely among metal atoms. As a result, the hydrogen conducts electricity and is called **liquid metallic hydrogen**. Flow patterns in this fluid conductor generate a magnetic field 10 times stronger than that of Earth.

Beneath the layer of metallic hydrogen, Jupiter's core is a sphere about 10 to 20 times as massive as Earth's. It is probably composed of metals and rock surrounded by lighter elements such as carbon, nitrogen, and oxygen.

The Galileo spacecraft was launched in October 1989 and rendezvoused with Jupiter in December 1995. Once in orbit around the gas giant, the spacecraft launched a suicide probe to parachute through the outer atmosphere and collect data until it heated up and eventually vaporized. As expected, the atmosphere was primarily hydrogen and helium, with smaller concentrations of ammonia, water, and methane.

More than 300 years ago, two European astronomers reported seeing what came to be called the **Great Red Spot** on the surface of Jupiter. Although its size, shape, and color have changed from year to year, the spot remains intact to this day ($\triangleleft\triangleright$ Figure 23.17). If the Earth's crust were peeled off like an orange rind and laid flat on the Jovian surface, it would fit entirely within the Great Red Spot. Measurements show that the Great Red Spot is a giant hurricanelike storm. Hurricanes on Earth dissipate after a week, yet this storm on Jupiter has existed for centuries. Other wind systems, rotating in linear bands around the planet, have also persisted for centuries ($\triangleleft\triangleright$ Figure 23.18). One important mission of the Galileo suicide probe was to measure atmospheric pressures, temperatures, and wind speeds below the visible outer layers.

In the outer atmosphere, where the pressure was 0.4 bar, the Galileo probe measured a wind speed of 360 kilometers per hour with a temperature of −140°C. After falling 130 kilometers, the probe reported that the wind speed had increased to 650 kilometers per hour, the pressure was 22 bar, and the temperature was about +150°C. Buffeted by winds, squeezed by intense pressure, and heated beyond the tolerance of its electronics, the spacecraft stopped transmitting. Scientists calculate that 40 minutes later, when the spacecraft had sunk to deeper levels of the Jovian atmosphere, the temperature had increased to 650°C, the pressure increased to 260 bar, and the aluminum shell liquified. As the probe continued to fall, pressure and temperature rose until the titanium hull melted; streaming metallic droplets flashed into vapor as the probe vanished.

On Earth, winds are driven by the Sun's heat. But Jupiter receives only about 4 percent of the solar energy that Earth receives. Moreover, strong winds rip through the Jovian atmosphere far below the deepest penetration of solar light and heat. For these reasons, scientists deduce that the Jovian weather is not driven by solar heat but by heat from deep within the planet. This heat accumulated when Jupiter first formed. As the heat slowly rises, it warms the lower atmosphere, which then transmits heat by convection to higher levels. The weather systems are stable because the planet's interior heat flux changes only over hundreds to thousands of years.

Saturn

Saturn, the second largest planet, is similar to Jupiter. It has the lowest density of all the planets, so low that the entire planet would float on water if there were a basin of water large enough to hold it. Such a low density implies that it, like Jupiter, must be composed primarily of hydrogen and helium, with a relatively small core of rock and metal. In fact, in many ways, Saturn and Jupiter are alike. For example, Saturn's atmosphere is similar to that of Jupiter. Dense clouds and great storm systems envelop the planet.

$\triangleleft\triangleright$ **FIGURE 23.16** Jupiter consists of four main layers: a turbulent atmosphere, a sea consisting mainly of liquid molecular hydrogen with smaller amounts of atomic helium, a layer of liquid metallic hydrogen, and a rocky metallic core. Earth is drawn to scale on the left.

Earth

Red spot

Atmosphere

Liquid metallic hydrogen

Rocky metallic core

Liquid molecular hydrogen and atomic helium

◇ **FIGURE 23.17** Jupiter's Great Red Spot dwarfs the superimposed image of the Earth (to scale) shown on the bottom right. The vivid colors were generated by computer enhancement.

Uranus and Neptune

Uranus and **Neptune** are so distant and faint that they were unknown to ancient astronomers. The Voyager II spacecraft, launched in 1977, flew by Jupiter in 1979 and by Saturn in 1981. It encountered Uranus by 1986 and Neptune in 1989. The journey from Earth to Neptune covered 7.1 billion kilometers and took 12 years. The craft passed within 4,800 kilometers of Neptune's cloud tops, only 33 kilometers from the planned path. The strength of the radio signals received from Voyager measured one ten-quadrillionth of a watt ($1/10^{16}$). It took 38 radio antennas on four continents to absorb enough radio energy to interpret the signals.

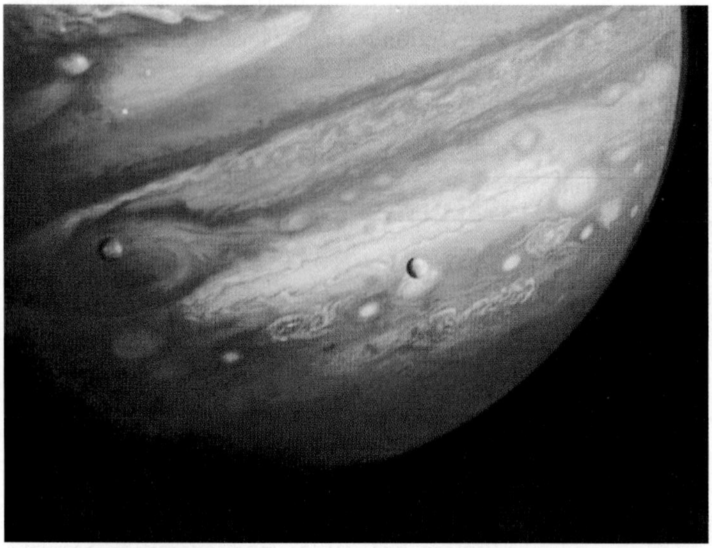

Both Uranus and Neptune are enveloped by thick atmospheres composed primarily of hydrogen and helium, with smaller amounts of carbon, nitrogen, and oxygen compounds. Beneath the atmosphere, their outer layers are molecular hydrogen, but neither body is massive enough to generate liquid metallic hydrogen. Their interiors are composed of methane, ammonia, and water, and the cores are probably a mixture of rock and metals. Uranus and Neptune are denser than Jupiter and Saturn, because these outermost giants contain relatively larger, solid cores.

Spacecraft and the Hubble Space Telescope have revealed rapidly changing weather on Uranus and Neptune. On Neptune, winds of at least 1,100

◇ **FIGURE 23.18** This colorful, turbulent complex cloud system of Jupiter was photographed by the Voyager spacecraft. The sphere on the left, in front of the Great Red Spot, is Io; Europa lies to the right against a white oval.

kilometers per hour rip through the atmosphere, clouds rise and fall, and one region is marked by a cyclonic storm system called the Great Dark Spot, similar to Jupiter's Great Red Spot. According to one controversial hypothesis, under the intense pressure near the core, methane decomposes into carbon and hydrogen and the carbon then crystallizes into diamond. Convection currents carry the heat released during the formation of diamond to the planet's surface to power the winds.

Voyager II recorded that the magnetic field of Uranus is tilted 58° from its axis. This was unexpected, as current explanations suggest that the magnetic fields of all planets should be roughly aligned with the spin axis. At first, scientists thought that Voyager II just happened to pass Uranus during a magnetic field reversal. However, Voyager II later recorded that the magnetic field on Neptune is tilted 50° from its axis. Because the probability of catching two planets during magnetic reversals is extremely low, there must be another explanation. However, at present no satisfactory hypothesis has been developed.

NASA

◆ **FIGURE 23.19** *Voyager I captured an image of a volcanic explosion on Io (shown on the horizon). The eruption is ejecting solid material to an altitude of about 200 kilometers.*

INTERACTIVE QUESTION: *The radius of Io (1,815 km) is similar to that of Earth's Moon (1,738 km). Discuss the differences between these moons and the reasons for these differences.*

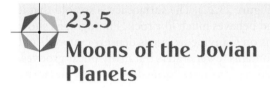

23.5
Moons of the Jovian Planets

The Moons of Jupiter

In 1610 Galileo discovered four tiny specks of light orbiting Jupiter. He reasoned that they must be satellites of the giant planet. By 1999, astronomers had identified 16 moons orbiting Jupiter. By 2005, 60 were known, but 52 of these are relatively small, with irregular orbits. The four discovered by Galileo are the largest and most widely studied.

Io

The innermost moon of Jupiter, **Io**, is about the size of Earth's Moon and is slightly denser. Because it is too small to have retained heat generated during its formation or by radioactive decay, many astronomers expected that it would have a cold, lifeless, cratered, Moon-like surface. However, images beamed to Earth from the Voyager spacecraft showed huge masses of gas and rock erupting to a height of 200 kilometers above the satellite's surface. This was the first evidence of active, extraterrestrial volcanism (◆ Figure 23.19). Images from the later Galileo spacecraft showed 100 volcanoes erupting simultaneously, making Io the most active volcanic body in the Solar System.

Recall that the gravitational field of our Moon causes the rise and fall of ocean tides on Earth. At the same time, Earth's gravity distorts lunar rock. Thus Earth's gravitation is responsible for deep-focus moonquakes. Jupiter is 300 times more massive than Earth, so its gravitational effects on Io are correspondingly greater. In addition, the three nearby satellites, Europa, Ganymede, and Callisto, are large enough to exert significant gravitational forces on Io, but these forces pull in different directions from that of Jupiter. This combination of oscillating and opposing gravitational forces causes so much rock distortion and frictional heating that volcanic activity is nearly continuous on Io.

Astronomers infer that meteorites bombarded Io and the other moons of Jupiter, as they did all other bodies in the Solar System. Yet the frequent lava flows have obliterated all ancient landforms, giving Io a smooth and nearly crater-free surface.

Europa

The second closest of Jupiter's moons, **Europa**, is similar to Earth in that much of its interior is composed of rock and much of its surface is covered with water. One major difference is that, on Europa, the water is frozen into a vast, planetary ice crust. The Galileo spacecraft transmitted images showing a fractured, jumbled, chaotic terrain resembling patterns created by Arctic ice on Earth (◆ Figure 23.20). ◆ Figure 23.21 shows a smooth region overlying an older, wrinkled surface. Data suggest that this smooth region is young ice formed when liquid water erupted to the surface and froze. Thus pools or oceans of liquid water or a

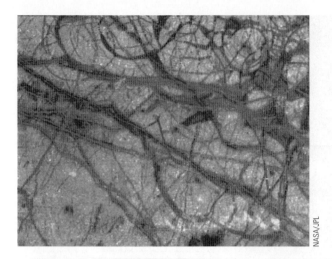

NASA/JPL

◇ **FIGURE 23.20** This jumbled terrain on Europa resembles Arctic pack ice as it breaks up in the spring. Scientists estimate that in this region, the ice crust is a few kilometers thick and is floating on subsurface water.

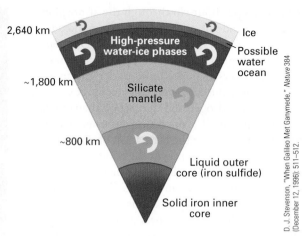

D. J. Stevenson, "When Galileo Met Ganymede," *Nature* 384 (December 12, 1996): 511–512.

◇ **FIGURE 23.22** Recent data suggest that Ganymede has a conducting, convecting core; a silicate mantle; and surface layers consisting of ice and water.

water/ice slurry probably lies beneath the surface ice. Astronomers estimate that this surface crust is 10 kilometers thick in many regions. Calculations show that the subterranean oceans are warmed by tidal effects similar to, though weaker than, those that cause Io's volcanism.

Scientists speculate that the chemical and physical environment in these subterranean oceans is favorable for life. There is no evidence whatsoever that life actually exists there, but the possibility is tantalizing.

Ganymede and Callisto

The Galileo spacecraft measured a magnetic field on **Ganymede**, indicating that this moon has a convecting, metallic core. Other measurements imply that the core is surrounded by a silicate mantle covered by a water/ice crust (◇ Figure 23.22). The surface ice is so cold that it is brittle and behaves much like rock. Photographs show two terrains on Ganymede: one is densely cratered, and the other contains fewer craters but many linear grooves. The cratered regions were formed by ancient meteorite storms. The grooved regions probably developed when

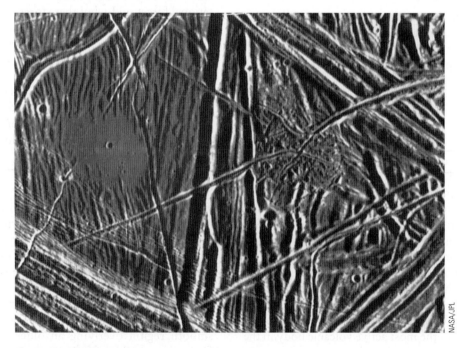

NASA/JPL

◇ **FIGURE 23.21** The smooth, circular region in the center-left of this photograph was formed when subsurface water rose to the surface of Europa and froze, covering older wrinkles and fractures in the crust.

NASA/JPL

◀▶ **FIGURE 23.23** A close-up of a young terrain on Ganymede shows numerous grooves less than a kilometer wide. One likely explanation is that these grooves were formed by recent tectonic activity.

the crust cracked and water from the warm interior flowed over the surface and froze, much as lava flowed over the surfaces of the terrestrial planets and the Moon. Lateral displacements of the grooves and ridges are indicative of Earth-like, horizontal, plate tectonic activity (◀▶ Figure 23.23).

Callisto, the outermost Galilean moon, is heavily cratered, indicating that its surface is very old. Its craters are shaped differently from those on either Ganymede or the Earth's Moon. Perhaps they have been modified by ice flowing slowly across its surface. Recent measurements indicate that a subterranean ocean may exist on Callisto, but the jury is still out on a similar feature on Ganymede.

Saturn's Moons

Thirty-one moons orbit Saturn. (Twelve of these were discovered in 2001, and more may be detected by the time you read this book.) Saturn's largest moon, **Titan**, is larger than the planet Mercury. Titan is unique because it is the only moon in the Solar System with an appreciable atmosphere. This atmosphere has been retained because Titan is relatively massive and extremely cold. The major

constituents are nitrogen, mixed with methane, CH_4, and smaller concentrations of trace gases. The average temperature on the surface of Titan is $-178°C$, and the atmospheric pressure is 1.5 times greater than that on Earth's surface. These conditions are close to the temperatures and pressures at which methane can exist as a solid, liquid, or vapor.

After a seven-year journey, the Cassini spacecraft reached Titan in December 2004, and released a probe named Huygens that parachuted through the outer clouds and landed on the surface. Images showed a surprisingly Earth-like landscape, with steep-sided hills and features that looked like riverbeds, eroded hillsides, coastlines, and sandbars. The best evidence indicates that these topographic features were formed by wind, tectonic activity, and flowing liquids (◀▶ Figure 23.24). During the extreme cold on Titan, water is permanently locked up as ice, but the temperature and pressure is such that methane in the Titan atmosphere could exist in the liquid, vapor, or solid states. Thus in the past, methane rain has fallen from the clouds and methane rivers flowed across the surface. Liquid methane may remain on the planetary surface today. This situation is analogous to Earth's environment, where water can exist as liquid, gas, or solid and frequently changes among those three states.

Methane, the simplest organic compound, reacts with nitrogen and other materials in Titan's environment to form more complex, organic molecules. These organic compounds do not decompose at low temperature, so the satellite's

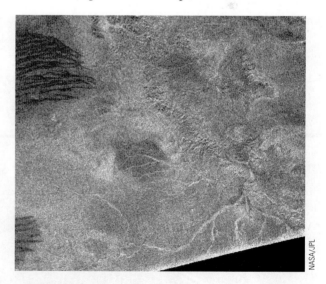

NASA/JPL

◀▶ **FIGURE 23.24** Radar images of Titan obtained in February 2005 show a well-developed drainage pattern in the lower right of the image, and apparent sand dunes. Previous images show features that may have formed by tectonic processes. Titan appears to have a young and dynamic surface that is modified by volcanism, tectonism, erosion, and impact cratering.

surface is likely to be covered by a tarlike organic goo. It is possible that a similar layer collected on early Earth and later underwent chemical reactions to form life. However, Titan is so cold that life probably has not formed there.

The Moons of Uranus and Neptune

Twenty-two moons orbit Uranus. Several of the moons are small and irregularly shaped, indicating that they may be debris from a collision with a smaller planet or moon. Neptune has 11 moons. The largest moon is Triton, which is about 75 percent rock and 25 percent ice. Like many other planets and moons, its surface is covered by impact craters, mountains, and flat, crater-free plains. While the maria on Earth's Moon are blanketed by lava, those on Triton are filled with ice or frozen methane.

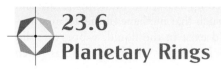

23.6 Planetary Rings

Although all the Jovian planets have one or more rings, by far the most spectacular are those of Saturn, which are visible from Earth even through a small telescope. Photographs from space probes show seven major rings, each containing thousands of smaller ringlets (◆ Figures 23.25 and 23.26). The entire ring system is only 10 to 25 meters thick, less than the length of a football field. However, the ring system is extremely

◆ **FIGURE 23.26** *Voyager I* took this color-enhanced close-up view of Saturn's rings and ringlets.

wide. The innermost ring is only 7,000 kilometers from Saturn's surface, whereas the outer edge of the most distant ring is 432,000 kilometers from the planet, a distance greater than that between Earth and our Moon. Thus the ring system measures 425,000 kilometers from its inner to its outer edge. A scale model of the ring system with the thickness of a compact disk would be 30 kilometers in diameter.

Saturn's rings are composed of dust, rock, and ice. The particles in the outer rings are only a few ten-thousandths of a centimeter in diameter (about the size of a clay particle), but the innermost rings contain chunks as large as a flying barn. Each piece orbits the planet independently.

Saturn's rings may be fragments of a moon that never coalesced. Alternatively, they may be the remnants of a moon that formed and was then ripped apart by Saturn's gravitational field. If a moon were close enough to its planet, the tidal effects would be greater than the gravitational attraction holding the moon together, and it would break up. Thus a solid moon cannot exist too close to a planet. Images from Cassini spacecraft show that gravitational forces from Saturn's moons have herded the ring particles into intricate spirals and twists.

◆ **FIGURE 23.25** An image of Saturn shows its spectacular ring system.

23.7
Pluto and Beyond

No spacecraft has ever visited **Pluto,** and our highest-resolution photographs are of poor quality compared with those of other planets (◆ Figure 23.27). Yet astronomers have deduced the properties of this planet from a variety of data.

In 1978, astronomers discovered a satellite, Charon, orbiting Pluto. When Pluto and Charon orbit each other, they periodically block one another from view. By precisely measuring these appearances and disappearances, astronomers measured Charon's orbit. Using mathematical laws derived by Kepler and Newton in the 1600s, they calculated the relative masses of Pluto and Charon from the radius of Charon's orbit and the time required for one complete orbital revolution. From these calculations, astronomers learned that Pluto is the smallest planet in the Solar System, smaller than Earth's Moon. It has a mass only ⅕₀₀ that of Earth.

From the diameter and mass, scientists calculated Pluto's density to be 2 g/cm³. The density of granite is about 2.6 g/cm³, and ice has a density of a little less than 1 g/cm³. Because Pluto's density is greater than ice but less than rock, planetary scientists infer that Pluto is a mixture of rock and ice.

If you could stand on Pluto, you would observe that the Sun is more than 1,000 times fainter than it appears from Earth. Infrared measurements show that Pluto's surface temperature is about −220°C. Spectral analysis of Pluto's bright surface shows that it contains frozen methane. Its atmosphere is extremely thin and composed mainly of carbon monoxide, nitrogen, and some methane. In 2005, astronomers observed two previously unrecorded moons, in addition to Charon, orbiting Pluto.

Pluto's orbit is more elliptical than that of any other planet. Pluto is also remarkably similar in size and density to Neptune's moon, Triton. The size and density similarities and orbital anomalies have led many astronomers to postulate that both Pluto and Charon were once moons

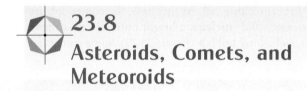
Space Telescope Science Institute

A Ground-based telescope   B Hubble Space Telescope

◆ **FIGURE 23.27**   **(A)** Ground-based image of Pluto and Charon and **(B)** a similar view from the Hubble Space Telescope.

of Neptune and were pulled out of their orbits by a close encounter with another object.

A few decades ago, astronomers detected a population of tiny ice dwarfs, similar to Pluto, in the outer reaches of the Solar System. According to calculations, tens of thousands of these ice dwarfs, called **Kuiper-belt objects,** may exist. Kuiper-belt objects are residual planetesimals left over from the formation of the Solar System. In 2005, astronomers at the California Institute of Technology reported an object in a highly elliptical orbit that almost crosses Neptune's path and then races far into the cold, frigid realm of the Kuiper-belt objects. This object, called 2003 UB313 and nicknamed Xena, is 1.5 times as large as Pluto and has been heralded as the 10th planet. But the definition of a planet has now become ambiguous. Many astronomers claim that Pluto is so anomalous that it should be a Kuiper-belt object and not a planet, while others claim that the largest Kuiper-belt objects should be upgraded to planets. The argument concerns nomenclature and doesn't really affect our understanding of the complex and wondrous Solar System we live in.

23.8
Asteroids, Comets, and Meteoroids

Asteroids

Astronomers have discovered a wide ring between the orbits of Mars and Jupiter that contains tens of thousands of small, orbiting bodies, called **asteroids.** The largest asteroid, Ceres, has a diameter of 930 kilometers. Three others are about half that size, and most are far smaller. The orbit of an asteroid is not permanent like that of a planet. If an asteroid passes near a planet without getting too close, the planet's gravity pulls the asteroid out of its current orbit and deflects it into a new orbit around the Sun. Thus an asteroid may change its orbit frequently and erratically. Many asteroids orbit near Earth, and some even cross Earth's orbit. If an asteroid passes too close to a planet, it will crash into its surface. As discussed previously, ancient asteroid impacts may have caused mass extinctions on Earth. Statistical analysis shows that there is a 1 percent chance that an asteroid with a diameter greater than 10 kilometers will strike Earth in the next 1,000 years. Such an impact would kill billions and could possibly cause the extinction of the human race.

One series of images of the asteroid Mathilde show an impact crater larger than the asteroid's mean radius. How could an object sustain such an impact without breaking apart? According to one hypothesis, Mathilde is not solid rock, as Earth is. Instead it is a compressed mass of fractured rock and rubble. Thus the bolide impact scattered the fractured rock without transmitting force through a solid, brittle body. However, craters, grooves, and surface rocks on nearby Eros indicate that it is a solid body.

Comets

Occasionally, a glowing object appears in the sky, travels slowly around the Sun in an elongated elliptical orbit, and then disappears into space (◇ Figure 23.28). Such an object is called a **comet**, after the Greek word for "long haired." Despite their fiery appearance, comets are cold, and their light is reflected sunlight.

Comets originate in the outer reaches of the Solar System, and much of the time they travel through the cold void beyond Pluto's orbit. A comet is composed mainly of water-ice mixed with frozen crystals of methane, ammonia, carbon dioxide, and other compounds. Smaller concentrations of dust particles, composed of silicate rock and metals, are mixed with the lighter ices.

When a comet is millions of kilometers from the Sun, it is a ball without a tail. As the comet approaches the Sun and is heated, some of its surface vaporizes. Solar wind blows some of the lighter particles away from the comet's head to form a long tail. At this time, the comet consists of a dense, solid **nucleus**, a bright outer sheath called a **coma**, and a long **tail** (◇ Figure 23.29). Some comet tails

◇ **FIGURE 23.28** Comet Hale–Bopp was the brightest comet seen from Earth in decades. It was brightest in March and April of 1997.

are more than 140 million kilometers long, almost as long as the distance from Earth to the Sun. As a comet orbits the Sun, the solar wind constantly blows the tail so that it always extends away from the Sun. By terrestrial standards, a comet tail would represent a good, cold laboratory vacuum, yet viewed from a celestial perspective it looks like a hot, dense, fiery arrow.

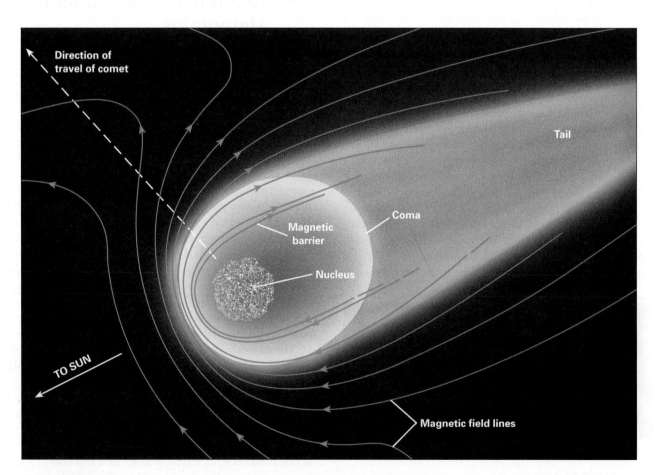

◇ **FIGURE 23.29** The nucleus of this comet has been enlarged several thousand times to show detail. When the comet interacts with the solar wind, magnetic field lines are generated as shown. Ions produced from gases streaming away from the nucleus are trapped within the field, creating the characteristically shaped tail.

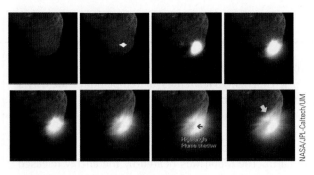

NASA/JPL-Caltech/UM

◀▶ **FIGURE 23.30** The eight images depict the development of the ejecta plume formed when Deep Impact's probe collided with comet Tempel 1 on July 3, 2005. The red arrows in images 3 and 7 highlight shadows cast by the ejecta. The yellow arrow on image 8 indicates the zone of avoidance in the up range direction. The 8 images were spaced 0.84 seconds apart.

Halley's comet passed so close to Earth in 1910 that its visit was a momentous event. When the comet returned to the inner Solar System in 1986, it was studied by six spacecraft as well as by several ground-based observatories. Its nucleus is a peanut-shaped mass approximately 16 by 8 by 8 kilometers, about the same size and shape as Manhattan Island. The cold, relatively dense coma of Halley's comet had a radius of about 4,500 kilometers when it passed by.

In one spectacular experiment conducted in 2005, astronomers fired a 372 kilogram metal probe into comet Tempel 1, while a mother ship recorded the effects of the impact. The probe triggered two quick flashes, the first when the collision heated the surface to thousands of degrees, and the second, a few milliseconds later, when the now-molten probe penetrated deeper into the nucleus and ejected a layer of volatile material (◀▶ Figure 23.30). Photographs taken just prior to the collision reveal a landscape sculpted by outgassing, melting, and natural impacts—all evidence of a complex geological history. Analysis of the impact ejecta suggest that comet Tempel 1 was composed of fine dust, more like talcum powder than beach sand. There were no large chunks, so the comet did not have a solid ice crust. In the near-surface interior, the space probe detected organic compounds, which could form the basis for living organisms.

Meteoroids

As tens of thousands of asteroids race through the Solar System in changing paths, many collide and break apart, forming smaller fragments and pieces of dust. A **meteoroid** is an asteroid or a fragment of a comet that orbits through the inner Solar System. If a meteoroid travels too close to Earth's gravitational field, it falls. Friction with the atmosphere heats it until it glows. To our eyes it is a fiery streak in the sky, which we call a **meteor** or, colloquially, a shooting star. Most meteors are barely larger than a grain of sand when they enter the atmosphere and vaporize completely during their descent. Larger ones, however, may reach Earth's surface. A meteor that strikes Earth's surface is called a **meteorite** (◀▶ Figure 23.31).

Most meteorites are **stony meteorites** and are composed of 90 percent silicate rock and 10 percent iron and nickel. The 90:10 mass ratio of rock to metal is similar to the mass ratio of the mantle to the core in Earth. Therefore, geologists think that meteorites reflect the primordial composition of the Solar System and are windows into our Solar System's past. Most stony meteorites contain small grains about 1 millimeter in diameter called **chondrules**, which are composed largely of olivine and pyroxene. Many chondrules contain organic molecules, including amino acids, the building blocks of proteins.

Some meteorites are metallic and consist mainly of iron and nickel, the elements that make up Earth's core, while the remainder are stony-iron, containing roughly equal quantities of silicates and iron-nickel. Some of our knowledge of Earth's mantle and core comes from studying meteorites, which may be similar to the mantles and cores of other planetary bodies. With the exception of rocks returned from the Apollo Moon missions, meteorites are the only physical samples we have from space.

NASA

◀▶ **FIGURE 23.31** Scientists believe that this meteorite is a fragment from the asteroid Vesta.

SUMMARY

Mercury is the closest planet to the Sun. It has virtually no atmosphere, and it rotates slowly on its axis; therefore, it experiences extremes of temperature. Mercury's surface is heavily cratered from meteorite bombardment that occurred early in the history of the Solar System. **Venus** has a hot, dense atmosphere as a result of greenhouse warming. Its surface shows signs of recent tectonic activity, probably resulting from vertical upwelling, called **blob tectonics.** Earth's **Moon** probably formed from the debris of a collision between a Mars-sized body and Earth. The Moon was heated by the energy released during condensation of the debris, by radioactive decay, and by meteorite bombardment. Evidence of ancient volcanism exists, but the Moon is cold and inactive today. **Mars** is a dry, cold planet with a thin atmosphere, but its surface bears signs of tectonic activity and ancient water erosion. Spirit and Opportunity detected clear signs that water once existed on Mars.

Jupiter, Saturn, Uranus, and **Neptune** are all large planets with low densities. Jupiter and Saturn have dense atmospheres, surfaces of liquid hydrogen, inner zones of liquid metallic hydrogen, and cores of rock and metal. The largest moons of Jupiter are **Io,** which is heated by gravitational forces and exhibits extensive volcanic activity; **Europa,** which is ice covered; and **Ganymede** and **Callisto,** which are large spheres of rock and ice. The rings of Saturn are made up of many small particles of dust, rock, and ice. They formed from a moon that was fragmented or from rock and ice that never coalesced to form a moon. **Titan,** the largest moon of Saturn, has an atmosphere rich in nitrogen and methane. Photos of the surface show evidence that landscape was sculpted by wind, flowing liquids, and tectonic activity. **Uranus** and **Neptune** have higher proportions of rock and ice than Jupiter and Saturn do, and their magnetic fields are not in line with their axes of rotation. **Pluto,** the most distant planet, has a low density and is a small planet composed of ice and rock.

Asteroids are small, planet-like bodies in solar orbits between those of Mars and Jupiter. A **comet** is an object that orbits the Sun in an elongated, elliptical orbit. Comets are composed mainly of water-ice and other frozen volatiles, mixed with smaller concentrations of dust, silicate rock, and metals. When a comet is in the inner part of the Solar System, it consists of a small, dense **nucleus** composed of ice and rock; an outer sheath or **coma** composed of gases, water vapor, and dust; and a long **tail** made up of particles blown outward by the solar wind. A **meteorite** is a fallen **meteoroid,** a piece of matter from the inner Solar System. Most meteorites are stony, and some contain organic molecules. About 10 percent of all meteorites are metallic.

Key Terms

For Review

1. List the nine planets in order of distance from the Sun, and distinguish between the terrestrial and Jovian planets.

2. Give a brief description of Mercury. Include its atmosphere, surface temperature, surface features, and speed of rotation about its axis.

3. Compare and contrast the atmospheres of Mercury, Venus, the Moon, Earth, and Mars.

4. Venus boiled, life evolved on Earth under moderate temperatures, and Mars froze. Discuss the evolution of the atmospheres and climates on these three planets. (Refer back to the beginning of Chapter 1 for additional information to answer this question.)

5. Discuss how the information from Spirit and Opportunity creates a picture of the climate history of Mars.

6. Discuss the evidence that the Martian atmosphere was once considerably different than it is today.

7. Why are there fewer meteorite craters visible on Venus than on Mercury?

8. Explain how Venusian tectonics differs from tectonics on Earth.

9. Compare and contrast the surface topography of Mercury, Venus, the Moon, Earth, and Mars.

10. What leads us to believe that the Moon was hot at one time in its history? How was the Moon heated?

11. Discuss evidence for blob tectonics and for horizontal tectonics on Mars.

12. Describe the composition of the planet Jupiter. How does it differ from that of Earth?

13. Discuss Jovian weather and what heat source drives it. Compare Jovian weather with that on Earth.

14. Compare and contrast the four Galilean moons of Jupiter.

15. Compare and contrast Saturn with Jupiter.

16. Compare and contrast Titan with Earth.

17. Compare and contrast Pluto with Earth and Jupiter.

18. Compare and contrast Pluto with Triton.

19. Is a comet hot, dense, and fiery? If so, what is the energy source? If not, why do comets look like burning masses of gas?

For Discussion

1. If Mercury rotated once every 24 hours as Earth does, would you expect daytime temperatures on that planet to be higher or lower than they are? Defend your answer.

2. Explain how we can learn about Earth's early history by studying the Moon.

3. At one time, Venus and Earth probably had similar climates, except that Venus was about 20°C warmer. If you could somehow cool the surface of Venus by 20°C, would conditions on that planet likely become similar to those on Earth? Explain.

4. Scientists speculate that even though the surface of Mercury is mostly hot, ice may exist in polar craters. The surface of Venus is also hot. Would you expect to find ice anywhere on Venus? If your answer is yes, where would you look? If your answer is no, why not?

5. Speculate on how a mantle plume could have generated faults on the surface of Venus even though this did not lead to rifting of tectonic plates.

6. Stephen Saunders, project scientist for NASA's Magellan mission, wrote, "Venus has been shaped by processes fundamentally similar to those that have taken place on Earth, but often with dramatically different results." Give examples to support or refute this statement.

7. Different minerals reflect radar differently, so radar data can be used to infer mineral structure. When a mineral weathers, it changes to a new mineral, and the radar reflectivity changes. Discuss how scientists could use radar reflectivity to age date Venusian landforms.

8. Refer to Figures 23.2 and 23.10. Imagine that you know nothing about Mars except that it is a planet and that these two photographs were taken of its surface. What information can you deduce from these pictures alone? Defend your conclusions.

9. Imagine that three new planets were found between Earth and Mars. What could you tell about the geologic history of each, given the following limited data: (a) Planet X: the entire surface of

this planet is covered with sedimentary rock. (b) Planet Y: this planet's atmosphere contains large quantities of water, ammonia, methane, and hydrogen sulfide. (c) Planet Z: about one-third of this planet is covered with numerous impact craters. Smaller craters can be seen within the largest ones. Another third of the surface is much smoother and is scattered with a few small craters. The remainder of the planet has no visible craters but is marked by great topographic relief, including mountain ranges and smooth plains but no canyons or river channels.

10. In his novel *2010: Odyssey II,* Arthur C. Clarke tells of a group of astronauts who traveled to the vicinity of Jupiter to retrieve a damaged spacecraft. At the end of the novel, Jupiter undergoes rapid changes and becomes a star. Is such a scenario plausible? Why did Mr. Clarke choose

Jupiter as the planet to undergo such a change? Would any other planet have been as believable?

11. About 4 billion years from now, the Sun will probably grow significantly larger and hotter. How will this affect the composition and structure of Jupiter?

12. Write a short, science-fiction story about space travel within the Solar System. You may make the plot fantastic and fictitious, but place the characters in scientifically plausible settings.

Stars, Space, and Galaxies

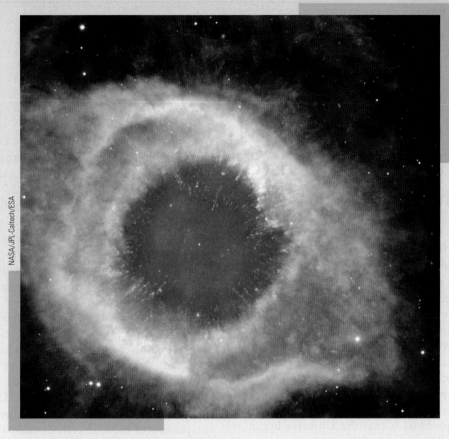

Six hundred and fifty light-years away in the constellation Aquarius, a dead star about the size of Earth and called a white dwarf, can be seen at the center of the image as a white dot. It is spewing out massive amounts of hot gas and intense ultraviolet radiation, creating a planetary nebula. In this false-color image, NASA's Hubble and Spitzer Space Telescopes have imaged the nebula. The colorful gaseous material seen was once part of the central star, but was lost in the death throes of the star as it became a white dwarf.

NASA/JPL-Caltech/ESA

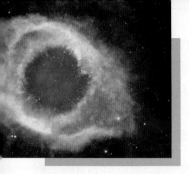

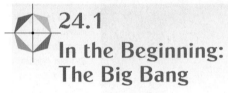

Earth is only one planet orbiting one star among roughly 100 billion stars in our Milky Way Galaxy. In turn, the Milky Way is only one galaxy of billions in the Universe. Looking beyond the Solar System into galactic and intergalactic space, we must stretch our minds to nearly unimaginable distances, look backward in time to events that occurred billions of years before Earth formed, and attempt to fathom energy sources powerful enough to create an entire Universe. ■

24.1

In the Beginning: The Big Bang

In our search for the origin and history of Earth, we have looked back more than 4.6 billion years to the time when a diffuse cloud of dust and gas coalesced to form the Solar System. But now, as we look into our galaxy and beyond, we ask, How did that cloud of dust and gas form? As our search for answers deepens, we finally ask, How and when did the Universe begin?

Before we explore the origin of the Universe, we must ask an even more fundamental question: Did it begin at all? One possibility is that the Universe has always existed and there was no beginning, no start of time. An alternative hypothesis is that the Universe began at a specific time and has been evolving ever since.

In 1929 Edwin Hubble observed that all galaxies are moving away from each other. By projecting the galactic motion backward in time, he reasoned that they must have started moving outward from a common center, and at the same time. Therefore, scientists calculated that in the beginning the entire Universe was compressed into a single, infinitely dense point. This point was so small that we cannot compare it with anything we know or can even imagine. According to modern theory, this point exploded. But it was no ordinary explosion. It cannot even be compared with a hydrogen bomb or supernova explosion. This explosion, called the **big bang**, instantaneously created the Universe. Matter, energy, and space came into existence with this single event. It was the start of space and time.

Astronomers calculate the timing of the big bang by measuring speeds of galaxies and the distances among them. They then calculate backward in time to determine when they were all joined together in a single point. In 2003, astronomers combined a variety of techniques to determine that the Universe is 13.7 billion years old.

Scientists generally start their discussions of the origin of the Universe when it was a trillionth of a trillionth of a billionth of a second old. How can we reach that far back in time with any degree of certainty? In one series of experiments, scientists study the collision behavior of particles at very high velocities in modern particle accelerators. These results are then compared with observations of deep space. In the 75 years since Hubble's pioneering work, the experiments, observations, and calculations have led to disagreements, paradoxes, and unsolved mysteries. Yet the preponderance of evidence is so persuasive that almost all astronomers agree with the fundamental premise of the big bang theory. Three lines of evidence and logic support the big bang theory. The first is the expansion of the Universe, discussed previously. To understand the second two, we must study the first 300,000 years in the life of the Universe.

Let us go back to that unimaginably distant time when the Universe was a trillionth of a trillionth of a billionth of a second old. At that time, the Universe was only about as big as a grapefruit and the temperature was about 100 billion degrees Kelvin (written 100 billion K).[1] But this primordial Universe was expanding rapidly as it was propelled outward by the force of the explosion that formed it. As the Universe expanded, it cooled. During the first second, the Universe cooled to about 10 billion degrees, 1,000 times the temperature in the center of the modern Sun (◀▶ Figure 24.1). At such high temperatures, atoms do not exist. Instead, the

---

[1]Astronomers generally report temperature on the Kelvin scale, which is based on fundamental thermodynamic properties ($0°C = 273$ Kelvin). At high temperatures, the 273-degree difference between the two is negligible.

Time	Description of Universe			Average temperature of Universe
0	Point sphere of infinite density			
Less than 0.01 second	Radiant energy	Electrons Neutrinos	Positrons	Other fundamental particles	100 billion °C
1 second	Radiant energy	Electrons Neutrinos		Protons and neutrons form	10 billion °C
1.5 to 10 minutes				Helium and deuterium nuclei	Below 1 billion °C
300,000 years				Atoms form	A few thousand °C
1 billion years		Proto galaxies			?
5 billion years		Primeval galaxies Quasars			?
Today, about 14 billion years		Today's galaxies			−275 °C

◈ **FIGURE 24.1**   A brief pictorial outline of the evolution of the Universe.

Universe consisted of a plasma of radiant energy, electrons, protons, neutrons, and extremely light particles called neutrinos.

Over the next few minutes, the most exciting action came from the behavior of protons and neutrons. Recall that protons are positively charged particles. At large distances (by subatomic standards) two positive protons repel each other, like two north poles of a magnet. However, if protons are hot enough, they collide with sufficient energy to overcome the repulsion. At very close distances, two protons attract and bond together. This bonding is called fusion. Hot protons also fuse with hot neutrons.

A hydrogen nucleus consists of a lone proton. At the right temperature, four protons (hydrogen nuclei) fuse together to form a nucleus of the next heavier element, helium (◈ Figure 24.2). This nuclear fusion is identical to reactions that occur in a star or in a hydrogen bomb.

When the Universe was less than 1.5 minutes old, it was so hot that helium nuclei were blasted apart almost as soon as they formed. After 10 minutes, the Universe was so cool that fusion could no longer occur. Thus, the formation of helium nuclei, called the **primordial nucle-**

osynthesis occurred over a time span of 8.5 minutes. Enough fusion had occurred during this time so 6 percent of the total nuclei in the Universe were helium, while 94 percent remained as hydrogen nuclei. Although primordial nucleosynthesis created a trace of lithium (the next heaviest element after helium), it did not produce any heavier elements. There was no carbon, nitrogen, or oxygen, so life could not have evolved, no silicon or metals, so solid planets could not have evolved. The Universe was still in its infancy.

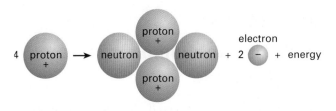

◈ **FIGURE 24.2**   In a series of reactions, four hydrogen nuclei fuse to form helium. Two electrons and tremendous quantities of energy are released.

When scientists calculate the expected ratio of hydrogen to helium in the modern Universe from the conditions of the primordial Universe, they arrive at a number almost exactly in agreement with the observed abundances. This agreement provides the second line of evidence and logic to support the big bang theory. It is also a testimonial to human genius. As the Universe continued to expand, it finally cooled to a few thousand degrees by about 300,000 years after the big bang. At this crucial temperature, electrons became attached to the hydrogen and helium nuclei. Thus the first atoms formed and, in a sense, the modern Universe was born.

Before atoms existed, the Universe was a chaotic plasma that scattered and absorbed photons. As a result, photons could never move far in any direction. The Universe was foglike or opaque. But when electrons combined with nuclei to form hydrogen and helium atoms, conditions changed radically. An atom absorbs light in only a relatively few specific wavelengths. All the rest of the light passes through unhindered. Thus a hydrogen helium-filled Universe is like a pair of sunglasses. Sunglasses filter out certain wavelengths of light but permit other wavelengths to pass through.

In the 1960s, an astrophysicist at Princeton named Robert Dicke predicted that we should be able to detect the primordial photons that began moving in a straight line as soon as atoms formed. He calculated that continued expansion of the Universe would have cooled the primordial photons to about 2.7 K (2.7°C above absolute zero). In 1964, Arno Penzias and Robert Wilson at Bell Laboratories detected a very faint photon radiation that was 2.7 K and was uniformly distributed throughout space. Thus experimental observation agreed precisely with Dicke's calculation.

Walk outside at any time of day at any place on Earth and turn your palm upward to the sky. Every second, a million billion low-energy photons will strike your palm. These photons are in the microwave band of the electromagnetic spectrum. The energy is so faint and weak that you could never feel it. Yet this **cosmic background radiation** began traveling through space when the Universe was only 300,000 years old. Today, the cosmic background radiation is called the echo of the big bang. The prediction and discovery of the cosmic background radiation provides the third convincing line of evidence to support the big bang theory.

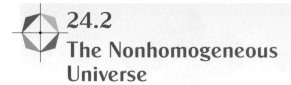

24.2
The Nonhomogeneous Universe

Penzias and Wilson observed that the cosmic background radiation was uniform throughout space. This measurement implied that the Universe was homogeneous in its infancy. However, the modern Universe is

clearly not homogeneous. Matter is concentrated into stars, stars are clumped into galaxies, and galaxies are grouped into clusters containing tens of thousands of galaxies. Even the clusters group into superclusters. Most of the space between the clusters and superclusters contains no galaxies at all (◀ Figure 24.3).

The question then arose: How and when did an initially uniform, homogeneous Universe concentrate into stars, galaxies, and clusters? According to one hypothesis, the original, grapefruit-sized Universe had to obey laws of quantum mechanics, the same laws that describe modern atoms. Calculations based on quantum mechanics showed that the earliest Universe contained tiny waves of energy, space, and time. These waves, like sound waves, contained alternating regions of higher and lower densities. If this model were correct, then these bands of varying energy densities would have created tiny temperature differences in the cosmic background radiation.

In the 1980s, physicists calculated that the temperature differences would be about one-ten-thousandth of a degree Celsius, far too small to have been detected by Penzias and Wilson's radio telescope. In order to search for these temperature differences, they had to build a much more precise radio telescope. Because the atmosphere interferes with microwave transmissions, they needed to mount this antenna on a satellite.

In 1989, astronomers launched the Cosmic Background Explorer (COBE) satellite that was capable of measuring 0.0001°C temperature differences in the cosmic background radiation. As it slowly scanned the Universe, COBE registered numerous fluctuations in the background temperature. After mapping the temperature of the Universe for 3 years, COBE scientists reported that cosmic background radiation varies by tiny amounts from one region to another. This data suggested that the primordial Universe was not homogeneous as Penzias and Wilson had inferred. Matter and energy had concentrated into clumps during the earliest infancy of the Universe, perhaps in the first billionth of a second. George Smoot, a researcher on the project, called these variations "the imprints of tiny ripples in the fabric of space-time put there by the primeval explosion." In 2003, data from a second spacecraft, called the WMAP satellite, confirmed and refined the COBE results.

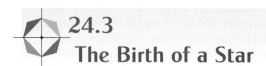

24.3
The Birth of a Star

The cosmic background radiation emanated from the primordial sea of atoms, particles, and energy of the big bang. This radiation is called **first generation energy**. When you look up into the sky on a clear, moonless night, you see stars sparkling in the blackness. If you owned a powerful telescope, you could peer deeper into space and detect distant galaxies, or even curious structures called quasars. All of these objects emit what astronomers call

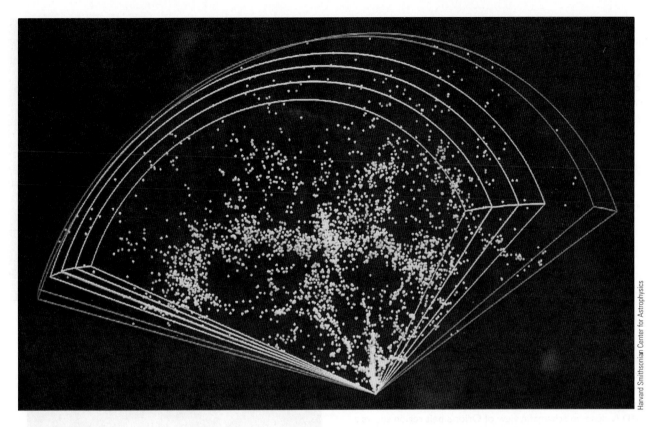

◀▶ **FIGURE 24.3** A three-dimensional, computer-generated drawing of a portion of the Universe. Notice that matter is unevenly distributed.

second generation energy. They use the term *second generation* because concentrated matter—stars, galaxies, or quasars—emit this light. To understand our universe, we must learn how matter in the subtly nonhomogeneous, early Universe clumped together to form concentrated bodies that emit light and energy of their own.

Over time, denser regions of the Universe drew matter inward by gravity to form huge clouds of hydrogen and helium, called **nebulae**. Within each cloud, matter further agglomerated into billions of smaller bodies. As the atoms in a nebula accelerated inward under the force of gravity, they collided rapidly with one another. Thus the center of each cloud became very dense and hot. Under the intense heat, electrons were stripped away from their atoms, leaving a plasma of positively charged nuclei and negatively charged electrons. If the cloud were originally large enough, the gravitational attraction accelerated the nuclei until they collided with enough energy to fuse. When fusion started, the collapsing portion of the original nebula became a star.

Fusion generates energetic photons and tremendous quantities of heat. If a dense sphere of hydrogen the size of a pinhead were to fuse completely, it would release as much energy as is released by burning several thousand tons of coal. Both the photons and the hot particles generated by fusion accelerate outward against the force of gravity. Thus two opposing processes occur in a star. Gravity pulls particles inward, but at the same time fusion energy drives them outward (◀▶ Figure 24.4). The

balance between these two processes determines the diameter and density of a star of a given mass. At equilibrium, a star with an average mass has a dense core surrounded by a less-dense shell.

A star exists for millions, billions, or tens of billions of years, and we can never observe one long enough to watch its birth, life, and death. We can, however, observe young, middle-aged, and old stars and thus piece together the story of stellar evolution. It is as if an alien came to Earth for one day to observe the life of a human being. It would be impossible to observe the birth, growth, and death of a single person, but the alien could observe babies, children, middle-aged people, and old people and thus infer the course of a human life.

One nebulae near the belt in the constellation Orion is a nursery for the birth of new stars (◀▶ Figure 24.5). The stars in one portion of the nebula are at least 12 million years old. Stellar ages decrease progressively toward the southwest until the stars are less than one million years old in the region near Orion's sword. This age progression implies that star formation moved steadily from one end of the nebula to the other.

Astronomers are uncertain how the first stars formed in a large nebula. However, once the first stars began to generate energy, they heated the surrounding gas in the nebula. The heated gas expanded so that its pressure and density increased. From the rate of stellar evolution, astronomers calculated that one or more of the initial stars were massive and short lived. These stars

◆ **FIGURE 24.4** The diameter and density of a star are determined by two opposing forces. Gravity pulls particles inward, while energy from fusion reactions in the core forces particles outward.

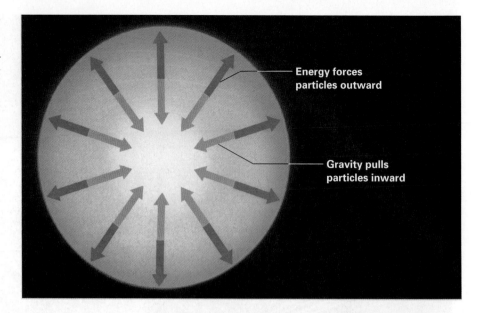

Energy forces particles outward

Gravity pulls particles inward

◆ **FIGURE 24.5** **(A)** The constellation Orion was named after a mythical Greek hunter. The three stars across the center of the constellation show the hunter's belt, and the points of light angling down from the belt show his sword. One of the objects in his belt is an emission nebula.
(B) A close-up schematic view of Orion's belt region shows a huge system of nebulae. Young stars are evolving within this giant stellar nursery.
Source: J. Bally, "Filaments in Creation's Heart," *Nature* 382 (July 11, 1996): 114.

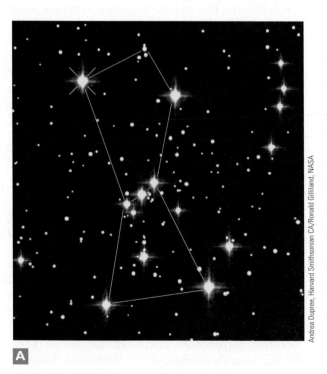

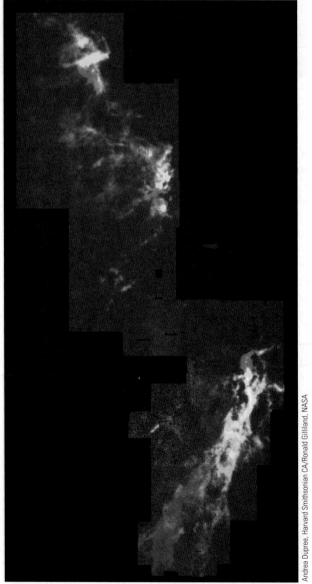

Andrea Dupree, Harvard Smithsonian CA/Ronald Gilliland, NASA

Andrea Dupree, Harvard Smithsonian CA/Ronald Gilliland, NASA

A

B

CHAPTER 24 • Stars, Space, and Galaxies

quickly exhausted their supply of fuel and exploded. Hot gases burst into the surrounding nebula, creating a shock wave that increased the pressure and density even more. When the pressure and gas density reached a critical value, gravitational forces between molecules were strong enough to pull the particles together. Matter sep-arated into discrete regions falling toward common centers. Recent observations by the Hubble telescope show that these regions of varying density appear as towering pillars of light and shadow (◁▷ Figure 24.6). Hundreds of young stars, only 8 million to 300,000 years old, are forming along the edges of the cloud.

◁▷ **FIGURE 24.6**   The Hubble Space Telescope has provided a close-up view of stellar nurseries near Orion's belt. The pillarlike structures are about 3 light-years tall and are composed of cold gas and dust. Energy from nearby stars and shock waves from ex-ploding stars have concentrated the gas in the pillars and initiated star formation.

J. Hester and P. Scowen (ASU) and NASA

Let us look at the structure of our own Sun and then return to study the lifecycle of a star.

24.4
The Sun

Nobody has ever traveled to the Sun—and nobody ever will. We have never sent spacecraft into the solar atmosphere to collect samples of the torridly hot gases that glow on its exterior. The core will be forever opaque and invisible because any material, any instrument, or any spacecraft would vaporize long before it penetrated into the solar interior. All of our information about the Sun is derived from indirect evidence, yet scientific reasoning is so powerful that we have a reasonably complete and accurate picture of this life-giving orb.

By studying the orbital properties of the planets, scientists can calculate the gravitational force of the Sun and hence its mass, approximately 2×10^{30} kg. Although this huge number is almost impossible to comprehend, our Sun is a star of average mass. The Sun's diameter is 1.4 million kilometers (109 Earth diameters).

Recall from Chapter 22 that astronomers can determine the composition of a star by analyzing its emission and absorption spectra. From these studies, we have learned that hydrogen accounts for 92 percent of the Sun's atoms, and helium is second in abundance at 7.8 percent. All the remaining elements make up only 0.2 percent. The Sun's structure is shown in ◆ Figure 24.7.

The Core

In the late 1800s, scientists calculated that the Sun emitted so much energy that it could not be powered by the burning of conventional fuels. But if the Sun was not a giant fire, what was it? In 1905, Einstein announced his theory of relativity and shortly thereafter other scientists began to unravel the mystery of nuclear reactions. These studies showed that the Sun is powered by hydrogen fusion. Every second, the Sun converts 4 million tons of hydrogen into energy and radiates this energy into space. Yet the Sun is so massive that there is no detectable change in mass from year to year, or even from century to century.

Our understanding of the solar **core** is derived from calculations of hydrogen fusion reactions and gravitational forces. From these studies, scientists are reasonably certain that the core of the Sun is extremely hot, more than 15 million K, and its density is 150 times that of water. The Sun's core is 140,000 kilometers in diameter and contains enough hydrogen to fuel its fusion reaction for another 5 billion years. When the hydrogen in the core is used up, the Sun will change drastically, as described in sections 24.5 and 24.6.

Most of the energy generated by fusion in the Sun's core is emitted as radiation of high-energy photons. When a photon flies out of the core, atoms almost immediately absorb it in a broad region surrounding the core called the **radiative zone**. Atoms reemit photons soon after they absorb

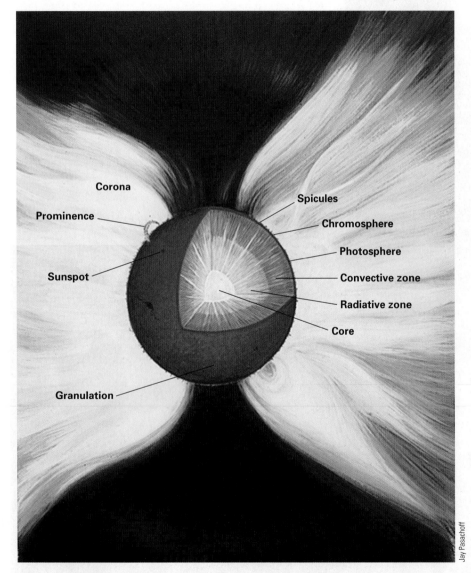

Jay Pasachoff

◆ **FIGURE 24.7** Structure of the Sun. Fusion occurs in the Sun's core. Energy escapes first by radiation, then by convection. The photosphere is the thin layer visible from Earth. Surface features are shown in more detail in Figures 24.8 through 24.10.

Corona
Prominence
Sunspot
Granulation

Spicules
Chromosphere
Photosphere
Convective zone
Radiative zone
Core

them, but the gas density in the region is so high that a photon is quickly reabsorbed again. This process of absorption and emission occurs repeatedly until the photon reaches the **convective zone**. The convective zone is much cooler on its surface than in its interior. As a result, hot gas rises, cools, and then sinks in huge convection cells. This convection transports heat to the outer layer that we see, called the **photosphere**.

If photons traveled directly from the core to the photosphere at the speed of light, they would complete the journey in about 4 seconds. However, the process of absorption, emission, and convection is so much slower that energy requires about a million years to make the journey.

The Photosphere

Even though the entire Sun is gaseous, the photosphere is called the solar atmosphere because it is cool and diffuse compared to the core. It is a thin surface veneer, only 400 kilometers thick. The pressure at the middle of the photosphere is about 1/100 that of Earth's atmosphere at sea level, and its average temperature is 5,700 K, about the same as that in Earth's core. Fusion does not occur at this relatively low temperature. Thus the core heats the photosphere, and the sunlight we see from Earth comes from this thin, glowing atmosphere of hydrogen and helium.

The photosphere has a granular structure, with each grain about 1,000 kilometers across, or about the size of Texas. Convection currents that carry energy to the surface form the granules. If you could watch the Sun's surface at close range, the granules would appear and disappear like bubbles in a pot of boiling water. The bright, yellow granules in ◀▶ Figure 24.8 are formed by hot, rising gas, and the darker, reddish regions are cooler areas of descending gas.

Large, dark spots called **sunspots** also appear regularly on the Sun's surface. A single sunspot may be small and last for only a few days, or it may be as large as 150,000 kilometers in diameter and remain visible for months. Because a sunspot is 1,000 degrees cooler than the surrounding area, it radiates only about one-half as much energy and hence appears dark compared to the rest of the photosphere. Since heat normally flows from a hot body to a cooler one, a large, cool region on the Sun's surface should quickly be heated and disappear. Yet some sunspots persist for long periods.

Astronomers believe that the Sun's magnetic fields restrict mixing of the photosphere and inhibit warming of sunspots. The magnetic fields associated with sunspots also produce flares, by accelerating glowing gases to velocities in excess of 900 km/hr and sending shock waves smashing through the solar atmosphere. The flares emit charged particles that interrupt radio communication and cause aurorae (northern and southern lights) on Earth.

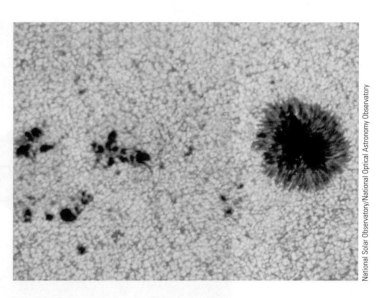

National Solar Observatory/National Optical Astronomy Observatory

◀▶ **FIGURE 24.8** A close-up of the Sun's surface shows granular structures and sunspots. The large, black structures are sunspots. Each small, bright-yellow dot is a rising column of hot gas about 1,000 km in diameter. The darker, reddish regions between the dots are cooler, descending gases.

The Sun's Outer Layers

The photosphere is not bounded by a sharp, outer surface. A turbulent, diffuse, gaseous layer called the **chromosphere** lies above the photosphere and is about 2,000 to 3,000 kilometers thick. The chromosphere can be seen with the naked eye only when the Moon blocks out the photosphere during a solar eclipse. Then it appears as a narrow, red fringe at the edge of the blocked-out Sun. Jets of gas called **spicules** shoot upward from the chromosphere, looking like flames from a burning log. An average spicule is about 700 kilometers across and 7,000 kilometers high and lasts about 5 to 15 minutes.

An even-more-diffuse region called the **corona** lies beyond the chromosphere. During a full solar eclipse, the corona appears as a beautiful halo around the Sun (◀▶ Figure 24.9). Its density is one-billionth that of the atmosphere at the Earth's surface. In a physics laboratory, that would be a good vacuum.

The corona is extremely hot, about 2 million K. How does the photosphere, at 5,700 degrees, heat the corona to 2 million degrees? According to one hypothesis, twisting magnetic fields accelerate particles in the corona. When particles are moving quickly, their temperature is high. **Prominences** are red, flame-like jets of gas that rise out of the corona and travel as much as 1 million kilometers into space (◀▶ Figure 24.10). Some prominences are held aloft for weeks or months by the Sun's magnetic fields.

The high temperature in the corona strips electrons from their atoms, reducing hydrogen and helium to bare nuclei in a sea of electrons. These nuclei and electrons are moving so rapidly that some fly off into space, forming the solar wind that extends outward toward the far reaches of the Solar System.

◇ **FIGURE 24.9** During a solar eclipse the photosphere is blocked by the Moon. The thin, red streaks beyond the Moon's outline are portions of the chromosphere, and the large, white zone is the corona.

◇ **FIGURE 24.10** This prominence rises approximately 505,000 kilometers above the photosphere.

FOCUS ON

Measuring Distances in Space

Distance on Earth is measured in millimeters, meters, or kilometers.

However, even a kilometer is too small to express distance in space. Light travels in a vacuum at a constant rate—3.0×10^8 m/sec—never faster and never slower. One **light-year** is the distance traveled by light in a year, 9.5 trillion (9.5×10^{12}) kilometers. The closest star to our Solar System is 38 trillion (38×10^{12}) kilometers, or 4 light-years, away. But an object 4 light-years away is a close neighbor in space. Some objects are more than 12 billion light-years from Earth.

Recall from Chapter 22 that astronomers detected Earth's motion around the Sun by observing the parallax shift of distant stars. The same approach can be used to calculate the distance to a star. The parallax angle is measured by observing the apparent shift in position of a star at a 6-month interval, after Earth has completed half of a revolution around the Sun (Figure 1). The angles even to the closest stars are small and are expressed in arc seconds; one arc second is equal to 1/3,600 of a degree. If you were to view a coin the size of a quarter from a distance of 5 kilometers, it would appear to have a diameter of 1 arc second. Parallax angles can be used to measure the distance to close stars but are not accurate for more distant stars or faraway galaxies.

A common unit of distance, the **parsec**, is the distance to a star with a parallax angle of 1 arc second. One parsec is equal to 3.2 light-years, or 3.1×10^{13} kilometers.

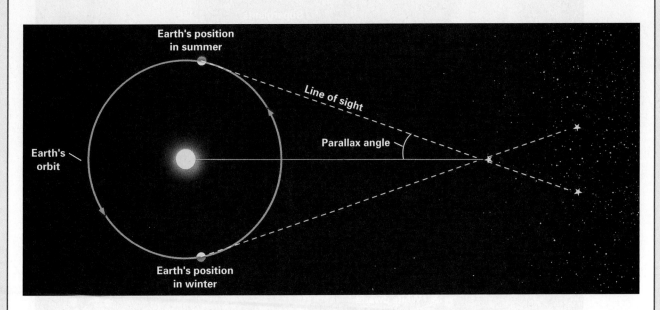

◆ **FIGURE 1** Scientists use simple trigonometry to calculate the distance to an object in space from a knowledge of the parallax angle.

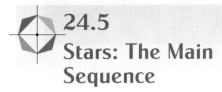

24.5
Stars: The Main Sequence

From our view on Earth, the Sun is the biggest and brightest object in the sky. However, the Sun looks so big and bright only because it is the closest star. When astronomers began to study stars in detail, one of their first efforts was to catalog them by their brightness. The **apparent brightness** or **apparent luminosity** of a star is its luminosity as seen from Earth. A star can appear luminous either because it is intrinsically bright or because it is close. (Think of a car headlight. The light appears to become brighter as the car approaches. The headlight bulb is not changing in luminosity; the car is just moving closer.) The **absolute brightness** or **absolute luminosity** is how bright a star would appear if it were a fixed distance away. Astronomers have chosen a standard distance of 10 parsec, or 32.6 light-years, to calculate absolute luminosities. (See Focus On box for definitions of parsec and light-year.) By convention, astronomers use a scale in which the absolute luminosity of a star is divided by the absolute luminosity of the Sun. Thus the Sun has a luminosity of 1.0, bright objects have a luminosity greater than 1, and faint objects have a value less than 1.

A second visible property of a star is its color. Different stars have different colors; some are reddish, while others are yellow or blue. The color of a star is a measure of its temperature. Blue is the hottest, yellow is intermediate, and red is the coolest. (The same relationship can be observed in a flame from a welding torch. The hottest, central portion of the flame is blue, and the cooler, outer edge is red.)

Between 1911 and 1913, Ejnar Hertzsprung and Henry Russell plotted the absolute luminosities of stars against their temperatures, as measured by the color. ◈ Figure 24.11 is a **Hertzsprung–Russell, or H–R, diagram.** In an H–R diagram, luminosity increases from bottom to top and temperature increases from right to

left. Thus hot, big, luminous stars are on the upper left, and cool, small, less-luminous stars are on the lower right.

The interesting observation about the H–R diagram is that about 90 percent of all stars fall along a sinuous band from upper left (bright and hot) to lower right (dim and cool). This band is called the **main sequence.** Thus luminosity increases with temperature in main-sequence stars. Notice that our Sun is more or less in the middle of the main sequence.

All main-sequence stars are composed primarily of hydrogen and helium and are fueled by hydrogen fusion. The major reason for differences in temperature and luminosity among main-sequence stars is that some are more massive than others. Because the force of gravity is stronger in massive stars than in less-massive stars, hydrogen nuclei are packed more tightly and move more rapidly. As a result, fusion is more rapid and intense in massive stars, so they are hotter and more luminous

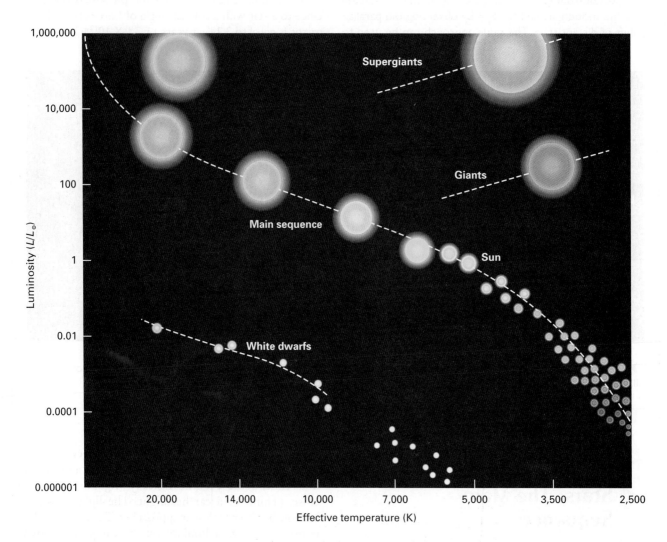

◈ **FIGURE 24.11**   A Hertzsprung–Russell (H–R) diagram. About 90 percent of all stars fall along the main sequence, which is marked by a white-dotted-line H–R diagram. Our Sun is an average-sized star with a surface temperature of about 6,000 K; it glows yellow.

INTERACTIVE QUESTION: **(A)** *What is the temperature range of blue, main-sequence stars?* **(B)** *What is the characteristic luminosity and temperature of a red giant? A white dwarf?*

(upper left of the H–R diagram). The least-massive stars, shown on the lower right, are cool and less luminous. Notice that our Sun is about halfway along the main sequence, in the yellow band.

To explain the stars that do not lie on the main sequence, we must consider the life and death of a star.

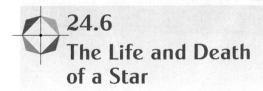

24.6
The Life and Death of a Star

Stars About the Same Mass As Our Sun

In a mature star such as our Sun, hydrogen nuclei fuse to form helium but the helium nuclei do not fuse to form heavier elements. For fusion to occur, two nuclei must collide so energetically that they overcome their nuclear repulsion. Hydrogen nuclei contain one proton, but helium contains two. Therefore the repulsion between two helium nuclei is much greater than that between two hydrogen nuclei. This stronger repulsion can be overcome if the nuclei are moving very rapidly. Because higher temperature causes nuclei to move more rapidly, helium fusion occurs only when a star becomes much hotter than the temperature at which hydrogen fusion occurs.

The lifecycle of a star with a mass similar to our Sun is shown in ◈ Figure 24.12. The Sun is now midway through its mature phase as a main-sequence star. It has been shining for about 5 billion years and will continue to shine much as it is today for another 5 billion years (◈ Figure 24.12A). During this entire period, the Sun produces energy by hydrogen fusion and remains on the main sequence.

After about 10 billion years, the outer shell of a star such as our Sun still contains large quantities of hydrogen, but most of the hydrogen in the core has fused to helium (◈ Figure 24.12B). The star's behavior now changes drastically. Because the hydrogen in the core is nearly

used up, hydrogen fusion slows down, less nuclear energy is produced, and the core cools. As it cools, the outward pressure of particles and energy decreases. Then the core starts to contract under the force of gravity. This gravitational contraction causes the core to grow hotter. It seems a paradox that when the nuclear reactions decrease, the core becomes hotter, but that is what happens.

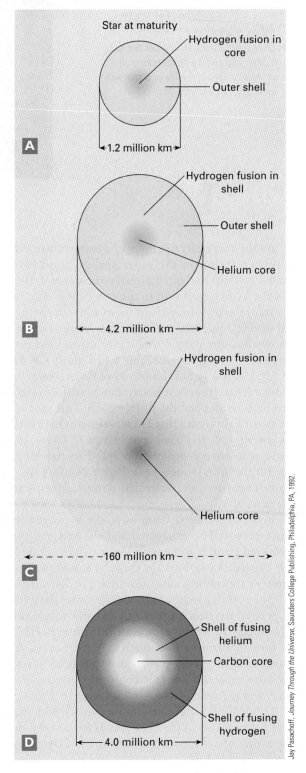

Jay Pasachoff, *Journey Through the Universe*, Saunders College Publishing, Philadelphia, PA, 1992.

◈ **FIGURE 24.12** Evolution of a star. **(A)** At maturity, the energy in a star the size of our Sun is derived from hydrogen fusion in the core. **(B)** When most of the hydrogen in the core is consumed, the star first contracts, then expands as hydrogen fusion starts in the outer shell. **(C)** In the red-giant phase, gravitational coalescence in the core and hydrogen fusion in the outer shell produce hundreds of times as much energy as was produced when the star was mature. **(D)** The star contracts again when helium fusion initiates in the core. The four parts cannot be drawn to scale as the star's diameter varies from 1.2 million kilometers to 160 million kilometers.

Also, in all cases the core has been drawn larger than scale to show detail.

FIGURE 24.13 A star the mass of our Sun passes through six major stages in its lifecycle: **(1)** after the original nebula condenses, **(2)** the star glows from the heat of gravitational coalescence; **(3)** the star enters the main sequence when hydrogen fusion starts in the core; **(4)** after hydrogen fusion ends in the core, the star leaves the main sequence and passes through the red-giant stage; **(5)** finally it explodes to produce a planetary nebula, and then **(6)** the remaining core glows as a white dwarf.

INTERACTIVE QUESTION: *List the solar luminosities and surface temperatures of the six major stages in the life of a star with the same mass as our Sun.*

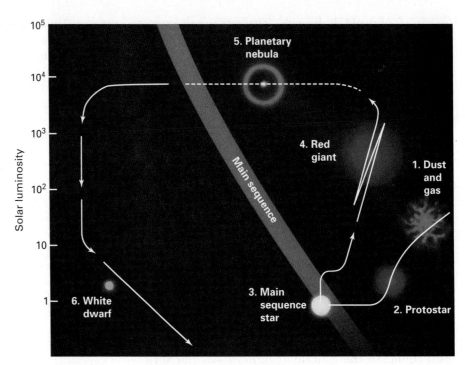

As the core heats up, the rising temperature initiates hydrogen fusion in the outer shell. The star is now heated by both the gravitational coalescence in the core and the hydrogen fusion in the outer shell. As a result, the star releases hundreds of times as much energy as it did when it was mature. This intense energy output now causes the outer parts of the star to expand and become brighter. The star has become a **red giant** (◆ Figure 24.12C). A red giant is hundreds of times larger than an ordinary star. Its core is hotter, but its surface is so large that heat escapes and the surface cools. This cool surface emits red light (recall that red wavelengths have the lowest energy of the visible spectrum). These sudden changes in energy production and diameter move the star off the main sequence (◆ Figure 24.13). Thus, a red giant is brighter and cooler than a main sequence star of the same mass.

Five billion years from now, the hydrogen in our Sun's core will be exhausted and the Sun will expand into a red giant. It will engulf Mercury, Venus, and Earth (◆ Figure 24.14). Perhaps the heat will blow much of Jupiter's atmosphere away, exposing a rocky surface.

The core of a red giant condenses under the influence of gravity and gets hotter until its temperature reaches 100 million K. At this temperature, helium nuclei begin to fuse to form carbon nuclei. When helium fusion starts, radiant energy pushes outward once again, and the core expands. The star cools, its outer layers contract, and it enters a second stable phase.

Gradually, as more helium fuses to carbon, the carbon accumulates in the core just as helium did during the earlier life of the star (◆ Figure 24.12D). When the

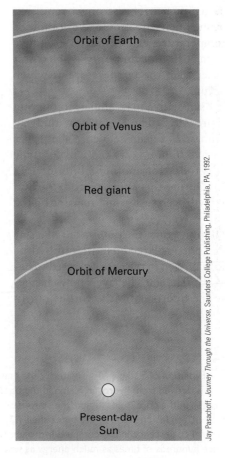

◆ **FIGURE 24.14** The Sun will engulf Earth when it expands to its red-giant stage.

Jay Pasachoff, Journey Through the Universe, Saunders College Publishing, Philadelphia, PA, 1992.

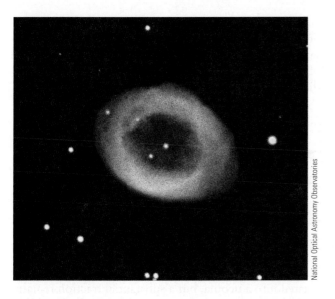

National Optical Astronomy Observatories

◆ **FIGURE 24.15** The Ring Nebula is a sphere of gas and dust expelled as a dying star exploded.

helium is used up, fusion ceases again and the carbon core contracts. This gravitational contraction causes the core to heat up again.

What happens next depends on the star's initial mass. Astronomers express the mass of a star relative to that of the Sun. One **solar mass** is the mass of the Sun. In a star as massive as our Sun, contraction of the carbon core is not intense enough to raise its temperature sufficiently to initiate fusion of the carbon nuclei. However, gravitational contraction of the carbon core does release enough energy to blow a shell of gas out into space. This shell is called a **planetary nebula** (◆ Figure 24.15). Meanwhile, the material remaining in the star contracts until atoms are squeezed so tightly together that only the pressure exerted by the electrons prevents further compression. A dying star as massive as our Sun will eventually shrink until its diameter is approximately that of Earth. Such a shrunken star no longer produces energy, and it glows solely from its residual heat produced during past eras. The star has become a **white dwarf** (◆ Figure 24.16). It will continue to cool slowly over tens of billions of years, but it will never change diameter again. No further nuclear reactions will occur. Its gravitational force is not strong enough to overcome the strength of the electrons, so it will never contract further.

Stars with a Large Mass

Some stars do not die so gently (◆ Figure 24.16). If the star is larger than 1.44 solar masses, a white dwarf does not form. Instead, as helium fusion ends, gravitational contraction produces enough heat to fuse carbon. Renewed fusion produces increasingly heavier elements in a stepwise process until iron forms. Iron is different from the lighter elements. When hydrogen or helium nuclei fuse, energy is released. In contrast, iron fusion absorbs energy and thus cools a star. When this happens, the thermal pressure that forced the stellar gases outward diminishes and the star collapses under the influence of gravity. This collapse releases large amounts of heat. Within a few seconds—a fantastically short time in the life of a star—the star's temperature reaches trillions of degrees and the star explodes to create a **supernova**. For a brief period, a supernova shines as brightly as hundreds of billions of normal stars and may even emit as much energy as an entire galaxy. To observers on Earth, it appears as though a new, brilliant star suddenly materialized in the sky, only to become dim and disappear to the naked eye within a few months.

On February 24, 1987, an astronomer named Ian Shelton was carrying out research unrelated to supernovas. When he developed one of his photographic plates, he saw a bright star where previously there was

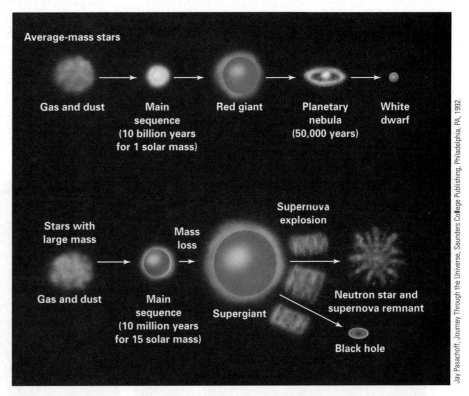

Jay Pasachoff, Journey Through the Universe, Saunders College Publishing, Philadelphia, PA, 1992.

◆ **FIGURE 24.16** A star with a mass about the same as our Sun passes through its lifecycle until it becomes a white dwarf. A more massive star follows a different route, ending as a neutron star or a black hole, depending on its initial mass.

only a dim one ($\diamondsuit$ Figure 24.17). He walked outside, looked into the sky, and saw the star with his naked eye. This was the first supernova explosion visible to the naked eye since 1604, five years before the invention of the telescope.

A supernova explosion is violent enough to send shock waves racing through the atmosphere of the star, fragmenting atomic nuclei and shooting subatomic particles in all directions. Many of the nuclear particles collide with sufficient energy both to fuse and to split apart. These processes form all the known elements heavier than iron. Thus in studying the evolution of stars, scientists learned how the heavy elements originated.

First- and Second-Generation Stars

Hydrogen and helium were the first elements to form when our Universe was 300,000 years old. Originally, all the stars in the Universe were composed entirely of these two lightest elements. These old, first-generation stars are called **population II stars**, and a few still exist.

Within the cores of population II stars, hydrogen fused to helium, changing the ratio of these two elements in the Universe. As stellar evolution continued, helium fused to carbon, and carbon fused to heavier elements, up to iron. Elements heavier than iron formed during supernova explosions of massive population II stars.

When a star dies, it blasts gas and dust into space to form a new nebula. Eventually, the nebulae condense once again into new, second-generation stars, called **population I stars**. Population I stars begin life with primordial hydrogen and helium mixed with small concentrations of heavy elements that were inherited from population II stars and from supernova explosions. The Sun is a population I star that condensed from a nebula containing all the natural elements.

Thus our Solar System was born from the debris of dying stars. Solar systems containing Earth-like planets and living organisms could never form around population II stars because these stars do not have all the necessary elements. As you can see, in studying the lifecycle of stars, we also learn about the origins of the elements that make up Earth and its life forms.

24.7
Neutron Stars, Pulsars, and Black Holes

Neutron Stars and Pulsars

In a supernova explosion, most of the matter in a star is blasted into a nebula, but a substantial fraction remains behind, compressed into a tight sphere. In the 1930s, scientists developed a hypothesis to explain what happens within this sphere. If it is between 2 and 3 solar masses, the gravitational force is so intense that the star cannot resist further compression the way a white dwarf does. Instead, the electrons and protons are squeezed together to form neutrons:

$$\text{Electrons} + \text{protons} = \text{neutrons}.$$

The neutrons then resist further compression and remain tightly packed. This ball of compressed neutrons is called a **neutron star**. A neutron star is extremely dense—approximately 10^{13} kg/cm^3 ($\diamondsuit$ Figure 24.18). If the entire Earth were as dense, it would fit inside a football stadium.

The first neutron star was discovered by accident in 1967, 30 years after scientists postulated their existence. Jocelyn Bell Burnell, a graduate student at the time, was studying radio emissions from distant galaxies. In one part of the sky, she detected a radio signal that switched

$\diamondsuit$ **FIGURE 24.17** During the supernova explosion of 1987, a relatively insignificant star **(A)** (see arrow) became the brightest object in that portion of the sky **(B)**.

on and off with a frequency of about one pulse every 1.33 seconds (◆ Figure 24.19). If such a signal were fed into the speaker of a conventional radio, you would hear a "click, click, click," evenly spaced, with one click every 1.33 seconds. Many radio signals arrive at Earth from outer space, but the emissions Burnell heard were unusual because they were sharp, regular, and spaced a little more than a second apart.

At first, astronomers considered the possibility that they might be a signal from intelligent life, so they called the signals LGM, for "little green men." But when Burnell found a similar pulsating source in a different region of the sky, scientists ruled out the possibility that two life forms in different parts of the Universe would send similar signals. Once it was established that the signals did not originate from intelligent beings, their sources were called **pulsars**. But naming the source did not explain it.

The first step toward identifying pulsars was to estimate their sizes. Not all parts of an object in space are equidistant from Earth (◆ Figure 24.20). If a large sphere emits a sharp burst of energy from its entire surface, some of the photons start their journey closer to Earth than others and therefore arrive sooner. A person on Earth listening to the radio noise hears not a sharp click but a more prolonged "cliiiiiick," because it takes a while for all the radio waves to arrive. Alternatively, a signal from a small source is much sharper. Pulsar signals are sharp, indicating that the source must be unusually small for an energetic object in space—30 kilometers in diameter or less. The smallest star previously recorded was a white dwarf 16,000 kilometers in diameter. Scientists then reasoned that the pulsar detected by Burnell might be the long-searched-for neutron star.

Astronomers suggested that the radio signals are emitted by an electromagnetic storm on the surface of a neutron star. Thus, a pulsar is a neutron star that emits intermittent, but regular, radio signals. According to this hypothesis, as the star rotates, so does the storm center. A receiver on Earth detects one click per revolution of the star, just as a lookout on a ship sees a lighthouse beam flash periodically as the beacon rotates (◆ Figure 24.21). Because a pulse is received once every 1.33 seconds, the pulsar must rotate that rapidly. White dwarfs

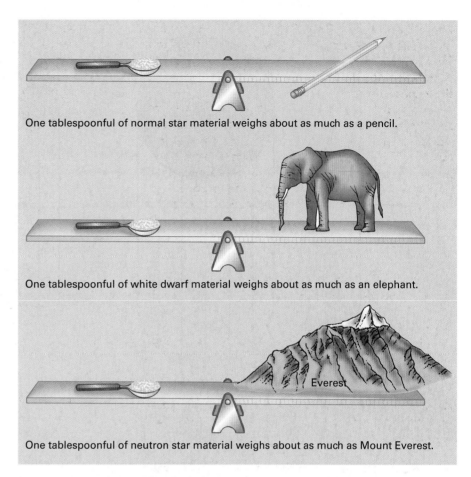

One tablespoonful of normal star material weighs about as much as a pencil.

One tablespoonful of white dwarf material weighs about as much as an elephant.

One tablespoonful of neutron star material weighs about as much as Mount Everest.

◆ **FIGURE 24.18** A normal star has a relatively low density; a white dwarf is more dense, and a neutron star has even higher density

can again be ruled out since they are too big to rotate so rapidly. But neutron stars are small enough to rotate that fast.

Astronomers searched for an example to test the hypothesis. They focused a radio telescope on the Crab Nebula, where a supernova explosion occurred during the Middle Ages. A pulsar signal was found exactly where the supernova had occurred about 950 years ago—precisely where a neutron star should be.

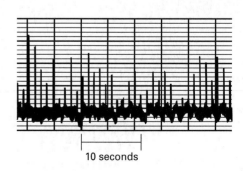

10 seconds

◆ **FIGURE 24.19** Pulsar signals appear as sharp spikes on the recording of a radio telescope.

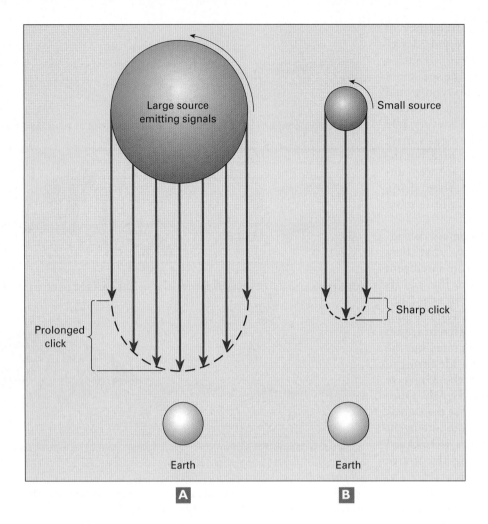

◆ **FIGURE 24.20** **(A)** A sharp signal from a larger sphere arrives over a longer time interval than **(B)** a sharp signal from a smaller sphere.

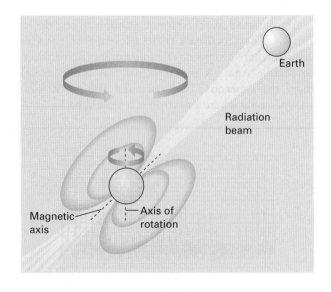

◆ **FIGURE 24.21** The radiation beam of a pulsar is detected only when it sweeps across Earth.

Black Holes

If a star with more than about 5 solar masses explodes, the core remnant remaining after the supernova explosion is thought to be more massive than 2 to 3 solar masses. When a sphere this massive contracts, the neutrons are not sufficiently strong to resist the gravitational force. Then the star shrinks to a diameter much smaller than a neutron star and becomes a **black hole**. Such a collapse is impossible to imagine in earthly terms. A tremendous mass, perhaps a trillion, trillion, trillion kilograms, shrinks to the size of a pinhead, then continues to shrink to the size of an atom, and then even smaller. Eventually it collapses to an infinitesimally small point of infinitely high density.

Such a small point of mass creates an extremely intense gravitational field. According to Einstein's theory of relativity, gravity affects photons. Thus starlight bends as it passes the Sun. If an object is massive and dense enough, its gravitational field becomes so intense that light and other radiant energy cannot escape. So, just as you cannot throw a ball from Earth to space because it falls back down, light cannot escape from a black hole because it is pulled back downward. Because no light can escape such an object, it is invisible; hence the name black hole. If you were to shine a flashlight beam, a radar beam, or any kind of radiation at a black hole, the energy would be absorbed. The beam could never be reflected back to your eyes; therefore, you would never see it again. It would be as though the beam just vanished into space. Similarly, if a spaceship flew too close to a black hole, it would be sucked in. No engine could possibly be powerful enough to accelerate the rocket back out, for no object can travel faster than the speed of light.

The search for a black hole is therefore even more difficult than the search for a neutron star. How do you find an object that is invisible and can neither emit nor reflect any energy? In short, how do you find a hole in space? Although it is theoretically impossible to see a black hole, astronomers can observe the effects of its gravitational field. Many stars exist in pairs or small clusters. If two stars are close together, they orbit around each other. Even if one becomes a black hole, the two still orbit about each other, but one is visible and the other invisible. The visible one appears to be orbiting an imaginary partner. Astronomers have studied several stars that seem to orbit in this unusual manner. In several cases, the invisible member of the pair has a mass equal to or greater than 3 solar masses. Because a normal star of 3 solar masses would be visible, the invisible partner may be a black hole. However, the simple observation that a star moves around an invisible companion does not prove that a black hole exists.

If a star were orbiting a black hole, great masses of gas from the star would be sucked into the black hole, to disappear forever (◆ Figure 24.22). As this matter started to fall into the hole, it would accelerate, just as a meteorite accelerates as it falls toward Earth. The gravitational field of a black hole is so intense that particles drawn into it would collide against each other with enough energy to emit X-rays. Thus just as a falling meteorite glows white-hot as it enters Earth's atmosphere,

anything tumbling into a black hole would glow even more energetically; that is, it would emit X-rays. These X-rays might then be detected here on Earth. But, you may ask, if light cannot escape from a black hole, how can the X-rays escape? The answer is that the X-rays are produced and escape from just outside the edge of the black hole. Thus, matter being sucked into a black hole sends off one final message before being pulled into the void from which no message can ever be sent.

An important experiment, therefore, was to focus an X-ray telescope on portions of the sky where a star appeared to orbit an invisible partner. Such telescopes must be located aboard space satellites because X-rays do not penetrate Earth's atmosphere. In the 1980s, an orbiting X-ray telescope detected X-ray sources adjacent to stars that appeared to orbit an unseen partner. More recent observations by ground-based observatories and the Hubble Space Telescope convince most astronomers that black holes exist.

Gamma Ray Bursts

If you could "see" electromagnetic radiation 10 million times more energetic than visible light, the Universe would appear to be a much more erratic, violent environment than we normally perceive. On average, once a day a flash of high-frequency radiation, called a **gamma ray burst**, bursts out of the heavens from a random direction, with a brief pulse containing more power than we receive from the Sun. Gamma ray bursts have a duration ranging from a few milliseconds to slightly more than two seconds. In 2002 astronomers determined that the longer bursts emanated from the death-throe explosion of a very massive star. But the origin of the short bursts proved more elusive. Imagine standing on the rim of a football stadium and trying to locate the source, intensity, and fine frequency structure of a flash from a camera flashbulb somewhere in the crowd. NASA scientists launched the Swift Space Observatory, equipped with an X-ray telescope, an ultraviolet/optical telescope, and a Burst Alert Telescope (BAT), all aligned and coordinated for instant observations over a wide range of frequencies. From these studies scientists determined that short, gamma ray bursts form when a neutron star is sucked into a black hole (◆ Figure 24.23) or when two neutron stars collide to form a black hole.

24.8 Galaxies

In the late 1700s, a French astronomer, Charles Messier, was studying comets. He recorded more than 100 fuzzy objects in the sky that clearly were not stars. When these objects were studied in the 1850s with more powerful telescopes, many were observed to have spiral structures like pinwheels. They were not comets, but what were

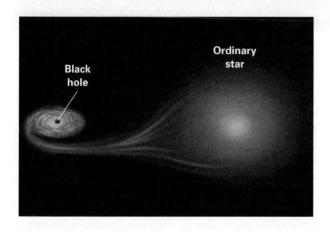

◆ **FIGURE 24.22** If a black hole and an ordinary star are in orbit around one another, gases from the star will eventually be sucked into the black hole.

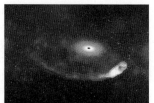

◀▶ **FIGURE 24.23** Short gamma ray bursts can be formed when a neutron star is sucked into a massive black hole. Source: http://www.nasa.gov/mission_pages/swift/bursts/short_burst_oct5.html.

they, and how far away were they? These questions were not answered until 1924 when Edwin Hubble determined that they were farther away than even the most distant known stars. In order for us to see them at all, they must be much more luminous than a star. Hubble concluded that each object is a **galaxy** composed of billions of stars. Today we recognize that galaxies and clusters of galaxies form the basic structure of the Universe.

About one-third of the galaxies in our region of the Universe are elliptical. Giant **elliptical galaxies** contain up to 10^{13} solar masses and may have a diameter larger than our Milky Way galaxy (which has a spiral shape, as discussed next). However, most ellipticals are dwarf and contain only a few million solar masses. Recent observations indicate that dwarf ellipticals may be the most common type of galaxy, far from our immediate galactic neighborhood.

Almost two-thirds of the bright galaxies in our region of the Universe are **spiral galaxies**.

The stars in a spiral are arranged in a thin disk, with arms radiating outward from a spherical center or nucleus (◀▶ Figure 24.24). The stars in the outer arms rotate around the nucleus like a giant pinwheel. A typical spiral galaxy contains about 100 billion stars. Nearly half of the spiral galaxies are **barred spirals**, having a straight bar of stars, gas, and dust extending across the nucleus (◀▶ Figure 24.25). A few percent of all galaxies are lens-shaped or irregular and show no obvious pattern.

Normally we think of galaxies as visible objects composed of collections of stars. Galaxy Virgo H 121 is a swirling cloud of hydrogen-rich gas, more massive than many visible galaxies, yet it contains no stars and emits no visible light, only radio-wave radiation. Astronomers hypothesize that dark galaxies have such a low density of matter that gravitational forces haven't been strong enough to initiate star formation.

Galactic Motion

Recall from Chapter 22 that astronomers measure the relative velocities of distant objects by measuring the frequency of light they emit. In 1929, 5 years after he had

Anglo-Australian Telescope Board

◀▶ **FIGURE 24.24** The dense nucleus of the spiral galaxy NGC 2997 is surrounded by spiraling arms composed of billions of stars.

described galaxies, Hubble noted that frequency of emitted light from almost every galaxy is shifted toward the red end of the spectrum. Hubble interpreted this red shift to mean that all the galaxies are flying away from us and from each other; the Universe is expanding. Moreover, he observed that the most-distant galaxies are moving outward at the greatest speeds, whereas the closer ones are receding more slowly. This relationship is known as **Hubble's law** (◀▶ Figure 24.26). Using Hubble's law, the distance from Earth to a galaxy can be calculated by measuring the galaxy's red shift.

◇▶ **FIGURE 24.25** A barred spiral galaxy, NGC 1365, displays a conspicuous bar running through its nucleus.

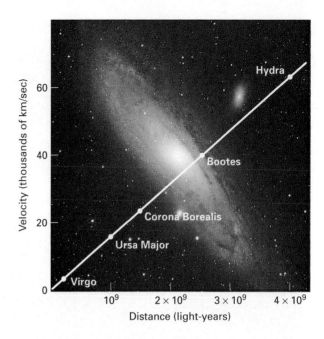

◇▶ **FIGURE 24.26** Hubble's law states that the velocity of a galaxy is directly proportional to its distance from Earth.

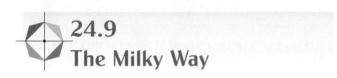

24.9
The Milky Way

Our Sun lies in a barred spiral galaxy called the **Milky Way** (◇▶ Figure 24.27). The Milky Way's disk is 2,000 light-years thick, and recent studies indicate that it is 200,000 light-years in diameter. The Milky Way contains about 400 billion stars. It has also swallowed up numerous, independent dwarf galaxies that now rotate in the outer-spiral arms. Because the disk is thin, like a phonograph record, an observer on Earth sees relatively few stars perpendicular to its plane. Thus most of the night sky contains a diffuse scattering of stars with large expanses of black space between them. However, if you look into the plane of the disk, you see a dense band of stars from horizon to horizon (◇▶ Figure 24.28).

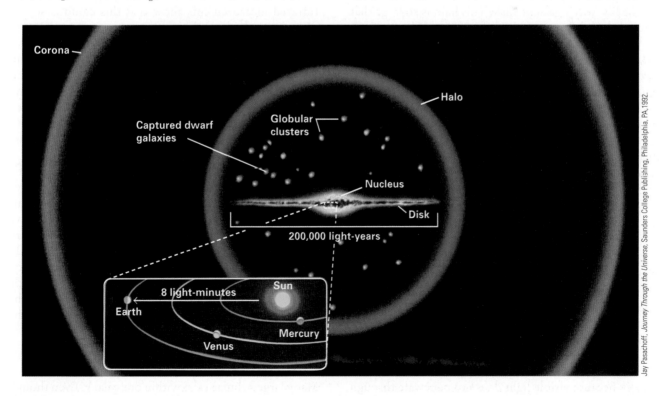

◇▶ **FIGURE 24.27** An artist's drawing of the Milky Way Galaxy shows the galactic disk, which is 200,000 light-years in diameter and 2,000 light-years thick. The distance from the Sun to Earth is only 8 light-minutes.

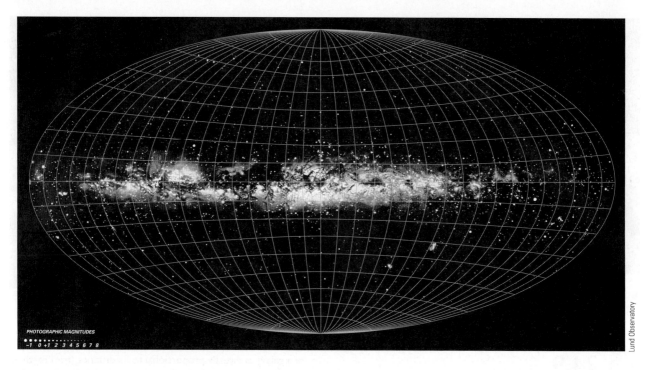

PHOTOGRAPHIC MAGNITUDES
● ● ● ● ● ● · · · · · · · ·
−1 0 +1 2 3 4 5 6 7 8

Lund Observatory

◈ **FIGURE 24.28**   The Milky Way appears as a solid band of light. However, this light is produced by billions of stars in the plane of the galactic disk.

This band is commonly called the Milky Way, although astronomers use the term to describe the entire galaxy. The galactic disk rotates about its center once every 200 million years, so in the 4.6-billion-year history of Earth, we have completed about 23 rotations.

A spherical **galactic halo** of dust and gas surrounds the Milky Way's galactic disk. This halo is so large that even though it is extremely diffuse, it contains as much as 90 percent of the mass of the galaxy. Many dim, and relatively old, stars exist within this halo. Some are concentrated in groups of 10,000 to 1 million stars. Each group is called a **globular cluster**.

The galactic halo and globular clusters are probably remnants of one or more protogalaxies that condensed to form the Milky Way. The spherical structure of the halo suggests that the entire galaxy was once spherical.

The Nucleus of the Milky Way

Photographs of other spiral galaxies show that the galactic nucleus shines much more brightly than the disk. The concentration of stars in the galactic nucleus is perhaps one million times greater than the concentration in the outer disk. If you could visit a planet orbiting one of these stars, you would never experience night, for the stars would light the planet from all directions. However, stable solar systems could not exist in this region because gravitational forces among nearby stars would rip planets from their orbits.

It is impossible to see into the nucleus of our own galaxy because visible light does not penetrate through the interstellar nebulae that lie in the way. Looking into the nucleus is like trying to see a ship on a foggy day.

However, just as sailors use radar to penetrate the fog, astronomers study the center of the galaxy by analyzing radio, infrared, and X-ray emissions that travel through the clouds. These studies provide a picture of the nucleus of the Milky Way.

A cloud of dust and gas orbits the galactic nucleus. Infrared measurements show that this cloud is so hot that it must be heated by an energy source with 10 to 80 million times the output of our Sun. The cloud is ring-shaped, with a hole in its center, like a doughnut. Relatively recently, 10,000 to 100,000 years ago, a giant explosion blew out the center of a larger cloud and created the hole. X-ray and gamma-ray emissions tell us that matter is now accelerating inward, back into the center at a rate of 1 solar mass every 1,000 years (◈ Figure 24.29).

By measuring the orbits of stars close to the galactic center, astronomers can measure the gravitational force of the galactic center, and hence its mass. Calculations show that an unseen object, 3.6 million times as massive as our Sun, lies at the heart of our galaxy. This object can only be a black hole. According to the most widely accepted hypothesis, early in the history of the galaxy, one huge star or many large ones formed in the galactic center. These stars were large enough to pass through the main sequence quickly and then collapse to become the black hole that now forms the center of our galaxy.

Galactic Nebula

Many large nebulae exist within our galaxy. Even though a nebula looks spectacular when viewed through a powerful telescope, the densest nebulae are only 10^{-13} as

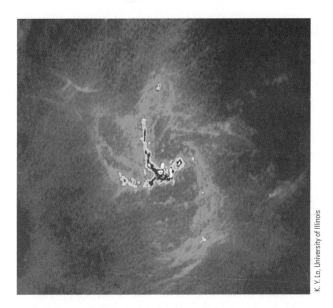

◆ **FIGURE 24.29** The bright region in the center of this image is a small but powerful radio source from the center of the galaxy. Gases falling toward the center produce the red, orange, and blue streamers.

ural elements. Astronomers have detected 100 different molecules in the Orion absorption nebula. Most are simple, such as hydrogen, water, carbon monoxide, and ammonia, but several, larger, organic molecules have also been detected. This nebula is the birthplace of stars.

24.10
Quasars

In the 1960s, astronomers studied many objects that look like stars but emit extremely large amounts of energy as radio waves. These objects were perplexing because normal stars, such as our Sun, emit mostly visible and ultraviolet light. During the following decade, astronomers discovered that the spectra of many of these objects coincide with that of hydrogen, except that they have a very large red shift. Such objects are now called **quasars** (◆ Figure 24.31).

Recall that a large, red shift indicates that an object is moving away from Earth at a high speed. If quasars obey Hubble's law, then they are very far away. In order to be visible when they are so far away, quasars must emit tremendous amounts of energy, 10 to 100 times more than an entire galaxy!

Quasars emit erratic bursts of energy. Recall from our discussion of pulsars that astronomers can estimate the diameter of a pulsar by the sharpness of a short burst

dense as the Earth's atmosphere at sea level, and less dense ones are only 10^{-18} as dense as our atmosphere! A nebula consists mainly of hydrogen and helium, with traces of heavier elements.

As explained in section 24.3, a huge system of nebulae lie in the constellation Orion. The entire structure is 100 light-years in length from north to south, and even though it is extremely diffuse, the nebula contains a mass equivalent to 200,000 of our Suns. ◆ Figure 24.30 shows a close-up of the Horsehead Nebula, which appears near the easternmost star of Orion's belt. The brightly colored region surrounding the horsehead is an **emission nebula**. This portion is so hot that its atoms and molecules glow like a neon light. A dark region, shaped like a horse's head in ◆ Figure 24.30, occupies the middle of one of the red emission nebulae. This dark region is an **absorption nebula**. The dust and gases are frigidly cold, about 13 K (−260°C), and absorb light from the emission nebula behind it. At these extremely low temperatures, most elements condense to form molecules. Although the two lightest elements, hydrogen and helium, constitute more than 99 percent of the cloud's total mass, the Orion absorption nebula contains all or most of the nat-

◆ **FIGURE 24.30** The Horsehead Nebula appears close to the bright star on Orion's belt shown in the left-center of this photograph. The bright, white lights in the photographs are other stars. The reddish glows are emission nebulae, and the horsehead-shaped dark area is an absorption nebula in front of an emission nebula.

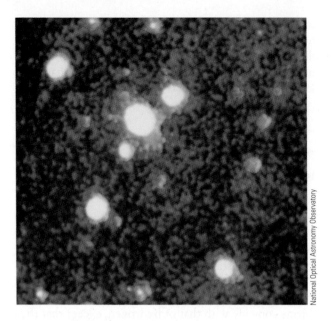

◁▷ **FIGURE 24.31** Quasar 3C 275.1 is the brightest object near the center of this photograph. An elliptical gas cloud surrounds the nucleus. This quasar is about 2 billion light-years from Earth. The smaller objects are normal galaxies.

National Optical Astronomy Observatory

of energy emitted from it. Similar techniques indicate that some quasars are several light-months in diameter. By comparison, the distance from Earth to the nearest star is 4 light-years.

Thus a quasar is much smaller than a galaxy but emits much more energy. Furthermore, it emits energy over a wide range of wavelengths. Most quasars are very distant from Earth. One common hypothesis for their origin is that a massive black hole, perhaps of 1 billion solar masses, lies at the center of a quasar and that the quasar emits energy as gas and dust accelerate into the hole.

Looking Backward into Time

Many quasars are 8 to 12 billion light-years away. If an object is 10 billion light-years away, the light we see today started its journey 10 billion years ago. Therefore, we see what was happening in the past but not what is happening today. If the object blew up and disappeared 9 billion years ago, we will not know about it for another billion years! Thus when we look at close objects, we see what happened recently, but when we look at distant objects, we see what happened in the distant past. One goal of building more-powerful telescopes is to study more-distant objects and therefore probe further back in time.

The oldest quasars must have formed when the Universe was young. Moreover, quasars contain heavy elements, and heavy elements form only in the supernova explosions of dying stars. Therefore, quasars are second-generation structures; they formed after earlier stars were born, evolved, exploded, and died. Hence, stars must have formed and passed through their lifecycle less than 2 billion years after the Universe formed.

24.11
Dark Matter

The Milky Way rotates slowly, like a giant pinwheel. Using laws of motion developed by Kepler and Newton in the seventeenth century, astronomers have attempted to calculate the mass of our galaxy from their knowledge of its rotational speed. However, these calculations produced a giant anomaly. The Milky Way is much more massive than we can account for if we add up all the known matter within it.

Satellite studies of the nonhomogeneity of galaxies (section 24.2) allow scientists to calculate the mass and gravitational force of the primordial Universe. These studies indicate that matter that we see is only 4 percent of the mass of the Universe! The other 96 percent is invisible and is called **dark matter**. What is it? A few years ago, many astrophysicists speculated that dark matter might be composed of planetlike objects that do not radiate energy and are therefore invisible, and black holes that cannot radiate energy. However, in 2003, results from the WMAP satellite convinced most scientists that the story is far more exotic, as explained next.

24.12
The End of the Universe

From the first instant of the big bang, the Universe has been affected by two opposing forces. Matter is expanding outward as a result of the big bang, but at the same time gravity is pulling all matter back inward toward a common center. If the gravitational force of the Universe is sufficient, all the galaxies will eventually slow down, reverse direction, and fall back to the center, forming another point of infinite density. This hypothesis is called a **closed Universe** (◁▷ Figure 24.32A). Some scientists have speculated that if the Universe is closed, it will collapse and then explode again to form a new Universe. In turn, the new Universe will expand and then collapse, creating a continuous chain of Universes. However, no mathematical model explains how another big bang would occur. Instead, current models predict that in a closed system, the Universe would collapse into a mammoth black hole.

Another possibility is that the gravitational force of the Universe is not sufficient to stop the expansion and that the galaxies will continue to fly apart forever. Within each galaxy, stars will eventually consume all their nuclear fuel and stop producing energy. As the stars fade and cool, the galaxies will continue to separate into the cold void. This scenario is called the **open Universe** (◁▷ Figure 24.32B).

In the mid to late 1990s, astronomers were attempting to measure the deceleration of the universe. If they could learn how fast it was slowing down, then they

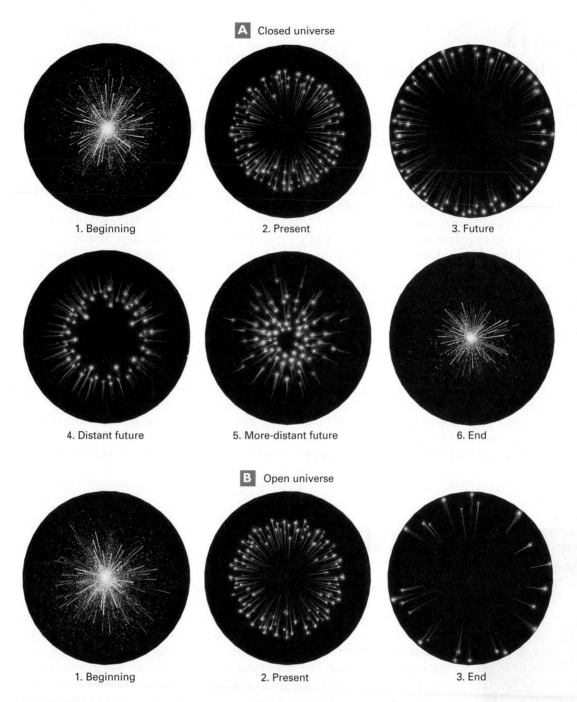

A Closed universe

1. Beginning 2. Present 3. Future

4. Distant future 5. More-distant future 6. End

B Open universe

1. Beginning 2. Present 3. End

◆ **FIGURE 24.32** A schematic representation of two possible cosmologies, a closed Universe cosmology and an open Universe cosmology. Recent data imply that the Universe may not only be open, but also accelerating.

reasoned that they could calculate whether the Universe is opened or closed. Astronomers made these measurements by studying the speed (red shift) of distant supernovae. The result was totally unexpected. The expansion of the Universe is not slowing down at all. Instead, it is accelerating!

The concept of an **accelerating Universe** is totally baffling. Objects accelerate only when they are acted on by a force. In this case, the force must be some sort of repulsion that acts against gravity and pushes the galaxies apart.

By 2003, WMAP data had convinced most astrophysicists that dark matter is composed of two components. **Dark mass**, which makes up 23 percent of the Universe, is an as-yet-undetected particle that exerts a conventional gravitational attraction. Another 73 percent of the Universe is **dark energy**—the poorly understood repulsive force that is causing the modern Universe to accelerate. As explained earlier, only 4 percent of the Universe is "ordinary" matter that makes stars shine, rivers flow, and people walk through grassy meadows on a spring day.

24.13
Why Are We So Lucky?

Planets and living organisms are composed of midmass and heavy elements. But, as we have learned, primordial population II stars were originally formed from the two lightest elements: hydrogen and helium. Thus in order for life to have evolved, population II stars had to pass through their lifecycles and produce carbon and other midmass elements. Then, some of these stars exploded in supernovae that created heavier elements.

Furthermore, life can only exist if atoms in the resultant population I stars fuse at an intermediate rate—not too fast and not too slow—so solar systems have time and the proper temperature for planets to evolve with favorable environments. These conditions are met only if the starting conditions of the big bang fit within an extraordinarily narrow window.

The conditions for life in the Universe would only be possible under the following conditions:

- If the initial temperature and expansion rate of the Universe were greater or less than what actually happened, the ratio of hydrogen to helium would have been different. In either case, stellar evolution would not have produced the existing conditions—and life as we know it would not have evolved.

- If the mass ratio of neutrons to protons were more nearly equal than they were in the primordial Universe, almost all the hydrogen would have converted to helium. Stellar evolution would have proceeded differently—and life as we know it would not have evolved.

- If gravity were considerably stronger relative to electromagnetism, stars would be more massive—and life as we know it would not have evolved.

- If nuclear forces were stronger or weaker than they are, primordial nucleosynthesis and stellar evolution would have been so different—and life as we know it would not have evolved.

- If electrons were more massive than they are, complex chemistry would not be possible—and life as we know it would not have evolved.

There is no fundamental reason why all the universal constants should agree exactly so that our Sun and our Earth formed exactly as they did. There is no fundamental reason why we live on this wonderful planet with green trees, azure seas, and other sentient beings to share our love and joys. Why are we so lucky?

No one knows, and perhaps no one will know. We have formulated a question whose answer transcends science and ventures into the unknowable.

SUMMARY

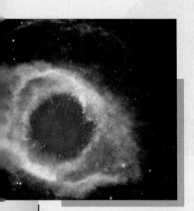

The **big bang** theory states that the Universe began with a cataclysmic explosion that instantly created space and time. Three main lines of reasoning support the big bang theory: (1) all the galaxies are flying away from one another, implying a common origin at a common time; (2) the calculated abundance of helium, formed during the primordial nucleosynthesis agrees with observation; and (3) the calculated, cosmic background radiation agrees with observation. The nonhomogeneity of the Universe was generated by quantum mechanical interactions very early in its infancy.

Stars form from condensing **nebulae**, clouds of gas and dust in interstellar space. When the condensing gases become hot enough, fusion begins. Two opposing forces occur in a star. Gravity pulls particles inward, but at the same time fusion energy generates gas pressure and drives them outward. The balance between these two forces determines the diameter and density of a star.

The central **core** of the Sun has a temperature of about 15 million K and a density 150 times that of water. Hydrogen fusion occurs in this region. The visible surface of the Sun, called the **photosphere**, is only 5,700 degrees K and has a pressure of 0.1 of the Earth's atmosphere at sea level. **Sunspots** are magnetic storms on the surface of the Sun. The outer layers of the Sun's atmosphere, called the **chromosphere** and the **corona**, are turbulent and diffuse.

The **apparent magnitude** or **apparent luminosity** of a star is its brightness as seen from Earth. A star can appear luminous either because it really is bright or because it is close. The **absolute magnitude** or **absolute luminosity** is how bright a star would appear if it were 32 light-years (10 parsecs) away. A **Hertzsprung–Russell, H–R, diagram** is a plot of luminosity versus stellar temperature. Most stars fall within the main sequence. All main-sequence stars are composed primarily of hydrogen and generate energy by hydrogen fusion.

Hydrogen nuclei in the core of a mature star the mass of our Sun fuse to form helium, with the release of large amounts of energy. When the hydrogen in the core is exhausted and hydrogen fusion ends in the core, gravity compresses the core and the temperature rises. Fusion then begins in the outer shell, and the star expands to become a **red giant**. In the following stage, helium nuclei fuse to form carbon. After helium fusion in an average-mass star ends, the star releases a **planetary nebula** and then shrinks to become a **white dwarf**. In a more-massive star, enough heat is produced to fuse heavier elements; the iron-rich core then explodes to become a **supernova**. The supernova explosion forms all elements heavier than iron. The remnant can contract to become a **neutron star**. If a neutron star emits pulses of radio waves, it is called a pulsar. If the mass of a dying star is great enough, it collapses into a **black hole**. **Gamma ray bursts** form when a neutron star falls into a black hole or when two neutron stars collide to form a black hole.

Population II stars formed from the primordial mixture of hydrogen and helium. In contrast, **population I** stars formed from nebulae that are the remains of exploding population II stars. Population I stars contain small concentrations of heavy elements.

Galaxies and clusters of galaxies form the basic structure of the Universe. Commonly, galaxies are **elliptical, spiral,** or **barred spiral**, although other shapes also exist. **Hubble's law** states that the most distant galaxies are moving away from Earth at the greatest speeds, while closer ones are receding more slowly.

Our Sun lies in the **Milky Way,** a barred spiral galaxy. In addition to stars in the main disk, a galaxy contains captured dwarf galaxies, diffuse clouds of gas and dust between the stars, and a **galactic halo** and **globular clusters** of stars surrounding the disk. The nucleus of the galaxy is dense and probably contains one or more massive black holes. Galactic nebulae are the birthing grounds of new population I stars.

Quasars are very far away, emit as much as 100 times the energy of an entire galaxy, and are small compared with a galaxy. A black hole may lie in the center of a quasar. Modern models indicate that today's galaxies evolved through collisions of smaller protogalaxies.

Up to 90 percent of the Universe is invisible and is called **dark matter**. In trying to determine whether the universe is **opened** or **closed**, astronomers learned, to their surprise, that it is **accelerating**. Life could have formed under only a very narrow range of fundamental physical constants.

Key Terms

For Review

1. Outline the three main forms of evidence to support the big bang theory.

2. Describe the universe when it was a trillionth of a trillionth of a billionth of a second old.

3. What happened between 1.5 minutes and 10 minutes after the Universe formed that greatly affected its evolution?

4. What is the cosmic background radiation? When and how did it form?

5. What is a light-year? How far is it in kilometers?

6. How do we know that the Universe is nonhomogeneous at present? How do we know when and how the nonhomogeneity evolved?

7. Briefly explain how a star is born.

8. What opposing factors determine the size and density of a star?

9. Draw a cutaway diagram of the Sun, labeling the core, the photosphere, sunspots, granules, the chromosphere, and the corona. Label the temperature and the relative density of each region.

10. What is the fundamental source of energy within the Sun? Is the Sun's chemical composition constant, or is it continuously changing?

11. How does energy travel from the core of the Sun to the surface? How does it travel from the surface of the Sun to Earth?

12. Compare and contrast the core of the Sun with its surface.

13. Compare and contrast the lifecycles of stars the mass of our Sun, 4 times as massive as our Sun, and 20 times as massive as our Sun. At what points do the lifecycles differ?

14. What is an H–R diagram? What conclusions do we draw from it?

15. What is a supernova? Do all stars eventually explode as supernovas? Explain.

16. What is a neutron star? Do all supernovas lead to the formation of neutron stars?

17. What is a black hole? How does it form, and how can we detect it?

18. Draw a picture of the Milky Way galaxy. Label the disk, the core, the halo, and globular clusters. Draw the approximate position of Earth.

19. What evidence indicates that the center of our galaxy was once the scene of a violent explosion?

20. List four characteristics of quasars that distinguish them from ordinary stars.

21. What is dark matter? How do we know that it exists?

22. Briefly outline the modern hypothesis for the end of the Universe.

For Discussion

1. According to a recent article in *Astronomy* (December 2002, pg. 34), 67 percent of Americans do not believe in the big bang theory. These people claim that because the big bang occurred so long ago and is impossible to observe, the scientific theory is merely a guess. In the same article, the authors point out that 30 percent of Americans do not believe in the Copernican theory of the heliocentric Solar System, and about half do not believe in Darwinian evolution. Discuss your ideas about the verity of these three theories in specific and about the scientific method in general.

2. What information (if any) could be gained by studying the absorption and emission spectra of the Moon and the planets?

3. Could a telescope be built to study a quasar as it exists today?

4. Hydrogen burns rapidly and explosively in air. Is this chemical combustion of hydrogen an important process within a star? Why or why not?

5. Explain why the density of the gases near the surface of the Sun is less than the density of the gases near the surface of Earth, even though the gravitational force of the Sun is much greater.

6. Using Figure 24.11, what would be the color of a star whose surface temperature is 10,000 K? 3,500 K?

7. What could you tell about the past history of a star if you knew that its core was composed primarily of carbon? Explain.

8. Compare and contrast a white dwarf with a red giant. Can a single star ever be both a red giant and a white dwarf during its lifetime?

9. Would you be likely to find life on planets that are orbiting around a population II star? Explain.

10. Certain stars that lie above and below the plane of the Milky Way contain fewer heavy elements than does our own Sun. From this information alone, what can you tell about the history of these stars?

11. Explain why the pulsar signals detected by Jocelyn Bell Burnell could not have originated from (a) an ordinary star, (b) an unknown planet in our Solar System, (c) a distant galaxy, or (d) a large, magnetic storm on a nearby star.

12. Could a black hole be hidden in our Solar System? Explain.

13. Would black holes represent a hazard to a rocket ship traveling to distant stars? Could the crew of such a rocket detect a black hole well in advance and avoid an encounter?

14. Is a quasar similar to a star, a galaxy, or neither? Discuss.

15. Arrange the following environments in the order of increasing densities: (a) intergalactic space, (b) the core of the Sun, (c) the corona of the Sun, (d) the region of space between planets in our Solar System, (e) galactic space, (f) the Earth's atmosphere at sea level.

ThomsonNOW™

Assess your understanding of this chapter's topics with additional comprehensive interactivities at **www.thomsonedu.com/login**, which also has current and up-to-date web links, additional readings, and exercises.

APPENDIX A

Identifying Common Minerals

About 3,500 minerals exist in Earth's crust. However, of this great number, only 30 or so are common. Therefore, when you pick up rocks and want to identify the minerals, you are most likely to be looking at the same small group of minerals over and over again. The following list includes the most common and abundant minerals in Earth's crust. A few important ore minerals, other minerals of economic value, and a few popular precious and semiprecious gems are included because they are of special interest.

The minerals in this table fall into four categories:

1. Rock-forming minerals are shown on pages A2 and A3. They are the most abundant minerals in the crust and make up the largest portions of all common rocks. The rock-forming minerals and mineral groups are feldspar, pyroxene, amphibole, mica, clay, olivine, quartz, calcite, and dolomite. If more than one mineral of a group is common, each mineral is listed under the group name. For example, three kinds of pyroxene are abundant: augite, diopside, and orthopyroxene. All three are described under pyroxene.

2. Accessory minerals are shown on pages A4 and A5. They are minerals that are common but that usually occur only in small amounts.

3. Ore minerals and other minerals of economic importance are shown on pages A6 and A7. They are minerals from which metals or other elements can be profitably recovered.

4. Gems are shown on pages A8 and A9. If the common gem name(s) is different from the mineral name, the gem name is given in parentheses following the mineral name. For example, emerald is the gem variety of the mineral beryl and is listed as Beryl (emerald).

Minerals are listed alphabetically within each of the four categories for quick reference. The physical properties most commonly used for identification of each mineral, and the kind(s) of rock in which each mineral is most often found, are listed to facilitate identification of these common minerals.

	Mineral Group or Mineral	Chemical Composition	Habit, Cleavage, Fracture	Usual Color
Amphibole	Actinolite	$Ca_2(MgFe)_5Si_8O_{22}(OH)_2$	Slender crystals, radiating, fibrous.	Blackish-green to black, dark green
	Hornblende	$(Ca,Na)_{2-3}(Mg,Fe,Al)_5Si_6(Si,Al)_2O_{22}(OH)_2$	Elongate crystals.	Blackish-green to black, dark green
Calcite		$CaCO_3$	Perfect cleavage into rhombs.	Usually white, but may be variously tinted
Clay Minerals	Illite	$K_{0.8}Al_2(Si_{3.2}Al_{0.8})O_{10}(OH)_2$		White
	Kaolinite	$Al_2Si_2O_5(OH)_4$		White
	Smectite	$Na_{0.3}Al_2(Si_{3.7}Al_{0.3})O_{10}(OH)_2$		White, buff
Dolomite		$CaMg(CO_3)_2$	Cleaves into rhombs; granular masses.	White, pink, gray, brown
Feldspar	Albite (sodium feldspar)	$NaAlSi_3O_8$ (sodic plagioclase)	Good cleavage in two directions, nearly 90°.	White, gray
	Orthoclase (potassium feldspar)	$KAlSi_3O_8$	Good cleavage in two directions at 90°.	White, pink, red, yellow-green, gray
	Plagioclase (feldspar containing both sodium and calcium)	$(Na,Ca)(Al,Si)_4O_8$	Good cleavage in two directions at 90°.	White, gray
Mica	Biotite	$K(Mg,Fe)_3AlSi_3O_{10}(OH)_2$	Perfect cleavage into thin sheets.	Black, brown, green
	Muscovite	$KAl_2Si_3O_{10}(OH)_2$	Perfect cleavage into thin sheets.	Colorless if thin
Olivine		$(MgFe)_2SiO_4$	Uneven fracture, often in granular masses.	Various shades of green
Pyroxene	Augite	$Ca(Mg,Fe,Al)(Al,Si_2O_6)$	Short stubby crystals have 4 or 8 sides in cross-section.	Blackish-green to light green
	Diopside	$CaMg(Si_2O_6)$	Usually short thick prisms; may be granular.	White to light green
	Orthopyroxene	$MgSiO_3$	Cleavage good at 87° and 93°; usually massive.	Pale green, brown, gray, or yellowish
Quartz		SiO_2	No cleavage, massive and as six-sided crystals.	Colorless, white, or tinted any color by impurities

Hardness	Streak	Specific Gravity	Other Properties	Type(s) of Rock in Which the Mineral Is Most Commonly Found
5–6	Pale green	3.2–3.6	Vitreous luster.	Low- to medium-grade metamorphic rocks
5–6	Pale Green	3.2	Crystals six-sided with 124° between cleavage faces.	Common in many granitic to basaltic igneous rocks, and many metamorphic rocks
3	White	2.7	Transparent to opaque. Rapid effervescence with HCl.	Limestone, marble, cave deposits
}	The clay minerals are so fine-grained that most physical properties cannot be identified.			Shale Shale, weathered bedrock, and soil Shale, weathered bedrock, and soil
3.5–4	White to pale gray	3.9–4.2	Effervesces slightly in cold dilute HCl.	Dolomite
6–6.5	White	2.6	Many show fine striations (twinning lines) on cleavage faces.	Granite, rhyolite, low-grade metamorphic rocks
6	White	2.6	Vitreous to pearly luster.	Granite, rhyolite, metamorphic rocks
6	White	2.6–2.7	May show striations as in albite.	Basalt, andesite, medium- to high-grade metamorphic rocks
2.2–2.5	White, gray	2.7–3.1	Vitreous luster, divides readily into thin flexible sheets.	Granitic to intermediate igneous rocks, many metamorphic rocks
2–2.5	White	2.7–3	Vitreous or pearly; flexible and elastic; splits easily.	Many metamorphic rocks, granite
6.5–7	White	3.2–3.3	Vitreous, glassy luster.	Basalt, peridotite
5.6	Pale green	5–6	Vitreous, distinguished from hornblende by the 87° angle between cleavage faces.	Basalt, peridotite, andesite, high-grade metamorphic rocks
5–6	White to greenish	3.2–3.6	Vitreous luster.	Medium-grade metamorphic rocks
5.5	White	3.2–3.5	Vitreous luster.	Periodotite, basalt, high-grade metamorphic rocks
7	White	2.6	Includes rock crystal, rose and milky quartz, amethyst, smoky quartz, etc.	Granite rhyolite, metamorphic rocks of all grades, sandstone, siltstone

Mineral Group or Mineral	Chemical Composition	Habit, Cleavage, Fracture	Usual Color
Apatite	$Ca_5(OH,F,Cl)(PO_4)_3$	Massive, granular.	Green, brown, red
Chlorite	$(Mg,Fe)_6(Si,Al)_4O_{10}(OH)_8$	Perfect cleavage as fine scales.	Green
Corundum	Al_2O_3	Short, six-sided barrel-shaped crystals.	Gray, light blue, and other colors
Epidote	$Ca_2(Al,Fe)Al_2O(SiO_4)(Si_2O_7)(OH)$	Usually granular masses; also as slender prisms.	Yellow-green, olive-green, to nearly black
Fluorite	CaF_2	Octahedral and also cubic crystals.	White, yellow, green, purple
Garnet — Almandine	$Fe_3Al_2(SiO_4)_3$	No cleavage, crystals 12- or 24-sided.	Deep red
Garnet — Grossular	$Ca_3Al_2(SiO_4)_3$	No cleavage, crystals 12- or 24-sided.	White, green, yellow, brown
Graphite	C	Foliated, scaly, or earthly masses.	Steel gray to black
Hematite	Fe_2O_3	Granular, massive, or earthy.	Brownish-red
Limonite	$2Fe_2O_3 \cdot 3H_2O$	Earthy fracture.	Brown or yellow
Magnetite	Fe_3O_4	Uneven fracture, granular masses.	Iron black
Pyrite	FeS_2	Uneven fracture cubes with striated faces, octahedrons.	Pale brass yellow (lighter than chalcopyrite)
Serpentine	$Mg_3Si_2O_5(OH)_4$	Uneven, often splintery fracture.	Light and dark green, yellow

Hardness	Streak	Specific Gravity	Other Properties	Type(s) of Rock in Which the Mineral Is Most Commonly Found
4.5–5	Pale red-brown	3.1	Crystals may have a partly melted appearance, glassy.	Common in small amounts in many igneous, metamorphic, and sedimentary rocks
2.0–2.5	Gray, white, pale green	2.8	Pearly to vitreous luster.	Common in low-grade metamorphic rocks
9	None	3.9–4.1	Hardness is distinctive.	Metamorphic rocks, some igneous rocks
6.7	Pale yellow to white	3.3	Vitreous luster.	Low- to medium-grade metamorphic rocks
4	White	3.2	Cleaves easily, vitreous, transparent to translucent.	Hydrothermal veins
6.5–7.5	White	4.2	Vitreous to resinous luster.	The most common garnet in metamorphic rocks
6.5–7.5	White	3.6	Vitreous to resinous luster.	Metamorphosed sandy limestones
1–2	Gray or black	2.2	Feels greasy; marks paper.	Metamorphic rocks
2.5	Dark red	2.5–5	Often earthy, dull appearance.	Common in all types of rocks. It can form by weathering of iron minerals, and is the source of color in nearly all red rocks.
1.5–4	Brownish-yellow	3.6	Earthy masses that resemble clay.	Common in all types of rocks. It can form by weathering of iron minerals, and is the source of color in most yellow-brown rocks.
5.5	Iron black	5.2	Metallic luster. Strongly magnetic.	Common in small amounts in most igneous rocks
6–6.5	Greenish-black	5	Metallic luster, brittle, very common.	The most common sulfide mineral. Igneous, metamorphic, and sedimentary rocks; hydrothermal veins
2.5	White	2.5	Waxy luster, smooth feel, brittle.	Alteration or metamorphism of basalt, peridotite, and other magnesium-rich rocks

Mineral Group or Mineral	Chemical Composition	Habit, Cleavage, Fracture	Usual Color
Anhydrite	$CaSO_4$	Granular masses, crystals with 2 good cleavage directions.	White, gray, blue-gray
Asbestos	$Mg_3Si_2O_5(OH)_4$	Fibrous.	White to pale olive-green
Azurite	$Cu_3(CO_3)_2(OH)_2$	Varied, may have fibrous crystals.	Azure blue
Bauxite	$Al(OH)_3$	Earthy masses.	Reddish to brown
Chalcopyrite	$CuFeS_2$	Uneven fracture.	Brass yellow
Chromite	$FeCr_2O_4$	Massive, granular, compact	Black
Cinnabar	HgS	Compact, granular masses.	Scarlet red to red-brown
Galena	PbS	Perfect cubic cleavage.	Lead or silver gray
Gypsum	$CaSO_4 \cdot 2H_2O$	Tabular crystals, fibrous, or granular.	White, pearly
Halite	$NaCl$	Granular masses, perfect cubic crystals.	White, also pale colors and gray
Hematite	Fe_2O_3	Granular, massive, or earthy.	Brownish-red, black
Malachite	$CuCO_3 \cdot Cu(OH)_2$	Uneven splintery fracture.	Bright green, dark green
Native copper	Cu	Malleable and ductile.	Copper red
Native gold	Au	Malleable and ductile.	Yellow
Native silver	Ag	Malleable and ductile.	Silver-white
Pyrolusite	MnO_2	Radiating or dendritic coatings on rocks.	Black
Sphalerite	ZnS	Perfect cleavage in 6 directions at 120°.	Shades of brown and red
Talc	$Mg_3Si_4O_{10}(OH)_2$	Perfect in one direction.	Green, white, gray

Hardness	Streak	Specific Gravity	Other Properties	Type(s) of Rock in Which the Mineral Is Most Commonly Found
3–3.5	White	2.9–3	Brittle; resembles marble but acid has no effect.	Sedimentary evaporite deposits
1–2.5	White	2.6–2.8	Pearly to greasy luster. Flexible, easily separated fibers.	A variety of serpentine, found in the same rock types
4	Pale blue	3.8	Vitreous to earthy, effervesces with HCl.	Weathered copper deposits
1.5–3.5	Pale reddish-brown	2.5	Dull luster, claylike masses with small round concretions.	Weathering of many rock types
3.5–4.5	Greenish-black	4.2	Metallic luster, softer than pyrite.	The most common copper ore mineral; hydrothermal veins, porphyry copper deposits
5.5	Dark brown	4.4	Metallic to submetallic luster.	Peridotites and other ultramafic igneous rocks
2.5	Scarlet red	8	Color and streak distinctive.	The most important mercury ore mineral; hydrothermal veins in young volcanic rocks
2.5	Gray	7.6	Metallic luster.	The most important lead ore; commonly also contains silver; hydrothermal veins
1–2.5	White	2.2–2.4	Thin sheets (selenite), fibrous (satinspar), massive (alabaster).	Sedimentary evaporite deposits
2.5–3	White	2.2	Pearly luster, salty taste, soluble in water.	Sedimentary evaporite deposits
2.5	Dark red	2.5–5	Often earthy, dull appearance, sometimes metallic luster.	Huge concentrations occur as sedimentary iron ore; the most important source of iron
3.5–4	Emerald green	4	Effervesces with HCl. Associated with azurite.	Weathered copper deposits
2.5–3	Copper red	8.9	Metallic luster.	Basaltic lavas
2.5–3	Yellow	19.3	Metallic luster.	Hydrothermal quartz-gold veins, sedimentary placer deposits
2.5–3	Silver-white	10.5	Metallic luster.	Hydrothermal veins, weathered silver deposits
1–2	Black	4.7	Sooty appearance.	Black stains on weathered surfaces of many rocks, manganese nodules on the sea floor
3.5	Reddish-brown	4	Resinous luster, may occur with galena, pyrite.	The most important ore mineral of zinc; hydrothermal veins
1–1.5	White	1–2.5	Greasy feel, occurs in foliated masses.	Low-grade metamorphic rocks

Mineral Group or Mineral	Chemical Composition	Habit, Cleavage, Fracture	Usual Color
Beryl (aquamarine, emerald)	$Be_3Al_2(SiO_3)_6$	Uneven fracture, hexagonal crystals.	Green, yellow, blue, pink
Chrysoberyl (cat's eye, alexandrite)	$BeAl_2O_4$	Tabular crystals.	Green, brown, yellow
Corundum (ruby, sapphire)	Al_2O_3	Short, six-sided barrel-shaped crystals.	Gray, red (ruby), blue (sapphire)
Diamond	C	Octahedral crystals.	Colorless or with pale tints
Garnet	$Fe_3Al_2(SiO_4)_3$	No cleavage, crystals 12- or 24-sided.	Deep red
Jadite (jade) (a pyroxene)	$NaAl(Si_2O_6)$	Compact fibrous aggregates.	Green
Olivine (peridot)	$(MgFe)_2SiO_4$	Uneven fracture, often in granular masses.	Various shades of green
Opal	$SiO_2 \cdot NH_2O$	Conchoidal fracture, amorphous, massive.	White and various colors
Quartz (rock crystal, amethyst, citrine, tiger eye, adventurine carneline, chrysoprase, agate, onyx, heliotrope, bloodstone, jasper)	SiO_2	No cleavage, massive and as six-sided crystals.	Colorless, white, or tinted any color by impurities
Spinel	$MgAl_2O_4$	No cleavage, rare octahedral crystals.	Black, dark green, or various colors
Topaz	$Al_2SiO_4(F,OH)_2$	Cleavage good in one direction; conchoidal fracture.	Colorless, white, pale tints of blue, pink
Tourmaline	$(Na,Ca)(Li,Mg,Al)(Al,Fe,Mn)_{6-}(BO_3)_3(Si_6O_{18})(OH)_4$	Poor cleavage, uneven fracture; striated crystals.	Black, brown, green, pink
Turquoise	$CuAl_6(PO_4)_4(OH)_8 \cdot 4H_2O$	Massive.	Blue-green
Zircon	$Zr(SiO_4)$	Cleavage poor, but often well-formed tetragonal crystals.	Colorless, gray, green, pink, bluish

Hardness	Streak	Specific Gravity	Other Properties	Type(s) of Rock in Which the Mineral Is Most Commonly Found
7.5–8.0	White	2.6–2.8	Vitreous luster.	Granite, granite pegmatite, mica schist
8.5	White	3.7–3.8	Vitreous luster.	Granite, granite pegmatite, mica schist
9	None	3.9–4.1	Ruby and sapphire are corundum varieties. Hardness is distinctive.	Metamorphic rocks, some igneous rocks
10	None	3.5	Adamantine luster. Hardness is distinctive.	Peridotite, kimberlite, sedimentary placer deposits
6.5–7.5	White	4.2	Vitreous to resinous luster.	Metamorphic rocks, igneous rocks, placer deposits
6.5–7	White, pale green	3.3	Vitreous luster.	High-pressure metamorphic rocks
6.5–7	White	3.2–3.3	Vitreous, glassy luster.	Basalt, peridotite
5.5–6.5	White	2.1	Vitreous, greasy, pearly luster. May show a play of colors.	Low-temperature hot springs and weathered near-surface deposits
7	White	2.6	Colors and other features differ among the varieties.	Quartz is found in nearly all rock types, although each of the gem varieties may form in special environments.
7.5–8.0	White	3.5–4.1	Vitreous luster. Hardness is distinctive.	High-grade metamorphic rocks, dark igneous rocks
8	Colorless	3.5–3.6	Vitreous luster.	Pegmatite, granite, rhyolite
7–7.5	White to gray	4.4–4.8	Vitreous, slightly resinous.	Pegmatite, granite, metamorphic rocks
6	Blue-green, white	2.6–2.8	Waxy luster.	Veins in weathered volcanic rocks in deserts
7.5	White	4.7	Adamantine luster.	Many types of igneous rocks

APPENDIX B

Systems of Measurements

I The SI System

In the past, scientists from different parts of the world have used different systems of measurement. However, global cooperation and communication make it essential to adopt a standard system. The International System of Units (SI) defines various units of measurement as well as prefixes for multiplying or dividing the units by decimal factors. Some primary and derived units important to geologists are listed here.

Time

The SI unit is the **second,** s or sec, which used to be based on the rotation of Earth but is now related to the vibration of atoms of cesium-133. SI prefixes are used for fractions of a second (such as milliseconds or microseconds), but the common words **minutes, hours,** and **days** are still used to express multiples of seconds.

Length

The SI unit is the **meter,** m, which used to be based on a standard platinum bar but is now defined in terms of wavelengths of light. The closest English equivalent is the **yard** (0.914 m). A **mile** is 1.61 kilometers (km). An inch is exactly 2.54 centimeters (cm).

Area

Area is length squared, as in **square meter, square foot,** and so on. The SI unit of area is the **are,** a, which is 100 sq m. More commonly used is the **hectare,** ha, which is 100 ares, or a square that is 100 m on each side. (The length of a U.S. football field plus one end zone is just about 100 m.) A hectare is 2.47 acres. An **acre** is 43,560 sq ft, which is a plot of 220 ft by 198 ft, for example.

Volume

Volume is length cubed, as in **cubic centimeter,** cm³, **cubic foot,** ft³, and so on. The SI unit is the **liter,** L, which is 1,000 cm³. A **quart** is 0.946 L; a U.S. liquid **gallon** (gal) is 3.785 L. A **barrel** of petroleum (U.S.) is 42 gal, or 159 L.

Mass

Mass is the amount of matter in an object. **Weight** is the force of gravity on an object. To illustrate the difference, an astronaut in space has no weight but still has mass. On Earth, the two terms are directly proportional and often used interchangeably. The SI unit of mass is the **kilogram,** kg, which is based on a standard platinum mass. A **pound** (avdp), lb, is a unit of weight. On the surface of Earth, 1 lb is equal to 0.454 kg. A **metric ton,** also written as **tonne,** is 1,000 kg, or about 2,205 lb.

Temperature

The Celsius scale is used in most laboratories to measure temperature. On the Celsius scale the freezing point of water is 0°C and the boiling point of water is 100°C.

The SI unit of temperature is the **Kelvin.** The coldest possible temperature, which is –273°C, is zero on the Kelvin scale. The size of 1 degree Kelvin is equal to 1 degree Celsius.

Celsius temperature (°C) = Kelvin temperature
(K) – 273 K

Fahrenheit temperature (°F) is not used in scientific writing, although it is still popular in English-speaking countries. The conversion between Fahrenheit and Celsius is shown.

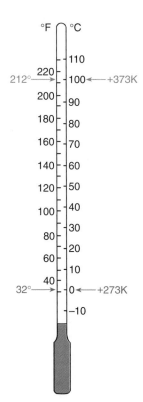

Energy

Energy is a measure of work or heat, which were once thought to be different quantities. Hence, two different sets of units were adopted and still persist, although we now know that work and heat are both forms of energy.

The SI unit of energy is the **joule**, J, the work required to exert a force of 1 newton through a distance of 1 m. In turn, a newton is the force that gives a mass of 1 kg an acceleration of 1 m/sec^2. In human terms, a joule is not much—it is about the amount of work required to lift a 100-g weight to a height of 1 m. Therefore, joule units are too small for discussions of machines, power plants, or energy policy. Larger units are

megajoule, $MJ = 10^6$ J (a day's work by one person)
gigajoule, $GJ = 10^9$ J (energy in half a tank of gasoline)

The energy unit used for heat is the **calorie**, cal, which is exactly 4.184 J. One calorie is just enough energy to warm 1 g of water 1°C. The more common unit used in measuring food energy is the **kilocalorie**, kcal, which is 1,000 cal. When **Calorie** is spelled with a capital C, it means kcal. If a cookbook says that a jelly doughnut has 185 calories, that is an error—it should say 185 Calories (capital C), or 185 kcal. A value of 185 calories (lowercase c) would be the energy in about one quarter of a thin slice of cucumber.

The unit of energy in the British system is the **British thermal unit,** or Btu, which is the energy needed to warm 1 lb of water 1°F.

1 Btu = 1,054 J = 1.054 kJ = 252 cal

The unit often referred to in discussions of national energy policies is the **quad,** which is 1 quadrillion Btu, or 10^{15} Btu.

Some approximate energy values are

1 barrel (42 gal) of petroleum = 5,900 MJ
1 ton of coal = 29,000 MJ
1 quad = 170 million barrels of oil, or 34 million tons of coal

II Exponential or Scientific Notation

Scientists all over the world use exponential or scientific notation. This system is based on exponents of 10, which are shorthand notations for repeated multiplications or divisions.

A positive exponent is a symbol for a number that is to be multiplied by itself a given number of times. Thus, the number 10^2 (read "ten squared" or "ten to the second power") is exponential notation for $10 \cdot 10 = 100$. Similarly, $3^4 = 3 \cdot 3 \cdot 3 \cdot 3 = 81$. The reciprocals of these numbers are expressed by negative exponents. Thus $10^{-2} = 1/10^2 = 1/(10 \cdot 10) = 1/100 = 0.01$.

To write 10^4 in longhand form you simply start with the number 1 and move the decimal four places to the right. Similarly, to write 10^{-4} you start with the number 1 and move the decimal four places to the left.

It is just as easy to go the other way—that is, to convert a number written in longhand form to an exponential expression. Thus, the decimal place of the number 1,000,000 is six places to the right of 1.

Similarly, the decimal place of the number 0.000001 is six places to the left of 1 and

$$0.000001 = 10^{-6}$$
6 places

What about a number like 3,000,000? If you write it $3 \cdot 1,000,000$, the exponential expression is simply $3 \cdot 10^6$. Thus, the mass of Earth, which, expressed in long numerical form is 3,120,000,000,000,000,000,000,000 kg, can be written more conveniently as $3.12 \cdot 10^{24}$ kg.

Table I   Prefixes for Use with Basic Units of the Metric System

Prefix	Symbol[†]	Power Equivalent			
geo*		10^{20}			
tera	T	10^{12}	=	1,000,000,000,000	Trillion
giga	G	10^9	=	1,000,000,000	Billion
mega	M	10^6	=	1,000,000	Million
kilo	k	10^3	=	1,000	Thousand
hecto	h	10^2	=	100	Hundred
deca	da	10^1	=	10	Ten
—	—	10^0	=	1	One
deci	d	10^{-1}	=	0.1	Tenth
centi	c	10^{-2}	=	0.01	Hundredth
milli	m	10^{-3}	=	0.001	Thousandth
micro	*m*	10^{-6}	=	0.000001	Millionth
nano	n	10^{-9}	=	0.000000001	Billionth
pico	p	10^{-12}	=	0.000000000001	Trillionth

*Not an official SI prefix but commonly used to describe very large quantities such as the mass of water in the oceans.

†The SI rules specify that its symbols are not followed by periods, nor are they changed in the plural. Thus, it is correct to write, "The tree is 10 m high," not "10 m. high" or "10 ms high."

Table 11 Handy Conversion Factors

To Convert From	To	Multiply By	
Centimeters	Feet	0.0328 ft/cm	
	Inches	0.394 in/cm	
	Meters	0.01 m/cm	(exactly)
	Micrometers (Microns)	1,000 μm/cm	(″)
	Miles (statute)	6.214×10^{-6} mi/cm	
	Millimeters	10 mm/cm	(exactly)
Feet	Centimeters	30.48 cm/ft	(exactly)
	Inches	12 in/ft	(″)
	Meters	0.3048 m/ft	(″)
	Micrometers (Microns)	304,800 μm/ft	(″)
	Miles (statute)	0.000189 mi/ft	
Grams	Kilograms	0.01 kg/g	(exactly)
	Micrograms	1×10^{6} μg/g	(″)
	Ounces (avdp.)	0.03527 oz/g	
	Pounds (avdp.)	0.002205 lb/g	
Hectares	Acres	2.47 acres/ha	
Inches	Centimeters	2.54 cm/in	(exactly)
	Feet	0.0833 ft/in	
	Meters	0.0254 m/in	(exactly)
	Yards	0.0278 yd/in	
Kilograms	Ounces (avdp.)	35.27 oz/kg	
	Pounds (avdp.)	2.205 lb/kg	
Kilometers	Miles	0.6214 mi/km	
Meters	Centimeters	100 cm/m	(exactly)
	Feet	3.2808 ft/m	
	Inches	39.37 in/m	
	Kilometers	0.001 km/m	(exactly)
	Miles (statute)	0.0006214 mi/m	
	Millimeters	1,000 mm/m	(exactly)
	Yards	1.0936 yd/m	
Miles (statute)	Centimeters	160,934 cm/mi	
	Feet	5,280 ft/mi	(exactly)
	Inches	63,360 in/mi	(exactly)
	Kilometers	1.609 km/mi	
	Meters	1,609 m/mi	
	Yards	1,760 yd/mi	(exactly)
Ounces (avdp.)	Grams	28.35 g/oz	
	Pounds (avdp.)	0.0625 lb/oz	(exactly)
Pounds (avdp.)	Grams	453.6 g/lb	
	Kilograms	0.454 kg/lb	
	Ounces (avdp.)	16 oz/lb	(exactly)

Periodic Table of Elements

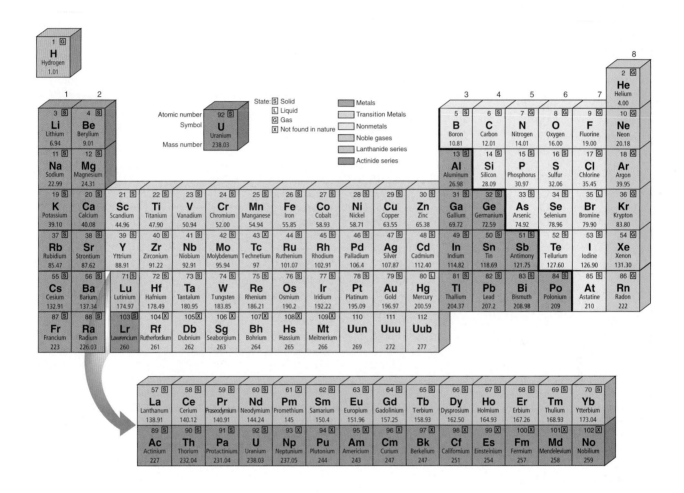

Element	Symbol	Atomic Number	Atomic Weight	Element	Symbol	Atomic Number	Atomic Weight
Actinium	Ac	89	(227)	Mercury	Hg	80	200.59
Aluminum	Al	13	26.9815	Molybdenum	Mo	42	95.94
Americium	Am	95	(243)	Neodymium	Nd	60	144.24
Antimony	Sb	51	121.75	Neon	Ne	10	20.1797
Argon	Ar	18	39.948	Neptunium	Np	93	(237)
Arsenic	As	33	74.9216	Nickel	Ni	28	58.69
Astatine	At	85	(210)	Niobium	Nb	41	92.9064
Barium	Ba	56	137.327	Nitrogen	N	7	14.0067
Berkelium	Bk	97	(247)	Nobelium	No	102	(259)
Beryllium	Be	4	9.01218	Osmium	Os	76	190.2
Bohrium	Bh	107	(262)	Oxygen	O	8	15.9994
Bismuth	Bi	83	208.980	Palladium	Pd	46	106.42
Boron	B	5	10.811	Phosphorus	P	15	30.9738
Bromine	Br	35	79.904	Platinum	Pt	78	195.08
Cadmium	Cd	48	112.411	Plutonium	Pu	94	(244)
Calcium	Ca	20	40.078	Polonium	Po	84	(209)
Californium	Cf	98	(251)	Potassium	K	19	39.0983
Carbon	C	6	12.011	Praseodymium	Pr	59	140.908
Cerium	Ce	58	140.115	Promethium	Pm	61	(145)
Cesium	Cs	55	132.905	Protactinium	Pa	91	(231)
Chlorine	Cl	17	35.4527	Radium	Ra	88	(226)
Chromium	Cr	24	51.9961	Radon	Rn	86	(222)
Cobalt	Co	27	58.9332	Rhenium	Re	75	186.207
Copper	Cu	29	63.546	Rhodium	Rh	45	102.906
Curium	Cm	96	(247)	Rubidium	Rb	37	85.4678
Dubnium	Db	105	(262)	Ruthenium	Ru	44	101.07
Dysprosium	Dy	66	162.50	Rutherfordium	Rf	104	(261)
Einsteinium	Es	99	(252)	Samarium	Sm	62	150.36
Erbium	Er	68	167.26	Scandium	Sc	21	44.9559
Europium	Eu	63	151.965	Seaborgium	Sg	106	(263)
Fermium	Fm	100	(257)	Selenium	Se	34	78.96
Fluorine	F	9	18.9984	Silicon	Si	14	28.0855
Francium	Fr	87	(223)	Silver	Ag	47	107.868
Gadolinium	Gd	64	157.25	Sodium	Na	11	22.9898
Gallium	Ga	31	69.723	Strontium	Sr	38	87.62
Germanium	Ge	32	72.61	Sulfur	S	16	32.066
Gold	Au	79	196.967	Tantalum	Ta	73	180.948
Hafnium	Hf	72	178.49	Technetium	Tc	43	(98)
Hassium	Hs	108	(265)	Tellurium	Te	52	127.60
Helium	He	2	4.00260	Terbium	Tb	65	158.925
Holmium	Ho	67	164.930	Thallium	Tl	81	204.383
Hydrogen	H	1	1.00794	Thorium	Th	90	232.038
Indium	In	49	114.82	Thulium	Tm	69	168.934
Iodine	I	53	126.904	Tin	Sn	50	118.710
Iridium	Ir	77	192.22	Titanium	Ti	22	47.88
Iron	Fe	26	55.847	Tungsten	W	74	183.85
Krypton	Kr	36	83.80	Ununbium	Uub	112	(277)
Lanthanum	La	57	139.906	Ununnilium	Uun	110	(269)
Lawrencium	Lr	103	(262)	Unununium	Uuu	111	(272)
Lead	Pb	82	207.2	Uranium	U	92	238.029
Lithium	Li	3	6.941	Vanadium	V	23	50.9415
Lutetium	Lu	71	174.967	Xenon	Xe	54	131.29
Magnesium	Mg	12	24.3050	Ytterbium	Yb	70	173.04
Manganese	Mn	25	54.9381	Yttrium	Y	39	88.9059
Meitnerium	Mt	109	(266)	Zinc	Zn	30	65.39
Mendelevium	Md	101	(258)	Zirconium	Zr	40	91.224

*The atomic weights given are from the 1987 IUPAC Atomic Weight table, to a maximum of six significant figures.

†Value in parentheses is the mass number of the isotope of longest known half-life.

Rock Symbols

The symbols used in this book for types of rocks are shown below:

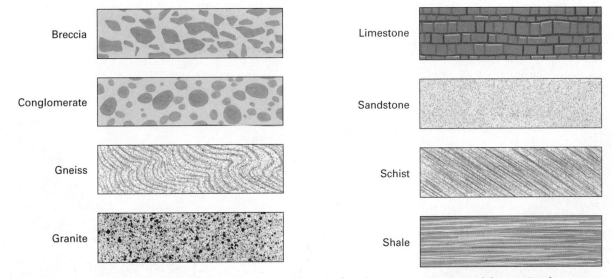

Breccia

Conglomerate

Gneiss

Granite

Limestone

Sandstone

Schist

Shale

In this book we have adopted consistent colors and style for depicting magma and layers in the upper mantle and crust.

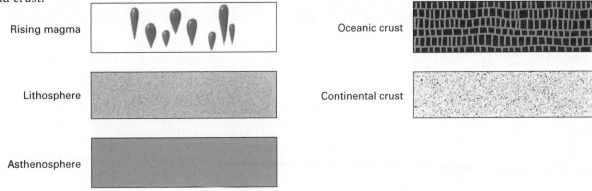

Rising magma

Lithosphere

Asthenosphere

Oceanic crust

Continental crust

SPRING SKY

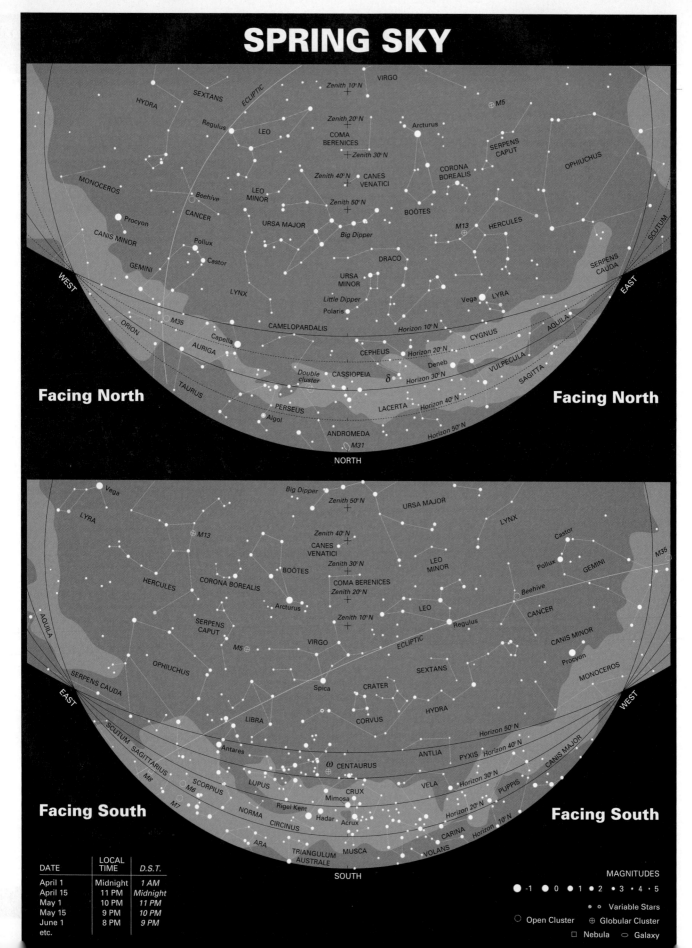

Facing North

Facing North

NORTH

Facing South

Facing South

SOUTH

DATE	LOCAL TIME	*D.S.T.*
April 1	Midnight	*1 AM*
April 15	11 PM	*Midnight*
May 1	10 PM	*11 PM*
May 15	9 PM	*10 PM*
June 1	8 PM	*9 PM*
etc.		

MAGNITUDES

● -1 ● 0 ● 1 ● 2 ● 3 · 4 · 5

○ · Variable Stars

○ Open Cluster ⊕ Globular Cluster

□ Nebula ○ Galaxy

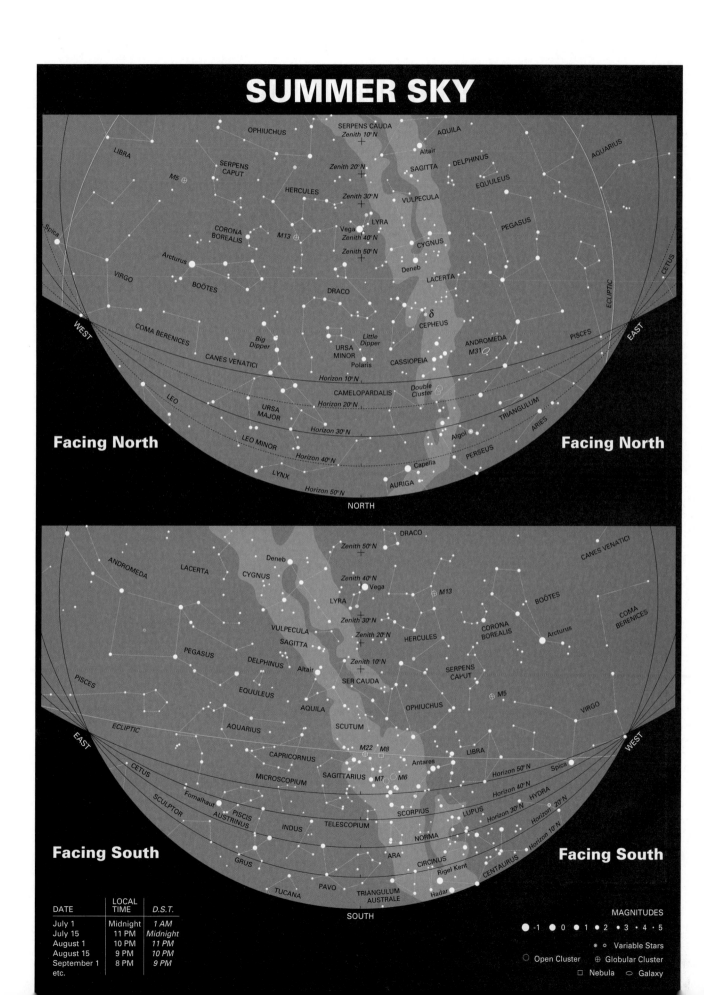

SUMMER SKY

Facing North

Facing North

Facing South

Facing South

DATE	LOCAL TIME	D.S.T.
July 1	Midnight	1 AM
July 15	11 PM	Midnight
August 1	10 PM	11 PM
August 15	9 PM	10 PM
September 1	8 PM	9 PM
etc.		

MAGNITUDES
● -1 ● 0 ● 1 ● 2 • 3 · 4 · 5

• ∘ Variable Stars
○ Open Cluster ⊕ Globular Cluster
□ Nebula ◠ Galaxy

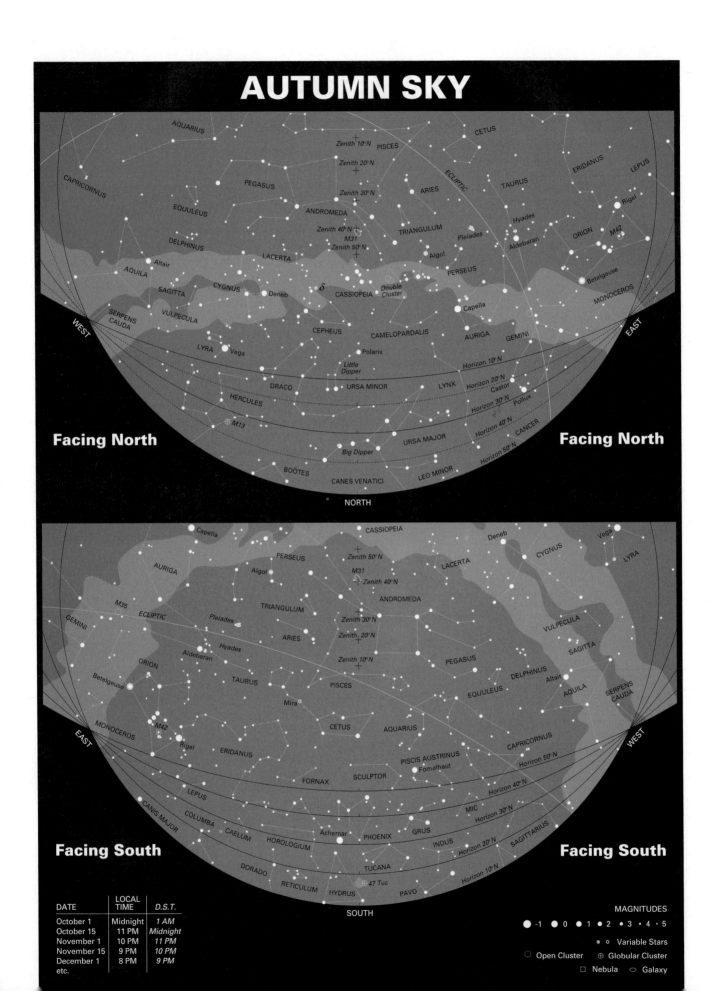

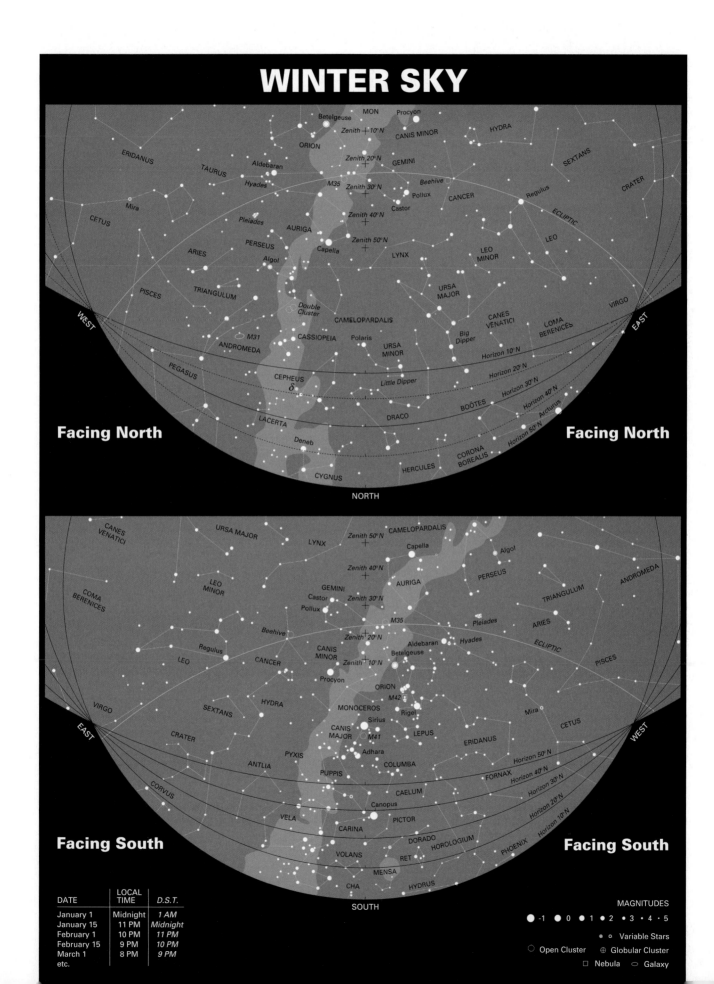

WINTER SKY

WINTER SKY

Facing North

Facing South

Glossary

A horizon   The uppermost layer of soil, composed of a mixture of organic matter and leached and weathered minerals (*syn:* topsoil).

aa   A lava flow that has a jagged, rubbly, broken surface.

ablation area   The lower portion of a glacier where more snow melts in summer than accumulates in winter so there is a net loss of glacial ice (*syn:* zone of wastage).

abrasion   The mechanical wearing and grinding of rock surfaces by friction and impact.

absolute age   Time measured in years.

absolute brightness of luminosity   The luminosity of a star as it would appear if it were a fixed distance away.

absolute humidity   *See* humidity.

absolute magnitude   The brightness that a star would appear to have if it were 32 light-years (10 parsecs) away.

absorption nebula   A cold nebula that absorbs light.

absorption of radiation   The process that occurs when energy is absorbed: the energy of a photon is converted to electrical, chemical, vibrational, or heat energy, and the photon disappears.

absorption spectrum   *See* spectrum.

abyssal fan   A large, fan-shaped accumulation of sediment deposited at the bases of many submarine canyons adjacent to the deep-sea floor (*syn:* submarine fan).

abyssal plain   A flat, level, largely featureless part of the ocean floor between the Mid-Oceanic Ridge and the continental rise.

accelerating universe   A model in which some force is overpowering gravity and causing the expansion of the universe to accelerate.

accreted terrain   A land mass that originated as an island arc or a microcontinent that was later added onto a continent.

accumulation area   The upper part of a glacier where accumulation of snow during the winter exceeds melting during the summer, causing a net gain of glacial ice.

acid precipitation   (often called *acid rain*) A condition in which natural precipitation becomes acidic after reacting with air pollutants.

acid rain   *See* acid precipitation.

active continental margin   A continental margin characterized by subduction of an oceanic lithospheric plate beneath a continental plate (*syn:* Andean margin).

adiabatic temperature changes   Temperature changes that occur without gain or loss of heat.

advection   In meteorology, the horizontal component of a convection current in air, i.e., the surface movement that is commonly called wind.

advection fog   Fog that forms when warm, moist air from the sea blows onto cooler land, where the air cools and water vapor condenses at ground level.

aerosol   Any small particle that is larger than a molecule and suspended in air.

air mass   A large body of air that has approximately the same temperature and humidity throughout.

air pressure   *See* barometric pressure.

albedo   The reflectivity of a surface. A mirror or bright, snowy surface reflects most of the incoming light and has a high albedo, whereas a rough, flat road surface has a low albedo.

alluvial fan   A fan-like accumulation of sediment created where a steep stream slows down rapidly as it reaches a relatively flat valley floor.

alpine glacier   A glacier that forms in mountainous terrain.

alternative energy resources   All energy resources other than fossil fuels and nuclear fission, including solar energy; hydroelectric power; geothermal energy; wind energy; biomass energy; tidal, wave, and heat energy from the seas; and nuclear fusion.

altimeter   A barometer with a scale calibrated in units of elevation rather than units of pressure.

altostratus cloud   A high-level stratus cloud.

amphibole   A group of double chain silicate minerals. Hornblende is a common amphibole.

anaerobic   Without oxygen; anaerobic bacteria are bacteria that live without oxygen.

Andean margin   A continental margin characterized by subduction of an oceanic lithospheric plate beneath a continental plate (*syn:* active continental margin).

andesite   A fine-grained gray or green volcanic rock intermediate in composition between basalt and granite, consisting of about equal amounts of plagioclase feldspar and mafic minerals.

angle of repose   The maximum slope or angle at which loose material remains stable.

angular unconformity   An unconformity in which younger sediment or sedimentary rocks rest on the eroded surface of tilted or folded older rocks.

anion   An ion that has a negative charge.

antecedent stream   A stream that was established before local uplift started and cut its channel at the same rate the land was rising.

anticline   A fold in rock that resembles an arch; the fold is convex upward, and the oldest rocks are in the middle.

anticyclone   A system of rotating winds that develop where descending air spreads over Earth's surface. In the Northern Hemisphere, the Coriolis effect deflects the diverging winds to the right, forming a pinwheel pattern with the air spiraling clockwise. In the Southern Hemisphere, the Coriolis effect deflects winds leftward and creates a counterclockwise spiral.

apparent brightness or luminosity   The luminosity of a star as seen from Earth.

apparent magnitude   The brightness of a star as seen from Earth.

aquifer   A porous and permeable body of rock that can yield economically significant quantities of groundwater.

Archean Eon   A division of geologic time 3.8 to 2.5 billion years ago. The oldest known rocks formed at the beginning of, or just prior to, the start of the Archean Eon.

arête   A sharp, narrow ridge between adjacent valleys that was formed by glacial erosion.

artesian aquifer   An inclined aquifer that is bounded top and bottom by layers of impermeable rock so the water is under pressure.

artesian well   A well drilled into an artesian aquifer, in which the water rises without pumping and in some cases spurts to the surface.

artificial channel   Any channel dredged to modify a natural channel or to alter the course of a stream.

artificial levee   A wall built along the banks of a stream to prevent rising floodwater from spilling out of the stream channel onto the flood plain.

aseismic ridge   A submarine mountain chain with little or no earthquake activity.

ash (volcanic)   Fine pyroclastic material less than 2 mm in diameter.

ash flow   A mixture of volcanic ash, larger pyroclastic particles, and gas that flows rapidly along Earth's surface as a result of an explosive volcanic eruption (*syn:* nuée ardente).

ash-flow tuff   A pyroclastic rock formed when an ash flow solidifies.

aspect   The orientation of a slope with respect to the Sun; the geographic orientation or exposure of a slope.

asteroid   One of the many small celestial bodies in orbit around the Sun. Most asteroids orbit between Mars and Jupiter.

asthenosphere   The portion of the upper mantle beneath the lithosphere. It consists of weak, plastic rock where magma may form and extends from a depth of about 100 kilometers to about 350 kilometers below the surface of Earth.

atmosphere   A mixture of gases, mostly nitrogen and oxygen, with smaller amounts of argon, carbon dioxide, and other gases. The atmosphere is held to Earth by gravity and thins rapidly with altitude.

atmospheric inversion   *See* inversion (atmospheric).

atmospheric pressure   The pressure of the atmosphere at any given location and time.

atoll   A circular reef that surrounds a lagoon and is bounded on the outside by deep water of the open sea.

atom   The fundamental unit of elements consisting of a small, dense, positively charged center called a nucleus surrounded by a diffuse cloud of negatively charged electrons.

B horizon   The soil layer just below the A horizon, where ions leached from the A horizon accumulate.

back arc basin   A sedimentary basin on the opposite side of the magmatic arc from the trench, either in an island arc or in an Andean continental margin.

backshore   The upper zone of a beach that is usually dry but is washed by waves during storms.

bajada   A broad depositional surface extending outward from a mountain front and that is formed by the merging of alluvial fans.

banded iron formation   Iron-rich, layered sedimentary rocks precipitated from the seas mostly between 2.6 and 1.9 billion years ago, as a result of rising atmospheric oxygen concentrations.

bank   The rising slope bordering the side of a stream channel.

bar   An elongate mound of sediment, usually composed of sand or gravel, in a stream channel or along a coastline.

barchan dune   A crescent-shaped dune, highest in the center, with the tips facing downwind.

barometer   A device used to measure barometric pressure.

barometric pressure   The pressure exerted by Earth's atmosphere.

barred spiral galaxy   A galaxy with a straight bar of stars, gas, and dust extending across the nucleus.

barrier island   A long, narrow, low-lying island that extends parallel to the shoreline.

barrier reef   A reef separated from the coast by a deep, wide lagoon.

basal slip   Movement of the entire mass of a glacier along the bedrock.

basalt   A dark-colored, very-fine-grained, mafic, volcanic rock composed of about half calcium-rich plagioclase feldspar and half pyroxene.

base level   The deepest level to which a stream can erode its bed. The ultimate base level is usually sea level, but this is seldom attained.

basement rock   The older granitic and related metamorphic rocks of Earth's crust that make up the foundations of continents.

basin   A low area of Earth's crust of tectonic origin, commonly filled with sediment.

batholith   A large, plutonic mass of intrusive rock with more than 100 square kilometers of surface exposed.

bauxite   A gray, yellow, or reddish-brown rock composed of a mixture of aluminum oxides and hydroxides. It is the principle ore of aluminum.

baymouth bar   A bar that extends partially or completely across the entrance to a bay.

beach   Any strip of shoreline washed by waves or tides.

beach drift   Sediment (usually sand) carried parallel to a beach when waves strike the beach obliquely. The waves push sand up and along the beach in the direction that the wave is traveling. When water recedes, the sand flows straight down the beach. Thus, at the end of one wave cycle, the sand has moved a short distance parallel to the beach. The next wave transports the sand a little farther, until, over time, sediment moves long distances.

beach terrace   A level portion of old beach elevated above the modern beach by uplift of the shoreline or fall of sea level.

bed   The floor of a stream channel. Also the thinnest layer in sedimentary rocks, commonly ranging in thickness from a centimeter to a meter or two.

bed load   That portion of a stream's load that is transported on or immediately above the streambed.

bedding (stratification)   Layering that develops as sediments are deposited.

bedrock   The solid rock that underlies soil or regolith.

Benioff zone   An inclined zone of earthquake activity that traces the upper portion of a subducting plane in a subduction zone.

big bang   An event thought to mark the beginning of our universe. The big bang theory postulates that 10 to 20 billion years ago, all matter exploded from an infinitely compressed state.

bioclastic sedimentary rock Sedimentary rocks such as most limestone, which are composed of broken shell fragments and similar remains of living organisms. The fragments are clastic, but they are of biological origin.

biodegradable pollutants A pollutant that is consumed and destroyed in a reasonable amount of time by organisms that live naturally in soil or water.

biome A community of plants living in a large geographic area characterized by a particular climate.

bioremediation The use of microorganisms to decompose an environmental contaminant.

biosphere The zone inhabited by life.

biotite Black, rock-forming mineral of the mica group.

bitumen A general name for solid and semi-solid hydrocarbons that are fusible and soluble in carbon bisulfide. Petroleum, asphalt, natural mineral wax, and asphaltites are all bitumens.

black hole A small region of space that contains matter packed so densely that light cannot escape from its intense gravitational field.

black smoker A jet of black water spouting from a fracture in the sea floor, commonly near the Mid-Oceanic Ridge. The black color is caused by precipitation of very-fine-grained metal sulfide minerals as hydrothermal solutions cool by contact with seawater.

blob tectonics Tectonic activity dominated by rising and sinking of the mantle and crust, believed to predominate on Venus.

blowout A small depression created by wind erosion.

blue shift The frequency shift of light waves observed in the spectrum of an object traveling toward an observer. This shift is caused by the Doppler effect.

body wave A seismic wave that travels through the interior of Earth.

bolide A large piece of space debris, such as an asteroid, that crashes into a planet.

braided stream A stream that divides into a network of branching and reuniting shallow channels separated by midchannel bars.

branching chain reaction A nuclear fission reaction in which the initial reaction releases two or three neutrons, each of which triggers the fission of additional nuclei.

break As a wave approaches a beach, the front of the wave rises over the base and the wave steepens until it collapses forward, or "breaks." Waves breaking along a shore are called surf.

breccia A coarse-grained sedimentary rock composed of angular, broken fragments cemented together.

brittle fracture Rupture that occurs when a rock breaks sharply.

burial metamorphism Metamorphism that results from deep burial of rocks in a sedimentary basin. Rocks metamorphosed in this way are usually unfoliated.

butte A flat-topped mountain, with several steep cliff faces. A butte is smaller and more tower-like than a mesa.

C horizon The lowest soil layer, composed of partly weathered bedrock grading downward into unweathered parent rock.

calcite A common rock-forming mineral, $CaCO_3$.

caldera A large circular depression caused by an explosive volcanic eruption.

caliche A hard soil layer formed when calcium carbonate precipitates and cements the soil.

calving A process in which large chunks of ice break off from tidewater glaciers to form icebergs.

cap rock An impermeable rock, usually shale, that prevents oil or gas from escaping upward from a reservoir.

capacity The maximum quantity of sediment that a stream can carry.

capillary action The action by which water is pulled upward through small pores by electrical attraction to the pore walls.

carbonate A mineral such as calcite ($CaCO_3$), containing CO_3^{-2} (the "carbonate anion") as a major part of its chemical composition.

carbonate platform An extensive accumulation of limestone, such as the Florida Keys and the Bahamas, formed on a continental shelf in warm regions where sediment does not muddy the water and reef-building organisms thrive.

carbonate rocks Rocks such as limestone and dolomite made up primarily of carbonate minerals.

catastrophism The principle that states that infrequent catastrophic events alter the course of Earth history and modify the path of slow geologic change.

cation A positively charged ion.

cavern An underground cavity or series of chambers created when groundwater dissolves large amounts of rock, usually limestone (*syn*: cave).

celestial spheres A hypothetical series of concentric spheres centered at the center of Earth. Aristotle postulated that the Sun, Moon, planets, and stars are imbedded in the spheres.

cementation The process by which clastic sediment is lithified by precipitation of a mineral cement among the grains of the sediment.

Cenozoic Era The latest of the four eras into which geologic time is subdivided, 65 million years ago to the present.

chalk A very-fine-grained, soft, earthy, white to gray bioclastic limestone made of the shells and skeletons of marine microorganisms.

channel The trough or groove through which a stream flows, defined by its bed and banks.

channel characteristics Channel characteristics refer to the shape and roughness of a stream channel.

chemical bond The linkage between atoms in molecules and between molecules and ions in crystals.

chemical remediation Treatment of a contaminated groundwater aquifer by injecting a chemical compound that reacts with the pollutant to produce harmless products. Common compounds used in chemical remediation include oxygen and dilute acids and bases.

chemical sedimentary rock Rocks such as rock salt, which form by direct precipitation of minerals from solution.

chemical weathering The chemical decomposition of rocks and minerals by exposure to air, water, and other chemicals in the environment.

chemosynthesis A process in which an organism uses energy from chemical reactions and thus is not dependent on photosynthesis.

chert A hard, dense, sedimentary rock composed of microcrystalline quartz (*syn*: flint).

chlorofluorocarbons (CFCs) Organic compounds containing chlorine and fluorine. These materials often become air pollutants.

chondrule A small grain about 1 millimeter in diameter, embedded in a meteorite, which is composed largely of olivine and pyroxene. Many chondrules contain organic molecules, including amino acids, the building blocks of proteins.

chromosphere A turbulent, diffuse, gaseous layer of the Sun that lies above the photosphere.

cinder cone A small volcano, as high as 300 meters, made up of loose, pyroclastic fragments blasted out of a central vent.

cinders (volcanic) Glassy, pyroclastic, volcanic fragments 4 to 32 millimeters in size.

cirque A steep-walled semicircular depression eroded into a mountain peak by a glacier.

cirrus cloud A wispy, high-level cloud.

clastic sediment Sediment composed of fragments of weathered rock that have been transported some distance from their place of origin.

clastic sedimentary rock Rock composed of lithified clastic sediment.

clay Any clastic mineral particle less than 1/256 millimeter in diameter. Also a group of layered silicate minerals.

claystone A fine-grained, clastic sedimentary rock composed predominantly of clay minerals and small amounts of quartz and other minerals of clay size.

Clean Water Act A federal law mandating the cleaning of the nation's rivers, lakes, and wetlands, and forbidding the discharge of pollutants into waterways.

cleavage The tendency of some minerals to break along certain crystallographic planes.

climate The composite pattern of long-term weather conditions that can be expected in a given region. Climate refers to yearly cycles of temperature, wind, rainfall, etc., and not to daily variations. *See* weather.

closed universe A model in which the gravitational force of the universe is so great that all the galaxies will eventually slow down, reverse direction, and fall back to the center, forming another point of infinite density.

cloud A collection of minute water droplets or ice crystals in air.

coal A flammable, organic, sedimentary rock formed from partially decomposed plant material and composed mainly of carbon.

coal bed methane Methane that is chemically bonded to coal. It can be released when pressure is lowered by drilling wells into a coal bed to remove the ground water. The decrease in pressure allows the methane to separate from the coal as a gas, enabling it to be recovered. Seven percent of the methane (natural gas) produced in the United States comes from this source.

cobbles Rounded rock fragments in the size range of 64 to 256 mm, larger than pebbles and smaller than boulders.

cold front A front that forms when moving cold air collides with stationary or slower-moving warm air. The dense, cold air distorts into a blunt wedge and pushes under the warmer air, creating a narrow band of violent weather commonly accompanied by cumulus and cumulonimbus clouds.

column A dripstone or speleothem formed when a stalactite and a stalagmite meet and fuse together.

columnar joint A regularly spaced crack that commonly develops in lava flows, forming five- or six-sided columns.

coma The bright outer sheath of a comet.

comet An interplanetary body composed of loosely bound rock and ice, which forms a bright head and extended fuzzy tail when it enters the inner potion of the solar system.

compaction Tight packing of sedimentary grains causing weak lithification and a decrease in porosity, usually resulting from the weight of overlying sediment.

competence A measure of the largest particles that a stream can transport.

composite cone A volcano that consists of alternate layers of unconsolidated pyroclastic material and lava flows (*syn:* stratovolcano).

compound A pure substance composed of two or more elements whose composition is constant.

compressive stress Stress that acts to shorten an object by squeezing it.

conduction The transport of heat by atomic or molecular motion.

cone of depression A cone-like depression in the water table, formed when water is pumped out of a well more rapidly than it can flow through the aquifer.

confining pressure *See* confining stress.

confining stress A form of stress that develops when rock or sediment is buried. Confining stress compresses rocks but does not distort them because the compressive force acts equally in all directions.

conformable The condition in which sedimentary layers were deposited continuously without interruption.

conglomerate A coarse-grained, clastic sedimentary rock, composed of rounded fragments larger than 2 mm in diameter and cemented in a fine-grained matrix of sand or silt.

constellation One of 88 groups of stars that astronomers refer to for convenience in referring to the positions of objects in the night sky.

consumption Any process that uses water and then returns it to Earth far from its source.

contact A boundary between two different rock types or between rocks of different ages.

contact metamorphic ore deposit An ore deposit formed by contact metamorphism.

contact metamorphism Metamorphism caused by heating of country rock, and/or addition of fluids, from a nearby igneous intrusion.

continental crust The predominantly granitic portion of the crust, 20 to 70 kilometers thick, that makes up the continents.

continental drift The theory proposed by Alfred Wegener that continents were once joined together and later split and drifted apart. The continental drift theory has been replaced by the more complete plate tectonics theory.

continental glacier A glacier that forms a continuous cover of ice over areas of 50,000 square kilometers or more and spreads outward in all directions under the influence of its own weight (*syn:* ice sheet).

continental margin The region between the shoreline of a continent and the deep ocean basins, including the continental shelf, continental slope, and continental rise. Also the region where thick, granitic continental crust joins thinner, basaltic oceanic crust.

continental rifting The process by which a continent is pulled apart at a divergent plate boundary.

continental rise An apron of sediment between the continental slope and the deep-sea floor.

continental shelf A shallow, nearly level area of continental crust covered by sediment and sedimentary rocks that is submerged below sea level at the edge of a continent between the shoreline and the continental slope.

continental slope The relatively steep (3°–6°) underwater slope between the continental shelf and the continental rise.

control rod A column of neutron-absorbing alloys that are spaced among fuel rods to fine-tune nuclear fission in a reactor.

convection The transport of heat by the movement of currents. In meteorology, horizontal air flow is called advection, whereas convection is reserved for vertical air flow.

convective zone The subsurface zone in a star where energy is transmitted primarily by convection.

convergent boundary A boundary where two lithospheric plates collide head-on.

coquina A bioclastic limestone consisting of coarse shell fragments cemented together.

core The innermost region of Earth, probably consisting of iron and nickel.

Coriolis effect The deflection of air or water currents caused by the rotation of Earth.

corona The luminous, irregular envelope of highly ionized gas outside the chromosphere of the Sun.

correlation Demonstration of the equivalence of rocks' or geologic features' ages from different locations.

cosmic background radiation Low-energy, microwave radiation that began traveling through space when the universe was only 300,000 years old and now pervades all space in the universe.

cost–benefit analysis A system of analysis that attempts to weigh the cost of an act or policy, such as pollution control, directly against the economic benefits.

country rock The older rock intruded by a younger igneous intrusion or mineral deposit.

crater A bowl-like depression at the summit of the volcano.

craton A segment of continental crust, usually in the interior of a continent, that has been tectonically stable for a long time, commonly a billion years or longer.

creep The slow movement of unconsolidated material downslope under the influence of gravity.

crescent moon A position in which a moon appears as a thin crescent, approximately four days after a new moon.

crest (of a wave) The highest part of a wave.

crevasse A fracture or crack in the upper 40 to 50 meters of a glacier.

cross-bedding An arrangement of small beds at an angle to the main sedimentary layering.

cross-cutting relationship Any relationship in which younger rocks or geological structures interrupt or cut across older rocks or structures.

crust Earth's outermost layer, about 7 to 70 kilometers thick and composed of relatively low-density silicate rocks.

crystal A solid element or compound whose atoms are arranged in a regular, orderly, periodically repeated array.

crystal face A planar surface that develops if a crystal grows freely in an uncrowded environment.

crystal habit The shape in which individual crystals grow and the manner in which crystals grow together in aggregates.

crystal settling A process in which the crystals that solidify first from a cooling magma settle to the bottom of a magma chamber because the solid minerals are more dense than liquid magma.

crystalline structure The orderly, repetitive arrangement of atoms in a crystal.

cumulonimbus cloud A cumulus cloud from which precipitation is falling.

cumulus cloud A column-like cloud with a flat bottom and a billowy top.

Curie point The temperature below which rocks can retain magnetism.

current A continuous flow of water in a concerted direction.

cyanobacteria Blue-green photosynthetic bacteria that were among the earliest photosynthetic life-forms on Earth.

cyclone A low-pressure region with its accompanying surface wind. Also a synonym for a tropical cyclone or hurricane.

dark energy The poorly understood repulsive force that causes the modern universe to accelerate.

dark mass A collection of as-yet-undetected particles, large enough to make up 23 percent of the universe, that exert conventional gravitational attraction.

dark matter Mysterious invisible matter that may comprise up to 90 percent of the mass of the universe.

daughter isotope An isotope formed by radioactive decay of another isotope.

debris flow A type of mass wasting in which particles move as a fluid and more than half of the particles are larger than sand.

deep-sea current Vertical and horizontal water flow below a depth of 400 meters in the oceans.

deflation Erosion by wind.

deformation Folding, faulting, and other changes in the shape of rocks or minerals in response to mechanical forces, such as those that occur in tectonically active regions.

degeneracy pressure The strength of the atomic particles that holds a white dwarf star from further collapse.

delta The nearly flat, alluvial, fan-shaped tract of land at the mouth of a stream.

deposition The laying down of rock-forming materials by any natural agent.

desert Any region that receives less than 25 centimeters (10 inches) of rain per year and consequently supports little or no vegetation.

desert pavement A continuous cover of stones created as wind erodes fine sediment, leaving larger rocks behind.

desertification A process by which semiarid land is converted to desert, often by improper farming or by climate change.

dew Moisture condensed onto objects from the atmosphere, usually during the night, when the ground and leaf surfaces become cooler than the surrounding air.

dew point The temperature at which the relative humidity of air reaches 100 percent and the air becomes saturated. If saturated air cools below the dew point, some of the water vapor generally condenses into liquid droplets (although sometimes the relative humidity can rise above 100 percent).

differential weathering The process by which certain rocks weather more rapidly than adjacent rocks, usually resulting in an uneven surface.

dike A sheetlike igneous rock that cuts across the structure of country rock.

directed stress Stress that acts most strongly in one direction.

discharge The volume of water flowing downstream per unit time. It is measured in units of cubic meters per second (m^3/sec).

disconformity A type of unconformity in which the sedimentary layers above and below the unconformity are parallel.

discordant Pertaining to a dike or other feature that cuts across sedimentary layers or other kinds of layering in country rock.

disseminated ore deposit A large, low-grade ore deposit in which generally fine-grained metal-bearing minerals are widely scattered throughout a rock body in sufficient concentration to make the deposit economical to mine.

dissolution A chemical weathering process in which mineral or rock passes into solution.

dissolved load The portion of a stream's sediment load that is carried in solution.

distributary A channel that flows outward from the main stream channel, such as is commonly found in deltas.

divergent boundary The boundary or zone where lithospheric plates separate from each other (*syn:* spreading center).

diversion system Any infrastructure used to transfer ground or surface water from its natural place and path in the hydrologic cycle to a new place and path to serve human needs.

docking The accretion of island arcs or microcontinents onto a continental margin.

doldrums A region of Earth near the equator in which hot, humid air moves vertically upward, forming a vast low-pressure region. Local squalls and rainstorms are common, and steady winds are rare.

dolomite A common, rock-forming mineral, $CaMg(CO_3)_2$.

dome A circular or elliptical anticlinal structure.

Doppler effect The observed change in frequency of light or sound that occurs when the source of the wave is moving relative to the observer.

downcutting Downward erosion by a stream.

drainage basin The region that is ultimately drained by a single river.

drift (glacial) Any rock or sediment transported and deposited by a glacier or by glacial meltwater.

drumlin An elongate hill formed when a glacier flows over and reshapes a mound of till or stratified drift.

dry adiabatic lapse rate The rate of cooling that occurs when dry air rises without gain or loss of heat.

dune A mound or ridge of wind-deposited sand.

earthflow A flowing mass of fine-grained soil particles mixed with water. Earthflows are less fluid than mudflows.

earthquake A sudden motion or trembling of Earth caused by the abrupt release of slowly accumulated elastic energy in rocks.

eccentricity The shape of the ellipse that constitutes Earth's orbit around the Sun.

echo sounder An instrument that emits sound waves and then records them after they reflect off the sea floor. The data is then used to record the topography of the sea floor.

eclipse A phenomenon that occurs when a heavenly body is shadowed by another and therefore rendered invisible. When the Moon lies directly between Earth and the Sun, the Moon blocks our view of the Sun and we observe a *solar eclipse*. When Earth lies directly between the Sun and the Moon, Earth's shadow falls on the Moon and we observe a *lunar eclipse*.

ecosystem A system formed by the interactions of a variety of individual organisms with each other and with their physical environments.

effluent stream A stream that receives water from ground water because its channel lies below the water table (*syn:* gaining stream).

Ekman transport The net effect of wind and the Coriolis effect on ocean current directions. In principle, Ekman transport in the Northern Hemisphere is 90° to the right of the wind direction.

El Niño An episodic weather pattern in which the trade winds slacken in the Pacific Ocean and warm water accumulates off the coast of South America and causes unusual rains and heavy snowfall in the Andes.

elastic deformation A deformation such that if the stress is removed, the material springs back to its original size and shape.

elastic limit The maximum stress that an object can withstand without permanent deformation.

electromagnetic radiation The transfer of energy by an oscillating electric and magnetic field; it travels as a wave and also behaves as a stream of particles.

electromagnetic spectrum The entire range of electromagnetic radiation from very-long-wavelength (low-frequency) radiation to very-short-wavelength (high-frequency) radiation.

electron A fundamental particle that forms a diffuse cloud of negative charge around an atom.

element A substance that cannot be broken down into other substances by ordinary chemical means. An element is made up of the same kind of atoms.

elliptical galaxy A galaxy with an elliptical appearance.

emergent coastline A coastline that was recently under water but has been exposed either because the land has risen or sea level has fallen.

emission nebula A glowing cloud of interstellar gas.

emission of radiation Energy, in the form of a photon, may be emitted, with the equivalent loss of electrical, chemical, or vibrational energy of the emitting substance.

emission spectrum *See* spectrum.

end moraine A moraine that forms at the end, or terminus, of a glacier.

energy resources Geologic resources, including petroleum, coal, natural gas, and nuclear fuels, used for heat, light, work, and communication.

entropy A measure of the disorder or homogeneity in a system.

eon The longest unit of geologic time. The most recent eon, the Phanerozoic Eon, is further subdivided into eras.

epicenter The point on Earth's surface directly above the focus of an earthquake.

epidemiology The study of the distribution of sickness in a population.

epoch   The smallest unit of geologic time. Periods are divided into epochs.

equatorial upwelling   Oceanic upwelling in which surface currents flowing westward on both sides of the equator are deflected poleward and are replaced by upward flow of deep, nutrient-rich waters.

equinox   Either of two times during a year when the Sun shines directly overhead at the equator. The equinoxes are the beginnings of spring and fall, when every portion of Earth receives 12 hours of daylight and 12 hours of darkness.

era   A geologic time unit. Eons are divided into eras, and, in turn, eras are subdivided into periods.

erosion   The removal of weathered rocks by moving water, wind, ice, or gravity.

erratic   A boulder that was transported to its present location by a glacier. Usually different from the bedrock in its immediate vicinity.

escape velocity   The velocity that an object must attain to escape the gravitational field of an object in space.

esker   A long, snake-like ridge formed by deposition in a stream that flowed on, within, or beneath a glacier.

estuary   A bay that formed when a broad river valley was submerged by rising sea level or a sinking coast.

eustatic sea-level change   A global sea-level change; it can be caused by three different processes: the growth or melting of glaciers, changes in water temperature, and changes in the volume of the Mid-Oceanic Ridge system.

eutrophic lake   A body of water characterized by abundant dissolved nitrates, phosphates, and other plant nutrients and by a seasonal deficiency of oxygen in bottom water. Such waterways are commonly shallow.

evaporation   The transformation of a liquid to a gas.

evaporation fog   Fog that forms when air is cooled by evaporation from a body of water, commonly a lake or river. The water evaporates, but the vapor cools and condenses to fog.

evaporite deposit   A chemically precipitated sedimentary rock that formed when dissolved ions were concentrated by evaporation of water.

evolution   The change in the physical and genetic characteristics of a species over time.

excited state   A relatively stationary state of higher energy than the lowest energy level (or ground state) of an electron in an atom or molecule.

exfoliation   Fracturing in which concentric plates or shells split from the main rock mass like the layers of an onion.

extensional stress   Tectonic stress in which rocks are pulled apart.

externalities   The cost of living in a degraded environment, including direct costs such as medical bills, lost work, and damage to waterways, crops, and livestock. It also includes indirect costs of environmental degradation, such as reduction in tourism and lowered land values.

extrusive igneous rock   An igneous rock formed from material that has erupted onto the surface of Earth.

eyepiece   The lens, closest to the eye, that focuses light in a refracting telescope.

fall   A type of mass wasting in which unconsolidated material falls freely or bounces down the face of a cliff.

fault   A fracture in rock along which displacement has occurred.

fault creep   A continuous, slow movement of solid rock along a fault, resulting from a constant stress acting over a long time.

fault zone   An area of numerous, closely spaced faults.

faunal succession (principle of)   *See* principle of faunal succession.

feedback mechanism   A feedback mechanism occurs when a small initial perturbation affects another component of Earth's systems, which amplifies the original effect, which perturbs the system even more, which leads to an even greater effect, and so on.

feldspar   A common group of aluminum silicate rock-forming minerals that contain potassium, sodium, or calcium.

felsic   A term describing any light-colored igneous rock containing large amounts of feldspar and silica.

fetch   The distance that the wind has traveled over the ocean without interruption.

firn   Hard, dense snow that has survived through one summer melt season. Firn is transitional between snow and glacial ice.

first-generation energy   Radiation emanated from the primordial sea of atoms, particles, and energy of the big bang.

First Great Oxidation Event   Approximately 2.4 billion years ago the oxygen concentration in Earth's atmosphere increased suddenly from trace amounts to appreciable quantities, probably because of a combination of biological and geochemical processes.

first law of thermodynamics   A fundamental law of nature that states that energy is always conserved.

fissility   Fine layering along which a rock splits easily.

fission   The breakdown of an atomic nucleus of an element of relatively high atomic number into two or more nuclei of lower atomic number, with conversion of part of its mass into energy.

fissure   An extensive break, crack, or fracture in rocks. Fissure eruptions are usually gentle volcanic eruptions in which lava flows from fissures in Earth's crust.

fjord   A long, deep, narrow arm of the sea bounded by steep walls, generally formed by submergence of a glacially eroded valley.

flash flood   A rapid, intense, local flood of short duration.

flood   Any relatively high stream flow that overtops the stream banks in any part of its course, covering land that is not usually under water.

flood basalt   Basaltic lava that erupts gently in great volume from cracks at Earth's surface to cover large areas of land and form basalt plateaus.

flood plain   That portion of a river valley adjacent to the channel; it is built by sediment deposited during floods and is covered by water during a flood.

flow   Mass wasting in which individual particles move downslope as a semifluid, not as a consolidated mass.

fly ash   Minerals that escape into the atmosphere, usually when coal burns and eventually settles as gritty dust.

focus   The initial rupture point of an earthquake.

fog   A cloud that forms at or very close to ground level.

fold   A bend in rock.

foliation   Layering in rock created by metamorphism.

footwall   The rock beneath an inclined fault.

forearc basin   A sedimentary basin between the trench and the magmatic arc either in an island arc or Andean continental margin.

foreshock Small earthquakes that precede a large quake by an interval ranging from a few seconds to a few weeks.

foreshore The zone that lies between the high and low tides; the intertidal region.

formation A lithologically distinct body of sedimentary, igneous, or metamorphic rock that can be recognized in the field and mapped.

fossil The preserved trace, imprint, or remains of a plant or animal.

fossil fuel Fuels formed from the partially decayed remains of plants and animals. The most commonly used fossil fuels are petroleum, coal, and natural gas.

fracture (1) The manner in which minerals break, other than along planes of cleavage. (2) A crack, joint, or fault in bedrock.

frequency The frequency of a wave is the number of complete wave cycles, from crest to crest, that pass by any point in a second.

front In meteorology, a line separating air masses of different temperature or density.

frontal weather system A weather system that develops when air masses collide.

frontal wedging A condition in which a moving mass of cool, dense air encounters a mass of warm, less-dense air; the cool, denser air slides under the warm air mass, forcing the warm air upward to create a weather front.

frost Ice crystals formed directly from vapor.

frost wedging A process in which water freezes in a crack in rock and the expansion wedges the rock apart.

fuel cell An electrochemical energy conversion device that produces electricity from an external supply of fuel, such as hydrogen and oxygen.

fuel rod A 2-meter-long column of fuel-grade uranium pellets used to fuel a nuclear reactor.

full moon A position in which a moon appears round and fully illuminated. A full moon appears when the Moon is on the opposite side of Earth from the Sun; the entire sunlit area is visible.

fusion (of atomic nuclei) The combination of nuclei of light elements (particularly hydrogen) to form heavier nuclei.

gabbro Igneous rock that is mineralogically identical to basalt but that has a medium- to coarse-grained texture because of its plutonic origin.

Gaia (Greek for *Earth*) In 1972 James Lovelock hypothesized that the oxygen produced by primitive organisms gradually accumulated to create the modern atmosphere. He used the term *Gaia* to emphasize the interconnectivity of all of Earth's systems.

gaining stream A stream that receives water from ground water because its channel lies below the water table (*syn:* effluent stream).

galactic halo A spherical cloud of dust and gas that surrounds the Milky Way's galactic disk.

galaxy A large volume of space containing many billions of stars, held together by mutual gravitational attraction.

gamma ray bursts Bursts of high-energy gamma rays.

gem A mineral that is prized for its rarity and beauty rather than for industrial use.

geocentric A model that places Earth at the center of the celestial bodies.

geologic column A composite, columnar diagram that shows the sequence of rocks at a given place or region, arranged to show their position in the geologic time scale.

geologic structure Any feature formed by deformation or movement of rocks, such as a fold or a fault. Also, the combination of all such features of an area or region.

geologic time scale A chronological arrangement of geologic time subdivided into units.

geology The study of Earth, the materials that it is made of, the physical and chemical changes that occur on its surface and in its interior, and the history of the planet and its life forms.

geosphere Solid Earth, consisting of the entire planet from the center of the core to the outer crust.

geostrophic gyres Surface current gyres controlled by the combination of wind, the Coriolis effect modified by Ekman transport, and sea surface topography (*geos* for Earth, and *strophe* for turning). Of the six great surface currents in the world oceans shown in Figure 16.13—two in the Northern Hemisphere and four in the Southern Hemisphere—five are geostrophic gyres.

geothermal energy Energy derived from the heat of Earth.

geothermal gradient The rate at which temperature increases with depth in Earth.

geyser A type of hot spring that intermittently erupts jets of hot water and steam. Geysers occur when ground water comes in contact with hot rock.

gibbous moon A bright waxing moon, with only a sliver of dark that appears approximately 10 days after a new moon.

glacial polish A smooth polish on bedrock created when fine particles transported at the base of a glacier abrade the bedrock.

glacial striation Parallel grooves and scratches in bedrock that form as rocks are dragged along at the base of a glacier.

glacier A massive, long-lasting accumulation of compacted snow and ice that forms on land and moves downslope or outward under its own weight.

glaze An ice coating that forms when rain falls on surfaces that are below freezing.

globular cluster A spherically symmetrical collection of stars that shared a common origin. Numerous globular clusters lie within the Milky Way's galactic halo.

gneiss A foliated rock with banded appearance formed by regional metamorphism.

Gondwanaland The southern part of Pangaea, consisting of what is now South America, Africa, Antarctica, India, and Australia.

graben A wedge-shaped block of rock that has dropped downward between two normal faults.

graded bedding A type of bedding in which larger particles are at the bottom of each bed and the particle size decreases toward the top.

graded stream A stream with a smooth, concave profile. A graded stream is in equilibrium with its sediment supply; it transports all the sediment supplied to it with neither erosion nor deposition in the streambed.

gradient The vertical drop of a stream over a specific distance.

gradualism Originally, gradualism was used to describe geological change that occurs as a consequence of slow or gradual accumulation of small events, such as the deposition of sand in a river delta to eventually form a large landform. More recently, scientists studying biological evolution use the term to describe a theory of evolution that proposes that species change gradually in small increments.

granite A medium- to coarse-grained, sialic, plutonic rock made predominately of potassium feldspar and quartz.

gravel Unconsolidated sediment consisting mainly of rounded particles with a diameter greater than 2 millimeters.

greenhouse effect An increase in the temperature of a planet's atmosphere caused when infrared-absorbing gases are introduced into the atmosphere.

groin A narrow wall built perpendicular to the shore to trap sand transported by currents and waves.

ground moraine A moraine formed when a melting glacier deposits till in a relatively thin layer over a broad area.

ground water Water contained in soil and bedrock. All subsurface water.

Gulf Stream One of many oceanic surface currents. It is 80 kilometers wide and 650 meters deep near the east coast of Florida and moves northward at approximately 5 kilometers per hour. Farther to the northeast the current widens and slows; east of New York it is more than 500 kilometers wide and travels at less than 8 kilometers per day.

guyot A flat-topped seamount.

gypsum A mineral with the formula, ($CaSO_4 2H_2O$). It commonly forms in evaporite deposits.

gyre A closed, circular current either in water or air.

Hadean Eon The earliest time in Earth's history.

hail Ice globules varying from 5 millimeters to a record 14 centimeters in diameter that fall from cumulonimbus clouds.

half-life The time it takes for half of the nuclei of a radioactive isotope in a sample to decompose.

halons Organic compounds containing bromine and chlorine. These materials often become air pollutants.

hanging valley A tributary glacial valley whose mouth lies high above the floor of the main valley.

hanging wall The rock above an inclined fault.

hardness The resistance of the surface of a mineral to scratching.

heat Heat is a measure of the total energy in a sample, associated with the motions of atoms and molecules.

heliocentric A model that places the Sun at the center of the Solar System.

Hertzsprung–Russell diagram (H–R diagram) A plot of absolute stellar magnitude (or luminosity) against temperature.

horn A sharp, pyramid-shaped rock summit where three or more cirques intersect near the summit.

hornblende A rock-forming mineral, the most common member of the amphibole group.

hornfels A fine-grained rock formed by contact metamorphism.

horse latitudes A region of Earth lying at about 30° north and south latitudes, in which air is falling, forming a vast, high-pressure region. Generally dry conditions prevail, and steady winds are rare.

horst A block of rock that has moved relatively upward and is bounded by two faults.

hot spot A persistent volcanic center thought to be located directly above a rising plume of hot mantle rock.

hot spring A spring formed where hot ground water flows to the surface.

Hubble's law A law that states that the velocity of a galaxy is proportional to its distance from Earth. Thus the most distant galaxies are traveling at the highest velocities.

humid continental climate A climate type characterized by hot summers, cold winters, and rainfall throughout the year.

humid subtropical climate A climate type characterized by hot, humid summers but cooler winters and rainfall that falls throughout the year.

humidity A measure of the amount of water vapor in the air. *Absolute humidity* is the amount of water vapor in a given volume of air. *Relative humidity* is the ratio of the amount of water vapor in a given volume of air divided by the maximum amount of water vapor that can be held by that air at a given temperature.

humus The dark, organic component of soil composed of litter that has decomposed sufficiently so that the origin of the individual pieces cannot be determined.

hurricane *See* tropical cyclone.

hydration The chemical combination of water with another substance.

hydraulic action The mechanical loosening and removal of material by flowing water.

hydride A compound of hydrogen and one or more metals. Hydrides can be heated to release hydrogen gas for use as a fuel.

hydrogen economy An energy economy in which electricity is used to dissociate water into hydrogen and oxygen, and the hydrogen is used as a fuel.

hydrologic cycle The constant circulation of water among the sea, the atmosphere, and the land.

hydrolysis A decomposition reaction involving water.

hydrosphere All of Earth's water, which circulates among oceans, continents, and the atmosphere.

hydrothermal metamorphism Changes in rock that are primarily caused by migrating hot water and by ions dissolved in the hot water.

hydrothermal ore deposit An ore deposit formed by precipitation of dissolved metals from hot water solutions.

hydrothermal processes Processes involving interactions between hot water or steam and rocks and minerals; processes in which ore forms as a result of interactions between hot water or steam, dissolved metals, and rocks and minerals.

hydrothermal vein A sheet-like mineral deposit that fills a fault or other fracture, precipitated from hot water solutions.

hypothesis A tentative or preliminary conclusion built on strong supporting evidence from scientific experiment or observation.

ice age A time of extensive glacial activity, when alpine glaciers descended into lowland valleys and continental glaciers spread over the higher latitudes.

ice fall A section of a glacier that flows down a steep gradient, so that the ice fractures into numerous crevasses and towering ice pinnacles.

ice sheet A glacier that forms a continuous cover of ice over areas of 50,000 square kilometers or more and spreads outward under the influence of its own weight (*syn:* continental glacier, ice cap).

iceberg A large chunk of ice that breaks from a glacier into a body of water.

igneous rock Rock that solidified from magma.

index fossil A fossil that identifies and dates the layers in which it is found. Index fossils are abundantly preserved in rocks, widespread geographically, and existed as a species or genus for only a relatively short time.

industrial mineral Any rock or mineral of economic value exclusive of metal ores, fuels, and gems.

inertia The tendency of an object to resist a change in motion.

influent stream A stream that lies above the water table. Water percolates from the stream channel downward into the saturated zone (*syn:* losing stream).

intensity (of an earthquake) A measure of the effects an earthquake has on buildings and people at a particular place.

intermediate rock Igneous rock with chemical and mineral composition between those of granite and basalt.

internal processes Earth processes that are initiated by movements within Earth and those internal movements themselves. For example, formation of magma, earthquakes, mountain building, and tectonic plate movement.

intertidal zone The part of a beach that lies between the high- and low-tide lines.

intracratonic basins A sedimentary basin located within a craton.

intrusive igneous rock A rock formed when magma solidifies within bodies of preexisting rock.

inversion (atmospheric) A meteorological condition in which the lower layers of air are cooler than those at higher altitudes. This cool air can remain relatively stagnant and allows air pollutants to concentrate in urban areas.

ion An atom with an electrical charge.

ionic substitution The replacement of one or more kinds of ions in a mineral by other kinds of ions of similar size and charge.

island arc A gently curving chain of volcanic islands in the ocean formed by convergence of two plates, each bearing ocean crust, and the resulting subduction of one plate beneath the other.

isobar Lines on a weather map connecting points of equal pressure.

isostasy The condition in which the lithosphere floats on the asthenosphere as an iceberg floats on water.

isostatic adjustment The rising and settling of portions of the lithosphere to maintain equilibrium as they float on the plastic asthenosphere.

isotherm A line on a weather map connecting points of equal temperature.

isotopes Atoms of the same element that have the same number of protons but different numbers of neutrons.

jet stream A relatively narrow, high altitude, fast-moving air current.

joint A fracture that occurs without movement of rock on either side of the break.

Jovian planets The outer planets, Jupiter, Saturn, Uranus, and Neptune, which are massive with a high proportion of the lighter elements.

kame A small mound or ridge of layered sediment deposited by a stream that flows on top of, within, or beneath a glacier.

karst topography A type of topography formed over limestone or other soluble rock and characterized by caverns, sinkholes, and underground drainage.

kerogen The solid bitumen in oil shales that yields oil when the shales are heated and distilled; the precursor of liquid petroleum.

kettle lake A lake that forms in a depression in glacial drift that was created by melting of a large chunk of ice left buried in the drift by a receding glacier. The ice prevents sediment from collecting; when the ice melts a lake or swamp may fill the depression.

key bed A thin, widespread, easily recognized sedimentary layer that can be used for correlation.

Koeppen Climate Classification A climate classification system used by climatologists throughout the world.

Kuiper-belt objects Tiny ice dwarfs, similar to Pluto, in the outer reaches of the Solar System.

lagoon A protected body of water separated from the sea by a reef or barrier island.

lake A large, inland body of standing water that occupies a depression in the land surface.

landslide A general term for the downslope movement of rock and regolith under the influence of gravity.

latent heat The heat released or absorbed by a substance during a change in state, i.e., melting, freezing, vaporization, condensation, or sublimation.

lateral erosion The action of a stream as it cuts into and erodes its banks. A meandering stream commonly swings from one side of its channel to the other, causing lateral erosion.

lateral moraine A ridge-like moraine that forms on or adjacent to the sides of a mountain glacier.

laterite A highly weathered soil rich in oxides of iron and aluminum that usually develops in warm, moist, tropical or temperate regions.

Laurasia The northern part of Pangaea, consisting of what is now North America and Eurasia.

lava Fluid magma that flows onto Earth's surface from a volcano or fissure. Also, the rock formed by solidification of the same material.

lava plateau A sequence of horizontal basalt lava flows that were extruded rapidly to cover a large region of Earth's surface (*syns:* flood basalt, basalt plateau).

law In the interpretation of scientific principles, a law is a statement of how events always occur under given conditions. It is considered to be factual and correct.

layers 1, 2, 3 The three layers of oceanic crust. The uppermost layer, layer 1, consists of mud. Layer 2 consists of pillow basalt. Layer 3 directly overlies the mantle and consists of basalt dikes and gabbro.

leaching The dissolution and downward movement of soluble components of rock and soil by percolating water.

light-year The distance traveled by light in one year, approximately 9.5×10^{12} kilometers.

limb The side of a fold in rock.

limestone A sedimentary rock consisting chiefly of calcium carbonate.

liquefaction A process in which a soil loses its shear strength during an earthquake and becomes a fluid.

liquid metallic hydrogen A form of hydrogen under extreme temperature and pressure. Pressure forces the atoms together so tightly that the electrons move freely throughout the packed nuclei, much as electrons travel freely among metal atoms.

lithification The conversion of loose sediment to solid rock.

lithologic correlation Correlation based on the lithologic continuity of a rock unit.

lithosphere The cool, rigid, outer layer of Earth, about 100 kilometers thick, which includes the crust and part of the upper mantle.

litter Leaves, twigs, and other plant or animal material that have fallen to the surface of the soil but have not decomposed.

loam Soil that contains a mixture of sand, clay, and silt and a generous amount of organic matter.

loess A homogenous, unlayered deposit of windblown silt, usually of glacial origin.

longitudinal dune A long, symmetrical dune oriented parallel with the direction of the prevailing wind.

longshore current A current flowing parallel and close to the coast that is generated when waves strike a shore at an angle.

longshore drift Sediment carried by longshore currents.

losing stream A stream that lies above the water table. Water percolates from the stream channel downward into the saturated zone (*syn:* influent stream).

low velocity layer A region of the upper mantle where seismic waves travel relatively slowly, approximately the same as the asthenosphere.

lunar eclipse *See* eclipse.

luster The quality and intensity of light reflected from the surface of a mineral.

mafic Referring to rocks and minerals with high magnesium and iron contents.

mafic rock Dark-colored igneous rock with high magnesium and iron content, and composed chiefly of iron- and magnesium-rich minerals.

magma Molten rock generated within Earth.

magmatic arc A narrow, elongate band of intrusive and volcanic activity associated with subduction.

magmatic processes Any processes having to do with magma; processes that form ore as liquid magma solidifies to form an igneous rock.

magnetic reversal A change in Earth's magnetic field in which the north magnetic pole becomes the south magnetic pole and vice versa.

magnetometer An instrument that measures Earth's magnetic field.

magnitude (of an earthquake) A measure of the strength of an earthquake, determined from seismic recordings. *See also* Richter scale.

main sequence A band running across a Hertzsprung–Russell diagram that contains most of the stars. Hydrogen fusion generates the energy in a main sequence star.

manganese nodule A manganese-rich, potato-shaped rock found on the ocean floor.

mantle A mostly solid layer of Earth lying beneath the crust and above the core. The mantle extends from the base of the crust to a depth of about 2,900 kilometers.

mantle convection The convective flow of solid rock in the mantle.

mantle plume A rising vertical column of mantle rock.

marble A metamorphic rock consisting of fine- to coarse-grained, recrystallized calcite and/or dolomite.

maria Dry, barren, flat expanses of volcanic rock on the Moon first thought to be seas.

marine, West-Coast climate A temperate climate type characterized by relatively mild but wet winters and cool summers.

mass extinction A rapid event during which a significant part of all life forms on Earth become extinct.

mass wasting The movement of earth material downslope primarily under the influence of gravity.

meander One of a series of sinuous curves or loops in the course of a stream.

mechanical weathering (physical) The disintegration of rock into smaller pieces by physical processes.

medial moraine A moraine formed in or on the middle of a glacier by the merging of lateral moraines as two glaciers flow together.

Mediterranean climate A climate type characterized by dry summers, rainy winters, and moderate temperatures.

Mercalli scale A scale of earthquake intensity that measures the strength of an earthquake in a particular place by its effect on buildings and people. The Richter scale has replaced it.

mesa A flat-topped mountain or a tableland that is smaller than a plateau and larger than a butte.

mesosphere The layer of air that lies above the stratopause and extends from about 55 km upward to 80 km above Earth's surface. Temperature decreases with elevation in the mesosphere.

Mesozoic era The part of geologic time roughly from 245 to 65 million years ago. Dinosaurs rose to prominence and became extinct during this era.

metallic mineral resources Ore; valuable concentrations of metals.

metamorphic facies A set of all metamorphic rock types that formed under similar temperature and pressure conditions.

metamorphic grade The intensity of metamorphism that formed a rock; the maximum temperature and pressure attained during metamorphism.

metamorphic rock A rock formed when igneous, sedimentary, or other metamorphic rocks recrystallize in response to elevated temperature, increased pressure, chemical change, and/or deformation.

metamorphism The process by which rocks and minerals change in response to changes in temperature, pressure, chemical conditions, and/or deformation.

meteor A falling meteoroid.

meteorite A fallen meteoroid.

meteoroid A small interplanetary body in an irregular orbit. Many meteoroids are fragments of asteroids formed during collisions.

mica A layered silicate mineral with a distinctive, platy, crystal habit and perfect cleavage. Muscovite and biotite are common micas.

microcontinent A small mass of continental crust, or an island arc, similar to Japan, New Zealand, and the modern islands of the southwest Pacific Ocean. The first supercontinent is thought to have formed when many microcontinents joined together about 2 billion years ago. Modern continents consist of many microcontinents that welded together over geologic time.

microwave radar Instruments and processes using reflections of electromagnetic radiation in the microwave range.

The data measures subtle swells and depressions on the sea surface, which reflect sea-floor topography.

Mid-Atlantic Ridge The portion of the Mid-Oceanic Ridge system that lies in the middle of the Atlantic Ocean, halfway between North America and South America to the west, and Europe and Africa to the east.

mid-channel bar An elongate lobe of sand and gravel formed in a stream channel.

Mid-Oceanic Ridge system The continuous submarine mountain chain that forms at the boundary between divergent tectonic plates within oceanic crust. It circles Earth like the seam on a baseball, forming Earth's longest mountain chain.

migmatite A rock composed of both igneous and metamorphic-looking materials. It forms at very high metamorphic grades, when rock begins to partially melt to form magma.

mineral A naturally occurring inorganic solid with a definite chemical composition and a crystalline structure.

mineral deposit A local enrichment of one or more minerals.

mineral reserve The known supply of ore in the ground.

mineral resources All useful rocks and minerals.

mineralization A process of fossilization in which the organic components of an organism are replaced by minerals.

Mohorovicic discontinuity (Moho) The boundary between the crust and the mantle identified by a change in the velocity of seismic waves.

Mohs hardness scale A standard, numbered from 1 to 10, to measure and express the hardness of minerals based on a series of 10 fairly common minerals, each of which is harder than those lower on the scale.

moment magnitude An earthquake scale based on the amount of movement and the surface area of a fault. The moment magnitude scale closely reflects the total amount of energy released during an earthquake.

monocline A fold with only one limb.

monsoon A continental wind system caused by uneven heating of land and sea. Monsoons generally blow from the sea to the land in the summer, when the continents are warmer than the ocean, and bring rain. In winter, when the ocean is warmer than the land, the monsoon winds reverse.

moraine A mound or ridge of till deposited directly by glacial ice.

mud Wet silt and clay.

mud cracks Irregular, usually polygonal fractures that develop when mud dries. The patterns may be preserved when the mud is lithified.

mudflow Mass wasting of fine-grained soil particles mixed with a large amount of water.

mudstone A nonfissile rock composed of a mixture of clay and silt.

native elements Chemical elements that occur in the native (not chemically bonded to other elements) state. With the exception of atmospheric gases, only about 20 elements occur as native elements.

natural gas A mixture of naturally occurring light hydrocarbons composed mainly of methane, CH_4.

natural levee A ridge or embankment of flood-deposited sediment along both banks of a stream channel.

neap tide A relatively small tide that forms when the Moon is 90° out of alignment with the Sun and Earth.

nebula An interstellar cloud of gas and dust. A *planetary nebula* is created when a star the size of our Sun explodes.

neutron A subatomic particle with the mass of a proton but no electrical charge.

neutron star A small, extremely dense star composed almost entirely of neutrons. (*See also* pulsar.)

new moon When the Moon moves to a position between Earth and the Sun, the Moon's sunlit side faces away from us. In this position the Moon is dark (called the new moon) because the surface that faces Earth is not illuminated.

nimbo A prefix or suffix added to cloud types to indicate precipitation.

nimbostratus cloud A stratus cloud from which precipitation is falling.

nonbiodegradable A pollutant that is not consumed and destroyed in a reasonable amount of time by organisms that live naturally in soil or water.

nonconformity A type of unconformity in which layered sedimentary rocks lie on an erosion surface cut into igneous or metamorphic rocks.

nonfoliated The lack of layering in metamorphic rock.

nonmetallic mineral resources Any economically useful rock or mineral that is not a metal, such as salt, building stone, sand, and gravel.

nonpoint source pollution Pollution generated over a broad area, such as fertilizers and pesticides spread over agricultural fields.

nonrenewable resource A resource in which formation of new deposits occurs much more slowly than consumption.

normal fault A fault in which the hanging wall has moved downward relative to the footwall.

normal lapse rate The decrease in temperature with elevation in air that is neither rising nor falling.

normal magnetic polarity A magnetic orientation the same as that of Earth's modern magnetic field.

nuclear fission The breakdown of an atomic nucleus of an element or relatively high atomic number into two or more nuclei of lower atomic number, with conversion of part of its mass into energy.

nuclear fuel Radioactive isotopes, such as those of uranium, used to generate electricity in nuclear reactors.

nucleus The small, dense, central portion of an atom composed of protons and neutrons. Nearly all of the mass of an atom is concentrated in the nucleus.

nuée ardente A swiftly flowing, often red-hot cloud of gas, volcanic ash, and other pyroclastics formed by an explosive volcanic eruption (*syn:* ash flow).

O horizon The uppermost layer of soil, named for its organic component and consisting mostly of litter and humus with a small proportion of minerals.

objective lens The lens, farthest from the eyepiece, that collects light from a distant object in a refracting telescope.

obsidian A black or dark-colored, glassy volcanic rock usually of rhyolitic composition.

occluded front A front that forms when a faster-moving cold mass traps a warm air mass against a second mass of cold air. The faster-moving cold air mass slides beneath the warm air, lifting it off the ground. Precipitation occurs along both frontal boundaries, resulting in a large zone of inclement weather.

oceanic crust The 7-to-10-kilometer-thick layer of sediment and basalt that underlies the ocean basins.

oceanic island A volcanic island formed at a hot spot above a mantle plume.

oceanic trench A long, narrow, steep-sided depression of the sea floor formed where a subducting oceanic plate sinks into the mantle.

Ogallala aquifer The aquifer that extends for almost 1,000 kilometers from the Rocky Mountains eastward beneath portions of the Great Plains.

oil A naturally occurring liquid or gas composed of a complex mixture of hydrocarbons (*syn:* petroleum).

oil shale A kerogen-bearing sedimentary rock that yields liquid or gaseous hydrocarbons when heated.

oil trap Any rock barrier that accumulates oil or gas by preventing its upward movement.

oligotrophic lake A body of water characterized by nearly pure water with low concentrations of nitrates, phosphates, and other plant nutrients. Oligotrophic waterways have low productivities, meaning that they sustain relatively few living organisms, although lakes of this type typically contain a few huge trout or similar game fish and are commonly deep.

olivine A common rock-forming mineral in mafic and ultramafic rocks with a composition that varies between Mg_2SiO_4 and Fe_2SiO_4.

open universe A model in which the gravitational force of the universe is not sufficient to stop the current expansion and the galaxies will continue to fly apart forever.

ore A natural material that is sufficiently enriched in one or more minerals to be mined profitably.

ore mineral A mineral from which metals or other elements can be profitably recovered.

organic sedimentary rock Sedimentary rocks such as coal, which consist of the lithified organic remains of plants or animals.

original horizontality (principle of) *See* principle of original horizontality.

orogeny The process of mountain building; all tectonic processes associated with mountain building.

orographic lifting Lifting of air that occurs when air flows over a mountain.

orthoclase A common rock-forming mineral; a variety of potassium feldspar ($KAlSi_3O_8$).

outgassing The release of gases from the interior of Earth.

outwash Sediment deposited by streams flowing from the terminus of a melting glacier.

outwash plain A broad, level surface formed where glacial sediment is deposited in front of or beyond a glacier.

oxbow lake A crescent-shaped lake formed where a meander is cut off from a stream and the ends of the cut-off meander become plugged with sediment.

oxidation The loss of electrons from a compound or element during a chemical reaction. In the weathering of common minerals, oxidation usually occurs when a mineral reacts with molecular oxygen.

ozone hole An unusually low ozone concentration in the stratosphere.

P wave *See* primary (P) wave.

pahoehoe A basaltic lava flow with a smooth, billowy, or ropy surface.

paleoclimatology The study of ancient climates.

paleontology The study of life that existed in the past.

Paleozoic era The part of geologic time occurring during a period from 570 to 245 million years ago. During this era invertebrates, fishes, amphibians, reptiles, ferns, and cone-bearing trees were dominant.

Pangaea A supercontinent, first identified and named by Alfred Wegener, that existed from about 300 to 200 million years ago.

parabolic dune A crescent-shaped dune with tips pointing into the wind.

parallax The apparent displacement of an object caused by the movement of the observer.

parent isotope A radioactive isotope that decays to produce a daughter isotope.

parent rock Any original rock before it is changed by weathering, metamorphism, or other geological processes.

parsec The distance to an object that would have a stellar parallax angle of 1 arc second. One parsec is about 3.2 light-years or 3×10^{13} km.

partial melting The process in which a silicate rock only partly melts as it is heated and forms magma that is more silica rich than the original rock.

particle In air pollution terminology, any pollutant larger than a molecule.

particulate *See* particle.

passive continental margin A margin characterized by a firm connection between continental and oceanic crust, where little tectonic activity occurs.

paternoster lake One of a series of lakes, strung out like beads and connected by short streams and waterfalls, created by glacial erosion.

peat A loose, unconsolidated, brownish mass of partially decayed plant matter; a precursor to coal.

pebble A sedimentary particle between 2 and 64 millimeters in diameter, larger than sand and smaller than a cobble.

pedalfer A common soil type that forms in humid environments, characterized by abundant iron and aluminum oxides and a concentration of clay in the B horizon.

pediment A gently sloping erosional surface that forms along a mountain front uphill from a bajada, usually covered by a patchy veneer of gravel only a few meters thick.

pedocal A soil formed in arid and semiarid climates characterized by an accumulation of calcium carbonate.

pegmatite An exceptionally coarse-grained igneous rock, usually with the same mineral content as granite.

pelagic sediment Muddy ocean sediment that consists of a mixture of clay and the skeletons of microscopic marine organisms.

penumbra A wide band outside of the umbra, where only a portion of the Sun is eclipsed during a solar eclipse.

perched water table The top of a localized lens of ground water that lies above the main water table, formed by a layer of impermeable rock or clay.

peridotite A coarse-grained plutonic rock composed mainly of olivine; it may also contain pyroxene, amphibole, or mica but little or no feldspar. The mantle is thought to be made of peridotite.

period A geologic-time unit longer than an epoch and shorter than an era.

permafrost   A layer of permanently frozen soil or subsoil that lies from about a half meter to a few meters beneath the surface in arctic environments.

permeability   A measure of the speed at which fluid can travel through a porous material.

persistent bioaccumulative toxic chemical (PBT)   A non-biodegradable toxin that is released into and accumulates in the environment.

petroleum   A naturally occurring liquid composed of a complex mixture of hydrocarbons.

pH scale   A logarithmic scale that measures the acidity of a solution. A solution with a pH of 7 is neutral, neither acidic nor basic. Numbers lower than 7 represent acidic solutions, and numbers higher than 7 represent basic ones.

Phanerozoic Eon   The most recent 570 million years of geologic time, represented by rocks that contain evident and abundant fossils.

phenocryst   A large, early formed crystal in a finer matrix in igneous rock.

photon   The smallest particle or packet of electromagnetic energy, such as light, infrared radiation, etc.

photosphere   The surface of the Sun visible from Earth.

photosynthesis   The process by which chlorophyll-bearing plant cells synthesize carbohydrates from carbon dioxide and water using sunlight as an energy source. Oxygen is released in the process.

phyllite   A metamorphic rock with a silky appearance and commonly wrinkled surface, intermediate in grade between slate and schist.

phytoplankton   Plankton that conduct photosynthesis like land-based plants and that are the base of the food chain for aquatic animals.

pillow basalt   Lava that solidified under water, forming spheroidal lumps like a stack of pillows.

placer deposit   A surface mineral deposit formed by the mechanical concentration of mineral particles (usually by water) from weathered debris.

planet   A celestial body that revolves in a fixed orbit around a star.

planetary nebula   A nebula created when a star the size of our Sun explodes.

plankton   Small marine organisms that live mostly within a few meters of the sea surface, where sunlight is available. Plankton conduct most of the photosynthesis and nutrient consumption in the ocean and form the base of the marine food web. Many plankton are single-celled and microscopic; others are more complex and are up to a few centimeters long. *See also* phytoplankton and zooplankton.

plastic deformation   A permanent change in the original shape of a solid that occurs without fracture.

plastic flow   *See* plastic deformation.

plate   A relatively rigid, independent segment of the lithosphere that can move independently of other plates.

plate boundary   A boundary between two lithospheric plates.

plate tectonics theory   A theory of global tectonics in which the lithosphere is segmented into several plates that move about relative to one another by floating on and gliding over the plastic asthenosphere. Seismic and tectonic activity occur mainly at the plate boundaries.

plateau   A large, elevated area of comparatively flat land.

platform   The part of a continent covered by a thin layer of nearly horizontal sedimentary rocks overlying older igneous and metamorphic rocks of the craton.

playa   A dry desert lake bed.

playa lake   An intermittent desert lake.

Pleistocene Ice Age   A span of time from roughly 2 or 3 million to 8,000 years ago characterized by several advances and retreats of glaciers.

plucking   A process in which glacial ice erodes rock by loosening particles and then lifting and carrying them downslope.

plume (of pollution)   The three-dimensional zone of an aquifer or of surface water affected by a dispersing pollutant.

pluton   An igneous intrusion.

plutonic rock   An igneous rock that forms deep (a kilometer or more) beneath Earth's surface.

pluvial lake   A lake formed during a time of abundant precipitation. Many pluvial lakes formed as continental ice sheets melted.

point bar   A stream deposit located on the inside of a growing meander.

point source pollution   Pollution which arises from a specific site such as a septic tank or a factory.

polar easterlies   Persistent polar surface winds in the Northern Hemisphere that flow from east to west.

polar front   The low-pressure boundary at about 60° latitude formed by warm air rising at the convergence of the polar easterlies and prevailing westerlies.

polar jet stream   Persistent, high-altitude air that flows along the polar front.

polarity   The magnetically positive (north) or negative (south) character of a magnetic pole.

pollution   The reduction of the quality of a resource by the introduction of impurities.

population I star   A relatively young star, formed from material ejected by an older, dying star, composed mainly of hydrogen and helium, with 1 percent heavier elements. Our Sun is a population I star.

population II star   An old star with lower concentration of heavy elements than a population I star.

pore space   The open space between grains in rock, sediment, or soil.

porosity   The proportion of the volume of a material that consists of open spaces.

porphyry   Any igneous rock containing larger crystals (phenocrysts) in a relatively fine-grained matrix.

porphyry copper   A large body of porphyritic igneous rock that contains disseminated copper sulfide minerals and that is usually mined by open pit methods.

Precambrian   All of geologic time before the Paleozoic era, encompassing approximately the first 4 billion years of Earth's history. Also, all rocks formed during that time.

precession   The circling or wobbling of the tilt of Earth's orbit.

precipitation   (1) A chemical reaction that produces a solid salt, or precipitate, from a solution. (2) Any form in which atmospheric moisture returns to Earth's surface—rain, snow, hail, and sleet.

pressure gradient   The change in air pressure over distance.

pressure-release fracturing   Fracturing of rock that occurs when pressure decreases and the rock expands as tectonic forces raise it from a depth of several kilometers and erosion removes overlying rocks.

pressure-release melting The melting of rock and the resulting formation of magma caused by a drop in pressure at constant temperature.

prevailing westerlies Persistent midlatitude surface winds that flow from the southwest in the Northern Hemisphere and from the northwest in the Southern Hemisphere.

primary (P) wave A seismic wave formed by alternate compression and expansion of rock. P waves travel faster than any other seismic waves.

primary air pollutant A pollutant released directly into the atmosphere.

primordial nucleosynthesis The formation of helium nuclei that occurred over a time span of the first 8.5 minutes in the life of the universe.

principle of crosscutting relationships The principle that a rock or feature must first exist before anything can happen to it or before another rock cuts across it.

principle of faunal succession The principle that fossil organisms succeed one another in a definite and recognizable sequence, so that sedimentary rocks of different ages contain different fossils and rocks of the same age contain identical fossils. Therefore, the relative ages of rocks can be identified from their fossils.

principle of original horizontality The principle that most sediment is deposited as nearly horizontal beds, and therefore most sedimentary rocks started out with nearly horizontal layering.

principle of superposition The principle that states that in any undisturbed sequence of sediment or sedimentary rocks, the age becomes progressively younger from bottom to top.

prominence A flame-like jet of gas rising from the Sun's corona.

Proterozoic Eon The portion of geological time occurring during a period from 2.5 billion to 570 million years ago.

proton A dense, massive, positively charged particle found in the nucleus of an atom.

protoplanets The planets in their earliest, incipient stage of formation.

protosun The Sun in its earliest, incipient stage of formation. The protosun was a condensing agglomeration of dust and gas.

pulsar A neutron star that emits a pulsating radio signal.

pumice Frothy, usually rhyolitic magma solidified into a rock so full of gas bubbles that it can float on water.

pyroclastic rock Any rock made up of material ejected explosively from a volcanic vent.

pyroxene A rock-forming silicate mineral group that consists of many similar minerals. Members of the pyroxene group are major constituents of basalt and gabbro.

quartz A rock-forming silicate mineral, SiO_2. Quartz is a widespread and abundant component of continental rocks but is rare in the oceanic crust and mantle.

quartzite A metamorphic rock composed mostly of quartz and formed by recrystallization of sandstone.

quasar An object, less than one light-year in diameter and very distant from Earth, that emits an extremely large quantity of energy.

radiation fog Radiation fog occurs when Earth's surface and the air near the surface cools by radiation during the night, and water vapor in the air condenses because it cools below the dew point.

radiative zone The zone of a star surrounding the core where energy is transmitted by absorption and radiation.

radioactivity The natural spontaneous decay of unstable nuclei.

radiometric dating The process of measuring the absolute age of geologic material by measuring the concentrations of radioactive isotopes and their decay products.

rain-shadow desert A desert formed on the lee side of a mountain range.

recessional moraine A moraine that forms at the terminus of a glacier as the glacier stabilizes temporarily during retreat.

recharge The replenishment of an aquifer by the addition of water.

red giant A stage in the life of a star when its core is composed of helium that is not undergoing fusion. A hot shell of hydrogen around this core is fusing at a rapid rate, producing enough energy to cause the star to expand.

red shift The frequency shift of light waves observed in the spectrum of an object traveling away from an observer. This shift is caused by the Doppler effect.

reef A wave-resistant ridge or mound built by corals or other marine organisms.

reflecting telescope A type of telescope that uses a mirror or mirrors to form the primary image.

reflection The return of a wave that strikes a surface.

refracting telescope A type of telescope where the primary image is formed by a lens or lenses.

refraction The bending of a wave that occurs when the wave changes velocity as it passes from one medium to another.

regional dynamothermal metamorphism Metamorphism accompanied by deformation affecting an extensive region of Earth's crust.

regional metamorphism Metamorphism that is broadly regional in extent, involving very large areas and volumes of rock. Includes both regional dynamothermal and regional burial metamorphism.

regolith The loose, unconsolidated, weathered material that overlies bedrock.

relative age Time expressed as the order in which rocks formed and geological events occurred, but not measured in years.

relative humidity *See* humidity.

remediation The treatment of a contaminated substance to remove or decompose a pollutant.

remote sensing The collection of information about an object by instruments that are not in direct contact with it.

reserves Known geological deposits that can be extracted profitably under current conditions.

reservoir Porous and permeable rock in which liquid petroleum or gas accumulates.

residual deposit A mineral deposit formed when water dissolves and removes most of the soluble ions from soil and rock near Earth's surface, leaving only relatively insoluble ions in the soil. Bauxite, the principal source of aluminum, forms as a residual deposit, and in some instances iron also concentrates by this process to become ore.

resolution The ability of an optical telescope to distinguish details.

retrograde motion The apparent motion of the planets where they appear to move backwards (westward) with respect to the stars.

reverse fault A fault in which the hanging wall has moved up relative to the footwall.

reversed magnetic polarity Magnetic orientations in rock which are opposite to the present orientation of Earth's field. Also, the condition in which Earth's magnetic field is opposite to its present orientation.

revolve To orbit a central point. For example, a satellite revolves around Earth. *See* rotate.

rhyolite A fine-grained, extrusive igneous rock compositionally equivalent to granite.

Richter scale A numerical scale of earthquake magnitude measured by the amplitude of the largest wave on a standardized seismograph.

rift valley An elongate depression that develops at a divergent plate boundary. Examples include continental rift valleys and the rift valley along the center of the Mid-Oceanic Ridge system.

rift zone A zone of separation of tectonic plates at a divergent plate boundary.

ring of fire The belt of subduction zones and major tectonic activity including extensive volcanism that borders the Pacific Ocean along the continental margins of Asia and the Americas.

rip current A current created when water flows back toward the sea after a wave breaks against the shore (*syn:* undertow).

ripple marks Small, parallel ridges and troughs formed in sediment by wind or water currents and waves. They may then be preserved when the sediment is lithified.

rock A naturally formed solid that is an aggregate of one or more different minerals.

rock avalanche A type of mass wasting in which a segment of bedrock slides over a tilted bedding plane or fracture. The moving mass usually breaks into fragments (*syn:* rockslide).

rock cycle The sequence of events in which rocks are formed, destroyed, altered, and reformed by geological processes.

rock dredge An open-mouthed steel net dragged along the sea floor behind a research ship for the purpose of sampling rocks from submarine outcrops.

rock flour Finely ground, silt-sized rock fragments formed by glacial abrasion.

rock-forming minerals The most abundant minerals in Earth's crust; the nine minerals or mineral groups that make up most rocks of Earth's crust. They are olivine, pyroxene, amphibole, mica, the clay minerals, quartz, feldspar, calcite, and dolomite.

rockslide A type of slide in which a segment of bedrock slides along a tilted bedding plane or fracture. The moving mass usually breaks into fragments (*syn:* rock avalanche).

rotate To turn or spin on an axis. Tops and planets rotate on their axes. *See* revolve.

rubble Angular particles with a diameter greater than 2 millimeters.

runoff Water that flows back to the oceans in surface streams.

S wave A seismic wave consisting of a shearing motion in which the oscillation is perpendicular to the direction of wave travel. S waves travel slower than P waves (*syn:* secondary wave).

salinity The saltiness of seawater.

salinization A process whereby salts accumulate in soil that is irrigated heavily.

salt cracking A weathering process in which salts that are dissolved in water in the pores of rock crystallize. This process widens cracks and pushes grains apart.

saltation Sediment transport in which particles bounce and hop along the surface.

salt-water intrusion A condition in coastal regions in which excessive pumping of fresh ground water causes salty ground water to invade an aquifer.

San Andreas Fault zone A zone of strike-slip faults, extending from San Francisco to San Diego, which forms the transform boundary between the Pacific plate and the North American plate. Frequent earthquakes occur along this fault zone.

sand Sedimentary grains that range from 1/16 to 2 millimeters in diameter.

sandstone Clastic sedimentary rock comprised primarily of lithified sand.

saturated zone The region below the water table where all the pores in rock or regolith are filled with water.

saturation The maximum amount of water vapor that air can hold; the maximum amount of a substance that can dissolve in another substance.

schist A strongly foliated metamorphic rock that has a well-developed parallelism of minerals such as micas.

scientific method A set of guidelines used by scientists to organize the systematic collection, classification, and interpretation of data. This system starts with observation, followed by forming a preliminary conclusion or hypothesis. If this preliminary conclusion is extensively tested and well substantiated, the hypothesis evolves into a theory.

sea arch An opening created when a cave is eroded all the way through a narrow headland.

sea stack A pillar of rock left when a sea arch collapses or when the inshore portion of a headland erodes faster than the tip.

sea-floor drilling A process in which drill rigs mounted on offshore platforms or on research vessels cut cylindrical cores from both sediment and rock of the sea floor, which are then brought to the surface for study.

sea-floor spreading The hypothesis that segments of oceanic crust are separating at the Mid-Oceanic Ridge.

seamount A submarine mountain, usually of volcanic origin, that rises 1 kilometer or more above the surrounding sea floor.

second generation energy Radiation emitted by stars, galaxies, quasars, and other concentrated forms of matter.

Second Great Oxidation Event About 600 million years ago, the oxygen concentration in Earth's atmosphere increased abruptly a second time in response to interactions among chemical, physical, and biological processes.

second law of thermodynamics A fundamental law of nature that states that a closed system always shifts toward disorder or homogeneity.

secondary air pollutant A pollutant generated by chemical reactions within the atmosphere.

secondary and tertiary recovery Production of oil or gas by artificially augmenting the reservoir energy, as by injection of water, detergent, or other fluid.

secondary wave   A seismic wave consisting of a shearing motion in which the oscillation is perpendicular to the direction of wave travel (*syn:* S wave). S waves travel slower than P waves.

sediment   Solid rock or mineral fragments transported and deposited by wind, water, gravity, or ice, precipitated by chemical reactions, or secreted by organisms, and that accumulate as layers in loose, unconsolidated form.

sedimentary rock   A rock formed when sediment is lithified.

sedimentary structure   Any structure formed in sedimentary rock during deposition or by later sedimentary processes—for example, bedding.

seismic profiler   A device used to construct a topographic profile of the ocean floor and to reveal layering in sediment and rock beneath the sea floor.

seismic tomography   A technique whereby seismic data from many earthquakes and recording stations are analyzed to provide a three-dimensional view of the interior of Earth.

seismic wave   All elastic waves that travel through rock, produced by an earthquake or explosion.

seismogram   The record made by a seismograph.

seismograph   An instrument that records seismic waves.

seismology   The study of earthquake waves and the interpretation of this data to elucidate the structure of the interior of Earth.

shale   A fine-grained clastic sedimentary rock with finely layered structure composed predominantly of clay minerals.

shear strength   The resistance of materials to being pulled apart along their cross section.

shear stress   Stress that acts in parallel but opposite directions.

sheet flood   A broad, thin sheet of flowing water that is not concentrated into channels, typically in arid regions.

shield volcano   A large, gently sloping volcanic mountain formed by successive flows of basaltic magma.

sialic rock   A rock such as granite and rhyolite that contains large proportions of silicon and aluminum.

silica   Silicon dioxide, SiO_2. Includes quartz, opal, chert, and many other varieties.

silicate   All minerals whose crystal structures contain silica tetrahedra. All rocks composed principally of silicate minerals.

silicate tetrahedron   A pyramid-shaped structure of a silicon ion bonded to four oxygen ions, $(SiO_4)^{4-}$.

sill   A tabular or sheetlike igneous intrusion parallel to the grain or layering of country rock.

silt   All sedimentary particles from 1/256 to 1/16 millimeter in size.

siltstone   Lithified silt.

sinkhole   A circular depression in karst topography caused by the collapse of a cavern roof or by dissolution of surface rocks.

660-kilometer discontinuity   A boundary in the mantle, at a depth of about 660 kilometers, where seismic wave velocities increase because pressure is great enough that the minerals in the mantle recrystallize to form denser minerals.

slate   A compact, fine-grained, low-grade metamorphic rock with slaty cleavage that can be split into slabs and thin plates. Intermediate in grade between shale and phyllite.

slaty cleavage   A parallel metamorphic foliation in a plane perpendicular to the direction of tectonic compression.

sleet   Small spheres of ice that develop when raindrops form in a warm cloud and freeze as they fall through a layer of cold air at lower elevation.

slide   All types of mass wasting in which the rock or regolith initially moves as a consolidated unit over a fracture surface.

slip   The distance that rocks on opposite sides of a fault have moved.

slip face   The steep lee side of a dune that is at the angle of repose for loose sand so that the sand slides or slips.

slump   A type of mass wasting in which the rock and regolith move as a consolidated unit with a backward rotation along a concave fracture.

smog   Smoky fog. The word is used loosely to define visible air pollution.

snow line   The altitude on an alpine glacier above which there is permanent snow. The boundary between the zone of accumulation and the zone of ablation.

soil   Soil scientists define soil as the upper layers of regolith that support plant growth. Engineers see soil as synonymous with regolith.

soil horizon   A layer of soil that is distinguishable from other horizons because of differences in appearance and in physical and chemical properties.

soil-moisture belt   The relatively thin, moist surface layer of soil directly above the unsaturated zone.

solar cell   A device that produces electricity directly from sunlight.

solar eclipse   *See* eclipse.

solar mass   The mass of our Sun.

solar wind   A stream of atomic particles shot into space by violent storms occurring in the outer regions of the Sun's atmosphere.

solifluction   A type of mass wasting that occurs when water-saturated soil moves slowly over permafrost.

solstice   Either of two times per year when the Sun shines directly overhead and is furthest from the equator. The solstices mark the beginnings of summer and winter. One solstice occurs on or about June 21 and marks the longest day of the year in the northern hemisphere and the shortest day in the southern hemisphere. The other solstice occurs on or about December 22, marking the longest day in the southern hemisphere and the shortest day in the northern hemisphere.

source rock   The geologic formation in which oil or gas originates.

specific gravity   The weight of a substance relative to the weight of an equal volume of water.

specific heat   The heat required to raise the temperature of one gram of a substance 1° Celsius.

spectrum (electromagnetic)   A pattern of wavelengths into which a beam of light or other electromagnetic radiation is separated. The spectrum is seen as colors, is photographed, or is detected by an electronic device. An *emission spectrum* is obtained from radiation emitted from a source. An *absorption spectrum* is obtained after radiation from a source has passed through a substance that absorbs some of the wavelengths.

speleothems   Any mineral deposit formed in caves by the action of water.

spheroidal weathering   Weathering in which the edges and corners of a rock weather more rapidly than the flat faces, giving rise to a rounded shape.

spicule   A small jet of gas at the edge of the Sun. An average spicule is about 700 kilometers across and 7,000 kilometers high and lasts about 5 to 15 minutes.

spiral galaxy   A type of galaxy characterized by arms that radiate out from the center like a pinwheel.

spit   A small point of sand or gravel extending from shore into a body of water.

spreading center   The boundary or zone where lithospheric plates rift or separate from each other (*syn:* divergent plate boundary).

spring   A place where ground water flows out of Earth to form a small stream or pool.

spring tide   A relatively large tide that forms when the Sun, Moon, and Earth are aligned.

stable air   A parcel of air with a temperature and humidity profile so that it does not rise rapidly, does not ascend to high elevations, and does not lead to cloud formation and precipitation.

stalactite   An icicle-like dripstone, deposited from drops of water, which hangs from the ceiling of a cavern.

stalagmite   A deposit of mineral matter that forms on the floor of a cavern by the action of dripping water.

star   A self-luminous celestial body of great mass that emits energy through nuclear fusion.

stationary front   A boundary between two stationary air masses of different temperatures.

steppe   A vast, semiarid, grass-covered plain, as found in southeast Europe, central Asia, and parts of central North America.

stock   An igneous intrusion with an exposed surface area of less than 100 square kilometers.

stony meteorite   A meteorite composed of approximately 90 percent silicate rock and 10 percent iron and nickel.

storm surge   An abnormally high coastal water level created by a combination of strong onshore winds and the low atmospheric pressure of a storm.

storm track   A path repeatedly followed by storms.

strain   The deformation (change in size or shape) that results from stress.

stratified drift   Sediment that was transported by a glacier and then sorted, deposited, and layered by glacial meltwater.

stratocumulus   Low, sheet-like clouds with some vertical structure.

stratopause   The boundary between the stratosphere and the mesosphere.

stratosphere   The layer of air extending from the tropopause to about 55 kilometers. Temperature remains constant and then increases with elevation in the stratosphere.

stratovolcano   A steep-sided volcano formed by an alternating series of lava flows and pyroclastic eruptions (*syn:* composite volcano).

stratus cloud   A horizontally layered, sheet-like cloud.

streak   The color of a fine powder of a mineral, usually obtained by rubbing the mineral on an unglazed, porcelain streak plate.

stream   A moving body of water, confined in a channel and flowing downslope.

stress   The force per unit area exerted against an object.

striations   Parallel scratches in bedrock caused by rocks embedded in the base of a flowing glacier.

strike-slip fault   A fault on which the motion is parallel with its strike and is primarily horizontal.

subarctic   The northernmost edge of the humid continental climate zone, lying just south of the arctic.

subduction   The process in which a lithospheric plate descends beneath another plate and dives into the asthenosphere.

subduction complex   In a subduction zone, sea-floor sediment, basaltic oceanic crust, and occasionally slices of the upper mantle are scraped from the upper layers of the subducting slab and added to the opposite plate to form a subduction complex.

subduction zone (or subduction boundary)   The region or boundary where a lithospheric plate descends into the asthenosphere.

sublimation   The process by which a solid transforms directly to a vapor or a vapor transforms directly to a solid without passing through the liquid phase.

submarine canyon   A deep, V-shaped, steep-walled trough eroded into a continental shelf and slope.

submarine delta   A submarine delta is that portion of a river delta that lies underwater.

submarine fan   A large, fan-shaped accumulation of sediment deposited at the bases of many submarine canyons adjacent to the deep-sea floor (*syn:* abyssal fan).

submarine hydrothermal ore   Ore that forms when hot seawater dissolves metals from sea-floor rocks and then, as it rises through the upper layers of oceanic crust, cools and precipitates the metals.

submergent coastline   A coastline that was recently above sea level but has been drowned either because the land has sunk or sea level has risen.

subsidence   Settling of Earth's surface that can either occur as petroleum or groundwater is removed by natural processes.

subtropical jet stream   Persistent, high-altitude air that flows between the trade winds and the westerlies.

sulfide   A mineral such as pyrite (FeS_2), containing sulfur (S) bonded to a metal ion.

sunspot   A cool, dark region on the Sun's surface formed by an intense magnetic disturbance.

supercontinent   A continent, such as Alfred Wegener's Pangaea, consisting of all or most of Earth's continental crust joined together to form a single, large landmass. At least three supercontinents are thought to have existed during the past 2 billion years, and each broke apart after a few hundred million years.

supercooling   A condition in which water droplets in air do not freeze even when the air cools below the freezing point.

Superfund   The Comprehensive Environmental Response, Compensation, and Liability Act (CERCLA). An act that provides an emergency fund to clean up chemical hazards and imposes fines for polluting the environment.

supernova   An exploding star that is releasing massive amounts of energy.

superposition (principle of)   *See* principle of superposition.

supersaturation   A condition in which the concentration of a solution exceeds the saturation point; a condition in which the relative humidity of air exceeds 100 percent.

surf   The chaotic turbulence created when a wave breaks near the beach.

surface current   Horizontal flow of water in the upper 400 meters of the oceans.

surface mine   A hole excavated into Earth's surface for the purpose of recovering mineral or fuel resources.

surface processes   All processes that sculpt Earth's surface, such as erosion, transport, and deposition.

surface wave   An earthquake wave that travels along the surface of Earth, or along a boundary between layers within Earth (*syn:* L wave).

suspended load   That portion of a stream's load that is carried for a considerable time in suspension, free from contact with the streambed.

suture   The junction created when two continents or other masses of crust collide and weld into a single mass of continental crust.

syncline   A fold that arches downward and whose center contains the youngest rocks.

system   Any combination of interrelated, interacting components.

taiga   A biome of conifers and muskeg that survives the extreme winters of the subarctic.

tail (of a comet)   *See* comet.

talus   An accumulation of loose, angular rocks at the base of a cliff from which they have been cleared by mass wasting.

tar sands   Sand deposits permeated with heavy oil and an oil-like substance called bitumen, which are too thick to be pumped. Tar sands are dug up and heated with steam to make the bitumen fluid enough to separate from the sand. The bitumen is then treated chemically and heated to convert it to crude oil.

tarn   A small lake at the base of a cirque.

tectonic plate   A large segment of Earth's outermost, cool, rigid shell, consisting of the lithosphere and overlying crust. Tectonic plates float on the weak, plastic asthenosphere.

tectonics   A branch of geology dealing with the broad architecture of the outer part of Earth, specifically the relationships, origins, and histories of major structural and deformational features.

temperate rainforest   Common along the northwest coast of North America, from Oregon to Alaska, temperate rainforests grow in marine, West-Coast climates where rainfall is more than 100 centimeters per year and is relatively constant throughout the year.

temperature   Temperature is proportional to the average speed of atoms and molecules in a sample.

temperature inversion   *See* inversion (atmospheric).

tensional stress   Stress that pulls rock apart and is the opposite of tectonic compression.

terminal moraine   An end moraine that forms when a glacier is at its greatest advance.

terminus   The end, or foot, of a glacier.

terrestrial planets   The four Earth-like planets closest to the sun—Mercury, Venus, Earth, and Mars—which are composed primarily of rocky and metallic substances.

terrigenous sediment   Sea-floor sediment derived directly from land.

texture   The size, shape, and arrangement of mineral grains, or crystals, in a rock.

theory   A well-substantiated scientific conclusion based on an original hypothesis that is not substantively contradicted and is generally accepted by the vast majority of the scientific community.

thermal expansion and contraction   Fracturing of rock that occurs when temperature changes rapidly, causing the surface of the rock to heat or cool faster than its interior and to expand or contract faster than the interior.

thermocline   A layer of ocean water between 0.5 and 2 kilometers deep where the temperature drops rapidly with depth.

thermodynamics   *See* first and second laws of thermodynamics.

thermohaline circulation   Ocean circulation caused by differences in water temperature and salinity (and therefore density).

thermosphere   An extremely high and diffuse region of the atmosphere lying above the mesosphere. Temperature remains constant and then rises rapidly with elevation in the thermosphere.

three-cell model   A model of global wind systems that depicts three convection cells in each hemisphere.

threshold effect   A threshold effect occurs when the environment initially changes slowly (or not at all) in response to a small perturbation, but after the threshold is crossed, an additional small perturbation causes rapid change.

thrust fault   A type of reverse fault with a dip of 45° or less over most of its extent.

tidal current   A current caused by the tides.

tide   The cyclic rise and fall of ocean water caused by the gravitational force of the Moon and, to a lesser extent, of the Sun.

tidewater glacier   A glacier that flows directly into the sea.

till   Sediment deposited directly by glacial ice.

tillite   Solid rock composed of lithified till.

time correlation   Correlation based on age equivalence of rock units.

time-travel curve   A curve that records arrival times of P and S earthquake waves, used to measure the distance from a recording station to an earthquake epicenter.

topsoil   The fertile, dark-colored surface soil, or A horizon.

tornado   A small, intense, short-lived, funnel-shaped storm that protrudes from the base of a cumulonimbus cloud.

trade winds   The winds that blow steadily from the northeast in the Northern Hemisphere and from the southeast in the Southern Hemisphere between 5° and 30° north and south latitudes.

transform boundary   A boundary between two lithospheric plates where the plates are sliding horizontally past one another.

transform fault   A strike-slip fault between two offset segments of a mid-oceanic ridge.

transpiration   Direct evaporation from the leaf surfaces of plants.

transport   The movement of sediment by flowing water, ice, wind, or gravity.

transverse dune   A relatively long, straight dune with a gently sloping windward side and a steep lee face that is orientated perpendicular to the prevailing wind.

trench   A long, narrow depression of the sea floor formed where a subducting plate sinks into the mantle.

tributary   Any stream that contributes water to another stream.

Tropic of Cancer   The latitude 23.5° north of the equator. On June 21, the summer solstice in the Northern Hemisphere, sunlight strikes Earth from directly overhead at noon at this latitude.

Tropic of Capricorn   The latitude 23.5° south of the equator. On December 22, the summer solstice in the Southern Hemisphere, sunlight strikes Earth from directly overhead at noon at this latitude.

tropical cyclone   A broad, circular storm with intense low pressure that forms over warm oceans (also called a hurricane, typhoon, or cyclone).

tropical monsoon   A climate type characterized by large seasonal variations in rainfall, with high rainfall during the wet months.

tropical savannah   Climate type characterized by large seasonal variations in rainfall, but with a relatively low annual rainfall.

tropopause   The boundary between the troposphere and the stratosphere.

troposphere   The layer of air that lies closest to Earth's surface and extends upward to about 17 km. Temperature generally decreases with elevation in the troposphere.

trough   The lowest part of a wave.

truncated spur   A triangular-shaped rock face that forms when a valley glacier cuts off the lower portion of an arête.

tsunami   A large sea wave, produced by a submarine earthquake or a volcano, characterized by long wavelength and great speed.

tundra   A biome dominated by low-lying sedges, grasses, mosses, lichen, and small bushes that grow in the arctic and high alpine regions.

turbidity current   A rapidly flowing submarine current laden with suspended sediment, which results from mass wasting on the continental shelf or slope.

turnover   A process, usually occurring in fall and spring in temperate-climate lakes, in which convection mixes lake water so that it becomes of uniform temperature from top to bottom of the lake.

typhoon   The name used in the western Pacific to describe a tropical cyclone.

ultimate base level   The lowest possible level of downcutting of a stream, usually sea level.

ultramafic   Referring to rocks and minerals with magnesium and iron contents that are even higher than those of mafic rocks and minerals.

ultramafic rock   Rock composed mostly of minerals containing iron and magnesium—for example, peridotite.

umbra   The part of an eclipse shadow where the Sun is completely eclipsed.

unconformity   A gap in the geological record, such as an interruption of deposition of sediments or a break between eroded igneous and overlying sedimentary strata, usually of long duration.

underground mine   A mine consisting of subterranean passages that commonly follow ore veins or coal seams.

underthrusting   The process by which one continent may be forced beneath the other during a continent–continent collision.

uniformitarianism   The principle that states that geological processes and scientific laws operating today also operated in the past and that past geologic events can be explained by forces observable today. "The present is the key to the past." The principle does not imply that geo-

logic change goes on at a constant rate, nor does it exclude catastrophes such as impacts of large meteorites.

unit cell   The smallest group of atoms that perfectly describes the arrangement of all atoms in a crystal and that repeats itself to form the crystal structure.

unsaturated zone   A subsurface zone above the water table that may be moist but is not saturated; it lies above the zone of saturation (*syn:* zone of aeration).

unstable air   A parcel of air with a temperature and humidity profile so that it rises rapidly, ascends to high elevations, and leads to cloud formation and precipitation.

upper mantle   The part of the mantle that extends from the base of the crust downward to about 670 kilometers beneath the surface.

upslope fog   Fog that forms when air cools as it rises along a land surface.

upwelling   A rising ocean current that transports water from the depths to the surface.

urban heat island effect   A local change in climate caused by a city.

U-shaped valley   A glacially eroded valley with a characteristic U-shaped cross section.

valley train   A long and relatively narrow strip of outwash deposited in a mountain valley by the streams flowing from an alpine glacier.

vent   A volcanic opening through which lava and rock fragments erupt.

vesicle   A bubble formed by expanding gases in volcanic rocks.

volatile   A compound that evaporates rapidly and therefore easily escapes into the atmosphere.

volcanic ash   *See* ash (volcanic).

volcanic neck   A vertical, pipe-like intrusion formed by the solidification of magma in the vent of a volcano.

volcanic rock   A rock that formed when magma erupted, cooled, and solidified within a kilometer or less of Earth's surface.

volcano   A hill or mountain formed from lava and rock fragments ejected through a volcanic vent.

waning moon   The 13 days after a full moon and before a new moon when the visible portion of a moon decreases every day.

warm front   A front that forms when moving warm air collides with a stationary or slower-moving cold air mass. The moving warm air rises over the denser cold air as the two masses collide, cools adiabatically, and the cooling generates clouds and precipitation.

wash   An intermittent stream channel found in a desert.

water table   The upper surface of a body of ground water at the top of the zone of saturation and below the zone of aeration.

wave height   The vertical distance from the crest to the trough of a wave.

wave period   The time interval between two crests (or two troughs) as a wave passes a stationary observer.

wave-cut cliff   A cliff created when a rocky coast is eroded by waves.

wave-cut platform   A flat or gently sloping platform created by erosion of a rocky shoreline.

wavelength   The distance between successive wave crests (or troughs).

waxing moon The 13 days after a new moon and before a full moon when the visible portion of a moon increases every day.

weather The state of the atmosphere on a particular day as characterized by temperature, wind, cloudiness, humidity, and precipitation. *See* climate.

weathering The decomposition and disintegration of rocks and minerals at Earth's surface by mechanical and chemical processes.

welded tuff A hard, tough, glass-rich pyroclastic rock formed by cooling of an ash flow that was hot enough to deform plastically and partly melt after it stopped moving; it often appears layered or streaky.

well A hole dug or drilled into Earth, generally for the production of water, petroleum, natural gas, brine, sulfur, or for exploration.

wet adiabatic lapse rate The rate of cooling that occurs when moist air rises and condensation occurs, without gain or loss of heat.

wetlands Areas known variously as swamps, bogs, marshes, sloughs, mud flats, flood plains, and many other names, wetlands are the boundary between land and water. Some are water soaked or flooded throughout the entire year; others are dry for much of the year and wet only during times of high water. Some are wet only during exceptionally wet years and may be dry for several years at a time.

white dwarf A stage in the life of a star when fusion has halted and the star glows solely from the residual heat produced during past eras. White dwarfs are very small stars.

wind Horizontal airflow caused by the pressure differences resulting from unequal heating of Earth's atmosphere. Winds near Earth's surface always flow from a region of high pressure toward a low-pressure region.

wind shear A condition in which air currents rise and fall simultaneously within the same cloud.

withdrawal Any process that uses water and then returns it to Earth locally.

zone of ablation The lower-altitude part of an alpine glacier where more snow melts in summer than accumulates in winter. When the snow melts, a surface of old, hard glacial ice is left behind.

zone of accumulation The higher-altitude section of an alpine glacier where more snow falls in winter than melts in summer and where snow accumulates from year to year. There the glacier's surface is covered by snow year-round.

zone of aeration A subsurface zone above the water table that may be moist but is not saturated; it lies above the zone of saturation (*syn:* unsaturated zone).

zone of saturation A subsurface zone below the water table in which the soil and bedrock are completely saturated with water.

zone of wastage The lower portion of a glacier where more snow melts in summer than accumulates in winter so there is a net loss of glacial ice (*syn:* ablation area).

zooplankton Tiny marine animals that live mostly within a few meters of the sea surface and feed on phytoplankton.

Photo Credits

This page constitutes an extension of the copyright page. We have made every effort to trace the ownership of all copyrighted material and to secure permission from copyright holders. In the event of any question arising as to the use of any material, we will be pleased to make the necessary corrections in future printings. Thanks are due to the following authors, publishers, and agents for permission to use the material indicated.

Chapter 1. **1:** © John M. Roberts/CORBIS **3:** top, JPL/USGS; inset, NASA **3:** bottom, NASA/JPL **6:** bottom, Courtesy of Graham R. Thompson/Jonathan Turk **6:** top, Courtesey of Graham R. Thompson/Jonathan Turk **8:** Courtesy NASA **9:** Loiuse K. Broman/Photo Researchers **12:** Meteor Crater National Monument **14:** bottom, Courtesy of Graham R. Thompson/Jonathan Turk **19:** © Royalty-Free Photodisc/Getty Images

Chapter 2. **20:** center, Jose Manuel Sanchis Calvete/Corbis **21:** right, Courtesy Esoteric Hydroponics/Guildford, UK **21:** left, MII/USGS **22:** Environmental Protection Agency **24:** top, Courtesy of Graham R. Thompson/Jonathan Turk **24:** bottom, Courtesy of Graham R. Thompson/Jonathan Turk **25:** bottom right, © American Museum of Natural History **28:** center, A.E. Seaman Mineral Museum, Michigan Technological University. Photo by G. Robinson **28:** bottom, Courtesy of Graham R. Thompson/Jonathon Turk **29:** top left, Arkansas Geological Commission, J.M. Howard, Photographer **29:** top, Geoffrey Sutton **29:** center, Geoffrey Sutton **29:** bottom, Geoffrey Sutton **29:** top right, Geoffrey Sutton **30:** bottom center, Arthur R. Hill/Visuals Unlimited **30:** bottom right, Arthur R. Hill/Visuals Unlimited **30:** top, Geoffrey Sutton **30:** bottom left, © Arthur R. Hill/Visuals Unlimited **31:** top right, American Museum of Natural History **31:** top left, Breck P. Kent **35:** center left, Breck P. Kent **35:** bottom right, Breck P. Kent **35:** center right, Breck P. Kent **35:** top left, Geoffrey Sutton **35:** center left, Geoffrey Sutton **35:** top right, Jeffrey Scovill **35:** center right, Jeffrey Scovill **35:** top center, Jeffrey Scoville **35:** bottom left, Mark A. Scneider/Dembinsky Photo **36:** center right, American Museum of Natural History **36:** bottom left, Courtesy of Graham R. Thompson/Jonathon Turk **36:** top right, Smithsonian Institution **36:** top left, Ward's Natural Science Establishment **37:** top, © 1995 Kent Wood **37:** bottom, © 1995 Kent Wood **39:** top, Courtesy of Graham R. Thompson/Jonathon Turk

Chapter 3. **44:** Courtesy of Graham R. Thompson/Jonathan Turk **46:** top right, Courtesy of Graham R. Thompson/Jonathan Turk **46:** bottom, Courtesy of Graham R. Thompson/Jonathan Turk **46:** top left, USGS **47:** center left, Courtesy of Graham R. Thompson/Jonathan Turk **47:** top left, Courtesy of Graham R. Thompson/Jonathan Turk **47:** top right, Courtesy of Graham R. Thompson/Jonathan Turk **47:** center right, Courtesy of Graham R. Thompson/Jonathan Turk **47:** bottom center, Courtesy of Graham R. Thompson/Jonathan Turk **48:** top left, Courtesy of Graham R. Thompson/Jonathan Turk **48:** bottom right, Courtesy of Graham R. Thompson/Jonathan Turk **49:** top left, Geoffrey Sutton **49:** bottom right, Geoffrey Sutton **49:** bottom left,

Geoffrey Sutton **49:** top right, Geoffry Sutton **51:** top right, Photo by Steve Self **51:** top left, Steven Sheriff **54:** bottom right, Courtesy of Graham R. Thompson/Jonathan Turk **54:** top left, Courtesy of Graham R. Thompson/Jonathan Turk **55:** bottom left, Courtesy of Graham R. Thompson/Jonathan Turk **55:** bottom right, Courtesy of Graham R. Thompson/Jonathan Turk **55:** top right, Courtesy of Graham R. Thompson/Jonathan Turk **56:** top left, Courtesy of Graham R. Thompson/Jonathan Turk **56:** top right, Courtesy of Graham R. Thompson/Jonathan Turk **57:** top right, Breck P. Kent **57:** bottom right, Breck P. Kent **57:** top left, Courtesy of Graham R. Thompson/Jonathan Turk **58:** top left, David Schwimmer **59:** bottom left, Courtesy of Graham R. Thompson/Jonathan Turk **59:** top, Donovan Reese/Getty Images **60:** bottom right, Breck P. Kent **60:** top left, Courtesy of Graham R. Thompson/Jonathan Turk **60:** top right, Courtesy of Graham R. Thompson/Jonathan Turk **62:** bottom left, Breck P. Kent **62:** bottom right, © Terri Wright **63:** bottom left, Courtesy of Graham R. Thompson/Jonathan Turk **64:** bottom left, Courtesy of Graham R. Thompson/Jonathan Turk **64:** bottom right, Courtesy of Graham R. Thompson/Jonathan Turk **65:** Courtesy of Graham R. Thompson/Jonathan Turk **66:** top center, Breck P. Kent **66:** top left, Courtesy of Graham R. Thompson/Jonathan Turk **66:** top center, Courtesy of Graham R. Thompson/Jonathan Turk **66:** top right, Courtesy of Graham R. Thompson/Jonathan Turk **66:** bottom left, Courtesy of Graham R. Thompson/Jonathan Turk **66:** bottom right, Courtesy of Graham R. Thompson/Jonathan Turk **67:** top right, Breck P. Kent **67:** top left, Courtesy of Graham R. Thompson/Jonathan Turk **67:** center right, Courtesy of Graham R. Thompson/Jonathan Turk **67:** bottom right, Courtesy of Graham R. Thompson/Jonathan Turk **67:** center right, Courtesy of Graham R. Thompson/Jonathan Turk **67:** bottom left, Courtesy of Graham R. Thompson/Jonathan Turk

Chapter 4. **73:** © Dirk Wiersma /Photo Researchers, Inc. **79:** Courtesy of Graham R. Thompson/Jonathan Turk **80:** top left, Courtesy of Graham R. Thompson/Jonathan Turk **80:** top right, Courtesy of Graham R. Thompson/Jonathan Turk **80:** bottom left, Courtesy of Graham R. Thompson/Jonathan Turk **81:** bottom, © Corbis/CORBIS SYGMA **82:** top, Courtesy of Graham R. Thompson/Jonathan Turk **83:** left, Courtesy of Graham R. Thompson/Jonathan Turk **84:** Courtesy of Graham R. Thompson/Jonathan Turk **91:** top right, Museum of the Rockies **91:** bottom, National Museum of Natural History, J.H. Maternes mural, with permission **91:** top left, Wards Natural Science Environment

Chapter 5. **96:** Courtesy of Graham R. Thompson/Jonathan Turk **98:** top, Erich Lessing /Art Resource, NY **98:** bottom, Montana Historical Society **99:** Courtesy of Graham R. Thompson/Jonathan Turk **102:** top left, Dudley Foster, Woods Hole Oceanographic Institution **102:** bottom right, Institute of Oceanographic Sciences/NERC/Science Photo Library/Photo Researchers **104:** top left, Maud Davis Barker/Montana Historical Society **105:** MII/USGS **106:** bottom, Agricultural Stabilization and Conservastion Service/USDA **106:** top right, H.E. Malde/USGS **106:** top left, Peabody Coal Company **107:** bottom, Adapted

Index

FOR THOMPSON AND TURK'S
Earth Science and the Environment

Take charge of your learning with **ThomsonNOW**™ for *Earth Science and the Environment!* This powerful and interactive resource will help you gauge your own unique study needs. Then, it gives you a *Personalized Study* plan that helps you focus your study time on the concepts you most need to master.

This unmatched resource enhances the text, providing you with a seamless, integrated learning system. Which means you'll spend less time flipping through pages or navigating websites and more time learning exactly what you need to!

HOW DOES IT WORK?

Just log on to **ThomsonNOW for *Earth Science and the Environment*** by using the access code packaged with your text.* You will immediately notice that **ThomsonNOW** is as easy to use as surfing the web. Just a click of the mouse gives you the freedom to build a complete, *Personalized Study* plan for yourself by taking advantage of all three of **ThomsonNOW's** powerful components.

WHAT DO YOU KNOW?

A *Pre-Test* is the first step. *Pre-Test* questions are designed to help you identify what you need to study more—you receive your score automatically. Once you've completed the *Pre-Test,* a detailed *Personalized Study* plan (based on your test results) outlines the concepts you need to review and presents links to specific pages in your textbook that discuss that concept.

WHAT DO YOU NEED TO LEARN?

Working from your *Personalized Study* plan, you will be guided to media-enhanced activities—such as animated *Active Figures* taken from the text—to help you master the material. This section helps you gain a full understanding of the chapter material.

WHAT HAVE YOU LEARNED?

An optional *Post-Test* ensures that you've mastered the concepts in each chapter. As with the *Pre-Test,* your individual results may be e-mailed to your instructor to help you both assess your progress. If you need to improve your score, **ThomsonNOW for *Earth Science and the Environment*** will work with you as you continue to build your knowledge.

Get started today at **www.thomsonedu.com/login**

* Access to **ThomsonNOW** for *Earth Science and the Environment* is web based. Your instructor may have chosen to package the access code card with your new text. If your instructor did not order the access code card to be packaged with your text—or if you have a used copy of the text—you can still purchase an access code. Just visit **www.thomsonedu.com**, where you'll find easy-to-follow instructions to help you purchase your access code.

Surface and Deep-Water Currents of the Atlantic Ocean

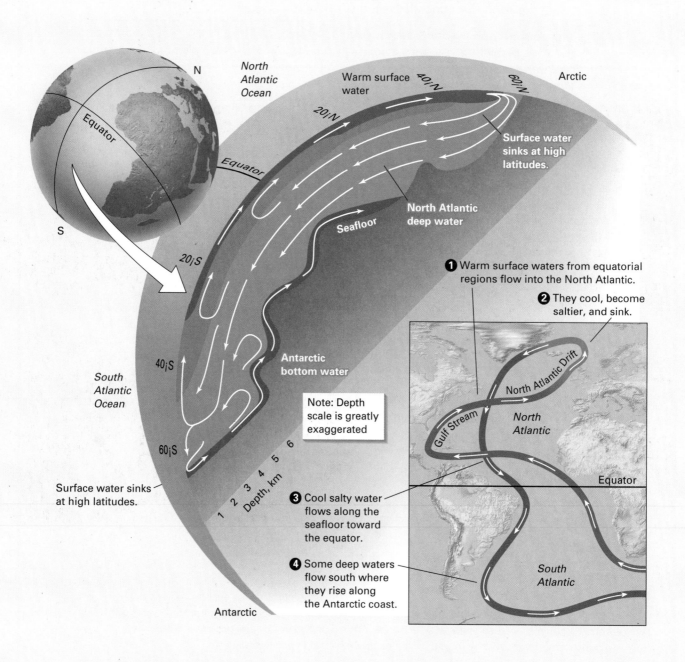